Operator Theory: Advances and Applications
Vol. 121

Problems and Methods in Mathematical Physics

The Siegfried Prössdorf Memorial Volume

Proceedings of the 11th TMP, Chemnitz (Germany), March 25–28, 1999

J. Elschner
I. Gohberg
B. Silbermann
Editors

Springer Basel AG

Editors:

Johannes Elschner
Weierstrass Institute for Applied Analysis and Stochastics
Mohrenstrasse 39
10117 Berlin
Germany

Bernd Silbermann
Faculty of Mathematics
Technical University of Chemnitz
09107 Chemnitz
Germany

I. Gohberg
Department of Mathematical Sciences
Raymond and Beverly Sackler
Faculty of Exact Sciences
Tel Aviv University
69978 Ramat Aviv
Israel

© 2000 Mathematics Subject Classification 35-xx, 41-xx, 42-xx, 45-xx, 47-xx, 65Nxx, 65Rxx, 65Txx

A CIP catalogue record for this book is available from the
Library of Congress, Washington D.C., USA

Deutsche Bibliothek Cataloging-in-Publication Data
Problems and methods in mathematical physics : the Siegfried Prössdorf memorial volume ; proceedings of the 11th TMP, Chemnitz (Germany), March 25 - 28, 1999 / J. Elschner ..., ed.. - Basel ; Boston ; Berlin : Birkhäuser, 2001
(Operator theory ; Vol. 121)
ISBN 978-3-0348-9500-2 ISBN 978-3-0348-8276-7 (eBook)
DOI 10.1007/978-3-0348-8276-7

Originally published by Birkhäuser Verlag in 2001
Softcover reprint of the hardcover 1st edition 2001
Printed on acid-free paper produced from chlorine-free pulp. TCF ∞
Cover design: Heinz Hiltbrunner, Basel

ISBN 978-3-0348-9500-2

9 8 7 6 5 4 3 2 1

Siegfried Prößdorf

1939 – 1998

Table of contents

Operator Theory:
Advances and Applications, Vol. 121

Obituary

Professor Dr. rer. nat. habil. Siegfried Prößdorf

January 2, 1939 – July 19, 1998

For me, as for many colleagues in Germany and abroad, the news of Siegfried Prößdorf's death was so shocking as to be unfathomable. Only now, little by little, are its implications becoming clearer.

In the Spring of 1997 preparations had begun for the conference in honor of Siegfried Prößdorf's 60th birthday. His sudden and untimely death on July 19, 1998 brought these plans to a halt. Nevertheless, many of his friends and colleagues were of the opinion that the conference, the 11th TMP, should take place anyway, now to honor the life and work of Siegfried Prößdorf.

We have lost one of the most prolific mathematicians that ever worked in the area of integral equations and their numerical analysis. It is not possible to fully describe his life and his scientific work in the few broad strokes to which I am restricted here. The course of his career was decidedly influenced by his undergraduate studies at the Mathematics Department of Leningrad University from 1958 to 1963 and subsequent graduate work there from 1963 to 1966. Fichtenholz, Natanson, Smirnow, and other excellent mathematicians were his lecturers. In particular, the influence of Professor Salomon Michlin on Siegfried Prößdorf's work can not be overestimated. In the 1960s, various classes of convolution functions whose generating functions have zeroes and thus violate normalization conditions became the object of intense study. Michlin steered Prößdorf's interest towards this area, became his graduate mentor and remained his lifelong friend and collaborator. Upon his graduation in 1966, Prößdorf accepted a position at the Mathematics Institute at the then-called Technical College of Karl-Marx-Stadt (now Chemnitz University of Technology). It was my good fortune to become an assistant in his group in 1967.

His extraordinary talents were demonstrated in his habilitation thesis in 1967 and finally in his advancement to tenure in 1969. In addition, Siegfried Prößdorf showed himself to be an exemplary teacher and organizer. He built up a flourishing research team and enthusiastically involved himself in teaching. Prößdorf's lectures in Analysis are well-remembered to this day. They set the high standards that the mathematics faculty of the Technical University in Chemnitz feels its duty to maintain.

In the late 1960s and early 1970s tensions rose in the Mathematics Institute due to contrary interpretations of the role of Mathematics within the natural sciences and within society. Siegfried Prößdorf, with characteristic determination and integrity, maintained the position that Mathematics is indivisible and that its course of progress is determined by its own innate logic and dynamics. Nevertheless, he certainly did not deny the importance of external influences as stimulating and enriching. For example, he considered computer technology as being a means to raise the applicability of mathematical methods to completely new levels.

In the early 1970s, although the potential for conflict at the Mathematics Institute in Karl-Marx-Stadt kept increasing, Siegfried Prößdorf's reputation outside of Karl-Marx-Stadt grew steadily. These events engendered his notion to leave Karl-Marx-Stadt. As I know it, he received various offers from the then GDR Academy of Sciences and finally, in 1975, moved reluctantly to Berlin. Prior to that, in 1972-73, he was a guest professor at the Mathematics Institute of the Academy of Sciences in Kishinev (Moldavia). This period of his life brought two important developments: For one, Prößdorf wrote his first book, which was later quite well respected and widely distributed. The second is his acquaintance and developing friendship with Israel Gohberg, whose scientific work was for Prößdorf's research group as significant as that of Salomon Michlin. Stimulated by Israel Gohberg's suggestions, in 1973-74 Siegfried Prößdorf and I began working on our own problems in numerical analysis. The scope of these investigations grew steadily and influenced all his scientific activities at the Weierstraß Institute in Berlin. In the mid 1970s he had his first contacts with Professors Meister and Wendland who at the time were still working in Darmstadt. They, too, proved to be good friends and colleagues in difficult times. Wolfgang Wendland suggested to Prößdorf that he explore applications of spline approximations to boundary integral equations. Together with his coworkers in Berlin, he earned lasting honors for his work in this area, of which a representative example is the application of Mellin techniques in numerical analysis.

As I already mentioned, respect for Siegfried Prößdorf's contributions grew with each passing year. His list of publications comprises 130 refereed papers and five books. He gave innumerable lectures at international conferences and other occasions and was on the editorial boards of four respected mathematical periodicals. His receiving the National Prize in 1980 cannot be viewed as a mere episode.

Siegfried Prößdorf leaves behind a school in which not only his own students but also their progeny continues to work. The course that he laid is clear and leads to the future. Despite his personal importance as a scientist, he remained modest, tolerant, and supportive.

I recall with pleasure my many visits to Roswitha and Siegfried Prößdorf, our discourses, and the sincere empathy shown by him for the manifold problems that we, his friends, colleagues and students, shared with him. His open and congenial manner, and his commitment, won him many friends.

A glance at the conference program and the list of participants of the 11th TMP will confirm what is stated here. I believe that Siegfried Prößdorf would have enjoyed this conference immensely, since it was a rare occasion for so many of his good friends and nearest colleagues to collect in one place, to honor him, and to discuss mathematical problems that were important to his work.

His early death leaves an emptiness that, in my mind and heart, can never be filled.

We are reaping today that which Siegfried Prößdorf sowed many years ago. His life and work tell us that this process needs continuous renewal. We honor him best when we follow these ideas.

Bernd Silbermann
Chemnitz, March 2000

Operator Theory:
Advances and Applications, Vol. 121

In memory of Siegfried Prößdorf

VLADIMIR MAZ'YA

I was honoured to chair the memorial session in the 11th TMP Conference in Klaffenbach dedicated to Siegfried Prößdorf. He was my dearest friend and his untimely passing away left an incurable wound in me. I was the first among the participants of this conference to meet Siegfried. This happened when he was a freshman at the Department of Mathematics of Leningrad University in 1958, 19 years old, being one year younger than myself. He was handsome and had an attractive disposition which he retained during his entire life. It seems that we felt mutual sympathy, although we did not speak much during his undergraduate years. He graduated in 1963 and went back to Leipzig, where he succeeded in getting permission to start post-graduate studies in Leningrad. I remember, as if it were today, how we met at V.I. Smirnov's seminar in the autumn of the same year. Since then we saw each other more often because S.G. Mikhlin, to whom I was very close, became his supervisor. For Siegfried's thesis Mikhlin proposed to investigate one-dimensional singular integral operators with degenerate symbols. Soon, under Mikhlin's influence, I also became interested in degenerate symbols, but for the quite different multidimensional case. So, we had some mathematics to speak about.

The life of foreign exchange students at the Leningrad University in the 1960s was far from being comfortable. The dormitory on Detskaya Street was in poor condition, and the quality of food in the student restaurant left much to be desired. This probably caused Siegfried's ulcer, which was operated on in Leningrad. Afterwards, he never felt completely healthy, but judging by his flourishing appearance this was impossible to guess. His wife Roswitha, at that time a student of a medical institute in Leningrad, cared for him selflessly.

After Prößdorf returned home we met rather rarely, mostly during visits by him and his wife to Leningrad. Before the Perestroika I had never been allowed to go abroad as a mathematician. For private visits to countries of the eastern block the regulations were milder, and when in 1980 Siegfried and Roswitha invited my wife Tatyana and myself to Berlin, we were lucky to get permission for the trip. We spent two weeks in the GDR giving talks in East Berlin and other places, which was actually not allowed by the Soviet authorities for a scientist travelling privately.

I remember clearly one day when Siegfried took Roswitha, Tatyana and me to the village Caputh. We intended to visit the summer house of Einstein, which turned out to be closed. The owner of a neighbouring villa invited us for a glass of wine, while he himself was sitting in his swimming pool. Later, because neither of us could remember his name, Siegfried called him Zweistein when recalling that day. The visit had an exciting continuation. When we rented a boat and were in the middle of the lake, a heavy rainfall began. Thoroughly wet, we found a nearby restaurant where our wives had some warming cognac. Siegfried (our driver) was unable to enjoy anything of this kind and I kept him company out of solidarity. On the way back, we saw large trees struck down by the storm. Clearly, we were lucky to have got inside just in time.

There were other meetings in the USSR, Germany and Sweden. I mention only our unforgettable trip to Schwarzwald and the Bodensee in June 1997 and the last meeting in February 1998 in Darmstadt on the occasion of our friend E. Meister's retirement. With Siegfried we discussed plans for future collaboration and nothing seemed to foretell the tragic end which was soon to come.

My dominant memory of Siegfried is that of an extraordinarily warm and kind person, "a real gentleman", as Paolo Ricci once called him speaking to me. I am especially grateful to him for the help he readily gave me during the Brezhnev time, when there were difficulties with publications for some categories of Soviet mathematicians, including myself. It is to his credit that my first books were published by the Teubner and Akademie–Verlag in the GDR.

We never wrote joint articles. However, we are the authors of two parts of a volume in "Encyclopaedia of Mathematical Sciences" dedicated to integral equations (Vol. 27, 1998, Springer–Verlag). I did my best to convince the publishers that Prößdorf's name should appear on the cover, but I failed since according to their policy, the cover was exclusively designated for the editors.

⋆⋆⋆

Prößdorf was a mathematician of high international reputation. I followed his research from the beginning of his scientific career. In more than 130 publications, among them numerous books and survey articles, he dealt with problems in operator theory, in the theory and numerical analysis of integral equations, pseudodifferential equations and boundary value problems, in approximation theory and boundary element methods. In the following I will try to describe some of his work.

Prößdorf was the first who characterized a class of linear operators in Banach spaces that admit an unbounded regularization [7].* Simultaneously he explicitly constructed unbounded regularizers for systems of one-dimensional singular integral equations with a degenerate symbol matrix [8], [9], [14], [15], [17]. These results became a part of his dissertation [1] (University of Leningrad 1966). On the basis of these results, he then created a complete theory of non-elliptic one-dimensional singular integral and Wiener–Hopf equations, with symbols vanishing at a finite number of points, in various spaces of functions and distributions (Fredholm property, index formulas, as well as existence, regularity and effective

*The references correspond to the list of Siegfried Prößdorf's publications.

construction of solutions) [2], [11], [13], [19]–[24]. A detailed presentation of this theory can be found in his monograph "Einige Klassen singulärer Gleichungen", which was published in 1974 and has later been translated into English and Russian. This work has been extended and generalized in various directions (see, e.g., the more recent monographs by Dybin, Roozemond, and Böttcher/Silbermann).

Together with Mikhlin, Prößdorf wrote the monograph "Singular Integral Operators" (Akademie–Verlag 1980, Springer–Verlag 1986, and Academia Sinica Publ. House, Beijing 1990), which became a standard reference and a fundamental textbook in the field of one and multidimensional singular integral operators. The book presents the basic aspects of the theory and leads the reader into this area, which is crucial for many applications.

In a series of papers (partly joint with Gohberg and Silbermann) he dealt with the convergence and error analysis of projection methods for the approximate solution of non-elliptic singular integral and Wiener–Hopf equations (e.g., finite section methods, polynomial and trigonometric Galerkin and collocation methods, mechanical quadrature) (see, e.g., [32]–[34], [36], [38]–[40], [42], [44]). These results were systematically presented in the monograph "Projektionsverfahren und die näherungsweise Lösung singulärer Gleichungen" by Prößdorf and Silbermann. This book is essentially based on a perturbation theory of projection methods with unbounded projectors developed by the authors. In recent years it has strongly influenced the work of many other mathematicians (e.g., Sloan, Atkinson, Wendland, McLean, Kress, Stephan, and their students).

Nowadays the results of Prößdorf on the asymptotic convergence order of Fourier series and their Cesaro means in Hölder norms are well known in approximation theory as "Prößdorf's Theorems". They play an important role in the convergence analysis of projection methods in these norms. Recently they have been generalized by many mathematicians (Leindler, Prestin et. al.) into different directions (e.g., for more general summation and averaging procedures).

Prößdorf also made interesting contributions to the factorization of matrix functions and to Wiener–Hopf methods for the explicit solution of boundary value problems of mathematical physics.

During the last 10–15 years his research activities were mainly devoted to numerical analysis, in particular, to stability and error analysis of spline and wavelet methods including compression techniques for the efficient approximate solution of boundary integral and pseudodifferential equations (the so-called boundary element methods). Some of his results in this very active and rapidly developing domain are the following:

Necessary and sufficient conditions for the stability of various classes of spline and wavelet approximation methods (e.g., Galerkin, collocation and quadrature methods) for singular integral and pseudodifferential equations (see, e.g., [51], [53], [62], [70], [71], [78], [88], [89], [97], [100], [105], [117], [130], [131]). These investigations by Prößdorf underscored the importance of the notion of the symbol (called by him "numerical symbol"), which occured in the eighties in papers of Silbermann and his co-workers and which now plays a role in numerical analysis similar to the symbol in the classical theory of pseudodifferential operators (an approximation method is stable if and only if its symbol is elliptic).

The notion of the symbol is closely connected with basic principles in modern numerical analysis, the so-called local principles. In his papers [58], [59], [62], [69], Prößdorf considered spline approximation methods with the help of the abstract local principle of Gohberg and Krupnik. This approach allows the reduction of the stability problem to the case of a convolution operator with "frozen" symbol (the latter, as a rule, is treated by Toeplitz matrix techniques). Later on Prößdorf (partly in collaboration with other mathematicians) successfully applied this principle to the numerical analysis of singular integral equations on curves with corners and to integral equations with Mellin convolution kernels (with Rathsfeld [74], [79], [81], [83]) as well as to the convergence analysis of spline and wavelet collocation and Galerkin methods for multidimensional pseudodifferential equations (with Schneider [88], [91], [97], and with Dahmen, Schneider [105]–[107]). These results and methods opened new perspectives in the analysis of boundary element methods, which today are among the standard methods of engineering mathematics.

He contributed to the development of efficient quadrature methods for pseudodifferential equations and for integral equations with Mellin convolution kernels and logarithmic difference kernels [74], [81], [90], [95], [96], [101], [103], [112], [113], [116]. Subsequently these methods have been used with great success for the numerical solution of integral equations appearing in fracture mechanics and Prandtl's airfoil theory ([83], [85], [87], [104], [123], [125]).

Prößdorf's joint monograph [6] with Silbermann is at the forefront of the research in this important field. A major part of this book has been shaped by a fruitful team-work with the authors' (former) students and (subsequent) coworkers Böttcher, Elschner, Hagen, Junghanns, Rathsfeld, Roch, and Schmidt.

The fast solvers for general Galerkin–Petrov methods applied to boundary integral and pseudodifferential equations [102], [106], [107], developed by Prößdorf in cooperation with Dahmen and Schneider during the years 1992–1994, are of significant practical and theoretical interest. They rely on an appropriate compression of the stiffness matrix with respect to the corresponding wavelet bases. In combination with the general convergence theory from [105], this leads to efficient numerical algorithms of almost linear computational complexity, which have the same optimal convergence order as in the non-compressed case. The numerical experiments presented in [111] for the double layer potential equation on a polyhedron confirm the applicability of these methods to practically relevant equations and boundary value problems.

Another important field of research in Prößdorf's work was that of ill-posed problems. In collaboration with Bruckner, Cheng, Pereverzev, Vainikko and Yamamoto he obtained interesting regularization results for Symm's equation [118], [133], Volterra equations [126] and integral equations of the first kind with analytic kernels [132]. In particular, the effect of self-regularization, i.e., regularization by discretization, has been studied in [134].

The above comments show that Siegfried Prößdorf has made fundamental contributions to a very broad field of analysis and numerics, including their applications.

Department of Mathematics
Linköping University
S-58183 Linköping
Sweden

Submitted: 1.11.1999

Operator Theory:
Advances and Applications, Vol. 121

Siegfried Prößdorf – A short review of his scientific career

JÜRGEN SPREKELS

Dear colleagues, Ladies and Gentlemen,

We have met here today in order to bid farewell to Professor Dr. Siegfried Prößdorf who has left us, on July 19, 1998, much too early. As the Director of the Weierstrass Institute, the institution to which Siegfried Prößdorf belonged almost 23 years long in leading positions, but also as a scientist, it is a sad duty and – at the same time – a real concern and great honour to me to stand here to pay tribute to the merits and to the achievements of the life of this outstanding researcher, colleague and human being.

I will confine myself to talking about his activity in the Weierstrass Institute. Other colleagues are in a better position to give a comprehensive appraisal of his scientific work; Prof. Silbermann will take the floor after me in order to do so.

First I would like to give a short summary of the developing of Siegfried Prößdorf into the outstanding scientist that he has been. Born on January 2, 1939, in Baldenhain in Thuringia, he went to primary school until 1953, then to secondary school. In 1957 he passed his school-leaving examination in Ronneburg in the district of Gera.

After that he took up the study of mathematics, first at the University of Leipzig, and afterwards from 1958 to 1963 at the University of Leningrad where he graduated with the mark "summa cum laude". From 1963 to 1966, he got a training for university lectureship in the field of mathematical physics at the Leningrad University in the chair of W. I. Smirnov. His tutor was Professor Mikhlin. In May 1966, he received his doctor's degree (Dr. rer. nat.).

From 1966 to 1968, he worked as a head scientific assistant at the Mathematical Institute of the Technical University of Karl Marx Stadt. It was there that he habilitated in July 1967 and was appointed university lecturer for mathematics on February 1, 1968, and full professor for analysis on September 1, 1969.

On September 1, 1975, he started working at the then Karl Weierstrass Institute for Mathematics of the GDR's Academy of Sciences in Berlin. From 1975 to 1984, he lead the section for applied analysis and numerical mathematics. Since 1984 he was head of a department and later, after a restructuration, head of a research group. In 1980, he received the GDR's National Award for Science and Technology for his scientific work.

After the reunification of Germany the Weierstrass institute was newly founded and he became one of the leading scientists of the institute and head of the group "Integral Equations and Pseudodifferential Equations" right from the beginning on January 1st, 1992.

His extraordinary many-sidedness and his high productivity made Siegfried Prößdorf one of the internationally leading scientists in his fields of research. A total of 152 publications in the fields of operator theory, integral equations, pseudodifferential equations, boundary value problems, numerical analysis, boundary element methods and approximation theory are his scientific legacy for us. I would like to mention here only his monographs some of which have been translated into several languages:

- Einige Klassen singulärer Gleichungen (1974) (Some classes of singular equations),
- Projektionsverfahren und die näherungsweise Lösung singulärer Gleichungen (1977, together with Prof. Silbermann) (Projection methods and the approximative solution of singular equations),
- Singuläre Integraloperatoren (1980, together with Prof. Mikhlin) (Singular integral operators),
- Linear integral equations (1991),
- Numerical analysis for Integral and Related Operator Equations (1991, together with Prof. Silbermann).

More than 100 of these publications were written during his time at the WIAS and at its predecessor institute. In addition, he leaves more than 20 scientists who graduated with him and six who habilitated under his guidance. He was co-editor of a number of internationally renowned mathematical journals and he gave invited plenary talks at many international conferences.

His international contacts were manifold. He co-operated with outstanding colleagues in Germany and abroad; colleagues like Hsiao, Wendland, Meister, Gohberg, Maz'ya, Sloan, Plamenevskii, Koshelev, Gilbert, Saranen, to name only a few, answered his invitation to the Weierstrass Institute, and the reputation of the institute in the international community profited enormously from their names.

The conferences organized by Siegfried Prößdorf were very important for the Weierstrass Institute, too. I only want to mention

- the International Symposium "Operator Equations and Numerical Analysis" (1992), held in Gosen near Berlin,
- the Oberwolfach Conference "Singuläre Integral- und Pseudodifferential-Operatoren und ihre Anwendungen" (1994) ("Singular integral operators and pseudodiffenential operators and their applications"),
- the Workshop on "Multiscale Methods in Numerical Analysis" (1995), at WIAS,

- the Conference "Recent approximation theory results in the numerical solution of differential and integral equations", Cortona, Italy, 1995.

Unfortunately, he did not live to see the International Congress of Mathematicians in 1998 in Berlin or the Conference on Functional Analysis, Partial Differential Equations and Applications held in Rostock from August 31 to September 4, 1998, in honour of Professor Maz'ya, in the organization of which he had taken part.

The outstanding scientific position of Siegfried Prößdorf in the international community is also reflected in the large number of invitations that he got to visit foreign research institutions. He visited Leningrad, Moscow, Kishinev, Amsterdam, Rome, Florence, Pavia, Turin, Helsinki, Paris, Calgary, Toronto, Peking, Tokyo and Canberra, to name only a few stages. It may be characteristic of the situation in Germany before the reunification that I personally met Siegfried Prößdorf for the first time in 1986 at the University of Delaware in the United States.

Ladies and gentlemen,

What I just told you underlines clearly the importance that the outstanding researcher Siegfried Prößdorf had for the scientific reputation of the Weierstrass Institute. But also inside the institute, Siegfried Prößdorf played an important irretrievable part, namely as the committed head of his research group who always spoke up for his collaborators, who promoted their scientific development with a lot of understanding, and – last but not least – as a member of the institute's management. I would like to thank him here personally above all for his efforts for the general interests of the institute, for his readiness to co-operate straightforwardly and constructively in the solving of the institute's problems and for some good advice that he gave to me personally.

Beside all these things we will miss above all the human being Siegfried Prößdorf; his sense of humour, his wit, his charm and his personality will not be forgotten. With Professor Siegfried Prößdorf, the Weierstrass Institute has lost one of its scientific top performers. He left a rich intellectual legacy. We will keep him in lasting remembrance and we will honour his memory.

Thank you very much.

Weierstrass Institute for Applied Analysis and Stochastics
Mohrenstrasse 39
D-10117 Berlin
Germany

Submitted: 28.9.1999

Publications of Siegfried Prößdorf

1. One–dimensional singular integral equations and systems of equations with degenerating symbol (Russian), Dissertation, Leningrad State University, 1966.

2. Eindimensionale singuläre Integralgleichungen und Faltungsgleichungen nicht normalen Typs in lokalkonvexen Räumen, Habilitationsschrift, TU Chemnitz, 1967.

Books

3. *Einige Klassen singulärer Gleichungen*, Akademie–Verlag, Berlin, 1974, Birkhäuser Verlag, Basel, 1974; English transl.: *Some Classes of Singular Equations*, North–Holland, Amsterdam, 1978; Russian transl.: Mir, Moscow, 1979.

4. (with B. Silbermann)
Projektionsverfahren und die näherungsweise Lösung singulärer Gleichungen, Teubner–Texte zur Mathematik, Teubner–Verlag, Leipzig, 1977.

5. (with S.G. Mikhlin)
Singuläre Integraloperatoren, Akademie–Verlag, Berlin, 1980; English transl.: *Singular Integral Operators*, Akademie–Verlag, Berlin, 1986, Springer–Verlag, Berlin, 1986.

6. (with B. Silbermann)
Numerical Analysis for Integral and Related Operator Equations, Akademie–Verlag, Berlin, 1991, Birkhäuser Verlag, Basel, 1991.

Articles

7. Operators admitting an unbounded regularization (Russian), *Vestnik Leningrad Univ.* **13** (1965), 59–67.

8. On the theory of systems of singular integral equations with degenerating symbol matrix. I (Russian), *Vestnik Leningrad Univ.* **19** (1965), 58–73.

9. On the theory of systems of singular integral equations with degenerating symbol matrix. II (Russian), *Vestnik Leningrad Univ.* **7** (1966), 68–75.

10. On the stability of the index of a one–dimensional singular integral operator with degenerating symbol (Russian), *Probl. Math. Analysis*, Leningrad State Univ., 1966, pp. 70–79.

11. On linear equations in spaces of test functions and distributions (Russian), *Dokl. Akad. Nauk SSSR* **166**, No. 4 (1966), 802–805.

12. The index of the one–dimensional singular operator with vanishing symbol on the space $C^\infty(\Gamma)$ (Russian), *Vestnik Leningrad Univ.* **7** (1966), 154–156.

13. Die Indexformel für ein System eindimensionaler singulärer Integralgleichungen nicht normalen Typs, *Math. Z.* **106** (1968), 73–80.

14. Über eine Verallgemeinerung des Ableitungsbegriffes und die Regularisierung singulärer Integralgleichungen, *Wiss. Z. Tech. Hochschule Karl–Marx–Stadt* **10**, No. 5 (1968), 551–554.

15. Ein Satz über die äquivalente Regularisierung abgeschlossener Operatoren, *Wiss. Z. Tech. Hochschule Karl–Marx–Stadt* **10**, No. 5 (1968), 555–556.

16. Zur Theorie singulärer Integralgleichungen nicht normalen Typs, *Wiss. Z. Tech. Hochschule Karl–Marx–Stadt* **11**, No. 3 (1969), 353–357.

17. (with E. Teichmann)
Über die äquivalente rechte Regularisierung einer singulären Integralgleichung nicht normalen Typs, *Wiss. Z. Tech. Hochschule Karl–Marx–Stadt* **11**, No. 3 (1969), 361–365.

18. Über eine Klasse singulärer Integralgleichungen nicht normalen Typs, *Math. Ann.* **183** (1969), 130–150.

19. Zur Theorie der Faltungsgleichungen nicht normalen Typs, *Math. Nachr.* **42** (1969), 103–131.

20. Über koerzitive und nichtkoerzitive Probleme bei singulären Integralgleichungen, *Mitteilungen Math. Gesellschaft der DDR* **3–4** (1970), 61–78.

21. Zur Lösung eines Systems singulärer Integralgleichungen mit entartetem Symbol, *Elliptische Differentialgleichungen*, Vol. 1, Akademie–Verlag, Berlin, 1970, pp. 111–118.

22. (with L. v. Wolfersdorf)
Zur Theorie der Noetherschen Operatoren und einiger singulärer Integral– und Integrodifferentialgleichungen, *Wiss. Z. Tech. Hochschule Karl–Marx–Stadt* **13**, No. 1 (1971), 103–116.

23. The singular integral equation with a symbol vanishing at a finite number of points (Russian), *Math. Issled.* **7**, No. 1 (1972), 116–132.

24. On systems of singular integral equations with degenerating symbol (Russian), *Math. Issled.* **7**, No. 2 (1972), 129–142.

25. Über eine Algebra von Pseudodifferentialoperatoren im Halbraum, *Math. Nachr.* **52** (1972), 113–139.

26. Über Pseudodifferentialoperatoren im Halbraum, *Beiträge Anal.* **4** (1972), 53–60.

27. Über einige Klassen singulärer Integralgleichungen mit entartetem Symbol, *Vorträge der Wiss. Haupttagung Math. Ges. DDR*, Vol. 3, Dresden, 1972, pp. 175–176.

28. (with B. Silbermann)
Über die normale Auflösbarkeit des singulären Integraloperators vom nicht normalen Typ, *Math. Nachr.* **55** (1973), 73–88.

29. (with B. Silbermann)
Über die normale Auflösbarkeit von Systemen singulärer Integralgleichungen vom nicht normalen Typ, *Math. Nachr.* **56** (1973), 131–144.

30. Singuläre Integralgleichungen nicht normalen Typs, *Entwicklung der Mathematik in der DDR*, Deutscher Verlag der Wissenschaften, Berlin, 1974, pp. 476–479.

31. (with B. Silbermann)
Ein Projektionsverfahren zur Lösung singulärer Gleichungen vom nicht normalen Typ, *Wiss. Z. Tech. Hochschule Karl-Marx-Stadt* **16**, No. 2 (1974), 367–376.

32. (with B. Silbermann)
Ein Projektionsverfahren zur Lösung abstrakter singulärer Gleichungen vom nicht normalen Typ und einige seiner Anwendungen, *Math. Nachr.* **61** (1974), 133–155.

33. (with I. Gohberg)
Ein Projektionsverfahren zur Lösung entarteter Systeme von diskreten Wiener–Hopf–Gleichungen, *Math. Nachr.* **65** (1975), 19–45.

34. Systeme einiger singulärer Gleichungen vom nicht normalen Typ und Projektionsverfahren zu ihrer Lösung, *Studia Math.* **53** (1975), 225–252.

35. Zur Konvergenz der Fourierreihen hölderstetiger Funktionen, *Math. Nachr.* **69** (1975), 7–14.

36. (with B. Silbermann)
Verallgemeinerte Projektionsverfahren zur Lösung singulärer Gleichungen vom nicht normalen Typ, *Math. Nachr.* **68** (1975), 7–28.

37. Entartete Integralgleichungen, *5. Tagung über Probleme und Methoden der Mathematischen Physik*, Vol. 2, Wiss. Schriftenreihe Tech. Hochschule Karl–Marx–Stadt, 1975, pp. 263–287.

38. (with B. Silbermann)
Einige allgemeine Sätze zur Theorie der Projektionsverfahren für lineare Operatorgleichungen in Banachräumen, *Math. Nachr.* **75** (1976), 61–72.

39. (with B. Silbermann)
On the convergence of the reduction and collocation methods for systems of singular integral equations (Russian), *Dokl. Akad. Nauk SSSR* **226**, No. 3, (1976), 516–519; English transl.: *Sov. Math. Dokl.* **17** (1976), 140–143.

40. (with B. Silbermann)
General convergence theorems of projection methods for operator equations in Banach spaces (Russian), *Dokl. Akad. Nauk SSSR* **230**, No. 3 (1976), 527–529; English transl.: *Sov. Math. Dokl.* **17** (1976), 1347–1349.

41. Allgemeine Konvergenztheoreme zu Projektionsverfahren für lineare Operatorgleichungen und einige Anwendungen, *Theory of Nonlinear Operators*, Proceedings International Summer School, Berlin, 1975, Akademie–Verlag, Berlin, 1977, pp. 217–227.

42. (with B. Silbermann)
Projektionsverfahren zur Lösung von Systemen singulärer Gleichungen vom nicht normalen Typ, *Rev. Roum. Math. Pures Appl.* **22** (1977), 965–991.

43. (with G. Unger)
Zur Faktorisierung von Matrixfunktionen in Algebren mit zwei Normen, *Math. Nachr.* **79** (1977), 37–47.

44. (with B. Silbermann)
Zur Kollokations– und Reduktionsmethode für Systeme singulärer Integralgleichungen, *7. Internat. Kongr. Anwendungen der Mathematik in den Ingenieurwiss.*, Weimar, 1975, Verlag für Bauwesen, Berlin, 1977, pp. 289–293.

45. (with B. Silbermann)
Gestörte Projektionsverfahren und einige ihrer Anwendungen, *Theory of Nonlinear Operators*, Proceedings International Summer School, Berlin, 1977, Akademie–Verlag, Berlin, 1978, pp. 229–237.

46. On approximation methods for the solution of one–dimensional singular integral equations, *Appl. Anal.* **7** (1978), 259–270.

47. (with G. Schmidt)
Notwendige und hinreichende Bedingungen für die Konvergenz des Kollokationsverfahrens bei singulären Integralgleichungen, *Math. Nachr.* **89** (1979), 203–215.

48. (with B. Silbermann)
Über Näherungsverfahren zur Lösung singulärer Gleichungen, *7. Tagung über*

Probleme und Methoden der Mathematischen Physik, Tagungsberichte, Vol. 2, Tech. Hochschule Karl–Marx–Stadt, 1979, pp. 96–114.

49. On the convergence of two collocation methods for the approximate solution of one–dimensional singular integral equations, *Proceedings International Conference "Functions, Series, Operators"*, Budapest, Aug. 1980, Janos Bolyai Math. Soc., Budapest, 1980.

50. On the method of mechanical quadratures for singular integro–differential equations, *Proceedings 2nd Symposium on Integral Equations and Their Applications* (ed. J. Wolska–Bochenek), Warsaw, Dec. 1979, Politechnika Warszawska, Inst. Mat., Warsaw, 1980.

51. (with G. Schmidt)
A finite element collocation method for singular integral equations, *Math. Nachr.* **100** (1981), 33–60.

52. New results on the approximate solution of singular integral equations, *International Conference on Complex Analysis and Applications*, Summaries, Varna, Sept. 1981, pp. 171–173.

53. A finite element method for the solution of singular integral equations, *Komplexe Analysis und ihre Anwendung auf partielle Differentialgleichungen* (ed. W. Tutschke), Kongreß– und Tagungsberichte, Vol. 4, Martin–Luther–Univ. Halle–Wittenberg, Halle, 1981.

54. Zur Splinekollokation für lineare Operatoren in Sobolewräumen, *Recent Trends in Mathematics*, Teubner–Texte zur Mathematik, Vol. 50, Teubner–Verlag, Leipzig, 1983, pp. 251–262.

55. On the degenerate Riemann–Hilbert problem (Russian), *Partial Differential Equations*, Banach Center Publ., Vol. 10, 1983, pp. 301–310.

56. Approximation methods for solving singular integral equations, *Complex Analysis: Methods, Trends and Applications* (eds. E. Lanckau and W. Tutschke), Akademie–Verlag, Berlin, 1983, pp. 131–141.

57. Starke Elliptizität singulärer Integraloperatoren und Spline–Approximation, *Linear and Complex Analysis Problem Book - 199 Research Problems* (eds. V.P. Havin, S.V. Hruščev and N.K. Nikol'ski), Lecture Notes in Math., Vol. 1043, Springer–Verlag, Berlin, Heidelberg, 1984, pp. 298–302.

58. A localization principle in the theory of finite element methods, *Probleme und Methoden der Mathematischen Physik*, Teubner–Texte zur Mathematik, Vol. 63, Teubner–Verlag, Leipzig, 1984, pp. 169–177.

59. Ein Lokalisierungsprinzip in der Theorie der Spline–Approximationen und einige Anwendungen, *Math. Nachr.* **119** (1984), 239–255.

60. A finite element collocation method for systems of singular integral equations, *Complex Analysis and Applications*, Publishing House of the Bulgarian Acad. Sci., Sofia, 1984, pp. 428–439.

61. (with J. Elschner)
Finite element methods for singular integral equations on an interval, *Engineering Analysis* **1** (1984), 83–87.

62. (with A. Rathsfeld)
A spline collocation method for singular integral equations with piecewise continuous coefficients, *Integral Equations Operator Theory* **7** (1984), 536–560.

63. (with A. Rathsfeld)
Strongly elliptic singular integral equations with piecewise continuous coefficients and the convergence of spline Galerkin and collocation methods, *Constructive Theory of Functions*, Publishing House of the Bulgarian Acad. Sci., Sofia, 1984, pp. 710–717.

64. Strongly elliptic singular integral equations with piecewise continuous coefficients, *Function Theoretic Methods in Partial Differential and Integral Equations*, Tagungsbericht 49/1985, Mathem. Forschungsinstitut Oberwolfach, 1985, pp. 17–18.

65. (with A. Rathsfeld)
On strongly elliptic singular integral operators with piecewise continuous coefficients, *Integral Equations Operator Theory* **8** (1985), 825–841.

66. (with J. Elschner, A. Rathsfeld and G. Schmidt)
Spline approximation of singular integral equations, *Demonstratio Math.* **18** (1985), 661–672.

67. (with A. Rathsfeld)
On spline Galerkin methods for singular integral equations with piecewise continuous coefficients, *Numer. Math.* **48** (1986), 99–118.

68. On the approximate solution of singular integral equations and Wiener–Hopf equations, *Wiener–Hopf Problems with Applications*, Tagungsbericht 31/1986, Mathem. Forschungsinstitut Oberwolfach, 1986, pp. 11–12.

69. A localization principle in the theory of finite element methods and some applications (Russian), *Proceedings International Conference on Partial Differential Equations*, Novosibirsk, Oct. 1983, Nauka, Novosibirsk, 1986, pp. 160–167.

70. (with A. Rathsfeld)
Stabilitätskriterien für Näherungsverfahren bei singulären Integralgleichungen in L^p, *Z. Anal. Anw.* **6** (1987), 539–558.

71. (with A. Rathsfeld)
On quadrature methods and spline approximation of singular integral equations, *Boundary Elements IX* (eds. C.A. Brebbia, W.L. Wendland, G. Kuhn), Vol. 1: Mathematical and Computational Aspects, Springer–Verlag, Berlin, 1987, pp. 193–211.

72. (with U. Szyszka)
On spline collocation of singular integral equations on nonuniform meshes, *Seminar Analysis: Operator equations and numerical analysis 1986/87*, Karl–Weierstraß–Institut Math., Akad. Wiss. DDR, Berlin, 1987, pp. 123–137.

73. Einige Entwicklungsrichtungen der Theorie der linearen Integralgleichungen, *Mitteilungen Math. Gesellschaft der DDR* **3** (1987), 21–36.

74. (with A. Rathsfeld)
Mellin techniques in the numerical analysis for one–dimensional singular integral equations, Report R–MATH–06/88, Karl–Weierstraß–Institut Math., Akad. Wiss. DDR, Berlin, 1988.

75. (with G.C. Hsiao)
On the stability of the spline collocation method for a class of integral equations of the first kind, *Appl. Anal.* **30** (1988), 249–261.

76. Prof. Dr. Dr. h.c. Solomon G. Michlin zum 80. Geburtstag, *Z. Anal. Anw.* **7** (1988), 97–98.

77. Linear integral equations (Russian), *Analysis IV* (eds. V.G. Maz'ya and S.M. Nikol'ski), VINITI, Moscow, 1988, pp. 5–130; English transl.: *Encyclopaedia of Mathematical Sciences*, Vol. 27, Springer–Verlag, Berlin, 1991, pp. 1–125.

78. Numerische Behandlung singulärer Integralgleichungen, *ZAMM* **69** (1989), T5–T13.

79. (with A. Rathsfeld)
Quadrature and collocation methods for singular integral equations on curves with corners, *Z. Anal. Anw.* **8** (1989), 197–220.

80. (with W. McLean and W.L. Wendland)
Pointwise error estimates for the trigonometric collocation method applied to singular integral equations and periodic pseudodifferential equations, *J. Integral Equations Appl.* **2** (1989), 125–146.

81. (with A. Rathsfeld)
Quadrature methods for strongly elliptic Cauchy singular integral equations on an interval, *The Gohberg–Anniversary Collection* (eds. H. Dym et al.), Vol. 2, Oper. Theory: Advances and Applications, Vol. 41, Birkhäuser Verlag, Basel, 1989, pp. 435–471.

82. Recent results in numerical analysis for singular integral equations, *Problems and Methods in Mathematical Physics*, Teubner–Texte zur Mathematik, Teubner–Verlag, Leipzig, 1989, pp. 224–234.

83. (with A. Rathsfeld)
On an integral equation of the first kind arising from a cruciform crack problem, *Integral Equations and Inverse Problems* (eds. Bl. Sendov and V. Petkov), Proceedings International Conference Varna, 1989, Longman, Coventry, 1990, pp. 210–219.

84. (with F.–O. Speck)
A factorization procedure for two by two matrix functions on the circle with two rationally independent entries, *Proc. Roy. Soc. Edinburgh* **115 A** (1990), 119–138.

85. (with G. Monegato)
On the numerical treatment of an integral equation arising from a cruciform crack problem, *Math. Methods Appl. Sci.* **12** (1990), 489–502.

86. (with J. Prestin)
Error estimates in generalized trigonometric Hölder–Zygmund norms, *Z. Anal. Anw.* **9** (1990), 343–349.

87. (with D. Tordella)
On an extension of Prandtl's lifting line theory to curved wings, *Impact Comput. Sci. Engrg.* **3** (1991), 192–212.

88. (with R. Schneider)
A spline collocation method for multidimensional strongly elliptic pseudodifferential operators of order zero, *Integral Equations Operator Theory* **14** (1991), 399–435.

89. On the super–approximation property of Galerkin's method with finite elements, *Numer. Math.* **59** (1991), 711–722.

90. Approximation methods for integral equations of Cauchy and Mellin type, *Atti Sem. Mat. Fis. Univ. Modena* **34** (1991), 103–114.

91. (with R. Schneider)
Pseudodifference operators – A symbolic calculus for approximation methods for periodic elliptic pseudodifferential equations. Part I: Calculus, Preprint No. 1438, Technische Hochschule Darmstadt, Fachbereich Mathematik, 1991.

92. Solomon G. Mikhlin (1908–1990), *Z. Anal. Anw.* **10** (1991), 121–122.

93. (with G. Mastroianni)
On quadrature methods of Gauss type for singular integral equations and the airfoil equation, *Proceedings of the Seventh GAMM–Seminar* (ed. W. Hackbusch), Kiel, 1991, Vieweg, Braunschweig, 1992, pp. 129–140.

94. (with R. Schneider)
Convergence of spline approximation methods for periodic elliptic pseudodifferential equations, *Proceedings of the Seventh GAMM-Seminar* (ed. W. Hackbusch), Kiel, 1991, Vieweg, Braunschweig, 1992, pp. 141–152.

95. (with I.H. Sloan)
Quadrature method for singular integral equations on closed curves, *Numer. Math.* **61** (1992), 543–559.

96. (with G. Mastroianni)
A quadrature method for Cauchy integral equations with weakly singular perturbation kernel, *J. Integral Equations Appl.* **4** (1992), 205–228.

97. (with R. Schneider)
Spline approximation methods for multidimensional periodic pseudodifferential equations, *Integral Equations Operator Theory* **15** (1992), 626–672.

98. Spline collocation methods for boundary integral equations in $\mathbb{R}^n$, Rend. Sem. Mat. Univ. Politecn. Torino , Special Issue, 1992, pp. 245–261.

99. Spline approximation methods for multidimensional singular integral and pseudodifferential equations, *Continuum Mechanics and Related Problems of Analysis* (eds. M. Balavadze, I. Kiguradze, V. Kokilashvili), Proceedings of the International Symposium Dedicated to the Centenary of Academician N. Muskhelishvili, Tbilisi, 6–11 June 1991, Metsniereba, Tbilisi, 1993, pp. 290–302.

100. (with G.C. Hsiao)
A generalization of the Arnold–Wendland lemma to a modified collocation method for boundary integral equations in R^3, *Math. Nachr.* **163** (1993), 133–144.

101. (with J. Saranen and I.H. Sloan)
A discrete method for the logarithmic–kernel integral equation on an open arc, *J. Austral. Math. Soc.* **B 34** (1993), 401–418.

102. (with W. Dahmen and R. Schneider)
Wavelet approximation methods for pseudodifferential equations II: Matrix compression and fast solution, *Adv. Comput. Math.* **1** (1993), 259–335.

103. (with W. McLean and W.L. Wendland)
A fully–discrete trigonometric collocation method, *J. Integral Equations Appl.* **5** (1993), 103–129.

104. (with G. Monegato)
Uniform convergence estimates for a collocation and a discrete collocation method for the generalized airfoil equation, *Contributions in Numerical Mathematics* (ed. R.P. Agarwal), Vol. 2, World Scientific Series in Applicable Analysis, 1993, pp. 285–299.

105. (with W. Dahmen and R. Schneider)
Wavelet approximation methods for pseudodifferential equations I: Stability and convergence, *Math. Zeitschrift* **215** (1994), 583–620.

106. (with W. Dahmen and R. Schneider)
Multiscale methods for pseudodifferential equations, *Recent Advances in Wavelet Analysis* (eds. L.L. Schumaker and G. Webb), Academic Press, 1994, pp. 191–235.

107. (with W. Dahmen and R. Schneider)
Multiscale methods for pseudodifferential equations on smooth closed manifolds, *Wavelets: Theory, Algorithms, and Applications* (eds. C.K. Chui, L. Montefusco, and L. Puccio), Academic Press, 1994, pp. 385–424.

108. (with W. Dahmen and R. Schneider)
Wavelets zur schnellen Lösung von Randintegralgleichungen und angewandte harmonische Analysis, *ZAMM* **74** (1994), T 505–T 507.

109. Wavelet approximation methods for integral and pseudodifferential equations, *Problems and Methods in Mathematical Physics* (eds. L. Jentsch and F. Tröltzsch), Teubner–Texte zur Mathematik, Vol. 134, Teubner–Verlag, Stuttgart, Leipzig, 1994, pp. 148–164.

110. (with G. Mastroianni)
Some nodes matrices appearing in the numerical analysis for singular integral equations, *BIT* **34** (1994), 120–128.

111. (with W. Dahmen, B. Kleemann and R. Schneider)
A multiscale method for the double layer potential equation on a polyhedron, *Advances in Computational Mathematics* (eds. H.P. Dikshit and C.A. Micchelli), World Scientific, Singapore, 1994, pp. 15–57.

112. (with J. Saranen)
A fully discrete approximation method for the exterior Neumann problem of the Helmholtz equation, *Z. Anal. Anw.* **13** (1994), 683–695.

113. (with D. Elliott)
An algorithm for the approximate solution of integral equations of Mellin type, *Numer. Math.* **70** (1995), 427–452.

114. (with R. Duduchava)
On the approximation of singular integral equations by equations with smooth kernels, *Integral Equations Operator Theory* **21** (1995), 224–237.

115. (with B.N. Khoromskij)
Multilevel preconditioning on the refined interface and optimal boundary solvers for the Laplace equation, *Adv. Comput. Math.* **4** (1995), 331–355.

116. (with J. Elschner and I.H. Sloan)
The qualocation method for Symm's integral equation on a polygon, *Math. Nachr.* **177** (1996), 81–108.

117. (with W. McLean)
Boundary element collocation methods using splines with multiple knots, *Numer. Math.* **74** (1996), 419–451.

118. (with G. Bruckner and G. Vainikko)
Error bounds of discretization methods for boundary integral equations with noisy data, *Appl. Anal.* **63** (1996), 25–37.

119. (with B. Kleemann, W. Dahmen and R. Schneider)
Multiscale methods for pseudodifferential equations, *ZAMM* **76** (1996), Suppl. 1, 7–10.

120. (with Yu. Demjanovich, A. Koshelev and G. Leonov)
S. G. Mikhlin (1908–1990), *Math. Nachr.* **177** (1996), 5–7.

121. (with W. McLean)
Boundary element collocation methods with reduced inter–element smoothness, *IABEM Symposium on Boundary Integral Methods for Nonlinear Problems* (eds. L. Morino and W.L. Wendland), Kluwer Academic Publishers, 1997, pp. 155–160.

122. Boundary element methods using splines and multiwavelets, *Proceedings of the Annual Meeting of the Japan Mathematical Society*, Applied Mathematics, Matsumoto, 1997, pp. 94–103.

123. (with G. Chiocchia and D. Tordella)
The lifting line equation for a curved wing in oscillatory motion, *ZAMM* **77** (1997), 295–315.

124. (with W. Dahmen, B. Kleemann and R. Schneider)
Multiscale methods for the solution of the Helmholtz and Laplace equations, *Boundary Element Topics* (ed. W.L. Wendland), Springer, Berlin, 1997, 189–219.

125. (with S. Okada)
On the solution of the generalized airfoil equation, *J. Integral Equations Appl.* **9** (1997), 71–98.

126. (with S.V. Pereverzev)
A discretization of Volterra integral equations of the third kind with weakly singular kernels, *J. Inverse and Ill–Posed Problems* **5** (1997), 565–577.

127. (with D. Eidus, R. Khvoles, G. Kresin, E. Merzbach, T. Shaposhnikova, P. Sobolevskij and M. Solomiak)
Mathematical work of Vladimir Maz'ya (On the occasion of his 60th birthday), *Funct. Differ. Equ.* **4** (1997), 3–11.

128. Tricomi's composition formula and the analysis of multiwavelet approximation methods for boundary integral equations, *Proceedings of the International Conference "Tricomi's Ideas and Contemporary Applied Mathematics"*, Rome and Turin, 1997, Atti dei Convegni Lincei, Vol. 147, Accademia Nazionale dei Lincei, Rome, 1998, pp. 73–91.

129. (with B.N. Khoromskij)
Fast computations with the harmonic Poincaré–Steklov operators on nested refined meshes, *Adv. Comput. Math.* **8** (1998), 111–135.

130. (with J. Schult)
Multiwavelet approximation methods for pseudodifferential equations on curves. Stability and convergence analysis, *Adv. Comput. Math.* **9** (1998), 145–171.

131. (with J. Schult)
Approximation and commutator properties of projections onto shift–invariant subspaces and applications to boundary integral equations, *J. Integral Equations Appl.* **10** (1998), 417–444.

132. (with J. Cheng and M. Yamamoto)
Local estimation for an integral equation of first kind with analytic kernel, *J. Inverse and Ill–Posed Problems* **6** (1998), 115–126.

133. (with S.V. Pereverzev)
On the characterization of self–regularization properties of a fully discrete projection method for Symm's integral equation, Preprint No. 394, Weierstrass Institute Berlin, 1998, *J. Integral Equations Appl.*, to appear.

134. (with M. Yamamoto)
Discretization methods for the Lavrent'ev regularization, this Volume.

Dissertations directed by Siegfried Prößdorf

1. Bernd Silbermann (TU Chemnitz 1970):
   Einige Fragen zur Theorie singulärer Integraloperatoren nicht normalen Typs.

2. Walter Mach (TU Chemnitz 1971):
   Zur numerischen Lösung von Wiener–Hopf–Gleichungen.

3. Jürgen Schulz (TU Chemnitz 1972):
   Ein Näherungsverfahren zur Lösung eindimensionaler singulärer Integralgleichungen nicht normalen Typs.

4. Sybille Meyer (TU Chemnitz 1973):
   Einige Beiträge zur Theorie singulärer Integralgleichungen nicht normalen Typs.

5. Michael Lorenz (TU Chemnitz 1974):
   Über mehrdimensionale singuläre Integraloperatoren und Pseudomultiplikationsoperatoren nicht normalen Typs.

6. Wolfgang Sprößig (TU Chemnitz 1974):
   Über die Regularisierung eines Systems zweidimensionaler singulärer Integralgleichungen, dessen Symbol endlich viele Nullstellen ganzzahliger Ordnung besitzt.

7. Johannes Elschner (TU Chemnitz 1976):
   Entartete gewöhnliche Differentialoperatoren in Räumen differenzierbarer Funktionen.

8. Achim Voigtländer (TU Chemnitz 1977):
   Über Systeme paariger Wiener–Hopf–Gleichungen nicht normalen Typs.

9. Gunther Schmidt (Weierstraß–Institut Berlin 1981):
   Über die Konvergenz von Kollokationsverfahren zur näherungsweisen Lösung singulärer Integralgleichungen.

10. Andreas Pomp (Weierstraß–Institut Berlin 1982):
    Zur Konvergenz des Reduktionsverfahrens für Wiener–Hopfsche Gleichungen.

11. Andreas Rathsfeld (Weierstraß–Institut Berlin 1990):
Stark elliptische singuläre Integralgleichungen vom Cauchy–Typ und Splineverfahren zur numerischen Lösung.

12. Kathrin Bühring (FU Berlin 1995):
Quadrature methods for the Cauchy singular integral equation on curves with corner points and for the hypersingular integral equation on the interval.

13. Jörg Schult (TU Chemnitz 1998):
Multiwavelet–Approximationsmethoden für periodische Pseudodifferentialgleichungen.

Operator Theory:
Advances and Applications, Vol. 121

On the theory of a singular Vekua system

HEINRICH BEGEHR, DAO-QING DAI

We study the solvability of the Riemann-Hilbert problem for a singular Vekua system. For the number of continuous solutions, we shall show that it depends not only on the index but also on the location and type of the singularities, moreover it does not depend continuously on the coefficients of the equation. These suggest essential difficulties to obtain a general theory for singular Vekua systems.

1. Introduction

For the first order elliptic system of equations

$$u_{\bar{z}} = A(z)u + B(z)\bar{u}, \tag{1.1}$$

with regular coefficients $A, B \in L^p(\overline{\Omega})(p > 2)$ when Ω is a bounded domain in the complex plane $\mathbb{C}$ or $A, B \in L^{p,2}(\mathbb{C})$, function theory, boundary value problems and their generalizations were investigated extensively over the past years, see, e.g. [1], [7].

In shell theory, elliptic systems of equations with singular coefficients were encountered [8]. Many results appeared afterwards, for more references, see [2, 3, 4, 5, 6] and the references therein.

Let Ω be the unit disk $\{z : |z| < 1\}$ in the complex plane. In this paper we consider a singular Vekua system

$$w_{\bar{z}} = \frac{\mu}{\bar{z} - \bar{z}_0} w + aw + b\bar{w}, \tag{1.2}$$

where μ is a constant, $z_0 \in \overline{\Omega}$. We shall consider the following Riemann–Hilbert problem for equation (1.2):

Problem RH Find a regular solution $w(z)$ of (1.2) in Ω, continuous in $\overline{\Omega}$ and $w_{\bar{z}}, w_z \in L^p(\overline{\Omega})$ for some $p > 2$, satisfying the condition

$$\operatorname{Re}[\overline{G(z)}w(z)] = g(z), \quad z \in \partial\Omega, \tag{1.3}$$

on the boundary $\partial\Omega$, where G and g are given functions on $\partial\Omega$.

Throughout this paper, we assume that
1) $a, b \in L^p(\Omega)(p > 2)$;
2) $G, g \in C^\alpha(\partial\Omega)(1/2 < \alpha < 1)$, $G \neq 0$ and $\kappa = \text{Ind}_{\partial\Omega} G = \frac{1}{2\pi}\Delta_{\partial\Omega} \arg G$ is an

integer;
3) μ is a real constant.

The motivation for using the model equation (1.2) is that:
1) It can be used to reveal for the first time several phenomena that do not occur in the regular coefficients case.
2) Our results can be used to display flaws of results in the literature.

We shall show that the number of continuous solutions of the problem **RH** depends on the constant μ and the location of the singular point z_0.

The organization of this paper is as follows: in Section 2, we give some preliminary results, in Section 3 we study the case where $z_0 \in \Omega$. The results are applied to several models existing in literature, which show necessity for us to study the model (1.2). In Section 4 we assume $z_0 \in \partial\Omega$. For the solvability of the problem **RH**, we have the necessary condition

$$g(z_0) = 0,$$

see (4.4) below, which is not needed when $z_0 \in \Omega$.

2. Preliminaries

In this section we collect some results which will be used later. Consider the normal boundary condition

$$\operatorname{Re}[\overline{G(z)}u(z)] = g(z), \quad z \in \partial\Omega, \tag{2.1}$$

and the non–normal Riemann–Hilbert boundary condition

$$\operatorname{Re}[\overline{G(z)}(z - z_0)w(z)] = g(z), \quad z \in \partial\Omega. \tag{2.2}$$

For the problem (1.1), (2.1), we seek a solution $u \in W^1_{p_2}(\Omega)$ for some $p_2 > 2$. We have

Lemma 2.1. *For the problem (1.1), (2.1), we have*
1) When $\kappa \geq 0$, it is always solvable and the corresponding homogeneous problem, i.e. $g(z) = 0$, has $2\kappa + 1$ linearly independent solutions over the field of real numbers;
2) When $\kappa < 0$, it is solvable if the function g satisfies $2|\kappa| - 1$ solvability conditions.

Proof. See for example [1] and [7]. □

For the problem (1.2),(2.2), we seek a solution $u \in C^0(\overline{\Omega}\backslash\{z_0\})\cap W^1_{p_3}(\Omega_\epsilon)(p_3 > 2)$ such that $u(z) = O(|z - z_0|^{-\beta}), \beta < 1$, where $\Omega_\epsilon = \Omega\backslash\{z : |z - z_0| \leq \epsilon\}$ for a small fixed constant ϵ.

Lemma 2.2. *1) If the problem (1.2) (2.2) is solvable then*

$$g(z_0) = 0. \tag{2.3}$$

2) If the condition (2.3) holds then we have
i) if $\kappa > 0$, the homogeneous problem has 2κ linearly independent solutions over the field of real numbers, the non–homogeneous problem is always solvable.
ii) if $\kappa = 0$, it has a unique solution.
iii) if $\kappa < 0$, the homogeneous problem has only the trivial solution, the non–homogeneous problem is solvable if g satisfies $2|\kappa|$ real solvability conditions.

3) Let u be a solution. We have $u \in C^0(\overline{\Omega}\backslash\{z_0\}) \cap W^1_{p_4}(\Omega_\epsilon)$ for some $p_4 > 2$ and $u(z) = O(|z - z_0|^{\alpha-1})$.

Proof. See [2]. □

We denote by n the integral part of μ

$$\mu = n + \lambda,$$

with $0 \leq \lambda < 1$ and set

$$w_0(z) = \begin{cases} \dfrac{(\overline{z} - \overline{z_0})^n}{(z - z_0)^{n-1}}, & \lambda = 0, \\ \dfrac{(\overline{z} - \overline{z_0})^n |z - z_0|^{2\lambda}}{(z - z_0)^n}, & 0 < \lambda \leq \dfrac{1}{2}, \\ \dfrac{(\overline{z} - \overline{z_0})^n |z - z_0|^{2\lambda}}{(z - z_0)^{n+1}}, & \dfrac{1}{2} < \lambda < 1. \end{cases}$$

For the regularity of $w_0(z)$, through direct calculation we have the following results.

Lemma 2.3. *Let the function w_0 be defined as above. Then w_0 is continuous on $\overline{\Omega}$. Moreover, we have*
1) $w_0 \in W^1_q(\Omega), \forall q \geq 2$ if $\lambda = 0$,
2) $w_0 \in W^1_q(\Omega), q < 2/(1 - 2\lambda)$ if $0 < \lambda \leq 1/2$,
3) $w_0 \in W^1_q(\Omega), q < 1/(1 - \lambda)$ if $1/2 < \lambda < 1$.

3. The case $|z_0| < 1$

In this section, we assume that the singularity point $z_0 \in \Omega$. We introduce a function $\sigma : \mathbb{R} \to \mathbb{Z}$ by

$$\sigma(\mu) = \begin{cases} 2n - 1, & \lambda = 0, \\ 2n, & 0 < \lambda \leq \frac{1}{2}, \\ 2n + 1, & \frac{1}{2} < \lambda < 1. \end{cases}$$

Then $\sigma(\mu)$ is not continuous at integers and half integers.

For the solvability of the problem **RH**, we have

Theorem 3.1. *Let $\kappa_0 = \kappa + \sigma$. Then we have*
1) when $\kappa_0 \geq 0$, the problem **RH** *is always solvable and the corresponding homogeneous problem has $2\kappa_0 + 1$ linearly independent solutions over the field of real numbers;*
2) when $\kappa_0 < 0$, the problem **RH** *is solvable if the function g satisfies $2|\kappa_0| - 1$ solvability conditions.*

Proof. By virtue of Lemma 2.3 we have $w_0 \in W^1_{p_1}(\Omega)$ for some $p_1 > 2$, which depends on λ as shown in Lemma 2.3.

By means of the transform

$$w(z) = w_0(z)\varphi(z), \tag{3.1}$$

the problem **RH** becomes

$$\begin{cases} \varphi_{\bar{z}} = a\varphi + b_0\overline{\varphi} \text{ in } \Omega, \\ \mathrm{Re}[\overline{G_0(z)}\varphi(z)] = g(z) \text{ on } \partial\Omega, \end{cases} \tag{3.2}$$

where $b_0(z) = b(z)\overline{w_0(z)}/w_0(z), G_0(z) = \overline{w_0(z)}G(z)$.

We look for solutions of (3.2) such that $\varphi \in C^0(\overline{\Omega}\backslash\{z_0\}) \cap W^1_{p_6}(\Omega_\epsilon)$ for some $p_6 > 2$ and $\varphi(z) = O(|z - z_0|^{-\alpha_0})$, where $\alpha_0 < 1$ if $\lambda = 0$, $\alpha_0 < 2\lambda$ if $0 < \lambda \leq 1/2$, $\alpha_0 < 2\lambda - 1$ if $1/2 < \lambda < 1$.

Noticing that when $w \in C^0(\overline{\Omega})$, $w_{\bar{z}} \in L^{p_7}(\Omega)$ for $p_7 > 2$, from the equation (1.2) we must have $w(z_0) = 0$. Then it can be shown that the problem **RH** is equivalent to the problem (3.2). Furthermore, by virtue of similarity principle, the solution of (3.2) in the above sence has only removable sigularity in $\overline{\Omega}$. Hence continuous solutions can be sought and Lemma 2.1 can be applied to (3.2).

Since $b_0(z) \in L^p(\Omega)$ and $G_0 \in C^\alpha(\partial\Omega)$, the problem (3.2) is a normal problem and its index is

$$\kappa_0 = \mathrm{Ind}_{\partial\Omega} G_0 = \kappa + \sigma.$$

By virtue of Lemma 2.1, the problem (3.2) is always solvable in $W^1_{p_8}(\Omega)(p_8 > 2)$ when $\kappa_0 \geq 0$ and the corresponding homogeneous problem has $2\kappa_0 + 1$ linearly independent solutions in $W^1_{p_8}(\Omega)$, over the field of real numbers. Moreover, when $\kappa_0 < 0$ the problem (3.2) is solvable if and only if the function g satisfies $2|\kappa_0| - 1$ solvability conditions.

Hence the function $w(z)$ defined by (3.1) solves the problem **RH**. □

Remark 3.2. From Theorem 3.1, it follows that the number of solutions to the problem **RH** does not depend continuously on μ when $\mu \to n^+, (n + \frac{1}{2})^+, n \in \mathbb{Z}$. This shows a difference between the singular case and the regular case, cf. [7], (ch. 3, section 12).

Remark 3.3. In [6] (p. 72), the following problem

$$\begin{cases} U_{\bar{z}} + [b(0)/2\bar{z}]\,U = 0, |z| < R & (4.3) \\ \mathrm{Re}(ie^{i\varphi}U) = 0, |z| = R & (4.4) \end{cases}$$

was introduced as a conjugate problem. The solutions were sought in the class of functions continuous outside $z = 0$ and admitting no more than a first order singularity at $z = 0$. The author found that (4.3), (4.4) had only the non-trivial solution $U(z) = \frac{1}{z}|z|^{-b(0)}$.

In fact, the number of solutions depends on the constant $b(0)$. In Theorem 3.1 we are dealing with continuous solutions. This can be modified in seeking solutions in the above sense also. For example, when $-2 < b(0) < -1$, the problem (4.3), (4.4) has three linearly independent solutions

$$i|z|^{-b(0)}(1 - z^{-2}), z^{-1}|z|^{-b(0)}, |z|^{-b(0)}(1 + z^{-2}),$$

having at most a first order singularity at $z = 0$.

Remark 3.4. In [4] the following boundary value problem was investigated

$$\begin{cases} w_{\bar{z}} + \dfrac{A(z)}{|z|}w + \dfrac{B(z)}{|z|}\overline{w} = F, |z| < 1, \\ \mathrm{Re}[z^{-\kappa}w(z)] = g(z), |z| = 1, \end{cases} \tag{3.3}$$

where $A, B, F \in L^q(\Omega)(q > 2)$.

Under the assumption that the norms of $A(z)$ and $B(z)$ are sufficiently small ([4] (p. 357) condition (1.2)), Theorem 3 of [4] says that the homogeneous problem has $2\kappa - 1$ linearly independent solutions, over the field of real numbers when $\kappa > 0$.

In [5], problem (3.3) was generalized to several singularities.

Let

$$A(z) = \frac{\mu|z|}{\bar{z}}, B(z) = 0.$$

We choose $1/2 > \mu > 0$ such that $A(z)$ satisfies the condition (1.2) of [4]. Then $n = 0$, $0 < \lambda(< \frac{1}{2})$, and $\sigma = 0$. By virtue of Theorem 3.1, $\kappa_0 = \kappa$. Hence the homogeneous problem (3.3) has $2\kappa + 1$ linearly independent solutions. However, if we let $-1/2 < \mu < 0$, then $n = -1, \frac{1}{2} < \lambda$ and $\sigma = -1$. By virtue of Theorem 3.1, $\kappa_0 = \kappa - 1$ and the corresponding homogeneous problem (3.3) has $2\kappa - 1$ linearly independent solutions.

This shows that it does not suffice to consider modules of its coefficients when studying singular equations.

4. The case $|z_0| = 1$

In this section, we assume that the singularity point z_0 is located on the boundary $\partial\Omega$, i.e., $|z_0| = 1$.

With the transform

$$w(z) = w_0(z)\varphi(z), \tag{4.1}$$

the problem **RH** becomes

$$\begin{cases} \varphi_{\bar{z}} = a\varphi + b_1\overline{\varphi}, & |z| < 1, \\ \text{Re}[\overline{G(z)}w_0(z)\varphi(z)] = g(z), & |z| = 1, \end{cases} \tag{4.2}$$

where $b_1(z) = b(z)\,\overline{w_0(z)}/w_0(z) \in L^p(\Omega)$.

Since we are seeking continuous solutions of the Problem **RH**, through the transform (4.1), the function φ can have a singularity at z_0

$$\varphi(z) = O(|z - z_0|^{-\beta}), \tag{4.3}$$

where β satisfies

$$\begin{array}{ll} \beta < 1, & \text{if} \quad \lambda = 0, \\ \beta < 2\lambda, & \text{if} \quad 0 < \lambda \le \frac{1}{2}, \\ \beta < 2\lambda - 1, & \text{if} \quad \frac{1}{2} < \lambda < 1. \end{array}$$

From (4.2) and (4.3), we have

Lemma 4.1. *If the problem (3.3) is solvable then*

$$g(z_0) = 0. \tag{4.4}$$

We now consider three cases.

case (i): $\lambda = 0$

From (4.2) we get

$$\begin{cases} \varphi_{\bar{z}} = a\varphi + b_1\overline{\varphi}, |z| < 1, \\ \text{Re}[\overline{G_1(z)}(z - z_0)\varphi(z)] = g(z), |z| = 1, \end{cases} \tag{4.5}$$

where $G_1(z) = (-1)^n \overline{z_0}^n z^n G(z)$.

Since $G_1(z) \neq 0$ on $\partial\Omega$, its index is given by

$$\kappa_1 = \kappa + n.$$

From Lemma 2.2 we get

Lemma 4.2. *Let the condition (4.4) hold. Then we have*
1) When $\kappa_1 > 0$, the homogeneous problem (4.5) has $2\kappa_1$ linearly independent solutions over the field of real numbers, the non-homogeneous problem is always solvable.
2) When $\kappa_1 = 0$, it has a unique solution.
3) When $\kappa_1 < 0$, the homogeneous problem has only the trivial solution, the problem (4.5) is solvable if the function g satisfies $2|\kappa_1|$ real solvability conditions.

Moreover, for the solution we have

$$\varphi(z) = O(|z - z_0|^{\alpha-1}).$$

case (ii): $0 < \lambda \leq \frac{1}{2}$

We have from (4.2) the normal Riemann–Hilbert problem

$$\begin{cases} \varphi_{\bar{z}} = a\varphi + b_1\overline{\varphi}, |z| < 1, \\ \mathrm{Re}[\overline{G_2(z)}\varphi(z)] = g_2(z), |z| = 1, \end{cases} \tag{4.6}$$

where $G_2(z) = (-1)^n z_0^n z^n G(z)$, $g_2(z) = g(z)\,|z - z_0|^{-2\lambda}$. Its index is

$$\kappa_2 = \kappa + n.$$

From properties of Cauchy type integrals we have

Lemma 4.3. *Let the condition (4.4) hold. Then we have*
1) When $\kappa_2 \geq 0$, *the homogeneous problem (4.6) has* $2\kappa_2 + 1$ *linearly independent solutions over the field of real numbers, the non-homogeneous problem is always solvable.*
2) When $\kappa_2 < 0$, *the homogeneous problem (4.6) has only the trivial solution; the non-homogeneous problem is solvable if the function* g *satisfies* $1 - 2\kappa_2$ *solvability conditions.*

Moreover, for the solution we have when $\alpha - 2\lambda < 0$

$$\varphi(z) = O(|z - z_0|^{\alpha-2\lambda});$$

otherwise φ *is continuous at* z_0.
case (iii): $\frac{1}{2} < \lambda < 1$

From (4.2) we get

$$\begin{cases} \varphi_{\bar{z}} = a\varphi + b_1\overline{\varphi}, |z| < 1, \\ \mathrm{Re}[\overline{G_3(z)}(z - z_0)\varphi(z)] = g_3(z), |z| = 1, \end{cases} \tag{4.7}$$

where $G_3(z) = (-1)^{n+1} z_0^{n+1} z^{n+1} G(z)$, $g_3(z) = g(z)\,|z - z_0|^{-2(\lambda-1)}$. Its index is

$$\kappa_3 = \kappa + n + 1.$$

From Lemma 2.2 we have

Lemma 4.4. *Let the condition (4.4) hold. Then we have*
1) When $\kappa_3 > 0$, *the homogeneous problem (4.7) has* $2\kappa_3$ *linearly independent solutions over the field of real numbers, the non-homogeneous problem is always solvable.*
2) When $\kappa_3 = 0$, *it has a unique solution.*

3) When $\kappa_3 < 0$, the homogeneous problem has only the trivial solution, the problem (4.7) is solvable if the function g satisfies $2|\kappa_3|$ real solvability conditions.
Moreover, for the solution we have when $\lambda > (1+\alpha)/2$

$$\varphi(z) = O(|z - z_0|^{1+\alpha-2\lambda}),$$

otherwise φ is continuous at z_0.

For the Problem **RH**, from Lemmas 4.1 – 4.4 and (4.1) we finally get

Theorem 4.5. *The Problem* **RH** *is solvable only if the condition (4.4) holds.*
Conversely if (4.4) is satisfied then we have the following:

case (i) If $\kappa_1 > 0$, the homogeneous problem has $2\kappa_1$ linearly independent solutions over the field of real numbers, the non–homogeneous problem is always solvable; if $\kappa_1 = 0$, it has a unique solution; when $\kappa_1 < 0$, the homogeneous problem has only the trivial solution, the non–homogeneous problem is solvable when the function g satisfies $2|\kappa_1|$ real solvability conditions.

case (ii) If $\kappa_2 \geq 0$, the homogeneous problem has $2\kappa_2 + 1$ linearly independent solutions over the field of real numbers, the non–homogeneous problem is always solvable; if $\kappa_2 < 0$, the homogeneous problem has only the trivial solution; the non–homogeneous problem is solvable if the function g satisfies $1 - 2\kappa_2$ solvability conditions.

case (iii) If $\kappa_3 > 0$, the homogeneous problem has $2\kappa_3$ linearly independent solutions over the field of real numbers, the non–homogeneous problem is always solvable; if $\kappa_3 = 0$, it has a unique solution; when $\kappa_3 < 0$, the homogeneous problem has only the trivial solution, the non–homogeneous problem is solvable when the function g satisfies $2|\kappa_3|$ real solvability conditions.

Acknowledgements

The second–named author is partly supported by the Alexander von Humboldt Stiftung and the National Science Foundation of China.

References

[1] H. Begehr, Complex analytic methods for partial differential equations. An Introductory Text. World Scientific, Singapore etc., 1994.

[2] H. Begehr & D. Q. Dai, On continuous solutions of a generalized Cauchy-Riemann system with more than one singularity, Preprint, 1999.

[3] M. Reissig & A. Timofeev, Special Vekua equations with singular coefficients, Preprint, Bergakademie Freiburg, 1998.

[4] A. Tungatarov, On the theory of Carleman-Vekua equation with a singular point, Russian Acad. Sci. Sb. Math. 78(1994), 357-365.

[5] A. Tungatarov, Continuous solutions of the generalized Cauchy-Riemann system with a finite number of singular points, Mat. Zametki 56(1994), 105-115 (in Russian).

[6] Z. D. Usmanov, Generalized Cauchy-Riemann systems with a singular point, Longman, Harlow, 1997.

[7] I. N. Vekua, Generalized analytic functions, Pergamon, Oxford, 1962.

[8] I. N. Vekua, On the class of elliptic systems with singularities, Proc. of the International Conference on Functional Analysis and Related Topics, Tokyo, April 1969.

Freie Universität Berlin
I. Math. Institut
Arnimallee 3
14195 Berlin, Germany

Zhongshan University
Department of Mathematics
510275 Guangzhou
China

1991 Mathematics Subject Classification: Primary 30G20; Secondary 30E25, 35Q15

Submitted: 17.4.2000

Operator Theory:
Advances and Applications, Vol. 121

Singular integral operators with complex conjugation from the viewpoint of pseudodifferential operators

A. Böttcher, Yu.I. Karlovich, V.S. Rabinovich

To the Memory of Siegfried Prössdorf

We study the algebra $\mathcal{A}$ generated by singular integral operators $aI + bS$ and the operator of complex conjugation V on the weighted Lebesgue space $L^p(\Gamma, w)$. Our approach is based on transforming the operators in $\mathcal{A}$ locally into Mellin pseudodifferential operators. By having recourse to the Fredholm and index theory of the latter class of operators, we can establish Fredholm criteria and index formulas for operators in $\mathcal{A}$ with slowly oscillating coefficients in the case of slowly oscillating composed curves Γ and slowly oscillating Muckenhoupt weights w. We are in particular able to consider curves with whirl points, in which case massive local spectra may emerge even for constant coefficients and weights.

1. Introduction

Let Γ be an oriented and rectifiable curve in the plane and let w be a weight on Γ, that is, suppose $w := \Gamma \to [0, +\infty]$ is a measurable function such that $w^{-1}(\{0, +\infty\})$ has zero length measure. For $1 < p < \infty$, we denote by $L^p_N(\Gamma, w)$ the Banach space of all measurable vector-functions $f : \Gamma \to \mathbf{C}^N$ such that

$$\|f\| := \left(\int_\Gamma \|f(t)\|^p_{\mathbf{C}^N} w(t)^p \, |dt| \right)^{1/p} < \infty;$$

the scalar field of $L^p_N(\Gamma, w)$ is $\mathbf{C}$. The same space $L^p_N(\Gamma, w)$ considered over the scalar field $\mathbf{R}$ will be denoted by $RL^p_N(\Gamma, w)$.

In this paper we will assume that Γ is a slowly oscillating composed curve and that w is a slowly oscillating Muckenhoupt weight on Γ. Precise definitions are in Section 2. We remark that the class of curves we admit includes piecewise Lyapunov curves with angles (but not with cusps) and also non-Lipschitz curves with certain whirl points, for example, curves which locally look like the logarithmic star

$$\{0\} \cup \bigcup_{j=1}^{n} \left\{ re^{i(\log r + \mu_j)} : r \in (0, \varepsilon) \right\}, \quad 0 \le \mu_1 < \mu_2 < \ldots < \mu_n < 2\pi.$$

Our assumptions on the curve Γ and the weight w guarantee that the Cauchy singular integral operator S,

$$(Sf)(t) := \lim_{\varepsilon \to 0} \frac{1}{\pi i} \int_{\{\tau \in \Gamma : |\tau - t| \geq \varepsilon\}} \frac{f(\tau)}{\tau - t} \, d\tau \quad (t \in \Gamma),$$

is bounded on $L^p_N(\Gamma, w)$ (as usual, we let the operator S act componentwise on vector-functions).

We let $PSO_{N \times N}(\Gamma)$ stand for the class of all piecewise slowly oscillating matrix-functions $a : \Gamma \to \mathbf{C}^{N \times N}$ (again see Section 2 for the precise definition). Given an essentially bounded matrix function $a : \Gamma \to \mathbf{C}^{N \times N}$, we denote by aI the operator of multiplication by a on $L^p_N(\Gamma, w)$, and in case aI follows another operator, B say, we abbreviate aIB to aB.

Let $\mathcal{L}(X)$ be the Banach algebra of all bounded linear operators on a Banach space X. We denote by $\mathcal{C}_{N \times N}$ the smallest closed subalgebra of $\mathcal{L}(L^p_N(\Gamma, w))$ which contains the set

$$\{aI : a \in PSO_{N \times N}(\Gamma)\} \cup \{S\}.$$

The Fredholm and index theory of operators in $\mathcal{C}_{N \times N}$ is now well understood: see Gohberg and Krupnik's book [11], Roch and Silbermann's report [24] for piecewise continous coefficients, piecewise Lyapunov curves, and power weights and see the recent works [1], [2], [3], [4], [5], [22] (and the many references in these sources) for more general coefficients, curves, and weights.

Now let V be the operator of complex conjugation:

$$(Vf)(t) := \overline{f(t)} \quad (t \in \Gamma).$$

Again we think of V as acting componentwise on vector-functions. Clearly, V is sesquilinear on $L^p_N(\Gamma, w)$ and linear on $RL^p_N(\Gamma, w)$. We denote by $\mathcal{A}_{N \times N}$ the smallest closed subalgebra of $\mathcal{L}(RL^p_N(\Gamma, w))$ which contains the algebra $\mathcal{C}_{N \times N}$ and the operator V. Operators in $\mathcal{A}_{N \times N}$ emerge in many applications (see, e.g., [8], [13], [14], [18], [25]), and the Fredholm and index theory for these operators is much more difficult and therefore less complete than the corresponding theory for operators in $\mathcal{C}_{N \times N}$. In the case of piecewise continuous coefficients, piecewise Lyapunov curves Γ without cusps, and power weights w, Fredholm criteria and index formulas for operators in $\mathcal{A}_{N \times N}$ were established by Costabel [7], Nyaga [16], [17], and Duduchava and Latsabidze [9], [12]; the theory of these operators was completed in Roch and Silbermann's book [24]. Only recently were Duduchava, Latsabidze, and Saginashvili [10] able to extend these results to piecewise Lyapunov curves with so-called cusps of the first order.

In the present paper, we do not admit cusps and their twisted analogues. However, we allow the curve, the weight, and the coefficients to be slowly oscillating, and we will establish Fredholm and index results for operators in $\mathcal{A}_{N \times N}$ under these quite general assumptions.

Following the common approach, we put

$$J = \frac{1}{2}\begin{pmatrix} I & V \\ iI & -iV \end{pmatrix}, \quad J^{-1} = \begin{pmatrix} I & -iI \\ V & -iV \end{pmatrix},$$

and, for $A \in \mathcal{L}(RL^p_N(\Gamma, w))$, we set

$$\hat{A} = J^{-1}\begin{pmatrix} A & 0 \\ 0 & A \end{pmatrix} J =: \begin{pmatrix} A_{11} & A_{12} \\ A_{21} & A_{22} \end{pmatrix}.$$

Since

$$\hat{aI} = \begin{pmatrix} aI & 0 \\ 0 & \overline{a}I \end{pmatrix}, \quad \hat{S} = \begin{pmatrix} S & 0 \\ 0 & VSV \end{pmatrix}, \quad \hat{V} = \begin{pmatrix} 0 & I \\ I & 0 \end{pmatrix},$$

we see that $\hat{A} \in \mathcal{L}(L^p_{2N}(\Gamma, w))$ for every $A \in \mathcal{A}_{N\times N}$ and that $A \in \mathcal{A}_{N\times N}$ is Fredholm on $RL^p_N(\Gamma, w)$ if and only if $\hat{A}$ is Fredholm on $L^p_{2N}(\Gamma, w)$, in which case $\operatorname{Ind} A = \operatorname{Ind} \hat{A}$. Consequently, if VSV would belong to $\mathcal{C}_{N\times N}$, the Fredholm theory of operators $\mathcal{A}_{N\times N}$ could be settled by having recourse to the Fredholm theory of operators in $\mathcal{C}_{2N\times 2N}$. In case Γ is a piecewise Lyapunov curve without cusps and w is a power weight, one can show that VSV indeed lies in $\mathcal{C}_{N\times N}$ (see, e.g., [7] or [24, Proposition 10.1]). The real problems arise if $VSV \notin \mathcal{C}_{N\times N}$, and this is in general the case for the curves and weights studied in this paper.

We here proceed as follows. Given $A \in \mathcal{A}_{N\times N}$, consider the linear operator $\hat{A}$ on $L^p_{2N}(\Gamma, w)$. The operator entries A_{jk} $(j, k = 1, 2)$ of $\hat{A}$ belong to the smallest closed subalgebra $\mathcal{B}_{N\times N}$ of $\mathcal{L}(L^p_N(\Gamma, w))$ containing the set

$$\{aI : a \in PSO_{N\times N}(\Gamma)\} \cup \{S, VSV\}.$$

Let $t_0 \in \Gamma$ and suppose that, in a neighborhood of t_0, Γ is comprised of n simple arcs that have only the point t_0 in common. We associate a Mellin pseudodifferential operator $OP(\sigma_B)$ on $L^p_{nN}(\mathbf{R}_+, dr/r)$ with each operator $B \in \mathcal{B}_{N\times N}$, where

$$\big(OP(\sigma_B)f\big)(r) := \int_{\mathbf{R}} \frac{d\lambda}{2\pi} \int_{\mathbf{R}_+} \sigma_B(r, \lambda) f(\varrho) \Big(\frac{r}{\varrho}\Big)^{i\lambda} \frac{d\varrho}{\varrho} \quad (r \in \mathbf{R}_+).$$

If $B = aI$, then $OP(\sigma_B)$ is nothing but a multiplication operator. The explicit form of the essential part of $\sigma_S(r, \lambda)$ is known from previous work, and because we will see that

$$OP(\sigma_{VSV}) = V\, OP(\sigma_S)\, V,$$

it follows immediately from the definition of $OP(\sigma_B)$ that

$$\sigma_{VSV}(r, \lambda) = \overline{\sigma_S(r, -\lambda)}.$$

Thus, although VSV need not belong to $\mathcal{C}_{N\times N}$, the operator $OP(\sigma_{VSV})$ is a Mellin pseudodifferential operator together with $OP(\sigma_S)$.

We construct the correspondence $B \mapsto OP(\sigma_B)$ so that $A \in \mathcal{A}_{N\times N}$ is locally Fredholm at $t_0 \in \Gamma$ if and only if the Mellin pseudodifferential operator

$$\begin{pmatrix} OP(\sigma_{A_{11}}) & OP(\sigma_{A_{12}}) \\ OP(\sigma_{A_{21}}) & OP(\sigma_{A_{22}}) \end{pmatrix} \in \mathcal{L}(L^p_{2nN}(\mathbf{R}_+, dr/r))$$

is locally Fredholm at the origin. Our assumptions on the coefficients aI, the curve Γ, and the weight w ensure that the Mellin pseudodifferential operators we obtain have slowly oscillating symbols. The theory of such operators was elaborated by one of the authors in [19], [20], [21], [22], [23]. Using these results and applying standard localization and index separation techniques, we arrive at Fredholm criteria and index formulas for operators in $\mathcal{A}_{N\times N}$ on $RL^p_N(\Gamma, w)$.

2. Curves, weights, and coefficients

A set $\gamma \subset \mathbf{C}$ is called a *simple smooth arc* if there exists a homeomorphism $\varphi : [0,1] \to \gamma$ such that $\varphi \in C^\infty(0,1)$ and $\varphi'(r) \neq 0$ for all $r \in (0,1)$. The points $\varphi(0)$ and $\varphi(1)$ are called the *endpoints* of γ. We refer to a set $\Gamma \subset \mathbf{C}$ as a *composed curve* if $\Gamma = \bigcup_{k=1}^K \Gamma_k$ where $\Gamma_1, \ldots, \Gamma_K$ are oriented and rectifiable simple smooth arcs each pair of which has at most endpoints in common. A *node* of Γ is a point which is endpoint of at least one of the arcs $\Gamma_1, \ldots, \Gamma_K$. The set of all nodes of Γ will be denoted by F.

A C^∞ function $f : (0, \varepsilon) \to \mathbf{C}$ is said to be *slowly oscillating at the origin* if

$$\sup_{r\in(0,\varepsilon)} \left| \left(r\frac{d}{dr} \right)^k f(r) \right| < \infty \ \text{ for } \ k = 0, 1, 2, \ldots \tag{2.1}$$

and

$$\lim_{r\to 0} |rf'(r)| = 0. \tag{2.2}$$

We remark that (2.1) and (2.2) imply that actually

$$\lim_{r\to 0} \left| \left(r\frac{d}{dr} \right)^k f(r) \right| = 0 \ \text{ for } \ k = 1, 2, 3, \ldots$$

To have an example, notice that if $f(r) = g(\log(-\log r))$ where $g \in C^\infty(\mathbf{R})$ and g as well as all derivates of g are bounded, then f is slowly oscillating at the origin.

Suppose Γ is a composed curve and $t_0 \in F$. We say that Γ is *slowly oscillating at t_0* if there is an $\varepsilon > 0$ such that the portion

$$\Gamma(t_0, \varepsilon) := \{\tau \in \Gamma : |\tau - t| < \varepsilon\}$$

is of the form

$$\Gamma(t_0, \varepsilon) = \{t_0\} \cup \gamma_1 \cup \ldots \cup \gamma_n$$

where

$$\gamma_j = \{t_0 + re^{i(\theta(r)+\theta_j(r))} : r \in (0, \varepsilon)\} \quad (j = 1, \ldots, n),$$

$\theta, \theta_1, \ldots, \theta_n$ are real-valued C^∞ functions subject to the additional constraint

$$0 \le m_1 < \theta_1(r) < M_1 < m_2 < \theta_2(r) < M_2 < \ldots < m_n < \theta_n(r) < M_n < 2\pi$$

for all $r \in (0, \varepsilon)$ with certain constants m_j, M_j (that may depend on t_0) and where $r\theta'(r), r\theta_1'(r), \ldots, r\theta_n'(r)$ are slowly oscillating at $r = 0$. Under these assumptions, the functions θ_j $(j = 1, \ldots, n)$ are also slowly oscillating at $r = 0$. For example, if

$$\theta(r) + \theta_j(r) = \delta \log r + \mu_j \quad (j = 1, \ldots, n)$$

with $0 \le \mu_1 < \mu_2 < \ldots < \mu_n < 2\pi$, then the above assumptions are all met. A composed curve that is slowly oscillating at each of its nodes will be referred to as a *slowly oscillating composed curve.*

Let $w : \Gamma \to [0, +\infty]$ be a function that assumes values in $(0, +\infty)$ on $\Gamma \setminus F$ and is C^∞ on $\Gamma \setminus F$. We call w a *slowly oscillating weight at* $t_0 \in F$ if, with the above notation, w is of the form

$$w(t_0 + re^{i(\theta(r)+\theta_j(r))}) = e^{v(r)}, \quad r \in (0, \varepsilon), \ j \in \{1, \ldots, n\},$$

where $rv'(r)$ is slowly oscillating at $r = 0$. If, in addition,

$$-1/p < \liminf_{r\to 0} rv'(r) \le \limsup_{r\to 0} rv'(r) < 1/q,$$

where $1/p + 1/q = 1$, then w is referred to as a slowly oscillating Muckenhoupt weight at t_0. For instance, the weight w arising from

$$v(r) = f\big(\log(-\log r)\big) \log r, \quad r \in (0, \varepsilon)$$

with a bounded function $f \in C^\infty(\mathbf{R})$ all derivates of which are also bounded is slowly oscillating at t_0; in this case we have

$$\liminf_{r\to 0} / \sup rv'(r) = \quad \liminf_{x\to +\infty} / \sup \big(f(x) + f'(x)\big).$$

In case w is a slowly oscillating Muckenhoupt weight at each node of Γ we simply call w a *slowly oscillating Muckenhoupt weight.*

Finally, a function $a : \Gamma \to \mathbf{C}$ is said to be *piecewise slowly oscillating on* Γ, $a \in PSO(\Gamma)$, if a is C^∞ on $\Gamma \setminus F$ and if for each node $t_0 \in F$ we have

$$a(t_0 + re^{i(\theta(r)+\theta_j(r))}) = a_{t_0,j}(r), \ r \in (0, \varepsilon), \ j \in \{1, \ldots, n\}$$

where $a_{t_0,1}(r), \ldots, a_{t_0,n}(r)$ are slowly oscillating at $r = 0$.

Given a set E, we let E_M and $E_{M\times M}$ stand for the column-vectors of length M with components from E and the $M \times M$ matrix-functions with entries in E, respectively.

Throughout what follows we assume that Γ is a slowly oscillating composed curve and that w is a slowly oscillating Muckenhoupt weight on Γ. Taking into account that

$$|d(re^{i\varphi(r)})| = \sqrt{1+(r\varphi'(r))^2}\,dr,$$

it is easily seen that slowly oscillating composed curves are Carleson curves (= Ahlfors-David curves), and it is well known that slowly oscillating Muckenhoupt weights satisfy the so-called Muckenhoupt A_p condition (see, e.g., [2]). Consequently, the Cauchy singular integral operator S is a bounded linear operator on $L^p(\Gamma, w)$.

3. Mellin pseudodifferential operators

For $1 < p < \infty$, let $L^p(\mathbf{R}_+, d\mu)$ be the usual Lebesgue space on $\mathbf{R}_+ = (0, +\infty)$ with respect to the measure $d\mu(r) = dr/r$. Notice that $d\mu$ is the normalized invariant measure on the multiplicative group $\mathbf{R}_+$.

We denote by $\mathcal{E}$ the collection of all functions $a \in C^\infty(\mathbf{R}_+ \times \mathbf{R})$ with the following properties: for each pair (j,k) of nonnegative integers j,k there exists a constant $c_{jk} < \infty$ such that

$$|(r\partial_r)^j \partial_\lambda^k a(r,\lambda)| \le c_{jk}(1+|\lambda|)^{-k} \text{ for all } (r,\lambda) \in \mathbf{R}_+ \times \mathbf{R}$$

and for each nonnegative integer k we have

$$\lim_{r\to 0} \sup_{\lambda\in\mathbf{R}} |(r\partial_r)\partial_\lambda^k a(r,\lambda)|(1+|\lambda|)^k = 0,$$
$$\lim_{r\to \infty} \sup_{\lambda\in\mathbf{R}} |(r\partial_r)\partial_\lambda^k a(r,\lambda)|(1+|\lambda|)^k = 0.$$

For $a \in \mathcal{E}_{M\times M}$, the Mellin pseudodifferential operator $OP(a)$ is formally defined by

$$(OP(a)f)(r) = \int_{\mathbf{R}} \frac{d\lambda}{2\pi} \int_{\mathbf{R}_+} a(r,\lambda) f(\varrho) \left(\frac{r}{\varrho}\right)^{i\lambda} \frac{d\varrho}{\varrho} \quad (r \in \mathbf{R}_+);$$

this makes sense for all f in $[C_0^\infty(\mathbf{R}_+)]_M$, the vector-functions in $[C^\infty(\mathbf{R}_+)]_M$ with compact support. One can show (see, e.g., [15] and [6, Theorem 9]) that $OP(a)$ extends to a bounded operator on $L^p_M(\mathbf{R}_+, d\mu)$. Furthermore, if $a, b \in \mathcal{E}_{M\times M}$ then

$$OP(a)OP(b) - OP(ab) \in \mathcal{K}\big(L^p_M(\mathbf{R}_+, d\mu)\big) \tag{3.1}$$

(see [22]), where $\mathcal{K}(X)$ denotes the set of all compact linear operators on X.

An operator $B \in \mathcal{L}(L^p_M(\mathbf{R}_+, d\mu))$ is said to be *locally Fredholm at the origin* if there exists an $\varepsilon > 0$ and operators $C', C'' \in \mathcal{L}(L^p_M(\mathbf{R}_+, d\mu))$ such that

$$C'B\chi_{(0,\varepsilon)}I - \chi_{(0,\varepsilon)}I,\ \chi_{(0,\varepsilon)}BC'' - \chi_{(0,\varepsilon)}I \in \mathcal{K}\big(L^p_M(\mathbf{R}_+, d\mu)\big),$$

where $\chi_{(0,\varepsilon)}$ is the characteristic function of the interval $(0, \varepsilon)$.

Theorem 3.1. [22] *Let $a \in \mathcal{E}_{M\times M}$. The operator $OP(a)$ is locally Fredholm at the origin if and only if*

$$\liminf_{r\to 0} \inf_{\lambda\in\mathbf{R}} |\det a(r,\lambda)| > 0.$$

Recall that a bounded linear Banach space operator is called a *Fredholm operator* if B is invertible modulo compact operators. In that case B has finite kernel and cokernel dimensions, and the *index* $\operatorname{Ind} B$ is defined as the kernel dimension minus the cokernel dimension. For $R \in (0, \infty)$, let $Q(R)$ be the rectangle

$$Q(R) := \big\{(r,\lambda) \in \mathbf{R}_+ \times \mathbf{R} : 1/R < r < R,\ |\lambda| < R\big\}$$

and let $\partial Q(R)$ be the boundary of $Q(R)$. Given a continuous function

$$\varphi : \partial Q(R) \to \mathbf{C} \setminus \{0\},$$

we denote by $\operatorname{wind} \varphi(\partial Q(R))$ the winding number about the origin of the closed curve described by $\varphi(t)$ as t traces out $\partial Q(R)$ once in the counter-clockwise direction.

Theorem 3.2. [22] *Let $a \in \mathcal{E}_{M\times M}$. The operator $OP(a)$ is Fredholm on the space $L^p_M(\mathbf{R}_+, d\mu)$ if and only if*

$$\lim_{R\to\infty} \inf_{(r,\lambda)\notin Q(R)} |\det a(r,\lambda)| > 0.$$

If $OP(a)$ is Fredholm, then

$$\operatorname{Ind} OP(a) = -\lim_{R\to\infty} \operatorname{wind} \det a\big(\partial Q(R)\big).$$

4. Symbol construction

Let Γ and w be as in Section 2 and fix a point $t_0 \in \Gamma$. Denote the characteristic function of the portion $\Gamma(t_0, \varepsilon)$ by χ_ε. An operator $B \in \mathcal{L}(L^p_M(\Gamma, w))$ is called *locally Fredholm* at t_0 if there exist an $\varepsilon > 0$ and operators $C', C'' \in \mathcal{L}(L^p_M(\Gamma, w))$ such that

$$C'B\chi_\varepsilon I - \chi_\varepsilon I,\ \chi_\varepsilon BC'' - \chi_\varepsilon I \in \mathcal{K}\big(L^p_M(\Gamma, w)\big).$$

The *operators of local type* on $L^p_M(\Gamma, w)$ are the operators $B \in \mathcal{L}(L^p_M(\Gamma, w))$ for which

$$B\varphi I - \varphi B \in \mathcal{K}\big(L^p_M(\Gamma, w)\big) \text{ for all } \varphi \in C(\Gamma).$$

The local principle of Simonenko (see[26] or [2, Section 8.2]) states that an operator of local type is Fredholm on $L^p_M(\Gamma, w)$ if and only if it is locally Fredholm at each point of Γ. The operators aI ($a \in L^\infty_{M\times M}(\Gamma)$) and S are known to be of local type. Since

$$VSV\varphi I - \varphi VSV = V(S\overline{\varphi} I - \overline{\varphi} S)V;$$

it follows that all operators in the algebra $\mathcal{B}_{M\times M}$ are also of local type. We therefore arrive at the following conclusion (recall Section 1).

Theorem 4.1. *An operator $A \in \mathcal{A}_{N\times N}$ is Fredholm on $RL^p_N(\Gamma, w)$ if and only if the operator $\hat{A} \in \mathcal{B}_{2N\times 2N}$ is locally Fredholm on $L^p_{2N}(\Gamma, w)$ at each point $t_0 \in \Gamma$.*

At the points of $\Gamma \setminus F$, criteria for the local Fredholmness of the operators $\hat{A}$ ($A \in \mathcal{A}_{N\times N}$) are well known. Thus, fix $t_0 \in \Gamma \setminus F$. Let I_N be the $N \times N$ identity matrix. For

$$A \in \{aI : a \in PSO_{N\times N}(\Gamma)\} \cup \{S, V\}$$

we define the $2N \times 2N$ matrices $\mathbf{Sym}^{\pm}_{t_0} A$ as follows:

$$\begin{aligned}
\mathbf{Sym}^{+}_{t_0} aI &= \mathbf{Sym}^{-}_{t_0} aI = \begin{pmatrix} a(t_0) & 0 \\ 0 & \overline{a(t_0)} \end{pmatrix}, \\
\mathbf{Sym}^{+}_{t_0} S &= \begin{pmatrix} I_N & 0 \\ 0 & -I_N \end{pmatrix}, \quad \mathbf{Sym}^{-}_{t_0} S = \begin{pmatrix} -I_N & 0 \\ 0 & I_N \end{pmatrix}, \\
\mathbf{Sym}^{+}_{t_0} V &= \mathbf{Sym}^{-}_{t_0} V = \begin{pmatrix} 0 & I_N \\ I_N & 0 \end{pmatrix}.
\end{aligned}$$

Theorem 4.2. *Let $t_0 \in \Gamma \setminus F$. The two maps $A \mapsto \mathbf{Sym}^{\pm}_{t_0} A$ extend to Banach algebra homomorphisms of $\mathcal{A}_{N\times N}$ into $\mathbf{C}^{2N\times 2N}$. If $A \in \mathcal{A}_{N\times N}$, then $\hat{A} \in \mathcal{B}_{2N\times 2N}$ is locally Fredholm on $L^p_{2N}(\Gamma, w)$ at t_0 if and only if*

$$\det \mathbf{Sym}^{+}_{t_0} A \neq 0 \text{ and } \det \mathbf{Sym}^{-}_{t_0} A \neq 0.$$

Now suppose that $t_0 \in F$. We fix a sufficiently small $\varepsilon > 0$ and change Γ and w to $\tilde{\Gamma}$ and $\tilde{w}$ as follows. Let

$$\tilde{\Gamma} = \{t_0\} \cup \tilde{\gamma}_1 \cup \ldots \cup \tilde{\gamma}_n$$

where

$$\tilde{\gamma}_j = \{t_0 + re^{i(\tilde{\theta}(r) + \tilde{\theta}_j(r))} : r \in (0, \infty)\} \quad (j = 1, \ldots, n)$$

with C^∞ functions $\tilde{\theta}, \tilde{\theta}_1, \ldots, \tilde{\theta}_n$ such that

$$\tilde{\theta}(r) + \tilde{\theta}_j(r) = \begin{cases} \theta(r) + \theta_j(r) & \text{for} \quad r \in (0, \varepsilon/2), \\ \text{constant} & \text{for} \quad r \in (\varepsilon, \infty), \end{cases}$$

$$0 \leq m_1 < \tilde{\theta}_1(r) < M_1 < m_2 < \tilde{\theta}(r) < M_2 < \ldots < m_n < \tilde{\theta}_n(r) < M_n < 2\pi$$

for $r \in (0, \infty)$, and define $\tilde{w}$ on $\tilde{\Gamma}$ by

$$\tilde{w}\big(t_0 + re^{i(\tilde{\theta}(r) + \tilde{\theta}_j(r))}\big) = e^{\tilde{v}(r)} \quad (j = 1, \ldots, n)$$

where $\tilde{v}$ is a real-valued C^∞ function such that

$$\tilde{v}(r) = \begin{cases} v(r) & \text{for} \quad r \in (0, \varepsilon/2), \\ 0 & \text{for} \quad r \in (\varepsilon, \infty). \end{cases}$$

Let $\mathcal{B}^0_{M\times M}$ be the set of all operators on $L^p_M(\Gamma, w)$ which can be represented as finite sums of finite (ordered) products of the form

$$B = \sum_j \prod_k B_{jk}$$

where the operators B_{jk} belong to

(4.1) $$\{aI : a \in PSO_{M\times M}(\Gamma)\} \cup \{S, VSV\}.$$

Clearly, $\mathcal{B}^0_{M\times M}$ is dense in $\mathcal{B}_{M\times M}$. The analogues of $\mathcal{B}^0_{M\times M}$ and $\mathcal{B}_{M\times M}$ for $L^p_M(\tilde{\Gamma}, \tilde{w})$ will be denoted by $\tilde{\mathcal{B}}^0_{M\times M}$ and $\tilde{\mathcal{B}}_{M\times M}$. For an operator B in (4.1), we define the operator $\Psi(B) \in \tilde{\mathcal{B}}^0_{M\times M}$ as follows: if B is the Cauchy singular integral operator on Γ, then $\Psi(B)$ is the Cauchy singular integral operator on $\tilde{\Gamma}$ (thus, $\Psi(S_\Gamma) = S_{\tilde{\Gamma}}$); if $B = VSV$, then $\Psi(B) = V\Psi(S)V$; if $B = aI$, then $\Psi(B) = \tilde{a}I$ where $\tilde{a}I \in PSO_{M\times M}(\tilde{\Gamma})$ is any matrix-function such that

$$\tilde{a}(\tau) = \begin{cases} a(\tau) & \text{for} \quad |\tau - t_0| < \varepsilon/2, \\ 1 & \text{for} \quad |\tau - t_0| > \varepsilon. \end{cases}$$

The extension of Ψ from the set (4.1) to all of $\mathcal{B}^0_{M\times M}$ is handicaped by the circumstance that in general $(ab)^\sim \neq \tilde{a}\tilde{b}$. Let therefore $\tilde{J}_{t_0}$ be the smallest two-sided ideal of $\tilde{\mathcal{B}}^0_{M\times M}$ which contains the set

$$\Big\{\varphi I : \varphi \in PSO_{M\times M}(\tilde{\Gamma}),\ \varphi|\tilde{\Gamma}(t_0, \varepsilon/2) = 0\Big\}.$$

The map

$$\tilde{\Psi} : \mathcal{B}^0_{M\times M} \to \tilde{\mathcal{B}}^0_{M\times M}/\tilde{J}_{t_0},\ \tilde{\Psi}\Big(\sum_j \prod_k B_{jk}\Big) = \sum_j \prod_k \Psi(B_{jk}) + \tilde{J}_{t_0}$$

is a well-defined algebra homomorphism, and it is not difficult to prove the following.

Proposition 4.3. *Let* $B \in \mathcal{B}^0_{M\times M}$ *and* $\tilde{B} \in \tilde{\Psi}(B)$. *Then* B *is locally Fredholm on* $L^p_M(\Gamma, w)$ *at* t_0 *if and only if* $\tilde{B} \in \tilde{\mathcal{B}}^0_{M\times M}$ *is locally Fredholm on* $L^p_M(\tilde{\Gamma}, \tilde{w})$ *at* t_0.

The map $\Phi_{t_0} : L^p_M(\tilde{\Gamma}, \tilde{w}) \to L^p_{nM}(\mathbf{R}_+, d\mu)$ given by

$$(\Phi_{t_0} f)(r) = r^{1/p} e^{\tilde{v}(r)} \text{ column } \left(f(t_0 + re^{i(\tilde{\theta}(r)+\tilde{\theta}_1(r))}), \ldots, f(t_0 + re^{i(\tilde{\theta}(r)+\tilde{\theta}_n(r))})\right)$$

is obviously a Banach space isomorphism. This easily implies the following result.

Proposition 4.4. *If* $\tilde{B} \in \tilde{\mathcal{B}}^0_{M\times M}$, *then for* $\tilde{B}$ *to be locally Fredholm on* $L^p_M(\tilde{\Gamma}, \tilde{w})$ *at* t_0 *it is necessary and sufficient that*

$$\Phi_{t_0} \tilde{B} \Phi_{t_0}^{-1} \in \dot{\mathcal{L}}(L^p_{nM}(\mathbf{R}_+, d\mu))$$

be locally Fredholm at the origin.

For B in the set (4.1), define $\Sigma_{t_0}(B) \in \mathcal{L}(L^p_{nM}(\mathbf{R}_+, d\mu))$ by

$$\Sigma_{t_0}(B) = \Phi_{t_0} \Psi(B) \Phi_{t_0}^{-1}.$$

We now show that the operators $\Sigma_{t_0}(B)$ are Mellin pseudodifferential operators. Things are easy for multiplication operators.

Theorem 4.5. *If* $a \in PSO_{M\times M}(\Gamma)$, *then*

$$\Sigma_{t_0}(aI) = OP(\sigma_{aI})$$

where $\sigma_{aI} \in \mathcal{E}_{nM\times nM}$ *is the matrix function*

$$\sigma_{aI}(r, \lambda) = \operatorname{diag}\left(\tilde{a}_1(r), \ldots, \tilde{a}_n(r)\right), \quad \tilde{a}_j(r) = \tilde{a}(t_0 + re^{i(\tilde{\theta}(r)+\tilde{\theta}_j(r))}) \quad (j = 1, \ldots, n).$$

We remark that $OP(\sigma_{aI})$ is actually nothing else than the operator of multiplication by $\operatorname{diag}(\tilde{a}_1, \ldots, \tilde{a}_n)$ on $L^p_{nM}(\mathbf{R}_+, d\mu)$.

To get $\Sigma_{t_0}(S)$ and $\Sigma_{t_0}(VSV)$, we need some more notation. Put $\varepsilon_k = 1$ if t_0 is the starting point of the oriented arc γ_k and let $\varepsilon_k = -1$ if t_0 is the terminating point of the oriented arc γ_k. Define

$$\nu : [0, 2\pi) \times (\mathbf{C} \setminus i\mathbf{Z}) \to \mathbf{C}$$

by

$$\nu(\delta, z) = \begin{cases} \coth(\pi z) & \text{for } \delta = 0, \\ \dfrac{e^{(\pi-\delta)z}}{\sinh(\pi z)} & \text{for } \delta \in (0, 2\pi). \end{cases}$$

For $j, k \in \{1, \dots, n\}$, let

$$s_{jk} : \mathbf{R}_+ \times (\mathbf{C} \setminus i\mathbf{Z}) \to \mathbf{C}, \quad \overline{s}_{jk} : \mathbf{R}_+ \times (\mathbf{C} \setminus i\mathbf{Z}) \to \mathbf{C}$$

be the functions

$$s_{jk}(r,z) = \begin{cases} \varepsilon_k \nu(2\pi + \tilde{\theta}_j(r) - \tilde{\theta}_k(r), z) & \text{if } j < k, \\ \varepsilon_k \nu(0,z) & \text{if } j = k, \\ \varepsilon_k \nu(\tilde{\theta}_j(r) - \tilde{\theta}_k(r), z) & \text{if } j > k, \end{cases} \tag{4.2}$$

$$\overline{s}_{jk}(r,z) = \begin{cases} -\varepsilon_k \nu(\tilde{\theta}_k(r) - \tilde{\theta}_j(r), z) & \text{if } j < k, \\ -\varepsilon_k \nu(0,z) & \text{if } j = k, \\ -\varepsilon_k \nu(2\pi + \tilde{\theta}_k(r) - \tilde{\theta}_j(r), z) & \text{if } j > k, \end{cases} \tag{4.3}$$

and put

$$s(r,z) = \big(s_{jk}(r,z)\big)_{j,k=1}^n, \quad \overline{s}(r,z) = \big(\overline{s}_{jk}(r,z)\big)_{j,k=1}^n. \tag{4.4}$$

Theorem 4.6. *We have*

$$\Sigma_{t_0}(S) = OP(\sigma_S), \quad \Sigma_{t_0}(VSV) = OP(\sigma_{VSV})$$

where $\sigma_S, \sigma_{VSV} \in \mathcal{E}_{nM \times nM}$ *are given by*

$$\begin{aligned} \sigma_S(r,\lambda) &= s\left(r, \frac{\lambda + i(1/p + r\tilde{v}'(r))}{1 + ir\tilde{\theta}'(r)}\right) \otimes I_M + q_S(r,\lambda), \\ \sigma_{VSV}(r,\lambda) &= \overline{s}\left(r, \frac{\lambda + i(1/p + r\tilde{v}'(r))}{1 - ir\tilde{\theta}'(r)}\right) \otimes I_M + q_{VSV}(r,\lambda) \end{aligned}$$

with $q_S, q_{VSV} \in \mathcal{E}_{nM \times nM}$ *satisfying*

$$\lim_{r \to 0} \sup_{\lambda \in \mathbf{R}} \|q_S(r,\lambda)\|_{\mathbf{C}^{nM \times nM}} = \lim_{r \to 0} \sup_{\lambda \in \mathbf{R}} \|q_{VSV}(r,\lambda)\|_{\mathbf{C}^{nM \times nM}} = 0. \tag{4.5}$$

Proof. For the operator S, this was established in [4] (also see [3] and [22] for additional details of the proof). Because

$$\begin{aligned} &\Sigma_{t_0}(VSV) = \Phi_{t_0}\Psi(VSV)\Phi_{t_0}^{-1} = \Phi_{t_0} V \Psi(S) V \Phi_{t_0}^{-1} \\ &= \Phi_{t_0} V \Phi_{t_0}^{-1} \Phi_{t_0} \Psi(S) \Phi_{t_0}^{-1} \Phi_{t_0} V \Phi_{t_0}^{-1} = V \Sigma_{t_0}(S) V = V OP(\sigma_S) V \end{aligned}$$

we deduce that

$$\begin{aligned} &\big(\Sigma_{t_0}(VSV)f\big)(r) = \int_{\mathbf{R}} \frac{d\lambda}{2\pi} \int_{\mathbf{R}_+} \overline{\sigma_S(r,\lambda)} f(\varrho) \Big(\frac{r}{\varrho}\Big)^{-i\lambda} \frac{d\varrho}{\varrho} \\ &= \int_{\mathbf{R}} \frac{d\lambda}{2\pi} \int_{\mathbf{R}_+} \overline{\sigma_S(r,-\lambda)} f(\varrho) \Big(\frac{r}{\varrho}\Big)^{i\lambda} \frac{d\varrho}{\varrho}, \end{aligned}$$

which gives $\Sigma_{t_0}(VSV) = OP(\sigma_{VSV})$ with

$$\begin{aligned}
\sigma_{VSV}(r,\lambda) &= \overline{\sigma_S(r,-\lambda)} \\
&= \left(\overline{s_{jk}\left(r, \frac{-\lambda + i(1/p + r\tilde{v}'(r))}{1+i\tilde{\theta}'(r)}\right)}\right)_{j,k=1}^{n} \otimes I_M + \overline{q_S(r,-\lambda)} \\
&= \left(s_{jk}\left(r, \frac{-\lambda - i(1/p + r\tilde{v}'(r))}{1-i\tilde{\theta}'(r)}\right)\right)_{j,k=1}^{n} \otimes I_M + \overline{q_S(r,-\lambda)} \\
&= \left(\overline{s}_{jk}\left(r, \frac{\lambda + i(1/p + r\tilde{v}'(r))}{1-i\tilde{\theta}'(r)}\right)\right)_{j,k=1}^{n} \otimes I_M + \overline{q_S(r,-\lambda)}.
\end{aligned}$$

□

We are now in a position to state a final result for the local Fredholmness of operators in $\mathcal{B}^0_{M\times M}$ at $t_0 \in F$. For $r \in (0,\varepsilon)$ and $z \in \mathbf{C}\setminus i\mathbf{Z}$, define $s(r,z)$ and $\overline{s}(r,z)$ by (4.2), (4.3), (4.4) with $\tilde{\theta}$ replaced by θ, and put

$$\begin{aligned}
(\mathrm{Sym}_{t_0}\, aI)(r,\lambda) &= \mathrm{diag}(a_1(r),\dots,a_n(r)) \quad (a \in PSO_{M\times M}(\Gamma)) \\
&\quad \text{where } a_j(r) = a(t_0 + re^{i(\theta(r)+\theta_j(r))}), \\
(\mathrm{Sym}_{t_0}\, S)(r,\lambda) &= s\left(r, \frac{\lambda + i(1/p + rv'(r))}{1+ir\theta'(r)}\right) \otimes I_M, \\
(\mathrm{Sym}_{t_0} VSV)(r,\lambda) &= \overline{s}\left(r, \frac{\lambda + i(1/p + rv'(r))}{1-ir\theta'(r)}\right) \otimes I_M.
\end{aligned}$$

Theorem 4.7. *Let $t_0 \in F$. For $B = \Sigma_j \Pi_k B_{jk} \in \mathcal{B}^0_{M\times M}$, define $\mathrm{Sym}_{t_0} B$ by*

$$(\mathrm{Sym}_{t_0} B)(r,\lambda) = \sum_j \prod_k (\mathrm{Sym}_{t_0} B_{jk})(r,\lambda).$$

An operator $B \in \mathcal{B}^0_{M\times M}$ is locally Fredholm on $L^p_M(\Gamma,w)$ at t_0 if and only if

$$\liminf_{r\to 0} \inf_{\lambda\in\mathbf{R}} |\det(\mathrm{Sym}_{t_0} B)(r,\lambda)| > 0.$$

Proof. Propositions 4.3 and 4.4 imply that B is locally Fredholm at t_0 if and only if the operator

$$\Sigma_{t_0}(B) := \sum_j \prod_k \Sigma_{t_0}(B_{jk})$$

is locally Fredholm at the origin. Theorems 4.5 and 4.6 in conjunction with (3.1) shows that

$$\Sigma_{t_0}(B) = OP\Big(\sum_j \prod_k \sigma_{B_{jk}}\Big) + \text{ compact operator,}$$

and hence Theorem 3.1 yields that B is locally Fredholm at t_0 if and only if

$$\liminf_{r\to 0} \inf_{\lambda\in\mathbf{R}} \Big|\det\Big(\sum_j \prod_k \sigma_{B_{jk}}(r,\lambda)\Big)\Big| > 0. \tag{4.6}$$

If $r > 0$ is sufficiently small, then, again by Theorems 4.5 and 4.6,

$$\begin{aligned}
\sigma_{aI}(r,\lambda) &= \operatorname{diag}(a_1(r),\dots,a_n(r)) = (\mathrm{Sym}_{t_0} aI)(r,\lambda),\\
\sigma_S(r,\lambda) &= s\left(r, \frac{\lambda + i(1/p + rv'(r))}{1 + ir\theta'(r)}\right) \otimes I_N + q_S(r,\lambda)\\
&= (\mathrm{Sym}_{t_0} S)(r,\lambda) + q_S(r,\lambda),\\
\sigma_{VSV}(r,\lambda) &= \overline{s}\left(r, \frac{\lambda + i(1/p + rv'(r))}{1 + ir\theta(r)}\right) \otimes I_N + q_{VSV}(r,\lambda)\\
&= (\mathrm{Sym}_{t_0} VSV)(r,\lambda) + q_{VSV}(r,\lambda).
\end{aligned}$$

Thus, (4.5) shows us that (4.6) is satisfied if and only if

$$\liminf_{r\to 0} \inf_{\lambda\in\mathbf{R}} \Big|\det\Big(\sum_j \prod_k (\mathrm{Sym}_{t_0} B_{jk})(r,\lambda)\Big)\Big| > 0.$$

□

Let $BC_{M\times M}(\mathbf{R}_+ \times \mathbf{R})$ denote the Banach algebra of all bounded and continuous functions of $\mathbf{R}_+ \times \mathbf{R}$ into $\mathbf{C}^{M\times M}$, and let $\mathcal{Z}_0^M$ be the closed two-sided ideal of $BC_{M\times M}(\mathbf{R}_+ \times \mathbf{R})$ which consists of the matrix functions a satisfying

$$\lim_{r\to 0} \sup_{\lambda\in\mathbf{R}} \|a(r,\lambda)\|_{\mathbf{C}^{M\times M}} = 0.$$

Using the estimates for local norms established in [22] one can show that the map

$$\mathrm{Sym}_{t_0} : \mathcal{B}^0_{M\times M} \to BC_{M\times M}(\mathbf{R}_+ \times \mathbf{R})/\mathcal{Z}_0^M$$

is a continuous algebra homomorphism. Consequently, Sym_{t_0} extends to a Banach algebra homomorphism

$$\mathrm{Sym}_{t_0} : \mathcal{B}_{M\times M} \to BC_{M\times M}(\mathbf{R}_+ \times \mathbf{R})/\mathcal{Z}_0^M.$$

Again employing the estimates of [22] we arrive at the following consequence of Theorem 4.7.

Theorem 4.8. *Let $t_0 \in F$. An operator $B \in \mathcal{B}_{M\times M}$ is locally Fredholm on $L^p_M(\Gamma, w)$ at t_0 if and only if*

$$\liminf_{r\to 0} \inf_{\lambda\in\mathbf{R}} |\det b(r,\lambda)| > 0$$

where $b \in BC_{M\times M}(\mathbf{R}_+ \times \mathbf{R})$ is any element of the coset $\mathrm{Sym}_{t_0} B$.

We finally return to the operators in $\mathcal{A}_{N\times N}$. For $A \in \mathcal{A}_{N\times N}$ and $t_0 \in F$, define

$$\mathbf{Sym}_{t_0} A = \mathrm{Sym}_{t_0} \hat{A} \quad \left(\in BC_{2nN\times 2nN}(\mathbf{R}_+ \times \mathbf{R})/\mathcal{Z}_0^{2nN} \right).$$

In particular, for $a \in PSO_{N\times N}(\Gamma)$ we have, modulo $\mathcal{Z}_0^{2nN}$,

$$(\mathbf{Sym}_{t_0} aI)(r,\lambda) = \begin{pmatrix} \mathrm{diag}\,(a_1(r),\ldots,a_n(r)) & 0 \\ 0 & \mathrm{diag}\,(\overline{a_1(r)},\ldots,\overline{a_n(r)}) \end{pmatrix}$$

$$\text{where } a_j(r) = a(t_0 + e^{i(\theta(r)+\theta_j(r))}), \quad r \in (0,\varepsilon)$$

and, again modulo $\mathcal{Z}_0^{2nN}$,

$$(\mathbf{Sym}_{t_0} S)(r,\lambda) = \begin{pmatrix} s\left(r, \frac{\lambda+i(1/p+rv'(r))}{1+ir\theta'(r)}\right) \otimes I_N & 0 \\ 0 & \overline{s}\left(r, \frac{\lambda+i(1/p+rv'(r))}{1-ir\theta'(r)}\right) \otimes I_N \end{pmatrix},$$

$$(\mathbf{Sym}_{t_0} V)(r,\lambda) = \begin{pmatrix} 0 & I_{nN} \\ I_{nN} & 0 \end{pmatrix}.$$

The following theorem is a consequence of Theorem 4.8 and of what was said in Section 1.

Theorem 4.9. *Let $t_0 \in F$. An operator $A \in \mathcal{A}_{N\times N}$ is locally Fredholm on $RL^p_N(\Gamma, w)$ at t_0 if and only if*

$$\liminf_{r\to 0} \inf_{\lambda\in\mathbf{R}} |\det a(r,\lambda)| > 0$$

where $a \in BC_{2nN\times 2nN}(\mathbf{R}_+ \times \mathbf{R})$ is any element of the coset $\mathbf{Sym}_{t_0} A$.

Combining Theorems 4.1, 4.2, and 4.9 we obtain a Fredholm criterion for operators in $\mathcal{A}_{N\times N}$.

5. Index formula

Standard separation techniques (see [22]) and Theorem 3.2 allow us to compute $\mathrm{Ind}\,A$ for $A \in \mathcal{A}_{N\times N}$. The result is as follows.

If $A \in \mathcal{A}_{N\times N}$ is Fredholm, then, by Theorems 4.1 and 4.2, $\det \mathbf{Sym}_t^{\pm} A \neq 0$ for all $t \in \Gamma \setminus F$. The set $\Gamma \setminus F$ is the union of K smooth simple arcs Γ_k $(k = 1,\ldots,K)$ whose endpoints are removed. Since $\det \mathbf{Sym}_t^{\pm} A \neq 0$ for all $t \in \Gamma \setminus F$, we can choose a continuous argument

$$\Gamma_k \to \mathbf{R}, \; t \mapsto \arg \frac{\det \mathbf{Sym}_t^+ A}{\det \mathbf{Sym}_t^- A} \tag{5.1}$$

of $\det \mathbf{Sym}_t^+ A / \det \mathbf{Sym}_t^- A$ on Γ_k. It is clear that (5.1) has a finite increment

$$\left[\arg \frac{\det \mathbf{Sym}_t^+ A}{\det \mathbf{Sym}_t^- A}\right]_{l_k}$$

as t traces out an arbitrary arc segment l_k of Γ_k in the positive direction.

For every node $t_0 \in F$, which is the center of the star $\Gamma(t_0, \varepsilon) = \{t_0\} \cup \gamma_1 \cup \ldots \cup \gamma_{n(t_0)}$, and for every sufficiently small $r > 0$, we consider the points

$$\tau_j(t_0, r) = t_0 + re^{i(\theta(r)+\theta_j(r))} \in \gamma_j \quad (j = 1, \ldots, n(t_0)).$$

Then for each arc Γ_k $(k = 1, \ldots, K)$, there exist exactly two points

$$\tau_k^{\pm}(r) \in \Gamma_k \cap \left(\bigcup_{t_0 \in F} \bigcup_{j=1}^{n(t_0)} \{\tau_j(t_0, r)\} \right).$$

Let $\Gamma_k^{(r)}$ be the arc of Γ_k whose endpoints are $\tau_k^{\pm}(r)$ and which is oriented as Γ_k. For $t_0 \in F$, choose any representative $a_{t_0} \in \mathbf{Sym}_{t_0} A$. By Theorems 4.1 and 4.9, we can choose a continuous argument $\mathbf{R} \to \mathbf{R}$, $\lambda \mapsto \arg \det a_{t_0}(r, \lambda)$ if only $r > 0$ is sufficiently small. The increment $\left[\arg \det a_{t_0}(r, \lambda)\right]_{-\infty}^{\infty}$ of $\arg \det a_{t_0}(r, \lambda)$ as λ moves from $-\infty$ to $+\infty$ can be shown to be finite whenever $r > 0$ is small enough.

The Fredholm index of A on $RL_N^p(\Gamma, w)$ is

$$\operatorname{Ind} A = -\frac{1}{2\pi} \lim_{r \to 0} \left(\sum_{k=1}^{K} \left[\arg \frac{\det \mathbf{Sym}_t^+ A}{\det \mathbf{Sym}_t^- A}\right]_{\Gamma_k^{(r)}} + \sum_{t_0 \in F} \left[\arg \det a_{t_0}(r, \lambda)\right]_{-\infty}^{\infty} \right).$$

6. The double layer potential

Let Γ be a slowly oscillating composed curve. If $t \in \Gamma \setminus F$, then t is an inner point of one of the simple smooth arcs $\Gamma_1, \ldots, \Gamma_K$ whose union is Γ. Let $t \in \Gamma_j$. We denote by $n(t)$ the normalized normal vector to Γ_j at t which points to the right-hand side of the oriented arc Γ_j. A straightforward computation gives

$$((S + VSV)f)(t) = \lim_{\varepsilon \to 0} \frac{2}{\pi} \int_{\{\tau \in \Gamma : |\tau - t| \geq \varepsilon\}} \frac{(n(\tau), \tau - t)}{|\tau - t|^2} f(\tau) |d\tau| \quad (t \in \Gamma),$$

and hence $S + VSV$ is nothing but the double layer potential operator on Γ.

For a point $t_0 \in \Gamma$, the local spectrum $\operatorname{sp}_{t_0}(S + VSV)$ is defined as the set of all $\lambda \in \mathbf{C}$ for which $S + VSV - \lambda I$ is not locally Fredholm on $RL^p(\Gamma, w)$ at t_0. At the points $t_0 \in \Gamma \setminus F$ we have $\operatorname{sp}_{t_0}(S + VSV) = \{0\}$. In this section we determine the local spectrum $\operatorname{sp}_{t_0}(S + VSV)$ in some more interesting cases.

Example 1. Let Γ be the logarithmic spiral

$$\Gamma = \{t_0\} \cup \{t_0 + re^{i\delta \log r} : r \in (0,1)\}$$

and suppose

$$w(t_0 + re^{i\delta \log r}) = r^\eta$$

with $\eta \in (-1/p, 1/q)$. We put $\varepsilon = +1$ and $\varepsilon = -1$ in dependence on whether t_0 is the starting or terminating point of Γ. Also let $\beta = 1/p + \eta$. We have, modulo $\mathcal{Z}_0^2$,

$$\begin{aligned}
(\mathbf{Sym}_{t_0} S)(r,\lambda) &= \begin{pmatrix} \varepsilon \coth\left(\pi \frac{\lambda + i\beta}{1+i\delta}\right) & 0 \\ 0 & -\varepsilon \coth\left(\pi \frac{\lambda + i\beta}{1-i\delta}\right) \end{pmatrix}, \\
(\mathbf{Sym}_{t_0} VSV)(r,\lambda) &= \begin{pmatrix} -\varepsilon \coth\left(\pi \frac{\lambda + i\beta}{1-i\delta}\right) & 0 \\ 0 & \varepsilon \coth\left(\pi \frac{\lambda + i\beta}{1+i\delta}\right) \end{pmatrix}, \\
(\mathbf{Sym}_{t_0} \zeta I)(r,\lambda) &= \begin{pmatrix} \zeta & 0 \\ 0 & \overline{\zeta} \end{pmatrix}.
\end{aligned}$$

Notice that these matrices do not depend on r. Theorem 4.9 implies that $\zeta \in \mathrm{sp}_{t_0}(S + VSV)$ if and only if

$$\det(\mathbf{Sym}_{t_0}(S + VSV - \zeta I))(\lambda) = 0$$

for some $\lambda \in \overline{\mathbf{R}} := \mathbf{R} \cup \{\pm\infty\}$. What results is that

$$\text{(6.1)} \quad \mathrm{sp}_{t_0}(S + VSV) = \varepsilon \left\{ \coth\left(\pi \frac{\lambda + i\beta}{1+i\delta}\right) - \coth\left(\pi \frac{\lambda + i\beta}{1-i\delta}\right) : \lambda \in \overline{\mathbf{R}} \right\}.$$

If $\delta = 0$, we obtain the well known result that $\mathrm{sp}_{t_0}(S + VSV) = \{0\}$. Figure 1 shows plots of the set on the right of (6.1) and thus of $\mathrm{sp}_{t_0}(S + VSV)$ for six choices of the parameters δ and p (with $\varepsilon = 1$ and $\eta = 0$).

Example 2. Let $t_0 \in F$ and suppose

$$\Gamma(t_0, \varepsilon) = \{t_0\} \cup \{t_0 + re^{i(\delta \log r + \mu_1)} : r \in (0,\varepsilon\} \cup \{t_0 + re^{i(\delta \log r + \mu_2)} : r \in (0,\varepsilon\}$$

with constants μ_1, μ_2 such that $0 \le \mu_1 < \mu_2 < 2\pi$. Further, define

$$w(t_0 + re^{i(\delta \log r + \mu_j)}) = r^\eta \quad (j = 1,2),$$

where $\eta \in (-1/p, 1/q)$, and assume the orientation is so that $\varepsilon_1 \varepsilon_2 = -1$. Put

$$\beta = 1/p + \eta, \quad a(\lambda) = \frac{\lambda + i\beta}{1+i\delta}, \quad b(\lambda) = \frac{\lambda + i\beta}{1-i\delta}, \quad \gamma = \pi + \mu_1 - \mu_2,$$
$$\operatorname{ch} z = \cosh z, \quad \operatorname{sh} z = \sinh z, \quad \operatorname{cth} z = \coth z.$$

We have, modulo $\mathcal{Z}_0^4$,

$$(\mathbf{Sym}_{t_0}(S+VSV-\zeta I))(r,\lambda)=\begin{pmatrix} A(\lambda)-\zeta I_2 & 0 \\ 0 & \overline{A(-\lambda)}-\overline{\zeta}I_2 \end{pmatrix}$$

where I_2 is the 2×2 identity matrix and

$$A(\lambda)=\begin{pmatrix} \varepsilon_1\Big(\big(\mathrm{cth}\,(\pi a(\lambda))-\mathrm{cth}\,(\pi b(\lambda))\Big) & \varepsilon_2\left(\dfrac{e^{-\gamma a(\lambda)}}{\mathrm{sh}\,(\pi a(\lambda))}-\dfrac{e^{\gamma b(\lambda)}}{\mathrm{sh}\,(\pi b(\lambda))}\right) \\ \varepsilon_1\left(\dfrac{e^{\gamma a(\lambda)}}{\mathrm{sh}\,(\pi a(\lambda))}-\dfrac{e^{-\gamma b(\lambda)}}{\mathrm{sh}\,(\pi b(\lambda))}\right) & \varepsilon_2\Big(\mathrm{cth}\,(\pi a(\lambda))-\mathrm{cth}\,(\pi b(\lambda))\Big) \end{pmatrix}.$$

The symbol is again independent of r, and it follows that

$$\det(\mathbf{Sym}_{t_0}(S+VSV-\zeta I))(\lambda)=\det(A(\lambda)-\zeta I_2)\det(\overline{A(-\lambda)}-\overline{\zeta}I_2),$$

whence, by Theorem 4.9,

$$\mathrm{sp}_{t_0}(S+VSV)=\{\zeta\in\mathbf{C}:\det(A(\lambda)-\zeta I_2)=0\ \text{ for some }\ \lambda\in\overline{\mathbf{R}}\}.$$

A direct calculation gives

$$\det(A(\lambda)-\zeta I_2)=\zeta^2+2\frac{\mathrm{ch}\,\big(\pi(a(\lambda)-b(\lambda))\big)-\mathrm{ch}\,\big(\gamma(a(\lambda)+b(\lambda))\big)}{\mathrm{sh}\,\big(\pi a(\lambda)\big)\mathrm{sh}\,\big(\pi b(\lambda)\big)}. \tag{6.2}$$

If $\delta=0$, this can be simplified to

$$\det(A(\lambda)-\zeta I_2)=\zeta^2-\left(2\frac{\mathrm{sh}\,(\gamma a(\lambda))}{\mathrm{sh}\,(\pi a(\lambda))}\right)^2.$$

If, in addition, $\gamma=\pi+\mu_1-\mu_2=0$, the last formula yields that $\mathrm{sp}_{t_0}(S+VSV)=\{0\}$.

Figure 2 and 3 show $\mathrm{sp}_{t_0}(S+VSV)$ for some cases in which $\gamma\neq 0$ (and $\eta=0$). The upper two pictures of Figure 2 correspond to the case where $\mu_2-\mu_1=0.1$, that is, to a case "close to a cusp". Notice the different scales on the real and imaginary axes. The four lower pictures of Figure 2 indicate the evolution of the local spectrum in dependence of δ, while the six pictures of Figure 3 provide an idea of the metamorphosis of the local spectrum in dependence on p.

The upper left picture of Figure 4 shows a positively oriented Jordan curve with three exceptional points (nodes): a whirl point A, a corner point B, and a cusp C which is supposed to have the order 1. In the other three pictures in the two upper rows of Figure 4 we see the local spectrum of $S+VSV$ on $RL^2(\Gamma)$ at the points A,B,C. For the points A and B, the result follows from (6.2); we remark that the result for the point B is well known. The approach of this paper is not applicable to the point C. The fact that in this case the local spectrum is the segment $[-2, 2]$ was established only recently by Duduchava, Latsabidze, and Saginashvili [10]. The essential spectrum of $S+VSV$, that is, the set of all $\zeta \in \mathbf{C}$ for which $S+VSV-\zeta I$ is not Fredholm, is the union of the three sets plotted in the right upper and the two middle pictures of Figure 4.

Example 3. Finally, suppose $t_0 \in F$ and, for some $\varepsilon < 1$,

$$\Gamma(t_0, \varepsilon) = \{t_0\} \cup \bigcup_{j=1,2} \{t_0 + re^{i\mu_j + ih(\log(-\log r))\log r} : r \in (0, \varepsilon)\}$$

where $0 \le \mu_1 < \mu_2 < 2\pi$ and $h(x) = \delta_0 + \alpha \sin x$ with real constants δ_0, α. We consider the operator $S+VSV$ on $RL^2(\Gamma)$. As in Example 2 we get

$$(\mathbf{Sym}_{t_0}(V+VSV-\zeta I))(r,\lambda) = \begin{pmatrix} A(r,\lambda)-\zeta I_2 & 0 \\ 0 & \overline{A(r,-\lambda)}-\overline{\zeta} I_2 \end{pmatrix}$$

where now $A(r, \lambda)$ is the matrix on the right of the definition of the matrix $A(\lambda)$ with the replacements

$$a(\lambda) \mapsto a(r,\lambda) = \frac{\lambda + i/2}{1 + ir\theta'(r)}, \quad b(\lambda) \mapsto b(r,\lambda) = \frac{\lambda + i/2}{1 - ir\theta'(r)},$$

$$\gamma = \pi + \mu_1 - \mu_2, \quad \theta(r) = h(\log(-\log r))\log r.$$

Again

$$\det(\mathbf{Sym}_{t_0}(S+VSV-\zeta I))(r,\lambda) = \det(A(r,\lambda)-\zeta I_2)\det(\overline{A(r,-\lambda)}-\overline{\zeta} I_2),$$

and proceeding as in Example 2 we deduce from Theorem 4.9 that $\mathrm{sp}_{t_0}(S+VSV)$ is the set of all $\zeta \in \mathbf{C}$ for which there exist a $\lambda \in \overline{\mathbf{R}}$ and a sequence $\{r_n\} \subset \mathbf{R}_+$ such that $r_n \to 0$ and

$$\zeta^2 = -2 \lim_{n\to\infty} \frac{\mathrm{ch}\left(\pi(a(r_n,\lambda) - b(r_n,\lambda))\right) - \mathrm{ch}\left(\gamma(a(r_n,\lambda) + b(r_n,\lambda))\right)}{\mathrm{sh}\left(\pi a(r_n,\lambda)\right)\mathrm{sh}\left(\pi b(r_n,\lambda)\right)}.$$

Equivalently, on denoting by Δ the set of the partial limits of $r\theta'(r)$ as $r \to 0$, we can describe $\mathrm{sp}_{t_0}(S+VSV)$ as the set of all $\zeta \in \mathbf{C}$ for which

$$\zeta^2 \in \left\{-2\frac{\mathrm{ch}\left(\pi(a(\delta,\lambda) - b(\delta,\lambda))\right) - \mathrm{ch}\left(\gamma(a(\delta,\lambda) + b(\delta,\lambda))\right)}{\mathrm{sh}\left(\pi a(\delta,\lambda)\right)\mathrm{sh}\left(\pi b(\delta,\lambda)\right)} : \delta \in \Delta, \lambda \in \overline{\mathbf{R}}\right\}.$$

It is easily seen that

$$\Delta = [\delta_0 - |\alpha|\sqrt{2}, \delta_0 + |\alpha|\sqrt{2}].$$

Thus, $\mathrm{sp}_{t_0}(S+VSV)$ is the union over $\delta \in \Delta$ of the local spectra described in Example 2. In this way we obtain what we call heavy local spectra. Two examples are indicated in the lower two pictures of Figure 4, which correspond to the cases $\Delta = [0.1, 1.5]$ and $\Delta = [1.8, 2.2]$.

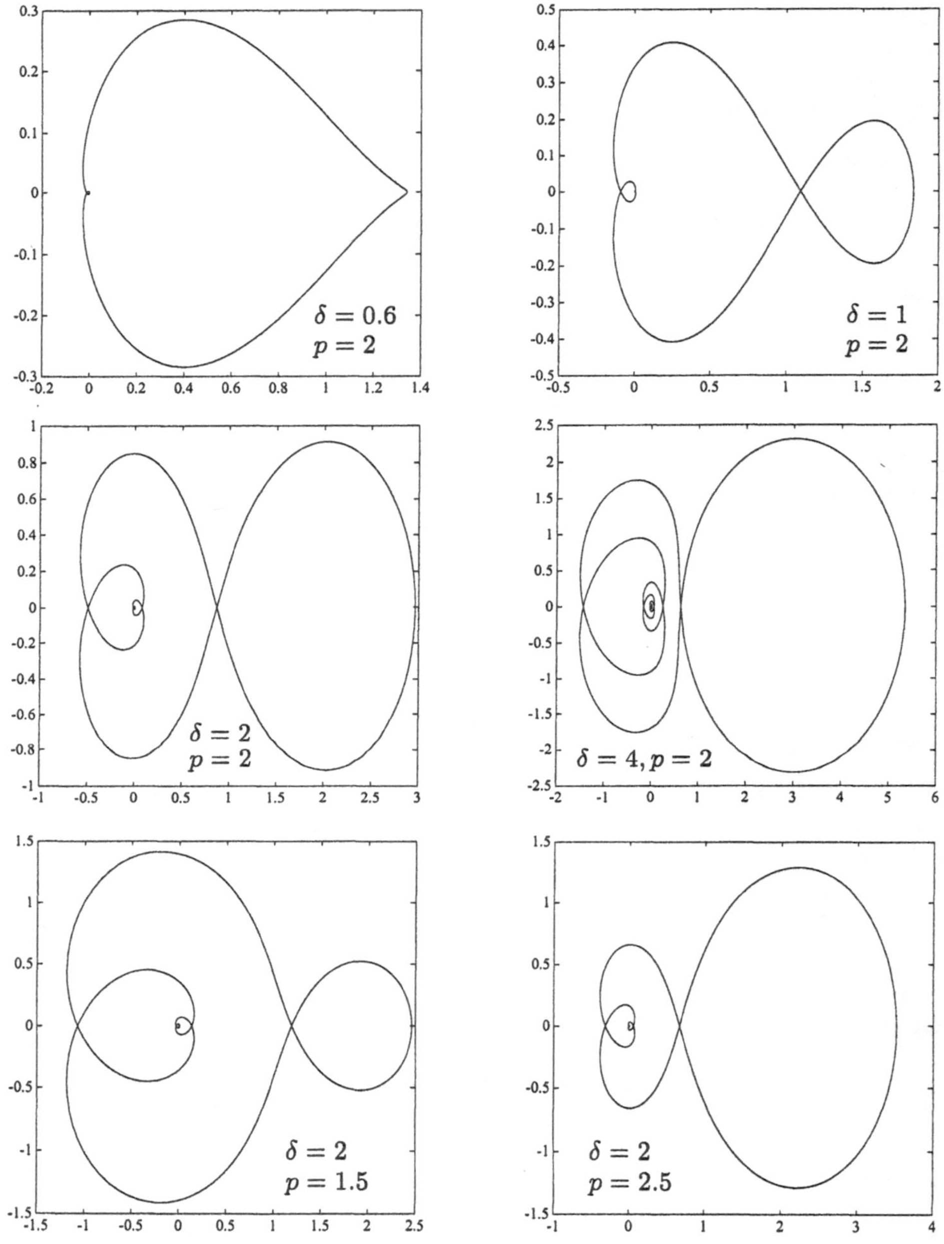

Figure 1: The local spectra $\mathrm{sp}_{t_0}(S + VSV)$ at the starting point t_0 of a simple arc for various values of δ and p.

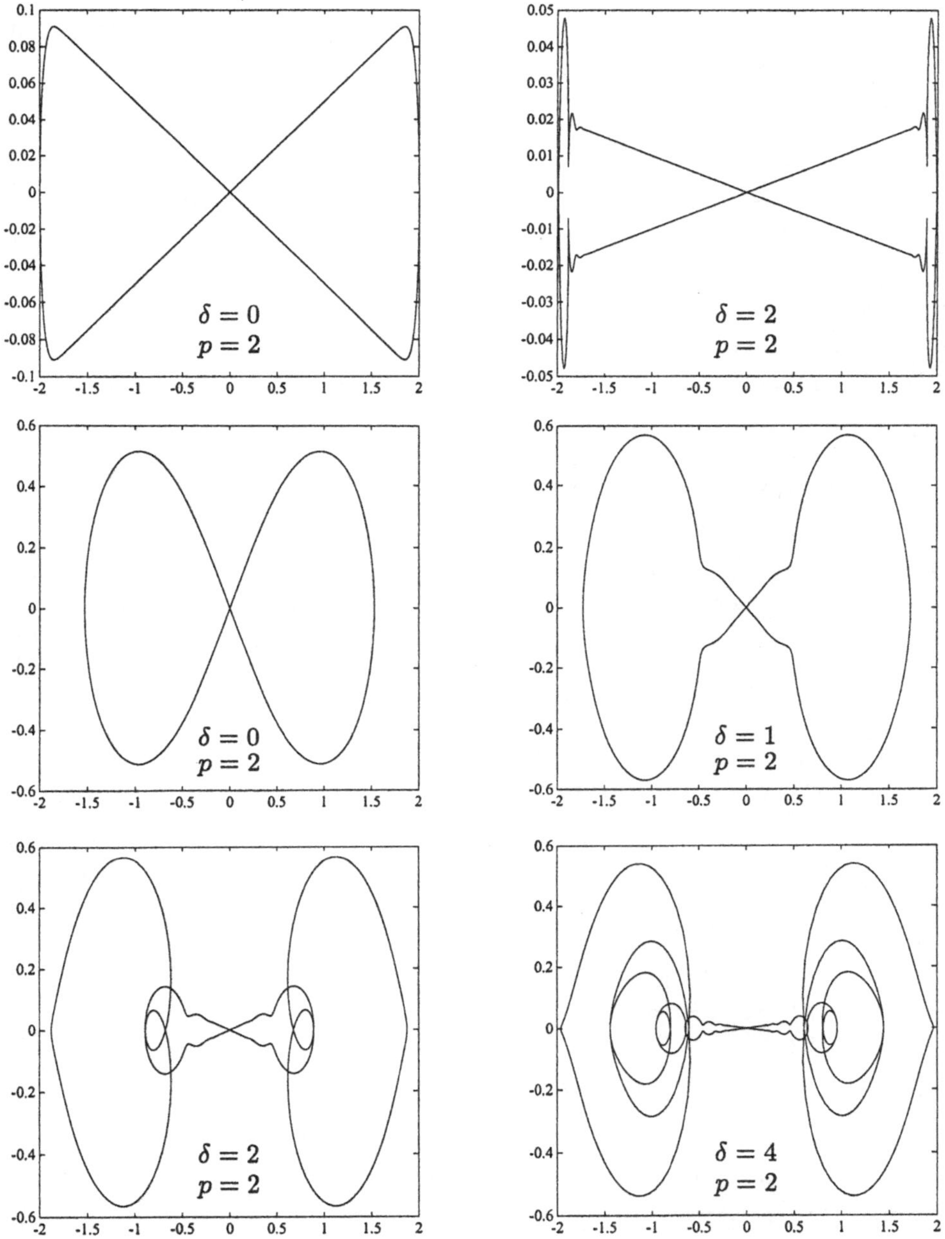

Figure 2: The local spectra $\mathrm{sp}_{t_0}(S + VSV)$ at a point of a Jordan curve with $\mu_2 - \mu_1 = 0.1$ (upper two pictures) for two values of δ and at a point of a Jordan curve with $\mu_2 - \mu_1 = 1.4$ (lower four pictures) for four values of δ.

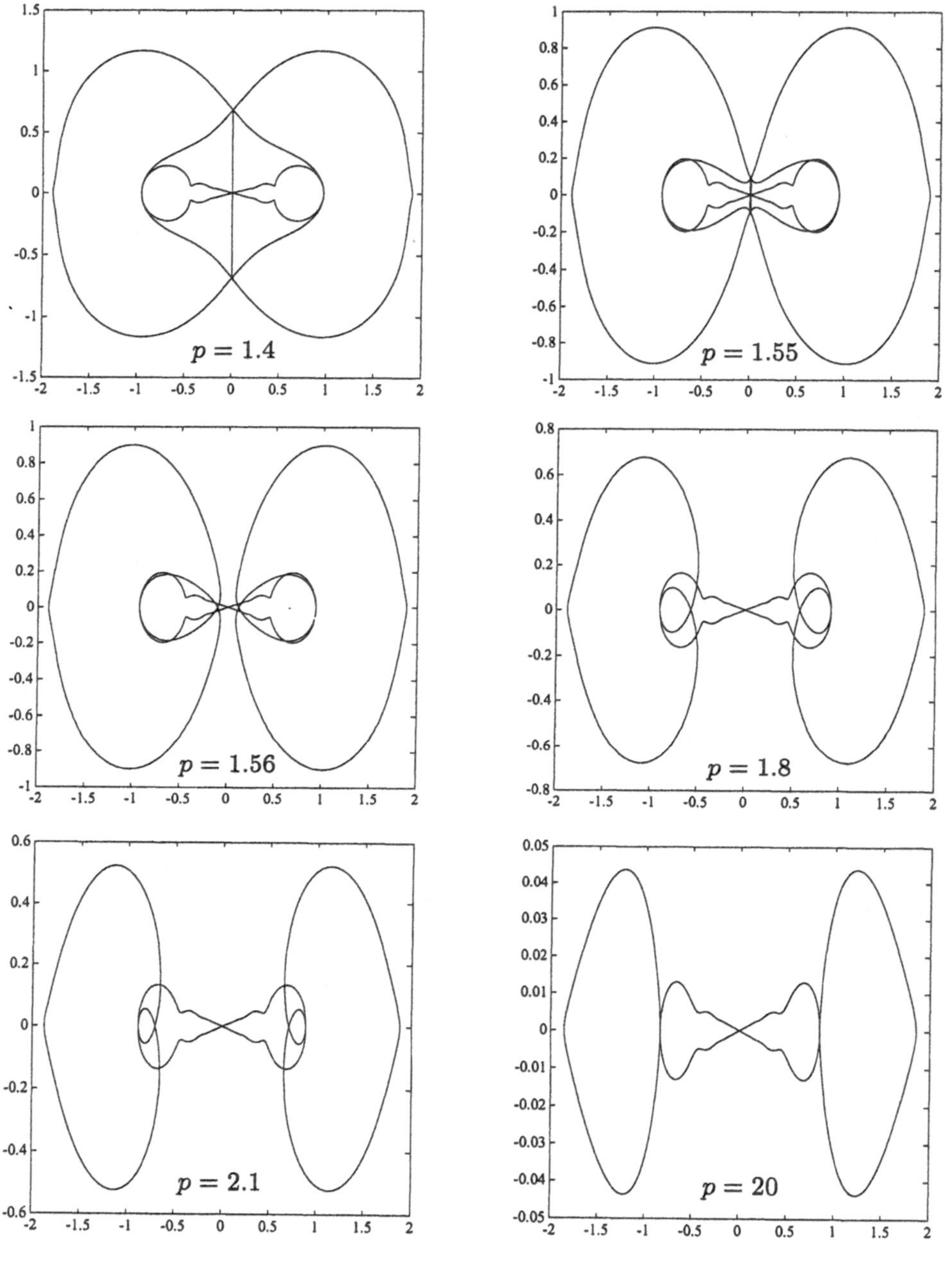

Figure 3: The local spectra $\mathrm{sp}_{t_0}(S + VSV)$ at a point of a Jordan curve with $\mu_2 - \mu_1 = 1.4$ and $\delta = 2$ for six values of p.

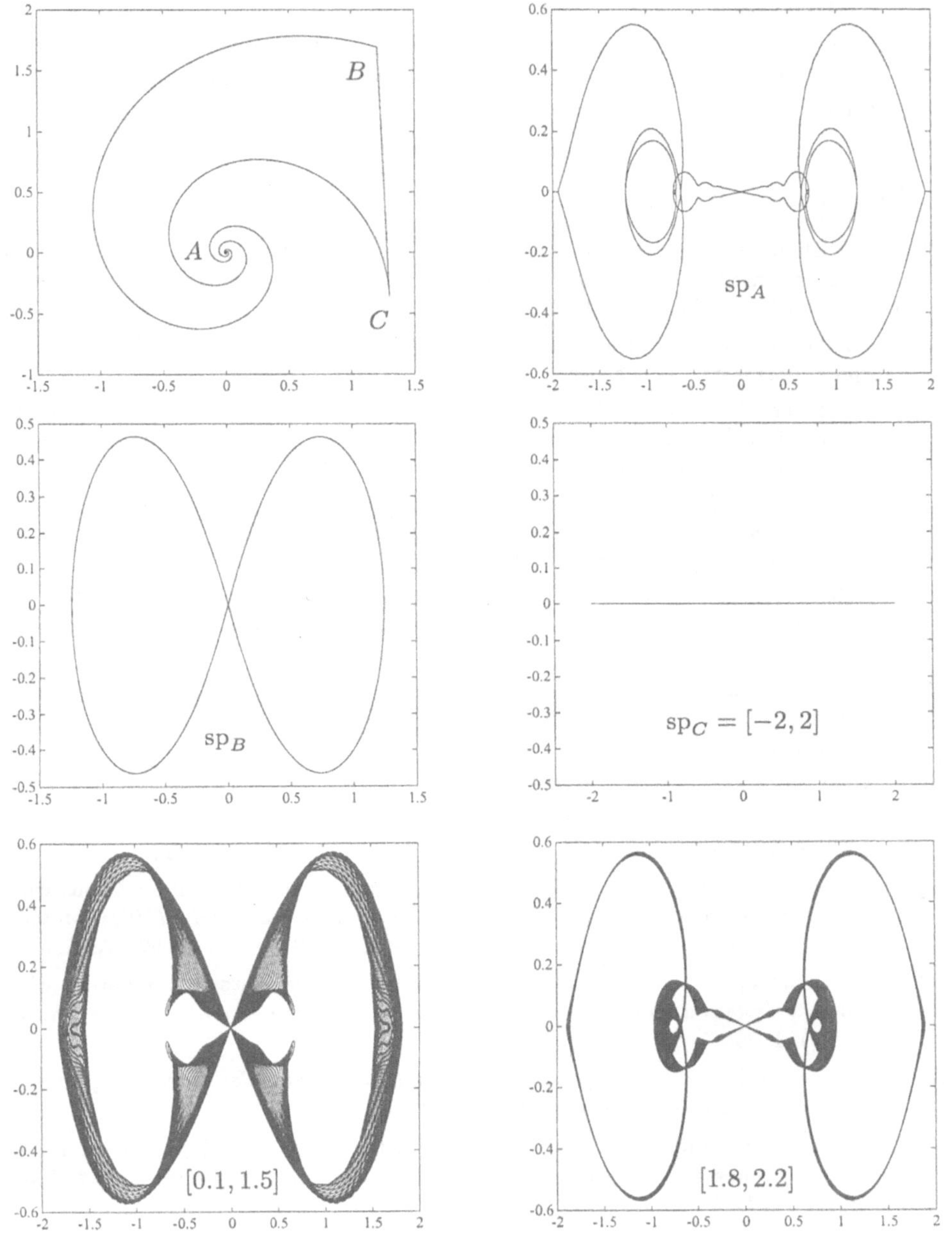

Figure 4: The upper left picture shows a curve three points of which are labeled. The next three pictures show the local spectra of $S + VSV$ at these points. Heavy local spectra are indicated in the lower two pictures.

Acknowledgements

Karlovich was partially supported by CONACYT grant, Cátedra Patrimonial, No. 99017-EX, and CONACYT project 32726-E, México. Rabinovich was partially supported by CONACYT project 32424-E, México.

References

[1] C.J. Bishop, A. Böttcher, Yu.I. Karlovich, I. Spitkovsky: Local spectra and index of singular integral operators with piecewise continuous coefficients on composed curves. *Math. Nachrichten* **206** (1999), 5-83.

[2] A. Böttcher and Yu.I. Karlovich: *Carleson Curves, Muckenhoupt Weights, and Toeplitz Operators*. Birkhäuser Verlag, Basel 1997.

[3] A. Böttcher, Yu.I. Karlovich, V.S. Rabinovich: Emergence, persistence, and disappearance of logarithmic spirals in the spectra of singular integral operators. *Integral Equations and Operator Theory* **25** (1996), 406–444.

[4] A. Böttcher, Yu.I. Karlovich, V.S. Rabinovich: Mellin pseudodifferential operators with slowly varying symbols and singular integrals on Carleson curves with Muckenhoupt weights. *Manuscripta Mathematica* **95** (1998), 363–376.

[5] A. Böttcher, Yu.I. Karlovich, V.S. Rabinovich: The method of limit operators for one-dimensional singular integrals with slowly oscillating data. *J. Operator Theory* **43** (2000), 171-198.

[6] R. Coifman and Y. Meyer: Au delà des opérateurs pseudo-différentiels. *Astérisque* **57** (1978), 1-185.

[7] M. Costabel: Singular integral operators on curves with corners. *Integral Equations and Operator Theory* **3** (1980), 323–349.

[8] R. Duduchava: On general singular integral operators of the plane theory of elasticity. *Rendiconti Sem. Mat. Univ. e Politecn. Torino* **42** (1984), 15–41.

[9] R. Duduchava and T. Latsabidze: On the index of singular integral equations with complexly conjugated functions on piecewise smooth curves. *Trudy Tbiliss. Matem. Inst. im. A.M. Razmadze Akad. Nauk Gruzinsk. SSR* **76** (1985), 40–59 [Russian].

[10] R. Duduchava, T. Latsabidze, A. Saginashvili: Singular integral operators with complex conjugation on curves with cusps. *Integral Equations and Operator Theory* **22** (1995), 1–36.

[11] I. Gohberg and N. Krupnik: *One-Dimensional Linear Singular Integral Equations*. Vols. I and II. Birkhäuser Verlag, Basel 1992.

[12] T. Latsabidze: Singular integral operators with complex conjugation on piecewise smooth curves. *Trudy Tbiliss. Matem. Inst. im. A.M. Razmadze Akad. Nauk Gruzinsk. SSR* **76** (1985), 107–122 [Russian].

[13] V.G. Maz'ya and S. Prössdorf: *Linear and Boundary Integral Equations*. Vol. 27 (Analysis IV) of the Encyclopaedia of Math. Sciences, Springer-Verlag, Berlin, Heidelberg, New York 1991.

[14] N. Muskhelishvili: *Singular Integral Equations*. Nordhoff, Groningen 1963.

[15] M. Nagase: The L^p-boundedness of pseudo-differential operators with non-regular symbols. *Comm. PDE* **2** (1977), 1045-1061.

[16] V. Nyaga: On the singular integral operator with conjugation on contours with angular points. *Matem. Issled.* **73** (1983), 47–50 [Russian].

[17] V. Nyaga: Conditions for the Fredholmness of singular integral operators with conjugation in the case of a piecewise Lyapunov contour. In: *Issled. po Funkts. Analizu i Diff. Uravneniyam*, pp. 90–101, Shtiintsa, Kishinev 1984 [Russian].

[18] V. Parton and P. Perlin: *Integral Equations of Elasticity Theory.* Nauka, Moscow 1977 [Russian].

[19] V.S. Rabinovich: Singular integral operators on a composed contour with oscillating tangent and pseudodifferential Mellin operators. *Soviet Math. Dokl.* **44** (1992), 791–796.

[20] V.S. Rabinovich: Singular integral operators on composed contours and pseudodifferential operators. *Math. Notes* **58** (1995), 722–734.

[21] V.S. Rabinovich: Pseudodifferential operators with analytic symbols and some of their applications. In: *Proceeding of the Seminar on Linear Topological Spaces and Complex Analysis*, pp.79–98, Metu-Tübitak, Ankara 1995.

[22] V.S. Rabinovich: Algebras of singular integral operators on composed contours with nodes that are logarithmic whirl points. *Izv. Ross. Akad. Nauk, Ser. Mat.* **60** (1996), 169–200 [Russian].

[23] V.S. Rabinovich: Mellin pseudodifferential operators techniques in the theory of singular integral operators on some Carleson curves. *Operator Theory: Advances and Applications* **102** (1998), 201–218.

[24] S. Roch and B. Silbermann: *Algebras of Convolution Operators and Their Image in the Calkin Algebra.* Report R–Math–05/90, Karl-Weierstrass-Inst. f. Math., Berlin 1990.

[25] H.S. Shapiro: *The Schwarz Function and Its Generalization to Higher Dimensions.* John Wiley and Sons, Inc., New York 1992.

[26] I.B. Simonenko and Chin Ngok Minh: *The Local Method in the Theory of One-Dimensional Singular Integral Equations with Piecewise Continuous Coefficients. Noethericity.* Rostov-on-Don Univ. Press 1986 [Russian].

TU Chemnitz
Fak. f. Mathematik
09107 Chemnitz
Germany

CINVESTAV del I.P.N.
Departamento de Matemáticas
Apartado Postal 14-740
México D.F. 07000, México

ESIME-Zacatenco
National Polytechnic Inst.
Dept. of Telecommunication
Ed. 1, Av. IPN,
México D.F. 07738, México

1991 Mathematics Subject Classification: Primary 45E05; Secondary 30E25, 31A10, 47G30

Submitted: 27.3.2000

Operator Theory:
Advances and Applications, Vol. 121

Biorthogonal wavelet approximation methods for the heat equation

CHRISTIAN BOURGEOIS, SERGE NICAISE

We consider the integral formulation of the heat equation in a smooth domain of $\mathbb{R}^2$ with Dirichlet and Neumann boundary conditions. The unknown solutions of the integral equations belong to anisotropic Sobolev spaces and are approximated by the Galerkin method using an appropriate wavelet basis. This allows to compress the stiffness matrix from $O(N^2)$ to $O(N)$, and to obtain a uniformly bounded condition number. Finally, we show that the compressed scheme converges as fast as the Galerkin method based on B-splines basis.

1. Introduction

The boundary element method applied to the heat equation on a smooth domain Ω of $\mathbb{R}^2$ leads to the resolution of a two-dimensional problem [4, 9]. Unfortunately, the stiffness matrix is full, and in general ill-conditioned. Many authors (see [5, 6, 7, 12, 13]) have introduced new bases made of wavelets to overcome these difficulties, but only, to our knowledge, for elliptic problems. Therefore our goal is to adapt the strategy to a parabolic case.

In this paper, we extend some previous results (see [1]) into biorthogonal wavelets and all integral representations. The Dirichlet and the Neumann problems will be solved using direct or indirect formulations (see [4, 9]). Their kernels involve an exponential function and the first step is then to check that this function satisfies a decay property in the sense of [13] to fit well in our compression procedure. The second step consists in the characterization of the anisotropic Sobolev spaces $\tilde{H}^{s,p}(\Sigma_T)$ with $s \neq p$ in terms of biorthogonal wavelets, since the integral formulation of the heat equation is given in those spaces. This will be done using tensor products of 1-dimensional wavelets. Moreover, for some of the formulations, the associated bilinear form is no more coercive. We overcome this difficulty by using compact perturbation arguments (see [2, 10]).

The schedule of the paper is the following. In section 2 we recall the integral formulations of the problem in a general setting and define the anisotropic Sobolev spaces which are characterized by a multiscale basis in section 3. The Galerkin method is also presented and, due to the above characterization, we obtain a well-conditioned system after rescaling the basis functions. The next part is devoted to the decay of some coefficients of the stiffness matrix and a compression procedure is retailed to construct a sparse representation. We conclude by giving an error estimate between the exact solution and the solution of the compressed Galerkin scheme in section 5.

2. Integral formulation

In this section, we write the different integral formulations of the heat equation in a general form, in order to treat Dirichlet or Neumann problems by direct or indirect methods as well. Only few parameters will change. The integral formulations can be found in [4, 9].

The heat equation reads as

$$\begin{cases} -\Delta\Phi + \partial_t\Phi = 0 & \text{in } Q_T = \Omega \times (0,T), \\ B\Phi = g & \text{on } \Sigma_T = \Gamma \times (0,T), \\ \Phi(x,0) = 0 & \forall x \in \Omega, \end{cases} \tag{2.1}$$

with

$$B\Phi = \Phi_{|\Sigma_T}, \tag{2.2}$$

for the Dirichlet problem and

$$B\Phi = \partial_n \Phi_{|\Sigma_T}, \tag{2.3}$$

for the Neumann case.

For a given function Φ on Q_T, we introduce the useful notation Φ^+ (resp. Φ^-) for the exterior (resp. interior) limit of Φ on Σ_T.

Let V and W denote the single- and double-layer heat potentials :

$$(Vu)(x,t) = \int_0^t \int_\Gamma u(y,\tau)E(x-y,t-\tau)d\Gamma_y d\tau, \tag{2.4}$$

$$(Wu)(x,t) = \int_0^t \int_\Gamma u(y,\tau)\partial_{n_y}E(x-y,t-\tau)d\Gamma_y d\tau, \tag{2.5}$$

for all $(x,t) \in Q_T \cup Q_T^c$, with $Q_T^c = \Omega^c \times (0,T)$, where E is the fundamental solution of the heat equation:

$$E(x,t) = \frac{H(t)}{4\pi t} \exp\left(-\frac{|x|^2}{4t}\right), \tag{2.6}$$

$H(t)$ being Heaviside's function. We also recall the definitions of the single-layer operator S, the double-layer operator D, its spatial adjoint D', and the hypersingular operator H :

$$(Su)(x,t) = \int_0^t \int_\Gamma u(y,\tau)E(x-y,t-\tau)d\Gamma_y d\tau, \tag{2.7}$$

$$(2.8) \qquad (Du)(x,t) = \int_0^t \int_\Gamma u(y,\tau)\partial_{n_y} E(x-y,t-\tau)d\Gamma_y d\tau,$$

$$(2.9) \qquad (D'u)(x,t) = \int_0^t \int_\Gamma u(y,\tau)\partial_{n_x} E(x-y,t-\tau)d\Gamma_y d\tau,$$

$$(2.10) \qquad (Hu)(x,t) = -\partial_{n_x} \int_0^t \int_\Gamma u(y,\tau)\partial_{n_y} E(x-y,t-\tau)d\Gamma_y d\tau,$$

for $(x,t) \in \Sigma_T$.

There exist several integral representations for the solution Φ of the problem (2.1) which can be written in the form

$$(2.11) \qquad \Phi = Ru,$$

where u is the solution of an integral equation

$$(2.12) \qquad Au = f.$$

A is a pseudodifferential operator bounded on anisotropic Sobolev spaces. These spaces are defined as follows. For $s, p \geq 0$, we set

$$(2.13) \qquad H^{s,p}(\Sigma_T) = L^2((0,T); H^s(\Gamma)) \bigcap L^2(\Gamma; H^p(0,T)),$$

$$(2.14) \qquad \tilde{H}^{s,p}(\Sigma_T) = \{u = U_{|\Sigma_T} \colon U \in H^{s,p}(\Sigma)\ ,\ U(.,t) = 0, t < 0\},$$

if $\Sigma = \Gamma \times \mathbb{R}$ and $H^s(\Gamma)$ is the classic Sobolev space because the boundary Γ is smooth. In the following, the norm of $\tilde{H}^{s,p}(\Sigma_T)$ will be denoted by $\|\cdot\|_{s,p}$.

The solution u of the integral equation (2.12) belongs to the space $\tilde{H}^{s,s/2}(\Sigma_T)$ with $s = \pm\frac{1}{2}$.

For instance, the direct method for the Dirichlet problem gives

$$A = S, \quad u = \partial_n \Phi^-, \quad \text{and } f = (\frac{1}{2}I + D)g.$$

The unknown $u = \partial_n \Phi^-$ belongs to the space $\tilde{H}^{-1/2,-1/4}(\Sigma_T)$ and Φ admits the representation formula

$$\Phi = Vu - Wg.$$

As a second example, the double-layer representation of the Dirichlet problem is given by

$$\Phi = -Wu,$$

where u is solution of

$$(\frac{1}{2}I - D)u = g,$$

with $u = \Phi^- - \Phi^+ \in \tilde{H}^{1/2,1/4}(\Sigma_T)$.

For all the formulations, the operator A is of order $r \in \{r_1, r_2, r_3\} = \{-1, 0, 1\}$ and its principal part A' can be written as :

$$(A'u)(x,t) = \int_0^t \int_\Gamma u(y,\tau) K_i(x,y,t,\tau) d\Gamma_y d\tau, \quad (x,t) \in \Sigma_T, \tag{2.15}$$

with the different kernels

$$K_1(x,y,t,\tau) = E(x-y,t-\tau) \quad \text{if } A' = S, \tag{2.16}$$

$$K_2(x,y,t,\tau) = \begin{cases} \partial_{n_y} E(x-y,t-\tau) & \text{if } A' = D, \\ \partial_{n_x} E(x-y,t-\tau) & \text{if } A' = D', \end{cases} \tag{2.17}$$

$$K_3(x,y,t,\tau) = -\partial_{n_x}\partial_{n_y} E(x-y,t-\tau) \quad \text{if } A' = H. \tag{2.18}$$

The advantage of the boundary element method is the reduction of the complexity, passing from a 3D-problem to a 2D-case. But the discretization of (2.12) leads to densely populated matrices, which are ill-conditioned in general. We avoid these drawbacks by a construction of adapted wavelet basis.

3. Multiscale bases

In this section, we extend the results of [1] to the construction of a biorthogonal wavelet basis which characterizes the anisotropic Sobolev space $\tilde{H}^{s,s/2}(\Sigma_T)$, with $s \geq -1/2$. The idea is to use tensor products of wavelets in one dimension but with two uniform meshes different in space and time.

First of all, there exists a smooth parametrisation of Σ_T by

$$\begin{aligned} \phi\colon [0,1]^2 &\to \Gamma \times (0,T) \\ (\theta, s) &\to (x(\theta), Ts). \end{aligned}$$

This mapping ϕ yields an isomorphism between $\tilde{H}^{s,s/2}(\Sigma_T)$ and $\tilde{H}^{s,s/2}([0,1]^2)$.

Therefore, we subdivide $[0,1]^2$ into two uniform meshes, dividing $[0,1]$ by 2^j subintervals (space-part) and 2^{2j} subintervals (time-part) with $j \geq 0$.

We take V_j as the following approximation space

$$\begin{aligned} V_j = \{&f(x,t) \in C([0,1]^2) \text{ s.t. } f(x,.) \\ &\text{and } f(.,t) \text{ are piecewise linear on } [0,1]\}. \end{aligned} \tag{3.1}$$

As usual, for any $\lambda = (j,k) \in \mathbb{Z}^{\not k}$, we set

$$\psi_{j,k}(x) = 2^{j/2}\psi(2^j x - k).$$

We will denote by $\tilde{\Psi} = \{\tilde{\psi}_{j,k} : (j,k) \in \nabla\}$ the dual system, where $\nabla = \{(j,k) : k \in \nabla_j, j = -1, 0, 1, \ldots\}$, i.e.

$$\langle \psi_{j,k}, \tilde{\psi}_{j',k'} \rangle = \delta_{(j,k),(j',k')}, \quad (j,k),(j',k') \in \nabla.$$

The space-part basis is constructed by taking a periodized biorthogonal wavelet basis of $[0,1]$. This set is denoted by

$$\Psi_1 = \{\psi_\lambda : \lambda \in \nabla_1\}, \tag{3.2}$$

and the wavelets are exact of order d_1 and have $\tilde{d}_1$ vanishing moments.

For the time-part, we define the space :

$$\tilde{H}^s([0,1]) = \{u = U_{|[0,1]} : U \in H^s(\mathbb{R}) \text{ and } U(t) = 0, t < 0\}.$$

Introducing the set $Z \subseteq \{0,1\}$ to indicate the location of the Dirichlet boundary condition (here $Z = \{0\}$), we infer from [8] the existence of a biorthogonal wavelet basis which characterizes the Sobolev space $\tilde{H}^s([0,1])$. The resulting system is written

$$\Psi^Z = \{\psi_\lambda^Z : \lambda \in \nabla^Z\} \tag{3.3}$$

$$= \{\psi_\lambda^L\} \cup \{\psi_\lambda^I\} \cup \{\psi_\lambda^R\}, \tag{3.4}$$

and

$$\tilde{\Psi}^Z = \{\tilde{\psi}_\lambda^{\tilde{Z}} : \lambda \in \nabla^Z\}$$

for the dual system. The set Ψ^Z is composed of interior wavelets Ψ_λ^I which are the "classical" biorthogonal wavelets whose support is strictly included in $(0,1)$. The end-point wavelets Ψ^L, Ψ^R are constructed by integrating on the left lower order wavelets (which raises the number of zero boundary conditions) and differentiating on the right. This process preserves the stability of the system and a biorthogonalization leads to the final result. We refer to [8] for more details.

Once the set Ψ^Z is obtained, we set for $j \geq 0$ and $Z = \{0\}$:

$$\{\Psi_{2;j}\} = \{\Psi_{2j}^Z\} \cup \{\Psi_{2j-1}^Z\}.$$

We define a sequence of spaces W_j by $V_{j+1} = V_j \overset{\not\perp}{\bigoplus} W_j$, where

$$W_j = Span\{\Psi_{1;j,k_1}(x)\Psi_{2;j,k_2}(t)\}_{\substack{k_1 \in \nabla_{1j},\\ k_2 \in \nabla_{2j}}} \tag{3.5}$$

$$:= Span\{\Xi_{j,K}(x,t)\}_{K\in\nabla}, \tag{3.6}$$

for $j \geq j_0$. Now we can prove the following result :

Theorem 3.1. *Let $\{N_{j_0;k_1,k_2}\}_{k_1,k_2}$ be a basis of V_{j_0}. If*

$$u(x,t) = \sum_{k_1,k_2=0}^{k_0} d_{j_0;k_1,k_2} N_{j_0;k_1,k_2}(x,t) + \sum_{j\geq j_0}\sum_{K\in\nabla} c_{j,K}\Xi_{j,K}(x,t)$$

belongs to $\tilde{H}^{s,s/2}([0,1]^2)$ *for* $s \in [-\frac{1}{2}, \frac{5}{2})$ *(with the degree of exactness of the space-part basis* $d_1 \geq 2$ *if* $s < 3/2$*;* $d_1 \geq 3$ *if* $s \in [3/2, 5/2)$ *and the degree of exactness of the time-part basis* $d_2 \geq 2$*), then we have*

$$(3.7) \qquad \|u\|^2_{\tilde{H}^{s,s/2}([0,1]^2)} \sim \sum_{k_1,k_2=0}^{k_0} |d_{j_0;k_1,k_2}|^2 + \sum_{j,K} 2^{2sj} |c_{j,K}|^2.$$

Proof. We recall that

$$\begin{aligned} \tilde{H}^{s,s/2}([0,1]^2) &= L^2([0,1]; H^s([0,1])) \cap L^2([0,1]; \tilde{H}^{s/2}([0,1])) \\ &= H_1 \cap H_2. \end{aligned}$$

We write

$$\begin{aligned} u(x,t) &= \sum_{k_1,k_2=0}^{k_0} d_{j_0;k_1,k_2} \varphi_{j_0,k_1}(x) \cdot \varphi_{j_0,k_2}(t) \\ &+ \sum_{\substack{j \geq j_0, \\ k_1=0,\ldots,k_0, k_2 \in \nabla_{2j}}} c^1_{j;k_1,k_2} \varphi_{j_0,k_1}(x) \cdot \Psi_{2;j_0,k_2}(t) \\ &+ \sum_{\substack{j \geq j_0, \\ k_1 \in \nabla_{1j}, k_2=0,\ldots,k_0}} c^2_{j;k_1,k_2} \Psi_{1;j,k_1}(x) \cdot \varphi_{j_0,k_2}(t) \\ &+ \sum_{\substack{j \geq j_0, \\ k_1 \in \nabla_{1j}, k_2 \in \nabla_{2j}}} c^3_{j;k_1,k_2} \Psi_{1;j,k_1}(x) \cdot \Psi_{2;j,k_2}(t) \\ &:= \sum_{i=0}^{3} u^i(x,t). \end{aligned}$$

Each term of the sum is estimated in the following way. We write

$$u^1(x,t) = \sum_{\substack{j \geq j_0, \\ k_1=0,\ldots,k_0}} c^1_{j,k_1}(t) \varphi_{j_0,k_1}(x),$$

with

$$c^1_{j,k_1}(t) := \sum_{k_2 \in \nabla_{2j}} c^1_{j;k_1,k_2} \Psi_{2;j,k_2}(t) \in W_j.$$

Due to the characterization of $\tilde{H}^{s/2}([0,1])$ by the set $\{\Psi_{2;j,k_2}(t)\}$, we get

$$\|u^1\|^2_{H_2} \sim \sum_{j,k_1,k_2} 2^{2sj} |c^1_{j;k_1,k_2}|^2.$$

In the same way, the set $\{\Psi_{1;j,k_1}\}$ characterizes $H^s([0,1])$ and we have

$$\|u^2\|^2_{H_1} \sim \sum_{j,k_1,k_2} 2^{2sj} |c^2_{j;k_1,k_2}|^2,$$

where

$$u^2(x,t) = \sum_{\substack{j \geq j_0, \\ k_1=0,\ldots,k_0}} c^2_{j,k_1,k_2} \Psi_{1;j,k_1}(x) \varphi_{j_0,k_2}(t).$$

We therefore obtain the following equivalences :

$$\begin{aligned} \|u\|^2_{H_1} \sim & \sum_{k_1,k_2} |d_{j_0;k_1,k_2}|^2 + \sum_{j,k_1,k_2} |c^1_{j;k_1,k_2}|^2 \\ & + \sum_{j,k_1,k_2} 2^{2sj} |c^2_{j;k_1,k_2}|^2 + \sum_{j,k_1,k_2} 2^{2sj} |c^3_{j;k_1,k_2}|^2, \end{aligned}$$

and

$$\begin{aligned} \|u\|^2_{H_2} \sim & \sum_{k_1,k_2} |d_{j_0;k_1,k_2}|^2 + \sum_{j,k_1,k_2} 2^{2sj} |c^1_{j;k_1,k_2}|^2 \\ & + \sum_{j,k_1,k_2} |c^2_{j;k_1,k_2}|^2 + \sum_{j,k_1,k_2} 2^{2sj} |c^3_{j;k_1,k_2}|^2, \end{aligned}$$

which leads to (3.7) by definition of the space $\tilde{H}^{s,s/2}([0,1])$. □

We now write the variational formulation of problem (2.12): we seek the solution $u \in \tilde{H}^{s/2,s/4}(\Sigma_T)$, with $s \in \{-1,1\}$ of

$$a(u,v) = \langle Au, v\rangle_{\Sigma_T} = \langle f, v\rangle_{\Sigma_T}, \forall v \in \tilde{H}^{r/2,r/4}(\Sigma_T). \tag{3.8}$$

The coercivity of the bilinear form a is known (see [4]) if $A = S$ or $A = H$. The main properties of the single-layer operator S and hypersingular operator H are the following :

Proposition 3.2.

- *The operator* $S : \tilde{H}^{s,s/2}(\Sigma_T) \longrightarrow \tilde{H}^{s+1,\frac{s+1}{2}}(\Sigma_T)$ *is an isomorphism for all* $s \geq -1/2$ *and* S *is strongly coercive :*

$$(u, Su)_{\Sigma_T} \geq C\|u\|^2_{-\frac{1}{2},-\frac{1}{4}}, \forall u \in \tilde{H}^{-1/2,-1/4}(\Sigma_T).$$

- *The operator* $H : \tilde{H}^{s,s/2}(\Sigma_T) \longrightarrow \tilde{H}^{s-1,\frac{s-1}{2}}(\Sigma_T)$ *is an isomorphism for all* $s \geq 1/2$ *and*

$$(u, Hu)_{\Sigma_T} \geq C\|u\|^2_{\frac{1}{2},\frac{1}{4}}, \forall u \in \tilde{H}^{1/2,1/4}(\Sigma_T).$$

In the case $A = \frac{1}{2}I \pm D$, a is not coercive on $L^2(\Sigma_T)$ and we use the compact perturbation arguments of [10] to overcome this difficulty.

Theorem 3.3. *Let* H *be a separable Hilbert space with a norm denoted by* $\|.\|$, *let* A *be a one-to-one continuous operator on* H *and let* H_n, $n \in \mathbb{N}$ *be a family of* *finite-dimensional subspaces of H. Denote by P_n the orthogonal projection on H_n and assume that*

$$\|(I - P_n)w\| \to 0, \quad as\ n \to \infty, \forall w \in H. \tag{3.9}$$

If there exists a positive constant α and an integer N such that

$$\|P_n A x_n\| \geq \alpha \|x_n\|, \forall x_n \in H_n, \forall n > N, \tag{3.10}$$

then for all $f \in H$, the problem

$$Ay = f \tag{3.11}$$

has a unique solution $y \in H$. Moreover, for all $n > N$, there exists a unique solution $y_n \in H_n$ of the approximate problem

$$P_n A y_n = P_n f \tag{3.12}$$

satisfying

$$\|y - y_n\| \leq c\|y - P_n y\|. \tag{3.13}$$

In our setting, the assumption (3.10) is satisfied thanks to the following result (see [2]):

Lemma 3.4. *If $A = \frac{I}{2} \pm D$ is an isomorphism from H into itself with a compact operator D, then A satisfies (3.10).*

Proof. This is a direct consequence of the Lemma in [10]. □

Since the compactness of D is shown in [4], Theorem 3.3 and Lemma 3.4 guarantee existence and uniqueness for the approximation problem: find the solution $u_L \in V_L$ of

$$a(u_L, v_L) = \langle f, v_L \rangle_{\Sigma_T}, \forall v_L \in V_L. \tag{3.14}$$

Consequently, the subsequent result is a direct consequence of Theorems 3.1 and 3.3 and Proposition 3.2 :

Theorem 3.5. *Let $u \in \tilde{H}^{t,t/2}(\Sigma_T)$ be the solution of (2.12) with $t \in [-\frac{1}{2}, \frac{5}{2})$. Then its Galerkin approximation $u_L \in V_L$ satisfies*

$$\|u - u_L\|_{r/2,r/4} \leq C 2^{-(t-\frac{r}{2})L} \|u\|_{t,t/2}. \tag{3.15}$$

If we denote by A_L the corresponding stiffness matrix in the basis $\{\Xi_{j;K}\}$, a rescaling of the basis functions allows to obtain a well-conditioned stiffness matrix.

Proposition 3.6. *Let D_s be the diagonal matrix with entries*

$$2^{sj}\delta_{(j,k),(j',k')}, \quad (j,k),(j',k') \in \nabla.$$

Then

$$Cond\,(D_{-r/2} A_L D_{-r/2}) \leq C.$$

4. Compression of the stiffness matrix

In this section, we retail the compression procedure of the stiffness matrix. When the operator A is of the form $A = \frac{1}{2}I \pm D$, we only compress the nonlocal operator D. Denoting by A' the "nonlocal" part of A (namely $A' = \pm D$ in this case), we introduce the bilinear form

$$a'(u,v) := \langle A'u, v\rangle_{\Sigma_T}. \tag{4.1}$$

For $k_1 \in \nabla_1, k_2 \in \nabla_2, k_1' \in \nabla_1', k_2' \in \nabla_2'$, we introduce the notation:

$$a'(\Xi_{j;k_1,k_2}, \Xi_{j';k_1',k_2'}) = A'_{j,j';K,K'}.$$

For two continuous functions ϕ, ψ with a compact support on Γ, let $S(\phi)$ be the interior of the support of ϕ and let $\delta(\phi,\psi)$ be the distance between $S(\phi)$ and $S(\psi)$.

We also introduce the "pseudo-distance":

$$dist_{j,j';K,K'} = \delta(\Psi_{1;j,k_1}, \Psi_{1;j',k_1'})^2 + \delta(\Psi_{2;j,k_2}, \Psi_{2;j',k_2'}). \tag{4.2}$$

The decay property of $A'_{j,j';K;K'}$ can be expressed as follows:

Proposition 4.1. *For all k_1, k_1', k_2, k_2' as above, one has*

$$|A'_{j,j';K,K'}| \leq C \frac{2^{-b_i(j+j')}}{(dist_{j,j';K,K'})^{b_i}}, \tag{4.3}$$

with $b_i = \tilde{d}_1 + 2\tilde{d}_2 + \frac{r_i+3}{2}$, for $i = 1,2,3$, where $r_1 = -1$ if $A = S$, $r_2 = 0$ if $A = \frac{1}{2}I \pm D$, and $r_3 = +1$ if $A = H$.

Proof. By Fubini's theorem, we have

$$\begin{aligned} A'_{j,j';K,K'} &= \int_{\Gamma^2\times(0,T)\times(0,\tau)} \Psi_{2;j',k_2'}(t)\Psi_{1;j',k_1'}(x)\Psi_{2;j,k_2}(\tau)\Psi_{1;j,k_1}(y) \\ &\quad K_i(x,y,t,\tau)d\Gamma_x d\Gamma_y dt d\tau, \end{aligned}$$

with K_i, $i \in \{1,2,3\}$ beeing defined by (2.16)-(2.18). By definition of the moment property of Ψ_1 and Ψ_2, there exists functions θ_1, θ_2 such that

$$\Psi_{1;j,k_1} = \theta^{(\tilde{d}_1)}_{1;j,k_1}, \quad \Psi_{2;j,k_2} = \theta^{(\tilde{d}_2)}_{2;2j,\hat{k}_2}, \quad \Psi_{2;j,k_2} = \theta^{(\tilde{d}_2)}_{2;2j-1,\tilde{k}_2}.$$

Therefore, after integrating by parts, we may write

$$\begin{aligned} |A'_{j,j';K,K'}| &\leq C \int_{[0,1]^2\times(0,T)\times(0,t)} \left| \frac{\partial^{2(\tilde{d}_1+\tilde{d}_2)} K_i(x,y,t,\tau)}{\partial x^{\tilde{d}_1}\partial y^{\tilde{d}_1}\partial t^{\tilde{d}_2}\partial\tau^{\tilde{d}_2}} \right| \\ &\quad \times \left| \theta_{2;2j,\hat{k}_2}(\tau)\theta_{2;2j',\hat{k}_2'}(t)\theta_{1;j,k_1}(y)\theta_{1;j',k_1'}(x) \right| dxdydtd\tau. \end{aligned}$$

We get the conclusion thanks to the two following estimates

$$\begin{aligned}\left|\frac{\partial^{2(\tilde d_1+\tilde d_2)}K_i(x,y,t,\tau)}{\partial x^{\tilde d_1}\partial y^{\tilde d_1}\partial t^{\tilde d_2}\partial\tau^{\tilde d_1}}\right| &\leq C\frac{1}{(|x-y|^2+|t-\tau|)^{b_i}}, \quad \forall i\in\{1,\dots,3\},\\ \|\theta_{i;j,k}\|_{L^1([0,1])} &\leq C2^{-(\tilde d_i+\frac{1}{2})j}, i\in\{1,2\}.\end{aligned}$$

□

For $j,j'\in\{j_0,\dots,L\}$, we can define the compressed subblocks

$$\tilde A'_{j,j'}=\left(\tilde A'_{j,j';K,K'}\right)_{\substack{K\in\nabla,\\ K'\in\nabla'}}$$

where we set

$$\tilde A'_{j,j';K,K'}=\begin{cases}0 & \text{if } dist_{j,j';K,K'}\geq \mathcal{B}_{j,j'}\\ A'_{j,j';K,K'} & \text{else.}\end{cases} \tag{4.4}$$

We associate with A'_L (resp. $\tilde A'_L$, the global compressed matrix) the operator $\mathcal{A}'_L$ (resp. $\tilde{\mathcal{A}}'_L$) from V_L into its dual $\tilde V_L=\mathrm{Span}(\tilde\Psi_j)$. The next result gives an estimate for the consistancy part of the error.

Theorem 4.2. *Let $t,\tilde t\in[-\frac{1}{2},\frac{5}{2})$ and assume that the parameters $\mathcal{B}_{j,j'}$ satisfy*

$$\mathcal{B}_{j,j'}\geq\max\{2^{-2j},2^{-2j'},a2^{\alpha(L-j)}2^{\tilde\alpha(L-j')}2^{-2L}\}, \tag{4.5}$$

for $\alpha,\tilde\alpha,a>0$ such that

$$\alpha>\frac{t+b_i+\tau}{b_i-\frac{r_i+3}{2}},\quad \tilde\alpha>\frac{\tilde t+b_i-r_i-3+\tau}{b_i-\frac{r_i+3}{2}},\quad \forall i\in\{1,2,3\}, \tag{4.6}$$

and for some $\tau>0$. The parameter a is arbitrary. Then for any $u_L,\tilde u_L\in V_L$, we have

$$\left|\left\langle(\mathcal{A}'_L-\tilde{\mathcal{A}}'_L)u_L,\tilde u_L\right\rangle\right|\leq Ca^{3/2-b_i}2^{-L(t+\tilde t-r_i)}\|\mu_L\|_{t,t/2}\|\tilde\mu_L\|_{\tilde t,\tilde t/2}. \tag{4.7}$$

Proof. By Theorem 3.1 and the definition of $\tilde A'_L$, we have

$$\left|\left\langle(\mathcal{A}'_L-\tilde{\mathcal{A}}'_L)u_L,\tilde u_L\right\rangle\right|\leq C2^{-L(t+\tilde t-r_i)}\|E_L\|_2\|\mu_L\|_{t,t/2}\|\tilde\mu_L\|_{\tilde t,\tilde t/2},$$

where we define the matrix E_L by

$$E_L=2^{(L-j)t}2^{(L-j')\tilde t}\left(A'_{j,j';K,K'}-\tilde A'_{j,j';K,K'}\right)_{j,j';K,K'}.$$

We estimate $\|E_L\|_2$ by the Schur lemma (see [12]) with the sequence $\gamma_{j,K}=2^{-\tau j}$, for a real number $\tau>0$. Using the estimate (see [1] for the proof)

$$\|A'_{j,j'}-\tilde A'_{j,j'}\|_\infty\leq C2^{-b_ij}2^{-(b_i-3)j'}\mathcal{B}_{j,j'}^{-b_i}\max\{\mathcal{B}_{j,j'}^{-3/2},2^{-3j},2^{-3j'}\},$$

and the hypothesis (4.5), we obtain

$$\sum_{j',K'} |E_{L;j,j';K,K'}|\gamma_{j',K'} \leq Ca^{3/2-b_i}\gamma_{j,K}2^{(L-j)[t+(\frac{3+r_i}{2}-b_i)(\alpha-\kappa_1)]} \times \sum_{j'=j_0}^{L-1} 2^{(L-j')[\tilde{t}+(\frac{3+r_i}{2}-b_i)(\tilde{\alpha}-\kappa_1')]},$$

for two parameters κ_1, κ_1' satisfying $\kappa_1 + \kappa_1' = 2$, $\kappa_1(b_i - \frac{3+r_i}{2}) = b_i - \tau$ and $\kappa_1'(b_i - \frac{3+r_i}{2}) = b_i - r_i - 3 + \tau$. If (4.6) holds, we deduce from the above estimate that

$$\sum_{j',K'} |E_{L;j,j';K,K'}|\gamma_{j',K'} \leq Ca^{3/2-b_i}\gamma_{j,K}.$$

We similarly show, under the assumption (4.6), that

$$\sum_{j,K} |E_{L;j,j';K,K'}|\gamma_{j,K} \leq Ca^{3/2-b_i}\gamma_{j',K'}.$$

□

Concerning the number of non-zero entries of the truncated matrix $\tilde{A}_L'$, we can prove the next Lemma (see [1])

Lemma 4.3. *Assume that $\alpha, \tilde{\alpha} < 2$ for all $\{d_i\}_{i=1,2}$ with $d_i \in \{2,3\}$. Suppose that the parameters $\mathcal{B}_{j,j'}$ satisfy*

$$\mathcal{B}_{j,j'} = \max\{2^{-2j}, 2^{-2j'}, a2^{\alpha(L-j)}2^{\tilde{\alpha}(L-j')}2^{-2L}\}. \tag{4.8}$$

Then the number of non-zero elements of $\tilde{D}_L$ is of order $N_L \log N_L$, when $N_L = 2^{3L}$ is the size of the matrix.

Remark 4.4. For the sake of brevity, we do not retail the second compression defined for matrix entries corresponding to wavelets of different scales with overlapping support. This second compression allows to reduce the number of non-zero elements from $O(N^2)$ to $O(N)$ (see [13]).

5. Convergence

The compressed Galerkin scheme is defined by $\tilde{A}_L\vec{\tilde{u}}_L = f$ or equivalently by

$$\langle\tilde{A}_L\tilde{u}_L, v_L\rangle = \langle f, v_L\rangle, \forall v_L \in V_L,$$

when the components of $\vec{\tilde{u}}_L$ are the coefficients of $\tilde{u}_L$ with respect to the wavelet basis. To estimate the error between u and $\tilde{u}_L$, we first treat the cases where the bilinear form a is coercive. For these cases, we apply directly the first Strang Lemma. When a is not coercive, we state an equivalent of this Lemma (see [3]).

Theorem 5.1. *Under the assumption of Theorem 3.3, assume that there exists a sequence of operators A_n from H_n to H_n such that A_n is injective. Then the error between the exact solution $y \in H$ of $Ay = f$ and the approximation $z_n \in H_n$ determined by*

$$A_n z_n = P_n f,$$

is estimated as follows:

$$\|y - y_n\| \leq C\left(\|y - P_n y\| + \|P_n A z_n - A_n z_n\|\right).$$

It remains to check that the compressed scheme is invertible.

Lemma 5.2. *Assume that (4.5) and (4.6) are satisfied, for all $d_i \in \{2,3\}, i = 1,2$ with $t = \tilde{t} = \frac{r}{2}$. Then there exists $a_\star > 0$ (independent of L) such that for all $a \geq a_\star$, the operator $\tilde{\mathcal{A}}_L$ is injective and satisfies*

$$\|\tilde{\mathcal{A}}_L x_n\| \geq \beta \|x_n\|,$$

for an appropriate constant $\beta > 0$.

Proof. By Theorem 4.2 and assumption (3.10) satisfied by $\mathcal{A}_L$ for L large enough, we get that

$$\|\mathcal{A}_L - \tilde{\mathcal{A}}_L\| \leq \alpha \leq \|\mathcal{A}_L^{-1}\|^{-1}.$$

This yields the desired injectivity of $\tilde{\mathcal{A}}_L$ with the help of Neumann's series. □

Corollary 5.3. *Under the assumptions of the previous Lemma, for all $a \geq a_\star$, we have*

$$\kappa(\tilde{\mathcal{A}}_L) \leq C.$$

Finally, we have the following error estimate.

Theorem 5.4. *Let the assumptions of Theorem 3.5 be satisfied. Assume that (4.6) holds, for all $d_i \in \{2,3\}, i = 1,2$, $t \in [-\frac{1}{2}, \frac{5}{2})$ and $\tilde{t} = \frac{r}{2}$. Assume that the compression parameters satisfy (4.5). Then there exists an $a_\star > 0$ (independent of L) such that for all $a \geq a_\star$, we have*

$$\|u - \tilde{u}_L\|_{r/2,r/4} \leq C 2^{-(t-\frac{r}{2})L} \|\mu\|_{t,t/2}. \tag{5.1}$$

Proof. We denote by Q_L the projection on V_L along W_L. By the first Strang Lemma or by Theorem 5.1, we get

$$\|u - \tilde{u}_L\|_{r/2,r/4} \leq C\left(\|u - Q_L u\|_{r/2,r/4} + \|P_L \mathcal{D}_L \tilde{u}_L - \tilde{\mathcal{D}}_L \tilde{u}_L\|_{r/2,r/4}\right).$$

The first term on the right hand side is estimated by Theorem 3.5 while the second one is estimated by Theorem 4.2. □

References

[1] C. Bourgeois and S. Nicaise, *Prewavelet analysis of the heat equation,* Num. Math., 2000, to appear.

[2] M. Bourlard, S. Nicaise and L. Paquet, *An adapted Galerkin method for the resolution of Dirichlet and Neumann problems in a polygonal domain,* Math. Meth. Appl. Sci., **12**, 1990, 251-265.

[3] P.G. Ciarlet, *The finite element method for elliptic problems,* St. Math. Appl., **4**, North-Holland, 1978.

[4] M. Costabel, *Boundary integral operators for the heat equation,* Int. Eq. Op. Th., **13**, 1990, 498-552.

[5] W. Dahmen and A. Kunoth, *Multilevel preconditioning,* Num. Math., **63**, 1992, 315-344.

[6] W. Dahmen, S. Prößdorf and R. Schneider, *Wavelet approximation methods for pseudodifferential equations I : Stability and convergence,* Math. Z., **215**, 1994, 583-620.

[7] W. Dahmen, S. Prößdorf and R. Schneider, *Wavelet approximation methods for pseudodifferential equations II: Matrix compression and fast solution,* Advances in Comp. Math., **1**, 1993, 259-335.

[8] W. Dahmen and R. Schneider, *Wavelets with complementary boundary conditions-function spaces on the cube,* Result. Math. 34, 1998, 255-293.

[9] G.C. Hsiao and J. Saranen, *Boundary integral solution of the two-dimensional heat equation,* Math. Meth. Appl. Sc., **16**, 1993, 87-114.

[10] S. Hildebrandt and E. Wienholtz, *Constructive proofs of representation theorems in separable Hilbert space,* Comm. Pure Appl. Math., **17**, 1964, 369-373.

[11] J.L. Lions and E. Magenes, *Problèmes aux limites non homogènes et applications,* T.**1**, Dunod, 1968.

[12] Y. Meyer, *Ondelettes et opérateurs 2: Opérateur de Calderon-Zygmund,* Hermann, Paris, 1990.

[13] R. Schneider, *Multiscalen- und Wavelet- Matrixkompression: Analysisbasierte Methoden zur effizienten Lösung grosser vollbesetzter Gleichungssystem,* Advances in Numerical Mathematics, Teubner, Stuttgart, 1998.

Christian Bourgeois
FB Mathematik, TU Chemnitz,
D-09107 Chemnitz, Germany

Serge Nicaise
Université de Valenciennes, MACS, I.S.T.V.,
B.P. 311, F-59304 Valenciennes Cedex, France

1991 Mathematics Subject Classification: Primary 65 N 38; Secondary 45 L 10

Submitted: 28.4.2000

Operator Theory:
Advances and Applications, Vol. 121

Orientable and nonorientable Riemann–Hilbert problems

MESSOUD A. EFENDIEV, WOLFGANG L. WENDLAND

Dedicated to the memory of Siegfried Prössdorf

We introduce orientable and nonorientable Riemann-Hilbert problems. We prove that the number of connected components of solutions differs significantly for the orientable and nonorientable cases; that is, countably many to two, respectively. Moreover we analyze CW–structures of connected components. We use the degree of quasilinear Fredholm maps to prove global existence of solutions to Riemann–Hilbert problems.

1. Introduction

We consider here nonlinear Riemann–Hilbert problems (abbr. RHP) of the following form:

Let G_q be a given q–connected domain, with boundary $\partial G_q = \bigcup_{j=1}^{q} \Gamma_j$ consisting of q separated Jordan curves Γ_j with parametric representations $\zeta = t_j(s)$ for $0 \le s_j \le 2\pi$ and $j = 1, \dots, q$.

Find a function $w(z) = u(z) + iv(z)$ holomorphic in G_q and continuous in the closure $\overline{G}_q$, satisfying the boundary condition

$$F(\zeta, u(\zeta), v(\zeta)) = 0 \quad \text{for } \zeta \in \partial G_q \qquad (H_q) \tag{1.1}$$

where $F(\zeta, u, v)$ is a given real–valued function.

For the special case when $F(\zeta, u, v)$ on Γ_j is independent of ζ, i.e. $F_{|\Gamma_j} = F_j(u, v)$ for $j = 1, \dots, q$, this Riemann–Hilbert problem reduces to a problem of conformal mapping where G_q is mapped onto a domain in the (u, v)–plane, whose boundary is given by the equations: $F_j(u, v) = 0$ for $j = 1, \dots, q$. Here, non–univalued conformal mappings must also be admitted.

On the other hand, if the function $F(\zeta, u, v)$ is linear in u and v, i.e.

$$F(\zeta, u, v) = A(\zeta) \cdot u + B(\zeta) \cdot v + C(\zeta), \quad \text{for } \zeta \in \partial G_q, \tag{1.2}$$

then (H_q) becomes the well–known classical linear Riemann–Hilbert boundary value problem.

With his celebrated thesis "Grundlagen für eine allgemeine Theorie der Funktionen einer veränderlichen Größe", B. Riemann was in 1851 the first mathematician who formulated and analyzed such conformal mapping problems. The linear Riemann–Hilbert problem for the simply connected case was considered in the pioneering papers by D. Hilbert [11] and F. Noether [17]. The complete solution of the linear Riemann- Hilbert problem for the simply connected case can e.g. be found in the book [9] (see also [15], [18]). The conformal mapping for multiply connected domains is treated in [13] and the linear Riemann–Hilbert problems in such domains in [9].

In recent years, one can also observe much activity concerning the nonlinear Riemann–Hilbert problem for simply connected domains.

However most of these papers (for complete references see [5, 10, 20, 22]) are based on "analytic" conditions, such as growth, monotonicity, and coerciveness. In contrast to these concepts, here we are more interested in problems (H_q), where the familiy of curves $\{F(\zeta, u, v) = 0 \text{ for } \zeta \in \partial G_q\}$ is characterized by topological conditions only, which we will call **global cases**. Correspondingly, we will speak of global existence theorems for solutions to (H_q). In this context we shall consider two basically different cases of families of curves in the (u, v)–plane, $\{F(\zeta, u, v) = 0$ for $\zeta \in \partial G_q\}$, namely

(a) *it is a family of closed curves or*

(b) *it is a family of curves where every one is non–closed and going off to infinity.*

Correspondingly, we will speak of nonlinear Riemann–Hilbert problems with closed and non–closed boundary data. Throughout this paper we are only interested in global existence theorems for solutions of the nonlinear (RHP).

The first global existence theorems for solutions of nonlinear (RHPs) in the case $G_1 = \{z||z| < 1\}$ were obtained in [20, 2, 3] for the cases a) and b), respectively. These results were extended to weaker smoothness in [22]. Some other problems concerning global existence of holomorphic solutions for the nonlinear (RHP) can be found in [1, 4, 5, 6, 14, 21, 22], (see also further references therein).

Here we present a new global existence theorem for solutions of the nonlinear (RHP) which is based on a degree theory for quasilinear Fredholm mappings between Banach spaces.

Setting of the problem: Let $G_1 = \{z||z| < 1\}$ be the unit disk in the complex z-plane. We denote the angular coordinate on $\Gamma = \partial G_1$ by $\tau, 0 \leq \tau < 2\pi$.

Definition 1.1. We will say, that a meromorphic function $Z(z)$ in G_1 belongs to the class $\mathcal{M}$, if for every $z_0 \in \partial G_1$ there exists a neighbourhood $U(z_0, \delta), \delta > 0$, such that for all $z \in U(z_0, \delta) \cap \bar{G}_1$, either $Z(z)$ or $1/Z(z)$ is smooth up to the boundary.

Functions in $\mathcal{M}$ can be considered as mappings from $G_1 = \{z \mid |z| < 1\}$ onto the Riemann sphere S^2. It is not difficult to see that the class $\mathcal{M}$ is closed with respect to the mappings

$$Z(z) \mapsto \frac{R_1(z) \cdot Z(z) + R_2(z)}{R_3(z) \cdot Z(z) + R_4(z)}, \tag{1.3}$$

where $R_j(z)$ are rational functions, $j = \overline{1,4}$. For functions in the class $\mathcal{M}$, the formulation of the Riemann–Hilbert problem below is natural. To this end, we assume the following condition to be satisfied.

Condition 1.2. Let γ_τ be a family of closed, nonselfintersecting and smooth curves on S^2. Moreover, they depend smoothly on $\tau \in [0, 2\pi)$.

Statement of the problem. *Suppose $z \in G_1$ and Condition 1.2 be satisfied. Find a function $Z(z) \in \mathcal{M}$, such that*

$$Z(e^{i\tau}) \in \gamma_\tau . \tag{H_1}$$

Let

$$B = \{(Z, \tau) \in S^2 \times S^1 \mid Z \notin \gamma_\tau\} \tag{1.4}$$

Definition 1.3. Suppose Condition 1.2 is satisfied. Then the family of γ_τ is called orientable, if the domain B consists of two connected components; and called nonorientable, if B is connected.

In other words, in the orientable case, each component of $S^2 \setminus \gamma_\tau$ returns to its original place, when τ varies from 0 to 2π, whereas in the nonorientable case there is a change of the original place.

An example of a nonorientable family is:

$$\gamma_\tau : \cos\frac{\tau}{2} \cdot X + \sin\frac{\tau}{2} \cdot Y = 0, \ 0 \le \tau < 2\pi \tag{1.5}$$

An example of an orientable family is:

$$\gamma_\tau : \cos\tau \cdot X + \sin\tau \cdot Y = 0, \ 0 \le \tau < 2\pi \tag{1.6}$$

Correspondingly, we will speak of an *orientable or nonorientable Riemann–Hilbert problem*, respectively.

2. The orientable Riemann–Hilbert problem

We begin with the formulation of our main result for orientable Riemann-Hilbert problems.

Theorem 2.1. *Let γ_τ be an orientable family on S^2. Then the manifold of solutions of the nonlinear Hilbert problem is nonempty and consists of a countable number of connected components. Moreover, each component has the homotopy type of a CW-complex, with the cells of odd dimension.*

The proof of Theorem 2.1 will be based on several lemmata, formulated below.

Lemma 2.2. *Let γ_τ be an orientable family on S^2. Then there exist $\Gamma_j : S^1 \to S^2$, $j = 1, 2$, which satisfy the following conditions:*

1. *$\Gamma_j : S^1 \to S^2$ are continuous, $j = 1, 2$,*
2. *$\Gamma_j(e^{i\tau}) \neq 0, \infty$ for all $\tau \in [0, 2\pi]$ and $\Gamma_j(e^{i\tau}) \notin \gamma_\tau$, $j = 1, 2$,*
3. *$\Gamma_1(e^{i\tau})$ and $\Gamma_2(e^{i\tau})$ lie in different components of $S^2 \setminus \gamma_\tau$.*

Proof. Let B be given by (1.4). B is a locally trivial bundle over S^1 with projection $p : (Z, \tau) \mapsto \tau$. Let $Z_{1,0}, Z_{2,0}$ be points, which lie in different components of $S^2 \setminus \gamma_0, Z_{j,0} \neq 0, \infty$, $j = 1, 2$.

From the homotopy covering theorem (see [12]) it follows that there exist continuous curves $Z_j(\tau)$, $0 \leq \tau < 2\pi$, $j = 1, 2$, such that $Z_j(\tau) \notin \gamma_\tau$ for every $\tau \in [0, 2\pi)$. Without loss of generality, we assume that $Z_j(\tau) \neq 0, \infty$ for each $\tau \in [0, 2\pi)$. Indeed, let $\tilde{\Gamma}_j(\tau)$ be some smooth approximation of $Z_j(\tau)$, so that also $\tilde{\Gamma}_j(\tau) \notin \gamma_\tau$ for each $\tau \in [0, 2\pi), j = 1, 2$. It is clear, that $\text{meas}(\tilde{\Gamma}_j([0, 2\pi])) = 0$, $j = 1, 2$, since the image of a set of measure zero under a smooth mapping also has measure zero. Let $0 \in \tilde{\Gamma}_j([0, 2\pi])$. Then we can make an arbitrarily small rotation of S^2, so that $0 \notin \tilde{\tilde{\Gamma}}_j([0, 2\pi])$, where by $\tilde{\tilde{\Gamma}}_j([0, 2\pi])$ we denote the image of the curve $\tau \mapsto \tilde{\Gamma}_j(\tau)$ under the rotation of S^2.

In the same way, we can also guarantee, that $\tilde{\tilde{\Gamma}}_j(\tau) \neq \infty$ for each $\tau \in [0, 2\pi]$. In general, however, $\tilde{\tilde{\Gamma}}_j(0) \neq \tilde{\tilde{\Gamma}}_j(2\pi)$, $j = 1, 2$.

Since the family of curves γ_τ is orientable, the points $\tilde{\tilde{\Gamma}}_j(0)$ and $\tilde{\tilde{\Gamma}}_j(2\pi)$ lie in the same component of $S^2 \setminus \gamma_0$. Therefore there exist continuous connecting curves $d_j(\sigma), 0 \leq \sigma \leq 1, j = 1, 2$, with $d_j(0) = \tilde{\tilde{\Gamma}}(2\pi - \delta)$, $d_j(1) = \tilde{\tilde{\Gamma}}_j(0)$ for any sufficiently small $\delta > 0$. Let

$$\Gamma_j(\tau) := \begin{cases} \tilde{\tilde{\Gamma}}_j(\tau) & \text{for } 0 \leq \tau < 2\pi - \delta\,, \\ d_j\left(\dfrac{\tau - 2\pi + \delta}{\delta}\right) & \text{for } 2\pi - \delta \leq \tau \leq 2\pi\,. \end{cases} \tag{2.1}$$

Then $\Gamma_j(0) = \Gamma_j(2\pi)$, $j = 1, 2$ and $\Gamma_j(\tau) \notin \gamma_\tau$ for every sufficiently small $\delta > 0$. ∎

Lemma 2.3. *For an arbitrary 2π-periodic continuous function $f(\tau)$ and for arbitrary $\varepsilon > 0$, there exists a rational function $R(z)$, such that*

$$|R(e^{i\tau}) - f(\tau)| < \varepsilon \tag{2.2}$$

for every τ with $0 \leq \tau < 2\pi$.

Proof. Indeed, let

$$f_k := \frac{1}{2\pi}\int_0^{2\pi} f(\tau)e^{-ik\tau}\,d\tau \tag{2.3}$$

and

$$\sigma_N(f) := \sum_{k=-N}^{N}\left(1-\frac{|k|}{N+1}\right) f_k e^{ik\tau}\,;$$

be Cesaro's sum; then we can choose the rational function as

$$R(z) = \sum_{k=-N}^{N}\left(1-\frac{|k|}{N+1}\right)\cdot f_k\cdot z^k\,. \tag{2.4}$$

If N is sufficiently large then (2.2) is satisfied. ∎

Let $\varepsilon = \min\operatorname{dist}(\Gamma_j(\tau),\gamma_\tau)$, $j=1,2$, $0\le\tau<2\pi$, where $\Gamma_j(\tau)$ are the curves from Lemma 2.2.

Let $R_j(z)$ be rational functions, such that

$$|R_j(e^{i\tau})-\Gamma_j(\tau)| < \varepsilon \text{ for all } \tau\in[0,2\pi)\,,\ j=1,2\,.$$

Throughout this paper, the functions R_j , $j=1,2$, will be fixed.

Let $Z(z)$ be any solution of the orientable Riemann–Hilbert problem and define

$$Z_1(z) := \frac{Z(z)-R_1(z)}{Z(z)-R_2(z)}. \tag{2.5}$$

It is clear that $Z_1(z)$ is a meromorphic solution of the following (RHP):

$$Z_1(z)\in\mathcal{M} \quad\text{and}\quad Z_1(e^{i\tau})\in\gamma'_\tau\,, \tag{2.6} \qquad (H'_1)$$

where γ'_τ is the image of γ_τ under the mapping

$$Z\mapsto\frac{Z-R_1(e^{i\tau})}{Z-R_2(e^{i\tau})}\,. \tag{2.7}$$

It is not difficult to see, that the curves γ'_τ of this family are closed, nonselfintersecting, smooth and circumvent in their interior the point $X_1=Y_1=0$.

Let any points $z_1^+,\dots,z_{m_+}^+$, $z_1^-,\dots,z_{m_-}^-$ be fixed, $|z_j^+|<1$, $j=\overline{1,m_+}$, $|z_k^-|<1$, $k=\overline{1,m_-}$.

Theorem 2.4. *The meromorphic solutions of the nonlinear Riemann-Hilbert problem* (H_1), *i.e.* $Z_1\in\mathcal{M}$ *and* $Z_1(e^{i\tau})\in\gamma'_\tau$, *having fixed zeroes* $z_1^+,\dots,z_{m_+}^+$, $|z_j^+|<1$, $j=\overline{1,m_+}$, $Z(z)\neq 0$ *for* $z\neq z_j^+$ *and having fixed poles* $z_1^-,\dots,z_{m_-}^-$,

$|z_k^-| < 1$, $k = \overline{1, m_-}$ *(counted according to their multiplicities)*, $Z(z) \neq \infty$ *for* $z \neq z_k^-$*, form a manifold that is homeomorphic to a circle.*

Proof. Let $Z_1(z)$ be meromorphic function satsfying all the above conditions, i. e. $Z_1(z) \in \mathcal{M}$, $Z_1(e^{i\tau}) \in \gamma'_\tau$, $Z_1(z_j^+) = 0$, $Z_1(z_k^-) = \infty$ and $Z_1(z) \neq 0$, $Z_1(z) \neq \infty$ for $z \neq z_j^+$ and $z \neq z_k^-$, respectively.

Then

$$Z_2(z) = Z_1(z) \cdot \frac{\prod_{k=1}^{m_-}(z - z_k^-)}{\prod_{j=1}^{m_+}(z - z_j^+)} \tag{2.8}$$

is a holomorphic function in G_1 and $Z_2(z) \neq 0$, for all $z \in G_1$ and $Z_2(e^{i\tau}) \in \gamma''_\tau$, where γ''_τ is image of the family of curves γ'_τ under the mapping

$$Z_1 \mapsto Z_1 \cdot \frac{\prod_{k=1}^{m_-}(z - z_k^-)}{\prod_{j=1}^{m_+}(z - z_j^+)}. \tag{2.9}$$

Conversely, if $Z_2(z)$ is a holomorphic function in G_1 such that $Z_2(z) \neq 0$ for all $z \in G_1$ and $Z_2(e^{i\tau}) \in \gamma''_\tau$, then

$$Z_1(z) = Z_2(z) \cdot \frac{\prod_{j=1}^{m_+}(z - z_j^+)}{\prod_{k=1}^{m_-}(z - z_k^-)} \tag{2.10}$$

satisfies all the conditions of Theorem 2.4.

By $\mathcal{X}_l(S^1)$ we denote a class of all functions $Z(z) = X(z) + iY(z)$ holomorphic in G_1 with $Z(e^{i\tau}) \in H^l(S^1)$, where by $H^l(S^1)$ we denote the standard Sobolev space of order $l \geq 2$.

Our main goal is now to prove the existence of holomorphic solutions $Z_2(z) = X_2(z) + iY_2(z)$ of the nonlinear (RHP), that is $Z_2 \in \mathcal{X}_l(S^1)$ with $Z_2(e^{i\tau}) \in \gamma''_\tau$. To this end we consider the locally trivial bundle

$$\mathcal{G} := \{(e^{i\tau}, Z_2) \in S^1 \times \mathbb{C} | Z_2 \in \gamma''_\tau\} \tag{2.11}$$

over S^1 with fibers diffeomorphic to a circle. In what follows, the bundle $\mathcal{G}$ will be called the "boundary bundle" corresponding to the nonlinear (RHP). In other words, we want to characterize the set of solutions

$$\mathcal{G}_* = \{(Z_2(z) \in \mathcal{X}_l(S^1) | (e^{i\tau}, Z_2(e^{i\tau})) \in \mathcal{G}, \frac{1}{2\pi}\int_0^{2\pi} d \arg Z_2(e^{i\tau}) = 0\}.$$

Let $F(e^{i\tau}, X_2, Y_2)$ be a real–valued function with the following properties:

1) $F \in C^\infty(S^1 \times (\mathbb{R}^2 \setminus \{0\}))$;

2) $0 < C \leq |grad_{X_2,Y_2}\ F| \leq C^{-1}$ on $S^1 \times (\mathbb{R}^2 \setminus \{0\})$,

3) $\mathcal{G} = \{(e^{i\tau}, X_2 + iY_2) \in S^1 \times \mathbb{C} \mid F(e^{i\tau}, X_2, Y_2) = 0\}$.

We now introduce the new holomorphic function $\xi(z) = lnZ_2(z) = u + iv$. Since $Z_2(z) \neq 0$ in G_1, it is possible to choose a continuous branch of the logarithm. Then $\xi(z) \in \mathcal{X}_l(S^1)$ and satisfies

$$f(e^{i\tau}, u(e^{i\tau}), v(e^{i\tau})) = 0 \tag{2.12}$$

where $f(e^{i\tau}, u, v) := F(e^{i\tau}, e^u \cos v, e^u \sin v)$. Vice versa, if $\xi(z) = u + iv$ satisfies (2.12), then $Z_2(z) = exp\ \xi(z) = x_2 + y_2$ satisfies

$$F(e^{i\tau}, X_2(e^{i\tau}), Y_2(e^{i\tau})) = 0\,. \tag{2.13}$$

Note that in (2.13) only the curves $f(e^{i\tau}, u, v) = 0$ are important and the function f itself can be selected to be quite general, so that we assume the following conditions to be satisfied:

Condition 2.5.

1) $f(e^{i\tau}, u, v) = u$ for $|u| > C_1$,
2) $f(e^{i\tau}, u, v + 2\pi) = f(e^{i\tau}, u, v)$,
3) $0 < C \leq |grad_{u,v} f| \leq C^{-1}$ for all (τ, u, v).

Due to the one–to–one correspondence between (2.12) and (2.13) it suffices to describe the set of solutions (1.11). For this purpose we reduce the problem (2.12) to a nonlinear singular integral equation on $\Gamma = \partial G_1$. Indeed, it is well known that

$$v(e^{i\tau}) = \frac{1}{2\pi}\int_0^{2\pi} u(e^{i\sigma})\cot\frac{\tau - \sigma}{2}d\sigma + \lambda =: H_\lambda u(e^{i\tau}), \tag{2.14}$$

where λ is a real number and the integral (2.14) is taken in the sense of the Cauchy principal value (see [9]). For simplicity of presentation we will consider H_λ as acting on 2π-periodic functions on the real line, that is

$$H_\lambda u(\tau) := \frac{1}{2\pi}\int_0^{2\pi} u(\sigma)\cot\frac{\tau - \sigma}{2}d\sigma + \lambda\,, \quad \lambda \in \mathbb{R}^1\,; \tag{2.15}$$

and write $\xi(\tau), f(\tau, u, v)$ instead of $\xi(e^{i\tau}), f(e^{i\tau}, u, v)$, respectively (this will not lead to misunderstandings). Hence, condition (2.12) is equivalent to the following nonlinear singular integral equation:

$$f(\tau, u(\tau), H_\lambda u(\tau)) = 0\,, \tag{2.16}$$

or, equivalently,

$$A_\lambda u(\tau) = 0 \ \text{ with } A_\lambda : u(\tau) \mapsto f(\tau, u(\tau), H_\lambda u(\tau))\,. \tag{2.17}$$

For proving the global solvability of (2.17) we make use of the degree theory for quasilinear Fredholm mappings between Banach spaces (see [3, 5, 6, 7, 16, 20]). For the reader's convenience, we recall here briefly some basic aspects of this theory. Let X, Y be Banach spaces with X compactly embedded into another Banach space X_1. By $\Phi_0(X,Y)$ we denote the set of linear bounded Fredholm operators of index 0 equipped with the associated operator norm.

Definition 2.6. A mapping $A : X \to Y$ is called "quasilinear Fredholm" provided A has a representation of the form

$$Au = L_u(u) + C(u) \text{ for } u \in X \tag{2.18}$$

where

1) $L : X_1 \to \Phi_0(X,Y)$ is a continuous map,

2) $C : X_1 \to Y$ is continuous.

We will call $L_u : X_1 \to \Phi_0(X,Y)$ the **principal part** of A. Clearly, L_u is not uniquely determined by A, but its equivalence class in the Calkin algebra is determined modulo compact operators (see [19]). We denote the class of quasilinear Fredholm mappings between Banach spaces X and Y by $F - QL(X,Y)$.

Definition 2.7. Let $\Omega \subset X$. The mapping $A : X \to Y$ is called "proper on Ω" if the preimage $A^{-1}K \cap \Omega$ of any compact $K \Subset Y$ is a compact subset in Ω.

We will need the following lemma in what follows.

Lemma 2.8. *Let $A : X \to Y$ be quasilinear Fredholm. Then the mapping A is proper on any bounded closed set Ω in X.*

Proof. Without loss of generality we assume that $A : X \to Y$ has a representation of the form (see [7])

$$Au = \tilde{L}_u(u) + \tilde{C}(u) \tag{2.19}$$

where

1) $\tilde{L} : X_1 \to GL(X,Y)$ is continuous,

2) $\tilde{C} : X_1 \to Y$ is continuous.

By $GL(X,Y)$ we denote the set of invertible operators in $L(X,Y)$, the linear space of bounded linear operators from X to Y. Let $K \Subset Y$ be any compact set and $h \in K$ whith $Au = h$ where $u \in \Omega$. Then by (2.19) we have

$$u = \tilde{L}_u^{-1}(h) - \tilde{L}_u^{-1}\tilde{C}(u)\,.$$

Since Ω is bounded in X, it follows that $i(\Omega)$ is compact in X_1 where $i : X \to X_1$ is the compact embedding. The operators L_u depend continuously on $u \in X_1$, and therefore L_u^{-1} depends continuously on $u \in X_1$ as well. Consider

$$\tilde{X} = A^{-1}(K) \cap \Omega$$

and a sequence $\{u_n\} \in \tilde{X}$. By the compactness of K in Y and $\tilde{X}$ in X_1, we can choose a subsequence $\{u_{n_k}\}$ such that

$$u_{n_k} \to u_0 \text{ in } X_1 \text{ and } h_{n_k} \to h_0 \text{ in } Y.$$

Since $C(u)$ and L_u^{-1} depend continuously on $u \in X_1$, we obtain

$$u_{n_k} = \tilde{L}_{u_{n_k}}^{-1} h_{n_k} - \tilde{L}_{u_{n_k}}^{-1} \tilde{C}(u_{n_k}) \to \tilde{L}_{u_0}^{-1} h_0 - \tilde{L}_{u_0}^{-1} \tilde{C}(u_0) = u_0 \,.$$

Here the limit is in the norm in X, hence, $u_{n_k} \to u_0$ in X. Since $Au_0 = h_0 \in K$ and $u_0 \in \Omega$, then $u_0 \in \tilde{X}$. Therefore $\tilde{X}$ is compact. ∎

Corollary 2.9. *Let $A \in F{-}QL(X,Y)$ and suppose the following a priori estimate,*

$$||u||_X \leq \Phi(||Au||_Y)\,, \tag{2.20}$$

where $\Phi(x)$ is a positive monotonically increasing function on $0 \leq x < +\infty$. Then A is proper.

Assume for $A \in F-QL(X,Y)$ that an a priory estimate of the form (2.20) holds (or $0 \notin A(\partial\Omega)$, where $\Omega \subset X$ is bounded). Then, as was shown in [3, 5, 7, 20], one can construct a degree of mappings denote by $d(A)$ with the properties:

1) *$d(A) \neq 0$ implies that $Ax = y$ has at least one solution for any $y \in Y$,*
2) *$|d(A_t)| = \text{const}$, that is, the absolute value of degree is invariant under any admissible homotopy.*

A mapping $A : [0,1] \times X \to Y$ is called an "admissible homotopy" provided that A has a representation of the form

$$A_t(u) = L_{t,u}(u) + C(t,u)$$

with continuous $L : [0,1] \times X_1 \to \Phi_0(X,Y)$ and compact $C : [0,1] \times X \to Y$, for which a uniform a priori estimate of the form

$$||u|| \leq \Phi(||A_t u||)$$

holds or if $0 \notin A_t(\partial\Omega)$ for all $0 \leq t \leq 1$.

Lemma 2.10. *Let $f(\tau,u,v)$ satisfy Conditions* 2.5. *Then Equation* (2.17) *has a solution for every λ.*

Proof. It can be proved that $A_\lambda \in F{-}QL(H^l(S^1), H^l(S^1))$ with $l \geq 2$ and satisfies an a priori estimate of the form (2.20) (see [5, 20]). Hence, the degree $d(A_\lambda)$ is defined. It remains to compute $d(A_\lambda)$. For this purpose, consider a homotopy $f^t(\tau,u,v)$ with $0 \leq t \leq 1$ between $f^0(\tau,u,v) = f(\tau,u,v)$ and $f^1(\tau,u,v) = u$ in the space of functions satisfying Conditions 2.5. This homotopy can, for example, explicitly be given in the following way: Let $\gamma_\tau^0 := \{(u,v) | f(\tau,u,v) = 0\}$ and $\gamma_\tau^1 =$

$\{(u,v)|u=0\}$ be a given family of curves and the vertical line in the (u,v)–plane, respectively. We denote by $\tilde{G}_\tau$ the domain in the (u,v)–plane whose boundary consists of the curves $\tilde{\gamma}^1_\tau$ and γ^0_τ where $\tilde{\gamma}^1_\tau := \gamma^1_\tau + d$ with an appropriately chosen constant $d > \max\limits_{\gamma^0_\tau} u$. Consider the following Dirichlet problem: *Find a bounded solution* Φ *of*

$$\Delta\Phi = 0 \text{ in } \tilde{G}_\tau\,, \tag{2.21}$$

$$\Phi|_{\gamma^0_\tau} = 0 \text{ and } \Phi|_{\tilde{\gamma}^1_\tau} = 1\,. \tag{2.22}$$

The existence of a solution to (2.21)–(2.22) is evident. Then the function

$$f^t(\tau,u,v) := \Phi(\tau,u,v) - t, \quad 0 \le t \le 1\,, \tag{2.23}$$

is already a homotopy between γ^0_τ and $\tilde{\gamma}^0_\tau$. With an additional shift we obtain the desired homotopy.

According to the homotopy invariance property of the degree (its absolute value is invariant) we have $|d(f^0_\lambda)| = |d(f^1_\lambda)| = 1$. Hence the equation (1.16) has at least one solution for every λ. In other words, the set of solutions to (2.17) is nonempty. This proves Lemma 2.10. ∎

Our next task is to describe the set of all solutions of (2.13) such that

$$\frac{1}{2\pi}\int_0^{2\pi} d\ \arg Z_2(e^{i\tau}) = 0. \tag{2.24}$$

Lemma 2.11. *The set of all solutions* $Z_2(z)$ *of* (2.13) *satisfying* (2.24) *is homeomorphic to the circle.*

Proof. According to Corollary 2.9, the operator A_λ defined by (1.16) is proper for each $\lambda \in \mathbb{R}$. As a result we have that the set of all solutions of equation (1.16) for each λ is compact as well as the union of all sets of solutions for $\lambda \in [0,2\pi]$. Therefore, the set of all solutions $Z_2(z)$ of (2.13) with $\frac{1}{2\pi}\int_0^{2\pi} d\ \arg Z_2(e^{i\tau}) = 0$ is compact, too. In order to describe its structure, note that the Frechet derivative of the operator $A_{Z_2(\tau)} = F(\tau, X_2(\tau), Y_2(\tau))$ is defined by

$$A'_{Z_2(\tau)}h(\tau) = a(\tau)Re h(\tau) + b(\tau) Im h(\tau) \tag{2.25}$$

where $a(\tau) := F'_{X_2}(\tau, X_2(\tau), Y_2(\tau))$ and $b(\tau) := F'_{Y_2}(\tau, X_2(\tau), Y_2(\tau))$ Due to Conditions 2.5 it follows that $a(\tau) + ib(\tau) \neq 0$. Since $\frac{1}{2}\int_0^{2\pi} d\ \arg Z_2(e^{i\tau}) = 0$, then $\frac{1}{2\pi}\int_0^{2\pi} d\ \arg(a+ib) = 0$. Hence, the homogeneous problem $h_\tau = 0$ in G_1,

$$a(\tau)Re h(\tau) + b(\tau) Im h(\tau) = 0 \text{ on } \partial G_1$$

has a one–dimensional kernel (see [9]). From the surjective implicit function theorem it follows that the set Q of solutions of the problem (2.13) is a smooth and one–dimensional manifold. Since Q is compact, Q must consist of a finite number of closed smooth curves (see[20]). In fact, there is only one such curve, as we will see below, based on the so–called continuation principle. Indeed, let $F^t(\tau, X, Y)$ connect $F^0(\tau, X, Y) = F(\tau, X, Y)$ by a continuous chain with $F^1(\tau, X, Y) := \ln(X^2 + Y^2)$ in the class of functions satisfying Conditions 2.5. Note that for $F^1(\tau, X, Y) = ln(X^2 + Y^2)$ the set Q of all solutions consists just of all rotations and forms a circle. Let Q_t be the set of all solutions of problem (2.13) for F^t. According to the surjective implicit function theorem, Q_t is homeomorphic to Q_{t_0} for every $t_0 \in [0,1]$ and for all t sufficiently close to t_0. Hence, the number of components of Q_t is locally constant with respect to t, and thus, constant on all of $[0, 1]$. Hence, we proved that the set of solutions of (2.13) is nonempty and is homeomorphic to the circle. ■

Remark: *Let these solutions be parametrized by the parameter ω with $0 \leq \omega \leq 2\pi$. We denote these solutions by $Z_{2,\omega}(z)$. Then $Z_{2,\omega}(e^{i\tau})$, for each point $e^{i\tau} \in \partial G_1$, winds once around zero when ω varies from 0 to 2π.*

Consequently, we obtain the assertion of Theorem 2.4. Then the function

$$Z(z) = \frac{Z_1(z) \cdot R_2(z) - R_1(z)}{Z_1(z) - 1} \tag{2.26}$$

is a meromorphic function which satisfies the following conditions:

1) $Z(z) \in \mathcal{M}$,
2) $Z(e^{i\tau}) \in \gamma_\tau$. (H_1)

In other words, $Z(z)$ is a solution of the orientable nonlinear Riemann–Hilbert problem. Now we can characterize the structure of the set of all meromorphic solutions of the nonlinear Riemann-Hilbert problem in the orientable case. According to Theorem 2.4, the set of meromorphic functions $Z_1(z) \in \mathcal{M}$, having the fixed zeros $z_1^+, \ldots, z_{m_+}^+$, $j = \overline{1, m_+}$, $Z_1(z) \neq 0$ for $z \neq z_j^+$ and having the fixed poles $z_1^-, \ldots, z_{m_-}^-$, $|z_k^-| < 1$, $k = \overline{1, m_-}$, $Z_1(z) \neq \infty$ for $z \neq z_k^-$, $k = \overline{1, m_-}$ and satisfying $Z_1(e^{i\tau}) \in \gamma'_\tau$, form a manifold which is homeomorphic to a circle. By ζ_0 we now denote the set of all pairs (ζ^+, ζ^-), where $\zeta^+ = (z_1^+, \ldots, z_{m_+}^+)$, $\zeta^- = (z_1^-, \ldots, z_{m_-}^-)$, $|z_j^+| < 1, |z_k^-| < 1$, $j = \overline{1, m_+}$, $k = \overline{1, m_-}$, with the equivalence relation defined by

$$(\zeta^+, \zeta^-) \sim (\zeta^+, z_0, z_0, \zeta^-)\,,\ z_j^\pm \in G_1,\ z_0 \in G_1\,.$$

It is clear, that ζ_0 is a countable CW-complex (see [8]). By $\mathcal{A}^{(0)}$ we denote the topological space of meromorphic solutions of the nonlinear Riemann–Hilbert problem with the following topology: two meromorphic solutions $Z_1(z)$ and $Z_2(z)$ are close to each other, if they are close to each other as mappings of ∂G_1 into S^2 in some class $C^k(S^1, S^2)$ with some $k \geq 0$. Let $(\mathcal{A}^{(0)}, \pi, \zeta_0)$ be a triple where $\pi : \mathcal{A}^{(0)} \to \zeta_0$ is defined in the following way:

$$\pi(Z(z)) = (\zeta_+(Z_1(z)), \zeta_-(Z_1(z))) \tag{2.27}$$

where ζ^+ and (ζ^-) denote the collections of zeros and poles, respectively, of the function

$$Z_1(z) = \frac{Z(z) - R_1(z)}{Z(z) - R_2(z)} . \tag{2.28}$$

According to Theorem 2.4, the preimage of $(\zeta^+, \zeta^-) \in \zeta_0$ under the mapping π is a smooth curve, homeomorphic to a circle, depending continuously on $(\zeta^+, \zeta^-) \in \zeta_0$ which follows from the implicit function theorem. Consequently, $\mathcal{A}^{(0)}$ is a locally trivial bundle over ζ_0, with the fiber, which is homeomorphic to the circle. Since ζ_0 is a CW-complex with cells of even real dimension, which have a countable number of components. Hence, $\mathcal{A}^{(0)}$ is a CW-complex with cells of odd real dimension which has a countable number of connected components. The invariant which can distinguish connected components is the number

$$m_+ - m_- = \frac{1}{2\pi} \int_0^{2\pi} d \arg Z_1(e^{i\tau}) \quad . \tag{2.29}$$

■

3. The nonorientable Riemann–Hilbert problem

Let γ_τ be a nonorientable family of curves on S^2 and $Z(z)$ be any solution of the nonlinear Riemann-Hilbert problem, i. e. $Z(z) \in \mathcal{M}$ and $Z(e^{i\tau}) \in \gamma_\tau$ (RHP). Let $Z_1(z)$ be a meromorphic function, defined by

$$Z_1(z) = Z(z^2) . \tag{3.1}$$

Then

$$Z_1(-z) = Z_1(z) \text{ and } Z_1(e^{i\tau}) \in \gamma'_\tau , \qquad (RHP_1) \tag{3.2}$$

where $\gamma'_\tau = \gamma_{2\tau}$.

It is clear that $\{\gamma'_\tau\}$ is an orientable family on S^2. The curves $\{\gamma'_\tau\}$ and $\{\gamma'_{\tau+\pi}\}$ coincide for each τ. Let $\Gamma : S^1 \to S^2$ be a closed, continuous curve, such that $\Gamma(e^{i\tau}) \neq 0, \infty$ and $\Gamma(e^{i\tau}) \notin \gamma'_\tau$ as given in Lemma 2.8. Then $\Gamma(e^{i\tau}) \neq \Gamma(e^{i(\tau+\pi)})$, since these points lie in different components of $S^2 \setminus \gamma_\tau$. From Lemma 2.10 it follows, that there exists a rational function $R(z)$, such that

$$|R(e^{i\tau}) - \Gamma(e^{i\tau})| < \varepsilon \text{ with } \varepsilon = \min \operatorname{dist}(\Gamma(e^{i\tau}), \gamma'_\tau) . \tag{3.3}$$

Let

$$Z_2(z) := \frac{Z_1(z) - R(-z)}{Z_1(z) - R(z)} , \tag{3.4}$$

then

$$Z_1(z) = \frac{Z_2(z) \cdot R(z) - R(-z)}{Z_2(z) - 1} . \tag{3.5}$$

It is not difficult to see, that

$$(3.6)\qquad Z_2(-z) = \frac{1}{Z_2(z)} \quad\text{and}\quad Z_2(e^{i\tau}) \in \gamma''_\tau\,, \qquad (RHP_2)$$

where γ''_τ is the image of γ'_τ under the mapping

$$(3.7)\qquad Z_1 \mapsto \frac{Z_1 - R(e^{i(\tau+\pi)})}{Z_1 - R(e^{i\tau})}\,.$$

Moreover, the mapping

$$(3.8)\qquad inv: \quad Z_2 \mapsto \frac{1}{Z_2}$$

transforms the curve γ''_τ to $\gamma''_{\tau+\pi}$. The curves γ''_τ, $\tau \in [0, 2\pi)$ are closed, nonintersecting, smooth and circumvent in their interior the point $X_2 = Y_2 = 0$. It is obvious that if z_j is a pole of the meromorphic function $Z_2(z)$, then $-z_j$ is a zero of the meromorphic function $Z_2(z)$, since $Z_2(-z) = \frac{1}{Z_2(z)}$. Now we consider

$$(3.9)\qquad Z_3(z) := Z_2(z)\cdot \frac{\prod_{j=1}^m (z - z_j)}{\prod_{j=1}^m (z + z_j)}\,.$$

It is clear that $Z_3(z)$ is a holomorphic function in G_1 with

$$Z_3(z) \neq 0\,,\ Z_3(-z) = \frac{1}{Z_3(z)} \ \text{and}\ Z_3(e^{i\tau}) \in \gamma'''_\tau \qquad (RHP_3)$$

where γ'''_τ is the image of γ''_τ under the mapping

$$(3.10)\qquad Z_2 \mapsto Z_2\cdot \frac{\prod_{j=1}^m (e^{i\tau} - z_j)}{\prod_{j=1}^m (e^{i\tau} + z_j)}.$$

Moreover, the curves γ'''_τ, $\tau \in [0, 2\pi)$ are closed, nonselfintersecting, smooth and circumvent in their interior the point $X_3 = Y_3 = 0$. According to Lemma 2.11, the set of solutions $Z_3(z)$ of the nonlinear Riemann–Hilbert problem: $Z_3(e^{i\tau}) \in \gamma'''_\tau$ and $Z_3(z) \neq 0$ for all $z \in G_1$, forms a manifold homeomorphic to a circle. Let Ω be the set of these solutions. As before, these solutions are parametrized by the parameter ω, which takes values $0 \le \omega \le 2\pi$. We will denote the parametrized solutions by $Z_{3,\omega}(z)$. From (3.8) and the definition of γ''' it follows, that if $Z_3(z)$ is a solution of the nonlinear Riemann–Hilbert problem with $Z_3(e^{i\tau}) \in \gamma'''_\tau$ then $1/Z_3(-z)$ is also a solution. Hence, we have the following mapping $\Psi : \Omega \to \Omega$, defined by

$$(3.11)\qquad \Psi : Z_{3,\omega}(z) \mapsto \frac{1}{Z_{3,\omega}(-z)} = Z_{3,\omega'}(z)\,.$$

Due to the Remark, the point $Z_{3,\omega}(e^{i\tau})$ winds once around zero, when ω varies from 0 to 2π and the point $Z_{3,\omega}(e^{i\tau})$ moves monotonically along the curve γ'''_τ.

Consequently, we have a homeomorphism of Ω onto γ_τ''' for each $\tau \in [0, 2\pi]$. We denote this homeomorphism by $\varphi_\tau : \Omega \to \gamma_\tau'''$. The mapping φ_τ induces an orientation on each curve γ_τ''' and these orientations for different $\tau \in [0, 2\pi]$ are compatible, i. e.

$$\text{(3.12)} \qquad \frac{1}{2\pi}\int_0^{2\pi} d\,\arg Z_{3,\omega}(e^{i\tau}) = \varepsilon \text{ for every } \tau \in [0, 2\pi]\,; \quad \text{and } \varepsilon = \pm 1\,.$$

Hence, following diagram is commutative:

$$\text{(3.13)} \qquad \begin{array}{ccc} \Omega & \xrightarrow{\Psi} & \Omega \\ \varphi_\tau \downarrow & & \downarrow \varphi_{\tau+\pi} \\ \gamma_\tau''' & \xrightarrow{\text{inv}} & \gamma_{\tau+\pi}''' \end{array}$$

Since the mapping *"inv"* changes the orientation, the mapping $\Psi : \Omega \to \Omega$ must also change the orientation of Ω. Therefore, the homeomorphism $\Psi : \Omega \to \Omega$ has exactly two fixed points. Consequently, there exist exactly two holomorphic functions $Z_3^{(1)}(z)$ and $Z_3^{(2)}(z)$, which satisfy the following conditions:

1) $Z_3(e^{i\tau}) \in \gamma_\tau'''$,
2) $Z_3(z) \neq 0$ in G_1,
3) $Z_3(-z) = \dfrac{1}{Z_3(z)}$.

The last property is the condition for finding a solution of the reduced problem (RHP_2). Then the desired solution of the original nonlinear Riemann–Hilbert problem (RHP) can be obtained by formulae (3.9), (3.5), (3.1).

In the same manner as in the orientable case, we can define ζ_n and $\mathcal{A}^{(n)}$ and $\pi : \mathcal{A}^{(n)} \to \zeta_n$, $\pi(Z(z)) = (\zeta_+(Z_2(z)), \zeta_-(Z_2(z)))$, where $Z_2(z) = \frac{Z(z^2)-R(-z)}{Z(z^2)-R(z)}$. In the nonorientable case, we have $m_+ = m_- = m$. Moreover, in contrast to the orientable case, in the nonorientable case, ζ_n is contractible. Hence, the triple $(\mathcal{A}^{(n)}, \pi, \zeta_n)$ is a trivial bundle and consists of two connected components. Since in this case, $\pi^{-1}(\zeta_+, \zeta_-)$ consists of two points and ζ_n is the CW–complex we obtain: $\mathcal{A}^{(n)}$ consists of two connected components, which have the homotopy type of the CW–complex, with cells of even real dimension. Now we can state the main result of the nonorientable Riemann–Hilbert problem.

Theorem 3.1. *Let γ_τ be a nonorientable family on S^2. Then the manifold of solutions of the nonlinear Riemann–Hilbert problem (RHP) is non–empty and consists of two connected components, which have homotopy type of a CW-complex with the cells of even dimension.*

Acknowledgements

The authors gratefully acknowledge the substantial support by the collaborative research centre 'Sonderforschungsbereich 404' on Field Interaction Problems at the University of Stuttgart.

References

[1] M.A. Efendiev. *Asymptotics of the solutions of a nonlinear Hilbert problem.* Dokl. Akad. Nauk. SSSR. **270**(1983) N 1. 58–61. English translation in Soviet Math. Dokl. **27**(1983) 569–571.

[2] M.A. Efendiev. *On a property of the conjugate integral and a nonlinear Hilbert problem for non compact boundary data.* Izv.Akad.Nauk Azerb.SSR **3**, N.5 (1984) 9–12 (Russian).

[3] M.A. Efendiev. *On a property of the conjugate integral and a nonlinear Hilbert problem.* Soviet Math. Dokl. **35** (1987) 535-539.

[4] M.A. Efendiev and H. Begehr. *On the asymptotics of meromorphic solutions for nonlinear Riemann–Hilbert problems.* Z. Angew. Math. Mech. **76** (1996) 25–28.

[5] M.A. Efendiev and W. Wendland. *Nonlinear Riemann–Hilbert problems for multiply connected domains.* Nonlinear Analysis, Theory, Methods & Applications **27** (1996) 37–58.

[6] M.A. Efendiev and W. Wendland. *Nonlinear Riemann-Hilbert problems without transversality.* Math. Nachrichten **183** (1997) 73–89.

[7] P. M. Fitzpatrick and J. Pejsachowicz. *An extension of the Leray–Schauder degree for fully nonlinear elliptic problems.* In: Proceedings of Symposia in Pure Mathematics, Vol. **45**, Amer. Math. Soc. Providence, R. I. (1986) 425–439.

[8] D. B. Fuks and V. A. Rokhlin: *Beginner's Course in Topology.* Springer–Verlag, Berlin 1984.

[9] F. D. Gakhov. *Boundary Value Problems.* Pergamon Press, Oxford 1966.

[10] A. I. Guseinov. *On nonlinear boundary value problem theory of analytic functions.* Math. Sbornik USSR **2** (1950) 237–246.

[11] D. Hilbert. *Über eine Anwendung der Integralgleichungen auf ein Problem der Funktionentheorie.* Verhandl. des III. Internat. Mathematiker Kongresses, Heidelberg (1904).

[12] D. Husemoller. *Fiber Bundles.* Springer-Verlag, Berlin 1966.

[13] M. V. Keldysh. *On the problem of conformal mapping of multiply connected domains onto canonical domains.* Russian Math. Surveys **6** (1939) 90–119 (Russian).

[14] F.G. Maksudov and M.A. Efendiev. *A nonlinear Hilbert problem for a doubly connected domain.* Soviet Math. Dokl. **34** (1987) 349–351.

[15] E. Meister. *Randwertaufgaben der Funktionentheorie.* B.G. Teubner, Stuttgart 1983.

[16] L. Nirenberg. Personal communication, 1994-1996.

[17] F. Noether. *Über eine Klasse singulärer Integralgleichungen.* Math. Annalen **82** (1921) 42–63.

[18] S. Prößdorf. *On the degenerate Riemann-Hilbert problem.* In: Partial Differential Equations, Banach Centre Publ. **10** (1983) 301-310 (Russian).

[19] S. Prössdorf. Linear integral equations. Chap.I in *Analysis IV, Encyclopaedia of Mathematical Sciences*, Vol.27, Springer–Verlag, Berlin, 1991.

[20] A. I. Šnirel'man. *The degree of quasi–ruled mapping and a nonlinear Hilbert problem.* Math. USSR-Sbornik **18** (1972) 373–396.

[21] E. Wegert and M.A. Efendiev. *Nonlinear Riemann-Hilbert problems with Lipschitz continuous boundary conditions.* Preprint 94–2, Math. Inst. A, University Stuttgart, Germany.

[22] E. Wegert. *Nonlinear Boundary Value Problems for Holomorphic Functions and Singular Integral Equations.* Akademie–Verlag, Berlin, 1992.

M. A. Efendiev
Freie Universität Berlin
FB Mathematik und Informatik
Arnimallee 2-6
D–14195 Berlin
Germany

W. L. Wendland
Universität Stuttgart
Mathematisches Institut A
Pfaffenwaldring 57
D–70569 Stuttgart
Germany
e-mail:wendland@mathematik.uni-stuttgart.de

1991 Mathematics Subject Classification: 35Q15, 35J65, 45G05, 47H11

Submitted: 16.5.2000

Operator Theory:
Advances and Applications, Vol. 121

Diffraction in Periodic Structures and Optimal Design of Binary Gratings. Part II: Gradient Formulas for TM Polarization

JOHANNES ELSCHNER, GUNTHER SCHMIDT

Dedicated to the memory of our teacher Siegfried Prößdorf

This paper provides the mathematical foundation of analytic formulae for derivatives of TM reflection and transmission coefficients of diffraction gratings with respect to geometric parameters of non-smooth grating profiles and interfaces. This problem arises in optimal design problems for those optical devices studied in Part I. The derivatives can be expressed by contour integrals involving the direct and adjoint solutions of TM diffraction problems.

1. Introduction

Diffractive optics is a modern technology in which optical devices are micromachined with complicated structural features on the order of the length of light waves. Exploiting diffraction effects, those devices can perform functions unattainable with conventional optics. It is widely acknowledged that geometrical optics approximations to the underlying electromagnetic field equations are not accurate for these diffractive elements, hence, their mathematical modelling has to rely on Maxwell's equations or related partial differential equations. The simplest case, the scattering of time–harmonic waves from infinite periodic structures, is a classical problem, dating back to Rayleigh and Bloch. It can be transformed to two quasiperiodic transmission problems for the Helmholtz equation in the whole plane corresponding to the TE and TM polarisation of the incoming wave, respectively. Although various numerical methods have been developed to compute the solution for a given periodic grating (among them a highly accurate integral equation code by A. Pomp, J. Creutziger and B. Kleemann, realized during their work in the group of S. Prößdorf at the Karl–Weierstrass–Institute), rigorous results on the existence and uniqueness of solutions have been obtained only during the last decade; see the references given in part I of this paper [5].

Based on a variational approach to this problem, which goes back to Bonnet-Bendhia & Starling ([1]) and Bao & Dobson (see [2]), it was also possible to develop gradient type optimization methods for finding the optimal design of diffractive gratings with desired far–field patterns. In [5] we derived analytic formulae for derivatives of certain cost functionals involving the reflection and transmission coefficients of so called binary gratings. Roughly speaking, the surface of a binary grating can be given by a periodic step–function separating different optical materials, and the derivatives have to be taken with respect to the width or height of those steps. It turned out that these derivatives can be expressed as one–dimensional integrals over the part of the surface to be varied. In the TE case one has to integrate the product of the solutions of the direct and certain adjoint problem, whereas in the TM case the integrand is the product of their gradients. Unfortunately, due to the singularities of the solutions of TM problems near corners of the grating surface, the product of gradients might be non–integrable. So the formula for the derivatives has to be modified. In [5] we have given, without proof, one of these modifications.

The topic of the present paper is to study in more detail the dependence of the solution of TM diffraction problems with respect to variations of the (non–smooth) grating profile and interfaces between different optical materials. We prove the unique solvability of these problems for quite general small variations of grating profiles and interfaces and obtain different analytic formulae for the derivatives of the reflection and transmission coefficients with respect to these variations, which can be expressed as path–independent contour integrals.

The outline of the paper is as follows. In Section 2, we briefly describe the TE and TM diffraction problems and present their variational formulations and some basic results. In Section 3, we study the perturbation of TM problems arising after sufficiently smooth (piecewise C^1) variations of interfaces. We prove the unique solvability of these perturbed problems and show that the derivative of diffraction coefficients can be expressed as a certain domain integral. This formula is simplified in Section 4 in different ways to get contour integrals or, in the case of strong singularities of solutions, contour integrals plus point functionals. In Section 5 we apply these results to the special case of binary gratings, leading in particular to a simple proof of the above mentioned modified formula.

The authors are grateful to Prof. S. A. Nazarov for many fruitful discussions, especially concerning the topics of Section 3.

2. Variational formulation of TE and TM problems

Consider a diffractive grating with period d consisting of nonmagnetic materials (of permeability μ_0) with different dielectric constants ϵ. The coordinate system is chosen such that the grating is invariant in the x_3–direction and periodic in the x_1–direction. Thus the diffraction problem is determined by the function $\epsilon(x_1, x_2)$ which is d–periodic in x_1. This function is assumed to be piecewise constant and complex valued with $0 \le \arg \epsilon < \pi$. We assume that the material above and below the grating is homogeneous with $\epsilon = \epsilon^+ > 0$ and ϵ^- respectively.

Assume that an incoming plane wave with time dependence $\exp(-i\omega t)$ is incident in the (x_1, x_2)–plane upon the grating from the top with the angle of incidence $\theta \in (-\pi/2, \pi/2)$. Then the electromagnetic field does not depend on x_3. In either case of polarization, one of the fields $\mathbf{E}$ or $\mathbf{H}$ remains parallel to the x_3–axis and is therefore determined by a single scalar quantity $v = v(x_1, x_2)$ (equal to the transverse component of $\mathbf{E}$ in the TE case and to the transverse component of $\mathbf{H}$ in the TM case). The function v satisfies two–dimensional Helmholtz equations in the regions with constant permittivity, together with some radiation condition at infinity. At the material interfaces the solutions are subjected to well known transmission conditions. For TE polarisation the solution and its normal derivative $\partial_n v$ have to cross the set of interfaces Λ between different materials continuously, whereas in TM polarisation the product $\epsilon^{-1}\partial_n v$ has to be continuous (for more details cf. the classical monograph [7]) .

For notational convenience we will change the length scale by a factor of $2\pi/d$ so that the grating becomes 2π–periodic, $\epsilon(x_1 + 2\pi, x_2) = \epsilon(x_1, x_2)$. Let us introduce the piecewise constant function

$$k = \frac{\omega d}{2\pi}(\mu_0 \epsilon)^{1/2} = \frac{d}{\lambda}\nu ,$$

where λ is the length of the incoming plane wave and ν is the optical index of the corresponding material. The constant values of k above and below the grating are denoted by k^+ and k^-, respectively.

Then the incoming plane wave is of the form $(\mathbf{E}^i, \mathbf{H}^i) = (\mathbf{p}, \mathbf{q})\, e^{-i\omega t}\, e^{i(\alpha x_1 - \beta x_2)}$, where $\alpha = k^+ \sin\theta$, $\beta = k^+ \cos\theta$, and the total diffracted field can be obtained as superposition of solutions of the TE and TM polarisation cases.

In TE polarization only the x_3–component E_3 of the electric field is different from zero. It is α–quasiperiodic, $E_3(x_1 + 2\pi, x_2) = e^{2\pi i\alpha} E_3(x_1, x_2)$, and satisfies the Helmholtz equation

$$\Delta E_3 + k^2 E_3 = 0 \quad \text{in } \mathbf{R}^2 . \tag{2.1}$$

The radiation condition that must be imposed for $|x_2| \to \infty$ states that E_3 remains bounded and is representable as superposition of outgoing waves, i.e.

$$\begin{aligned} E_3 &= p_3 e^{-i\beta x_2} + \sum_{n\in\mathbf{Z}} E_n^+ e^{i(n+\alpha)x_1 + i\beta_n^+ x_2} && \text{for } x_2 \to \infty , \\ E_3 &= \sum_{n\in\mathbf{Z}} E_n^- e^{i(n+\alpha)x_1 - i\beta_n^- x_2} && \text{for } x_2 \to -\infty . \end{aligned} \tag{2.2}$$

where $E_n^\pm$ are complex numbers and

$$\beta_n^\pm = \beta_n^\pm(\alpha) := |(k^\pm)^2 - (n+\alpha)^2|^{1/2}\, e^{i\gamma_n^\pm/2} , \quad n \in \mathbf{Z} , \tag{2.3}$$

with

$$\gamma_n^\pm = \arg((k^\pm)^2 - (n+\alpha)^2) , \quad 0 \le \gamma_n^\pm < 2\pi .$$

Note that $\beta_0^+ = \beta$ and that, for real $k^\pm$,

$$\beta_n^\pm = \begin{cases} ((k^\pm)^2 - (n+\alpha)^2)^{1/2} , & k^\pm > |n+\alpha| , \\ i((n+\alpha)^2 - (k^\pm)^2)^{1/2} , & k^\pm < |n+\alpha| . \end{cases}$$

In TM polarization only the x_3–component H_3 of the electric field is different from zero. This α–quasiperiodic function satisfies the Helmholtz equation

$$\nabla \cdot \left(\frac{1}{k^2}\nabla H_3\right) + k^2 H_3 = 0 \quad \text{in } \mathbf{R}^2 , \tag{2.4}$$

together with the radiation condition

$$\begin{aligned} H_3 &= q_3 e^{-i\beta x_2} + \sum_{n\in\mathbf{Z}} H_n^+ e^{i(n+\alpha)x_1 + i\beta_n^+ x_2} && \text{for } x_2 \to \infty , \\ H_3 &= \sum_{n\in\mathbf{Z}} H_n^- e^{i(n+\alpha)x_1 - i\beta_n^- x_2} && \text{for } x_2 \to -\infty . \end{aligned} \tag{2.5}$$

The diffraction problems admit variational formulations in a bounded periodic cell which were introduced in [1], [2]. Define for example the 2π–periodic function $u = e^{-i\alpha x_1} E_3$. It satisfies the partial differential equation

$$\Delta_\alpha u + k^2 u = 0$$

where we use the notation

$$\nabla_\alpha = \nabla + i(\alpha, 0) \ , \ \Delta_\alpha = \nabla_\alpha \cdot \nabla_\alpha = \Delta + 2i\alpha\partial_{x_1} - \alpha^2 .$$

The outgoing wave conditions are equivalent to nonlocal boundary conditions on some artificial boundaries $\Gamma^\pm := \{x_2 = \pm b\}$ above and below the grating, respectively, of the form

$$\partial_n u|_{\Gamma^+} = -T_\alpha^+ u - 2p_3 i\beta e^{-i\beta b} , \quad \partial_n u|_{\Gamma^-} = -T_\alpha^- u , \tag{2.6}$$

where $T_\alpha^\pm u$ is the periodic pseudodifferential operators of order 1

$$(T_\alpha^\pm v)(x) := -\sum_{n\in\mathbf{Z}} i\beta_n^\pm \hat{v}_n e^{inx} , \quad \hat{v}_n = (2\pi)^{-1}\int_0^{2\pi} v(x) e^{-inx}\, dx , \tag{2.7}$$

acting on boundary values $u|_{\Gamma^\pm} \in H_p^{s-1/2}(\Gamma^\pm)$ of functions $u \in H_p^s(\Omega)$, $s \geq 0$. Here $H_p^s(\Omega)$ denotes restriction to the rectangular domain $\Omega = [0, 2\pi] \times [-b, b]$ of all functions in the Sobolev space $H_{loc}^s(\mathbf{R}^2)$ which are 2π–periodic in x_1. Integration by parts leads to the variational formulation for the TE diffraction problem

$$\begin{aligned} B_{TE}(u, \varphi) &= \int_\Omega \nabla_\alpha u \cdot \overline{\nabla_\alpha \varphi} - \int_\Omega k^2 u \bar{\varphi} + \int_{\Gamma^+} (T_\alpha^+ u)\, \bar{\varphi} + \int_{\Gamma^-} (T_\alpha^- u)\, \bar{\varphi} \\ &= -2ip_3\beta e^{-i\beta b} \int_{\Gamma^+} \bar{\varphi} , \quad \forall \varphi \in H_p^1(\Omega) . \end{aligned} \tag{2.8}$$

Analogously, the TM diffraction problem admits the variational formulation for the function $u = e^{-i\alpha x_1} H_3$:

$$
\begin{aligned}
B_{TM}(u,\varphi) &= \int\limits_{\Omega} \frac{1}{k^2} \nabla_\alpha u \cdot \overline{\nabla_\alpha \varphi} - \int\limits_{\Omega} u\,\bar\varphi + \frac{1}{(k^+)^2} \int\limits_{\Gamma^+} (T_\alpha^+ u)\,\bar\varphi + \frac{1}{(k^-)^2} \int\limits_{\Gamma^-} (T_\alpha^- u)\,\bar\varphi \\
&= -\frac{2iq_3\beta\, e^{-i\beta b}}{(k^+)^2} \int\limits_{\Gamma^+} \bar\varphi\,, \quad \forall \varphi \in H_p^1(\Omega)\,.
\end{aligned} \tag{2.9}
$$

In [5], the following properties have been proved under the assumption on the optical indices of the materials that

$$
\operatorname{Re} k(x_1,x_2) > 0\,,\ \operatorname{Im} k(x_1,x_2) \ge 0\,,\ k^+ > 0\,, \tag{2.10}
$$

which is satisfied for all practically relevant materials.

1. If Im $k > 0$ in some subdomain $\Omega_1 \subset \Omega$ then for any $\omega > 0$ there exists at most one solution $u \in H_p^1(\Omega)$.
2. For any $\theta_0 \in (0, \pi/2)$ there exists a frequency $\omega_0 > 0$ such that the variational problem (2.8) resp. (2.9) admits a unique solution $u \in H_p^1(\Omega)$ for all incidence angles θ with $|\theta| \le \theta_0$ and all frequencies ω with $0 < \omega \le \omega_0$.
3. The sesquilinear forms B_{TE} and B_{TM} are strongly elliptic over $H_p^1(\Omega)$, i.e., after multiplication by some complex number they satisfy a Gårding inequality.
4. (i) The diffraction problems (2.8) and (2.9) are always solvable in $H_p^1(\Omega)$. For all but a countable set of frequencies ω_j, $\omega_j \to \infty$, these solutions are unique.

   (ii) Introduce the set of Rayleigh frequencies

   $$\mathcal{R} = \left\{(\omega,\theta) : \exists n \in \mathbf{Z} \text{ s. th. } (k^\pm)^2 = (n+\alpha)^2\right\}.$$

   If for $(\omega^0, \theta^0) \notin \mathcal{R}$ the TE or TM diffraction problem is uniquely solvable, then the solution depends analytically on ω and θ in a neighbourhood of this point.

3. Variation of interfaces

Define the finite sets of indices $P^\pm = \{n \in \mathbf{Z} : \beta_n^\pm > 0\}$, where $\beta_n^\pm$ is given by (2.3). Then the Rayleigh amplitudes $E_n^\pm$ and $H_n^\pm$ $(n \in P^\pm)$, which are called the reflection resp. transmission coefficients for TE and TM polarization, correspond to the propagating modes in (2.2), (2.5) and are used to compute the so called efficiencies of the diffractive grating. Note that $P^- = \emptyset$ if Im $k^- \neq 0$.

We are interested in the solvability of the problems and the dependence of Rayleigh coefficients if parts of the interfaces Λ between different materials are varied. The variation of interfaces leads to a new piecewise constant function k_h, where we assume that meas $\Omega_h = O(h)$ with $\Omega_h = \{x \in \Omega : k(x) \neq k_h(x)\}$. Let B^h_{TE} denote the variational form of the TE problem for the perturbed geometry, then

$$\begin{aligned} \left|B^h_{TE}(u,\varphi) - B_{TE}(u,\varphi)\right| &= \left|\int_{\Omega_h} (k^2 - k_h^2)\, u\, \bar\varphi\right| \\ &\leq \|k^2 - k_h^2\|_{L_p(\Omega_h)} \|u\|_{L_q(\Omega_h)} \|\varphi\|_{L_r(\Omega_h)} \end{aligned}$$

for $p^{-1} + q^{-1} + r^{-1} = 1$. Hence, the variation of interfaces represents a compact and small perturbation of the form B_{TE} ensuring the unique solvability of B^h_{TE} for all sufficiently small h.

In the TM case the situation is more involved. The relation

$$\left|B^h_{TM}(u,\varphi) - B_{TM}(u,\varphi)\right| = \left|\int_{\Omega_h} \Big(\frac{1}{k^2} - \frac{1}{k_h^2}\Big) \nabla_\alpha u\, \overline{\nabla_\alpha \varphi}\right|$$

shows that the variation of interfaces is a strong perturbation of the TM diffraction problem. Therefore we consider a more regularly perturbed diffraction problem

$$B^h_{TM}(u,\varphi) = -\frac{2iq_3\beta\, e^{-i\beta b}}{(k^+)^2} \int_{\Gamma+} \bar\varphi\,, \quad \forall \varphi \in H^1_p(\Omega)\,, \tag{3.1}$$

assuming that, for sufficiently small $|h|$, the perturbed interface Λ_h is given by

$$\Lambda_h = \Phi_h(\Lambda)\,,\ \Phi_h(x) = x + h\,\chi(x)\,. \tag{3.2}$$

Here Φ_h is a C^1 diffeomorphism of Ω onto itself, and $\chi = (\chi_1, \chi_2)$ is 2π–periodic in x_1 and has compact support in $[0, 2\pi] \times (-b, b)$.

Then we can define the isomorphism $\Psi_h : H^1_p(\Omega) \to H^1_p(\Omega)$ which maps u to $u \circ \Phi_h^{-1}$. Moreover, $k_h = \Psi_h k$ and the change of variables $y = \Phi_h(x)$ provides

$$dy = |J(x)| dx$$

with

$$J(x) = 1 + h\Big(\frac{\partial \chi_1}{\partial x_1} + \frac{\partial \chi_2}{\partial x_2}\Big) + h^2\Big(\frac{\partial \chi_1}{\partial x_1}\frac{\partial \chi_2}{\partial x_2} - \frac{\partial \chi_1}{\partial x_2}\frac{\partial \chi_2}{\partial x_1}\Big)$$

and

$$\begin{aligned} \frac{\partial}{\partial y_1} &= \frac{1 + h\,\partial\chi_2/\partial x_2}{J(x)} \frac{\partial}{\partial x_1} - \frac{h\,\partial\chi_2/\partial x_1}{J(x)} \frac{\partial}{\partial x_2}\,, \\ \frac{\partial}{\partial y_2} &= -\frac{h\,\partial\chi_1/\partial x_2}{J(x)} \frac{\partial}{\partial x_1} + \frac{1 + h\,\partial\chi_1/\partial x_1}{J(x)} \frac{\partial}{\partial x_2}\,. \end{aligned}$$

Hence we obtain

$$\int_\Omega \Big(- \Psi_h u \overline{\Psi_h \varphi} + \frac{1}{k_h^2(y)} \nabla_\alpha \Psi_h u \cdot \overline{\nabla_\alpha \Psi_h \varphi} \Big)\, dy = - \int_\Omega u \overline{\varphi}\, J(x)\, dx$$
$$+ \int_\Omega \frac{((1+h\partial_2\chi_2)\partial_1 + i\alpha J(x) - h\partial_1\chi_2\partial_2)u\, ((1+h\partial_2\chi_2)\partial_1 - i\alpha J(x) - h\partial_1\chi_2\partial_2)\overline{\varphi}}{J(x)k^2(x)}$$
$$+ \int_\Omega \frac{(-h\partial_2\chi_1\partial_1 + (1+h\partial_1\chi_1)\partial_2)u\, (-h\partial_2\chi_1\partial_1 + (1+h\partial_1\chi_1)\partial_2)\overline{\varphi}}{J(x)k^2(x)}$$
$$= \int_\Omega \Big(- u\overline{\varphi} + \frac{1}{k^2} \nabla_\alpha u \overline{\nabla_\alpha \varphi}\Big) dx + hB_1(u,\varphi) + h^2 B_{2,h}(u,\varphi)\,,$$

where

$$\begin{aligned} B_1(u,\varphi) = & -\int_\Omega (\partial_1\chi_1 + \partial_2\chi_2) u\,\overline{\varphi} + \int_\Omega \frac{\partial_1\chi_1}{k^2}\big(\partial_2 u\,\overline{\partial_2\varphi} - \partial_1 u\,\overline{\partial_1\varphi} + \alpha^2 u\,\overline{\varphi}\big) \\ & + \int_\Omega \frac{\partial_2\chi_2}{k^2}\big(\partial_{1,\alpha} u\,\overline{\partial_{1,\alpha}\varphi} - \partial_2 u\,\overline{\partial_2\varphi}\big) \\ & - \int_\Omega \Big(\frac{\partial_1\chi_2}{k^2}\big(\partial_{1,\alpha} u\,\overline{\partial_2\varphi} + \partial_2 u\,\overline{\partial_{1,\alpha}\varphi}\big) + \frac{\partial_2\chi_1}{k^2}\big(\partial_1 u\,\overline{\partial_2\varphi} + \partial_2 u\,\overline{\partial_1\varphi}\big)\Big) \end{aligned} \tag{3.3}$$

and the remainder term satisfies

$$|B_{2,h}(u,\varphi)| \le c\|u\|_1\|\varphi\|_1\,,\; u,\varphi \in H_p^1(\Omega\,,\; |h| \le h_0\,.$$

Here we have used the notation $\partial_j = \partial/\partial x_j$, $\partial_{1,\alpha} = \partial_1 + i\alpha$ and the relation

$$J(x)^{-1} = 1 - h(\partial_1\chi_1 + \partial_2\chi_2) + O(h^2)\,,\; |h| \le h_0\,,$$

which holds uniformly in $x \in \Omega$. Since the boundary terms in the TM sesquilinear form remain unchanged, we have for $|h| \le h_0$

$$B_{TM}^h(\Psi_h u, \Psi_h \varphi) = B_{TM}(u,\varphi) + hB_1(u,\varphi) + h^2 B_{2,h}(u,\varphi)\,. \tag{3.4}$$

Theorem 3.1. *If the TM diffraction problem* (2.9) *has a unique solution and the perturbation of the grating geometry is given by the regular mapping* (3.2), *then for all sufficiently small h the perturbed problem* (3.1) *is also uniquely solvable. Moreover, the solution of this problem takes the form*

$$\Psi_h^{-1} u_h = u + hu_1 + h^2 u_{2,h}\,, \tag{3.5}$$

where u is the solution of the original problem (2.9), *$u_1 \in H_p^1(\Omega)$ solves the equation*

$$B_{TM}(u_1,\varphi) = -B_1(u,\varphi)\,,\quad \forall \varphi \in H_p^1(\Omega)\,, \tag{3.6}$$

and the remainder satisfies $\|u_{2,h}\|_1 \le c$ *for* $|h| \le h_0$.

Proof : Replacing u, φ with $\Psi_h^{-1}u, \Psi_h^{-1}\varphi$ in (3.4) and using the equivalence of norms $\|u\|_1 \sim \|\Psi_h u\|_1$ (uniformly in h), we obtain

$$\begin{aligned} B_{TM}^h(u,\varphi) &= B_{TM}(\Psi_h^{-1}u, \Psi_h^{-1}\varphi) + O(h)\|\Psi_h^{-1}u\|_1\|\Psi_h^{-1}\varphi\|_1 \\ &= B_{TM}(u,\varphi) + O(h)\|u\|_1\|\varphi\|_1 \,. \end{aligned}$$

Hence B_{TM}^h is a small perturbation of B_{TM}, which proves the unique solvability of (3.1).

Inserting the ansatz (3.5) for the solution u_h of (3.1) into (3.4) yields the following equation for $u_{2,h}$:

$$\begin{aligned} &B_{TM}(u_{2,h},\varphi) + hB_1(u_{2,h},\varphi) + h^2B_{2,h}(u_{2,h},\varphi) \\ &\quad = -B_1(u_1,\varphi) - B_{2,h}(u,\varphi) - hB_{2,h}(u_1,\varphi)\,, \quad \forall \varphi \in H_p^1(\Omega)\,. \end{aligned} \tag{3.7}$$

Recall that

$$\int_{\Gamma^+} \Psi_h\overline{\varphi} = \int_{\Gamma^+} \overline{\varphi}\,,$$

which implies $B_{TM}^h(u_h, \Psi_h\varphi) = B_{TM}(u,\varphi)$. Since the left–hand side of (3.7) takes the form $B_{TM}(u_{2,h},\varphi) + O(h)\|u_{2,h}\|_1\|\varphi\|_1$ and the right–hand side defines a (uniformly) bounded linear functional on $H_p^1(\Omega)$, we obtain a uniformly bounded solution $u_{2,h}$. ∎

Remark 3.2. It is not difficult to prove recursively that for any $N \ge 2$ the solution of (3.1) admits the expansion

$$\Psi_h^{-1}u_h = \sum_{j=0}^{N} h^j u_j + h^{N+1}u_{N+1,h}\,, \qquad \|u_{N+1,h}\|_1 \le c_N\,,$$

with $u_0 = u, u_1$ as above and certain functions $u_j \in H_p^1(\Omega)$, $j \ge 2$.

Now we are in the position to obtain a formula for the derivative of the Rayleigh coefficients $H_n^\pm$ with respect to the regular variations (3.2) of the interfaces Λ. These reflection and transmission coefficients are determined by the traces of the solution u of the problem (2.9) on the artificial boundaries $\Gamma^\pm$,

$$\begin{aligned} H_n^+ &= -q_3\,\delta_{0n}e^{-2i\beta b} + \frac{e^{-i\beta_n^+ b}}{2\pi}\int_{\Gamma^+} u\, e^{-inx_1}\,dx_1\,, \qquad && n \in P^+\,, \\ H_n^- &= \frac{e^{-i\beta_n^- b}}{2\pi}\int_{\Gamma^-} u\, e^{-inx_1}\,dx_1\,, && n \in P^-\,. \end{aligned} \tag{3.8}$$

Thus the derivative of $H_n^\pm$ is given by

$$DH_n^\pm(\chi) = \lim_{h\to 0} \frac{e^{-i\beta_n^\pm b}}{2\pi h}\int_{\Gamma^\pm} (u_h - u)\, e^{-inx_1}\,dx_1\,, \tag{3.9}$$

where u_h is the solution of the perturbed problem (3.1), (3.2). Let w denote the solution of the adjoint problem

$$B_{TM}(\varphi, w) = \frac{e^{-i\beta_n^\pm b}}{2\pi} \int_{\Gamma^\pm} \varphi\, e^{-inx_1}\, dx_1 , \quad \forall \varphi \in H_p^1(\Omega) . \tag{3.10}$$

Then

$$\frac{e^{-i\beta_n^\pm b}}{2\pi h} \int_{\Gamma^\pm} (u_h - u)\, e^{-inx_1}\, dx_1 = \frac{1}{h} B_{TM}(u_h - u, w) .$$

Since the right–hand side of equation (3.10) is a functional supported at the artificial boundary $\Gamma^\pm$, one has $B_{TM}(u_h, w) = B_{TM}(\Psi_h^{-1} u_h, w)$, and (3.5) then gives

$$\begin{aligned} h^{-1} B_{TM}(u_h - u, w) &= h^{-1} B_{TM}(\Psi_h^{-1} u_h - u, w) = B_{TM}(u_1, w) + h B_{TM}(u_{2,h}, w) \\ &= -B_1(u, w) + h B_{TM}(u_{2,h}, w) . \end{aligned}$$

Thus we have proved the following

Theorem 3.3. *The derivative of the reflection and transmission coefficients $H_n^\pm$ with respect to the variations* (3.2) *of the interface Λ is given by the formula*

$$DH_n^\pm(\chi) = -B_1(u, w) \tag{3.11}$$

where the sesquilinear form B_1 is defined by (3.3), *and u and w denote the solution of the direct and adjoint diffraction problems* (2.9), (3.10), *respectively.*

4. Derivatives of diffraction coefficients as contour integrals

Theorem 3.3 states that the derivative of the diffraction coefficients can be obtained from certain integrals with $\operatorname{supp} \nabla \chi$ as domain of integration. In the following formula (3.11) will be simplified by transforming these domain integrals to contour integrals. For the sake of simplicity we will consider in the following only the variation of interfaces between two different materials. This means the support of the function χ is divided by some part of the interface Λ into two subdomains, which will be denoted by Ω^+ and Ω^-. In each subdomain the function k takes constant values, denoted by k_+ and k_-, respectively.

Let $\Gamma \subset \Omega$ be a simple closed piecewise smooth curve enclosing the domain G such that $k = \mathit{const}$ in G. Let $\nu = (\nu_1, \nu_2)$ be the exterior normal to Γ, $\tau = (-\nu_2, \nu_1)$ the tangential vector, and introduce the weighted normal and tangential derivatives

$$\partial_{\nu,\alpha} = \nu_1 \partial_{1,\alpha} + \nu_2 \partial_2 , \; \partial_{\tau,\alpha} = -\nu_2 \partial_{1,\alpha} + \nu_1 \partial_2 .$$

We denote by $B_1(u,w;G)$ the right–hand side of (3.3) where the integrals are taken over G instead of Ω.

Lemma 4.1. *If* $\operatorname{supp}\chi\cap\Gamma$ *does not contain a corner point of* Λ, *then*

$$B_1(u,w;G) = \int_\Gamma \Big((\chi,\nu)\mathcal{J} + (\chi,\tau)\mathcal{K} + \chi_1\mathcal{L}\Big)\,, \tag{4.1}$$

where

$$\begin{aligned}
\mathcal{J} &= -u\overline{w} + \frac{1}{k^2}(\partial_{\tau,\alpha}u\,\overline{\partial_{\tau,\alpha}w} - \partial_{\nu,\alpha}u\,\overline{\partial_{\nu,\alpha}w})\\
\mathcal{K} &= -\frac{1}{k^2}(\partial_{\nu,\alpha}u\,\overline{\partial_{\tau,\alpha}w} + \partial_{\tau,\alpha}u\,\overline{\partial_{\nu,\alpha}w})\\
\mathcal{L} &= \frac{i\alpha}{k^2}(u\,\overline{\partial_{\nu,\alpha}w} - \partial_{\nu,\alpha}u\,\overline{w})\,.
\end{aligned}$$

Proof: We have

$$\begin{aligned}
&B_1(u,w;G)\\
&= \int_G \frac{1}{k^2}\Big(\partial_1\chi_1\big((\alpha^2-k^2)u\,\overline{w} + \partial_2 u\,\overline{\partial_2 w} - \partial_1 u\,\overline{\partial_1 w}\big) - \partial_2\chi_1\big(\partial_1 u\,\overline{\partial_2 w} + \partial_2 u\,\overline{\partial_1 w}\big)\Big)\\
&+ \int_G \frac{1}{k^2}\Big(\partial_2\chi_2\big(-k^2\,u\,\overline{w} + \partial_{1,\alpha}u\,\overline{\partial_{1,\alpha}w} - \partial_2 u\,\overline{\partial_2 w}\big) - \partial_1\chi_2\big(\partial_{1,\alpha}u\,\overline{\partial_2 w} + \partial_2 u\,\overline{\partial_{1,\alpha}w}\big)\Big)\\
&=: I_1 + I_2\,.
\end{aligned}$$

Let us start with the integral I_1. Green's formula yields

$$\begin{aligned}
&\int_G \partial_2\chi_1\big(\partial_1 u\,\overline{\partial_2 w} + \partial_2 u\,\overline{\partial_1 w}\big) = -\int_G \chi_1\Big(\partial_1 u\,\overline{\partial_2^2 w} + \partial_2^2 u\,\overline{\partial_1 w} + \partial_1\big(\partial_2 u\,\overline{\partial_2 w}\big)\Big)\\
&+ \int_\Gamma \chi_1\big(\partial_1 u\,\overline{\partial_2 w} + \partial_2 u\,\overline{\partial_1 w}\big)\nu_2 = -\int_G \chi_1\big(\partial_1 u\,\overline{\partial_2^2 w} + \partial_2^2 u\,\overline{\partial_1 w}\big)\\
&+ \int_G \partial_1\chi_1\partial_2 u\,\overline{\partial_2 w} + \int_\Gamma \chi_1\Big(\big(\partial_1 u\,\overline{\partial_2 w} + \partial_2 u\,\overline{\partial_1 w}\big)\nu_2 - \partial_2 u\,\overline{\partial_2 w}\,\nu_1\Big)\,,
\end{aligned}$$

$$\begin{aligned}
&\int_G \partial_1\chi_1\,\partial_1 u\,\overline{\partial_1 w} = -\int_G \chi_1\big(\partial_1^2 u\,\overline{\partial_1 w} + \partial_1 u\,\overline{\partial_1^2 w}\big) + \int_\Gamma \chi_1\partial_1 u\,\overline{\partial_1 w}\,\nu_1\\
&= -\int_G \chi_1\big((\partial_1^2 u + 2i\alpha\partial_1 u)\overline{\partial_1 w} + \partial_1 u(\overline{\partial_1^2 w + 2i\alpha\partial_1 w})\big) + \int_\Gamma \chi_1\partial_1 u\,\overline{\partial_1 w}\,\nu_1\,,
\end{aligned}$$

$$\int_G \partial_1\chi_1\,u\,\overline{w} = -\int_G \chi_1\big(\partial_1 u\,\overline{w} + u\,\overline{\partial_1 w}\big) + \int_\Gamma \chi_1\,u\,\overline{w}\,\nu_1\,,$$

which implies

$$\begin{aligned} I_1 = & \frac{1}{k^2} \int_G \chi_1 \partial_1 u \, (\partial_1^2 \overline{w} + \partial_2^2 \overline{w} - 2i\alpha \partial_1 \overline{w} + (k^2 - \alpha^2) \overline{w}) \\ & + \frac{1}{k^2} \int_G \chi_1 (\partial_1^2 u + \partial_2^2 u + 2i\alpha \partial_1 u + (k^2 - \alpha^2) u) \, \partial_1 \overline{w} \\ & - \frac{1}{k^2} \int_\Gamma \chi_1 \Big((\partial_1 u \, \partial_2 \overline{w} + \partial_2 u \, \partial_1 \overline{w}) \nu_2 + (\partial_1 u \, \partial_1 \overline{w} - \partial_2 u \, \partial_2 \overline{w}) \, \nu_1 \Big) \\ & - \frac{k^2 - \alpha^2}{k^2} \int_\Gamma \chi_1 \, u \, \overline{w} \, \nu_1 \, . \end{aligned}$$

Note that $u, w \in H^2(G \cap \operatorname{supp}\chi)$, and since

$$\Delta_\alpha u + k^2 u = \Delta_\alpha w + \overline{k}^2 w = 0 \quad \text{in } G \, ,$$

we obtain

$$\begin{aligned} I_1 = & - \frac{1}{k^2} \int_\Gamma \chi_1 \Big((\partial_1 u \, \partial_2 \overline{w} + \partial_2 u \, \partial_1 \overline{w}) \nu_2 + (\partial_1 u \, \partial_1 \overline{w} - \partial_2 u \, \partial_2 \overline{w}) \, \nu_1 \Big) \\ & - \frac{k^2 - \alpha^2}{k^2} \int_\Gamma \chi_1 \, u \, \overline{w} \, \nu_1 \, . \end{aligned}$$

Simple calculations show that

$$\begin{aligned} & (\partial_1 u \, \partial_2 \overline{w} + \partial_2 u \, \partial_1 \overline{w}) \nu_2 + (\partial_1 u \, \partial_1 \overline{w} - \partial_2 u \, \partial_2 \overline{w}) \, \nu_1 - \alpha^2 \, u \, \overline{w} \, \nu_1 \\ & = (\partial_{\nu,\alpha} u \, \overline{\partial_{\nu,\alpha} w} - \partial_{\tau,\alpha} u \, \overline{\partial_{\tau,\alpha} w}) \nu_1 - (\partial_{\nu,\alpha} u \, \overline{\partial_{\tau,\alpha} w} + \partial_{\tau,\alpha} u \, \overline{\partial_{\nu,\alpha} w}) \nu_2 \\ & \quad + i \, \alpha (\partial_{\nu,\alpha} u \, \overline{w} - u \, \overline{\partial_{\nu,\alpha} w}) \, . \end{aligned}$$

Therefore we have

$$I_1 = I_1(\chi_1) = \int_\Gamma \Big(\chi_1 \nu_1 \mathcal{J} - \chi_1 \nu_2 \mathcal{K} + \chi_1 \mathcal{L} \Big) \, .$$

Similarly one verifies that

$$I_2 = I_2(\chi_2) = \int_\Gamma \Big(\chi_2 \nu_2 \mathcal{J} + \chi_2 \nu_1 \mathcal{K} \Big) \, ,$$

which finishes the proof of (4.1). ∎

Remark 4.2. An inspection of the above proof shows that $I_1 = I_1(\chi_1) = 0$ if $\chi_1 \equiv 1$ in G. Moreover, since

$$\int_\Gamma (u \, \overline{\partial_{\nu,\alpha} w} - \partial_{\nu,\alpha} u \, \overline{w}) = 0$$

by the second Green formula, the third integral in (4.1) vanishes if χ_1 is constant. Thus, for $\chi_1 \equiv 1$ the integral $I_1(\chi_1)$ takes the form

$$\int_\Gamma (\nu_1 \mathcal{J} - \nu_2 \mathcal{K})$$

and is zero as long as $\Gamma \subset \overline{\Omega^+}$ (or $\Gamma \subset \overline{\Omega^-}$) does not contain a corner of the interface Λ. Analogously, $I_2 = I_2(\chi_2)$ with $\chi_2 \equiv 1$ always vanishes in that case.

Corollary 4.3. *If Λ has no corner points, then*

$$\text{(4.2)} \quad DH_n^\pm(\chi) = -B_1(u,w) = \int_\Lambda (\chi,\nu) \left[\frac{1}{k^2}\left(\partial_{\nu,\alpha} u\, \overline{\partial_{\nu,\alpha} w} - \partial_{\tau,\alpha} u\, \overline{\partial_{\tau,\alpha} w}\right)\right]_\Lambda .$$

Here ν denotes the normal to Λ pointing from Ω^+ into Ω^- and $[v]_\Lambda$ stands for the jump $v|_\Lambda^+ - v|_\Lambda^-$ across Λ, where $v|_\Lambda^\pm$ represents the limit as the interface is approached from the region $\Omega^\pm$.

Proof: Applying Lemma 4.1 with $G = \Omega^\pm$, we obtain

$$\begin{aligned} B_1(u,w) &= B_1(u,w;\Omega^+) + B_1(u,w;\Omega^-) \\ &= \int_\Lambda \left((\chi,\nu)\mathcal{J}|_\Lambda^+ + (\chi,\tau)\mathcal{K}|_\Lambda^+ + \chi_1 \mathcal{L}|_\Lambda^+\right) - \int_\Lambda \left((\chi,\nu)\mathcal{J}|_\Lambda^- + (\chi,\tau)\mathcal{K}|_\Lambda^- + \chi_1 \mathcal{L}|_\Lambda^-\right). \end{aligned}$$

Recall that $\operatorname{supp}\chi \cap \Gamma = \emptyset$ and the integrands are 2π–periodic in x_1. Using the transmission conditions for u and w then gives

$$[u\,\overline{w}]_\Lambda = [\mathcal{K}]_\Lambda = [\mathcal{L}]_\Lambda = 0 ,$$

hence

$$B_1(u,w) = \int_\Lambda (\chi,\nu) \left[\frac{1}{k^2}\left(\partial_{\tau,\alpha} u\, \overline{\partial_{\tau,\alpha} w} - \partial_{\nu,\alpha} u\, \overline{\partial_{\nu,\alpha} w}\right)\right]_\Lambda . \qquad \blacksquare$$

We now extend formula (4.2) to the case of corner points. Assume first that Λ has exactly one corner point at O, and denote by δ the angle at O seen from Ω^+. Without loss of generality we may assume that Ω^+ locally coincides with the sector $\{(r,\varphi) : 0 < r < \infty, |\varphi| < \delta/2\}$, where (r,φ) denote polar coordinates centered at O.

To describe the singularities of solutions to problem (2.9) near O, consider the transcendental equation

$$\text{(4.3)} \qquad \frac{\sin(\pi-\delta)\lambda}{\sin \pi\lambda} = \sigma \frac{k_-^2 + k_+^2}{k_-^2 - k_+^2} , \quad \sigma = \pm 1 .$$

Denote by λ_0 the unique zero of (4.3) in the strip $0 < \operatorname{Re}\lambda < 1$ if it exists. It was proved in [6, Lemma 4.2] that (4.3) has exactly one simple root in that strip if $|k_-| \neq k_+$ and no root there if $|k_-| = k_+$. Moreover (see [6, Thm. 4.1]), the solution $u \in H_p^1(\Omega)$ of the TM diffraction problem (2.9) satisfies

$$\xi u|_{\Omega^\pm} = C + C^\pm r^{\lambda_0} u_0^\pm + u_1^\pm , \tag{4.4}$$

where ξ is a smooth cut–off function near O, C and $C^\pm$ are certain complex constants, the remainder terms $u_1^\pm$ satisfy

$$u_1^\pm \in H^{2-\epsilon}(\Omega^\pm) \quad \text{for all} \quad \epsilon > 0 ,$$

and the functions $u_0^\pm$ take the form

$$u_0^+(\varphi) = \cos\lambda_0\varphi \, , \; u_0^-(\varphi) = \cos\lambda_0(\varphi - \pi) \tag{4.5}$$

or

$$u_0^+(\varphi) = \sin\lambda_0\varphi \, , \; u_0^-(\varphi) = \sin\lambda_0(\varphi - \pi) \tag{4.6}$$

corresponding to the case $\sigma = +1$ or $\sigma = -1$ in (4.3). For fixed $\varepsilon > 0$, let $O_{\pm\varepsilon}$ be the two points on Λ satisfying $\operatorname{dist}(O, O_{\pm\varepsilon}) = \varepsilon$ and set $\Lambda_\varepsilon = \Lambda \setminus (\overline{OO_{-\varepsilon}} \cup \overline{OO_\varepsilon})$.

Theorem 4.4. *With* $\mathcal{G} := (\chi, \nu) \left[\frac{1}{k^2} \left(\partial_{\nu,\alpha} u \, \overline{\partial_{\nu,\alpha} w} - \partial_{\tau,\alpha} u \, \overline{\partial_{\tau,\alpha} w} \right) \right]_\Lambda$ *we have*

$$DH_n^\pm(\chi) = \lim_{\varepsilon \to 0} \Big(\int\limits_{\Lambda_\varepsilon} \mathcal{G} + \frac{\varepsilon}{2\lambda_0 - 1} \big(\mathcal{G}(O_{-\varepsilon}) + \mathcal{G}(O_\varepsilon)\big) \Big) . \tag{4.7}$$

Remark 4.5. Since the function $\overline{w}$ also admits the representation (4.4)–(4.6) (with other constants C, $C^\pm$ and remainder terms), one obtains that $\mathcal{G}(x) = O(r^{2\lambda_0 - 2})$ as $r \to 0$. Thus (4.7) coincides with formula (4.2) if $\operatorname{Re}\lambda_0 > 1/2$. This is always true if $k_\pm$ are real; cf. [3]. Note that the case $0 < \lambda_0 \leq 1/2$ is excluded by our assumptions (2.10).

Proof of Theorem 4.4: Let $\Omega_\varepsilon^\pm = \Omega^\pm \setminus \{r \leq \varepsilon\}$ and denote by $S_\varepsilon^\pm$ the (clockwise oriented) circular arcs $\Omega_\varepsilon^\pm \cap \{r = \varepsilon\}$ with endpoints $O_{-\varepsilon}$, O_ε. Applying Lemma 4.1 with $G = \Omega_\varepsilon^\pm$ gives

$$B_1(u, w) = \lim_{\varepsilon \to 0} \Big(B_1(u, w; \Omega_\varepsilon^+) + B_1(u, w; \Omega_\varepsilon^-) \Big) = \lim_{\varepsilon \to 0} \Big(- \int\limits_{\Lambda_\varepsilon} \mathcal{G} + \int\limits_{S_\varepsilon^+} \mathcal{H} + \int\limits_{S_\varepsilon^-} \mathcal{H} \Big) ,$$

where $\mathcal{H} := (\chi, \nu)\mathcal{J} + (\chi, \tau)\mathcal{K} + \chi_1 \mathcal{L}$; cp. (4.1). It remains to show that, for $\operatorname{Re}\lambda_0 \leq 1/2$ and $\lambda_0 \neq 1/2$,

$$\int\limits_{S_\varepsilon^\pm} \mathcal{H} = \mp \frac{\varepsilon}{2\lambda_0 - 1} \Big((\mathcal{H}|_\Lambda^\pm)(O_\varepsilon) + (\mathcal{H}|_\Lambda^\pm)(O_{-\varepsilon}) \Big) + o(\varepsilon) \tag{4.8}$$

as $\varepsilon \to 0$. To prove this, it is enough to replace $\mathcal{H}$ by

$$\frac{(\chi(O),\nu)}{k^2}(\partial_\tau u\,\partial_\tau \overline{w} - \partial_\nu u\,\partial_\nu \overline{w}) - \frac{(\chi(O),\tau)}{k^2}(\partial_\nu u\,\partial_\tau \overline{w} + \partial_\tau u\,\partial_\nu \overline{w})$$

and to insert the principal asymptotic term

$$u^0(r,\varphi) = \begin{cases} r^{\lambda_0} u_0^+(\varphi)\,, & \varphi \in (-\delta/2, \delta/2), \\ r^{\lambda_0} u_0^-(\varphi)\,, & \varphi \in (\delta/2, 2\pi - \delta/2), \end{cases}$$

for u and $\overline{w}$, with $u_0^\pm$ defined in (4.5) or (4.6).

Thus (4.8) is proved provided we have shown that

$$\int_{S_\varepsilon^\pm} \mathcal{H}_0 = \mp \frac{\varepsilon}{2\lambda_0 - 1}\Big((\mathcal{H}_0|_\Lambda^\pm)(O_\varepsilon) + (\mathcal{H}_0|_\Lambda^\pm)(O_{-\varepsilon})\Big), \tag{4.9}$$

where

$$\mathcal{H}_0 = (\chi(O),\nu)\big((\partial_\tau u^0)^2 - (\partial_\nu u^0)^2\big) - 2(\chi(O),\tau)\partial_\nu u^0\,\partial_\tau u^0\,.$$

Consider, for example, (4.9) with the plus sign and $u_0^+ = \cos\lambda_0\varphi$. Since $\nu = -(\cos\varphi, \sin\varphi)$, $\tau = (\sin\varphi, -\cos\varphi)$, $\partial_\nu = -\partial_r$, $\partial_\tau = r^{-1}\partial_\varphi$ on S_ε^+, we then have

$$\begin{aligned}
\int_{S_\varepsilon^+} \mathcal{H}_0 &= \int_{S_\varepsilon^+} \lambda_0^2 r^{2\lambda_0-2}\big(\chi_1(O)\cos\varphi + \chi_2(O)\sin\varphi\big)\big(\cos^2\lambda_0\varphi - \sin^2\lambda_0\varphi\big) \\
&\quad + \int_{S_\varepsilon^+} 2\lambda_0^2 r^{2\lambda_0-2}\big(\chi_1(O)\sin\varphi - \chi_2(O)\cos\varphi\big)\sin\lambda_0\varphi\,\cos\lambda_0\varphi \\
&= \lambda_0^2\,\varepsilon^{2\lambda_0-1}\Big(\chi_1(O)\int_{-\delta/2}^{\delta/2}(\cos\varphi\,\cos 2\lambda_0\varphi + \sin\varphi\,\sin 2\lambda_0\varphi)\,d\varphi \\
&\qquad -\chi_2(O)\int_{-\delta/2}^{\delta/2}(\cos\varphi\,\sin 2\lambda_0\varphi - \sin\varphi\,\cos 2\lambda_0\varphi)\,d\varphi\Big) \\
&= \lambda_0^2\,\varepsilon^{2\lambda_0-1}\Big(\chi_1(O)\int_{-\delta/2}^{\delta/2}\cos(2\lambda_0-1)\varphi\,d\varphi - \chi_2(O)\int_{-\delta/2}^{\delta/2}\sin(2\lambda_0-1)\varphi\,d\varphi\Big) \\
&= \frac{2\chi_1(O)\lambda_0^2\,\varepsilon^{2\lambda_0-1}}{2\lambda_0-1}\sin(2\lambda_0-1)\delta/2\,.
\end{aligned}$$

On the other hand, since $\nu = (\sin\delta/2, \cos\delta/2)$, $\tau = (\cos\delta/2, -\sin\delta/2)$, $\partial_\nu = r^{-1}\partial_\varphi$, $\partial_\tau = \partial_r$ on $\{\varphi = -\delta/2\}$ and $\nu = (\sin\delta/2, -\cos\delta/2)$, $\tau = (-\cos\delta/2, -\sin\delta/2)$,

$\partial_\nu = -r^{-1}\partial_\varphi$, $\partial_\tau = -\partial_r$ on $\{\varphi = \delta/2\}$, we obtain

$$\begin{aligned}(\mathcal{H}_0|_\Lambda^+)(O_\varepsilon) &= \big(\chi_1(O)\sin\delta/2 + \chi_2(O)\cos\delta/2\big)\lambda_0^2\,\varepsilon^{2\lambda_0-2}\cos\lambda_0\delta \\ &\quad -\big(\chi_1(O)\cos\delta/2 - \chi_2(O)\sin\delta/2\big)\lambda_0^2\,\varepsilon^{2\lambda_0-2}\sin\lambda_0\delta\,, \\ (\mathcal{H}_0|_\Lambda^+)(O_{-\varepsilon}) &= \big(\chi_1(O)\sin\delta/2 - \chi_2(O)\cos\delta/2\big)\lambda_0^2\,\varepsilon^{2\lambda_0-2}\cos\lambda_0\delta \\ &\quad -\big(\chi_1(O)\cos\delta/2 + \chi_2(O)\sin\delta/2\big)\lambda_0^2\,\varepsilon^{2\lambda_0-2}\sin\lambda_0\delta\,,\end{aligned}$$

which implies

$$\begin{aligned}&\frac{\varepsilon}{2\lambda_0-1}\Big((\mathcal{H}_0|_\Lambda^+)(O_\varepsilon) + (\mathcal{H}_0|_\Lambda^+)(O_{-\varepsilon})\Big) \\ &= -2\chi_1(O)\frac{\lambda_0^2\,\varepsilon^{2\lambda_0-1}}{2\lambda_0-1}(\cos\delta/2\,\sin\lambda_0\delta - \sin\delta/2\,\cos\lambda_0\delta) \\ &= -\frac{2\chi_1(O)\lambda_0^2\,\varepsilon^{2\lambda_0-1}}{2\lambda_0-1}\,\sin(2\lambda_0-1)\delta/2\,,\end{aligned}$$

hence (4.9) for the plus sign. In the other cases the proof of (4.9) is analogous. ∎

Remark 4.6. The extension of (4.7) to the case of finitely many corners $O_1,\dots,O_r$ of Λ with angles $\delta_1,\dots,\delta_r$ is straightforward. Let $O_{j,\pm\varepsilon} \in \Lambda$ be the points with $\operatorname{dist}(O_j, O_{j,\pm\varepsilon}) = \varepsilon$. Then formula (4.7) holds with $\Lambda_\varepsilon = \Lambda \setminus \bigcup_{j=1}^{r}(\overline{O_jO_{j,-\varepsilon}} \cup \overline{O_jO_{j,\varepsilon}})$ and the correction terms replaced by the sum

$$\sum_{j=1}^{r}\frac{\varepsilon}{2\lambda_j-1}\Big(\mathcal{G}(O_{j,-\varepsilon}) + \mathcal{G}(O_{j,\varepsilon})\Big)\Big)\,,$$

where λ_j denotes the root of equation (4.3) (with $\delta = \delta_j$) in the strip $0 < \operatorname{Re}\lambda < 1$.

Note that formula (4.7) requires the knowledge of the zero λ_0 of the transcendental equation (4.3). An alternative expression for $DH_n^\pm(\chi)$ can be given by a path–independent contour integral.

Theorem 4.7. *Assume that Λ has only one corner point at O, and let $\Gamma = \partial G \subset \Omega$ be an arbitrary simple closed piecewise smooth curve around O. Then*

$$\begin{aligned}DH_n^\pm(\chi) &= \int_\Gamma \Big((\chi(O),\nu)\mathcal{J} + (\chi(O),\tau)\mathcal{K}\Big) \\ &\quad + \int_{G\cap\Lambda}(\chi - \chi(O),\nu)\left[\frac{1}{k^2}\big(\partial_{\nu,\alpha}u\,\overline{\partial_{\nu,\alpha}w} - \partial_{\tau,\alpha}u\,\overline{\partial_{\tau,\alpha}w}\big)\right]_\Lambda \\ &\quad + \int_{\Lambda\setminus G}(\chi,\nu)\left[\frac{1}{k^2}\big(\partial_{\nu,\alpha}u\,\overline{\partial_{\nu,\alpha}w} - \partial_{\tau,\alpha}u\,\overline{\partial_{\tau,\alpha}w}\big)\right]_\Lambda\,.\end{aligned} \tag{4.10}$$

To prove (4.10), we first extend Lemma 4.1 to the case where $\mathrm{supp}\chi \cap \Gamma$ contains a corner point of the interface Λ.

Lemma 4.8. *Let $\Gamma = \partial G$ be a simple closed piecewise smooth curve such that $k = const$ in G and that $\mathrm{supp}\chi \cap \Gamma$ contains exactly one corner point O of Λ. Then*

$$B_1(u,w;G) = \int_{\Gamma} \Big((\chi - \chi(O), \nu)\mathcal{J} + (\chi - \chi(O), \tau)\mathcal{K} + \chi_1 \mathcal{L} \Big). \tag{4.11}$$

Proof: Let $G_\varepsilon = G \setminus \{r \le \varepsilon\}$, $r = \mathrm{dist}(x, O)$ and $\Gamma_\varepsilon = \partial G_\varepsilon$. Replacing χ by $\chi - \chi(O)$, as in the proof of Lemma 4.1 one obtains by partial integration that

$$B_1(u,w;G) = \lim_{\varepsilon \to 0} B_1(u,w;G_\varepsilon) = \lim_{\varepsilon \to 0} \int_{\Gamma_\varepsilon} \Big((\chi - \chi(O), \nu)\mathcal{J} + (\chi - \chi(O), \tau)\mathcal{K} + \chi_1 \mathcal{L} \Big).$$

Recall that the integral of $\chi_1(O)\mathcal{L}$ vanishes; see Remark 4.2. Using the asymptotics (4.4) of u and $\overline{w}$, one can pass to the limit in the last expression giving formula (4.11). ∎

Proof of Theorem 4.7: Lemma 4.8 applied to $G = \Omega^\pm$ and the transmission conditions for u and w yield

$$\begin{aligned} B_1(u,w) &= \Big(B_1(u,w;\Omega^+) + B_1(u,w;\Omega^-) \Big) \\ &= \int_{\partial\Omega} \Big((\chi - \chi(O), \nu)\mathcal{J} + (\chi - \chi(O), \tau)\mathcal{K} + \chi_1 \mathcal{L} \Big) \\ &\quad - \int_{\Lambda} (\chi - \chi(O), \nu) \left[\frac{1}{k^2} \left(\partial_{\nu,\alpha} u \, \overline{\partial_{\nu,\alpha} w} - \partial_{\tau,\alpha} u \, \overline{\partial_{\tau,\alpha} w} \right) \right]_\Lambda \\ &= - \int_{\partial\Omega} \Big((\chi(O), \nu)\mathcal{J} + (\chi(O), \tau)\mathcal{K} \Big) \\ &\quad - \int_{\Lambda} (\chi - \chi(O), \nu) \left[\frac{1}{k^2} \left(\partial_{\nu,\alpha} u \, \overline{\partial_{\nu,\alpha} w} - \partial_{\tau,\alpha} u \, \overline{\partial_{\tau,\alpha} w} \right) \right]_\Lambda , \end{aligned}$$

which proves (4.10) for $\Gamma = \partial\Omega$. On the other hand, if $G_1 \subset \Omega$ is a simply connected domain such that $O \notin \overline{G_1}$, then Remark 4.2 implies

$$\int_{\partial G_1^\pm} \Big((\chi(O), \nu)\mathcal{J} + (\chi(O), \tau)\mathcal{K} \Big) = 0 ,$$

where $G_1^\pm = G_1 \cap \Omega^\pm$. Hence, by the transmission conditions for u and w,

$$\begin{aligned} &\int_{\partial G_1} \Big((\chi(O), \nu)\mathcal{J} + (\chi(O), \tau)\mathcal{K} \Big) \\ &\quad = - \int_{G \cap \Lambda} (\chi(O), \nu) \left[\frac{1}{k^2} \left(\partial_{\nu,\alpha} u \, \overline{\partial_{\nu,\alpha} w} - \partial_{\tau,\alpha} u \, \overline{\partial_{\tau,\alpha} w} \right) \right]_\Lambda \end{aligned} \tag{4.12}$$

so that the right–hand side of (4.10) is in fact independent of the contour Γ. ∎

Remark 4.9. Formula (4.10) easily extends to the case of finitely many corners $O_1, \ldots, O_r$ of the interface Λ. Let $\Gamma_j = \partial G_j$ be a simple piecewise smooth curve enclosing the corner point O_j only. Then the right–hand side of (4.10) has to be replaced by the sum

$$\sum_j \Big(\int_{\Gamma_j} \big((\chi(O_j), \nu)\mathcal{J} + (\chi(O_j), \tau)\mathcal{K} \big) + \int_{G_j \cap \Lambda} (\chi - \chi(O_j), \nu) \left[\frac{1}{k^2} \big(\partial_{\nu,\alpha} u \, \overline{\partial_{\nu,\alpha} w} - \partial_{\tau,\alpha} u \, \overline{\partial_{\tau,\alpha} w} \big) \right]_\Lambda \Big) + \int_{\Lambda \setminus (\cup G_j)} (\chi, \nu) \left[\frac{1}{k^2} \big(\partial_{\nu,\alpha} u \, \overline{\partial_{\nu,\alpha} w} - \partial_{\tau,\alpha} u \, \overline{\partial_{\tau,\alpha} w} \big) \right]_\Lambda . \tag{4.13}$$

Indeed, choosing cut–off functions ξ_j near O_j such that $\sum_j \xi_j \equiv 1$ in some neighbourhood of Λ, one applies formula (4.10) with χ replaced by $\xi_j \chi$ and summing over j then gives the result with sufficiently small discs G_j with centres O_j. Again, by virtue of (4.12), the resulting expression (4.13) is independent of the choice of the contours Γ_j.

Remark 4.10. Repeating the arguments used in the proofs of Theorem 3.3 and Corollary 4.3 one obtains the following formula for the derivative of the TE reflection and transmission coefficients $E_n^\pm$ with respect to the variations (3.2) of the interface Λ:

$$DE_n^\pm(\chi) = \int_\Lambda (\chi, \nu) \left[k^2 u \overline{w} \right]_\Lambda . \tag{4.14}$$

Here u is the solution of the direct TE problem (2.8), w solves the corresponding adjoint problem and Λ may be an arbitrary Lipschitz curve. A special case of (4.14) was first proved in [4].

5. Applications to binary gratings

For simplicity we restrict to a binary grating with two transition points t_1, $t_2 = 2\pi$ and the height t_3. Let $O_1 = (t_1, 0)$, $O_2 = (t_1, t_3)$, $O_3 = (2\pi, t_3)$ and $\Sigma_1 = \overline{O_1 O_2}$, $\Sigma_2 = \overline{O_2 O_3}$.

We first compute the derivative $D_1 H_n^\pm$ of the Rayleigh coefficients with respect to the variation of t_1. Then the mapping (3.2) takes the form

$$\Phi_h(x) = x + h\,\chi(x), \quad \chi(x) = (\chi_1(x), 0),$$

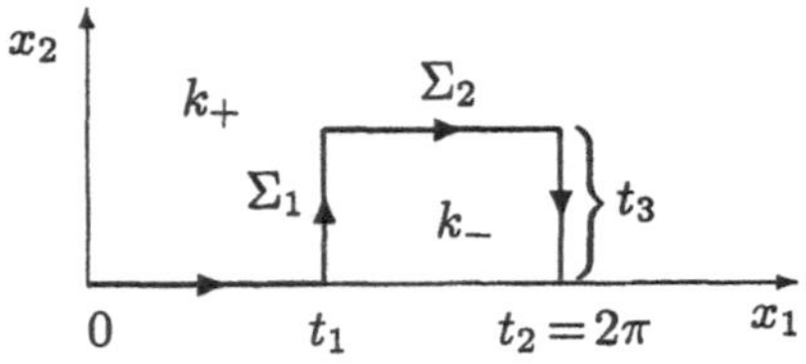

Figure 1: Cross section of a simple binary grating

where $\chi_1 \equiv 1$ in some neighbourhood of Σ_1 and $\chi \in C_o^\infty(U)$ for a somewhat larger neighbourhood U (not containing other corners of the profile curve Λ). Since

$$\chi - \chi(O_1) = \chi - \chi(O_2) = 0 \text{ on } \Sigma_1 \ , \ (\chi, \nu) = 0 \text{ on } \Lambda \setminus \Sigma_1 \ ,$$

from Remark 4.9 we easily obtain

Corollary 5.1. *Let Γ be an arbitrary simple closed piecewise smooth curve around Σ_1, which does not encircle corner points on Λ different from O_1, O_2. Then*

$$\begin{aligned} D_1 H_n^\pm &= \int\limits_\Gamma \Big(\nu_1 \big(-u\overline{w} + \frac{1}{k^2}(\partial_2 u \, \partial_2 \overline{w} - \partial_{1,\alpha} u \, \overline{\partial_{1,\alpha} w}) \big) \\ &\qquad + \frac{\nu_2}{k^2}(\partial_2 u \, \overline{\partial_{1,\alpha} w} + \partial_{1,\alpha} u \, \partial_2 \overline{w}) \Big), \end{aligned} \tag{5.1}$$

where u and w denote the solutions of the direct and adjoint diffraction problems (2.9), (3.10), *respectively.*

To prove (5.1), one may choose for example rectangles G_j $(j = 1, 2)$ around O_j with a common side such that $\Gamma = \partial(G_1 \cup G_2)$ encloses the segment Σ_1. Then formula (5.1) follows immediately from Remark 4.9. Note that (4.13) reduces to

$$\sum_{j=1,2} \int\limits_{\partial G_j} ((\chi(O_j), \nu)\mathcal{J} + (\chi(O_j), \tau)\mathcal{K}) = \int\limits_\Gamma (\nu_1 \mathcal{J} - \nu_2 \mathcal{K})$$

with $\mathcal{J}$, $\mathcal{K}$ defined in (4.1). The fact that the integral in (5.1) is path–independent is an easy consequence of Remark 4.2.

Define $O_{j,\pm\varepsilon}$ as in Sec. 3, and let $\Sigma_{1,\varepsilon} = \overline{O_{1,\varepsilon} O_{2,-\varepsilon}}$. Let further λ_0 be the root of equation (4.3) with $\delta = \pi/2$, lying in the strip $0 < \operatorname{Re} \lambda < 1$. Note that for all corner points of a binary grating the same transcendental equation occurs. Since $\nu = (1, 0), \tau = (0, 1)$ on Σ_1 and $(\chi, \nu) = 0$ on $\Lambda \setminus \Sigma_1$, Remark 4.6 implies immediately

Corollary 5.2. *With* $\mathcal{G} := \left[\frac{1}{k^2}(\partial_{1,\alpha} u \, \overline{\partial_{1,\alpha} w} - \partial_2 u \, \overline{\partial_2 w}) \right]_\Lambda$ *we have*

$$D_1 H_n^\pm = \lim_{\varepsilon \to 0} \Big(\int\limits_{\Sigma_{1,\varepsilon}} \mathcal{G} + \frac{\varepsilon}{2\lambda_0 - 1} (\mathcal{G}(O_{1,\varepsilon}) + \mathcal{G}(O_{2,-\varepsilon})) \Big) .$$

This result has been stated, without proof, in [5, Remark 4.3].

We now compute the derivative $D_2 H_n^{\pm}$ with respect to the height of the binary grating. In this case the mapping (3.2) is of the form

$$\Phi_h(x) = x + h\,\chi(x)\,, \quad \chi(x) = (0, \chi_2(x))\,,$$

where $\chi_2 \equiv 1$ near Σ_2 and $\chi_2 \in C_o^\infty(U)$ for a sufficiently small neighbourhood U of Σ_2. Note that $\nu = (0,-1), \tau = (1,0)$ on Σ_2 and $\chi - \chi(O_2) = \chi - \chi(O_3) = 0$ on Σ_2 and $(\chi, \nu) = 0$ on $\Lambda \setminus \Sigma_2$. As above we then obtain

Corollary 5.3. *Let Γ be an arbitrary simple closed piecewise smooth curve enclosing Σ_2 but no other corner points of Λ. Then*

$$\begin{aligned} D_2 H_n^{\pm} &= \int_\Gamma \Big(\nu_2 \big(-u\overline{w} + \frac{1}{k^2}(\partial_{1,\alpha} u\, \overline{\partial_{1,\alpha} w} - \partial_2 u\, \overline{\partial_2 w})\big) \\ &\qquad - \frac{\nu_1}{k^2}(\partial_2 u\, \overline{\partial_{1,\alpha} w} + \partial_{1,\alpha} u\, \partial_2 \overline{w})\Big) \\ &= \lim_{\varepsilon \to 0} \Big(\int_{\Sigma_{2,\varepsilon}} \mathcal{G} + \frac{\varepsilon}{2\lambda_0 - 1}\big(\mathcal{G}(O_{2,\varepsilon}) + \mathcal{G}(O_{3,-\varepsilon})\big)\Big)\,, \end{aligned}$$

where $\Sigma_{2,\varepsilon} = \overline{O_{2,\varepsilon} O_{3,-\varepsilon}}$ and $\mathcal{G}$ is defined as in Corollary 5.2.

Finally, we remark that for Re $\lambda_0 > 1/2$ we have

$$D_i H_n^{\pm} = \int_{\Sigma_i} \Big[\frac{1}{k^2}(\partial_{1,\alpha} u\, \overline{\partial_{1,\alpha} w} - \partial_2 u\, \overline{\partial_2 w}) \Big]_\Lambda\,,$$

which gives

$$\begin{aligned} D_1 H_n^{\pm} &= \frac{1}{k_+^2} \int_{\Sigma_1} (\partial_{1,\alpha} u|_\Lambda^+\, \overline{\partial_{1,\alpha} w|_\Lambda^+} - \partial_2 u\, \partial_2 \overline{w}) \\ &\quad - \frac{1}{k_-^2} \int_{\Sigma_1} (\partial_{1,\alpha} u|_\Lambda^-\, \overline{\partial_{1,\alpha} w|_\Lambda^-} - \partial_2 u\, \partial_2 \overline{w}) \\ &= \frac{k_+^2 - k_-^2}{(k_+ k_-)^2} \int_{\Sigma_1} (\partial_{1,\alpha} u|_\Lambda^+\, \overline{\partial_{1,\alpha} w|_\Lambda^-} + \partial_2 u\, \partial_2 \overline{w})\,, \end{aligned}$$

$$\begin{aligned} D_2 H_n^{\pm} &= \frac{1}{k_+^2} \int_{\Sigma_2} (\partial_{1,\alpha} u\, \overline{\partial_{1,\alpha} w} - \partial_2 u|_\Lambda^+\, \partial_2 \overline{w}|_\Lambda^+) \\ &\quad - \frac{1}{k_-^2} \int_{\Sigma_2} (\partial_{1,\alpha} u\, \overline{\partial_{1,\alpha} w} - \partial_2 u|_\Lambda^-\, \partial_2 \overline{w}|_\Lambda^-) \\ &= \frac{k_-^2 - k_+^2}{(k_+ k_-)^2} \int_{\Sigma_2} (\partial_{1,\alpha} u\, \overline{\partial_{1,\alpha} w} + \partial_2 u|_\Lambda^+\, \partial_2 \overline{w}|_\Lambda^-)\,. \end{aligned}$$

These formulas have been proved in [5, Sec. 4.3] using another approach.

References

[1] A.-S. Bonnet-Bendhia and F. Starling, Guided waves by electromagnetic gratings and non–uniqueness examples for the diffraction problem, *Math. Meth. Appl. Sci.* **17** (1994), 305–338.

[2] G. Bao, D. C. Dobson, and J. A. Cox, Mathematical studies in rigorous grating theory, *J. Opt. Soc. Amer.* **A 12** (1995), 1029–1042.

[3] M. Costabel and E. Stephan, A direct boundary integral equation method for transmission problems, *J. Math. Anal. Appl.* **106** (1985), 367–413.

[4] D. C. Dobson, Optimal shape design of blazed diffraction gratings, *Appl. Math. Opt.* **40** (1999), 61-78.

[5] J. Elschner and G. Schmidt, Diffraction in periodic structures and optimal design of binary gratings I. Direct problems and gradient formulas, *Math. Meth. Appl. Sci.* **21** (1998), 1297–1342.

[6] J. Elschner, R. Hinder, F. Penzel, and G. Schmidt, Existence, uniqueness and regularity for solutions of the conical diffraction problem, M^3AS (to appear).

[7] R. Petit (ed.), *Electromagnetic theory of gratings*, Topics in Current Physics, Vol. **22**, Springer-Verlag, Berlin, 1980.

Weierstrass Institute
for Applied Analysis and Stochastics
Mohrenstrasse 39
D – 10117 Berlin
elschner@wias-berlin.de
schmidt@wias-berlin.de

1991 Mathematics Subject Classification: Primary 78A45, 35J20, 49J20

Submitted: 15.3.2000

Operator Theory:
Advances and Applications, Vol. 121

A connection between the determinant and characteristic numbers of an operator pencil

ISRAEL GOHBERG AND NAUM KRUPNIK

Dedicated to the memory of Professor Siegfried Prößdorf

In this paper a connection between the determinant of a polynomial operator pencil $A(\lambda) = I + \lambda A_1 + ... + \lambda^n A_n$ and the characteristic numbers of this pencil are established. The coefficients A_m of the pencil $A(\lambda)$ belong to some algebras D of operators on Banach spaces $\mathcal{B}$. Some applications are suggested.

1. Introduction

Let H be a separable Hilbert space and let A_j be some trace class operators on H. Denote by $A(\lambda)$ the polynomial operator pencil

$$A(\lambda) = I + \lambda A_1 + ... + \lambda^n A_n \tag{1.1}$$

and by $\mu_j(A)$ the characteristic numbers of the pencil (1.1). The following statements are known for the pencil $A(\lambda)$ (see, for example, [1], Chap. 4, Section 10):

Let the operators A_m $(m = 1, ..., n)$ admit representations $A_m = A_{m1} ... A_{mm}$ where $A_{mj} (j = 1, ..., m)$ are of trace class. Then

$$\sum_n \frac{1}{|\mu_n(A)|} < \infty, \quad \sum_n \frac{1}{\mu_n(A)} = -\mathrm{tr} A_1,$$
$$\prod \left(1 - \frac{\lambda}{\mu_n(A)}\right) = \det A(\lambda). \tag{1.2}$$

In general (for trace class operators A_m) statements (1.2) fail, but a following equality holds:

$$\det A(\lambda) = \prod_j \left[\left(1 - \frac{\lambda}{\mu_j(A)}\right) \exp \sum_{k=1}^{n-1} \frac{1}{k} \left(\frac{\lambda}{\mu_j(A)}\right)^k\right] \exp \sum_{k=1}^{n-1} \frac{(-\lambda)^{k+1}}{k} \mathrm{tr}(A_0^k)$$
(1.3)

where $L_0(\lambda) = I + \lambda A_0$ is a linearization of the pencil $A(\lambda)$. This means that $A_0 \in L(H^N)$ for some N and the pencil $L_0(\lambda)$ admits a factorization $L(\lambda) = B(\lambda)D(\lambda)C(\lambda)$ where $B(\lambda)$ and $C(\lambda)$ are correspondingly lower and upper triangular matrices with identity operator on the main diagonals, and $D(\lambda)$ is a diagonal matrix $D(\lambda) = \text{diag}\,(I, I, ..., I, A(\lambda))$. For details see [1], Chap. IV, Section 10.

In this paper we extend these results to the case where the coefficients A_m of the pencil $A(\lambda)$ belong to some algebras D of operators on Banach spaces $\mathcal{B}$.

2. Algebras generated by companion block matrices

Let $L(\mathcal{B})$ denote the algebra of all linear bounded operators on a Banach space $\mathcal{B}$ and $\mathcal{F}(\mathcal{B})$ the subalgebra of all operators of finite rank. By D ($D \subset L(\mathcal{B})$) we denote a Banach algebra with the following properties:

a) there exists a constant c such that $\|A\|_{L(\mathcal{B})} \leq c\|A\|_D, A \in D$ (embedded algebra);

b) $\|AB\|_D \leq \|A\|_D\|B\|_D \quad (A, B \in D)$;

c) $\mathcal{F}(\mathcal{B})$ is dense in D (approximation property)

and

d) there exists a constant C such that $|\text{tr}F| \leq C\|F\|_D$ for all $F \in \mathcal{F}$.

These conditions provide (see [1], Theorem II.2.1) that the functionals $\text{tr}F$ and $\det(I+F)$ admit continuous extensions from $\mathcal{F}(\mathcal{B})$ to D. The extended trace and determinant we denote by $\text{tr}_D A$ and $\det_D(I + A)$ $(A \in D)$.

For given D we denote by $\mathcal{A}_n$ the free algebra generated by matrices

$$\begin{bmatrix} 0 & I & \dots & 0 \\ . & . & \dots & . \\ 0 & 0 & \dots & I \\ R_n & R_{n-1} & \dots & R_1 \end{bmatrix} \tag{2.1}$$

where $R_1, ..., R_n \in D$.

Let $R = [R_{ik}]_{ik=1}^n \in \mathcal{A}_n$. Each operator R_{ii} on the main diagonal belongs to the algebra D because it is a polynomial $f(R_1, ..., R_n)$ without the constant term. Hence the following functional

$$\text{tr diag}\,(R) := \sum_{i=1}^{n} \text{tr}_D R_{ii} \tag{2.2}$$

is well defined. Direct calculations show that

$$\text{tr diag}(R_1 + R_2) = \text{tr diag}(R_1) + \text{tr diag}(R_2) \tag{2.3}$$

and

$$\text{tr diag}(R_1R_2) = \text{tr diag}(R_2R_1) \tag{2.4}$$

for each $R_1, R_2 \in \mathcal{A}_n$.

Let

$$A(\lambda) := I + \lambda A_1 + ... + \lambda^n A_n \quad (A_i \in D)$$

be an operator pencil and $\mu_j(A)$ the characteristic numbers of the pencil $A(\lambda)$, where the multiplicities are taken into account. Denote by A_c the companion operator

$$A_c := \begin{bmatrix} 0 & -I & ... & 0 \\ \cdot & \cdot & ... & \cdot \\ 0 & 0 & ... & -I \\ A_n & A_{n-1} & ... & A_1 \end{bmatrix}. \tag{2.5}$$

It is well known that $I + \lambda A_c$ is a linearization of the pencil $A(\lambda)$ (see [6], §12).

3. Main results

Let D be a Banach algebra and $D^{N\times N}$ the algebra of all $N \times N$ matrices with entries from D. If the algebra D satisfies the conditons a), b), c), d) (Section 2) then the algebras $D^{N\times N}$ satisfy these conditions too. This fact can be directly checked. We say that the Lidskii theorem holds in the algebra $D^{N\times N}$ if for each $X \in D^{N\times N}$

$$\sum_i |\lambda_i(X)| < \infty \text{ and } \mathrm{tr}_{D^{N\times N}}(X) = \sum_i \lambda_i(X), \tag{3.1}$$

where $\lambda_i(X)$ ($|\lambda_1(X)| \geq |\lambda_2(X)| \geq ...$) are the eigenvalues of X, where the multiplicities are taken into account.

THEOREM 1. *Let an algebra D satisfy conditions* a), b), c), d) (Section 2) *and let $\mu_j(A)$ be the characteristic numbers of the pencil*

$$A(\lambda) = I + \lambda A_1 + ... + \lambda^n A_n,$$

where the operators $A_m (m = 1, ..., n)$ admit representations $A_m = A_{m1}...A_{mm}$ with $A_{mj} \in D$ $(j = 1, ..., m)$. If in the algebra $D^{N\times N}$ $(N = 1 + n(n-1)/2)$ the Lidskii theorem (3.1) *holds, then*

$$\sum_n \frac{1}{|\mu_n(A)|} < \infty, \quad \sum_n \frac{1}{\mu_n(A)} = -\mathrm{tr}_D A_1,$$
$$\prod_n \left(1 - \frac{\lambda}{\mu_n(A)}\right) = \det{}_D A(\lambda). \tag{3.2}$$

The proof of this theorem uses the same arguments as the proof of Theorem IV.10.1 from [1]. This theorem was first proved in [5]. Here we omit this proof.

THEOREM 2. *Let an algebra D satisfy conditions* a), b), c), d) (Section 2) *and let $\mu_j(A)$ be the characteristic numbers of the pencil*

$$A(\lambda) = I + \lambda A_1 + \dots + \lambda^n A_n \quad (A_i \in D).$$

If in the algebra $D^{n\times n}$ the Lidskii theorem (3.1) *holds, then*

$$\det\nolimits_D A(\lambda) =$$

$$= \prod_j \left[\left(1-\frac{\lambda}{\mu_j(A)}\right)\exp\sum_{k=1}^{n-1}\frac{1}{k}\left(\frac{\lambda}{\mu_j(A)}\right)^k\right]\exp\sum_{k=1}^{n-1}\frac{(-\lambda)^{k+1}}{k}\operatorname{tr}\operatorname{diag}(A_c^k). \tag{3.3}$$

The following theorem gives explicit formulas for tr diag (A_c^k).

THEOREM 3. *Let an algebra D satisfy conditions* a), b), c), d) (Section 2) *and let $A_1, \dots, A_n \in D$. Then for each $m < n$ the following equalities hold:*

$$\operatorname{tr}\operatorname{diag}(A_c^m) = \operatorname{tr}_D(A_1^m)$$

$$+2\sum_{p_1+\dots+p_{2k}=m}\operatorname{tr}_D\left[A_1^{p_1-1}A_{p_2+1}A_1^{p_3-1}A_{p_4+1}\dots A_1^{p_{2k-1}-1}A_{p_{2k}+1}\right](-1)^{p_2+p_4+\dots+p_{2k}}$$

$$+\sum_{p_1+\dots+p_{2k+1}=m}\operatorname{tr}_D[A_1^{p_1-1}A_{p_2+1}A_1^{p_3-1}A_{p_4+1}\dots A_1^{p_{2k-1}-1}(A_{p_{2k}+1}A_1^{p_{2k+1}}$$

$$+(-1)^{p_{2k+1}}A_{p_{2k}+p_{2k+1}+1})](-1)^{p_2+p_4+\dots+p_{2k}} \tag{3.4}$$

where $p_j \geq 1$ and $k \geq 1$.

The proofs of Theorems 2 and 3 are presented in the next two sections.

4. Proof of Theorem 2

Let us recall the definition of an n-regularized determinant. Assume that $B \in L(\mathcal{B})$ and $B^k \in D$ for each $k \geq n$. Denote by $w = w(B)$ the following operator

$$w(B) = (I+B)\exp\left(\sum_{m=1}^{n-1}\frac{(-1)^m}{m}B^m\right) - I. \tag{4.1}$$

It is clear that

$$w(B) = \alpha_n B^n + \alpha_{n+1}B^{n+1} + \dots \in D. \tag{4.2}$$

It follows from here that $\det_D(I + w(\lambda B))$ is well defined and by definition it is the n-regularized determinant of the operator $I + \lambda B$. Let, in particular, A_c be the operator defined by (2.5). Since $A_c^n \in D^{n\times n}$, we have

$$\det{}_n(I+\lambda A_c) := \det{}_D\left[(I+\lambda A_c)\exp\sum_{m=1}^{n-1}\frac{(-\lambda)^m}{m}A_c^m\right] =$$

$$= \prod_j \left[\left(1-\frac{\lambda}{\mu_j(A_c)}\right)\exp\sum_{m=1}^{n-1}\left(\frac{\lambda}{\mu_j(A_c)}\right)^m\frac{1}{m}\right], \tag{4.3}$$

where $\mu_j(A_c)$ are the characteristic numbers of the pencil $I + \lambda A_c$. The second equality in (4.3) follows from conditions (3.1) with $N = n$. Indeed, using Theorem 1 for the linear pencil $I + \lambda w(A_c)$ we obtain

$$\det{}_{D^{n\times n}}(I+\lambda w(A_c)) = \prod_j (I+\lambda\lambda_j w(A_c)) =$$

$$= \prod_j \left[(1+\lambda\lambda_j(A_c))\exp\sum_{m=1}^{n-1}(-\lambda\lambda_j(A_c))^m\frac{1}{m}\right] =$$

$$= \prod_j \left[\left(1-\frac{\lambda}{\mu_j(A_c)}\right)\exp\sum_{m=1}^{n-1}\left(\frac{\lambda}{\mu_j(A_c)}\right)^m\frac{1}{m}\right].$$

Hence

$$\begin{aligned}\det{}_n(I+\lambda A_c) &:= \det{}_{D^{n\times n}}(I+\lambda w(A_c)) = \\ &= \prod_j \left[\left(1-\frac{\lambda}{\mu_j(A_c)}\right)\exp\sum_{m=1}^{n-1}\left(\frac{\lambda}{\mu_j(A_c)}\right)^m\frac{1}{m}\right].\end{aligned}$$

For operators $F \in \mathcal{F}(\mathcal{B})$ the n-regularized determinant can also be represented in the form

$$\det{}_n(I+\lambda F) = \det(I+\lambda F)\exp\sum_{k=1}^{m-1}\frac{(-\lambda)^k}{k}\mathrm{tr}F^k. \tag{4.4}$$

Let $F_k \in \mathcal{F}(\mathcal{B})$ $(k = 1, ..., n)$ and $F(\lambda) = \sum_{k=0}^n \lambda^k F_k$. Consider two linearizations of the pencil $F(\lambda)$: $I + \lambda F_c$ and $I + \lambda \widetilde{F}$, where

$$F_c := \begin{bmatrix} 0 & -I & ... & 0 \\ . & . & ... & . \\ 0 & 0 & ... & -I \\ F_n & F_{n-1} & ... & F_1 \end{bmatrix}; \quad \widetilde{F} := \begin{bmatrix} 0 & -P & ... & 0 \\ . & . & ... & . \\ 0 & 0 & ... & -P \\ F_n & F_{n-1} & ... & F_1 \end{bmatrix}$$

and P is a finite dimensional projection such that $F_j P = F_j$ $(j = 1, ..., n)$.

Since F_c^n and $\widetilde{F}^n$ are of finite rank and

$$\mu_j(F_c) = \mu_j(\widetilde{F}) = \mu_j(F) \tag{4.5}$$

it follows from (4.3), (4.4), (4.5) that

$$\begin{aligned} \det_n(I+\lambda F_c) = det_n(I+\lambda\widetilde{F}) &= \det(I+\lambda\widetilde{F})\exp\sum_{k=1}^{m-1}\frac{(-\lambda)^k}{k}\mathrm{tr}(\widetilde{F}^k) = \\ &= \prod_j\left[\left(1-\frac{\lambda}{\mu_j(F)}\right)\exp\sum_{m=1}^{n-1}\frac{1}{m}\left(\frac{\lambda}{\mu_j(F)}\right)^m\right]. \end{aligned} \tag{4.6}$$

It can be directly checked that

$$\mathrm{tr}(\widetilde{F}^k) = \mathrm{tr\ diag}(F_c^k)$$

and it follows from (4.6) that

$$\begin{aligned} \det_n(I+\lambda F_c) &= \det F(\lambda)\exp\sum_{k=1}^{n-1}\frac{(-\lambda)^k}{k}\ \mathrm{tr\ diag}(F_c^k) = \\ &= \prod_j\left[\left(1-\frac{\lambda}{\mu_j(F)}\right)\exp\sum_{m=1}^{n-1}\frac{1}{m}\left(\frac{\lambda}{\mu_j(F)}\right)^m\right]. \end{aligned} \tag{4.7}$$

Now for a given pencil

$$A(\lambda) = \sum_{k=0}^{n}\lambda^k A_k \qquad (A_k \in D)$$

we consider n sequences $\{F_m^{(l)}\}$ $(m = 1, ..., n)$ of operators $F_m^{(l)} \in \mathcal{F}(\mathcal{B})$ such that $\|F_m^{(l)} - A_m\|_D \to 0$ $(l \to \infty)$, and a sequence $F^{(l)}(\lambda)$ of pencils

$$F^{(l)}(\lambda) = \sum_{k=0}^{n}\lambda^k F_k^{(l)}.$$

It is clear that $\|F^{(l)}(\lambda) - A(\lambda)\|_D \to 0$ for each λ, and

$$\det F^{(l)}(\lambda) \to \det_D A(\lambda). \tag{4.8}$$

For each integer k the elements $F_{jj}^{(l)}$ and A_{jj} on the main diagonals of the matrices $(F_c^{(l)})^k$ and A_c^k belong to the algebra D and $\|F_{jj}^{(l)} - A_{jj}\|_D \to 0$. It follows from here that

$$\mathrm{tr\ diag}\left((F_c^{(l)})^k\right) \to \mathrm{tr\ diag}(A_c^k). \tag{4.9}$$

Now we are going to show that

$$\det_n(I+\lambda F_c^{(l)}) \to \det_n(I+\lambda A_c). \tag{4.10}$$

Since

$$A_c - F_c^{(l)} = \begin{bmatrix} 0 & 0 & \dots & 0 \\ \cdot & \cdot & \dots & \cdot \\ 0 & 0 & \dots & 0 \\ A_n - F_n^{(l)} & A_{n-1} - F_{n-1}^{(l)} & \dots & A_1 - F_1^{(l)} \end{bmatrix}$$

and $\|A_k - F_k^{(l)}\|_D \to 0$ it follows that for given $\epsilon > 0$ there exists N such that

$$\|A_c - F_c^{(l)}\|_{D^{n\times n}} < \epsilon \quad (l > N).$$

Let $\gamma = \max(\|A_c\|, \|F_c^{(l)}\|)$. Using the general inequality

$$\|x^m - y^m\| \le \|x-y\| \sum_{q=1}^{m-1} \|x\|^q \|y\|^{m-1-q}, \tag{4.11}$$

we obtain

$$\|A_c^m - (F_c^{(l)})^m\|_{D^{n\times n}} \le \epsilon(m-1)\gamma^{m-1} \quad (l > N). \tag{4.12}$$

It follows from (4.2) and (4.11) that

$$\begin{aligned} &\|w(\lambda A_c) - w(\lambda F_c^{(l)})\|_{D^{n\times n}} \le \\ &\le \sum_{m=n}^{\infty} |\alpha_m| \|A_c^m - (F_c^{(l)})^m\|_{D^{n\times n}} |\lambda|^m \le \epsilon \sum_{m=n}^{\infty} |\alpha_m|(m-1)\gamma^{m-1}|\lambda|^m. \end{aligned} \tag{4.13}$$

Since $\sum \alpha_m z^m$ $(m \ge n)$ is an entire function the series in the right-hand side of (4.13) converges and hence $\|w(\lambda A_c) - w(\lambda F_c^{(l)})\|_{D^{n\times n}} \to 0$. Thus

$$\begin{aligned} \det_n(I+\lambda F_c^{(l)}) &= \det(I + w(\lambda F_c^{(l)})) \\ &\to \det(I + w(\lambda A_c)) = \det_n(I+\lambda A_c). \end{aligned} \tag{4.14}$$

Since $\mu_j(A_c) = \mu_j(A)$, it follows from (4.3), (4.7), (4.9) and (4.14) that

$$\begin{aligned} &\prod_j \left[\left(1 - \frac{\lambda}{\mu_j(A)}\right) \exp \sum_{m=1}^{n-1} \frac{1}{m} \left(\frac{\lambda}{\mu_j(A)}\right)^m \right] = \\ &= \det_n(I+\lambda A_c) = \lim_{l\to\infty} \det(I+\lambda F_c^{(l)}) = \\ &= \lim_{l\to\infty} \det F^{(l)}(\lambda) \lim_{l\to\infty} \exp \sum_{k=1}^{n-1} \frac{(-\lambda)^k}{k} \operatorname{tr} \operatorname{diag}((F_c^{(l)})^k) = \\ &= \det_D A(\lambda) \exp \sum_{k=1}^{n-1} \frac{(-\lambda)^k}{k} \operatorname{tr} \operatorname{diag}(A_c^k). \end{aligned} \tag{4.15}$$

This proves the theorem. ■

5. Proof of Theorem 3

Let us represent the matrix A_c as a sum $B+C$, where

$$(5.1)\quad B:=\begin{bmatrix} 0 & -I & \dots & 0 \\ \cdot & \cdot & \dots & \cdot \\ 0 & 0 & \dots & -I \\ 0 & 0 & \dots & 0 \end{bmatrix}; \quad C:=\begin{bmatrix} 0 & 0 & \dots & 0 \\ \cdot & \cdot & \dots & \cdot \\ 0 & 0 & \dots & 0 \\ A_n & A_{n-1} & \dots & A_1 \end{bmatrix}.$$

Then

$$\begin{aligned} A_c^m &= (B+C)^m = B^m + C^m + \\ &+ \sum_{p_1+\dots+p_{2k}=m} (C^{p_1}B^{p_2}\dots C^{p_{2k-1}}B^{p_{2k}} + B^{p_{2k}}C^{p_1}B^{p_2}\dots C^{p_{2k-1}}) + \\ &+ \sum_{p_1+\dots+p_{2k+1}=m} (C^{p_1}B^{p_2}\dots C^{p_{2k-1}}B^{p_{2k}}C^{p_{2k+1}} + B^{p_{2k+1}}C^{p_1}B^{p_2}\dots C^{p_{2k-1}}B^{p_{2k}}), \end{aligned}$$
(5.2)

where $p_j \geq 1$ and $k \geq 1$.

For each operator $R \in D$ we denote by $\{R\}$ any matrix of the form

$$(5.3)\qquad \{R\} := \begin{bmatrix} 0 & \dots & 0 & 0 \\ \cdot & \dots & \cdot & \cdot \\ 0 & \dots & 0 & 0 \\ R_1 & \dots & R_{n-1} & R \end{bmatrix} \in D^{n\times n},$$

where R_j are some (arbitrary) matrices from D.

It is directly checked that $\{R\}\{S\} = \{RS\}$ for any $R, S \in D$ and the following equalities hold:

$$(5.4)\qquad C^p = A_1^{p-1}C; \quad C^pB^q = \left\{(-1)^q A_1^{p-1}A_{q+1}\right\}.$$

It follows from here that

$$\begin{aligned} &C^{p_1}B^{p_2}\dots C^{p_{2k-1}}B^{p_{2k}} = \\ (5.5)\qquad &= \left\{(-1)^{p_2+p_4+\dots+p_{2k}} A_1^{p_1-1}A_{p_2+1}A_1^{p_3-1}A_{p_4+1}\dots A_1^{p_{2k-1}-1}A_{p_{2k}+1}\right\} \end{aligned}$$

and

$$\begin{aligned} &\operatorname{tr}\operatorname{diag}(C^{p_1}B^{p_2}\dots C^{p_{2k-1}}B^{p_{2k}}) = \\ (5.6)\qquad &= (-1)^{p_2+\dots+p_{2k}}\operatorname{tr}_D\left(A_1^{p_1-1}A_{p_2+1}A_1^{p_3-1}A_{p_4+1}\dots A_1^{p_{2k-1}-1}A_{p_{2k}+1}\right). \end{aligned}$$

Using equalities (5.2), (5.6) and (2.3), (2.4) we obtain (5.2). ■

In particular, we obtain the following examples from this theorem:

$$\begin{aligned}
\text{tr diag}A_c &= \text{tr}_D A_1; \quad \text{tr diag}A_c^2 = \text{tr}_D(A_1^2 - 2A_2);\\
\text{tr diag}A_c^3 &= \text{tr}_D(A_1^3 - 3A_1A_2 + 3A_3);\\
\text{tr diag}A_c^4 &= \text{tr}_D(A_1^4 + 4A_1A_3 - 4A_1^2A_2 + 2A_2^2 - 4A_4);\\
\text{tr diag}A_c^5 &= \text{tr}_D(A_1^5 - 5A_1A_4 + 5A_3A_1^2 + 5A_1A_2^2 - 5A_2A_3 + 5A_5).
\end{aligned}$$

In the next three sections we give examples of algebras D for which equality (3.3) holds.

6. Trace class operators in Hilbert spaces

Let H be a separable Hilbert space and $D := \mathcal{S}_1(H)$ the algebra of all trace class operators on H. It is well known that $\mathcal{S}_1(H)$ satisfies conditions a) - d) (Section 2) and condition (3.1) for any N. It follows from Theorem 1 that equality (3.3) holds for the pencils

$$A(\lambda) = I + \sum_{k=1}^{n} \lambda^k A_k \quad (A_k \in \mathcal{S}_1(H)). \tag{6.1}$$

In our case ($D = \mathcal{S}_1(H)$) the determinant $\det_D A(\lambda)$ coincides with the standard determinant for trace class operators.

Equality (3.3) for the pencil (6.1) was obtained earlier in [1], (Theorem IV.10.3). But in [1] a corresponding linearization $I + \lambda A_0$ of the pencil $A(\lambda)$ as well as the functional $\text{tr}_D(A_0)^k$ were considered in an exotic algebra D.

7. 2/3 - Nuclear operators in Banach spaces

Let $\mathcal{B}$ be a Banach space with approximation property. By definition [2, 3] this means that for every compact $K \subset \mathcal{B}$ and every $\epsilon > 0$ there exists an operator F of finite rank such that $\|x - Fx\| < \epsilon$ for each $x \in K$.

Denote $\mathcal{N}_p = \mathcal{N}_p(\mathcal{B}) \quad (0 < p \leq 1, \quad \mathcal{N}_p \subset L(\mathcal{B}))$ the algebra of all p-nuclear operators, i. e. operators which admit representations in the form

$$T = \sum_k u_k \otimes f_k \quad (u_k \in \mathcal{B}, f_k \in \mathcal{B}')$$

such that

$$\sum_k \|u_k\|_{\mathcal{B}}^p \|f_k\|_{\mathcal{B}'}^p < \infty. \tag{7.1}$$

It is well known (see [2], [3], [1], [8]) that the algebra $\mathcal{N}_{2/3}$ satisfies conditions a)-d) (Section 2). It can be directly checked that

$$(7.2)\qquad \left(\mathcal{N}_{2/3}(\mathcal{B})\right)^{n\times n} = \mathcal{N}_{2/3}(\mathcal{B}^n)$$

and by the well known Grothendieck's 2/3 nuclear theorem ([Gr2], Ch.2; see also [1], Theorem V.3.1), condition (3.1) holds for the algebra $\left(\mathcal{N}_{2/3}(\mathcal{B})\right)^{n\times n} = \mathcal{N}_{2/3}(\mathcal{B}^n)$. It follows from Theorem 2 that equality (3.3) is true for the pencil

$$(7.3)\qquad A(\lambda) = I + \sum_{k=1}^{n} \lambda^k A_k$$

with $A_k \in \mathcal{N}_{2/3}(\mathcal{B})$.

Let, in addition, the operators A_k admit representations $A_k = A_{k1}...A_{kk}$ ($A_{kj} \in \mathcal{N}_{2/3}$). Then (see Theorem 1) the relations (3.2) hold for the operator pencil (7.3). This was obtained earlier in [1], Corollary V.3.2.

8. Another variant of trace class operators in Banach spaces

Let $\mathcal{B}$ be a Banach space (without requiring the approximation property), $T \in L(\mathcal{B})$ and $s_n(T)$ $(n \geq 1)$ denote the approximation numbers of the operator T :

$$(8.1)\qquad s_n(T) := \inf\{\|T - F\| : \text{rank}F < n\}.$$

Denote by $\mathcal{T} = \mathcal{T}(\mathcal{B})$ the set of all operators $A \in L(\mathcal{B})$ with summable approximation numbers:

$$(8.2)\qquad \sum_n s_n(A) < \infty.$$

It is well known that the set $\mathcal{T}$ endowed with the norm

$$(8.3)\qquad \|A\|_{\mathcal{T}} := \sum_{n\geq 1} s_n(A)$$

is a Banach algebra and conditions a) - c) (Section 2) are fulfilled for the algebra $D = \mathcal{T}(\mathcal{B})$ (see [8] and [7], Ch. 8). Condition d) follows from the generalized Weil inequality (see, for example [4], page 129). H. König ([4]) proved that for operators $A \in \mathcal{T}$ the famous Lidskii Theorem holds:

$$(8.4)\qquad \sum_j |\lambda_j(A)| < \infty, \quad \text{tr}(A) = \sum_j \lambda_j(A).$$

Let $T = [T_{jk}] \in \mathcal{T}^{N\times N}$. We represent T as $\sum \widetilde{T_{jk}}$ where the matrix $\widetilde{T_{jk}} \in \mathcal{T}^{N\times N}$ has only one non-zero entry T_{jk}. Since $T_{ik} \in \mathcal{T}(\mathcal{B})$ it is clear that $\widetilde{T_{jk}} \in \mathcal{T}(\mathcal{B}^N)$.

It follows from here and (8.4) that condition (3.1) is fulfilled for operators $A \in \mathcal{T}(\mathcal{B})^{N\times N}$ for each N. Using Theorem 2 we obtain that equality (3.3) holds for operator pencils

$$A(\lambda) = I + \sum_{k=1}^{n} \lambda^k A_k \tag{8.5}$$

with $A_k \in \mathcal{T}(\mathcal{B})$.

Let, in addition, the operators A_k admit representations $A_k = A_{k1}...A_{kk}$ ($A_{kj} \in \mathcal{T}(\mathcal{B})$). Then (see Theorem 1) the relations (3.2) hold for the operator pencil (8.5).

References

[1] Gohberg, I., Goldberg, S., Krupnik, N., *Traces and Determinants of Linear Operators*, OT, Vol. **116**, Birkhäuser, 2000.

[2] Grothendieck, A., *La Theorie de Fredholm*, Bull. Soc. Math. France **84** (1956), 319-384.

[3] Grothendieck, A., *Produits Tensoriels Topologiques et Espaces Nucléaires*, Memoirs of the American Mathematical Society, No. 16, 1966.

[4] König, H., *s-numbers,eigenvalues and trace theorem in Banach spaces*, Studia Math. LXVII (1980), 157-172.

[5] Kaashoek, M. A., van de Ven, M. P. A., *A linearization for operator polynomials with coefficients in certain operator ideals*, Ann. Math. Pura Appl., IV, CXXV (1980), 329-336.

[6] Markus, A., *Introduction to the spectral theory of polynomial operator pencils*, Amer. Math. Soc., Providence, 1988.

[7] Pietsch, A., *Nukleare lokalkonvexe Räume*, Akademie-Verlag, Berlin, 1965.

[8] Pietsch, A., *Eigenvalues and s-numbers*, Cambridge Studies in Advanced Math., Vol. **13**, Cambridge Univ. Press, Cambridge, 1987.

Israel Gohberg
School of Mathematical Sciences
Raymond and Beverly Sackler Faculty
of Exact Sciences
Tel-Aviv University, Israel

Naum Krupnik
Dept. of Mathematics and Computer Science
Bar-Ilan University
Ramat - Gan, 52900, Israel

1991 Mathematics Subject Classification: 47A05, 47A99

Submitted: 16.8.2000

Operator Theory:
Advances and Applications, Vol. 121

Random walk models approximating symmetric space-fractional diffusion processes

RUDOLF GORENFLO AND FRANCESCO MAINARDI

For the symmetric case of space-fractional diffusion processes (whose basic analytic theory has been developed in 1952 by Feller via inversion of Riesz potential operators) we present three random walk models discrete in space and time. We show that for properly scaled transition to vanishing space and time steps these models converge in distribution to the corresponding time-parameterized stable probability distribution. Finally, we analyze in detail a model, discrete in time but continuous in space, recently proposed by Chechkin and Gonchar.

1. Introduction: concepts and notations

By a "space-fractional" diffusion process (or Lévy-Feller diffusion process) we mean a process of diffusion of an extensive quantity with density $u(x,t)$ governed by an evolution equation

$$\frac{\partial u(x,t)}{\partial t} = D_\theta^\alpha u(x,t) \qquad (x \in \mathbf{R},\ t > 0) \tag{1.1}$$

with an initial condition

$$u(x,0) = f(x) \qquad (f \in L_1(\mathbf{R})). \tag{1.2}$$

We interpret x as space, t as time variable. D_θ^α is a pseudo-differential operator acting with respect to the space variable x, its symbol being

$$\widehat{D_\theta^\alpha}(\kappa) = -|\kappa|^\alpha e^{i\,(\operatorname{sign}\kappa)\,\theta\pi/2} = -|\kappa|^\alpha\, i^{(\operatorname{sign}\kappa)\,\theta}\,, \tag{1.3}$$

and the real parameters α and θ are restricted by the inequalities

$$0 < \alpha \le 2\,, \qquad |\theta| \le \begin{cases} \alpha & \text{if } 0 < \alpha \le 1\,, \\ 2-\alpha & \text{if } 1 < \alpha \le 2\,. \end{cases} \tag{1.4}$$

For a sufficiently well-behaved function or generalized function ϕ defined on $\mathbf{R}$ we denote by $\hat{\phi}$ its Fourier transform:

$$\hat{\phi}(\kappa) = \int_{-\infty}^{+\infty} e^{i\kappa x}\,\phi(x)\,dx \qquad (\kappa \in \mathbf{R})\,. \tag{1.5}$$

Then, for a generic linear pseudo-differential operator A acting on these functions, its symbol $\hat{A}(\kappa)$ turns out to be defined through the Fourier representation of $(A\phi)(x)$, namely $\widehat{A\phi}(\kappa) = \hat{A}(\kappa)\,\hat{\phi}(\kappa)$. An often applicable practical rule is

$$\hat{A}(\kappa) = (Ae^{-i\kappa x})e^{i\kappa x}, \qquad (\kappa \in \mathbf{R}). \tag{1.6}$$

If B is another pseudo-differential operator, then we have $\widehat{AB}(\kappa) = \hat{A}(\kappa)\,\hat{B}(\kappa)$. Let us remark, that we chose (1.5) to define the Fourier transform in agreement with the common terminology of probability theory.

Introducing the stable probability density $p_\alpha(x;\theta)$ whose characteristic function (Fourier transform) is

$$\hat{p}_\alpha(\kappa;\theta) = \exp\left(-|\kappa|^\alpha e^{i(\operatorname{sign}\kappa)\theta\pi/2}\right) \qquad (\kappa \in \mathbf{R}) \tag{1.7}$$

and rescaling $p_\alpha(x;\theta)$ for $x \in \mathbf{R},\ t > 0$ by the similarity variable $x\,t^{-\frac{1}{\alpha}}$ we obtain the time-dependent stable probability density

$$g_\alpha(x,t;\theta) = t^{-\frac{1}{\alpha}} p_\alpha(xt^{-\frac{1}{\alpha}};\theta) \qquad (x \in \mathbf{R},\, t > 0) \tag{1.8}$$

with which we can write the solution to (1.1), (1.2) in the form

$$u(x,t) = \int_{-\infty}^{+\infty} g_\alpha(x-\xi,t;\theta) f(\xi)\,d\xi. \tag{1.9}$$

Then, for all $t > 0$, we have

$$\begin{cases} u(\cdot,t) \in C^\infty \cap L_1(\mathbf{R}), & \int_{-\infty}^{+\infty} u(x,t)\,dx = \int_{-\infty}^{+\infty} f(x)\,dx \\ f(x) \ge 0 \quad \text{for all} \quad x \in \mathbf{R} \Rightarrow u(x,t) \ge 0 & \text{for all} \quad x \in \mathbf{R}. \end{cases} \tag{1.10}$$

If the initial function is a probability density then so is also the function $u(x,t)$ and we have in (1.1), (1.2) the description of a Markov process.

For orientation on the general theory of stable probability distributions we recommend in particular [F71] and [F52], Feller's parameterization being close to ours. In our foregoing considerations we essentially have surveyed results of Feller's pioneering paper [F52]. For a few parameters pairs $(\alpha,\ \theta)$ representations of $p_\alpha(x;\theta)$ in terms of elementary or well-investigated special functions are available in the literature but in other parameterization (see e.g. [Zo], [Sc], [SaT]). We here content ourselves with recognizing the classical Gauss process and the Cauchy process, respectively, in

$$g_2(x,t;0) = \frac{1}{2\sqrt{\pi}} t^{-\frac{1}{2}} \exp\left(-\frac{x^2}{4t}\right), \qquad g_1(x,t;0) = \frac{1}{\pi}\frac{t}{x^2+t^2}.$$

Feller in [F52] has shown that D_θ^α (in case $\alpha \neq 1$) can be viewed as inverse to the (later in [SKM] so called) Feller potential operator which is a linear combination of two Weyl integral operators. In the sequel we will restrict attention to the symmetric case $\theta = 0$ retaining however, the index 0 in order to be in concordance with the notation of our previous papers [GM98], [GFM], [GM99]. In [GM98] we have discussed for $\alpha \neq 1$ a random walk model for the whole range of values θ; here we now concentrate on models not treated there, requiring however $\theta = 0$. This means that henceforth we will treat the evolution equation

$$\frac{\partial u(x,t)}{\partial t} = D_0^\alpha \, u(x,t) \qquad (x \in \mathbf{R},\, t > 0) \tag{1.11}$$

with an initial condition

$$u(x,0) = f(x) \qquad (f \in L_1(\mathbf{R})). \tag{1.12}$$

The symbol of the pseudo-differential operator D_0^α is

$$\widehat{D_0^\alpha}(\kappa) = -|\kappa|^\alpha, \tag{1.13}$$

and the fundamental solution to (1.11) (namely, for $u(x,0) = \delta(x) =$ Dirac's delta function) is the function $u(x,t) = g_\alpha(x,t;0)$ whose Fourier transform is

$$\widehat{g_\alpha}(\kappa,t;0) = \exp\big(-t|\kappa|^\alpha\big) \qquad (\kappa \in \mathbf{R}\,,\; t > 0). \tag{1.14}$$

Then, see (1.7) and (1.8), we get by the Fourier inversion formula

$$g_\alpha(x,t;0) = \frac{1}{2\pi}\int_{-\infty}^{+\infty} e^{-i\kappa x}\,\exp(-t|\kappa|^\alpha)\,d\kappa = t^{-\frac{1}{\alpha}}\,p_\alpha(xt^{-\frac{1}{\alpha}};0)\,. \tag{1.15}$$

We will denote by $S(t)$ for $t > 0$ the random variable whose probability density is given by $g_\alpha(x,t;0)$.

Remark 1.1. We call a process described by (1.1), (1.2) a "space fractional" diffusion process because it is a "fractional" generalization of the classical diffusion process which is recovered by taking $\alpha = 2$, $\theta = 0$. Writing (1.13) in the form $\widehat{D_0^\alpha}(\kappa) = -(\kappa^2)^{\alpha/2}$ and observing that the operator D^2, defined by $\left(D^2\,\phi\right)(x) = \frac{d^2\phi(x)}{dx^2}$, has the symbol $-\kappa^2$, we see that $D_0^\alpha = -\left(-D^2\right)^{\alpha/2}$, hence the operator D_0^α is the negative of a fractional power of the (positive definite) operator $-D^2$. By calling a process described by (1.1), (1.2) also a Lévy-Feller diffusion process we honour both Lévy and Feller for their essential contributions [L25], [L54], [F52].

Our aim in the following sections is to derive discrete-space discrete-time random walk models approximating the space-fractional diffusion process (henceforth considered as a Markov process) described by (1.11) with $u(x,0) = \delta(x)$. We shall show that for properly scaled transition to vanishing space and time steps there is convergence in distribution to the probability distribution whose density is $g_\alpha(x,t;0)$. We shall give heuristic motivations for choosing our concrete models, using in a formal way calculations with symbols. The lack of rigour in these derivations will hopefully not be too annoying to the pure analyst, it will be remedied in the final end by rigorous proofs of convergence.

2. Operators and symbols

In this section we give a survey on the relevant operators and their symbols thereby always assuming

$$0 < \alpha \leq 2. \tag{2.1}$$

For general orientation and more rigorous treatment we refer to [SKM], [R] and [F52]. We need the operator D of differentiation, the Weyl operators I_+^α, I_-^α and their (formal) inverses $I_+^{-\alpha}$, $I_-^{-\alpha}$, the Riesz potential operator I_0^α whose negative inverse $-I_0^{-\alpha}$ is our pseudo-differential operator D_0^α (if $\alpha \neq 1$), and the Hilbert transform operator H. For sufficiently well behaved functions ϕ defined on $\mathbf{R}$ and if required with appropriate understanding of the occurring integrals as Cauchy principal values we have (for $x \in \mathbf{R}$)

$$(D\,\phi)(x) = \frac{d}{dx}\phi(x) = \phi'(x), \tag{2.2}$$

$$\begin{cases} (I_+^\alpha\,\phi)(x) = \dfrac{1}{\Gamma(\alpha)} \int_{-\infty}^{x} (x-\xi)^{\alpha-1}\,\phi(\xi)\,d\xi\,, \\ (I_-^\alpha\,\phi)(x) = \dfrac{1}{\Gamma(\alpha)} \int_{x}^{+\infty} (\xi-x)^{\alpha-1}\,\phi(\xi)\,d\xi\,, \end{cases} \tag{2.3}$$

$$I_\pm^{-\alpha} = \begin{cases} \pm D I_\pm^{1-\alpha} & \text{if } 0 < \alpha < 1\,, \\ D^2 I_\pm^{2-\alpha} & \text{if } 1 < \alpha \leq 2\,, \end{cases} \tag{2.4}$$

$$(I_0^\alpha\,\phi)(x) = \frac{(I_+^\alpha\,\phi)(x) + (I_-^\alpha\,\phi)(x)}{2\cos(\alpha\pi/2)} = \frac{\int_{-\infty}^{+\infty} |x-\xi|^{\alpha-1}\,\phi(\xi)\,d\xi}{2\Gamma(\alpha)\,\cos(\alpha\pi/2)} \quad \text{for} \quad \alpha \neq 1, \tag{2.5}$$

$$(H\phi)(x) = \frac{1}{\pi}\int_{-\infty}^{+\infty} \frac{\phi(\xi)}{x-\xi}\,d\xi\,. \tag{2.6}$$

Recalling formula (1.6) for calculation of symbols we get by direct computation

$$\hat{D} = -i\kappa\,, \tag{2.7}$$

$$\hat{H} = i\,\mathrm{sign}\,\kappa\,. \tag{2.8}$$

We take the symbols of the Weyl operators and the Riesz potential operators from [R, Theorem 4.10] as

$$\widehat{I_\pm^\alpha}(\kappa) = (\mp i\kappa)^{-\alpha} = |\kappa|^{-\alpha} e^{\pm i\,(\mathrm{sign}\,\kappa)\,\alpha\pi/2} = |\kappa|^{-\alpha}\, i^{\pm\alpha\,\mathrm{sign}\,\kappa}\,, \tag{2.9}$$

from which by addition we get

$$\widehat{I_0^\alpha}(\kappa) = |\kappa|^{-\alpha}\,. \tag{2.10}$$

As already remarked and as used in [F52]

$$D_0^\alpha = -I_0^{-\alpha} \quad \text{for} \quad \alpha \neq 1 \tag{2.11}$$

in agreement with the property of symbols

$$\widehat{D_0^\alpha}(\kappa) = -|\kappa|^\alpha = -\left(\widehat{I_0^\alpha}(\kappa)\right)^{-1} \quad \text{for} \quad \alpha \neq 1\,. \tag{2.12}$$

In the special case $\alpha = 1$ we observe that by (2.7) and (2.8)

$$\widehat{D_0^1}(\kappa) = -|\kappa| = -\hat{D}(\kappa)\,\hat{H}(\kappa)\,, \tag{2.13}$$

hence as already observed in [F52] (with a sign-modified version of the Hilbert transform)

$$D_0^1 = -D\,H\,. \tag{2.14}$$

Let us yet exhibit another representation of our pseudo-differential operator D_0^α. Via the semi-group property (see [F52] or [SKM])

$$I_0^\alpha\, I_0^\beta = I_0^{\alpha+\beta} \qquad \text{if} \quad \alpha,\beta \in (0,1)\,, \quad \alpha+\beta < 1\,,$$

analytic continuation to negative exponents can be justified and thus from (2.5) and (2.11) the formula

$$D_0^\alpha = -\frac{1}{2\cos(\alpha\pi/2)}(I_+^{-\alpha} + I_-^{-\alpha}) \qquad \text{for} \quad \alpha \neq 1\,. \tag{2.15}$$

In Section 4 we shall find random walk schemes in the case $\alpha \neq 1$ by approximating in (2.15) the operators $I_+^{-\alpha}$ and $I_-^{-\alpha}$ with the Grünwald-Letnikov discretization. In Section 5 a second random walk scheme, for the whole range $0 < \alpha \leq 2$, will be obtained by a straightforward discrete approximation of hypersingular integrals for $I_0^{-\alpha}$ and DH. From [SKM, formula (12.1')] we take (for $0 < \alpha < 2\,,\ \alpha \neq 1$)

$$(I_0^{-\alpha}\phi)(x) = \frac{1}{2\Gamma(-\alpha)\,\cos(\alpha\pi/2)} \int_0^\infty \frac{\phi(x+\xi) - 2\phi(x) + \phi(x-\xi)}{\xi^{\alpha+1}}\,d\xi. \tag{2.16}$$

Quite formally we can obtain (2.16) by replacing in (2.5) the integrand by $|\xi|^{\alpha-1}\,\phi(x-\xi)$, then replacing α by $-\alpha$, splitting $\int_{-\infty}^{\infty} = \int_{-\infty}^{0} + \int_0^\infty$, here regularizing the right hand side hypersingular integrals by subtracting $\phi(x)$ in the numerators, finally substituting $-\xi$ for ξ in the first right hand side integral and then putting both integrals together.

For convenience we simplify the coefficient in (2.16), introducing

$$b(\alpha) := -\frac{1}{2\Gamma(-\alpha)\cos(\alpha\pi/2)} = \frac{1}{\pi}\Gamma(\alpha+1)\sin(\alpha\pi/2), \tag{2.17}$$

where for the latter equality we have used $\sin(\alpha\pi) = 2\sin(\alpha\pi/2)\cos(\alpha\pi/2)$ and the reflection formula for the gamma function $\Gamma(-\alpha)\Gamma(\alpha+1) = -\pi/\sin(\alpha\pi)$. Then, for $0 < \alpha < 2$, $\alpha \neq 1$,

$$(D_0^\alpha\phi)(x) = -(I_0^{-\alpha}\phi)(x) = b(\alpha)\int_0^\infty \frac{\phi(x+\xi) - 2\phi(x) + \phi(x-\xi)}{\xi^{\alpha+1}}\,d\xi. \tag{2.18}$$

Note that for $\phi \in C^2(\mathbf{R})$ and $\phi(x)$ bounded the integral is finite as an improper Riemann integral and observe that $b(\alpha) > 0$ for the admitted values of α.

Because (2.17) gives $b(1) = 1/\pi$ we are tempted to believe the formula

$$(D_0^1\phi)(x) = \frac{1}{\pi}\int_0^\infty \frac{\phi(x+\xi) - 2\phi(x) + \phi(x-\xi)}{\xi^2}\,d\xi. \tag{2.19}$$

We will use (2.18) and (2.19) as motivation for our second random walk scheme in which the parameter value $\alpha = 1$ does no longer play a special role. We can indeed obtain (2.19) by looking at (2.14), then formally differentiating (behind the integral sign) (2.6) and then splitting and regularizing the resulting hypersingular integral in the same way as we have done in the case $\alpha \neq 1$.

In view of (2.18) and (2.19) we now have, with $b(\alpha)$ given in (2.17),

$$(D_0^\alpha\phi)(x) = b(\alpha)\int_0^\infty \frac{\phi(x+\xi) - 2\phi(x) + \phi(x-\xi)}{\xi^{\alpha+1}}\,d\xi \quad \text{for} \quad 0 < \alpha < 2. \tag{2.20}$$

Unfortunately, this formula loses its meaning in the case $\alpha = 2$.

3. General structure of the random walk models

What do we mean by a *random walk model, discrete in space and discrete in time, for a Markov process*? Let us be given a random variable Y taking its values in the set $\mathbf{Z}$ of integers according to the probabilities

$$P(Y = k) = p_k \qquad (k \in \mathbf{Z}) \tag{3.1}$$

with

$$\text{all } p_k \geq 0 \text{ and } \sum_{k\in\mathbf{Z}} p_k = 1. \tag{3.2}$$

We discretize the space variable $x \in \mathbf{R}$ and the time variable $t \geq 0$ by grid points $x_j = jh$ and instants $t_n = n\tau$, with $h > 0,\ \tau > 0,\ j \in \mathbf{Z},\ n \in \mathbf{N}_0$. Then, defining random variables

$$(3.3) \qquad S_n = \sum_{m=1}^{n} (h\,Y_m) = h \sum_{m=1}^{n} Y_m \qquad (n \in \mathbf{N})$$

with the Y_m as *independent identically distributed* random variables, all having the same probability distribution as the random variable Y, we interpret S_n as the position at time t_n of a random walker starting in the point $x = 0$ at $t = 0$. Denoting by $y_j(t_n)$ the probability of sojourn of the walker in point x_j at instant t_n, the recursion $S_{n+1} = S_n + hY_n$ implies

$$(3.4) \qquad y_j(t_{n+1}) = \sum_{k \in \mathbf{Z}} p_k\, y_{j-k}(t_n) \qquad (j \in \mathbf{Z},\ n \in \mathbf{N}_0),$$

and the walker starting at point $x_0 = 0$ means $y_0(0) = 1$ and $y_j(0) = 0$ for $j \neq 0$. However, in the recursion scheme (3.4) it is legitimate to use a more general initial sojourn probability distribution $(y_j(0) \mid j \in \mathbf{Z})$.

There is yet another possible interpretation of (3.4), namely as a *scheme of redistribution of an extensive quantity* (e.g. mass, charge, in the random walk interpretation probability), $y_j(t_n)$ being considered as a clump of this extensive quantity sitting in point x_j at instant t_n. Then (3.4) describes a conservative and non-negativity preserving redistribution scheme. In fact, for all $n \in \mathbf{N}$ it follows from (3.2) that, in analogy to (1.10),

$$\sum_{j \in \mathbf{Z}} y_j(t_n) = \sum_{j \in \mathbf{Z}} y_j(0) \qquad \text{if} \quad \sum_{j \in \mathbf{Z}} |y_j(0)| < \infty,$$

$$\text{all} \quad y_j(t_n) \geq 0 \qquad \text{if all} \quad y_j(0) \geq 0.$$

Such discrete redistribution schemes have been used by one of the authors in discretization of diffusion processes governed by second order linear parabolic differential equations ([G70], [G78], [GN]) as they discretely imitate essential properties of the continuous process.

We come nearer to the Cauchy problem (1.11), (1.12) by intending $y_j(t_n)$ as approximation to

$$\int_{x_j - h/2}^{x_j + h/2} u(x, t_n)\, dx$$

which, if $u(\cdot, t_n)$ is continuous, is $\approx hu(x_j, t_n)$ for small h.

We want to show that for proper choice of the probability distribution of the random variable Y and well-scaled transition

$$(3.5) \qquad \tau = \sigma(h)\,, \quad \sigma \quad \text{strictly monotonic}, \quad \sigma(h) \to 0 \quad \text{as} \quad h \to 0$$

the random walk "converges" in some sense to the Markov process with density $u(x,t)$ described by (1.11), (1.12) in the case that the initial function is a probability density. More specifically, we will prove for *fixed* $t > 0$, $t = t_n = n\tau$ with $\mathbf{N} \ni n = t/\tau \to \infty$ (and proper scaling of h and τ) that the random variable S_n of (3.3) converges *in distribution* (other terminology: *in law*) to the random variable $S(t)$ whose density is $g_\alpha(\cdot, t; 0)$, the fundamental solution (1.15) of (1.11). Observing that the distribution function $G_\alpha(x,t;0) = \int_{-\infty}^{x} g_\alpha(\xi,t;0)\,d\xi$ is continuous in x (due to the fast decay in $|\kappa|$ of $\hat{g}_\alpha(\kappa,t;0) = \exp(-t|\kappa|^\alpha)$ the density $g_\alpha(\cdot,t;0)$ is in $C^\infty(\mathbf{R})$) and invoking the continuity theorem of probability theory (see, e.g., [B, Theorem 8.28]), all we have to do is to show that for all $\kappa \in \mathbf{R}$ the characteristic function $\hat{y}(\kappa,t;h)$ of the random variable S_n tends to $\exp(-t|\kappa|^\alpha)$ as $h \to 0$. Note the equivalences following from $t = t_n = n\tau$ and (3.5)

$$n \to \infty \Leftrightarrow h \to 0 \Leftrightarrow \tau \to 0\,. \tag{3.6}$$

The general form of the characteristic function $\hat{y}(\kappa,t;h)$ can be found via the generating functions

$$\tilde{p}(z) = \sum_{j\in\mathbf{Z}} p_j z^j, \qquad \tilde{y}(z,t_n) = \sum_{j\in\mathbf{Z}} y_j(t_n)\, z^j. \tag{3.7}$$

As probabilities both the p_j and $y_j(t_n)$ sum up to 1 if added over the index j, hence these series converge absolutely and uniformly on the periphery $|z| = 1$ of the unit circle, and so the functions $\tilde{p}$ and $\tilde{y}_n$ are there uniformly continuous. The random walk S_n starting at $x = 0$, we have (using the Kronecker symbol) $y_j(0) = \delta_{j0}$, and the recursion (3.4) being a discrete convolution we get

$$\tilde{y}(z) = (\tilde{p}(z))^n. \tag{3.8}$$

Replacing in (3.7) z^j by $e^{i\kappa x_j} = e^{i\kappa jh}$ we obtain the corresponding characteristic functions ($\kappa \in \mathbf{R}$)

$$\hat{p}(\kappa;h) = \tilde{p}(e^{i\kappa h}), \qquad \hat{y}(\kappa,t_n;h) = \tilde{y}(e^{i\kappa h},t_n) = \left(\tilde{p}(e^{i\kappa h})\right)^n. \tag{3.9}$$

Recalling our fixation of $t = t_n = n\tau = n\sigma(h) > 0$, the scaling relation (3.5) and the equivalences (3.6) we have to show that

$$\hat{y}(\kappa,t;h) \to \exp(-t|\kappa|^\alpha) \text{ for } n \to \infty, \tag{3.10}$$

or, equivalently

$$\frac{1}{\sigma(h)} \log \tilde{p}(e^{i\kappa h}) \to -|\kappa|^\alpha \text{ as } h \to 0\,. \tag{3.11}$$

In the following sections we shall exhibit (3.10) as true for specific choices of the probabilities p_j and scalings $\tau = \sigma(h)$. The fact that, strictly speaking, $\tau = t/n = \sigma(h)$ and h in (3.6) and (3.11) are running through discrete sets will turn out as irrelevant for the proof of (3.11).

4. The Grünwald-Letnikov random walk

An idea suggesting itself is to discretize in (1.11) the time derivative $\frac{\partial u}{\partial t}$ by a two-level difference quotient and the operators $I_+^{-\alpha}$ and $I_-^{-\alpha}$ (see (2.15)) by the Grünwald-Letnikov approximation (see, *e.g.* , [SKM], [P]). This idea leads to

$$(4.1)\qquad \frac{y_j(t_{n+1}) - y_j(t_n)}{\tau} = {}_hD_0^\alpha y_j(t_n) = -\frac{1}{2\cos(\alpha\pi/2)}\left({}_hI_+^{-\alpha} + {}_h I_-^{-\alpha}\right) y_j(t_n).$$

with the operators ${}_hI_\pm^{-\alpha}$ still to be specified. We must exclude the singular case $\alpha = 1$, hence will distinguish from now on the cases

$$\text{(a)}\quad 0 < \alpha < 1, \qquad \text{(b)}\quad 1 < \alpha \le 2.$$

We define

$$(4.2)\qquad {}_hI_\pm^{-\alpha}\, y_j(t_n) = h^{-\alpha} \sum_{k=0}^{\infty} (-1)^k \binom{\alpha}{k}\, y_{j\mp k}(t_n) \qquad \text{in case (a)},$$

$$(4.3)\qquad {}_hI_\pm^{-\alpha}\, y_j(t_n) = h^{-\alpha} \sum_{k=0}^{\infty} (-1)^k \binom{\alpha}{k}\, y_{j\pm 1\mp k}(t_n) \qquad \text{in case (b)}.$$

Note in case (b) the shift of the index, that is required in order to obtain non-negative values for all transition probabilities p_j.

Solving (4.1) for $y_j(t_{n+1})$, thereby scaling by

$$(4.4)\qquad \tau = \mu h^\alpha =: \sigma(h),$$

gives (remember (3.4))

$$(4.5)\qquad y_j(t_{n+1}) = \sum_{k\in\mathbf{Z}} p_k\, y_{j-k}(t_n)$$

with in case (a)

$$(4.6)\qquad p_0 = 1 - \frac{\mu}{\cos(\alpha\pi/2)}, \quad p_k = (-1)^{|k|+1} \frac{\mu}{2\cos(\alpha\pi/2)} \binom{\alpha}{|k|} \quad \text{for} \quad k \ne 0,$$

in case (b)

$$(4.7)\qquad \begin{cases} p_0 = 1 + \dfrac{\mu}{\cos(\alpha\pi/2)} \dbinom{\alpha}{1}, \\[2ex] p_{\pm 1} = -\dfrac{\mu}{2\cos(\alpha\pi/2)} \left[1 + \dbinom{\alpha}{2}\right], \\[2ex] p_{\pm k} = (-1)^k \dfrac{\mu}{2\cos(\alpha\pi/2)} \dbinom{\alpha}{k+1} \quad \text{for } k = 2, 3, \ldots. \end{cases}$$

Then, all $p_k \geq 0$ if

(4.8) $$0 < \mu \leq \cos(\alpha\pi/2) \qquad \text{in case (a)},$$

(4.9) $$0 < \mu \leq |\cos(\alpha\pi/2)|/\alpha \qquad \text{in case (b)}.$$

Note that $\cos(\alpha\pi/2) < 0$ in case (b). In both cases, by rearrangement of series,

$$\sum_{k \in \mathbf{Z}} p_k = 1 - \frac{\mu}{\cos(\alpha\pi/2)} \sum_{j=0}^{+\infty} (-1)^j \binom{\alpha}{j} = 1 - 0 = 1.$$

Remark 4.1. For all $\alpha > 0$ the series $\sum_{j=0}^{+\infty} (-1)^j \binom{\alpha}{j}$ is absolutely convergent because $\binom{\alpha}{j} = O(j^{-\alpha-1})$ for $j \to \infty$.

We see that, under the conditions (4.8) or (4.9), respectively, we can put

$$P(Y = k) = p_k \qquad (k \in \mathbf{Z})$$

for the random variable Y of (3.1). Using (4.6) and (4.7) we identify the generating function $\tilde{p}$ of (3.7) as

(4.10) $$\tilde{p}(z) = 1 - \frac{\mu}{2\cos(\alpha\pi/2)} \{(1-z)^\alpha + (1-z^{-1})^\alpha\} \qquad \text{in case (a)},$$

(4.11) $$\tilde{p}(z) = 1 - \frac{\mu}{2\cos(\alpha\pi/2)} \{z^{-1}(1-z)^\alpha + z(1-z^{-1})^\alpha\} \quad \text{in case (b)}.$$

Let us verify the limit relation (3.11) which implies (3.10). Because of the symmetry relation (for $z = e^{i\kappa h}$, $\kappa \in \mathbf{R}$)

$$\tilde{p}(z) = \tilde{p}(z^{-1}) \text{ implying } \tilde{p}(e^{i\kappa h}) = \tilde{p}(e^{-i\kappa h})$$

it suffices to verify (3.11) for $\kappa > 0$ (the special case $\kappa = 0$ being trivial).

Let be $\kappa > 0$. Then in case (a) we have

$$\tilde{p}(z) = 1 - \frac{\mu}{\cos(\alpha\pi/2)} \Re(1-z)^\alpha = 1 - \mu(\kappa h)^\alpha + o(h^\alpha) \quad \text{as} \quad h \to 0,$$

since $(1-z)^\alpha \sim (-i\kappa h)^\alpha = e^{-i\alpha\pi/2}(\kappa h)^\alpha$. With the scaling (4.4), namely $\sigma(h) = \mu h^\alpha$, follows (3.11).

In case (b) we have

$$\tilde{p}(z) = 1 - \frac{\mu}{\cos(\alpha\pi/2)} \Re(z^{-1}(1-z)^\alpha),$$

and an analogous calculation gives

$$\frac{1}{\sigma(h)} \log \tilde{p}(e^{i\kappa h}) \sim -\kappa^\alpha \frac{\cos[(\alpha\pi/2) + \kappa h]}{\cos(\alpha\pi/2)} \quad \text{as} \quad h \to 0,$$

hence again (3.11). As result we have

Theorem 4.2. *Distinguish the cases* (a) $0 < \alpha < 1$, (b) $1 < \alpha \leq 2$. *Define the probabilities* $p_k = P(Y = k)$ *in case* (a) *by* (4.6) *with restriction* (4.8), *in case* (b) *by* (4.7) *with restriction* (4.9). *Let the scaling relation* $\tau = \mu h^\alpha = \sigma(h)$ *hold and let for fixed* $t > 0$ *the index* $n = t/\tau$ *run through* $\mathbf{N}$ *towards* ∞. *Then the random variable* S_n *of* (3.3) *converges in distribution to the random variable* $S(t)$ *whose probability density is given by* (1.15) *as* $g_\alpha(x, t; 0)$.

Remark 4.3. In the special case $\alpha = 2$ the familiar explicit difference scheme

$$y_j(t_{n+1}) = (1 - 2\mu)\, y_j(t_n) + \mu\, y_j(t_{n-1}) + \mu\, y_j(t_{n+1})$$

is recovered from (4.7), and (4.9) goes over into the well-known stability condition $0 < \mu \leq 1/2$.

Remark 4.4. The case $\alpha = 1$ is singular. For $\alpha \to 1$ both upper bounds in (4.8) and (4.9) tend to 0, and the denominators occurring in the definitions of the probabilities p_k tend to zero.

Remark 4.5. A motivation for the Grünwald-Letnikov approximation of $I_+^{-\alpha}$ can be drawn from the fact that $z = e^{i\kappa h}$ is the symbol of the backward shift by a step h: With

$$(T_h\phi)(x) = \phi(x + h), \qquad (T_{-h}\phi)(x) = \phi(x - h),$$

we have

$$\widehat{T_h}(\kappa) = e^{-i\kappa(x+h)} e^{i\kappa x} = e^{-i\kappa h}, \quad \widehat{T_{-h}}(\kappa) = e^{i\kappa h}.$$

From the symbol $h^{-1}(1 - z) = {}_h\widehat{D_+}(\kappa)$ of the usual backward approximation

$$({}_hD_+\phi)(x) = h^{-1}(\phi(x) - \phi(x - h)) = h^{-1}\,({}_hI\phi)(x)$$

we arrive by analogy at the symbol $h^{-\alpha}(1 - z)^\alpha$ as a candidate for the symbol of the operator ${}_hI_+^{-\alpha}$. Analogously we get $h^{-\alpha}(1 - z^{-1})^\alpha$ as the symbol for the operator ${}_hI_-^{-\alpha}$. We use the corresponding approximations in case (a) $0 < \alpha < 1$.

In case (b) $1 < \alpha \leq 2$ we use the form $D^2\, I_+^{2-\alpha}$ of the Riemann-Liouville left inverse of the operator I_+^α, and put ${}_hI_+^{-\alpha} = {}_hD^2\, {}_hI_+^{2-\alpha}$. The corresponding symbol then is, with symmetrically $({}_hD^2\phi)(x) = h^{-2}\,(\phi(x+h) - 2\phi(x) + \phi(x-h))$,

$$\begin{aligned} {}_h\widehat{I_+^{-\alpha}}(\kappa) &= {}_h\widehat{D^2}(\kappa)\, {}_h\widehat{I_+^{2-\alpha}}(\kappa) = h^{-2}\,(z^{-1} - 2 + z)\, h^{2-\alpha}\,(1 - z)^{-(2-\alpha)} \\ &= h^{-\alpha}\, z^{-1}\,(1 - 2z + z^2)\,(1 - z)^{\alpha-2} = h^{-\alpha}\, z^{-1}\,(1 - z)^\alpha. \end{aligned}$$

The symbol $h^{2-\alpha}(1-z)^{\alpha-2}$ for $\widehat{{}_hI_+^{2-\alpha}}$ here has been derived by the formal stipulation ${}_hI_+^{2-\alpha} = {}_hD_+^{-(2-\alpha)}$ using $\widehat{{}_hD_+}(\kappa) = h^{-1}(1-z)$.

Analogously we get $h^{-\alpha} z (1-z^{-1})^{\alpha}$ as symbol of $\widehat{{}_hI_-^{-\alpha}}$.

Remark 4.6. In [GM98] and [GM99] we have exploited the Grünwald-Letnikov random walks in the more general setting of not necessarily symmetric Lévy-Feller diffusion (see Section 1). The proof for the case of symmetry ($\theta = 0$) given in the present paper is considerably simpler.

5. The Gillis-Weiss random walk

Gillis and Weiss in 1970 (see [GiW]) showed (we interpret one of their results in the language of probability theory) that every symmetric random variable Y with values in $\mathbf{Z}$ and asymptotically $P(Y = k) \sim c/|k|^{\alpha+1}$ (where $c > 0$) lies in the domain of attraction of the corresponding symmetric Lévy distribution, hence can be used for an approximating random walk in the sense of Section 3. Only assuming their asymptotics they naturally cannot describe precisely how the coefficients μ and λ of the scaling law appear in the transition probabilities. However, from their analysis we can deduce that the scaling law is of the form

$$\tau = \sigma(h) = \mu\, h^{\alpha} \text{ if } 0 < \alpha < 2, \qquad \tau = \sigma(h) = \lambda\, h^2\, |\log h| \text{ if } \alpha = 2.$$

Remarkably, the parameter value $\alpha = 1$ is not singular, but the scaling law becomes discontinuous at $\alpha = 2$, thus giving an example of a distribution with non-finite variance lying in the domain of attraction of the normal (Gauss) distribution.

We will now re-work and complement their analysis in the framework of our Section 3 for the special symmetric probability distribution $(p_k \,|\, k \in \mathbf{Z})$ with

$$p_0 = 1 - 2\lambda \sum_{k=1}^{\infty} k^{-(\alpha+1)}, \quad p_k = \lambda |k|^{-(\alpha+1)} \quad \text{for} \quad k \neq 0, \tag{5.1}$$

where (so that $p_0 \geq 0$) λ is restricted by

$$0 < \lambda \leq \left(2 \sum_{k=1}^{\infty} k^{-(\alpha+1)} \right)^{-1}. \tag{5.2}$$

The parameter α is only restricted as in (1.4) by $0 < \alpha \leq 2$. Differently from Gillis and Weiss we motivate this choice of probabilities by (2.20), where the special character of the value $\alpha = 2$ already becomes visible. So, assume meanwhile $0 < \alpha < 2$.

Discretizing $D_0^{\alpha} u$ via a straightforward quadrature formula for the right hand side of (2.20) as

$$ {}_hD_0^{\alpha} y_j(t_n) = b(\alpha)\, h \sum_{k=1}^{\infty} \frac{y_{j+k}(t_n) - 2y_j(t_n) + y_{j-k}(t_n)}{(kh)^{\alpha+1}} \tag{5.3}$$

and solving the equation

$$\frac{y_j(t_{n+1}) - y_j(t_n)}{\tau} = {}_hD_0^\alpha\, y_j(t_n)$$

for $y_j(t_{n+1})$ we identify the transition probabilities p_k in (3.4) as

$$p_0 = 1 - 2\mu\, b(\alpha)\,\zeta(\alpha+1),\; p_k = \mu b(\alpha)\,|k|^{-(\alpha+1)} \;\text{ for }\; k \neq 0\,, \tag{5.4}$$

with $\mu = h^{-\alpha}\tau$, $b(\alpha) = \Gamma(\alpha+1)\sin(\alpha\pi/2)/\pi$ and the Riemann ζ-function

$$\zeta(z) = \sum_{k=1}^{\infty} k^{-z} \;\text{ for }\; \Re z > 1. \tag{5.5}$$

Obviously $\sum_{k\in\mathbf{Z}} p_k = 1$, and the non-negativity condition in (3.2) requires

$$0 < \mu \le \frac{1}{2\,b(\alpha)\,\zeta(\alpha+1)} = \frac{\pi}{2\Gamma(\alpha+1)\,\sin(\alpha\pi/2)\,\zeta(\alpha+1)}\,. \tag{5.6}$$

We want to free the parameter value $\alpha = 2$ from its singular character. Recalling (2.17) we see that $b(2) = 0$, so that in (5.4) $p_0 = 1$ and all $p_k = 0$ for $k \neq 0$ whereas the upper bound for μ in (5.6) tends to ∞ as $\alpha \to 2-$. This degenerate random walk obtained in (5.4) by formally setting $\alpha = 2$ being neither interesting nor useful we replace $\mu\, b(\alpha)$ by λ and obtain the transition probabilities in the form (5.1) with restriction (5.2). In (5.1) the special value $\alpha = 2$ seems to be a quite regular value, and we shall see that we have a valid random walk model for all α obeying $0 < \alpha \le 2$. However a price must be paid. Whereas for $0 < \alpha < 2$ we can scale by $\tau = \mu\, h^\alpha$ we can no longer do so in the case $\alpha = 2$. So, assume henceforth (if not explicitly stated otherwise) the condition (5.2).

We have now the generating function

$$\tilde{p}(z) = 1 - 2\lambda\zeta(\alpha+1) + \lambda \sum_{k=1}^{\infty} k^{-(\alpha+1)}(z^k + z^{-k}) \tag{5.7}$$

with $z = e^{i\kappa h}$, $\kappa \in \mathbf{R}$. With the polylogarithmic function

$$\Phi(z,\beta) = \sum_{k=1}^{\infty} \frac{z^k}{k^\beta} \qquad (\beta \in \mathbf{R})$$

we can write

$$\tilde{p}(z) = 1 - 2\lambda\,\zeta(\alpha+1) + \lambda\,\{\Phi(z,\alpha+1) + \Phi(z^{-1},\alpha+1)\} \tag{5.8}$$

and could carry out the required asymptotic analysis by specializing some of the formulas in [T]. See also [EHTF] and [Le] for properties of the polylogarithmic function and the more general Lerch function. We prefer, however, the direct way to obtain (3.11). This asymptotic relation is trivial for $\kappa = 0$, and because of $\tilde{p}(e^{i\kappa h}) = \tilde{p}(e^{-i\kappa h})$, it suffices to treat the case $\kappa > 0$ what we now will do.

From the common integral representation of the gamma function we take

$$k^{-(\alpha+1)} = \frac{1}{\Gamma(\alpha+1)} \int_0^\infty u^\alpha e^{-ku} du$$

and using $z^{-1} = \bar{z}$ we get

$$\tilde{p}(z) = 1 - 2\lambda\, \Re\gamma(z) \tag{5.9}$$

with

$$\Gamma(\alpha+1)\,\gamma(z) = \int_0^\infty u^\alpha \sum_{k=1}^\infty e^{-ku}(1-z^k)\,du = \int_0^\infty \frac{u^\alpha\, e^{-u}}{1-e^{-u}}\,\frac{1-z}{1-e^{-u}\,z}\,du\,. \tag{5.10}$$

The last equality in (5.10) has been obtained by summing the two involved geometric series.

In the Appendix we have performed in detail the required asymptotic analysis of $\Re\gamma(z)$ for $\nu = \kappa\, h \to 0+$ (κ fixed), which is resumed in formulas (A.6) and (A.7). Insertion of these asymptotic behaviours into (5.9) yields

$$\log\tilde{p}(e^{i\kappa h}) \sim -\frac{\lambda\pi}{\Gamma(\alpha+1)\sin(\alpha\pi/2)}\,|\kappa|^\alpha h^\alpha \ \text{ if } \ 0<\alpha<2,\ \kappa\neq 0\,, \tag{5.11}$$

$$\log\tilde{p}(e^{i\kappa h}) \sim -\lambda\kappa^2 h^2 \,\log(1/(|\kappa|h)) \ \text{ if } \ \alpha = 2,\ \kappa\neq 0\,. \tag{5.12}$$

Recalling that it suffices to prove (3.11) for $\kappa \neq 0$ and observing, that there the parameter κ can be treated like a constant, we see that $\log(1/(|\kappa|h)) \sim \log(1/h)$, where, because $h \to 0$, we can assume $0 < h < 1$.

Hence we can replace (5.12) by

$$\log\tilde{p}(e^{i\kappa h}) \sim -\lambda\kappa^2 h^2 \log\frac{1}{h} \ \text{ if } \ \alpha = 2,\ \kappa\neq 0\,. \tag{5.13}$$

Then the limit relation (3.11) (equivalently (3.10)) holds if we scale by

$$\tau = \sigma(h) = \frac{\lambda\pi}{\Gamma(\alpha+1)\,\sin(\alpha\pi/2)}\,h^\alpha \quad \text{if} \quad 0<\alpha<2\,, \tag{5.14}$$

$$\tau = \sigma(h) = \lambda h^2 \log\frac{1}{h} \quad \text{if} \quad \alpha = 2\,. \tag{5.15}$$

Putting $\mu = \lambda\pi/(\Gamma(\alpha+1)\,\sin(\alpha\pi/2) = \lambda/b(\alpha)$ in (5.9) with $b(\alpha)$ defined in (2.17) we obtain from (5.4) the regular scaling law

$$\tau = \sigma(h) = \mu\, h^\alpha \ \text{ for } \ 0<\alpha<2 \tag{5.16}$$

with the restriction (5.6) for μ. As result we have

Theorem 5.1. *Distinguish the cases* (i) $0 < \alpha < 2$, (ii) $\alpha = 2$. *Define the probabilities* $p_k = P(Y = k)$ *in case* (i) *by* (5.4) *with restriction* (5.6), *in case* (ii) *by*

$$p_0 = 1 - 2\lambda\zeta(3), \quad p_k = \lambda|k|^{-3} \quad \textit{for} \quad k \neq 0$$

with restriction $0 < \lambda \leq 1/(2\zeta(3))$. *Let the scaling relation*

$$\tau = \mu h^\alpha \ \text{ in case (i)}, \qquad \tau = \lambda h^\alpha \log\frac{1}{h} \ \text{ in case (ii)} \tag{5.17}$$

hold and let for fixed $t > 0$ *the index* $n = t/\tau$ *run through* **N** *towards* ∞. *Then the random variable* S_n *of* (3.3) *converges in distribution to the random variable* $S(t)$ *whose probability density is given by* (1.15) *as* $g_\alpha(x, t; 0)$.

Remark 5.2. We can use throughout $0 < \lambda \leq 2$ the parameter λ and then have in (5.1) under the restriction (5.2) a unified representation of the transition probabilities. Here, in contrast to the Grünwald-Letnikov random walk, the value $\alpha = 1$ does no longer play a special role. With $\mu = \lambda/b(\alpha)$ we have for $0 < \alpha < 2$ the regular scaling law $\tau = \mu h^\alpha$. However, the price to be paid for this unified representation is the non-regular scaling $\tau = h^2\log(1/h)$ for $\alpha = 2$. Another price is that the generating function $\tilde{p}(z)$ in (5.8) is non-elementary, requiring considerable efforts in its asymptotic analysis.

6. A globally binomial random walk

The random walk model discussed in Section 4 has the disadvantage that the case $\alpha = 1$ is excluded and the representation of the transition probabilities p_k for $1 < \alpha \leq 2$ is different from that for $0 < \alpha < 1$. However, for all admissible values of α we have the regular scaling law $\tau = \mu h^\alpha$. The method treated in Section 5 has the advantage of a unified representation of the transition probabilities in the whole interval $0 < \alpha \leq 2$, but the scaling law $\tau = \mu\, h^\alpha$ holds only for $0 < \alpha < 2$, it breaks down at $\alpha = 2$. In this section we present a model that in the whole interval $0 < \alpha \leq 2$ admits a unified representation of the p_k via binomial coefficients and has there a scaling law of the form $\tau = \mu\, h^\alpha$. Moreover, the generating function $\tilde{p}(z)$ is elementary for all $\alpha \in (0, 2]$.

The use of the binomial coefficients $\binom{\alpha}{j}$ in the Grünwald-Letnikov random walk has caused singular behaviour for $\alpha = 1$. One reason for this sad fact is that $\binom{1}{j} = 0$ for integer $j \geq 2$. We can remove this singular behaviour by removing the factor $\alpha - 1$.

For $0 < \alpha \leq 2$, $\alpha \neq 1$ let us define

$$p_0 = 1 - 2\lambda, \ p_k = (-1)^{k+1}\frac{\lambda}{\alpha - 1}\binom{\alpha}{|k| + 1} \quad \text{for} \ k \neq 0. \tag{6.1}$$

Observing that here the singularity at $\alpha = 1$ is removable, let us for $\alpha = 1$ define (via $\alpha \to 1$ in (6.1))

$$(6.2) \qquad p_0 = 1 - 2\lambda\,, \quad p_k = \frac{\lambda}{|k|(|k|+1)} \quad \text{for} \quad k \neq 0\,.$$

In (6.1) and (6.2) $\sum_{k\in\mathbf{Z}} p_k = 1$ and if $0 < \lambda \le 1/2$ all $p_k \ge 0$. In the special case $\alpha = 2$ we get

$$p_0 = 1 - 2\lambda\,, \quad p_1 = p_{-1} = \lambda\,, \quad p_k = 0 \;\; \text{for} \;\; |k| \ge 2\,,$$

the familiar random walk for approximation of the classical process governed by the equation $\frac{\partial u}{\partial t} = \frac{\partial^2 u}{\partial x^2}$.

The generating function $\tilde{p}(z) = \sum_{k\in\mathbf{Z}} p_k z^k$ has in the case $\alpha \neq 1$ the form

$$(6.3) \qquad \tilde{p}(z) = 1 - \lambda\{q(z) + q(z^{-1})\}$$

with

$$q(z) = \frac{1}{\alpha - 1}\,(1 - z^{-1})\,\{(1-z)^{\alpha-1} - 1\}\,.$$

By passing here to the limit or directly from (6.2) we get for $\alpha = 1$ the representation

$$(6.4) \quad \tilde{p}(z) = 1 - \lambda\{(1 - z^{-1})\log(1-z) + (1-z)\log(1-z^{-1})\}\,, \quad \tilde{p}(1) = 1\,.$$

We have proposed and investigated the particular random walk so generated (its transition probabilities given in (6.2)) in [GM99, Section 5].

In the special case $\alpha = 2$ we find

$$(6.5) \qquad \tilde{p}(z) = 1 + \lambda(z - 2 + z^{-1})\,.$$

We will now show that for all $\alpha \in (0, 2]$ there exists a finite positive number $c(\alpha)$ so that, with

$$(6.6) \qquad \mu = c(\alpha)\,\lambda\,,$$

we arrive for $\kappa \in \mathbf{R} \setminus \{0\}$ at the small h asymptotics

$$(6.7) \qquad \tilde{p}(e^{i\kappa h}) = 1 - \mu(|\kappa| h)^\alpha + o\left((|\kappa|\, h)^\alpha\right)$$

which implies (3.11). As in Sections 4 and 5 we can ignore the value $\kappa = 0$ as trivial.

Referring to [GM99] for detailed treatment of the case $\alpha = 1$, let now be $0 \neq \kappa \in \mathbf{R}$ and $0 < \alpha \le 2$, $\alpha \neq 1$, $z = e^{i\kappa h}$. In view of (6.3) we investigate the asymptotics of $q(z) + q(z^{-1})$ for $h \to 0$. From $z^{-1} = \bar{z}$ and

$$(1-\alpha)q(z) = z^{-1}(1-z)^\alpha - z^{-1} + 1 = e^{-i\kappa h}(1 - e^{i\kappa h})^\alpha - e^{-i\kappa h} + 1\,,$$

we conclude on

$$\text{(6.8)}\quad \psi(z) := (1-\alpha)\{q(z)+q(z^{-1})\} = 2\Re\left\{e^{-i\kappa h}(1-e^{i\kappa h})^{\alpha}\right\} + 2(1-\cos(\kappa h)),$$

and here

$$\text{(6.9)}\qquad \Re\left\{e^{-i\kappa\, h}(1-e^{i\kappa\, h})^{\alpha}\right\} \sim \Re\left((-i\kappa\, h)^{\alpha}\right) = (|\kappa|\, h)^{\alpha}\cos(\alpha\pi/2),$$

$$\text{(6.10)}\qquad 1-\cos(\kappa h) \sim \frac{1}{2}(|\kappa| h)^2 .$$

We distinguish three cases: (i) $0<\alpha<1$, (ii) $1<\alpha<2$, (iii) $\alpha=2$.
In cases (i) and (ii) the leading term in the asymptotics of $\psi(z)$ turns out to be

$$\psi(z) \sim 2\,(|\kappa|\, h)^{\alpha}\cos(\alpha\pi/2).$$

In case (iii) where $\alpha = 2$ however, this term is matched in order of magnitude by (6.10) so that we obtain

$$\psi(z) \sim 2(|\kappa|\, h)^2(-1) + (|\kappa|\, h)^2 = -(|\kappa|\, h)^2 .$$

Collecting results and dividing (6.8) by $1-\alpha$ we get (with $z = e^{i\kappa h}$)

$$\text{(6.11)}\qquad \lambda\{q(z)+q(z^{-1})\} \sim \begin{cases} \lambda\dfrac{2\cos(\alpha\pi/2)}{1-\alpha}(|\kappa|\, h)^{\alpha} & \text{if } 0<\alpha<2,\ \alpha\neq 1,\\ \lambda(|\kappa|\, h)^2 & \text{if } \alpha=2. \end{cases}$$

Hence, in view of (6.3), we obtain (6.7) with (6.6) by putting

$$\text{(6.12)}\qquad c(\alpha) = \begin{cases} \dfrac{2\cos(\alpha\pi/2)}{1-\alpha} & \text{if}\quad 0<\alpha<2,\ \alpha\neq 1,\\ 1 & \text{if}\quad \alpha=2. \end{cases}$$

The scaling coefficient $c(\alpha)$ allows continuous extension to the value $\alpha = 1$, giving $\lim_{\alpha\to 1} c(\alpha) = \pi$ in accordance with [GM99, formula (5.1)]. At $\alpha = 2$, however, $c(\alpha)$ is discontinuous. In fact

$$\text{(6.13)}\qquad c(2) = 1 \neq 2 = \lim_{\alpha\to 2} c(\alpha).$$

Let us finally display the transition probabilities with μ instead of λ as parameter.
For $0<\alpha<2,\ \alpha\neq 1$:

$$\text{(6.14)}\qquad \begin{cases} p_0 = 1 - 2\mu\,\dfrac{1-\alpha}{2\,\cos(\alpha\pi/2)},\\ p_k = \dfrac{(-1)^k}{2\cos(\alpha\pi/2)}\dbinom{\alpha}{|k|+1} \quad \text{for } k\neq 0,\\ 0<\mu\le \dfrac{\cos(\alpha\pi/2)}{1-\alpha}, \end{cases}$$

for $\alpha = 1$ (see [GM99, formula (5.1)]):

$$p_0 = 1 - \frac{2\mu}{\pi}, \quad p_k = \frac{\mu}{\pi|k|(|k|+1)} \quad \text{for} \quad k \neq 0, \quad 0 < \mu \leq \pi/2, \tag{6.15}$$

for $\alpha = 2$:

$$p_0 = 1 - 2\mu, \quad p_1 = p_{-1} = \mu, \quad p_k = 0 \quad \text{for} \quad |k| \geq 2, \quad 0 < \mu \leq 1/2. \tag{6.16}$$

The discontinuity at $\alpha = 2$ has so been transferred to the upper bound for μ.
We comprise the result in

Theorem 6.1. *Take the probabilities $p_k = P(Y = k)$ and the restrictions for μ as in formulas* (6.14), (6.15), (6.16), *and use the scaling relation $\tau = \mu\, h^\alpha$. Let for fixed $t > 0$ the index $n = t/\tau$ run through* **N** *towards ∞. Then the random variable S_n of* (3.3) *converges in distribution to the random variable $S(t)$ whose probability density is given by* (1.15) *as $g_\alpha(x, t; 0)$.*

7. The Chechkin-Gonchar random walk

In this section we adopt to each other considerations of Chechkin and Gonchar [ChG] and the framework of our Section 3, restricting attention to the parameter range $0 < \alpha < 2$. So doing we exclude the well-known case of the classical Gaussian process. We will obtain a *random walk*, which is *discrete in time* but *continuous in space*, in more precise words: whose jumping width (in the instants $t_n = n\tau$) can assume any real number, having an everywhere positive probability density. We modify our theory of Section 3 by allowing the random variable Y to have a strictly monotonic continuous distribution function $W(x) = P\,(Y < x)$ $(x \in \mathbf{R})$, that we furthermore require to be symmetric in the sense $W(x) + W(-x) = 1$ $(x \in \mathbf{R})$, being only interested in the symmetric case $\theta = 0$ of Section 1. We then have in the sum

$$S_n = \sum_{m=1}^{n} (h\, Y_m) = h \sum_{m=1}^{n} Y_m \qquad (n \in \mathbf{N}) \tag{7.1}$$

a description of a random walk, starting in the point $x = 0$. Here $h > 0$ is a scaling width that we let depend on the time-step $\tau > 0$ via a strictly monotonic scaling relation $\tau = \sigma(h)$, with $\sigma(h) \to 0$ as $h \to 0$. We expect the scaling relation to have the form $\tau = \mu\, h^\alpha$ with the positive coefficient μ to be specified, by having found orientation in Gnedenko's theorem on normal attraction (see [GnK], §35). It should be noted, however, that in this theorem the scaling constant C appearing there is given with a wrong value as has been remarked in [Ba].

As previously, we let the Y_m be independent identically distributed random variables, all having their distribution common with Y. However, we now assume

Y to have an everywhere positive (not necessarily bounded) probability density $w = W'$ which is an even function $w(x) = w(-x)$ $(x \in \mathbf{R})$. We will use the fact that w is normalized, $\int_{-\infty}^{+\infty} w(x)\, dx = 1$.

Fixing a value $t > 0$ and again setting $t = t_n = n\tau$ (equivalent to $n = t/\tau$) with $n \in \mathbf{N}$ we want that the random variable S_n converges in distribution to the random variable $S(t)$ whose density is given by (1.15). To this purpose we introduce a condition on the asymptotic behaviour of the density w, namely

$$(7.2) \qquad w(x) = (b + \epsilon(|x|))\, |x|^{-(\alpha+1)}, \quad |\epsilon(|x|)| \le \min\left\{K, E\,|x|^{-\gamma}\right\} \quad (x \in \mathbf{R}),$$

with positive constants b, K, E and γ.

With $\hat{w}(\kappa) = \int_{-\infty}^{+\infty} e^{i\kappa x}\, w(x)\, dx$ as characteristic function of the density $w(x)$ we observe that the random variable hY has density $w(x/h)/h$, hence the characteristic function $\hat{w}(\kappa h)$, and proceeding in analogy to the general method described in Section 3, replacing $\hat{p}(\kappa, h) = \tilde{p}(e^{i\kappa h})$ in (3.9) by $\hat{w}(\kappa h)$, we will find a scaling function $\sigma(h)$ such that for all $\kappa \in \mathbf{R}$, in analogy to (3.11),

$$(7.3) \qquad \frac{1}{\sigma(h)} \log \hat{w}(\kappa h) \to -|\kappa|^{\alpha} \quad \text{as} \quad h \to 0\,.$$

Of course, (7.3) is trivial for $\kappa = 0$. Since $\hat{w}$ like w is an even function it suffices to consider (7.3) for (fixed) values $\kappa > 0$. We see that (7.3) is equivalent to

$$(7.4) \qquad \hat{w}(\kappa h) = 1 - |\kappa|^{\alpha}\, \sigma(h) + o(\sigma(h)) \quad \text{as} \quad h \to 0\,.$$

In view of the symmetry and normalization properties of $w(x)$ and abbreviating $\kappa h = \nu$ we find

$$\hat{w}(\nu) - 1 = \int_0^{\infty} \left(e^{i\nu x} + e^{-i\nu x} - 2\right)\, w(x)\, dx = -4 \int_0^{\infty} \left(\sin(\nu x/2)\right)^2\, w(x)\, dx$$

so, using (7.2),

$$\hat{w}(\nu) = 1 - 2^{-\alpha+2}\, b\, \nu^{\alpha} \int_0^{\infty} \xi^{-\alpha-1}\, (\sin\xi)^2\, d\xi - 4 \int_0^{\infty} \epsilon(x)\, x^{-\alpha-1}\, \left(\sin(\nu x/2)\right)^2\, dx\,.$$

The first integral can be evaluated in terms of the gamma function. In fact, from [GR, (3.823)] we take

$$\int_0^{\infty} \xi^{-\alpha-1}\, (\sin\xi)^2\, d\xi = -\frac{\Gamma(-\alpha)\, \cos(\alpha\pi/2)}{2^{1-\alpha}} = \frac{\pi}{2^{2-\alpha}\, \Gamma(\alpha+1)\, \sin(\alpha\pi/2)}\,.$$

The latter equality follows by the reflection formula for the gamma function.

We estimate the second integral via decomposition $\int_0^{\infty} \ldots = \int_0^{\eta} \ldots + \int_{\eta}^{\infty} \ldots$, taking $\eta = \nu^{-(2\alpha+\gamma)/(2\alpha+2\gamma)}$, using $|\sin\xi| \le \min\{\xi, 1\}$ for $\xi \ge 0$ and the condition on $\epsilon(|x|)$ of (7.2). By careful calculation we find that it behaves asymptotically as $o(\nu^\alpha) = |\kappa|^\alpha\, o(h^\alpha)$. Combining these results and recalling that $\hat{w}$ is an even function, we obtain

$$\hat{w}(\kappa h) = 1 - |\kappa|^\alpha \frac{b\,\pi}{\Gamma(\alpha+1)\,\sin(\alpha\pi/2)}\, h^\alpha + |\kappa|^\alpha\, o(h^\alpha) \quad (h \to 0) \tag{7.5}$$

as valid for all $\kappa \in \mathbf{R}$. In view of (7.4) and the theory developed in Section 3 we thus arrive at the scaling relation

$$\tau = \sigma(h) = \mu\, h^\alpha\,, \quad \text{with} \quad \mu = \frac{b\,\pi}{\Gamma(\alpha+1)\,\sin(\alpha\pi/2)}\,. \tag{7.6}$$

Now we are in the position to formulate

Theorem 7.1. *Let $0 < \alpha < 2$ and assume the random variable Y to have a probability density w of the form* (7.2). *Let the scaling relation* (7.6) *hold and let for fixed $t > 0$ the index $n = t/\tau$ run through $\mathbf{N}$ towards ∞. Then the random variable S_n of* (7.1) *converges in distribution to the random variable $S(t)$ whose probability density is given by* (1.15).

Remark 7.2. According to the well-known asymptotic expansions of the function $p_\alpha(x;0) = g_\alpha(x,1;0)$ (see [F52], [F71], [Zo]) we have

$$b = \frac{\Gamma(\alpha+1)\,\sin(\alpha\pi/2)}{\pi} \tag{7.7}$$

if we take $w(x) = p_\alpha(x;0$, hence in this case $t = 1$, $\mu = 1$ and $h = \tau^{1/\alpha} = n^{-1/\alpha}$. If we require in (7.1) the Y_m to have this special density, then S_n for all $t > 0$ has the same probability distribution as $S(t)$ whose characteristic function is $\exp(-t\,|\kappa|^\alpha)$. We can here obtain the scaling relation also via the convolution theorem.

Remark 7.3. For actual simulation a random variable Y having the required properties is particularly useful if its distribution function $W(x) = \int_{-\infty}^{x} w(\xi)\, d\xi$ is easily invertible. We can then generate a realization of Y by a standard Monte Carlo method (see [HH]). Generate a random number y uniformly distributed in the interval $[0,1)$. Then solve the equation $y = W(x)$ for x and take x as a realization of Y. Chechkin and Gonchar in [ChG] have proposed to use

$$W(x) = \begin{cases} \frac{1}{2}\,(1+|x|^\alpha)^{-1} & \text{for } x < 0\,, \\ 1 - \frac{1}{2}\,(1+x^\alpha)^{-1} & \text{for } x \geq 0\,, \end{cases} \tag{7.8}$$

a function easily invertible. The density

$$w(x) = W'(x) = \frac{\alpha|x|^{\alpha-1}}{2\,(1+|x|^\alpha)^2} \tag{7.9}$$

has the property (7.2) with $b = \alpha/2$, $\gamma = \alpha$, hence we get

$$\mu = \frac{\pi}{2\Gamma(\alpha)\,\sin(\alpha\pi/2)}. \tag{7.10}$$

The density (7.9) is unbounded at the origin if $0 < \alpha < 1$. To avoid this we propose

$$w(x) = \frac{\alpha}{2}\,(1+|x|)^{-(\alpha+1)}, \tag{7.11}$$

which again satisfies the asymptotic condition (7.2). Then

$$W(x) = \begin{cases} \frac{1}{2}\,(1+|x|)^{-\alpha} & \text{for } x < 0, \\ 1 - \frac{1}{2}\,(1+x)^{-\alpha} & \text{for } x \geq 0 \end{cases} \tag{7.12}$$

is also easily invertible, and (7.10) remains valid.

Remark 7.4. Among the symmetric densities $p_\alpha(x;0)$, only the Cauchy density $w(x) = p_1(x;0) = (1/\pi)\,(1+x^2)^{-1}$ offers easy invertibility of the corresponding distribution function, namely of the function $W(x) = 1/2 + (1/\pi)\arctan x$. Via random numbers y_m uniformly distributed in [0,1) we can get realizations of the Y_m in $\tau\tan(\pi(y_m - 1/2))$ (here $\tau = 1\,h^1 = h$) and so obtain in S_n a snapshot at instant $t_n = n\,\tau$ of a true Cauchy process.

8. Conclusions

Anomalous diffusion processes have in recent years gained revived interest among physicists, and methods of fractional calculus have shown their usefulness for purposes of modelling. In the space-fractional case one is naturally led to a generalization of the classical diffusion equation with respect to the second-order spatial operator. One arrives in a natural way at the processes of Lévy-Feller type in which stable probability distributions play the essential role. Also among physicists and mathematicians who have found it rewarding to work in theory of finance, such processes are becoming more and more popular (see e.g. [M], [BoP], [MS]). So, it is no wonder that also in pure mathematics such types of processes are now investigated in great generality and analytical sophistication (see e.g. [J], [Be], [S], [Za]). From the more practical point of view discrete models are esteemed. They not only show that very different microscopic behaviour of particles can result in the same macroscopic behaviour but offer also possible visualizations of what is happening in such processes. Furthermore such discrete models can be used for simulation purposes, be it for simulation of particle paths via Monte Carlo methods (the microscopic view) or via solution of the underlying Cauchy problem for a pseudo-differential equation (the macroscopic view). And, last but not least, such models are fascinating as seen from the mathematical standpoint (or, more specifically, from the position of probability theory).

In our present investigation we first have given a survey on and drawn motivations from basic theory of fractional calculus and Lévy-Feller diffusion processes. Then we have obtained and rigorously analyzed (with respect to their convergence in distribution for passing to the limit of infinitely fine discretization) three models of random walk occurring on a regular spatio-temporal grid. The first model is devised from the Grünwald-Letnikov discretization of the two Weyl operators, the composition of which gives the inverse of the Riesz potential operator. The second model is an adaptation of ideas of Gillis and Weiss [GiW] to our framework. We have provided it with a new motivation, namely as obtainable from straightforward discretization of the hypersingular integral representation of the spatial pseudo-differential operator. The third model's intention is to overcome peculiar deficiencies of the first two models. It is a modification and improvement of the first model, and again properties of the binomial coefficients are used.

Finally, to offer also a highly efficient method for numerical simulation, we have mutually adapted our theoretical frame to ideas of Chechkin and Gonchar [ChG]. We so obtain a random walk still proceeding in equidistant instants of time but allowing spatial jumps of arbitrary length in positive or negative direction.

Appendix A: Asymptotics of an integral

Abbreviating $\kappa h = \nu$ in $z = e^{i\kappa h}$ in the right hand side of (5.10), and keeping in mind $0 < \alpha \le 2$, elementary calculation yields the equation

$$(A.1) \qquad \Gamma(\alpha+1)\,\Re\gamma(z) = \int_0^\infty u^\alpha\, e^{-u}\, \frac{(1+e^{-u})\,(1-\cos\nu)}{(1-e^{-u})\,|1-e^{-u}\,e^{i\nu}|^2}\, du$$

which we will treat asymptotically for $0 < \nu \to 0+$ by the Laplace method for integrals (see [dB]), using the fact that the lower bound $u = 0$ is the critical one (the integrand tending to ∞ as $u \to 0$). We have $1 - \cos\nu = \nu^2/2 + O(\nu^4)$ and

$$|1-e^{-u}e^{i\nu}|^2 = (1-e^{-u})^2 + 2e^{-u}(1-\cos\nu) = (1-e^{-u})^2 + \nu^2 e^{-u} + O(\nu^4),$$

uniformly in $0 \le u < \infty$, hence

$$\Gamma(\alpha+1)\,\Re\gamma(z) \sim \frac{\nu^2}{2}\int_0^\infty \frac{u^\alpha e^{-u}(1+e^{-u})}{(1-e^{-u})\{(1-e^{-u})^2+\nu^2 e^{-u}\}}\, du\,.$$

Because this integral diverges for $\nu = 0$ we can simplify the integrand (for *small* u) which, for small ν, gives the essential contribution: $1 + e^{-u} \sim 2$, $1 - e^{-u} \sim u$, $e^{-u} \sim 1$. We obtain

$$\Gamma(\alpha+1)\,\Re\gamma(z) \sim \nu^2 \int_0^\infty u^{\alpha-1}\frac{e^{-u}}{u^2+\nu^2}\, du$$

and, by substituting $u = \nu w$,

$$(A.2) \qquad \Gamma(\alpha+1)\Re\gamma(z) = \nu^\alpha \int_0^\infty w^{\alpha-1}\frac{e^{-\nu w}}{w^2+1}\,dw := \nu^\alpha \rho(\nu)\,.$$

In the investigation of the integral

$$(A.3) \qquad \rho(\nu) = \int_0^\infty w^{\alpha-1}\frac{e^{-\nu w}}{w^2+1}\,dw$$

we distinguish the cases (i) $0 < \alpha < 2$, (ii) $\alpha = 2$.

In the case (i) simply

$$\rho(\nu) \to \int_0^\infty \frac{w^{\alpha-1}}{w^2+1}\,dw \quad \text{for} \quad \nu \to 0$$

and with $\beta = \alpha - 1$, hence $-1 < \beta < 1$, we have to determine the value of

$$q(\beta) = \int_0^\infty \frac{x^\beta}{x^2+1}\,dx\,.$$

Observing that $q(-\beta) = q(\beta)$ (substitute $\xi = 1/x$) we do this $0 \le \beta < 1$. Complementation by (integrate along the upper edge of the negative real semi-axis)

$$\int_{-\infty}^0 \frac{x^\beta}{x^2+1}\,dx = e^{i\beta\pi}\int_0^{+\infty}\frac{x^\beta}{x^2+1}\,dx$$

gives, via the residue theorem,

$$\left(1+e^{i\beta\pi}\right) q(\beta) = \int_{-\infty}^{+\infty}\frac{x^\beta}{x^2+1}\,dx = \pi\, i^\beta = \pi\, e^{i\beta\pi/2}\,.$$

So

$$q(\beta) = \frac{\pi}{2\cos(\beta\pi/2)} = \frac{\pi}{2\sin(\alpha\pi/2)}\,,$$

and hence

$$(A.4) \qquad \rho(\nu) \to \frac{\pi}{2\sin(\alpha\pi/2)} \quad \text{if} \quad 0<\alpha<2 \quad (\nu\to 0+).$$

In case (ii) the integral diverges for $\nu = 0$, so we must proceed in another way. Inserting $\alpha = 2$ in (A.3) and differentiating we obtain for $\nu > 0$

$$-\rho'(\nu) = \int_0^\infty \frac{w^2 e^{-\nu w}}{w^2+1}\,dw = \int_0^\infty e^{-\nu w}\left(1-\frac{1}{w^2+1}\right)dw = \frac{1}{\nu} - \frac{\pi}{2} + o(1)\,,$$

and then by integration

$$(A.5) \qquad \rho(\nu) \sim -\log\nu = \log\frac{1}{\nu} \quad (\nu \to 0+).$$

Now we can collect results. From (A.1) - (A.5), using $\nu = \kappa h$ which because of symmetry we can replace by $|\kappa|\, h$ (admitting also negative values of κ) we deduce

$$(A.6)\quad \Re\gamma(z) \sim \frac{\pi}{2\Gamma(\alpha+1)\sin(\alpha\pi/2)}|\kappa|^\alpha h^\alpha \ \text{ if } \ 0<\alpha<2,\ \kappa\neq 0, \quad \text{as} \quad h\to 0,$$

$$(A.7)\qquad \Re\gamma(z) \sim \kappa^2 h^2 \log\frac{1}{|\kappa|h} \ \text{ if } \ \alpha=2,\ \kappa\neq 0, \quad \text{as} \quad h\to 0.$$

Acknowledgements

We are grateful to the Italian Istituto Nazionale di Alta Matematica and to the Research Commission of Free University of Berlin for supporting the joint efforts of our research groups in Berlin and Bologna.

References

[Ba] R. Bartles, Generating non-normal stable variates using limit theorem properties, *J. Stat. Comp. Simulation* **7** (1978), 199 - 212.

[Be] J. Bertoin, *Lévy Processes*, Cambridge: Cambridge University Press 1996.

[BoP] J.-P. Bouchaud and M. Potters, *Theory of Financial Risks*, Cambridge: Cambridge University Press 1999.

[Br] L. Breiman, *Probability*, Philadelphia: SIAM 1992.

[dB] N.G. de Brujn, *Asymptotic Methods in Analysis*, 2nd ed. New York: Dover Publications 1981 (1st ed. Amsterdam: North Holland 1958).

[ChG] A.V. Chechkin and V.Yu. Gonchar, A model for ordinary Lévy motion, downloadable from: `http://xxx.lanl.gov/abs/cond-mat/9901064`

[EHTF] A. Erdélyi, W. Magnus, F. Oberhettinger, F.G. Tricomi, *Higher Transcendental Functions*, Vol. I, New York: McGraw-Hill 1953.

[ETIT] A. Erdélyi, W. Magnus, F. Oberhettinger, F.G. Tricomi, *Tables of Integral Transforms*, Vol. I, New York: McGraw-Hill 1953.

[F52] W. Feller, *On a generalization of Marcel Riesz' potentials and the semi-groups generated by them*, Meddelanden Lunds Universitets Matematiska Seminarium (Comm. Sém. Mathém. Université de Lund), Tome suppl. dédié à M. Riesz. Lund 1952, 73 - 81.

[F71] W. Feller, *An Introduction to Probability Theory and its Applications*, Vol. 2, 2nd ed. New York: Wiley 1971 (1st ed. 1966).

[GiW] J.E. Gillis and G.H. Weiss, Expected number of distinct sites visited by a random walk with an infinite variance, *J. Math. Phys.* **11** (1970), 1307 - 1312.

[GnK] B.V. Gnedenko and A.N. Kolmogorov, *Limit Distributions for Sums of Independent Random Variables*, Cambridge, Mass.: Addison-Wesley 1954. Translated from the Russian edition: Moscow 1949, with notes by K.L. Chung, revised 1968.

[G70] R. Gorenflo, Nichtnegativitäts- und substanzerhaltende Differenzenschemata für lineare Diffusionsgleichungen, *Numer. Math.* **14** (1970), 448 – 467.

[G78] R. Gorenflo, Conservative difference schemes for diffusion problems, In: *Intern. Ser. Numer. Math.*: Vol. **39**. Basel: Birkhäuser-Verlag 1978, pp. 101 – 124.

[GFM] R. Gorenflo, G. De Fabritiis and F. Mainardi, Discrete random walk models for symmetric Lévy-Feller diffusion processes, *Physica A* **269** (1999), 79 – 89.

[GM98] R. Gorenflo and F. Mainardi, Random walk models for space-fractional diffusion processes, *Fractional Calculus & Applied Analysis* **1** (1998), 167 – 191.

[GM99] R. Gorenflo and F. Mainardi, Approximation of Lévy-Feller diffusion by random walk, *Journal for Analysis and its Applications* **18** (1999), 231 – 246.

[GN] R. Gorenflo and M. Niedack, Conservative difference schemes for diffusion problems with boundary and interface conditions, *Computing* **25** (1980), 299 – 316.

[GR] I.S. Gradshteyn and I.M. Ryzhik, *Tables of Integrals, Series and Products*, New York: Academic Press 1980. Translated from the Russian.

[HH] J.M. Hammersley and D.C. Handscomb, *Monte Carlo Methods*, London: Methuen 1964.

[J] N. Jacob, *Pseudo-differential Operators and Markov Processes*, Berlin: Akademie-Verlag 1996.

[Le] M. Lerch, Note sur la fonction $K(w,x,s) = \sum_{k=0}^{\infty} e^{2k\pi i x}/(w+k)^s$, *Acta Mathematica* **11** (1887), 19 – 24.

[L25] P. Lévy, *Calcul des probabilités*, Paris: Gauthier-Villars 1925.

[L54] P. Lévy, *Théorie de l'addition des variables aléatoires*, 2nd ed. Paris: Gauthier-Villars 1954 (1st ed. 1937).

[M] B.B. Mandelbrot, *Fractals and Scaling in Finance*, New York: Springer 1997.

[MS] R.N. Mantegna and H.E. Stanley, *An Introduction to Econophysics*, Cambridge: Cambridge University Press 1999.

[P] I. Podlubny, *Fractional Differential Equations*, San Diego: Academic Press 1999.

[R] B. Rubin, *Fractional Integrals and Potentials*, Harlow: Longman 1996.

[SKM] S.G. Samko, A.A. Kilbas and O.I. Marichev, *Fractional Integrals and Derivatives: Theory and Applications*, Amsterdam: Gordon and Breach 1993. Translated from the Russian edition, Minsk: Nauka i Technika 1987.

[SaT] G. Samorodnitsky and M.S. Taqqu, *Stable non-Gaussian Random Processes*, New York: Chapman & Hall 1994.

[S] K. Sato, *Lévy Processes and Infinitely Divisible Distributions*, Cambridge: Cambridge University Press 1999.

[Sc] W.R. Schneider, Stable distributions: Fox function representation and generalization, in S. Albeverio, G. Casati and D. Merlini (Eds), *Stochastic Processes in Classical and Quantum Systems*, Berlin: Springer Verlag 1986, 497-511.

[T] C. Truesdell, On a function which occurs in the theory of the structure of polymers, *Annals of Mathematics* **46** (1945), 144 – 157.

[Za] P.A. Zanzotto, On solution of one-dimensional stochastic differential equations driven by stable Lévy motion. *Stoch. Process. Appl.* **68** (1997), 209 – 228.

[Zo] V.M. Zolotarev, *One-dimensional Stable Distributions*, Providence, R.I.: Amer. Math. Soc. 1986. Translated from the Russian.

R. Gorenflo
Free University of Berlin,
Department of Mathematics
and Computer Science,
Arnimallee 2-6, D-14195 Berlin,
e-mail: gorenflo@math.fu-berlin.de

F. Mainardi
University of Bologna,
Department of Physics,
Via Irnerio 46, I-40126, Bologna, Italy,
e-mail: mainardi@bo.infn.it

1991 Mathematics Subject Classification: Primary 26A33, 60E07, 60J15, 60J60; Secondary 44A20, 45K05

Submitted: 22.11.1999

Operator Theory:
Advances and Applications, Vol. 121

On Qualocation and Collocation Methods for Singular Integral Equations with Piecewise Continuous Coefficients, Using Continuous Splines on Quasi-uniform Meshes

R.D. Grigorieff, Ian H. Sloan

Dedicated to the memory of Siegfried Prößdorf

In this paper the qualocation method (which includes the collocation method as a special case) is applied to index-zero singular integral equations with piecewise-continuous coefficients, using continuous splines defined on a quasi-uniform mesh. Because the mesh is not diffeomorphic to a uniform mesh, Fourier series techniques are not available. Instead use is made of recent superapproximation results of Grigorieff, Sloan and Brandts for continuous splines on general meshes. The main result of the paper is that if a particular qualocation method is stable when applied to the identity operator, then the qualocation method is L_2 stable when applied to a singular integral equation if and only if the same method is L_2 stable when applied to all frozen-coefficient versions of the equation. The main theoretical tool is a local principle for splines in the form given by Prößdorf.

1. Introduction

In this paper we study the qualocation method (of which the collocation method is a special case) for continuous splines on a possibly irregular (but quasi-uniform) mesh, applied to singular integral equations of the form

$$(1.1)\qquad (Ax)(t) := a(t)x(t) + b(t)(Sx)(t) = y(t), \qquad t \in \Gamma,$$

where

$$(Sx)(t) := \frac{1}{i\pi}\int_\Gamma \frac{1}{\tau - t}x(\tau)d\tau,$$

with the integral being defined in the Cauchy principal-value sense, and with $\Gamma \in \mathbb{C}$ being a simple, smooth, closed curve. The problem is considered in the space $L_p(\Gamma)$, where throughout this paper $p \in (1, \infty)$ is a fixed number.

The complex-valued functions a and b are assumed to be in $\overline{PC}$, the L_∞-closure of PC, which denotes the space of piecewise continuous functions on Γ. The functions f in $\overline{PC}$ are known to have an at most countable number of discontinuities, and in each point $t \in \Gamma$ the one-sided limits $f_+(t) := f(t+0)$ and $f_-(t) := f(t-0)$ exist. We specify functions in PC and $\overline{PC}$ to be left-sided continuous. Obviously, the space $\overline{PC}$ is a Banach space under the uniform norm.

We assume that equation (1.1) is uniquely solvable for arbitrary $y \in L_p(\Gamma)$. Hence, A is a bounded index-zero Fredholm operator in $L_p(\Gamma)$, with bounded inverse.

For coefficients a and b in $\overline{PC}$, necessary and sufficient conditions for A to be an index-zero Fredholm operator can be found in e.g. [8], Chap. 9, Th. 3.1.

In order to introduce the approximation scheme, we first introduce a smooth, bijective, 1-periodic mapping $\nu : [0,1] \to \Gamma$ of the periodic unit interval onto Γ. Then with the change of variable $t = \nu(s)$ (and with some abuse of notation) we may write the singular integral equation as

$$a(s)x(s) + b(s)(Sx)(s) = y(s), \qquad s \in \mathbb{R}, \tag{1.2}$$

where now

$$Sx(s) := \frac{1}{i\pi} \int_0^1 \frac{1}{\nu(\sigma) - \nu(s)} \nu'(\sigma)x(\sigma)d\sigma, \qquad s \in \mathbb{R}, \tag{1.3}$$

or

$$Ax := (aI + bS)x = y, \tag{1.4}$$

where I is the identity operator.

Now we introduce a sequence of partitions π_h of the (periodic) interval $[0,1]$,

$$\pi_h : 0 \le \sigma_0 < \sigma_1 < \cdots < \sigma_{n_h-1} < 1.$$

We define

$$h_k = \sigma_{k+1} - \sigma_k, \qquad k = 0, \ldots, n_h - 1,$$

where here and elsewhere we use a periodic labelling convention, so that $\sigma_{k+n_h} = 1 + \sigma_k$. And we denote by $h := \max h_k$ the maximum mesh diameter for the given partition. We shall assume that the sequence of partitions is quasi-uniform, i.e. there exists $C > 0$ such that $h \ge h_k \ge Ch$.

On the partition π_h we define the approximation space V_h as the space of 1-periodic functions v_h which are continuous on $\mathbb{R}$, and which satisfy, for given $r \ge 2$,

$$v_h \Big|_{(\sigma_k, \sigma_{k+1})} \in \mathbb{P}^{r-1} \qquad \text{for} \quad k \in \mathbb{Z},$$

where $\mathbb{P}^{r-1}$ denotes the space of complex-valued polynomials of degree $\le r-1$. Thus V_h is a space of C^0 splines of order r. Specific examples are the continuous piecewise-linear functions ($r = 2$), continuous piecewise quadratics ($r = 3$), and so on.

The qualocation method is a generalisation of the well known collocation method, which in form is reminiscent of the Galerkin method, except that the exact inner products are replaced by a quadrature approximation. Thus the Galerkin method for the space V_h (if we take (1.4) as our starting point) may be written as: find $x_h \in V_h$ such that

$$(Ax_h, \chi_h) = (y, \chi_h) \qquad \forall \chi_h \in V_h,$$

where

$$(f,g) := \int_0^1 f(s)\overline{g(s)}ds,$$

whereas the qualocation method takes the analogous form: find $x_h \in V_h$ such that

$$(1.5) \qquad (Ax_h, \chi_h)_h = (y, \chi_h)_h \qquad \forall \chi_h \in V_h,$$

where $(f,g)_h$ is defined not by an exact integral, but by a well-chosen composite quadrature rule on the partition π_h. Specifically, we write

$$(1.6) \qquad (f,g)_h := Q_h(f\overline{g}),$$

where

$$(1.7) \qquad Q_h F := \sum_{k=0}^{n_h-1} h_k \sum_{j=1}^{J} w_j F(\sigma_k + h_k \xi_j),$$

and where

$$0 \le \xi_1 < \xi_2 < \cdots < \xi_J \le 1, \qquad w_j > 0 \quad \text{for} \quad j = 1, \ldots, J.$$

Clearly, the quadrature rule Q_h is the composite quadrature rule obtained by mapping a suitably scaled version of an underlying J-point quadrature rule Q,

$$(1.8) \qquad QF := \sum_{j=1}^{J} w_j F(\xi_j) \approx \int_0^1 F(\sigma)d\sigma$$

onto each subinterval $[\sigma_k, \sigma_{k+1}]$.

The main result in this paper is Corollary 4.5 below, which asserts that a qualocation method for (1.2) is stable and optimally convergent provided the same qualocation method (characterised by the choice of the underlying quadrature rule Q and the space V_h) is stable for all the "frozen-coefficient" operators A_t, $t \in \mathbb{R}$. These operators are given by

$$(1.9) \qquad (A_t x)(s) := a_t(s)x(s) + b_t(s)(Sx)(s), \quad s, t \in \mathbb{R},$$

where for a function $e \in \overline{PC}$ we define $e_t \in PC$ for $t \in \mathbb{R}$ to be the two-valued step function

$$(1.10) \qquad e_t(s) := e_+(t)\chi_t^+(s) + e_-(t)\chi_t^-(s), \quad s \in \mathbb{R},$$

with χ_t^+ and χ_t^- denoting the (periodic) characteristic functions of the interval $(t, t+\frac{1}{2}]$ and $(t-\frac{1}{2}, t]$, respectively. That is, A_t is an operator in which the coefficients in A are frozen at their values on the two sides of t. At the present state of our knowledge we do not know how to establish the necessary stability properties for the frozen-coefficient case, but the question has been reduced by the analysis in this paper to the question of stability for the smaller class of singular integral operators with step-function coefficients.

We note the result in [17] for the case $p = 2$ and $a, b \in C(\Gamma_0)$, where Γ_0 is the unit circle, that for the stability of breakpoint collocation with piecewise-linear splines on uniform grids the strong ellipticity of the operator A, i.e.

$$a(t) + \lambda b(t) \neq 0, \quad t \in \Gamma_0, \ \lambda \in [-1, 1],$$

is necessary and sufficient (see [16] for the extension to $a, b \in PC$). Hence it is likely that the mere assumption of invertibility of A, under which our Equivalence Theorem 4.4 already holds, has at least to be supplemented by assumptions of a strong ellipticity kind for A in order to ensure stability of the qualocation method.

In most of the previous discussions of the qualocation method [3, 7, 12, 13, 20, 21, 22, 23], and in [5] even for problems with coefficients in PC, the approximating space V_h has been taken to be a space of smoothest splines (i.e. $v_h \in V_h$ belongs to C^{r-2}, instead of C^0 as here, if r is the order of the splines). Moreover (and this is a major restriction) the meshes have been required to be uniform (apart, perhaps, from a smooth transformation), in order to allow Fourier analysis. The very precise tools of Fourier analysis have allowed special quadrature rules to be designed in a number of papers [3, 20, 21, 22], which achieve high rates of convergence in special Sobolev norms. Approximating spaces of order r splines with lower smoothness C^d, $d \in [0, r-2]$, have been considered in [11], as well as a special kind of non-equidistant meshes (see also [6]).

In the present analysis, with C^0 splines and non-uniform meshes, the tools of Fourier analysis are no longer available to us, so that other techniques must be used. In consequence we must content ourselves for the present with analysing the error only in the L_2 norm, rather than in negative Sobolev norms. The analysis leaves considerable freedom in the choice of the quadrature rule Q, so leaving scope, perhaps, for the future design of quadrature rules that have superior convergence properties in certain negative Sobolev norms.

We remark that qualocation includes the collocation method (one example of which is the uniform-mesh ε-collocation method of Schmidt [19] in the case $r = 2$). Indeed, the qualocation method with $J = r - 1$ will be seen to be equivalent to a collocation method with collocation at all the points of the quadrature rule. The corresponding discrete projection is stable, provided that a certain mesh condition is satisfied. (For instance, it is sufficient that the breakpoints be included in the set of collocation points).

As far as we are aware, even the collocation results included here are new, except in so far as they follow from the well-known uniform mesh results of Prößdorf and Schmidt [17], Prößdorf [16], Saranen and Wendland [18], and Arnold and Wendland [2], or alternatively from the breakpoint collocation results of Arnold and Wendland [1] for the case of splines of odd degree (i.e. r even).

2. The qualocation projection

Let $\mathbb{T}$ denote the 1-periodic real line (i.e. $\mathbb{R}$ with points that differ by 1 identified), and let $C(\mathbb{T})$ be the set of 1-periodic continuous functions. For the spline space V_h and the approximate inner product $(\cdot,\cdot)_h$ defined in Section 1, let $R_h : C(\mathbb{T}) \to V_h$ be the linear operator defined by

$$(2.1) \qquad R_h v \in V_h, \quad (R_h v, \chi_h)_h = (v, \chi_h)_h \qquad \forall \chi_h \in V_h.$$

We shall call R_h the "qualocation projection" if it is well defined. A recent paper [10] studied R_h in the periodic setting used here (thereby extending earlier work in [9] for the case of an interval), finding that R_h is well defined if and only if **either** $J \geq r$ (where J is the number of points in the quadrature rule Q and r the order of the spline space), **or** $J = r-1$ and

$$(2.2) \qquad |\phi(0)| \neq |\phi(1)|, \quad \text{where} \quad \phi(\xi) := \prod_{j=1}^{r-1}(\xi - \xi_j).$$

Accordingly, in the following we shall always assume that either $J \geq r$ or $J = r-1$ and (2.2) holds.

If $J = r-1$ the number of distinct quadrature points on $[0,1)$ is $n_h(r-1)$ (since (2.2) prevents us from choosing both $\xi_0 = 0$ and $\xi_J = 1$), which equals dim V_h. It is then an easy argument to show that in this case $R_h = K_h$, where K_h is the interpolatory projection defined by

$$K_h v \in V_h, \ (K_h v)(\sigma_k + h_k\xi_j) = v(\sigma_k + h_k\xi_j), k = 0,\ldots,n_h-1, j = 1,\ldots,J.$$

This means that in this situation the qualocation method reduces to the collocation method.

The next step, beyond the mere existence of R_h, is stability. Following [9] and [10], we say that the sequence of projections $\{R_h\}$ is p-stable if there exists $C > 0$ such that

$$(2.3) \qquad \|R_h f\|_p \leq C|f|_{h,p} \qquad \forall f \in C(\mathbb{T}).$$

Here $\|\cdot\|_p$ denotes the usual L_p norm, and $|\cdot|_{h,p}$ is the discrete semi-norm defined by

$$|f|_{h,p} = Q_h(|f|^p)^{1/p} \quad \text{for} \quad 1 < p < \infty.$$

Since $R_h f$ depends on the values of f only at the quadrature points of the rule Q_h, it follows from (2.3) that the same result holds for all real-valued functions f on $\mathbb{T}$. In particular (2.3) extends to the class $\overline{PC}$.

We note in passing an important fact for the later argument, that the norm $\|\cdot\|_p$ and seminorm $|\cdot|_{h,p}$ are equivalent on the space V_h : it is shown in Lemma 2.2 of [10] that for given $p \in [1,\infty]$ there exists $C > 0$ (independent of h) such that

$$(2.4) \qquad C^{-1}\|\psi_h\|_p \leq |\psi_h|_{h,p} \leq C\|\psi_h\|_p \qquad \forall \psi_h \in V_h.$$

The same property holds, by Lemma 3.1 of [10], for the first derivatives: it is shown there that there exists $C > 0$ such that

$$C^{-1}\|\psi_h'\|_p \leq |\psi_h'|_{h,p} \leq C\|\psi_h'\|_p \qquad \forall \psi_h \in V_h. \tag{2.5}$$

A long list of sufficient conditions for the sequence $\{R_h\}$ to be p-stable is given in [10], see Theorem 5.4. In the interests of simplicity, we state here just three sufficiently representative cases.

Theorem 2.1. ([10], Corollary 5.5) *The sequence $\{R_h\}$ is p-stable if one of the following holds:*

1. *$J \geq r$ and Q is symmetric; or*
2. *$J = r - 1$ and $\xi_1 = 0$, $\xi_J < 1$; or*
3. *$r = 2$, $J = 1$, $0 < \xi_1 < \frac{1}{2}$ and for some fixed $\epsilon > 0$*

$$\frac{h_k}{h_{k-1} + h_k} \geq \xi_1 + \epsilon, \quad k = 0, \ldots, n-1.$$

Note the modest nature of the condition if $J \geq r$: **every** symmetric positive-weight quadrature rule Q yields p-stability with no restriction on the local mesh. The second case corresponds to collocation, for the situation in which the collocation points include the breakpoints, and says that in this case $\{R_h\}$ is p-stable with no restrictions on the mesh, provided the spline breakpoints are among the collocation points. The third case explores collocation in the piecewise-linear case, but in the more delicate situation that the collocation points (here just one per interval) are not the breakpoints. Again p-stability holds for all p, but this time with a restriction on the local mesh ratio h_k/h_{k-1}.

The p-stability is the key to the approximation and superapproximation properties established in [10], which we now state in a form adapted for our needs.

Theorem 2.2. ([10], Theorem 1.1) *Let $f \in W_p^1(\mathbb{T})$. Suppose that $\{R_h\}$ is p-stable. Then we have,*

$$|R_h f - f|_{h,p} + \|R_h f - f\|_p \leq Ch\|f'\|_p.$$

In the next theorem G is the operator of multiplication by a fixed function $g \in W_\infty^r(\mathbb{T})$, that is

$$G : L_p(\mathbb{T}) \to L_p(\mathbb{T}) : v \mapsto gv.$$

Theorem 2.3. ([10], Theorem 1.2) *Assume that $\{R_h\}$ is p-stable. Then for all $\psi_h \in V_h$*

$$|(I - R_h)G\psi_h|_{h,p} + \|(I - R_h)G\psi_h\|_p \leq Ch\|g'\|_{r-1,\infty}|\psi_h|_{h,p}.$$

If $\{R_h\}$ is also q-stable, where $\frac{1}{p}+\frac{1}{q}=1$, then for all $f \in C(\mathbb{T})$

$$|R_hG(I-R_h)f|_{h,p} + \|R_hG(I-R_h)f\|_p \le Ch\|g'\|_{r-1,\infty}|(I-R_h)f|_{h,p}.$$

The essential feature in the estimates in the above theorems is the occurrence of the factor h on the right-hand side.

We also require a superapproximation theorem for the derivative, analoguous to the first part of the last theorem. Recall that our grids are quasi-uniform.

Theorem 2.4. ([10], Theorem 5.9) *Assume one of the assumptions 1–3 of Theorem 2.1 to hold. Then for all $\psi_h \in V_h$*

$$\|[(I-R_h)G\psi_h]'\|_p \le Ch\|g'\|_{r-1,\infty}(|\psi_h|_{h,p} + |\psi_h'|_{h,p}).$$

In the subsequent analysis we need a convergence result for the sequence $\{R_hf\}$ for functions $f \in \overline{PC}$. In this result $p<\infty$ is crucial.

Proposition 2.5. *Let $\{R_h\}$ be p-stable. Then as $h \to 0$*

$$\|R_hf - f\|_p \to 0 \qquad \forall f \in \overline{PC}. \tag{2.6}$$

Proof. From the p-stability property (2.3) of $\{R_h\}$ follows

$$\|R_hf\|_p \le C\|f\|_\infty, \qquad \forall f \in \overline{PC}, \tag{2.7}$$

and hence it is sufficient to prove (2.6) for $f \in PC$; thus we assume $f \in PC$ in the sequel. By f_h we denote the continuous piecewise-linear interpolant of f in the breakpoints of π_h. Since $R_hf_h = f_h$ we obtain, with the aid of the p-stability of $\{R_h\}$,

$$\begin{aligned}
\|R_hf - f\|_p^p &= \|(R_h - I)(f-f_h)\|_p^p \\
&\le C(|f-f_h|_{h,p} + \|f-f_h\|_p)^p \\
&\le C\left(\sum_{k=0}^{n_h-1} h_k \sum_{j=1}^{J} w_j|(f-f_h)(\sigma_k + h_k\xi_j)\,|^p + \|f-f_h\|_p^p\right) \\
&\le C\sum_{k=0}^{n_h-1}{}' h_k \max_{s\in[\sigma_k,\sigma_{k+1}]}|(f-f_h)(s)|^p + C\sum_{k=0}^{n_h-1}{}'' h_k\|f\|_\infty^p.
\end{aligned}$$

Here, $\sum'$ denotes the sum over all k such that f is continuous in $[\sigma_k,\sigma_{k+1}]$ and $\sum''$ is the remaining sum, in which the number of terms is at most the (finite) number of (simple) discontinuities in f. For a piecewise continuous f we have

$$\max_{k\in\sum'}\max_{s\in[\sigma_k,\sigma_{k+1}]}|(f-f_h)(s)| \to 0 \qquad \text{as} \quad h \to 0,$$

together with $\sum'' h_k \to 0$, thus the result follows. □

Finally, we need a result by Crouzeix and Thomée [4] for the L_p-stability of the L_2-orthogonal projection P_h onto V_h. The theorem in [4] is proved for the orthogonal projection on the subspace of V_h satisfying zero Dirichlet boundary conditions, but the proof can be easily extended. The theorem also follows as a limiting case from the results in [10].

Theorem 2.6.([4]) *For all* $p \in [1, \infty]$

$$\|P_h x\|_p \leq C\|x\|_p \qquad \forall x \in L_p(\mathbb{T}), \tag{2.8}$$

and

$$\|(P_h x)'\|_p \leq C\|x'\|_p \qquad \forall x \in W_p^1(\mathbb{T}). \tag{2.9}$$

3. A local principle

Our analysis relies on a slightly modified version of Theorem 2.1 in [16]. The modification concerns the condition $\|Q_n f(I-Q_n)\|_Y \to 0$ as $n \to \infty$ (i.e. condition III.1, p.241 in [16]), which appears to be not easily applicable to the case of unbounded projections Q_n. In this section we first state the local principle in the form we need it. In a remark we then state the modified version for the general case considered in [16]. The proof can be obtained by adapting the proof of Theorem 1 in [16] to the changed assumptions, which can be done easily.

The following setting is suitable for our purposes. Let X and Z be Banach spaces with $Z \subset X$, and Z continuously imbedded in X. Let $\{X_n\}_{n \in \mathbb{N}}$ be a sequence of finite dimensional subspaces $X_n \subset Z$, and for each $n \in \mathbb{N}$ let $P_n \in \mathcal{L}(X, X_n)$ and let Q_n and $R_n \in \mathcal{L}(Z, X_n)$ be projections on X_n. Here $\mathcal{L}(V, W)$ denotes the space of linear bounded operators between the Banach spaces V and W. Finally, let $\mathcal{S} \subset \mathcal{L}(X, X)$ be a linear subspace.

The following conditions are essential in the local principle. By $\mathcal{K}(X, Z)$ we denote the set of linear compact operators from $X \to Z$.

I. $\|P_n x - x\|_X \to 0$ for $x \in X$, and $\|Q_n z - z\|_X \to 0$ for $z \in Z$, as $n \to \infty$.

II. 1. $\mathcal{K}(X, Z) \subset \mathcal{S}$.
   2. $AX_n \subset Z$ for $A \in \mathcal{S}$.
   3. $Q_n A P_n \in \mathcal{L}(X, X)$ and $\|Q_n A P_n x - x\|_X \to 0$ as $n \to \infty$, for $A \in \mathcal{S}$ and $x \in X$.

III. Let $\mathcal{M} \subset \mathcal{L}(X, X)$ be a set of multipliers in $\mathcal{S}$ (i.e. if $f, g \in \mathcal{M}$ then $fg \in \mathcal{M}$) such that $\mathcal{M}X_n \subset Z$, $\mathcal{M}AX_n \subset Z$, $A\mathcal{M}X_n \subset Z$ for $A \in \mathcal{S}$, and
   1. $\sup_n \|Q_n f P_n\|_X < \infty$ for $f \in \mathcal{M}$,
   2. $\|Q_n f(I - Q_n)AP_n\|_X \to 0$ as $n \to \infty$ for $A \in \mathcal{S}$ and $f \in \mathcal{M}$,
   3. $\|Q_n A(I - R_n)fP_n\|_X \to 0$ as $n \to \infty$ for $A \in \mathcal{S}$ and $f \in \mathcal{M}$,
   4. $\|Q_n f(I - Q_n)gP_n\|_X \to 0$ as $n \to \infty$ for $f, g \in \mathcal{M}$.

IV. For an index set J there exists for each $t \in J$ a subset $\mathcal{M}_t \subset \mathcal{M}$ such that
1. $0 \notin \mathcal{M}_t$ and for any two elements $f_t^{(j)} \in \mathcal{M}_t$, $j = 1, 2$, there exists a third element $f_t \in \mathcal{M}_t$ with $f_t^{(j)} f_t = f_t f_t^{(j)} = f_t$, for $j = 1, 2$,
2. each set $\{f_t\}_{t \in J}$ of elements $f_t \in \mathcal{M}_t$ contains a finite subset $f_{t_1}, \dots, f_{t_m}$ such that $f_{t_1} + \cdots + f_{t_m}$ is invertible in $\mathcal{M}$.

V. 1. $Af - fA \in \mathcal{K}(X, Z)$ for $A \in \mathcal{S}, f \in \mathcal{M}_t$ and $t \in J$,
2. for all $A \in \mathcal{S}$ there exist operators $A_t \in \mathcal{S}, t \in J$ such that for all $t \in J$ and $\epsilon > 0$ one can find $T_t \in \mathcal{K}(X, Z)$, $f_t \in \mathcal{M}_t$ and $n_0 \geq 1$ satisfying

$$\|Q_n(A - A_t) f_t P_n - Q_n T_t P_n\|_X < \epsilon \quad \text{for} \quad n \in \mathrm{N} \text{ with } n \geq n_0.$$

The following definitions are adapted from [16].

Definition 3.1. For an operator $A \in \mathcal{S}$ the sequence $\{Q_n A P_n\}$ is said to be **stable** if the operators $Q_n A P_n$ are invertible in X_n for $n \geq n_0$ and

$$\sup_{n \geq n_0} \|(Q_n A P_n)^{-1}\|_{X_n} < \infty.$$

Definition 3.2. For an operator $A \in \mathcal{S}$ the sequence $\{Q_n A P_n\}$ is said to be **locally stable from the right** if for all $t \in J$ and $n \in \mathrm{N}$ with $n \geq n_0$ there exist operators $T_t \in \mathcal{K}(X, Z)$, $D_{t,n} \in \mathcal{L}(X_n, X_n)$ and an element $f_t \in \mathcal{M}_t$ such that

$$\|Q_n f_t (A_t + T_t) D_{t,n} - Q_n f_t P_n\|_{X_n} \to 0 \quad \text{as} \quad n \to \infty \tag{3.1}$$

and

$$\sup_{n \geq n_0} \|D_{t,n}\|_{X_n} < \infty. \tag{3.2}$$

We are now in the position to state our version of Prößdorf's Theorem 1 from [16].

Theorem 3.3. ([16]) *Assume the above conditions I–V to hold and assume $A^{-1} \in \mathcal{L}(X, X)$ to exist. Then the sequence $\{Q_n A P_n\}$ is stable if and only if it is locally stable from the right.*

Remark 3.4 In the more general setting [16] of mappings $A : X \to Y$ the conditions III.2 and III.3 in [16] can be replaced by the following generalisations.

III.2 $\|Q_n f (I - Q_n) A P_n\|_{X \to Y} \to 0$ as $n \to \infty$ for $A \in \mathcal{S}$ and $f \in \mathcal{M}$,
$\|Q_n f (I - Q_n) g S_n\|_{Y \to Y} \to 0$ as $n \to \infty$ for $f, g \in \mathcal{M}$,

III.3 $\|Q_n A (I - R_n) f P_n\|_{X \to Y} \to 0$ as $n \to \infty$ for $A \in \mathcal{S}$ and $f \in \mathcal{M}$,
$\|R_n f (I - R_n) g P_n\|_{X \to X} \to 0$ as $n \to \infty$ for $f, g \in \mathcal{M}$.

4. Main theorems

Recall our prevailing assumptions, that $p \in (1, \infty)$, that the qualocation projection R_h is well-defined, and that the sequence $\{\pi_h\}$ of grids is quasi-uniform.

Theorem 4.1. *Let $\{R_h\}$ be p-stable. Then with A given by (1.4),*

$$\|R_h A P_h x - Ax\|_p \to 0 \quad \text{as} \quad h \to 0 \quad \text{for} \quad x \in L_p(\mathbb{T}). \tag{4.1}$$

As preparation for the proof we provide the following lemma.

Lemma 4.2. *Let $\{R_h\}$ be p-stable and $a, b \in \overline{PC}$. Then*

$$\|R_h a P_h x\|_p \le C\|a\|_\infty \|x\|_p, \; x \in L_p(\mathbb{T}), \tag{4.2}$$

and

$$\|R_h b S P_h x\|_p \le C\|b\|_\infty (\|S\|_p + \|S\|_{1,p})\|x\|_p, \; x \in L_p(\mathbb{T}). \tag{4.3}$$

Proof. For $x \in L_p(\mathbb{T})$ we obtain with the aid of the p-stability of $\{R_h\}$ and $\{P_h\}$ expressed in (2.3) and (2.8)

$$\begin{aligned} \|R_h a P_h x\|_p &\le C|aP_h x|_{h,p} \le C\|a\|_\infty |P_h x|_{h,p} \\ &\le C\|a\|_\infty \|P_h x\|_p \le C\|a\|_\infty \|x\|_p. \end{aligned}$$

In the third step we used (2.4), as we shall frequently do in the sequel without further reference.

In the proof of (4.3) we make use of the boundedness of the operator S as a mapping in $L_p(\mathbb{T})$ and in $W_p^1(\mathbb{T})$ (see [14], Theorem 2.2 and [15], Corollary 6.1.3), finding in this way

$$\begin{aligned} \|R_h b S P_h x\|_p &\le C\|b\|_\infty |SP_h x|_{h,p} \\ &\le C\|b\|_\infty \big(|(I - R_h)SP_h x|_{h,p} + \|R_h S P_h x\|_p\big) \qquad (4.4)\\ &\le C\|b\|_\infty \big(|(I - R_h)SP_h x|_{h,p} + \|(I - R_h)SP_h x\|_p + \|SP_h x\|_p\big). \end{aligned}$$

Theorem 2.2 yields

$$\begin{aligned} |(I - R_h)SP_h x|_{h,p} + \|(I - R_h)SP_h x\|_p &\le Ch\|SP_h x\|_{1,p} \le Ch\|S\|_{1,p}\|P_h x\|_{1,p} \\ &\le C\|S\|_{1,p}\|P_h x\|_p \le C\|S\|_{1,p}\|x\|_p, \end{aligned}$$

where the second last step comes from the inverse assumption for V_h. Taking into account also $\|SP_h x\|_p \le C\|S\|_p \; \|x\|_p$, the result follows from (4.4). □

Proof of Theorem 4.1 Since $A = aI + bS$ and we know the boundedness of $\{R_h a P_h\}$ and $\{R_h b S P_h\}$ in $L_p(\mathbb{T})$, it is sufficient to prove (4.1) for x in the dense subspace $C^1(\mathbb{T})$. For such x it follows from Theorem 2.6 and the approximation power of V_h that $\|P_h x - x\|_{1,p} \to 0$ as $h \to 0$, and then also $\|SP_h x - Sx\|_{1,p} \to 0$

as $h \to 0$. As a consequence, invoking also the continuous imbedding of $W_p^1(\mathbb{T})$ in $C(\mathbb{T})$,

$$\|AP_hx - Ax\|_\infty \le \|aP_hx - ax\|_\infty + \|bSP_hx - bSx\|_\infty \to 0 \qquad \text{as} \quad h \to 0.$$

Now $x \in C^1(\mathbb{T})$ implies $x \in W_p^1(\mathbb{T})$, so that $Sx \in W_p^1(\mathbb{T}) \subset C(\mathbb{T})$, and hence $Ax \in \overline{PC}$. Thus with the aid of p-stability of $\{R_h\}$, Proposition 2.5 and the result above we find

$$\begin{aligned}\|R_hAP_hx - Ax\|_p &\le \|R_h(AP_hx - Ax)\|_p + \|R_hAx - Ax\|_p \\ &\le \|AP_hx - Ax\|_\infty + \|R_hAx - Ax\|_p \to 0 \quad \text{as} \quad h \to 0,\end{aligned}$$

completing the proof. □

For the formulation of our main theorem, we define for $t \in \mathbb{T}$ the set $\mathcal{M}_t$ of multipliers in $L_p(\mathbb{T})$ by

$$(4.5) \qquad \mathcal{M}_t := \{f_t \in C_0^\infty(t - \frac{1}{4}, t + \frac{1}{4}),\ f_t(t) = 1,\ 0 \le f_t(s) \le 1,\ s \in \mathbb{T}\}.$$

We adapt Definition 3.2 to our present setting.

Definition 4.3. The sequence $\{R_hAP_h\}$ is said to be **locally stable from the right** if for all $t \in [0,1)$ and $n \in \mathbb{N}$ with $n \ge n_0$, there exist operators $T_t \in \mathcal{K}(L_p(\mathbb{T}), \overline{PC})$, $D_{t,h} \in \mathcal{L}(V_h, V_h)$ and a function $f_t \in \mathcal{M}_t$ such that the frozen-coefficient operator A_t from (1.9) satisfies

$$(4.6) \qquad \|R_hf_t(A_t + T_t)D_{t,h} - R_hf_tP_h\|_{V_h} \to 0 \qquad \text{as} \quad h \to 0$$

and

$$(4.7) \qquad \sup_h \|D_{t,h}\|_{V_h} < \infty.$$

Theorem 4.4. *Assume that (1.2) has for each $y \in L_p(\mathbb{T})$ a unique solution $x \in L_p(\mathbb{T})$, and that one of the conditions 1–3 in Theorem 2.1 holds. Then the sequence of qualocation operator $\{R_hAP_h\}$ is stable if and only if it is locally stable from the right.*

Corollary 4.5. *Let the assumptions of Theorem 4.4, be satisfied and let $p = 2$. Then the qualocation method applied to A is stable if and only if it is stable when applied to A_t for $t \in \mathbb{T}$.*

The proofs will be given through some lemmas which establish the conditions needed for the local principle.

We identify the various quantities in the abstract setting in the following way:

$$\begin{aligned}
&(4.8) \qquad & X &= L_p(\mathbb{T}),\ Z = \overline{PC},\ X_n = V_h, \\
&(4.9) \qquad & P_n &= P_h,\ Q_n = R_n = R_h, \\
&(4.10) \qquad & \mathcal{S} &= \{aI + bS : a, b \in \overline{PC}\} + \mathcal{K}(L_p(\mathbb{T}), \overline{PC}), \\
&(4.11) \qquad & \mathcal{M} &= \{f \in C^\infty(\mathbb{T}) : 0 \le f(s) \le 1,\ s \in \mathbb{T}\},
\end{aligned}$$

$J = \mathbb{T}$, and $\mathcal{M}_t$ is as already defined in (4.5). Note that with these identifications the general setting of Prößdorf's local principle is met. For the rest of this section we assume that one of the assumptions 1–3 of Theorem 2.1 holds.

Lemma 4.6. *Assumption I is satisfied.*
Proof That $\|P_h x - x\|_p \to 0$ as $h \to 0$ for $x \in L_p(\mathbb{T})$ follows from (2.8), while the second property is (2.6).

Lemma 4.7. *Assumption II holds.*
Proof II.1 is clear from (4.10). It is known ([15], Corollary 6.1.3) that $SW_p^1(\mathbb{T}) \subset W_p^1(\mathbb{T})$, from which II.2 holds, since $V_h \subset W_p^1(\mathbb{T}) \subset C(\mathbb{T})$ and $a, b \in \overline{PC}$. For II.3 observe that $R_h A P_h$ is bounded in $L_p(\mathbb{T})$ due to Lemma 4.2.

For the second part of II.3, observe that if $A = aI + bS$ then the result to be proved is given by Theorem 4.1, so that it only remains to prove the result with $A = K$, where $K \in \mathcal{K}(L_p(\mathbb{T}), \overline{PC})$. Now

$$\begin{aligned} \|R_h K P_h x - Kx\|_p &\le \|R_h K(P_h x - x)\|_p + \|R_h Kx - Kx\|_p \\ &\le C\|K(P_h x - x)\|_\infty + \|(R_h - I)Kx\|_p \\ &\le C\|P_h x - x\|_p + \|(R_h - I)Kx\|_p, \end{aligned}$$

in which both terms converge to zero as $h \to 0$ by the already proved Assumption I. □

Lemma 4.8. *Assumption III.2 holds.*
Proof Using the p- and q-stability of $\{R_h\}$ (see (2.3)) together with the second part of Theorem 2.3 we obtain, for $x \in L_p(\mathbb{T})$,

$$\begin{aligned} &\|R_h f(I - R_h) A P_h x\|_p \le Ch\|f'\|_{r-1,\infty} |(I - R_h) A P_h x|_{h,p} \\ &\le Ch\|f'\|_{r-1,\infty} |A P_h x|_{h,p} \qquad (4.12) \\ &\le Ch\|f'\|_{r-1,\infty} \big(\|a\|_\infty |P_h x|_{h,p} + \|b\|_\infty |S P_h x|_{h,p} + \|K P_h x\|_\infty\big). \end{aligned}$$

With the aid of Theorem 2.2 and (4.3) we see that

$$\begin{aligned} |S P_h x|_{h,p} &\le |(I - R_h) S P_h x|_{h,p} + |R_h S P_h x|_{h,p} \\ &\le Ch\|S P_h x\|_{1,p} + C\|R_h S P_h x\|_p \\ &\le Ch\|P_h x\|_{1,p} + C\|x\|_p \\ &\le C(\|P_h x\|_p + \|x\|_p) \le C\|x\|_p. \qquad (4.13) \end{aligned}$$

Since also $|P_h x|_{h,p} \le C\|P_h x\|_p \le C\|x\|_p$, and $\|K P_h x\|_\infty \le C\|P_h x\|_p \le C\|x\|_p$, III.2 follows from (4.12). □

Lemma 4.9. *Assumptions III.1 and III.4 hold.*
Proof Since $\mathcal{M} \subset \overline{PC}$, Assumption III.1 follows from (4.2). Assumption III.4 is a special case of III.2.

Lemma 4.10. *Assumption III.3 holds.*
Proof The p-stability of $\{R_h\}$ and the first part of Theorem 2.3 give, for $x \in L_p(\mathbb{T})$,

$$
\begin{aligned}
\|R_h a(I-R_h)fP_h x\|_p &\le C\|a\|_\infty |(I-R_h)fP_h x|_{h,p} \\
&\le Ch\|a\|_\infty \|f'\|_{r-1,\infty}|P_h x|_{h,p} \\
&\le Ch\|a\|_\infty \|f'\|_{r-1,\infty}\|x\|_p. \qquad (4.14)
\end{aligned}
$$

Similarly, for $K \in \mathcal{K}(L_p(\mathbb{T}), \overline{PC})$,

$$
\begin{aligned}
\|R_h K(I-R_h)fP_h x\|_p &\le C\|K(I-R_h)fP_h x\|_\infty \\
&\le C\|(I-R_h)fP_h x\|_p \\
&\le Ch\|f'\|_{r-1,\infty}|P_h x|_{h,p} \\
&\le Ch\|f'\|_{r-1,\infty}\|x\|_p. \qquad (4.15)
\end{aligned}
$$

For the other term we use also Theorem 2.2, to obtain

$$
\begin{aligned}
&\|R_h bS(I-R_h)fP_h x\|_p \le C\|b\|_\infty |S(I-R_h)fP_h x|_{h,p} \\
&\quad \le C\|b\|_\infty \left(|(I-R_h)S(I-R_h)fP_h x|_{h,p} + \|R_h S(I-R_h)fP_h x\|_p\right) \\
&\quad \le C\|b\|_\infty \left(h\|S(I-R_h)fP_h x\|_{1,p} + \|S(I-R_h)fP_h x\|_p + \right. \\
&\qquad\quad \left. \|(I-R_h)S(I-R_h)fP_h x\|_p\right) \\
&\quad \le C\|b\|_\infty \left(h\|S(I-R_h)fP_h x\|_{1,p} + \|(I-R_h)fP_h x\|_p\right) \\
&\quad \le C\|b\|_\infty h(\|(I-R_h)fP_h x\|_{1,p} + \|f'\|_{r-1,\infty}|P_h x|_{h,p}).
\end{aligned}
$$

We have $|P_h x|_{h,p} \le C\|x\|_p$. For the first term, with the aid of Theorem 2.4 we derive

$$
\begin{aligned}
\|(I-R_h)fP_h x\|_{1,p} &\le C\|f'\|_{r-1,\infty}h(|P_h x|_{h,p} + |(P_h x)'|_{h,p}) \\
&\le C\|f'\|_{r-1,\infty}h(\|P_h x\|_p + \|(P_h x)'\|_p) \\
&\le C\|f'\|_{r-1,\infty}\|P_h x\|_p \le C\|f'\|_{r-1,\infty}\|x\|_p.
\end{aligned}
$$

Here we took into account the estimate (2.5). We have thus shown that

$$
\|R_h bS(I-R_h)fP_h x\|_p \le C\|b\|_\infty \|f'\|_{r-1,\infty}h\|x\|_p. \qquad (4.16)
$$

The assertion is an immediate consequence of (4.14), (4.15) and (4.16). □

Lemma 4.11. *Assumption V holds.*
Proof If $A = aI + bS$ then let A_t be as in (1.9). Then the assumption V.1 is a consequence of [15], Lemma 6.1.7. To see that V.2 holds, for $f_t \in \mathcal{M}_t$, we define

$$
T_t := (A - A_t)f_t - f_t(A - A_t),
$$

which belongs to $\mathcal{K}(L_p(\mathbb{T}),\overline{PC})$ from V.1. Then for $x \in L_p(\mathbb{T})$

$$\begin{aligned}
&\|(R_h(A-A_t)f_tP_h - R_hT_tP_h)x\|_p \\
&= \|R_hf_t(A-A_t)P_hx\|_p \le C|f_t(A-A_t)P_hx|_{h,p} \\
&\le C\left(\sup_{s\in\operatorname{supp} f_t}|a(s)-a_t(s)|\,|P_hx|_{h,p} + \sup_{s\in\operatorname{supp} f_t}|b(s)-b_t(s)|\,|SP_hx|_{h,p}\right) \\
&\le C\left(\sup_{s\in\operatorname{supp} f_t}|a(s)-a_t(s)| + \sup_{s\in\operatorname{supp} f_t}|b(s)-b_t(s)|\right)\|x\|_p, \qquad (4.17)
\end{aligned}$$

where we invoked (4.13) in the last step. By construction of a_t and b_t (see (1.10)), the functions $a-a_t$ and $b-b_t$ are continuous at the point t. Hence by choosing $f_t \in \mathcal{M}_t$ with sufficiently small support, the factor in front of $\|x\|_p$ in (4.17) can be made smaller than any given $\epsilon > 0$.

For $A = K$, with $K \in \mathcal{K}(L_p(\mathbb{T}),\overline{PC})$ V.1 is clear, while V.2 is satisfied by choosing $A_t = A$ and $T_t = 0$. □

Evidently, assumption IV holds for our choice of $\mathcal{M}$ and $\mathcal{M}_t$. Now we can apply Theorem 3.3, which is the same as Theorem 4.4 in our setting. Thus the proof of Theorem 4.4 is complete.

Proof of Corollary 4.5 By an application of Theorem 4.2 in [8] for the case $p = 2$ the assumption that the operator A in (1.2) is Fredholm yields that for all $t \in \mathbb{T}$ the frozen-coefficient operator A_t is Fredholm of index zero. Moreover, we see from Definition 3.2 that the sequence $\{R_hAP_h\}$ is locally stable from the right if and only if $\{R_hA_{t_0}P_h\}$ is locally stable from the right for each $t_0 \in \mathbb{T}$ (the frozen-coefficient operators for A_{t_0} are to be the same operator). Choose any $t_0 \in \mathbb{T}$. Apply Theorem 4.4 with A_{t_0} in place of A. Clearly, A_{t_0} satisfies all conditions that are required for A, and as a result we obtain the stability of the qualocation method applied to A_{t_0}. □

Acknowledgements

The support of the Australian Research Council is gratefully acknowledged, as in that of the University of New South Wales, where this work was principally carried out. R.D. Grigorieff also acknowledges the support of the German Science Foundation DFG, and I.H. Sloan the hospitality of the University of Bath and the support of the United Kingdom Engineering and Physical Science Research Council.

References

[1] Arnold, D.N., and Wendland, W.L., On the asymptotic convergence of collocation methods, Math. Comp. **41**, pp. 349–381 (1983).

[2] Arnold, D.N., and Wendland, W.L., The convergence of spline collocation for strongly elliptic equations on curves, Numer. Math. **47**, pp. 317–341 (1985).

[3] Chandler, G.A., and Sloan, I.H., Spline qualocation methods for boundary integral equations, Numer. Math. **58**, pp. 537–567 (1990).

[4] Crouzeix, M., and Thomée, V., The stability in L_p and W_p^1 of the L_2-projection onto finite element function spaces, Math. Comp. **48**, pp. 521–532 (1987).

[5] Didenko, V.D., and Pel'ts, G.L., On the stability of the spline-qualocation method for singular integral equations with conjugation, Differential Equations **29**, pp. 1383–1391 (1993).

[6] Didenko, V.D., and Silbermann, B., Stability of approximation methods on locally non-equidistant meshes for singular integral equations, J. Integral Eqns. Applics. **11**, pp. 317–349 (1999).

[7] Didenko, V.D., Roch, S., and Silbermann, B., Approximation methods for singular integral equations with conjugation on curves with corners, SIAM J. Numer. Anal. **32**, pp. 1910–1939 (1995).

[8] Gohberg, I., and Krupnik, N., Einführung in die Theorie der eindimensionalen singulären Integraloperatoren, Birkhäuser Verlag, Basel, 1979.

[9] Grigorieff, R.D., and Sloan, I.H., Stability of discrete orthogonal projections for continuous splines, Bull. Austral. Math. Soc. **58**, pp. 307–332 (1998).

[10] Grigorieff, R.D., Sloan, I.H., and Brandts, J., Superapproximation and commutator properties of discrete orthogonal projections for continuous splines, submitted for publication.

[11] Hagen, R., Roch, S., and Silbermann, B., Spectral theory of approximation methods for convolution equations, Birkhäuser Verlag, Basel, 1995.

[12] Hagen, R., and Silbermann, B., On the stability of the qualocation method, in Seminar Analysis, Operator Equations and Numerical Analysis 1987/1988, Karl-Weierstraß-Institut für Mathematik, Berlin, pp. 43–52.

[13] Hagen, R., and Silbermann, B., On the convergence of the qualocation method, preprint 207, Technische Universität Chemnitz (1991).

[14] Michlin, S.G., and Prößdorf, S., Singular Integral Operators, Springer-Verlag, Berlin, 1986.

[15] Prößdorf, S., Some Classes of Singular Equations, North Holland Publishing Company, Amsterdam, 1978.

[16] Prößdorf, S., Ein Lokalisierungsprinzip in der Theorie der Spline-Approximationen und einige Anwendungen, Math. Nachrichten **119**, pp. 239–255 (1984).

[17] Prößdorf, S., and Schmidt, G., A finite element collocation method for singular integral equations, Math. Nachr. **100**, pp. 33–60 (1981).

[18] Saranen, J., and Wendland, W.L., On the asymptotic convergence of collocation methods with spline functions of even degree, Math. Comp. **45**, pp. 93–108 (1985).

[19] Schmidt, G., On spline collocation methods for boundary integral equations in the plane, Math. Meth. Appl. Sci. **7**, pp. 74-89 (1985).

[20] Sloan, I.H., A quadrature-based approach to improving the collocation method. Numer. Math. **54**, pp. 41–56 (1988).

[21] Sloan, I.H., and Wendland, W.L., A quadrature-based approach to improving the collocation method for splines of even degree, Z. für Anal. und ihre Anw. **8**, pp. 361–376 (1989).

[22] Sloan, I.H., and Wendland, W.L., Qualocation methods for elliptic boundary integral equations, Numer. Math. **79**, pp. 451-483 (1998).

[23] Sloan, I.H., and Wendland, W.L., Spline qualocation methods for variable-coefficient elliptic equations on curves, Numer. Math. **83**, pp. 497–533 (1999).

Technische Universität Berlin
Straße des 17 Juni 135
10623 Berlin, Germany

School of Mathematics
University of New South Wales
Sydney 2052, Australia

1991 Mathematics Subject Classification: Primary 65R10; Secondary 45E05

Submitted: 11.5.2000

Operator Theory:
Advances and Applications, Vol. 121

Toeplitz operators and the modelling of oscillating discontinuities with the help of Blaschke products

SERGEI M. GRUDSKY

In the present paper we establish theorems about the representation of functions with given asymptotics of the argument in a neighborhood of a discontinuity in the form of a Blaschke product or, more general, in the form of a superposition of a continuous function and a Blaschke product. On this foundation, a theory of normal solvability for Toeplitz operators $T(a)$ on the unit circle whose symbols have oscillating discontinuities is constructed. In particular, the cases of symbols of the form

$$|\arg a(e^{i\theta})| \sim |\theta|^{-\lambda}(\lambda > 0), \ |\arg a(e^{i\theta})| \sim \ln^{\beta}|\theta^{-1}|(\beta > 1), \ |\arg a(e^{i\theta})| \sim e^{|\theta|^{-\lambda}}(\lambda > 0)$$

are considered.

1. Introduction

The theory of singular integral operators with discontinuous coefficients and the equivalent theory of Toeplitz operators with discontinuous symbols have been developed intensively since the beginning of the sixties (see [1]–[7] and the references therein). In these theories, much attention has been paid to several classes of operators whose symbols are strongly oscillating near their discontinuities. One peculiarity of such operators is the fact that they are usually not Fredholm although they are normally solvable in most cases. The beginning of the theory of operators with oscillating symbols are the papers of N.V.Govorov and his successors ([8]–[13]), in which the results were formulated in terms of the Riemann boundary value problem in classes of entire or bounded functions, and the papers [14]–[15], where the case of almost-periodic discontinuities in the spaces L_p was considered. The further development of the theory is reflected in [16]–[31]. In these works, operators with oscillating symbols are studied in spaces of summable functions, the classes of admissible discontinuities were essentially extended, and part of the results were transfered to the matrix case.

The foundation of the afore-mentioned classical papers [14]–[15] and as well of the majority of subsequent investigations in this direction is an appropriate generalization of Wiener-Hopf factorization. (The most general factorization of such a type was introduced and investigated by I.M. Spitkovsky in [31]–[32]). The construction of a generalized factorization is usually the key problem when passing to a new class of operators.

In this paper, we present a theory of normal solvability for a large class of operators whose symbols have arguments that tend to infinity monotonously in neighborhoods of the discontinuities. Our approach is based on appropriate generalized factorizations and on theorems about the representation of functions with given asymptotics of the argument in a neighborhood of the discontinuity in the form of Blaschke products, or, more general, in the form of a superposition of a continuous function and a Blaschke product. It should be noted that our theorems on representations via Blaschke products are of independent interest in connection with questions of the theory of analytic functions.

The present paper is a continuation and essential extension of [22]–[25], [17]. The proof of the main result concerning the normal solvability of the class of operators under consideration (Theorem 2.3) uses the theory of u-factorization worked out in [23]–[25].

The structure of this paper is as follows. In Section 2 the main results are formulated and some examples illustrating the generality of these results are given. The other sections are devoted to the proofs of the theorems from Section 2.

In conclusion we note that the results of this paper allow us not only to consider cases of arguments that tend monotonously to infinity, but also cases of generalized periodic, almost periodic, and semi-almost periodic discontinuities. The basic tool for such extensions is Theorem 2.2. Moreover in the case of generalized periodic discontinuities, matrix operators can be considered. Here the word "generalized" means that the functions $\exp(i\lambda x)$, $\lambda \in \mathbf{R}$, which generate the algebra of uniformly almost-periodic functions, are replaced by the functions $\exp(i\lambda f(x))$, where $f(x)$ is a homeomorphism of the real line onto itself satisfying (after conformally mapping $\mathbf{R}$ onto $\mathbf{T}$) the hypothesis of Theorem 2.2 and, in particular, the conditions of Examples 2.6–2.12. A separate paper will be devoted to this problem.

2. Main results

Denote by $\mathbf{T}$ the unit circle with the center at the origin and by $L_p(\mathbf{T})$ the usual space of measurable functions on $\mathbf{T}$ with the norm

$$\|f\|_p = \left(\int_{\mathbf{T}} |f(t)|^p dt\right)^{1/p}, \quad 1 \le p < \infty,$$

and

$$\|f\|_\infty = \operatorname{ess\,sup}_{t\in\mathbf{T}} |f(t)| < \infty, \quad p = \infty.$$

Let P denote the analytic projector on $\mathbf{T}$, that is, $P = \frac{1}{2}(I + S)$, where I is the identity operator and S is the operator of singular integration,

$$(Sf)(t) = \frac{1}{\pi i} \int_{\mathbf{T}} \frac{f(\tau)}{\tau - t} d\tau, \quad t \in \mathbf{T}.$$

Here the integral is understood in sense of the principal value. We consider Toeplitz operators of the kind

$$T(a) = PaP : H_p(\mathbf{T}) \to H_p(\mathbf{T}) \tag{2.1}$$

acting in the Hardy spaces $H_p(\mathbf{T}) \overset{Df}{=} PL_p(\mathbf{T})$. It is well known that if the symbol of a Toeplitz operator is a function $a(t)$ in $L_\infty(\mathbf{T})$, then $T(a)$ is bounded in $H_p(\mathbf{T})$ for $1 < p < \infty$.

An operator A acting in a Banach space B is said to be normally solvable if its image $\operatorname{im} A$ is a closed subspace of B.

An operator A is called left (right) invertible in a Banach space B if there exists an operator $A_l^{-1} : B \to B$ ($A_r^{-1} : B \to B$) such that $A_l^{-1}A = I$ ($AA_r^{-1} = I$). It is well known that if A is left invertible (right invertible), then A is normally solvable and, moreover, $\dim\ker A = \{0\}$ and the subspace $\operatorname{im} A$ has a direct complement in B ($\operatorname{im} A = B$ and the subspace $\ker A$ has a direct complement in B).

We denote by $C(\mathbf{T})$ the set of continuous functions on the unit circle, and we connect with each function $c(t) \in C(\mathbf{T})$ that does not vanish on $\mathbf{T}$ its winding number $\operatorname{wind} c$ about the origin.

An infinite product of the kind

$$B(t) = \prod_{j\in\mathbf{Z}} \frac{\overline{z}_j}{|z_j|} \cdot \frac{t - z_j}{1 - \overline{z}_j t}, \tag{2.2}$$

where $|z_j| < 1$ and $\mathbf{Z}$ is the set of all integers, is referred to as a Blaschke product. It is well known that the condition

$$\sum_{j\in\mathbf{Z}} (1 - |z_j|) < \infty \tag{2.3}$$

is necessary and sufficient for the convergence of $B(t)$ for almost all $t \in \mathbf{T}$. Moreover, $|B(t)| = 1$ almost everywhere, and the points of discontinuity (and only these points!) are the limit points of the sequence of zeros $\{z_j\}$ of the Blaschke product.

Let $H_\infty(\mathbf{T})$ denote the subspace of $L_\infty(\mathbf{T})$ consisting of the functions which are almost everywhere representable as nontangential limit values of functions that are analytic and bounded inside $\mathbf{T}$. An unimodular function $u(t) \in H_\infty(\mathbf{T})$, that is, a function for which $|u(t)| = 1$ almost everywhere, is called an inner function. It is obvious that Blaschke products are inner functions.

Let us consider an unimodular function $a_0(t)$ such that $a_0(t)$ is continuous on $\mathbf{T} \setminus \{1\}$ and the following representation holds:

$$a_0(\exp(i\theta)) = \exp(i2\pi f(\theta)), \quad \theta \in (-\pi, \pi) \setminus \{0\}, \tag{2.4}$$

where $f(\theta)$ is a real-valued function that is continuous and monotonously increasing on the intervals $(-\pi, 0)$ and $(0, \pi)$ and that satisfies

$$\lim_{\theta\to 0\pm 0} f(\theta) = \mp\infty. \tag{2.5}$$

Without loss of generality assume that $f(-\pi+0)=f(\pi-0)=0$. On $\mathbf{R}\setminus\{0\}$, consider the function

$$\theta(x)=f^{-1}(-x). \tag{2.6}$$

Clearly, $\theta(x)$ decreases monotonously on the rays $(-\infty,0)$ and $(0,\infty)$ and we have

$$\theta(\pm\infty)=0, \quad \theta(0\pm 0)\overset{Df}{=}\theta(\pm 0)=\pm\pi. \tag{2.7}$$

Put

$$\Delta(n)=\begin{cases}\theta(n)-\theta(n+1), & n=+0,1,2,\dots;\\ \theta(n-1)-\theta(n), & n=-0,-1,-2,\dots\end{cases}$$

and let us consider the sequence of functions

$$\psi_n(s)=\frac{\theta(n)-\theta(n+s)}{\Delta(n)}, \quad s\in I,$$

on the segment $I=[-1/2,1/2]$.

Assume that this sequence converges uniformly on I and that, moreover,

$$\left.\begin{aligned}\lim_{n\to+\infty}\psi_n(s)&=\psi(s);\\ \lim_{n\to-\infty}\psi_n(s)&=-\psi(-s),\end{aligned}\right\} \tag{2.8}$$

where $\psi(s)$ is a monotonously increasing continuous function on I.

We introduce the notation

$$\xi(n)=\begin{cases}\theta(n+1)/\theta(n), & n=+0,1,2,\dots;\\ \theta(n-1)/\theta(n), & n=-0,-1,-2,\dots;\end{cases} \tag{2.9}$$

$$\alpha(n)=1-\xi(n) \tag{2.10}$$

and we consider the auxiliary functions

$$A(\theta)=\sum_{\theta(j)>\theta}\arctan(\alpha(j))-\sum_{\theta(j)<-\theta}\arctan(\alpha(j)), \quad \theta\in(0,\pi), \tag{2.11}$$

and

$$C(n)=\sum_{j=m(n)}^{n-\sigma}\arctan\frac{\Delta(j)}{\theta(j)-\theta(n)}-\sum_{j=n+\sigma}^{M(n)}\arctan\frac{\Delta(j)}{\theta(n)-\theta(j)}, \tag{2.12}$$

where $\sigma=\operatorname{sign} n$, $m=m(n)$ is the number j for which $|\theta(j)|\le(3/2)|\theta(n)|$ and $|j|$ is minimal, while $M=M(n)$ is the number j for which $|\theta(j)|\ge(1/2)|\theta(n)|$ and $|j|$ is maximal. Note that the value $A(\theta)$ characterizes the connection between the behavior of the function $f(\theta)$ in right and left half-neighborhoods of zero, while the value $C(n)$ characterizes the behavior of $\theta(x)$ in a neighborhood of the point $x=n$.

The first main result of the present paper is the following.

Theorem 2.1. *Let the unimodular function $a_0(t)$ be continuous on $\mathbf{T} \setminus \{1\}$ and have the form (2.4) with a real-valued monotonously increasing function $f(\theta)$ on $(-\pi, 0)$ and $(0, \pi)$ satisfying (2.5) and the condition*

$$\lim_{n \to \pm\infty} \xi(n) = 1. \tag{2.13}$$

In addition, suppose (2.8) holds and the limits

$$\lim_{\theta \to 0 \pm 0} A(\theta) = a, \quad a \in \mathbf{R}, \tag{2.14}$$

$$\lim_{n \to \infty} C(n) = 0 \tag{2.15}$$

exist. Then we have a representation

$$a_0(t) = B(t) g(B(t)) d(t) \tag{2.16}$$

with $g(t), d(t) \in C(\mathbf{T})$. Here $g(t)$ has the winding number zero, and the zeros of the Blaschke product $B(t)$ are $z_j = r_j \exp\{i\theta(j)\}$ where $r_j = (1 - \Delta(j)/2)/(1 + \Delta(j)/2)$, $j = \pm 1, \pm 2, \ldots$, and $\theta(x)$ is defined by (2.6).

Under some additional conditions representation (2.16) may be simplified.

Theorem 2.2. *Let the function $a_0(t)$ satisfy the hypothesis of Theorem 2.1 and let $\psi(s)$ from (2.8) have the form*

$$\psi(s) = s. \tag{2.17}$$

Then we have a representation

$$a_0(t) = U(t) c(t) \tag{2.18}$$

where $c(t) \in C(\mathbf{T})$ and $U(t)$ is the inner function

$$U(t) = \frac{r_0 + B(t)}{1 + r_0 B(t)}, \quad r_0 = e^{-2}, \tag{2.19}$$

and where the Blaschke product $B(t)$ is the same as that in Theorem 2.1.

Theorems 2.1 and 2.2 allow us to establish a theory of normal solvability for Toeplitz operators whose symbols have oscillating discontinuities. For simplicity, let us restrict ourselves to one point of discontinuity $t_0 = 1$.

Theorem 2.3. *Let $a(t) \in L_\infty(\mathbf{T})$ be continuous on the set $\mathbf{T} \setminus \{1\}$, suppose that $(1/a(t)) \in L_\infty(\mathbf{T})$ and that*

$$a(t) = |a(t)| a_0^\varepsilon(t), \quad \varepsilon = \pm 1, \tag{2.20}$$

where the unimodular function $a_0(t)$ *satisfies the hypothesis of Theorem 2.1. If* $\varepsilon = 1$*, then the Toeplitz operator* $T(a)$ *given by (2.1) is left invertible in* $H_p(\mathbf{T})$*,* $1 < p < \infty$*, and*

$$\dim(H_p(\mathbf{T})/\operatorname{im} T(a)) = \infty. \tag{2.21}$$

If $\varepsilon = -1$*, then the Toeplitz operator (2.1) is right invertible in* $H_p(\mathbf{T})$*,* $1 < p < \infty$*, and*

$$\dim \ker T(a) = \infty. \tag{2.22}$$

Theorems 2.1 to 2.3 are very general. However it should be noted that often the test of conditions (2.8), (2.14), (2.15) is complicated, because these conditions are expressed in terms of the inverse function $\theta(x) = f^{-1}(-x)$. Therefore we give some sufficient conditions for (2.8), (2.14), (2.15). These conditions are easier to check, since they are in terms of the function $f(\theta)$ itself.

Theorem 2.4. *Let the function* $f(\theta)$ *be twice continuously differentiable on* $[-\pi, \pi]\backslash\{0\}$*; suppose* $f'(\theta)$ *is monotonously decreasing on* $(0, \pi)$ *and monotonously increasing on* $(-\pi, 0)$*. If*

$$\lim_{\theta \to 0} \frac{f''(\theta)}{(f'(\theta))^2} = 0, \tag{2.23}$$

then (2.8) is realized with the function $\psi(s) = s$*. If*

$$\lim_{\theta \to 0} \frac{f''(\theta)|\theta|^{1/2}}{(f'(\theta))^{3/2}} = 0, \tag{2.24}$$

then (2.15) holds.

Finally, if the function $\psi(\theta) \overset{Df}{=} 1/(\theta f'(\theta))$ *monotonously tends to the zero as* $\theta \to 0$ *and the integral*

$$I = \int_0^\pi \left| \frac{1}{(sf'(s))^2} - \frac{1}{(sf'(-s))^2} \right| \frac{ds}{s} < \infty \tag{2.25}$$

converges, then condition (2.14) is satisfied.

Remark 2.5. The conditions imposed on the function $f(\theta)$ in Theorems 2.1–2.4 are of a local nature, because they only concern a neighborhood of the point $\theta = 0$. Thus, only the behavior of $f(\theta)$ in a neighborhood of the point $\theta = 0$ plays a role. In particular, in Theorems 2.1–2.4 the set $(-\pi, \pi) \setminus \{0\}$ may be replaced by the set $(-\delta, \delta) \setminus \{0\}$ for any $\delta \in (0, \pi)$.

The functions listed in the following examples are covered by Theorem 2.4.

Example 2.6. Power-like growth.

$$f(\theta) = \begin{cases} -c_+\theta^{-\lambda_+}, & \theta > 0, \\ c_-|\theta|^{-\lambda_-}, & \theta < 0, \end{cases} \qquad c_\pm > 0, \quad \underline{\lambda_\pm \in (0, \infty)}.$$

Example 2.7. Power-logarithmic-like growth.

$$f(\theta)=\begin{cases} -c_+\theta^{-\lambda_+}(\log|\theta|^{-1})^{\beta_+}, & \theta>0,\\ c_-\theta^{-\lambda_-}(\log|\theta|^{-1})^{\beta_-}, & \theta<0,\end{cases} \quad c_\pm>0,\ \lambda_\pm\in(0,\infty),\ \beta_\pm\in\mathbf{R}.$$

Example 2.8. Exponential and hyperexponential growth.

$$f(\theta)=\begin{cases} -c_+\exp\{d_+\theta^{-\lambda_+}\}, & \theta>0,\\ c_-\exp\{d_-|\theta|^{-\lambda_-}\}, & \theta<0,\end{cases} \quad c_\pm>0,\quad d_\pm>0.$$

$$f(\theta)=\begin{cases} -c_+\exp\{g_+\exp(d_+\theta^{-\lambda_+})\}, & \theta>0,\\ c_-\exp\{g_-\exp(d_-|\theta|^{-\lambda_-})\}, & \theta<0,\end{cases} \quad c_\pm>0,\quad d_\pm>0,\quad g_\pm>0.$$

Here $\lambda_\pm\in(0,\infty)$.

Example 2.8 reveals that the hypotheses of our theorems are well adapted to fast growing arguments $f(\theta)$. Moreover one can give examples of functions $f(\theta)$ satisfying the hypothesis of Theorem 2.2 and growing faster than any given function.

Now we will consider some cases of slowly growing arguments $f(\theta)$.

Example 2.9. Logarithmic growth.

$$f(\theta)=\begin{cases} -c(\log\theta^{-1})^{-\beta}, & \theta>0,\\ c(\log|\theta|^{-1})^{\beta}, & \theta<0,\end{cases} \quad c>0,\quad \beta>1,\quad \text{and}$$

$$f(\theta)=\begin{cases} -c(\log\theta^{-1})(\log\log\theta^{-1})^{-\beta}, & \theta>0,\\ c(\log|\theta|^{-1})(\log\log|\theta|^{-1})^{\beta}, & \theta<0,\end{cases} \quad c>0,\quad \beta>0.$$

In connection with Example 2.9 note that in the case of "purely logarithmic" growth ($|f(\theta)|=c\log|\theta|^{-1}$) the hypothesis of Theorem 2.1 is not satisfied, because in this case

$$\lim_{n\to\infty}\xi(n)=d,\quad 0<d<1, \tag{2.26}$$

and hence condition (2.13) is violated. Note that (2.26) is a condition that describes some sort of "slow oscillations". This case may be considered too and we will do that in a forthcoming publication.

Example 2.10. Mixed growth. With the help of the above examples we can build new examples by choosing the function $f(\theta)$ in different ways in the two half-neighborhoods of the point of discontinuity. For example, the results we formulated are true for the following function $f(\theta)$ having very fast growth on one side and very slow growth on the other side of the point $\theta=0$:

$$f(\theta)=\begin{cases} -c_+\exp\{g\exp(\theta^{-\lambda})\}, & \theta>0,\\ c_-\log^{\beta}(|\theta|^{-1}), & \theta<0,\end{cases} \quad c_\pm>0,\quad g>0,$$

where $\lambda\in(0,\infty)$, $\beta\in(3/2,\infty)$.

Example 2.11. Perturbation by nonmonotonous summands.

$$f(\theta) = -\sigma c|\theta|^{-\lambda} + c_1 \sin|\theta|^{-\beta}, \quad \sigma = \operatorname{sign}\theta, \quad c > 0,$$

where $c_1 \in \mathbf{R}$, $0 < 2\beta < \lambda$.

Example 2.12. Infinite index of power order. In the theory of the Riemann boundary-value problem with infinite index one often encounters symbols with arguments of the kind

$$f(\theta) = \begin{cases} -c_+(\theta)\theta^{-\lambda_+}, & \theta > 0, \\ c_-(\theta)|\theta|^{-\lambda_-}, & \theta < 0, \end{cases}$$

where $\lambda_\pm \in \mathbf{R}$, $c_+(\theta) \in C[0,\pi)$, $c_-(\theta) \in C(-\pi, 0]$. This case is covered by Theorem 2.4 if we suppose that the functions $c_\pm(\theta)$ are twice continuously differentiable on $(0,\pi)$ and $(-\pi, 0)$ and, in addition,

$$\lim_{\theta\to 0} c'_\pm(\theta)\theta = 0 \quad \text{and} \quad \lim_{\theta\to 0} c''_\pm(\theta)\theta^2 = 0.$$

3. Modelling of oscillations by the superposition of continuous functions and Blaschke products

This section is devoted to the proof of Theorem 2.1.

We begin with a few auxiliary results.

Proposition 3.1. *If a function $\theta(x)$ is of the kind (2.6), (2.7) and $\alpha(n)$ is given by (2.10), then we have the following:*

i) The two series $\sum_{j=1}^{\infty} \alpha(j)$ and $\sum_{j=-\infty}^{-1} \alpha(j)$ diverge.

ii) If $k, n \in \mathbf{Z}$ are numbers such that $kn > 0$ and $|k| < |n|$, then

$$\sum_{j=k}^{n} \alpha_j < \ln \frac{\theta(k)}{\theta(n+\sigma)}, \quad \sigma = \operatorname{sign} n. \tag{3.1}$$

We omit the proof of this proposition because it is elementary.

Condition (2.8) is most essential for the approach suggested here. The following proposition summarizes some consequences of this condition (which include, in particular, the existence of the limit (2.13)).

Proposition 3.2. *Let the function $\theta(x)$ be of the form (2.6) and suppose conditions (2.8) are satisfied. Then*

$$\lim_{n\to\pm\infty} \frac{\Delta(n\pm 1)}{\Delta(n)} = d, \quad \textit{where} \quad 0 \le d \le 1. \tag{3.2}$$

In addition,

(3.3) $$\lim_{n\to\infty} \xi(n) = d, \quad \lim_{n\to\infty} \alpha(n) = 1 - d,$$

(3.4) $$\lim_{n\to\infty} \frac{\alpha(n+1)}{\alpha(n)} = 1, \quad \textit{for } d > 0,$$

(3.5) $$\psi(1/2) - d\psi(-1/2) = 1.$$

Proof. Let $n > 0$. Due to the definition of the functions $\psi_n(s)$, $\psi_{n+1}(s)$ and the value $\Delta(n)$ we have

$$\frac{\Delta(n+1)}{\Delta(n)} = -\frac{1 - \psi_n(1/2)}{\psi_{n+1}(-1/2)}.$$

By conditions (2.8), this equality implies that

(3.6) $$\lim_{n\to\infty} \frac{\Delta(n+1)}{\Delta(n)} = -\frac{1 - \psi(1/2)}{\psi(-1/2)} \overset{Df}{=} d.$$

As $(-\psi(-1/2)) > 0$ and $0 < \psi(1/2) \le 1$, we have $0 \le d$. The value d cannot be greater than 1, since in this case the series $\sum\limits_{j=1}^{\infty} \Delta(j)$ would be divergent, whereas

$$\lim_{n\to\infty} \sum_{j=1}^{n} \Delta(j) = \lim_{n\to\infty} (\theta(1) - \theta(n+1)) = \theta(1) < \infty.$$

This proves (3.2).

Let us pass to (3.3). Suppose this relation is not true. Then there exists a limit point d_0 of the bounded sequence $\{\xi_j\}_{j=0}^{\infty}$ such that $d_0 \ne d$. Put

(3.7) $$d_j \overset{Df}{=} \frac{\Delta(j+1)}{\Delta(j)} = \xi(j)\frac{1 - \xi(j+1)}{1 - \xi(j)}$$

and rewrite this formula in the form

(3.8) $$\xi(j+1) - d_j = (\xi(j) - d_j)\xi^{-1}(j).$$

Suppose $d_0 < d$. Then, due to (3.2), there exists a number j_0 such that $\xi(j_0) < d_j$ for all $j \ge j_0$. Substituting $j = j_0$, $j = j_0 + 1$, $j = j_0 + 2, \ldots$ in (3.8) and taking into account that $\xi^{-1}(j) > 1$, we obtain that the sequence of positive numbers $\{\xi(j)\}_{j=j_0}^{\infty}$ is monotonously decreasing and hence has a limit d_0 as $j \to \infty$.

Passing to the limit in (3.8) we get $d_0 - d = (d_0 - d)d_0^{-1}$. As $d_0 \ne d$, it follows that $d_0 = 1$, which is impossible due to the assumption $d_0 < d \le 1$.

Let now $d_0 > d$. Using (3.8) one can show analogously that the sequence $\{\xi(j)\}_{j=0}^{\infty}$ is monotonously increasing. Passage to the limit in (3.8) again gives $d_0 = 1$, which is impossible since in this case (3.7) implies that there exists a

$0 < \widetilde{d} < 1$ such that $1 - \xi(j+1) > \widetilde{d}(1 - \xi(j))$ for all large enough j. Hence there exists a $c > 0$ such that

$$(1 - \xi(j)) < c\widetilde{d}^j \Rightarrow \theta(j+1) > (1 - c\widetilde{d}^j)\theta(j) \Rightarrow$$

$$\Rightarrow \theta(j+1) > \prod_{k=1}^{j}(1 - c\widetilde{d}^k)\theta(1) \Rightarrow \lim_{j\to\infty} \theta(j+1) > 0,$$

which contradicts the properties of the function $\theta(x)$ (2.6).

Thus, the unique limit point of the sequence $\{\xi(j)\}_{j=1}^{\infty}$ is the point d. This completes the proof of equality (3.3).

To prove (3.4), we represent the number d_j in the form (see (3.7))

$$d_j = \xi(j)\alpha(j+1)/\alpha(j).$$

Passing to the limit $j \to \infty$ we obtain (3.4). Finally, (3.5) follows from (3.6). □

Proposition 3.3. *If* $u \in [-1, 1]$, *then*

$$\Xi(u) \stackrel{Df}{=} \arctan u + \sum_{j=1}^{\infty}\left(\arctan\frac{1}{j+u} - \arctan\frac{1}{j-u}\right) = \tag{3.9}$$

$$= \begin{cases} \arctan(\tan \pi u/\tanh 1) - \pi, & u \in [-1, -1/2), \\ \arctan(\tan \pi u/\tanh 1), & u \in [-1/2, 1/2], \\ \arctan(\tan \pi u/\tanh 1) + \pi, & u \in (-1/2, 1], \end{cases}$$

where $\arctan x \in [-\pi/2, \pi/2]$.

Proof. For $|u| \le 1/2$ see [33], pp. 749–750. For $u \in [-1, 1] \setminus [-1/2, 1/2]$ we obtain (3.9) by analytic continuation. □

The proof of Theorem 2.1 is based on the following representation of the argument of Blaschke products. We assume that the sequence of the zeros of the product (2.2) is subject to the condition

$$\lim_{j\to\pm\infty} z_j = 1. \tag{3.10}$$

Thus, we can write z_j $z_j = r_j \exp(i\theta_j)$ with $0 < r_j < 1$, $-\pi < \theta_j \le \pi$ and $\lim_{j\to\pm\infty} r_j = 1$, $\lim_{j\to\pm\infty} \theta_j = 0$.

Proposition 3.4. *Let the Blaschke product (2.2) satisfy (2.3) and (3.10). Then we can choose a branch of the argument* $\arg B(t)$ *which is continuous on* $\mathbf{T} \setminus \{1\}$, *monotonously increasing on* $\mathbf{T} \setminus \{0\} = (1 + \widehat{0, 1} - 0)$, *and satisfies*

$$\lim_{t\to 1+0} \arg B(t) = A_+ < 0, \quad \lim_{t\to 1-0} \arg B(t) = A_- > 0.$$

In addition, at least one of the limits A_- and A_+ is infinite and the following formula holds:

$$\arg B(\exp(i\theta)) = \begin{cases} -2\left(\sum_{\theta_k\geq\theta}(\pi+\varphi_k(\theta))+\sum_{\theta_k<\theta}\varphi_k(\theta)\right), & \theta>0, \\ 2\left(\sum_{\theta_k\leq\theta}(\pi-\varphi_k(\theta))-\sum_{\theta_k>\theta}\varphi_k(\theta)\right), & \theta<0, \end{cases} \tag{3.11}$$

where

$$\varphi_k(\theta) = \arctan[\varepsilon_k \cot((\theta-\theta_k)/(2))], \quad \varepsilon_k = (1-r_k)/(1+r_k),$$

and $\arctan x \in [-\pi/2, \pi/2]$.

A full proof of this proposition can be found in [17], for example.

We now begin the **Proof of Theorem 2.1.**

First of all note that the zeros z_j satisfy condition (2.3) and that therefore the product defining $B(t)$ converges. Indeed,

$$\sum_{j\in\mathbf{Z}\setminus\{0\}}(1-|z_j|) \leq \text{const} \sum_{j\in\mathbf{Z}\setminus\{0\}}\Delta(j) = \text{const} \lim_{n\to\infty}\sum_{j=-n,j\neq0}^{n}(\theta(j)-\theta(j+1)) =$$

$$= \text{const} \lim_{n\to\infty}((\theta(1)-\theta(n+1))-(\theta(-1)-\theta(1-n))) = \text{const}\,(\theta(1)-\theta(-1)) < \infty.$$

Put $f_B(x) = \arg B(\exp\{i\theta(x)\})$ where $x = -f(\theta)$. Consider first the case of positive $x = n+s$ where $n\in\mathbf{N}$ and $s\in[-1/2,1/2]$. Then $\theta=\theta(x)>0$ and by (3.11),

$$f_B(x) = -2\pi(n-1) - 2\sum_{j=1}^{n-1}\varphi_j(\theta(x)) - 2\widetilde{\varphi}_n(\theta(x)) - \tag{3.12}$$

$$-2\sum_{j=n+1}^{\infty}\varphi_j(\theta(x)) - 2\sum_{j=\infty}^{-1}\varphi_j(\theta(x)),$$

where

$$\varphi_j(\theta(x)) = \arctan\left(\frac{\varepsilon_j}{2}\cot\left(\frac{\theta(x)-\theta(j)}{2}\right)\right),$$

$$\widetilde{\varphi}_n(\theta(x)) = \begin{cases} \varphi_n(\theta(x)), & s\leq 0, \\ \pi+\varphi_n(\theta(x)), & s>0. \end{cases}$$

We want to show that the function $f_B(x)$ is of the form

$$f_B(x) = -2\pi x + p(x-[x+1/2]) + O(x), \tag{3.13}$$

where $p(s)$ is continuous on the segment $[-1/2,1/2]$ and that, moreover,

$$p(-1/2) = p(1/2). \tag{3.14}$$

Here $[y]$ denotes integral part of the number y and $O(x)$ stands for a function that is continuous on the real line and satisfies

$$\lim_{x\to\pm\infty} O(x) = 0. \tag{3.15}$$

The proof of this fact will be split into several steps. We will show that a certain group of summands in (3.12) (whose numbers j are situated in a "neighborhood" of $j = n$) generate the function p in (3.13) and that the other summands generate a function that is continuous on the one-point compactification of $\mathbf{R}$.

First step. Let us consider the following group of summands:

$$A_1(\theta(x)) \overset{Df}{=} -2\left(\sum_{\theta(j)\geq\delta_1(x)} \varphi_j(\theta(x)) + \sum_{\theta(j)\leq-\delta_1(x)} \varphi_j(\theta(x))\right),$$

where $\delta_1 = \delta_1(x) > 0$ is an arbitrary function vanishing as $x \to \infty$ and satisfying the condition

$$\lim_{x\to\infty} (\theta(x)/\delta_1(x)) = 0. \tag{3.16}$$

Using (2.14) we want to show that

$$\lim_{\theta\to 0+0} A_1(\theta) = a_1, \quad a_1 \in \mathbf{R}. \tag{3.17}$$

To do this, we introduce the function

$$A_2(\theta(x)) \overset{Df}{=} A_1(\theta(x)) + 2A(\delta_1(x)) = A_2^+(\theta) - A_2^-(\theta),$$

where

$$A_2^+(\theta) = 2\sum_{\theta_j\geq\delta_1(x)} (\arctan\alpha(j) + \varphi_j(\theta(x))),$$

$$A_2^-(\theta) = 2\sum_{\theta(j)\leq-\delta_1(x)} (\arctan\alpha(j) - \varphi_j(\theta(x))).$$

For $A_2^+(\theta)$ we have

$$A_2^+(\theta) = 2\sum_{\theta(j)\geq\delta_1(x)} \left[\left(\arctan\alpha(j) - \arctan\left(\frac{\Delta(j)}{2}\cot\frac{\theta(j)}{2}\right)\right) + \right.$$

$$\left. + \left(\arctan\left(\frac{\Delta(j)}{2}\cot\frac{\theta(j)}{2}\right) + \varphi_j(\theta(x))\right)\right] \overset{Df}{=} A_{2,1}^+(\theta) + A_{2,2}^+(\theta).$$

In order to estimate $A_{2,2}^+(\theta)$ we use the following elementary inequality:

$$|\arctan u - \arctan v| \leq \arctan(u-v)|, \quad u, v > 0. \tag{3.18}$$

We have

$$|A_{2,2}^+(\theta)| \le 2 \left| \sum_{\theta_j \ge \delta_1(x)} \arctan\left(\frac{\Delta(j)}{2}\left(\cot\left(\frac{\theta(j)-\theta(x)}{2}\right) - \cot\left(\frac{\theta(j)}{2}\right)\right)\right)\right| \le$$

$$\le \text{const} \sum_{\theta(j)\ge\delta_1(x)} \Delta(j)\frac{\sin(\theta(x)/2)}{\sin((\theta(j)-\theta(x))/2)\sin(\theta(j)/2)} \le$$

$$\le \text{const}\,\theta(x) \sum_{\theta(j)\ge\delta_1(x)} \frac{\Delta(j)}{(\theta(j)-\theta(x))\theta(j)} \le$$

$$\le \text{const}\,\theta(x) \sum_{\theta(j)\ge\delta_1(x)} \left(\frac{\Delta(j)}{\theta(j)\theta(j+1)}\xi(j)\frac{1}{1-\theta(x)/\theta(j)}\right).$$

Taking into account the inequality $\theta(j) \ge \delta_1(x)$, (3.16), and (2.13) we see that $|A_{2,2}^+(\theta)|$ does not exceed

$$\text{const}\,\theta(x) \sum_{\theta(j)\ge\delta_1(x)} \left(\frac{1}{\theta(j+1)} - \frac{1}{\theta(j)}\right) = \text{const}\,\theta(x)\left(\frac{1}{\theta(j_*+1)} - \frac{1}{\theta(1)}\right),$$

where j_* is the maximal number for which $\theta(j_*) \ge \delta_1(x)$. Thus, from (3.16) and (2.13) it follows that $\lim_{\theta\to 0} A_{2,2}^+(\theta) = 0$. Note that, for each $\theta > 0$, $A_{2,1}^+(\theta)$ is a partial sum of the series

$$S = 2\sum_{j=1}^{\infty}\left(\arctan\alpha(j) - \arctan\frac{\Delta(j)}{2}\cot\frac{\theta(j)}{2}\right),$$

the convergence of which can be verified with the help of the inequality

$$|\cot u - (1/u)| \le \text{const}\,|u|, \quad |u| \le u_0 < \pi/2. \tag{3.19}$$

Using (3.18) and (3.19) we obtain

$$S \le \text{const}\sum_{j=1}^{\infty}\arctan\left(\frac{\Delta(j)}{2}\left(\frac{2}{\theta(j)} - \cot\frac{\theta(j)}{2}\right)\right) \le \text{const}\sum_{j=1}^{\infty}\Delta(j)\theta(j) < \infty.$$

Hence

$$\lim_{\theta\to 0+0} A_{2,1}^+(\theta) = a_2^+ = \lim_{\theta\to 0+0} A_2^+(\theta), \quad a_2^+ \in \mathbf{R}.$$

In the same way we can show that an analogous equality holds for the function $A_2^-(\theta)$. This completes the proof of (3.17).

Second step. We show that for the other summands (that is, for those given by $|\theta(j)| < \delta_1(x)$) the functions $\varphi_j(\theta(x))$ in (3.12) may be replaced by functions of the kind

$$\varphi_j^0(\theta(x)) = \arctan\left(\frac{\Delta(j)}{\theta(x)-\theta(j)}\right)$$

up to an infinitesimal as $x \to \infty$. Indeed, let

$$O_1(x) \stackrel{Df}{=} \sum_{\delta_1(x)>\theta(j)>\theta(n)} (\varphi_j(\theta(x)) - \varphi_j^0(\theta(x))).$$

Then using (3.18)–(3.19) we obtain

$$|O_1(x)| \leq \left| \sum_{\delta_1(x)>\theta(j)>\theta(n)} \arctan\left(\frac{\Delta(j)}{2}\left(\cot\frac{\theta(x)-\theta(j)}{2} - \frac{2}{\theta(x)-\theta(j)}\right)\right)\right| \leq$$

$$\leq \text{const} \sum_{\delta_1(x)>\theta(j)>\theta(n)} \Delta(j)(\theta(j)-\theta(x)) \leq \text{const}\, \delta_1(x) \sum_{\delta_1(x)>\theta(j)>\theta(n)} \Delta(j),$$

which is at most a constant times $\delta_1(x)$. Consequently, $\lim_{x\to\infty} O_1(x) = 0$. In an analogous way one can show that condition (3.15) is realized for the functions $O_2(x)$ and $O_3(x)$ given by

$$\sum_{j=n+1}^{\infty} (\varphi_j(\theta(x)) - \varphi_j^0(\theta(x))) \quad \text{and} \quad \sum_{0>\theta(j)>-\delta_1(x)} (\varphi_j(\theta(x)) - \varphi_j^0(\theta(x))),$$

respectively, and for

$$O_4(x) \stackrel{Df}{=} \widetilde{\varphi}_n(\theta(x)) - \varphi_n^0(\theta(x)),$$

where

$$\widetilde{\varphi}_n(\theta(x)) = \begin{cases} \varphi_n^0(\theta(x)), & s \leq 0, \\ \pi + \varphi_n^0(\theta(x)), & s > 0. \end{cases} \tag{3.20}$$

Thus, from equality (3.12) we can pass to the following:

$$f_B(x) = -2\pi(n-1) - 2 \sum_{\delta_1(x)>\theta(j)\geq\theta(n-1)} \varphi_j^0(\theta(x)) - 2\widetilde{\varphi}_n^0(\theta(x)) -$$

$$-2\sum_{j=n+1}^{\infty} \varphi_j^0(\theta(x)) - 2 \sum_{0>\theta(j)>-\delta_1(x)} \varphi_j^0(\theta(x)) + a_1 + O_5(x), \tag{3.21}$$

where a_1 is taken from (3.17) and the function $O_5(x)$ satisfies condition (3.15).

Third step. We show that the group of central summands whose numbers j are "not very far from n" gives a function $\xi(s)$ that is continuous and monotonously decreasing on $[-1/2, 1/2]$ and satisfies the equality

$$\xi(-1/2) - \xi(1/2) = 2\pi. \tag{3.22}$$

To do this, we use condition (2.8) and condition (3.2) ($d = 1$). We have

$$\widetilde{\varphi}_n^0(\theta(x)) = \left\{ \begin{array}{ll} -\arctan(1/\psi(s)), & s \leq 0 \\ \pi - \arctan(1/\psi(s)), & s > 0 \end{array} \right\} + O_n(x); \tag{3.23}$$

$$\varphi^0_{n+j}(\theta(x)) = \arctan\left(\frac{1}{-\psi_n(s)\cdot\frac{\Delta(n)}{\Delta(n+j)}+\frac{\Delta(n+1)}{\Delta(n+j)}+\cdots+\frac{\Delta(n+j-1)}{\Delta(n+j)}}\right) =$$

$$= \arctan\left(\frac{1}{j-\psi(s)}\right) + O_{n+j}(x).$$

Analogously,

$$\varphi^0_{n-j}(\theta(x)) = -\arctan\left(\frac{1}{j+\psi(s)}\right) + O_{n-j}(x),$$

where the continuous functions $O_n(x)$, $O_{n\pm j}(x)$, $j = 1, 2, \ldots$, satisfy condition (3.15). From this it follows that there exists a function $l = l(n)$ with values in the natural numbers which is unbounded and monotonously increasing (may be very slow) as $n \to \infty$ and for which

$$O_6(x) = \sum_{j=1}^{l(n)}\left(\varphi^0_{n+j}(\theta(x)) - \arctan\left(\frac{1}{j-\psi(s)}\right)\right) + \tag{3.24}$$

$$+\sum_{j=1}^{l(n)}\left(\varphi^0_{n-j}(\theta(x)) + \arctan\left(\frac{1}{j+\psi(s)}\right)\right),$$

also satisfies condition (3.15).

Consider now the function $\xi(s)$ given by

$$\xi(s) = 2\sum_{j=1}^{\infty}\left(\arctan\frac{1}{j-\psi(s)} - \arctan\frac{1}{j+\psi(s)}\right) - 2\xi_0(s), \tag{3.25}$$

where $\xi_0(s)$ is the function in the braces of (3.23). As the series in (3.25) converges uniformly for $s \in [-1/2, 1/2]$, we obtain from (3.24) the following modification of (3.21):

$$f_B(x) = -2\pi(n-1) + \xi(s) - 2\sum_{\delta_1(x)>\theta(j)>\theta(n-l(n))}\varphi^0_j(\theta(x)) - \tag{3.26}$$

$$-2\sum_{j=n+l(n)+1}^{\infty}\varphi^0_j(\theta_j(x)) - \sum_{0>\theta_j>-\delta_1(x)}\varphi^0_j(\theta(x)) + a_1 + O_7(x),$$

where $O_7(x)$ satisfies (3.15).

It should be noted that the function $\xi_0(s)$ may be written in the form $\xi_0(s) = \frac{\pi}{2} + \arctan\psi(s)$, which is easily seen from the elementary identity

$$\arctan u + \arctan(1/u) = \pi/2, \quad u > 0. \tag{3.27}$$

Thus, since $|\psi(s)| < 1$, Proposition 3.3 implies that the function (3.25) may be represented in the form

$$\xi(s) = -2\Xi(\psi(s)) - \pi, \quad s \in [-1/2, 1/2]. \tag{3.28}$$

Obviously, $\xi(s)$ is continuous and monotonously decreasing on $[-1/2, 1/2]$, and this function satisfies (3.22).

Fourth step. We show that sum of all summands of (3.26) (under the sign $\sum$) we have not considered so far tends to a finite limit as $x \to \infty$. First of all, with the help of condition (2.15), we show that this is true for the sums over $m(n) \le j < n - l(n)$ and $n + l(n) < j \le M(n)$, where $m(n)$ and $M(n)$ are defined by (2.12). The following inequalities are obvious :

$$2 \sum_{j=m(n)}^{n-l(n)-1} \arctan \frac{\Delta(j)}{\theta(j) - \theta(n+1)} < -2 \sum_{j=m(n)}^{n-l(n)-1} \varphi_j^0(\theta(x)) <$$

$$< 2 \sum_{j=m(n)}^{n-l(n)-1} \arctan \frac{\Delta(j)}{\theta(j) - \theta(n-1)};$$

$$-2 \sum_{j=n+l(n)+1}^{M(n)} \arctan \frac{\Delta(j)}{\theta(j) - \theta(n-1)} < 2 \sum_{j=n+l(n)+1}^{M(n)} (\varphi_j^0 \theta(x)) <$$

$$< -2 \sum_{j=n+l(n)+1}^{M(n)} \arctan \frac{\Delta(j)}{\theta(j) - \theta(n+1)}.$$

Substraction of the second inequality from the first gives the estimate

$$C_1(n+1) \le -2 \left(\sum_{j=m(n)}^{n-l(n)-1} \varphi_j^0(\theta(x)) + \sum_{j=n+l(n)+1}^{M(n)} \varphi_j^0(\theta(n)) \right) < C_1(n-1),$$

where

$$C_1(n \pm 1) \overset{Df}{=} 2C(n \pm 1) - 2 \sum_{j=1}^{l(n)} \left(\arctan \left(\frac{\Delta(n \pm 1 - j)}{\theta(n \pm 1 - j) - \theta(n \pm 1)} \right) + \right.$$

$$\left. + \arctan \left(\frac{\Delta(n \pm 1 + j)}{\theta(n \pm 1 + j) - \theta(n \pm 1)} \right) \right), \tag{3.29}$$

and $C(n)$ is the function defined by equality (2.12). Due to condition (2.15), $\lim_{n \to +\infty} C(n+1) = \lim_{n \to +\infty} C(n-1) = 0$. In the third step we showed that the sums on the right of (3.29) tend to the value $(\xi(s)+2\xi_0(s))\big|_{s=0}=0$ as $n\to\infty$ (see (3.25), (3.28)). Therefore the group of summands

$$-2\left(\sum_{j=m(n)}^{n-l(n)-1}\varphi_j^0(\theta(x))+\sum_{j=n+l(n)+1}\varphi_j(\theta(x))\right)\overset{Df}{=}O_8(x) \tag{3.30}$$

satisfies conditions (3.15).

To estimate the other groups of summands in (3.26) we show that the functions $\varphi_j^0(\theta(x))$ may be replaced up to an infinitesimal by the functions

$$\varphi_j^1(\theta(x))=\sigma\log\left(1+\sigma\frac{\Delta(j)}{\theta(x)-\theta(j)}\right)=\sigma\log\left(\frac{\theta(x)-\theta(j+\sigma)}{\theta(x)-\theta(j)}\right),$$

where $\sigma=\operatorname{sign} j$. Indeed, let

$$O_9(x)\overset{Df}{=}\sum_{\delta_1(x)>\theta(j)>\frac{3}{2}\theta(x)}(\varphi_j^0(\theta(x))-\varphi_j^1(\theta(x))).$$

Taking into account the inequality

$$|\arctan u-\log(1+u)|<\text{const}\cdot u^2,\quad u\geq u_0>-1, \tag{3.31}$$

we get

$$|O_9(x)|\leq\text{const}\sum_{\delta_1(x)>\theta(j)>\frac{3}{2}\theta(x)}\left(\frac{\Delta(j)}{\theta(x)-\theta(j)}\right)^2=$$

$$=\sum_{\delta_1(x)>\theta(j)>\frac{3}{2}\theta(x)}\alpha^2(j)/\left(1-\frac{\theta(x)}{\theta(j)}\right)^2\leq 9\sum_{\delta_1(x)>\theta(j)>\frac{3}{2}\theta(x)}\alpha_j^2.$$

Put

$$\delta_2(x)\overset{Df}{=}\sup_{\delta_1(x)>\theta(j)>\frac{3}{2}\theta(x)}\alpha(j)$$

and use inequality (3.1). It results that

$$|O_9(x)|\leq 9\delta_2(x)\log\frac{\delta_1(x)}{\theta(x)}.$$

Note that $\lim_{x\to\infty}\delta_2(x)=0$ and that, moreover, a smaller choice of $\delta_1(x)$ has as a consequence that $\delta_2(x)$ does not become larger. As $\delta_1(x)$ may be an arbitrary function tending to zero as $x\to\infty$ and satisfying condition (3.16), we can choose it so that

$$\lim_{x\to\infty}\delta_2(x)\log\frac{\delta_1(x)}{\theta(x)}=0. \tag{3.32}$$

This choice may be realized in the following way. Suppose that there is a $\delta_1(x)$ satisfying (3.16) but that relation (3.32) is not true. Then choose a new function

$$\widetilde{\delta}_1(x) \stackrel{Df}{=} \theta(x)/\delta_2(x).$$

We claim that

$$\widetilde{\delta}_1(x) < \delta_1(x). \tag{3.33}$$

Indeed, if we suppose the contrary then

$$\delta_2(x)\log\frac{1}{\delta_2(x)} > \delta_2(x)\log\frac{\delta_1(x)}{\theta(x)}$$

and hence (3.32) holds. Thus, for $\widetilde{\delta}_1(x)$ we have

$$\widetilde{\delta}_2(x)\log\frac{\delta_1(x)}{\theta(x)} \le \delta_2(x)\log\frac{1}{\delta_2(x)}$$

and condition (3.32) is realized. Hence, the function $O_9(x)$ satisfies (3.15). It follows that

$$\sum_{\delta_1(x)>\theta(j)>\frac{3}{2}\theta(x)} \varphi_j^0(\theta(x)) = \sum_{\delta_1>\theta(j)>\frac{3}{2}\theta(x)} \varphi_j^1(\theta(x)) + 2O_9(x) = \tag{3.34}$$

$$= \log\prod_{\delta_1>\theta(j)>\frac{3}{2}\theta(x)} \frac{\theta(x)-\theta(j+1)}{\theta(x)-\theta(j)} = \log\left(\frac{\theta(x)-\theta(j_2)}{\theta(x)-\theta(j_1)}\right) + 2O_9(x),$$

where

$$\theta(j_1) < \delta_1(x) \le \theta(j_1-1), \quad \theta(j_2) < \frac{3}{2}\theta(x) < \theta(j_2-1).$$

We now make the change of $\varphi_j^0(\theta(x))$ to $\varphi_j^1(\theta(x))$ for the other group of summands. We first estimate the function

$$O_{10}(x) \stackrel{Df}{=} \sum_{\frac{\theta(x)}{2}>\theta(j)>0} (\varphi_j^0(\theta(x)) - \varphi_j^1(\theta(x))).$$

Using inequality (3.31) we obtain

$$|O_{10}(x)| \le \text{const} \sum_{\frac{\theta(x)}{2}>\theta(j)>0} \left(\frac{\Delta(j)}{\theta(x)-\theta(j)}\right)^2 \le \frac{\text{const}}{\theta^2(x)} \sum_{\frac{\theta(x)}{2}>\theta(j)>0} \Delta^2(j) \le$$

$$\le \frac{\text{const}}{\theta^2(x)} \left(\sup_{\frac{\theta(x)}{2}>\theta(j)>0} \Delta(j)\right)\left(\sum_{\frac{\theta(x)}{2}>\theta(j)>0} \Delta(j)\right) \le \text{const}\,\frac{\theta^2(j_3)}{\theta^2(x)}\delta_3(x),$$

where $\delta_3(x) = \sup\limits_{\frac{\theta(x)}{2}>\theta(j)>0} \alpha_j$, and $\theta(j_3-1) \geq \frac{\theta(x)}{2} > \theta(j_3)$. As $(\theta(j_3)/\theta(x))^2 \leq \frac{1}{4}$ and $\lim\limits_{x\to\infty} \delta_3(x) = 0$, the function $O_{10}(x)$ satisfies condition (3.15). Thus,

$$\sum_{\frac{\theta(x)}{2}>\theta(j)>0} \varphi_j^0(\theta(x)) = \sum_{\frac{\theta(x)}{2}>\theta(j)>0} \varphi_j^1(\theta(x)) + O_{10}(x) = \tag{3.35}$$

$$\log \prod_{\frac{\theta(x)}{2}>\theta(j)>0} \frac{\theta(x)-\theta(j+1)}{\theta(x)-\theta(j)} + O_{10}(x) = \log\left(\frac{\theta(x)}{\theta(x)-\theta(j_3)}\right) + O_{10}(x).$$

We finally pass to the last group of summands. The function

$$O_{11} \stackrel{Df}{=} \sum_{0>\theta(j)>-\delta_1(x)} (\varphi_j^0(\theta(x)) + \varphi_j^1(\theta(x)))$$

can be estimated with the help of inequality (3.31):

$$|O_{11}(x)| \leq \text{const} \sum_{0>\theta(j)>-\delta_1(x)} \left(\frac{\Delta(j)}{\theta(x)-\theta(j)}\right)^2 \leq$$

$$\leq \text{const}\left(\sum_{-\theta(x)>\theta(j)\geq-\delta_1(x)} \left(\frac{\Delta(j)}{\theta(j)}\right)^2 + \sum_{0>\theta(j)>-\theta(x)} \left(\frac{\Delta(j)}{\theta(j)}\right)^2\right).$$

The resulting sums both tend to zero (see the proofs for the functions $O_9(x)$ and $O_{10}(x)$, respectively). Thus, $O_{11}(x)$ satisfies condition (3.15) and we have

$$\sum_{0>\theta(j)>-\delta_1(x)} \varphi_j^0(\theta(x)) = -\sum_{0>\theta(j)>-\delta_1(x)} \varphi_j^1(\theta(x)) + O_{11}(x) = \tag{3.36}$$

$$-\log \prod_{0>\theta(j)>-\delta_1(x)} \frac{\theta(x)-\theta(j-1)}{\theta(x)-\theta(j)} + O_{11}(x) = \log\frac{\theta(x)-\theta(j_4)}{\theta(x)} + O_{11}(x),$$

where $\theta(j_4) > -\delta_1(x) > \theta(j_4-1)$.

Now that equalities (3.30), (3.34), (3.35), and (3.36) are at our disposal, we can pass from representation (3.26) to the representation

$$f_B(x) = -2\pi(n-1) + \xi(s) - 2\log\left(\frac{\theta(j_2)-\theta(x)}{\theta(x)-\theta(j_3)} \cdot \frac{\theta(j_4)-\theta(x)}{\theta(x)-\theta(j_1)}\right) + a_1 + O_{12}(x), \tag{3.37}$$

where $\xi(s)$ is given by formula (3.28) and the constant a_1 is from (3.17). In addition, notice that, due to the reasoning of the first step of the proof, $a_1 = a_2 - 2a$, where a has the form (2.14) and

$$a_2 = 2\sum_{j=1}^{\infty}[\arctan(\alpha(j)) - \arctan(\Delta(j))/(2)\cot(\theta(j)/2))] -$$

$$-2\sum_{j=-\infty}^{-1}[\arctan(\alpha(j))+\arctan(\Delta(j)/2)\cot(\theta(j)/2))].$$

Taking into account the definition of the numbers j_2, j_3 and making use of relation (2.13) we obtain

$$\lim_{x\to\infty}\frac{\theta(j_2)}{\frac{3}{2}\theta(x)}=1 \quad\text{and}\quad \lim_{x\to\infty}\frac{\theta(j_3)}{\frac{1}{2}\theta(x)}=1.$$

Analogously,

$$\lim_{x\to\infty}\frac{\theta(j_1)}{\delta_1(x)}=1 \quad\text{and}\quad \lim_{x\to\infty}\frac{\theta(j_4)}{\delta_1(x)}=-1.$$

Hence, the logarithmic summands in (3.37) tend to zero as $x\to\infty$. Now recall that $x=n+s$ and $s=x-[x+1/2]$. Thus, (3.37) takes the form

$$f_B(x)=-2\pi x+p(x-[x+1/2])+\pi+a_1+O_{13}(x), \tag{3.38}$$

where $p(s)=\xi(s)+2\pi s+\pi=-2\Xi(s)+2\pi s$, $s\in[-1/2,1/2]$, and $O_{13}(x)$ satisfies condition (3.15). Note that, by (3.22), $p(-1/2)=p(1/2)$. Hence, the function $p(x-[x+1/2])$ is continuous and 1-periodic.

Representation (3.38) is in essence the same as (3.13), however we have proved it for $x>0$ only. In order to show it for $x<0$, let us consider the Blaschke product $B_1(t)=\overline{B(t^{-1})}$. Obviously,

$$\arg B_1(\exp i\theta(x))=-\arg B(\exp(-i\theta(x))), \tag{3.39}$$

and the zeros of the function $B_1(t)$ are $z_j^1=\overline{z}_j=r_j\exp(-i\theta(j))$. Due to (3.38), we have the following representation for $x>0$:

$$f_{B_1}(x)=-2\pi x+p_1((x-[x+1/2]))+\pi-a_1+O_{14}(x), \tag{3.40}$$

where $p_1(s)=-2\Xi(\psi(-s))+2\pi s$.

Indeed, by virtue of formula (3.8),

$$\psi_1(x)=\lim_{n\to\infty}\frac{\theta_1(n)-\theta_1(n+s)}{\Delta_1(n)}=\lim_{n\to\infty}\frac{-\theta(-n)+\theta(-n-s)}{-\theta(-n)+\theta(-n-1)}=$$

$$=-\lim_{j\to-\infty}\frac{\theta(j)-\theta(j-s)}{\Delta(j)}=\psi(-s),$$

which gives the explanation for the replacement of the function $p(s)$ in (3.38) by $p_1(s)$ in (3.40). Since $\theta_1(j)=-\theta(-j)$, we can replace the constant a_1 by $-a_1$. Now, as $-x-[-x+1/2]=-(x-[x+1/2])$, we infer from (3.39) that (3.38) holds for all $x\in\mathbf{R}$ (if $x<0$, then in (3.40) we encounter $-\pi$ instead of π, which, however, is inessential because the difference of these two values is 2π). Finally note that without loss of generality we can omit the constants in (3.38). Thus, we obtain a representation of the kind (3.13),

$$f_B(x) = -2\pi x + p(x - [x + 1/2]) + O(x). \tag{3.41}$$

Last step. Consider the function $u(x) = -2\pi x + p(x - [x+1/2])$ and note that this function monotonously decreases on the real line. As the function $p(x - [x + 1/2])$ is 1-periodic, the inverse function takes the form

$$u^{-1}(y) = -y/(2\pi) + \beta(y), \tag{3.42}$$

where $\beta(y)$ is 2π-periodic on $\mathbf{R}$. Act by map u^{-1} on both sides of equality (3.41). It results that

$$u^{-1}(\arg B(\exp(i\theta(x)))) = x + d_1(x), \tag{3.43}$$

where the function

$$d_1(x) = u^{-1}(-2\pi x + p(x - [x + 1/2]) + O(x)) - u^{-1}(2\pi x + p(x - [x + 1/2]))$$

is continuous on $\mathbf{R}$ and (since $u^{-1}(y)$ is uniformly continuous) zero at infinity. Recall that $x = -f(\theta)$. Thus, from (3.43) we obtain

$$\exp(-2\pi i u^{-1}(\arg B(\exp(i\theta))) = \exp(2\pi i f(\theta)) \exp(-2\pi i d_1(-f(\theta))).$$

By (3.42), the last equality implies that

$$B(t) \exp(-2\pi i \beta(-i \log B(t))) = a_0(t)/d(t),$$

where $d(t) \stackrel{Df}{=} \exp(2\pi i d_1(-f(\theta)))$ is continuous on $\mathbf{T}$. The function $\beta(y)$ is 2π-periodic, and therefore the superposition $g(u) \stackrel{Df}{=} \exp(-2\pi i\beta(-i \log u))$ is continuous on $\mathbf{T}$. In addition, $\operatorname{wind} g(u) = \beta(2\pi) - \beta(0) = 0$ which gives representation (2.16). Thus, the proof of Theorem 2.1 is complete.

4. Modelling of oscillations by inner functions

This section is devoted to the proof of Theorem 2.2. Essentially, this theorem is a corollary of Theorem 2.1. Indeed, let us use representation (3.41) for the argument of a Blaschke product obtained in the proof of Theorem 2.1 (see also (3.38), (3.28) and (3.9)). Taking into account that $\psi(s) = s$, we get

$$f_B(x) = -2\pi n - 2 \arctan\left(\frac{\tan \pi s}{\tanh 1}\right) + O(x), \tag{4.1}$$

where $x = n + s$, $s \in (-1/2, 1/2]$, and the function $O(x)$ is continuous and vanishes at infinity. Recall that, due to the definitions $x = -f(\theta)$ and $f_B(x) = \arg B(\exp(i\theta(x)))$, we have $\theta(x) = f^{-1}(-x)$.

The continuous branch of the argument of $U(t)$ (see (2.19)) may be chosen so that

$$|\arg U(t) - \arg B(t)| < \pi. \tag{4.2}$$

Simple plane geometry gives

$$\arg \frac{r_0 + \exp(i\varphi)}{1 + r_0 \exp(i\varphi)} = -2\arctan\left(\frac{1 - r_0}{1 + r_0}\cot\left(\frac{\varphi - \pi}{2}\right)\right) = 2\arctan\left(\tanh 1 \tan\frac{\varphi}{2}\right),$$

where the branch of the argument is continuous on the interval $\varphi \in (-\pi, \pi)$.

Now taking into account (4.1) and (4.2) we can calculate the continuous branch of $\arg U(t)$ in the following way:

$$\arg U(\exp(i\theta(x))) = -2\pi n + 2\arctan\left(\tanh 1\left[\tan\left(-\arctan\frac{\tan \pi s}{\tanh 1}\right) + O(x)\right]\right) =$$

$$= -2\pi n - 2\pi s + \widetilde{O}(x) = -2\pi x + \widetilde{O}(x),$$

where $\lim_{x\to\infty} \widetilde{O}(x) = 0$. From this (2.18) follows. Indeed,

$$U(t) = U(\exp(i\theta)) = \exp(i2\pi f(\theta))\exp(i\widetilde{O}(x) = a_0(t)/c(t).$$

5. Theory of normal solvability

For the proof of Theorem 2.3 we need some auxiliary definitions and theorems.

A function $a(t) \in L_\infty(\mathbf{T})$ is said to admit a p-factorization on $\mathbf{T}$, $1 < p < \infty$, if there exists a representation

$$b(t) = b_-(t)t^\kappa b_+(t), \tag{5.1}$$

where κ is an integer (the so-called index of the factorization) and the factors $b_\pm(t)$ satisfy the following conditions:

i) $b_+(t) \in H_q(\mathbf{T})$, $b_+^{-1}(t) \in H_p(\mathbf{T})$, $b_-(t) \in \overline{H_p(\mathbf{T})}$, $b_-^{-1}(t) \in \overline{H_q(\mathbf{T})}$, where $1/p + 1/q = 1$;

ii) the operator $b_+^{-1}(t)Pb_-^{-1}(t)I$ is bounded in the space $H_p(\mathbf{T})$.

It is well known [1]–[4], [7] that the Toeplitz operator $T(b)$ is Fredholm in the space $H_p(\mathbf{T})$ if and only if its symbol $b(t)$ has a p-factorization.

A function $a(t)$ is said to admit an (h,p)-factorization, $1 < p < \infty$, if there exists a representation

$$a(t) = h^\varepsilon(t)b(t), \quad \varepsilon = \pm 1, \tag{5.2}$$

where $h(t) \in H_\infty(\mathbf{T})$, $(1/h(t)) \in L_\infty(\mathbf{T})$, and the function $b(t)$ has a p-factorization of the kind (5.1).

We remark that condition (5.2) always includes the requirement that, for any polynomial $p(t) = \sum_{j=-m}^{n} c_j t^j$, $c_j \in \mathbf{C}$,

$$(p(t)/h(t)) \notin H_\infty(\mathbf{T}). \tag{5.3}$$

Otherwise factorizations (5.1) and (5.2) were equivalent. Indeed if $(p(t)/h(t)) \in H_\infty(\mathbf{T})$ then $(1/h(t)) \in H_\infty(\mathbf{T})+C(\mathbf{T})$ and ([2]) Toeplitz operator T_h is Fredholm.

Theorem 5.1. [[22]–[25]] *Let the function $a(t)$ be in $L_\infty(\mathbf{T})$ and have a representation of the kind (5.2). Then if $\varepsilon = 1$ ($\varepsilon = -1$), the Toeplitz operator $T(a)$ is left (right) invertible in the space $H_p(\mathbf{T})$, $1 < p < \infty$. In addition,*

$$\dim(H_p(\mathbf{T})/\operatorname{im} T(a)) = \infty \quad (\dim\ker T(a) = \infty),$$

and one of the left (right) inverses takes the form

$$T_l^{-1}(a) = Pa_+^{-1}(t)h^{-1}(t)Pa_-^{-1}(t)I \quad (T_r^{-1}(a) = a_+^{-1}(t)h(t)Pa_-^{-1}(t)I).$$

For the proof of the existence of an (h, p)-factorization of the symbol (2.20) we need some facts from the theory of u-factorization [23]–[24].

A function $a(t) \in L_\infty(\mathbf{T})$ is said to be u-periodic if there exists a representation

$$a(t) = g(u(t)) \tag{5.4}$$

where $g(t) \in L_\infty(\mathbf{T})$ and $u(t)$ is some inner function.

Theorem 5.2. *Let the function $g(t)$ be in $C(\mathbf{T})$ and suppose $g(t) \neq 0$ for $t \in \mathbf{T}$ and $\operatorname{wind} g(t) = \kappa$. Then, for every $1 < p < \infty$, the u-periodic function (5.4) possesses a $(u^{|\kappa|}, p)$-factorization*

$$a(t) = u^\kappa(t) g_-(u(t)) g_+(u(t)),$$

where $g(t) = g_-(t)t^\kappa g_+(t)$ is a p-factorization of the kind (5.1). Moreover, if $g(t)$ is a rational function then

$$g_+^{\pm 1}(u(t)) \in H_\infty(\mathbf{T}), \quad g_-^{\pm}(t) \in \overline{H_\infty(\mathbf{T})}. \tag{5.5}$$

Now we turn to the proof of Theorem 2.3.

Due to Theorem 2.1, a function $a(t)$ of the form (2.20) has a representation

$$a(t) = |a(t)|(B(t)g(B(t))d(t))^\varepsilon, \quad \varepsilon = \pm 1, \tag{5.6}$$

where $B(t)$ is a Blaschke product, $g(t), d(t) \in C(\mathbf{T})$, and $\operatorname{wind} g = 0$.

The positive function $|a(t)|$ satisfies the condition $\operatorname{ess\,inf}_{t\in\mathbf{T}}|a(t)| > 0$. Hence ([1]–[4]) it admits a factorization

$$|a(t)| = h_+(t)h_-(t), \tag{5.7}$$

where

$$h_{+}^{\pm 1}(t) \in H_{\infty}(\mathbf{T}), \quad h_{-}^{\pm 1}(t) \in \overline{H_{\infty}(\mathbf{T})}. \tag{5.8}$$

Further, let $g_\delta(t)$, $d_\delta(t)$ be rational functions such that

$$\sup_{t\in\mathbf{T}} |g(t) - g_\delta(t)| < \delta \quad \text{and} \quad \sup_{t\in\mathbf{T}} |d(t) - d_\delta(t)| < \delta,$$

where $\delta > 0$ is small enough. From (5.6) we get the representation

$$a(t) = B^{\varepsilon}(t)(h_{+}(t)d^{\varepsilon}_{\delta+}(t)g^{\varepsilon}_{\delta+}(B(t)))(h_{-}(t)d^{\varepsilon}_{\delta-}(t)g^{\varepsilon}_{\delta-}(B(t)))t^{\kappa_d}(1+m(t)). \tag{5.9}$$

Here $d_\delta(t) = d_{\delta+}(t)t^{\kappa_d}d_{\delta-}(t)$ and $g_\delta(t) = g_{\delta+}(t)g_{\delta-}(t)$ are factorizations of rational functions with factors satisfying condition (5.8), and $\sup_{t\in\mathbf{T}\setminus\{1\}} |m(t)| < \delta_1$, where δ_1 can be made arbitrarily small by suitably choosing δ. It is well known [1]–[4] that, for sufficiently small δ_1, the function $(1 + m(t))$ has a p-factorization with zero factorization index . Thus, taking into account (5.5) and (5.8) it is easily seen that (5.9) is a (B,p)-factorization. Hence Theorem 2.3 follows immediately from Theorem 5.1.

6. Proof of Theorem 2.4

We begin with the proof of the implication

$$(2.23) \Rightarrow \psi(s) = s. \tag{6.1}$$

The derivatives of functions $f(\theta)$ and $\theta(x)$ of the kind (2.6) are connected by the formulas

$$f'(\theta) = -\frac{1}{\theta'(x)}, \quad f''(\theta) = -\frac{\theta''(x)}{(\theta'(x))^3}. \tag{6.2}$$

Let first $\theta > 0$ ($x = n + s > 0$). Applying three times Lagrange's theorem we obtain

$$\psi_n(s) = \frac{\theta(n) - \theta(n+s)}{\theta(n) - \theta(n+1)} = \frac{s\theta'(n_1)}{\theta'(n_2)} =$$

$$= s\left(1 + \frac{\theta'(n_1) - \theta'(n_2)}{\theta'(n_2)}\right) = s\left(1 + \frac{\theta''(n_3)}{\theta'(n_2)}(n_1 - n_2)\right),$$

where $n_1 \in [n, n+s]$, $n_2 \in [n, n+1]$, $n_3 \in [\min(n_1,n_2), \max(n_1,n_2)]$.

If $n_1 \geq n_2$ then $n_3 \geq n_2$, and since the function $|\theta'(x)|$ is monotonously decreasing, we obtain from (6.2) that

$$\left|\frac{\theta''(n_3)}{\theta'(n_2)}\right| < \left|\frac{\theta''(n_3)}{\theta'(n_3)}\right| = \left|\frac{f''(\theta_3)}{(f'(\theta_3))^2}\right|,$$

where $\theta_3 = \theta(n_3)$.

If $n_1 \leq n_2$ then $n_1 \leq n_3$, and slightly transforming the representation of the function $\psi_n(s)$ we get

$$\psi_n(s) = s\left(\frac{\theta'(n_2)}{\theta'(n_1)}\right)^{-1} = s\left(1 + \frac{\theta''(n_3)}{\theta'(n_1)}(n_2 - n_1)\right)^{-1}.$$

Since (as earlier)

$$\left|\frac{\theta''(n_3)}{\theta'(n_1)}\right| \leq \left|\frac{\theta''(n_3)}{\theta'(n_3)}\right| = \left|\frac{f''(\theta_3)}{(f'(\theta_3))^2}\right|$$

and $\lim_{\theta\to 0} \theta_3 = 0$, we infer from (2.23) that $\lim_{n\to\infty} \psi_n(s) = \psi(s) = s$.

Analogously the case $\theta < 0$ can be settled.

We now pass to the proof of the implication (2.24) $\Rightarrow$ (2.15).

First of all note that (2.23) implies

$$\lim_{\theta\to 0} 1/(\theta f'(\theta)) = 0. \tag{6.3}$$

Further, due to (2.12), the function $C(n)$ has the form

$$C(n) = \sum_{j=m(n)}^{n-1} \arctan\frac{\Delta(j)}{\theta(j) - \theta(n)} - \sum_{j=n+1}^{M(n)} \arctan\frac{\Delta(j)}{\theta(n) - \theta(j)}$$

($n > 0$). Now recall that a function $O_6(n)$ of the kind (3.24) tends to zero as $n \to \infty$. If $x = n$ then $s = 0$ and $\psi(0) = 0$, whence

$$O_6(n) = \sum_{j=1}^{l(n)} \left(\arctan\left(\frac{\Delta(n+j)}{\theta(n-j) - \theta(n)}\right) - \arctan\left(\frac{\Delta(n+j)}{\theta(n) - \theta(n+j)}\right)\right),$$

where, recall, $\lim_{n\to\infty} l(n) = \infty$. Therefore instead of $C(n)$ we can consider the truncated expression

$$C_1(n) \stackrel{Df}{=} \sum_{j=m(n)}^{n-l(n)-1} \arctan\frac{\Delta(j)}{\theta(j) - \theta(n)} - \sum_{j=n+l(n)+1}^{M(n)} \arctan\frac{\Delta(j)}{\theta(n) - \theta(j)}.$$

Just as in the proof of Theorem 2.1, we represent $C_1(n)$ in the following form ("separation of logarithms"):

$$C_1(n) = \sum_{j=m(n)}^{n-l(n)-1} \left[\arctan\frac{\theta(j) - \theta(j+1)}{\theta(j) - \theta(n)} + \log\left(1 - \frac{\theta(j) - \theta(j+1)}{\theta(j) - \theta(n)}\right)\right] -$$

$$- \sum_{j=n+l(n)+1}^{M(n)} \left[\arctan\frac{\theta(j) - \theta(j+1)}{\theta(n) - \theta(j)} - \log\left(1 + \frac{\theta(j) - \theta(j+1)}{\theta(n) - \theta(j)}\right)\right] -$$

$$(6.4) \qquad -\log \frac{\theta(n)-\theta(M(n)+1)}{\theta(m(n))-\theta(n)} \cdot \frac{\theta(n-l(n))-\theta(n)}{\theta(n)-\theta(n+l(n)+1)}.$$

Because of the choice of the numbers $m(n)$ and $M(n)$,

$$\lim_{n\to\infty} \frac{\theta(n)-\theta(M(n)+1)}{\theta(m(n))-\theta(n)} = 1.$$

Further, the choice of $l(n)$ (see (3.24)) guarantees that

$$\theta(n)-\theta(n+l(n)+1) = (l(n)+1)\Delta(n)+o(1) \text{ and } \theta(n-l(n))-\theta(n) = l(n)\Delta(n)+o(1).$$

Hence

$$\lim_{n\to\infty} \frac{\theta(n-l(n))-\theta(n)}{\theta(n)-\theta(n+l(n)+1)} = 1.$$

Therefore the last logarithmic summand in equality (6.4) tends to zero, and taking into account that

$$|\arctan u \mp \log(1\pm u)| \le \text{const} \cdot u^2, \quad u \in (u_0, 1), \quad u_0 > 0,$$

it is sufficient to show sums of the kind

$$C_2(n) = \sum_{j=m(n)}^{n-l(n)-1} \left(\frac{\Delta(j)}{\theta(j)-\theta(n)}\right)^2 \quad \text{and} \quad C_3(n) = \sum_{j=n+l(n)+1}^{M(n)} \left(\frac{\Delta(j)}{\theta(n)-\theta(j)}\right)^2$$

also approach zero. We estimate $C_3(n)$ using Lagrange's theorem and the fact that the function $|\theta'(x)|$ tends to zero monotonously as $x \to \infty$:

$$|C_3(\theta)| = \sum_{j=n+l(n)+1}^{M(n)} \left(\frac{\theta'(j_1)}{\theta'(j_2)}\right)^2 \frac{1}{(n-j)^2} \le \sum_{j=n+l(n)+1}^{M(n)} \frac{1}{(j-n-1)(j-n)} =$$

$$= \sum_{j=n+l(n)+1}^{M(n)} \left(\frac{1}{j-n-1} - \frac{1}{j-n}\right) = \frac{1}{l(n)} - \frac{1}{M(n)-n},$$

where $j_1 \in [j, j+1]$, $j_2 \in [n, j]$.

Since the functions $l(n)$ and $M(n)-n$ tend to infinity as $n \to \infty$, we see that $\lim_{n\to\infty} C_3(n) = 0$.

We now proceed to the estimation of the function $C_2(n)$. Extract from it the sum $C_{2,1}(n)$ of all summands for which $\Delta(j) \le g\Delta(n)$, where $g > 1$ is any fixed number. Then (note that $m_1(n)$ is the smallest number for which $\Delta(j) \le g\Delta(n)$)

$$C_{2,1}(n) \overset{Df}{=} \sum_{j=m_1(n)}^{n-l(n)-1} \frac{\Delta^2(j)}{(\theta(j)-\theta(n))^2} \le$$

$$\le g\Delta(n)\sum_{j=m_1(n)}^{n-l(n)-1}\left(\frac{1}{\theta(j+1)-\theta(n)}-\frac{1}{\theta(j)-\theta(n)}\right)=$$

$$=g\left(\frac{\theta(n)-\theta(n+1)}{\theta(n-l(n))-\theta(n)}-\frac{\theta(n)-\theta(n+1)}{\theta(m_1(n))-\theta(n)}\right).$$

Since the functions $l(n)$ and $n-m_1(n)$ go to infinity as $n\to\infty$ ($m(n)\le m_1(n)\le n-l(n)-1$), we obtain $\lim_{n\to\infty} C_{2,1}(n)=0$. Thus, it remains to consider the sum

$$C_{2,2}(n)\overset{Df}{=}\sum_{j=m(n)}^{m_1(n)-1}\left(\frac{\Delta(j)}{\theta(j)-\theta(n)}\right)^2.$$

We can estimate this sum by the integral

$$I(n)=\int_{\frac{3}{2}\theta_n\le\theta(s)\le\theta_n+\delta}\left(\frac{\theta'(u)}{\theta(u)-\theta(n)}\right)^2 du,$$

where $\theta_n \overset{Df}{=} \theta(n)$ and $\delta>0$ is chosen so that $\theta'(u_g)=g\theta'(n)$ whenever $\theta(u_g)=\theta_n+\delta$. Indeed, since $|\theta'(j+1)|\le\Delta(j)\le|\theta'(j)|$ and $\lim_{j\to\infty}\theta'(j+1)/\theta'(j)=1$ we have

$$\left(\frac{\Delta(j)}{\theta(j)-\theta(n)}\right)^2\le \text{const}\int_j^{j+1}\left(\frac{\theta'(u)}{\theta(u)-\theta(n)}\right)^2 du.$$

The change of variables $s=\theta(u), u=-f(s)$ in the integral $I(n)$ gives

$$I(n)=\int_{\theta_n+\delta}^{\frac{3}{2}\theta_n}\frac{1}{(s-\theta_n)^2}\cdot\frac{ds}{f'(s)}. \tag{6.5}$$

We now calculate δ. Due to Lagrange's theorem,

$$f'(\theta_n+\delta)-f'(\theta_n)=\delta f''(\theta_n^*),\quad \theta_n^*\in\left(\theta,\frac{3}{2}\theta\right).$$

Taking into account that $f'(\theta_n+\delta)=f'(\theta_n)/g$ we obtain $\delta=\gamma|f'(\theta_n)/f''(\theta_n^*)|$ and $\gamma=(g-1)/g$.

Now we estimate the maximum of the integrand. For this purpose, we denote the denominator in (6.5) by $F(s)$ and calculate its derivative,

$$F'(s)=(s-\theta_n)(2f'(s)+(s-\theta_n)f''(s)).$$

If $F'(s_0)=0$, then

$$s_0-\theta_n=-2f'(s_0)/f''(s_0). \tag{6.6}$$

Thus, the maximum of the integrand is attained at the point (6.6) or at the endpoints of the interval $(\theta_n+\delta,\frac{3}{2}\theta)$. If the maximum is at the point (6.6), then

$$1/F(s_0)=(f''(s_0))^2/(4(f'(s_0))^3), \tag{6.7}$$

if the integrand has its maximum at the left endpoint of the interval, then

$$(6.8)\qquad \frac{1}{F(\theta_n+\delta)}=\frac{1}{\delta^2 f'(\theta_n+\delta)}=\frac{g(f''(\theta^*))^2}{\gamma^2(f'(\theta_n))^3}\leq\frac{g}{\gamma^2}\cdot\frac{(f''(\theta^*))^2}{(f'(\theta^*))^3},$$

and in case the maximum is at the right endpoint of the interval we have

$$(6.9)\qquad 1/F((3/2)\theta_n)=4/(\theta_n^2 f'((3/2)\theta_n)).$$

Thus in the cases (6.7), (6.8) we get

$$I(n)\leq \text{const}\,\theta_n \sup_{s\in(\theta_n,\frac{3}{2}\theta_n)}\frac{(f''(s))^2}{(f'(s))^3}\leq \text{const}\sup_{s\in(\theta_n,\frac{3}{2}\theta_n)}\left(\frac{f''(s)s^{1/2}}{f^{3/2}(s)}\right)^2,$$

while in the case (6.9) we have

$$I(\theta)\leq \text{const}\,1/(((3/2)\theta_n)f'((3/2)\theta_n)).$$

Consequently, (2.24) and (6.3) show that $\lim_{\theta\to 0} I(\theta)=0$, which gives (2.15).

Finally, the implication (2.25) ⇒ (2.14) is a consequence of the following proposition.

Proposition 6.1. *Let $f(\theta)$ be a continuously differentiable function on the set $(-\pi,\pi)\setminus\{0\}$ and suppose the function $q(\theta)=(\theta f(\theta))^{-1}$ decreases monotonously to zero as $\theta\to 0$. Then (2.14) holds if only the following integral converges:*

$$(6.10)\qquad I_0=\int_0^{\pi}[f'(s)\arctan(sf'(s))^{-1}-f'(-s)\arctan(sf'(-s))^{-1}]ds.$$

Proof. For the sake of definiteness, let us consider the case $\theta>0$. Due to (2.11), the function $A(\theta)$ admits a representation $A(\theta)=A_+(\theta)-A_-(\theta)$ with

$$A_+(\theta)=\sum_{j=1}^{N_+}\arctan\frac{\theta(j)-\theta(j+1)}{\theta(j)},\quad A_-(\theta)=\sum_{j=1}^{N_-}\arctan\frac{\theta(-j)-\theta(-j-1)}{\theta(-j)}.$$

Here N_+ and N_- are the largest natural numbers for which $\theta(N_+)<\theta$ and $\theta(-N_-)>-\theta$, respectively. We show that sum $A_+(\theta)$ may be replaced (up to summands tending to a finite limit as $\theta\to 0$) by the integral

$$I_+(\theta)=\int_0^{x_+}\arctan(\theta'(x)/\theta(x))dx,\quad \theta(x_+)=\theta.$$

(recall that $\theta(x)=f^{-1}(-x)$). Indeed, using Lagrange's theorem and the mean value theorem for integrals we obtain

$$(6.11)\qquad I_+(\theta)-A_+(\theta)=\sum_{j=0}^{N_+-1}\left(\arctan\frac{\theta'(j_1)}{\theta(j_1)}-\arctan\frac{\theta'((j+1)_2)}{\theta(j+1)}\right)+$$

$$+\int_{N_+-1}^{x_+} \arctan(\theta'(x)/\theta(x))dx,$$

where $j_1 \in [j, j+1]$, $(j+1)_2 \in (j+1, j+2)$. Since the last summand in (6.11) goes to zero as $\theta \to 0$ (because $x_+ - (N_- - 1) < 2$ and $q(\theta)$ tends to zero), it is sufficient to consider the sum in (6.11). Due to the monotonicity of the functions $q(\theta)$ and $\theta'(x)$, the summands of that sum are nonnegative. Therefore it suffices to show that the sum is uniformly bounded for $\theta \in (0, \pi)$.

Once again using the monotonicity of the functions $\theta'(x)$ and $\psi(\theta)$ we get

$$\sum_{j=0}^{N_+-1} \left(\arctan\frac{\theta'(j_1)}{\theta(j_1)} - \arctan\frac{\theta'((j+1)_2)}{\theta(j+1)}\right) \leq$$

$$\leq \sum_{j=0}^{N_+-1} \left(\arctan\frac{\theta'(j_1)}{\theta(j_1)} - \arctan\frac{\theta'((j+1)_2)}{\theta((j+1)_2)}\right) \leq$$

$$\leq \sum_{j=0}^{N_+-1} \left(\arctan\frac{\theta'(j)}{\theta(j)} - \arctan\frac{\theta'(j+2)}{\theta(j+2)}\right) =$$

$$= \left(\arctan\frac{\theta'(0)}{\theta(0)} + \arctan\frac{\theta'(1)}{\theta(1)} - \arctan\frac{\theta'(N_+)}{\theta(N_+)} -\right.$$

$$\left. - \arctan\frac{\theta'(N_+ + 1)}{\theta(N_+ + 1)}\right) \leq \arctan\frac{\theta'(0)}{\theta(0)} + \arctan\frac{\theta'(1)}{\theta(1)}.$$

Thus, instead of the sum $A_+(\theta)$ we can consider the integral $I_+(\theta)$. The change of variables $\theta(x) = s$, $x = -f(s)$ gives

$$I_+(\theta) = -\int_\theta^\pi f'(s)\arctan(1/(sf'(s)))ds.$$

Analogously we can show that the sum $A_-(\theta)$ may be replaced by the integral

$$I_-(\theta) = -\int_\theta^\pi f'(-s)\arctan(1/(sf'(-s)))ds.$$

Hence, the function $A(\theta)$ can be substituted by the integral

$$-\int_\theta^\pi [f'(s)\arctan(1/(sf'(s))) - f'(-s)\arctan(1/(sf'(-s)))]ds.$$

Now it is obvious that $A(\theta)$ tends to a finite limit as $\theta \to 0$ if (and only if) the integral I_0 given by (6.10) converges. □

We now show the implication (2.25) $\Rightarrow$ (6.10). Consider the function $F(v) = v^{-1}\arctan v$, and put $v_+ = 1/(sf'(s))$ and $v_- = 1/(sf'(-s))$. We have

$$|I_0| \leq \int_0^\pi |F(v_+) - F(v_-)|(ds/s) = \int_0^\pi |v_+ - v_-|\,|F'(v^*)|(ds/s),$$

where $v^* \in [v_+, v_-]$, $F'(v) = (v - (1+v^2)\arctan v)/(v^2(1+v^2))$.

Obviously, $|F'(v)| \le \text{const} \cdot v$. Therefore (2.25) implies that

$$|I_0| \le \text{const} \int_0^\pi |v_+ - v_-| v^*(ds/s) \le \text{const} \int_0^\pi |v_+ - v_-|\,|v_+ + v_-|(ds/s) =$$

$$= \text{const} \int_0^\pi |v_+^2 - v_-^2|(ds/s) < \infty.$$

Thus, (2.25) ⇒ (6.10) is true. Proposition 6.1 completes the proof of Theorem 2.4.

Acknowledgements

1. *The author is very grateful to Albrecht Böttcher for several useful discussions and for his help in the preparation of this work for publication.*
2. *This research was supported by the Russian fund of fundamental investigations (grant N 98-01-01023).*

References

[1] Böttcher A., Silbermann B. *Analysis of Toeplitz operators.* Akademie Verlag, Berlin, and Springer-Verlag, Heidelberg, New York (1990).

[2] Douglas R.G. *Banach algebra techniques in operator theory.* Academic Press, New York (1972).

[3] Douglas R.G. *Banach algebra techniques in the theory of Toeplitz operators.* CBMS Lecture Notes, **15**, Amer. Math. Soc., Providence, R.I. (1973).

[4] Gohberg I., Krupnik N.Ya. *Introduction to the theory of one-dimensional singular integral operators.* Shtiintsa, Kishinev (1973) (Russian); German transl.: Birkhäuser Verlag, Basel (1979).

[5] Khvedelidze B.V. *The method of Cauchy type integrals in discontinuous boundary-value problems in the theory of holomorphic functions of one complex variable.* Itogi nauki i tehniki, Sovremennye problemy matematiki, **7**, 5–162 (1975).

[6] Litvinchuk G.S., Spitkovsky I.M. *Factorization of measurable matrix functions.* Akademie-Verlag, Berlin, and Birkhäuser Verlag, Basel (1987).

[7] Simonenko I.B. *Some general questions of the theory of the Riemann boundary value problem.* Izv. Akad. Nauk SSSR, Mat., **32**, 5, 1138–1146 (1968) (Russian); also in: Math. USSR Izv, **2**, 1091–1099 (1968).

[8] Govorov N.V. *On the Riemann boundary value problem with infinite index.* DAN SSSR, **54**, 6, 1247–1249 (1964).

[9] Govorov N.V. *The Riemann boundary value problem with infinite index.* Moscow, Nauka, Gl. red. fiz.-mat. lit. (1986).

[10] Alehno A.G. *The nonhomogeneous Riemann boundary-value problem with infinite index in the case of multilateral vorticity.* Dokl. AN Belorus., **41**, 5, 35–43 (1997).

[11] Alekna P.Yu. *Necessary and sufficient conditions for the solvability of the nonhomogeneous Riemann boundary-value problem with minus-infinite index of logarithmic order* $\min(\alpha,\beta) > 1$ *for the half-plane.* Litovskii matematicheskii sbornik, **18**, 3, 5–14 (1978).

[12] Ostrovskii I.V. *Conditions for the solvability of the homogeneous Riemann boundary-value problem with infinite index.* Teoriya funktsii, funk. analiz i ih prilozh., 55, 3–23 (1991).

[13] Ostrovskii I.V. *Conditions for the solvability of the homogeneous Riemann boundary-value problem with infinite index.* Teoriya funktsii, funk. analiz i ih prilozh., 56, 95–105 (1991).

[14] Gohberg I., Feldman I.A. *On Wiener-Hopf integral difference equations.* Dokl. Akad. Nauk SSSR, **183**, 25–28 (1968) (Russian); Soviet Math. Dokl., **9**, 1312–1316 (1968).

[15] Coburn L.A., Douglas R.G. *Translation operators on half-line.* Proc. Nat. Acad. Sci. USA, **62**, 1010–1030 (1969).

[16] Abrahamse M.B. *The spectrum of Toeplitz operator with a multiplicative periodic symbol.* J. Funct. Anal., **31**, 2, 224–233 (1979).

[17] Böttcher A., Grudsky S.M. *Toeplitz operators with discontinuous symbols: phenomena beyond piecewise continuity.* Operator Theory: Advances and Applications, **90**, 55–118, Birkhäuser Verlag, Basel (1996).

[18] Saginashvili A.I. *Singular integral equations with coefficients having discontinuities of semi-almost periodic type.* Trudy Tbilissk. Mat. Inst. Razmadze, **66**, 84–95 (1980) (Russian); Engl. transl.: Amer. Math. Soc. Transl., **127**, 2(1986).

[19] Dybin V.B. *On the singular integral operator on the real line with almost-periodic coefficients.* Teoriya funktsii; diff. uravneniya i ih pril., Elista, 98–108 (1976).

[20] Grudsky S.M., Dybin V.B. *The Riemann boundary value problem with a coefficient having discontinuities of almost-periodic type.* DAN SSSR, **237**, 1, 21–24, (1977); AMS, Soviet Math. Dokl. **18**, 6, 1383–1387 (1977).

[21] Grudsky S.M. *The Riemann boundary value problem with semialmost-periodic discontinuities in* $L_p(\Gamma,\varrho)$. Differents. i integr. uravneniya i pribl. resheniya, Elista, 54–68 (1985).

[22] Grudsky S.M. *Singular integral equations and Riemann boundary value problems with infinite index in* $L_p(\Gamma,\omega)$. Izv. AN SSSR. Matematika, **49**, 1, 55–80 (1985).

[23] Grudsky S.M. *Singular integral operators with infinite index and Blaschke products.* Math. Nachr., **129**, 313–331 (1987).

[24] Grudsky S.M. *Factorization of u-periodic matrix functions and problems with infinite index.* DAN SSSR, **295**, 6, 1298–1302 (1987).

[25] Grudsky S.M. *Toeplitz operators with symbols with discontinuities of finite index type.* DAN Rossii, **242**, 3, 307–309 (1995).

[26] Monahov V.N., Semenko E.V. *Classes of correctness of the boundary-value conjugation problem of analytic functions with infinite index.* DAN SSSR, **286**, 1, 27–30 (1986).

[27] Sementzul A.A. *On singular integral equations with coefficients having discontinuities of almost-periodic type.* Mat. Issledovaniya, Kishinev, **6**, 3(21), 92–114 (1971).

[28] Sarason D. *Toeplitz operators with semi-almost periodic symbols.* Duke Math. J., **44**, 2, 357–364 (1977).

[29] Xia J. *Piecewise continuous almost-periodic functions and mean motion.* Trans. Amer. Math. Soc., **288**, 2, 801–811 (1985).

[30] Xia J. *Wiener-Hopf operators with piecewise continuous almost-periodic symbol.* J. Operator Theory, **14**, 1, 147–171 (1985).

[31] Spitkovsky I.M. *Generalized factorization of matrix functions and the Riemann boundary-value problem with infinite partial indices.* DAN SSSR, **286**, 3, 559–562 (1986).

[32] Spitkovsky I.M. *On the vector-valued Riemann boundary problem with infinite defect numbers and the factorization of matrix function connected with this problem.* Mat. Sb. (N.S.), **135**, 533–550 (1988) (Russian).

[33] Prudnikov A.P., Brychkov Yu.A., Marichev O.I. *Integrals and series.* Moscow, Nauka (1981).

Department of Mathematics and Mechanics
Rostov State University
344711 Rostov-on-Don, B. Sadovaya, 105
Russia

1991 Mathematics Subject Classification: Primary 47B35; Secondary 30D50, 30D55, 42A75, 45E05, 47A53

Submitted: 17.4.2000

Operator Theory:
Advances and Applications, Vol. 121

Towards $\mathcal{H}$–Matrix Approximation of Linear Complexity

WOLFGANG HACKBUSCH AND BORIS N. KHOROMSKIJ

In preceding papers [10]–[15], a class of matrices ($\mathcal{H}$-matrices) has been analysed which are data-sparse and allow an approximate matrix arithmetic of almost linear complexity. Several types of $\mathcal{H}$-matrices were shown to provide good approximation of nonlocal (integral) operators in FEM and BEM applications.

In the present paper, we develop special classes of $\mathcal{H}$-matrices with improved data sparsity to approximate elliptic problems posed in $\mathbb{R}^d$, $d = 1, 2, 3$. For the evalution of integral operators on spatial domains in $\mathbb{R}^d$, the idea is to apply degenerate kernel expansions supported only on the boundaries of geometrical clusters. This results in an algorithm of linear storage expenses, $O(n)$, which includes one call for an optimal Dirichlet solver (e.g., multi-grid method) on the involved cluster.

In the BEM application, our approach reduces the order of expansion from $O(\log^d n)$ down to $O(\log^{d-1} n)$ for rather general elliptic operators[1], where n is the problem size.

1. Introduction

In preceding papers [10]–[15], a class of hierarchical matrices ($\mathcal{H}$-matrices) has been analysed which are data-sparse and allow an approximate matrix arithmetic of almost linear complexity. Several types of $\mathcal{H}$-matrices were shown to provide good approximation of nonlocal (integral) operators in FEM and BEM applications. In the present paper, we develop special classes of $\mathcal{H}$-matrix approximations for elliptic problems which have linear memory requirements and allow a fast matrix-vector product of linear-logarithmic complexity under standard assumptions on the smoothness of fundamental solutions.

For evaluation of potential fields as well as for the calculation of particular solutions in elliptic problems, we are interested in the data–sparse approximation of integral operators on a spatial domain in $\mathbb{R}^d$, $d = 1, 2, 3$ whose kernels satisfy the property (2.2). The idea is to build the degenerate kernel expansions only on the boundaries of certain geometrical clusters and to use their elliptic extension into the corresponding interior product domain. As a result, the domain integrals are transformed into equivalent surface integrals. This leads to an algorithm of linear complexity. It requires one call of an optimal Dirichlet solver (say, multi-grid

[1]In the present paper, we call by the order of expansion the total number of terms in a separable approximation of the kernel.

method) on each geometrical cluster. We call the new approach the *wire–basket $\mathcal{H}$–matrix approximation*. For a general class of kernels, this method essentially reduces the amount of data (memory requirements) for each $n_\tau \times n_\sigma$ matrix block from $O(n_\tau \log^d n)$ for the standard $\mathcal{H}$–matrices down to $O(n_\tau^{1-1/d} \log^{d-1} n + n_\tau)$, where $n_\tau \sim n_\sigma$ is the corresponding blocksize and n is the global dimension.

As a by–product, we derive degenerate expansions of the order $O(\log^{d-1} n)$ for kernels of boundary integral operators. The efficiency of the corresponding expansions in BEM does not depend on the smoothness of the computational boundary Γ. However, the constructive way to avoid this dependence and to retain expansions of the optimal order $O(\log^{d-1} n)$, needs to compute the restriction of elliptic extension onto pieces of the boundary Γ located within associated "volume cluster" covering a part of the surface in the spatial product domain.

2. $\mathcal{H}$–Matrix Approximation Revisited

2.1. Integral Operators in Elliptic Problems

We consider the case of scalar elliptic operators. Let

$$\mathcal{L} = -\sum_{j,k=1}^{d} \partial_j a_{jk} \partial_k + \sum_{j=1}^{d} b_j \partial_j + c_0 \tag{2.1}$$

be a second order elliptic operator with real constant coefficients a_{jk}, b_j and c_0, defined in a polyhedral domain $\Omega \in \mathbb{R}^d$. Here $\partial_j = \frac{\partial}{\partial x_j}$. We assume that $\mathcal{L}$ has a fundamental solution S satisfying $\mathcal{L}S = \delta_0$, where δ_0 is the Dirac distribution supported at the origin. The associated weakly singular kernel $s(x,y) = S(x-y)$ defines a class of integral operators specified by their domain of definition. The function $(x,y) \to s(x,y)$ is C^∞ outside the diagonal of $\mathbb{R}^d \times \mathbb{R}^d$ and satisfies Assumption 1 from Section 2.3. Moreover,

$$\mathcal{L}_x s(x,y) = -\mathcal{L}_y^* s(x,y) = \delta(x-y), \tag{2.2}$$

where $\mathcal{L}^*$ is the formal adjoint of $\mathcal{L}$ with respect to the L^2 inner product. We further assume that $\mathcal{L}$ is $H_0^1(\Omega)$–elliptic which implies that for the bilinear form

$$a_\Omega(u,v) := \int_\Omega \Big(\sum_{j,k=1}^{d} a_{jk} \partial_k u \partial_j v + \sum_{j=1}^{d} b_j \partial_j u v + c_0 u v\Big) dx$$

there holds the inequality

$$a_\Omega(u,u) \geq c\|u\|_{1,\Omega}^2, \qquad \forall u \in H_0^1(\Omega), \text{ with } c > 0.$$

Note that the bilinear form $a_\Omega(\cdot,\cdot)$ is continuous on $H^1(\Omega) \times H^1(\Omega)$.

We consider the h-version of the Galerkin FE method for approximating the continuous integral operator $A : V \to V'$ defined in the Sobolev space $V = H^r(\Sigma)$ and deal with integral operators of the form

$$(2.3) \qquad (Au)(x) = \int_\Sigma s(x,y)u(y)dy, \qquad x \in \Sigma,$$

with s being the kernel function mentioned above or with s replaced by a suitable directional derivatives Ds of s. We distinguish two different cases:

(**A**) Σ is a bounded $(d-1)$-dimensional manifold (BEM applications);

(**B**) Σ is a polyhedron in $\mathbb{R}^d$, $d = 2, 3$ (FEM applications).

In the latter case, we confine ourselves to the case of the unit cube $\Sigma = (0,1)^{d_\Sigma}$ and an ansatz space $V_h := span\{\varphi_i\}_{i\in I} \subset V$ of piecewise constant/linear basis functions with respect to a quasiuniform tensor–product mesh. Therefore, we specify $V = H^r(\Sigma)$, $r \in [-1/2, 1/2]$, and $d_\Sigma = d - 1$ in Case (**A**), while $V = H^{-1}(\Sigma)$, $d_\Sigma = d$ for evaluation of the volume integral operators in Case (**B**). An extension of our approach to more general domains in $\mathbb{R}^d$, $d = 2, 3$, is possible.

2.2. Construction of $\mathcal{H}$-Matrices

We construct a data–sparse $\mathcal{H}$-matrix approximation for the integral operators A with asymptotically smooth kernels, see (2.11), and with locally analytic kernels (2.10). For the ease of presentation, we first consider the case of the piecewise constant Galerkin ansatz space $V_h \subset V$. Modifications for piecewise linear/bilinear elements will be discussed in Remark 2.2 below.

For the sake of completeness, we recall the important definition of an admissible block partitioning. Let I be the index set of unknowns (e.g., the FE-nodal points). For each $i \in I$, the support of the corresponding basis function φ_i is denoted by $X(i) := supp(\varphi_i)$. The *cluster tree* $T(I)$ is characterised by the following properties: (i) all vertices of $T(I)$ are subsets of I, (ii) $I \in T(I)$ is the root; (iii) if $\tau \in T(I)$ contains more than one element, the set $S(\tau)$ of sons of τ consists of at least 2 disjoint subsets satisfying $\tau = \dot\bigcup_{\sigma\in S(\tau)} \sigma$; (iv) the leaves of the tree are $\{i\}$ for all $i \in I$. For $\tau \in T(I)$ we extend the definition of the supports $X(\cdot)$ by $X(\tau) = \bigcup_{i\in\tau} X(i)$.

In the standard quasiuniform FE application, the cluster tree $T(I)$ is obtained by a recursive division of I into subsets of almost equal size having a diameter as small as possible. In the quasiuniform case, the term "almost equal size" can be understood in a geometrical sense (i.e., $diam(X(\tau')) \approx diam(X(\tau''))$ as well as with respect to the cardinality $\#\tau' \approx \#\tau''$. An appropriate construction of $T(I)$ will fulfil both criteria.

The matrix entries belong to the index set $I \times I$. In a canonical way (cf. [11]), a block-cluster tree $T(I \times I)$ can be constructed from $T(I)$, where all vertices $b \in T(I \times I)$ are of the form $b = \tau \times \sigma$ with $\tau, \sigma \in T(I)$. Given a matrix $M \in \mathbb{R}^{I \times I}$, the block-matrix corresponding to $b \in T(I \times I)$ is denoted by $M^b = (m_{ij})_{(i,j)\in b}$. A *block partitioning* $P_2 \subset T(I \times I)$ is a set of disjoint blocks $b \in T(I \times I)$, whose union equals $I \times I$. A block partitioning P_2 determines the $\mathcal{H}$-matrix format. We use the following explicit definition of $\mathcal{H}$-matrices.

Definition 2.1. Let a block partitioning P_2 of $I \times I$ and $k << n$ be given. The set of real $\mathcal{H}$-matrices induced by P_2 and k is

(2.4) $\mathcal{M}_{\mathcal{H},k}(I \times I, P_2) := \{M \in \mathbb{R}^{I \times I} : \forall b \in P_2 \text{ there holds } rank(M^b) \leq k\}.$

An admissibility condition is used to balance the size of matrix–blocks b and the distance between τ and σ (see [12] for more details). It takes into consideration the singularity location of the kernel function $s(x, y)$, $(x, y) \in \Sigma \times \Sigma$. We assume that the following *admissibility condition*

$$min\{diam(\sigma), diam(\tau)\} \leq 2\eta \, dist(\sigma, \tau) \tag{2.5}$$

holds for all $\sigma \times \tau \in P_2$, where $\eta \leq 1$ is a fixed threshold parameter. Here both *dist* and *diam* are defined with respect to the given norm $||\cdot||$ in $\mathbb{R}^d$. In the case of non-homogeneous coefficients, this norm will depend on the coefficient matrix of the elliptic operator. Specifically, assume that $\mathbf{L} = (a_{ij})_{i,j=1}^d \in \mathbb{R}^{d\times d}$ is the symmetric and positive definite coefficient matrix from (2.1). Define the matrix dependent scalar product and norm in $\mathbb{R}^d$ by $\langle u, v\rangle = (\mathbf{L}^{-1}u, v)$ and $||u|| = \langle u, u\rangle^{1/2}$, respectively, where $(\cdot,\cdot)$ is the Euclidean scalar product. In Section 5, we apply this construction to anisotropic elliptic equations.

For computational needs, we further use the splitting $P_2 = P_{far} \cup P_{near}$, where

$$P_{far} := \{\sigma \times \tau \in P_2 : dist(X(\tau), X(\sigma)) > 0\,\}. \tag{2.6}$$

Due to our assumption on a piecewise constant FE basis, the index set I is isomorphic to the disjoint supports X_i. While $\sigma \times \tau \in T(I \times I)$ indicates a matrix block, $X(\sigma) \times X(\tau) \subset \Sigma \times \Sigma$ is called the corresponding geometrical block. Denote by $\langle\cdot,\cdot\rangle = \langle\cdot,\cdot\rangle_\Sigma$ the L^2 inner product on the domain Σ. The standard $\mathcal{H}$-matrix approximation of the integral operator consists of three essential steps:

(a) the *admissible* block-partitioning $P_2 = P_{far} \cup P_{near}$ of the tensor product index set $I \times I$.

(b) the construction of an approximate integral operator $A_\mathcal{H} \in \mathcal{L}(V, V')$ with the kernel $s_\mathcal{H}(\cdot,\cdot)$ defined for each geometrical block $X(\sigma) \times X(\tau)$ with $\sigma \times \tau \in P_{far}$, by a *separable expansion* $s_{\tau,\sigma} = \sum_{i=1}^k a_i(x)c_i(y)$ of the order $k << n = dimV_h$; In the near-field area, the kernel function is unchanged.

(c) the setup of the Galerkin $\mathcal{H}$-matrix $\mathbf{A}_\mathcal{H} = \langle A_\mathcal{H}\varphi_i, \varphi_j\rangle_{i,j\in I}$ for the operator $A_\mathcal{H}$, where $\{\varphi_i\}$ is the FE basis of V_h.

Remark 2.2. In the case of piecewise linear/bilinear elements there is a minor difference in the definition of local separable expansions at Step (**b**). Since now the supports $X(\sigma) \times X(\tau)$ from different $\sigma \times \tau$ may overlap, the kernel function $s_{\mathcal{H}} : \Sigma \times \Sigma \to \mathbb{R}$ is defined by a multi-valued mapping in the overlap. This allows to use a smooth modification of $s_{\mathcal{H}}$ for all a_{ij}, $(i,j) \in \sigma \times \tau$.

2.3. Complexity and Approximation

A bound of the solution error caused by Step (**b**) as well as the computational complexity of the $\mathcal{H}$-matrix formats for quasiuniform meshes were considered in [11]. An almost linear complexity bound was proven in [12], which is valid in both cases (**A**) and (**B**). In the following statement we use the parametrization $\eta = \frac{\sqrt{d}}{2\mu}$, $\mu = 1, 2, \dots$ for the threshold parameter η.

Proposition 2.3. *Let $d \in \{1, 2, 3\}$, $A \in \mathcal{M}_{\mathcal{H},k}(I \times I, P_2)$ and $n = 2^{pd}$. Then the storage and matrix-vector multiplication expenses are bounded by*

$$\mathcal{N}_{st} \le (2^d - 1)(\sqrt{d}\eta^{-1} + 1)^d \, p\, k\, n, \qquad \mathcal{N}_{MV} \le \mathcal{N}_{st}, \tag{2.7}$$

where the cost unit of $\mathcal{N}_{MV}$ is one addition and one multiplication. Both estimates are asymptotically sharp.

The main goal of the present paper is the essential improvement of the bound (2.7) in Case (**B**) described in Sections 3 and 4.

The perturbation of the matrix induced by $A_{\mathcal{H}} - A$ yields a perturbed discrete solution of the original variational equation

$$\langle (\lambda I + A)u, v\rangle_\Sigma = \langle f, v\rangle_\Sigma \qquad \forall v \in V := H^r(\Sigma),\ r \le 1, \tag{2.8}$$

where $\lambda \in \mathbb{R}$ is a given parameter. For the given ansatz space $V_h \subset V$ of piecewise constant/linear FEs, consider the perturbed Galerkin equation for $u_{\mathcal{H}} \in V_h$,

$$\langle (\lambda I + A_{\mathcal{H}})u_{\mathcal{H}}, v\rangle_\Sigma = \langle f, v\rangle_\Sigma \qquad \forall v \in V_h. \tag{2.9}$$

Our error analysis will be based on different smoothness prerequisites. The first one requires analyticity of the kernel for $x \ne y$.

Assumption 1 *For any $x_0, y_0 \in \Omega$, $x_0 \ne y_0$, the kernel function $s(x,y)$ is analytic with respect to x and y at least in the domain*

$$|x - x_0| + |y - y_0| < |x_0 - y_0|. \tag{2.10}$$

A similar condition was used in [20] for the analysis of the Galerkin wavelet approximations in BEM.

The alternative assumption requires that the singularity function s is asymptotically smooth, i.e.,

Assumption 2 *For all $x, y \in \mathbb{R}^d, x \neq y$, and all multi-indices α, β with $|\alpha| = \alpha_1 + \ldots + \alpha_d$, $d = 2, 3$, let*

$$|\partial_x^\alpha \partial_y^\beta s(x,y)| \leq c(|\alpha|, |\beta|)|x-y|^{1-|\alpha|-|\beta|-d-2r}, \tag{2.11}$$

for all $|\alpha|, |\beta| \leq m$ such that $|\alpha| + |\beta| > 0$.

Here $2r \in \mathbb{R}$ is the order of the integral operator $A : H^r(\Sigma) \to H^{-r}(\Sigma)$ in Case (**A**) with the possible choice $r \in \{-\frac{1}{2}, 0, \frac{1}{2}\}$. In Case (**B**), we specify $r = -\frac{1}{2}$, such that the operator $A : H^{-1}(\Omega) \to H^1(\Omega)$ is continuous. Similar smoothness prerequisites are common in the wavelet or multi-resolution techniques [2, 3, 22], in the multipole expansion method (cf. [6] and references therein) as well as in the related mosaic–skeleton approach (cf. [25]).

Theorem 2.4. *Assume that (2.11) is valid and that V_h allows the standard inverse inequality. Suppose that the operator $\lambda I + A \in \mathcal{L}(V, V')$ is V-elliptic and let $r \in [-1/2, 1/2]$. Then there holds*

$$||u - u_{\mathcal{H}}||_V \lesssim \inf_{v_h \in V_h} ||u - v_h||_V + \frac{c(0,m)}{m!} \eta^m N^{d_\Sigma/2} ||u||_V, \tag{2.12}$$

where $c(0, m)$ is defined in (2.11).

Proof. The proof is a minor modification of the arguments from [12]. □

2.4. Cluster Tree on the Tensor-Product Index Set

In particular situations, we will be interested in the $\mathcal{H}$–matrix approximation of an integral operator A defined in (2.3) with $\Sigma = (0,1)^d$ and $d = 1, 2, 3$. Here we briefly recall the recursive construction of $\mathcal{H}$–matrices (cf. [12]). Consider the regular grid

$$I = \{\mathbf{i} = (i_1, ..., i_d) : 1 \leq i_k \leq N,\ k = 1, ..., d\}, \quad N = 2^p, \tag{2.13}$$

and define the norm $|\mathbf{i}|_\infty = \max_{1 \leq k \leq d} |i_k|$. The cardinality of I is $n = N^d = 2^{pd}$.

The cluster tree $T_1 = T(I)$ of I is based on a division of the underlying cubes into 2^d subcubes. The blocks

$$t_{\mathbf{j}}^\ell = \{\mathbf{i} :\ 2^{p-\ell} j_1 + 1 \leq i_1 \leq 2^{p-\ell}(j_1+1), ...,\ 2^{p-\ell} j_d + 1 \leq i_d \leq 2^{p-\ell}(j_d+1)\}$$

for $\mathbf{j} \in \{0, \ldots, 2^\ell - 1\}^d$ belong to level ℓ. $S_1(t_{\mathbf{j}'}^{\ell-1}) := \{t_{\mathbf{j}}^\ell : 0 \leq 2j'_k - i_k \leq 1, 1 \leq k \leq d\}$ defines the set of sons of the cluster $t_{\mathbf{j}'}^{\ell-1}$. Hence, the tree T_1 consisting of all blocks at all levels $\ell \in \{0, \ldots, p\}$ is a binary, quad- or octree for $d = 1, 2, 3$, respectively. The number of clusters on level ℓ equals $O(2^{d\ell})$.

Each index $\mathbf{i} \in I$ is associated with the d-dimensional cube

(2.14) $X_{\mathbf{i}} := \{(x_1, ..., x_d) : (i_1 - 1)h \le x_1 \le i_1 h, ..., (i_d - 1)h \le x_d \le i_d h\},$

which is the support of a piecewise constant function for the index $\mathbf{i}$ with $h = 2^{-p}$. Using the Euclidean norm, we obtain the diameter $diam(\tau) = \sqrt{d}\,2^{p-\ell}h = \sqrt{d}/2^{\ell}$ for blocks of level ℓ. Let $\tau = t^{\ell}_{\mathbf{j}}, \sigma = t^{\ell}_{\mathbf{j}'}$ be two blocks of level ℓ characterised by $\mathbf{j}$ and $\mathbf{j}'$. Then

$$dist(\tau, \sigma) = 2^{-\ell}\sqrt{\delta(j_1 - j_1')^2 + ... + \delta(j_d - j_d')^2} \tag{2.15}$$

with $\delta(\xi) := \max\{0, |\xi| - 1\}$. Let $T_2 = T(I \times I)$ be the block-cluster tree corresponding to the cluster tree $T_1 = T(I)$. The definition of T_2 implies the following remark (cf. [10]).

Remark 2.5. Let $\tau \times \sigma \in T(I \times I)$. Then $\tau, \sigma \in T(I)$ belong to the same level $\ell \in \{0, \ldots, p\}$.

The set of clusters $\tau \in T(I)$ from level ℓ is called T_1^{ℓ}. In view of Remark 2.5, for $\ell \in \{0, ..., p\}$, we denote by T_2^{ℓ} the set of clusters $\tau \times \sigma \in T_2$ such that blocks τ, σ belong to level ℓ. In particular, $T_2^0 = \{I \times I\}$ is the root of T_2 and $T_2^p = \{\{(x, y)\} : x, y \in I\}$ is the set of leaves. Correspondingly, we define $P_2^{\ell} := P_2 \cap T_2^{\ell}$.

3. $\mathcal{H}$–Matrices via Wire–Basket Expansions

3.1. Basic Idea: Description on the Continuous Level

The basic idea of the wire–basket approach is the interface representation of the scalar product $\langle A_{\mathcal{H}}u, v\rangle$, $u, v \in V_h$, for the hierarchical approximation to the operator A from (2.3) in Case (**B**). By definition, there holds

$$\langle A_{\mathcal{H}}v, u\rangle = \sum_{\tau\times\sigma\in P_2} \int_{X(\tau)\times X(\sigma)} s_{\tau,\sigma}(x, y)u(x)v(y)dydx. \tag{3.1}$$

First, we consider the exact Galerkin ansatz $\langle Av, u\rangle$ with the kernel–function $s(x, y)$ instead of $s_{\tau,\sigma}$ and transform each domain integral for $\tau \times \sigma \in P_{far}$ into its boundary form. For notational convenience, define a set of geometrical blocks (product subdomains) $X_{far} := \{X(\tau) \times X(\sigma) : \tau \times \sigma \in P_{far}\}$. In view of (2.2), we have

$$\mathcal{L}_x s(x, y) = \mathcal{L}_y^* s(x, y) = 0, \qquad (x, y) \in X(\tau) \times X(\sigma) \in X_{far}. \tag{3.2}$$

In the following, the symbol Ω is used as variable for a domain. Let $a_{\Omega}(\cdot, \cdot)$ be the bilinear form associated with the operator $\mathcal{L}$ as above. Then the first Green formula holds:

$$\langle v, \mathcal{L}u\rangle_{\Omega} = a_{\Omega}(u, v) - \langle \partial_{\nu}u, v\rangle_{\partial\Omega} \qquad \forall u \in H^1(\Omega, \mathcal{L}),\ v \in H^1(\Omega), \tag{3.3}$$

where $\langle\cdot,\cdot\rangle_\Sigma$ with $\Sigma = \partial\Omega$ is the $L^2(\Sigma)$ scalar product and ∂_ν is the conormal derivative

$$\partial_\nu := \sum_{j,k=1}^{d} n_j a_{jk}\partial_k - \sum_{j=1}^{d} n_j b_j$$

with $\partial_j = \partial/\partial x_j$ and n_j being the components of the outward unit normal vector.

Denote $W^0 = H^2(\Omega)\cap H_0^1(\Omega)$. For any $z \in L^2(\Omega)$, introduce a function $g_z \in W^0$ (note that convexity of the domain Ω implies the full elliptic regularity, $||g_z||_{2,\Omega} \le c||z||_{0,\Omega}$) such that

$$a_\Omega(g_z,\eta) = \langle z,\eta\rangle_\Omega \qquad \forall\eta \in H_0^1(\Omega). \tag{3.4}$$

The continuous operator $\mathcal{L}_\Omega^{-1} : L^2(\Omega) \to W^0$ is defined by $\mathcal{L}_\Omega^{-1}z = g_z$, which has a continuous extension to a mapping $\mathcal{L}_\Omega^{-1} : H^{-1}(\Omega) \to H_0^1(\Omega)$. In the case of a hierarchical cluster tree T_1 of depth p, the far-field component P_{far} from (2.6) may be specified by the choice of parameter $p_0 \in \mathbb{N}$, $p_0 = O(1)$, yielding the alternative definition

$$P_{far} = \cup_{\ell=2}^{p-p_0} P_2^\ell, \qquad P_{near} = \cup_{\ell=p-p_0+1}^{p} P_2^\ell. \tag{3.5}$$

We recall that on block–clusters from P_{far}, we approximate the kernel function by degenerate expansions, while matrix entries corresponding to P_{near} are computed as usual (we assume that the computation of one matrix entry costs $O(1)$ arithmetical operations). Remark that (3.2) implies the Galerkin orthogonality

$$a_\tau(g_u(x), s(x,y)) = a_\sigma(g_v(y), s(x,y)) = 0 \tag{3.6}$$

for $(x,y) \in X(\tau)\times X(\sigma) \in X_{far}$.

Lemma 3.1. *For any $u, v \in L^2(\Omega)$ there holds*

$$\langle Av,u\rangle_\Omega = \sum_{\tau\times\sigma\in P_{far}} \int_{\partial X(\tau)}\int_{\partial X(\sigma)} s(x,y)\partial_\nu g_v \partial_\nu g_u dxdy + \sum_{\tau\times\sigma\in P_{near}} \int_{X(\tau)\times X(\sigma)} s(x,y)u(x)v(y)dxdy. \tag{3.7}$$

Proof. (3.3) and (3.6) with $u = \mathcal{L}_x g_u$ and $v = \mathcal{L}_y g_v$ lead to

$$\int_{X(\tau)\times X(\sigma)} s(x,y)u(x)v(y)dxdy$$

$$= \int_{X(\tau)} \left\{ a_\sigma(g_v(y), s(x,y)) - \int_{\partial X(\sigma)} s(x,y)\partial_\nu g_v(y)dy \right\} \mathcal{L}_x g_u(x)dx$$

$$= \int_{\partial X(\sigma)} \left\{ -a_\tau(g_u(x), s(x,y)) + \int_{\partial X(\tau)} s(x,y)\partial_\nu g_u(x)dx \right\} \partial_\nu g_v(y)dy$$

$$= \int_{\partial X(\sigma)} \int_{\partial X(\tau)} s(x,y)\partial_\nu g_v(y)\partial_\nu g_u(x)dxdy$$

for all $\tau \times \sigma \in P_{far}$. Hence, (3.7) follows. □

Assume, we are given a degenerate expansion

$$s_{\partial\tau,\partial\sigma} = \sum_{\alpha=1}^{k_1} a_\alpha(x)c_\alpha(y), \qquad (x,y) \in \partial X(\tau) \times \partial X(\sigma) \tag{3.8}$$

for each $\tau \times \sigma \in P_{far}$ such that

$$|s(x,y) - s_{\partial\tau,\partial\sigma}(x,y)| \le c\eta^m \{dist(\tau,\sigma)\}^{2-d}, \qquad (x,y) \in \partial X(\tau) \times \partial X(\sigma),$$

where $k_1 = O(m^{d-1})$. Then we introduce the *wire-basket representation* of the operator $A_{\mathcal{H}}$ by

$$\langle A_{\mathcal{H}} v, u\rangle_\Omega = \sum_{\tau\times\sigma\in P_{near}} \int_{X(\tau)\times X(\sigma)} s(x,y)u(x)v(y)dxdy \tag{3.9}$$

$$+ \sum_{\tau\times\sigma\in P_{far}} \sum_\alpha \int_{\partial X(\tau)} a_\alpha(x)\partial_\nu g_u(x)dx \cdot \int_{\partial X(\sigma)} c_\alpha(y)\partial_\nu g_v(y)dy, \qquad u,v \in L^2(\Omega).$$

The second sum will be abbreviated by $\langle v, A_{\mathcal{H}} u\rangle^{far}$. Particular constructions of $s_{\partial\tau,\partial\sigma}$ will be considered in Section 4.

Below, we introduce the Galerkin approximation to the second sum in (3.9). To this end, we represent each integral over $\partial X(\tau)$ (or $\partial X(\sigma)$) in terms of domain integrals using an easily computable extension of $a_\alpha(x)$ (resp. $c_\alpha(y)$) into the interior of $X(\tau)$ (resp. $X(\sigma)$).

Remark 3.2. Let $E_x : H^{1/2}(\partial X(\tau)) \to H^1(X(\tau))$ be any continuous extension operator defined for each $\tau \in T_1$. Applying Green formula (3.3) and taking into account (3.4) we obtain

$$\int_{\partial X(\tau)} a_\alpha(x)\partial_\nu g_u(x)dx = -a_\tau(g_u, E_x a_\alpha) + \langle u, E_x a_\alpha\rangle_{X(\tau)}. \tag{3.10}$$

The same extension with respect to the y-variable is denoted by E_y.

With the given ansatz space $W_h \subset H^1(X(\tau))$, let $g_{h,z} \in W^0_{h,\tau} := W_h \cap H^1_0(X(\tau))$ be the Ritz projection of g_z defined by

$$a_\tau(g_{h,z}, \eta) = \langle z, \eta\rangle_{X(\tau)} \qquad \forall \eta \in W^0_{h,\tau}. \tag{3.11}$$

The FE Galerkin approximation to the far–field contribution in (3.9) is defined for any $u, v \in V_h$ by substituting $g_{h,u}$ (resp. $g_{h,v}$) into the right-hand side of (3.10) and choosing E_x (resp. E_y) as the extension into W_h by FE functions with minimal support inside $X(\tau)$.

Corollary 3.3. *For any $u, v \in V_h$, the FE Galerkin approximation $A_{\partial\mathcal{H},h} = A_h$ for the operator $A_{\mathcal{H}}$ is defined by*

$$(3.12) \qquad \langle A_h v, u\rangle_\Omega = \sum_{\tau\times\sigma\in P_{near}} \int_{X(\tau)\times X(\sigma)} s(x,y)u(x)v(y)dxdy$$

$$+ \sum_{\tau\times\sigma\in P_{far}} \sum_{\alpha=1}^{k_1} \left(\langle u, E_x a_\alpha\rangle_{X(\tau)} - a_\tau(g_{h,u}, E_x a_\alpha)\right)\left(\langle v, E_y c_\alpha\rangle_{X(\sigma)} - a_\sigma(g_{h,v}, E_y c_\alpha)\right).$$

Remark 3.4. Note that the Galerkin ansatz space $V_h \subset L^2(\Omega)$ restricted to the geometrical clusters $X(\tau)$ may differ from W_h defined above. However, for the ease of presentation, we further assume $V_h = W_h$ for each $\tau \in T_1(I)$.

3.2. Matrix Representation and Complexity Bound

The representation (3.12) defines the generalised $\mathcal{H}$–matrices which inherit the standard hierarchical block structure from the P_2-partitioning, but now the rank-k structure of blocks $b \in P_2^\ell$ is given implicitly based on the factorisation by local Schur–complement matrices. Such a factorisation allows to reduce the amount of data for the storage and matrix arithmetic essentially. To build the explicit representation of the matrix block $\mathbf{A}_h^{\tau\times\sigma}$ of the resulting $\mathcal{H}$–matrix, we introduce the local Schur–complement operator

$$S_{\tau,h} : W_{h,\tau}^0 \to \Gamma_{h,\tau} := (W_h)_{|\partial X(\tau)},$$

for $z \in W_{h,\tau}^0$ by

$$(3.13) \qquad \langle S_{\tau,h} z, w\rangle_{\partial X(\tau)} := \langle z, E_x w\rangle_{X(\tau)} - a_\tau(g_{h,z}, E_x w), \qquad w \in \Gamma_{h,\tau},$$

which, in fact, provides a FE variational approximation of the operator $S = \partial_\nu \mathcal{L}_{X(\tau)}^{-1} : L^2(X(\tau)) \to H^{1/2}(\partial X(\tau))$. The construction is independent of the extension operator E_x due to the Galerkin orthogonality, see (3.11).

The L^2–projection operators $\mathcal{Q}_h$ onto $\Gamma_{h,\tau}$ is given by

$$\langle \mathcal{Q}_h u, v\rangle_{\partial X(\tau)} = \langle u, v\rangle_{\partial X(\tau)} \qquad \forall v \in \Gamma_{h,\tau}.$$

Let $\mathbf{a}_{h,\alpha}, \mathbf{c}_{h,\alpha}$ be the vector representations of $\mathcal{Q}_h a_\alpha$ and $\mathcal{Q}_h c_\alpha$, while $\mathbf{S}_{\tau,h}$ and $\mathbf{S}_{\sigma,h}$ be the matrix representations of $S_{\tau,h}$ and $S_{\sigma,h}$, respectively. The matrix block $\mathbf{A}_h^{\tau\times\sigma}$ defined by (3.12) for the product index–set $\tau\times\sigma \in P_{far}$, has the factorisation

$$(3.14) \mathbf{A}_h^{\tau\times\sigma} = \sum_{\alpha=1}^{k_1} (\mathbf{S}_{\tau,h}^T \cdot \mathbf{a}_{h,\alpha}) \otimes (\mathbf{c}_{h,\alpha}^T \cdot \mathbf{S}_{\sigma,h}) = \mathbf{S}_{\tau,h}^T \left(\sum_{\alpha=1}^{k_1} \mathbf{a}_{h,\alpha} \otimes \mathbf{c}_{h,\alpha}^T\right) \mathbf{S}_{\sigma,h},$$

where $\mathbf{S}_{\tau,h} : \mathbb{R}^{n_\tau} \to \mathbb{R}^{n_{\partial\tau}}$, $\mathbf{a}_{h,\alpha} \in \mathbb{R}^{n_{\partial\tau}}$, $n_\tau = dim W_{h,\tau}$, $n_{\partial\tau} = dim \Gamma_{h,\tau}$ and the same for $\mathbf{S}_{\sigma,h}, \mathbf{c}_{h,\alpha}$. Clearly, (3.14) defines a matrix block of the rank $\leq k_1$.

Assumption 3 *Denote by $E_{\tau,h}$ the discrete $\mathcal{L}$–harmonic extension operator in $X(\tau)$. Assume that the Ritz projection $g_{h,u}$ in (3.12) and the extension operator $E_{\tau,h}$ (the same for $g_{h,v}$ and $E_{\sigma,h}$) can be evaluated on each geometrical cluster $X(\tau)$ with linear cost $c_{RP}n_\tau$ (say, by the multi–grid method).*

The advantage of the presented method is the reduction of the order of expansion, on the one hand, and the linear bound for $Q_{\tau,\sigma}$ with respect to the block-size n_τ, on the other hand. The latter is due to reduction of the volume integrals to the boundary of clusters. Moreover, the constant in the asymptotical complexity is essentially dominated by c_{RP} (specifying the cost of the Ritz projection) which may be smaller than the corresponding constant in the $\mathcal{H}$–matrix arithmetic, especially for $d = 3$.

Lemma 3.5. *Let our construction be based on a hierarchical cluster tree of the depth p such that $n_\tau \le c2^{d(p-\ell)}$ for $\tau \in T_1^\ell$, $\#P_2^\ell \le c2^{d\ell}$ and $|\tau| \le 2^{-d\ell}$, where $|\tau| = meas X(\tau)$. For the variable order approximation with $k_1(\ell) = (a_1(p-\ell)+b_1)^{d-1}$ the storage and matrix–vector multiplication expenses are dominated by*

$$\mathcal{N}_{st} = O(pn), \qquad \mathcal{N}_{MV} = O(n) + c_{RP}\, pn.$$

Proof. Let $p_0 > 0$ be defined by (3.5). Then the storage $\mathcal{N}_1$ for the implementation of the Schur–complement operator from (3.13) is dominated by the corresponding memory needs for solving problem (3.10) for all τ such that $\tau \times \sigma \in P_2^{far}$. Therefore,

$$\mathcal{N}_1 \le \sum_{\tau\times\sigma\in P_2^{far}} \#\tau = O((p-p_0)n).$$

On the other hand, a simple estimate $n_\tau = O(2^{d(p-\ell)}) = 2^{p-\ell}O(n_{\partial\tau})$ implies that the coefficients of rank-k matrix blocks from (3.14) need only a storage size of $O(k_1(n_{\partial\tau} + n_{\partial\sigma})) = O(k_1(n_\tau + n_\sigma)2^{\ell-p})$. Thus, the overall storage $\mathcal{N}_2$ for the rank-k blocks is estimated by

$$\text{(3.15)}\quad \mathcal{N}_2 = \sum_{\tau\times\sigma\in P_2^{far}} k_1(\ell)2^{\ell-p}(n_\tau + n_\sigma) \le \sum_{\ell} \sum_{\tau\times\sigma\in P_2^\ell} k_1(\ell)2^{\ell-p}2^{d(p-\ell)}$$

$$\le c\,2^{dp}\sum_{\ell} k_1(\ell)2^{\ell-p} \sum_{\tau:\tau\times\sigma\in P_2^\ell} 1 \le cn.$$

Since the matrix entries corresponding to $\tau\times\sigma \in P_2^{near}$ need only a storage size of $O(n)$, this proves the linear–logarithmic bound for $\mathcal{N}_{st} = \mathcal{N}_1 + \mathcal{N}_2$.

The matrix–vector product for each block $\tau\times\sigma \in P_2^{far}$ has the complexity

$$\text{(3.16)}\qquad Q_{\tau,\sigma} = ck_1(n_{\partial\tau} + n_{\partial\sigma}) + 2c_{RP}(n_\tau + n_\sigma).$$

In fact, due to Assumption 3 and (3.13), the implementation of $\mathbf{S}_{\sigma,h}$ is of linear cost. Furthermore, the matrix–vector product by $\mathbf{S}^T_{\tau,h}$ is equivalent to the implementation of the elliptic extension $E_{\tau,h}$ due to the relation (3.13) and the Galerkin orthogonality,

$$(3.17)\quad \langle z, S^T_{\tau,h} w\rangle_{X(\tau)} = \langle z, E_{\tau,h} w\rangle_{X(\tau)} - a_\tau(g_{h,z}, E_{\tau,h} w) = \langle z, E_{\tau,h} w\rangle_{X(\tau)}$$

for $z \in W^0_{h,\tau}$, $w \in \Gamma_{h,\tau}$. Thus, the matrix–vector product for each block on level ℓ has the complexity

$$(3.18)\qquad Q_{\tau,\sigma} = ck_1 n_\tau 2^{\ell-p} + 2c_{RP} n_\tau \leq c n_\tau .$$

Summation over all the blocks $\tau \times \sigma \in P_2$ completes our proof. □

To complete this section, we briefly discuss an optimised construction of the wire-basket scheme based on the reuse of particular solutions, see [14] for more details. Assume we are given a balanced hierarchical cluster tree T_1 of the depth p. The idea is that subdomain solvers are used only on the few coarse levels $\ell = 2, ..., \ell_0$ while the restriction of these solutions to the smaller domains corresponding to the levels $\ell \geq \ell_0$ yield particular solutions (with wrong Dirichlet data). The correction of the Dirichlet data is reduced to evaluation of the Dirichlet-Neumann map (Poincaré-Steklov operator) defined only on the boundary of geometrical clusters $X(\tau)$. Assuming a correction scheme for the Dirichlet data of the cost $O(n_{\partial\tau} \log^q n_{\partial\tau})$, $q = O(1)$, see Assumption 3 and Remark 3.6 below, linear complexity of the overall scheme holds.

Remark 3.6. For the most common elliptic operators of the form (2.1), the action of FE approximation $\mathbf{S}_\tau$ for the Poincaré-Steklov operator $S_{\mathcal{L}} = \partial_\nu E_\Omega$: $H^{1/2}(\partial\Omega) \to H^{-1/2}(\partial\Omega)$ defined for $\Omega = X(\tau)$ by

$$(3.19)\qquad \langle \mathbf{S}_\tau z, w\rangle_{\partial X(\tau)} := a_\tau(E_{\tau,h} z, E_x w), \qquad \forall\, z, w \in \Gamma_{h,\tau}$$

can be evaluated with the complexity and storage expense $O(n_{\partial\tau} \log^q n_{\partial\tau})$, $q = O(1)$. In fact, for the Laplace, biharmonic, Stokes and Lamé operators on rectangular domains, sparse approximations of linear–logarithmic complexity $O(n_{\partial\tau} \log^2 n_{\partial\tau})$ are known (cf. [17, 18, 19]). The extension of these results to the case of polygonal domains is possible [18].

3.3. Approximation Error

In the following, we estimate the approximation error of the scheme defined above. The optimal error bound $O(h)$ is based on the full elliptic regularity of the local problems (3.4) on $\Omega = X(\tau)$, $\tau \in T_1(I)$, as well as on the technical assumption concerning the "stability" of expansion coefficients (see (3.20) below). In the case of non-convex clusters, we arrive at an accuracy $O(h^\beta)$, $0 < \beta < 1$, depending on the elliptic regularity of the subproblems.

Assumption 4 *There exist positive constants C_0, C_1 and $\gamma \in \mathbb{R}$, such that for each $\tau \times \sigma \in P_{far}$ the coefficients $a_\alpha(x)$, $c_\alpha(y)$ for $(x,y) \in \partial X(\tau) \times \partial X(\sigma)$ from (3.8) satisfy the estimates*

$$(3.20) \qquad \sum_{\alpha=1}^{k_1} \left(||a_\alpha||_{1/2,\partial X(\tau)} ||c_\alpha||_{1/2,\partial X(\sigma)} + C_0 \gamma^\alpha_{0,1/2} + C_1 \gamma^\alpha_{1/2,-1/2} \right)$$

$$\leq c\,(|\tau|\,|\sigma|)^\gamma,$$

where

$$\gamma^\alpha_{0,1/2} = ||a_\alpha||_{1/2,\partial X(\tau)} ||c_\alpha||_{0,\partial X(\sigma)} + ||a_\alpha||_{0,\partial X(\tau)} ||c_\alpha||_{1/2,\partial X(\sigma)},$$

$$\gamma^\alpha_{1/2,-1/2} = ||a_\alpha||_{1/2,\partial X(\tau)} ||c_\alpha||_{-1/2,\partial X(\sigma)} + ||a_\alpha||_{-1/2,\partial X(\tau)} ||c_\alpha||_{1/2,\partial X(\sigma)}.$$

Assumption 5 *The operator $\mathcal{L}$ satisfies the Maximum-Minimum Principle ([9]).*

Lemma 3.7. *Let $k_1 = O(p^{d-1})$ and $r = -1/2$. Under Assumptions 4 and 5 there holds*

$$|\langle v, (A - A_{\partial\mathcal{H}})u\rangle_\Omega|$$

$$\leq c\eta^m n^q + c_\gamma \left((C_0^{-1} h^{3/2} + C_1^{-1} h) ||u||_0 ||v||_0 + +ch(||u||_0 ||v||_{-1} + ||u||_{-1} ||v||_0) \right),$$

with $q = O(1)$, for all $u, v \in L^2(\Omega)$, where $c_\gamma = \sum_{\ell=0}^{p} 2^{-d\ell(1+2\gamma)}$.

Proof. First, we use the representation

$$(3.21) \qquad \langle v, (A - A_{\partial\mathcal{H}})u\rangle_\Omega = \langle v, (A - A_{\mathcal{H}})u\rangle_\Omega + \langle v, (A_{\mathcal{H}} - A_{\partial\mathcal{H}})u\rangle_\Omega$$

indicating that the total error contains the standard consistency error $\langle v, (A - A_{\mathcal{H}})u\rangle_\Omega$, as well as the error involved by the local Ritz projections. The first term in the right-hand side in (3.21) is estimated using the stability of problem (3.2) with respect to the Dirichlet data on $\partial X(\tau) \times \partial X(\sigma)$. In fact, the Maximum-Minimum Principle with respect to both the x- and y-variables leads to

$$\max_{(x,y)\in X(\tau)\times X(\sigma)} \left| s(x,y) - \sum_\alpha E_\tau a_\alpha(x) E_\sigma c_\alpha(y) \right|$$

$$\leq \max_{(x,y)\in \partial X(\tau)\times \partial X(\sigma)} \left| s(x,y) - \sum_\alpha a_\alpha(x)\, c_\alpha(y) \right| \leq c\eta^m dist(\tau,\sigma)^{2-d}.$$

Then, similarly to the proof of Theorem 2.4, see [12], the following estimate holds

$$|\langle v, (A - A_{\mathcal{H}})u\rangle_\Omega| \leq c\eta^m n^q ||v||_{0,\Omega} ||u||_{0,\Omega}.$$

To estimate the second term in (3.21), we first note that the choice $image(E_x) = image(E_y) \subset W_h$ implies that the terms $\langle u, E_x a_\alpha\rangle_{X(\tau)}$ and $\langle v, E_y c_\alpha\rangle_{X(\sigma)}$ arising in (3.12) can be evaluated exactly. Then, it is sufficient to consider the bound of

$$\langle v, (A_{\mathcal{H}} - A_{\partial\mathcal{H}})u\rangle_\Omega = \sum_{\ell=2}^{p-p_0} \sum_{\tau\times\sigma\in P_2^\ell} \sum_{\alpha\le k_1} (e_\alpha^1 + e_\alpha^2 + e_\alpha^3 + e_\alpha^4), \tag{3.22}$$

where

$$\begin{aligned}
e_\alpha^1 &= a_\tau(g_{h,u} - g_u, E_x a_\alpha)\cdot a_\sigma(g_{h,v}, E_y c_\alpha),\\
e_\alpha^2 &= a_\tau(g_{h,u}, E_x a_\alpha)\cdot a_\sigma(g_{h,v} - g_v, E_y c_\alpha),\\
e_\alpha^3 &= -\langle u, E_x a_\alpha\rangle_{X(\tau)}\cdot a_\sigma(g_{h,v} - g_v, E_y c_\alpha),\\
e_\alpha^4 &= -a_\tau(g_{h,u} - g_u, E_x a_\alpha)\cdot \langle v, E_y c_\alpha\rangle_{X(\sigma)}.
\end{aligned}$$

Using the Galerkin orthogonality for $g_{h,z}$ and g_z, the a priori estimate for the Dirichlet boundary value problem in $X(\tau)$ and the standard H^1-error bound for the Ritz projection yield

$$|a_\tau(g_{h,u} - g_u, E_x a_\alpha)| = |a_\tau(g_{h,u} - g_u, E_{\tau,h} a_\alpha)| \le c\,h||u||_{0,X(\tau)}||a_\alpha||_{1/2,\partial X(\tau)}.$$

The L^2-estimate of the discrete $\mathcal{L}$-harmonic function

$$||E_{\tau,h} c_\alpha||_{0,X(\sigma)} \le c||c_\alpha||_{-1/2,\partial X(\sigma)} \tag{3.23}$$

is valid in the case of full elliptic regularity, the proof is similar to the case of Laplace equation, considered in [1]. This implies

$$\begin{aligned}
|a_\sigma(g_{h,v}, E_y c_\alpha) &\le c(||g_{h,v}||_{1,X(\sigma)}||E_{\sigma,h}c_\alpha||_{1,X(\sigma)} + ||v||_{0,X(\sigma)}||E_{\sigma,h}c_\alpha||_{0,X(\sigma)})\\
&\le c||v||_{-1,X(\sigma)}||c_\alpha||_{1/2,\partial X(\sigma)} + c||v||_{0,X(\sigma)}||c_\alpha||_{-1/2,\partial X(\sigma)},
\end{aligned}$$

where $||\cdot||_{-1,\Omega}$ is the norm of $H^{-1}(\Omega) = \big(H_0^1(\Omega)\big)'$. Therefore,

$$|e_\alpha^1| \le c\,h||u||_{0,X(\tau)}||a_\alpha||_{1/2,\partial X(\tau)}\big(||v||_{-1,X(\sigma)}||c_\alpha||_{1/2,\partial X(\sigma)} + ||v||_{0,X(\sigma)}||c_\alpha||_{-1/2,\partial X(\sigma)}\big),$$

$$|e_\alpha^2| \le c\,h||v||_{0,X(\sigma)}||c_\alpha||_{1/2,\partial X(\sigma)}\big(||u||_{-1,X(\tau)}||a_\alpha||_{1/2,\partial X(\tau)} + ||u||_{0,X(\tau)}||a_\alpha||_{-1/2,\partial X(\tau)}\big).$$

Using similar arguments and applying the bound

$$||E_y c_\alpha||_{0,X(\sigma)} \le c h^{1/2}||c_\alpha||_{0,\partial X(\sigma)}, \tag{3.24}$$

but now to both a_α and c_α, we obtain

$$|e_\alpha^3| \le c\,h^{3/2}||u||_{0,X(\tau)}||v||_{0,X(\sigma)}||a_\alpha||_{0,\partial X(\tau)}||c_\alpha||_{1/2,\partial X(\sigma)},$$

$$|e_\alpha^4| \le c\,h^{3/2}||u||_{0,X(\tau)}||v||_{0,X(\sigma)}||a_\alpha||_{1/2,\partial X(\tau)}||c_\alpha||_{0,\partial X(\sigma)}.$$

Finally, the substitution of these estimates into (3.22) yields

$$
\begin{aligned}
&|\langle v, (A_{\mathcal{H}} - A_{\partial\mathcal{H}})u\rangle_\Omega| \\
&\leq c\,h \sum_{\tau\times\sigma\in P_2} (\|u\|_{0,X(\tau)}\|v\|_{-1,X(\sigma)} + \|u\|_{-1,X(\tau)}\|v\|_{0,X(\sigma)}) \\
&\quad\cdot\sqrt{|\tau||\sigma|}\sum_{\alpha\leq k_1} \|a_\alpha\|_{1/2,\partial X(\tau)}\|c_\alpha\|_{1/2,\partial X(\sigma)} \\
&\quad + c\,h \sum_{\tau\times\sigma\in P_2} \|u\|_{0,X(\tau)}\|v\|_{0,X(\sigma)}\sqrt{|\tau||\sigma|}\sum_{\alpha\leq k_1}\gamma^{\alpha}_{1/2,-1/2} \\
&\quad + c\,h^{3/2} \sum_{\tau\times\sigma\in P_2} \|u\|_{0,X(\tau)}\|v\|_{0,X(\sigma)}\sqrt{|\tau||\sigma|}\sum_{\alpha\leq k_1}\gamma^{\alpha}_{0,1/2} \\
&\leq c\,h(\|u\|_{0,\Omega}\sum_{\ell=2}^{p-p_0}|\sigma|^{1+2\gamma}\,\|v\|_{-1,X(\sigma)}\sum_{\sigma:\tau\times\sigma\in P_2^\ell} 1 \\
&\quad + \|v\|_{0,\Omega}\sum_{\ell=2}^{p-p_0}|\tau|^{1+2\gamma}\,\|u\|_{-1,X(\tau)}\sum_{\tau:\tau\times\sigma\in P_2^\ell} 1) \\
&\quad + (C_0^{-1}\,h^{3/2} + C_1^{-1}h)\|u\|_{0,\Omega}\|v\|_{0,\Omega}\sum_{\ell=2}^{p-p_0}(|\tau||\sigma|)^{1/2+\gamma}\sum_{\tau:\tau\times\sigma\in P_2^\ell} 1.
\end{aligned}
$$

Hence, the assertion follows. □

Remark 3.8. Lemma 3.7 guarantees a hierarchical approximation of the exact Galerkin stiffness matrix with an error $O(h)$ with respect to the spectral norm. The approximation result remains valid even for $\gamma = 0$ in Assumption 4.

4. Construction of Kernel Expansions (Case B)

4.1. Polynomial Approximation of Multivariate Functions

We assume that our kernel function $s(x, y)$ satisfies Assumption 1, see Section 2.3. For deriving the desired low order expansions, we use classical approximation results for functions which are analytic in the interval $I_1 = [-1, 1]$.

Definition 4.1. A function $f \in C^\infty(I_1)$ has Bernstein's regularity ellipse $\mathcal{E}_H(I_1)$ if it admits an analytic extension to the closed ellipse $\mathcal{E}_H(I_1) \subset \mathbb{C}$ with foci in $z = \pm 1$ and the sum of semi–axes equal to $H > 1$.

The definition of $\mathcal{E}_H(I_1)$ for other intervals than $[-1, 1]$ is obvious.

The following statement goes back to the classical result of S.N. Bernstein (see also [24] for more details). In particular, we apply the result from [23].

Proposition 4.2. *Assume a function $f \in C^\infty(I_1)$ has the regularity ellipse $\mathcal{E}_H(I_1)$, according to Definition 4.1. Let $[I_N^1 f](x) \in P_N[I_1]$ on $[-1,1]$ be the interpolation polynomial with respect to the Chebyshev-Gauss-Lobatto nodes $\xi_j = \cos\frac{\pi j}{N}$, $j = 0,\dots,N$. Then the following approximation property holds:*

$$||f - I_N f||_{L^\infty(I_1)} \le cN \frac{H^{-N}}{H-1} \max_{z\in\mathcal{E}_H(I_1)} |f(z)|. \tag{4.1}$$

For multivariate functions $f = f(x_1,...,x_d) : \mathbb{R}^d \to \mathbb{R}$, we use the tensor product interpolant

$$\mathbf{I}_N f = I_N^1 ... I_N^d f \in P_N[I_1^d],$$

where $I_N^i f$ denotes the interpolation polynomial with respect to the variable x_i, $i = 1,...,d$, at the Chebyshev-Gauss-Lobatto nodes. The interpolation points ξ_α, $\alpha = (i_1,...,i_d) \in \mathbb{N}_0^d$, in I_1^d are obtained by the Cartesian product of the one-dimensional nodes,

$$\xi_\alpha := \left(\cos\frac{\pi i_1}{N}, ..., \cos\frac{\pi i_d}{N}\right).$$

Denote by X_{-i} the subset $X_{-i} := \{x_1,...,x_{i-1},x_{i+1},...,x_d\}$ of $d-1$ spatial variables. Related to Definition 4.1, we are interested in polynomial approximations of the following class of functions.

Assumption 6 *For given function $f \in C^\infty(I_1^d)$, assume there is $H_0 > 1$ such that for each of the subset $z_i \in X_{-i}$, $i = 1,...,d$ there exists an analytic extension with respect to $x_i \in \mathcal{E}_{H_0}(I_i) \subset \mathbb{C}$.*

Proposition 4.3. *Let Assumption 6 be valid. Then, for $1 < H < H_0$ there holds*

$$||f - \mathbf{I}_N f||_{L^\infty(I_1^d)} \le cN \log^{d-1} N \frac{H^{-N}}{H-1} M_H(f), \tag{4.2}$$

$$M_H(f) = \max_{j\le d}\{\max_{X_{-j}} \max_{x_j\in\mathcal{E}_H(I_j)} |f(x_1,...,x_d)|\}.$$

Proof. The multiple use of (4.1) and the triangle inequality lead to

$$|f - \mathbf{I}_N f| \le |f - I_N^1 f| + |I_N^1 (f - I_N^2 \dots I_N^d f)|$$

$$\le |f - I_N^1 f| + |I_N^1 (f - I_N^2 f)| + |I_N^1 I_N^2 (f - I_N^3 f)| +$$

$$\dots + |I_N^1 I_N^2 \dots I_N^{d-1} (f - I_N^d f)|$$

$$\le c \Big(\max_{X_{-1}} \max_{x_1 \in \mathcal{E}_H(I_1)} |f(x)| + \log N \max_{X_{-2}} \max_{x_2 \in \mathcal{E}_H(I_2)} |f(x)| +$$

$$\dots + \log^{d-1} N \max_{X_{-d}} \max_{x_d \in \mathcal{E}_H(I_d)} |f(x)| \Big) N \frac{H^{-N}}{H-1},$$

where, similar to [20], we apply the L^∞–estimate of the scalar interpolant I_N^i with respect to each space variable x_i, $i = 1, ..., d$,

$$||I_N^i f||_{L^\infty(I_i)} \le c \log N ||f||_{L^\infty(I_i)}, \qquad \forall f \in C^0(I_i).$$

Hence (4.2) follows. □

Remark 4.4. In the case of scaled domain I_δ^d, $I_\delta = [-\delta, \delta]$, $\delta > 0$ the factor H^{-N} in the error estimates (4.2), (4.3) is changed by $(H/\delta)^{-N}$.

4.2. Application to Volume Integral Operators (Case B)

Consider the kernel function $s(x_\tau, x_\sigma) = S(x_\tau - x_\sigma)$, $(x_\tau, x_\sigma) \in X(\tau) \times X(\sigma)$ associated with the fundamental solution S of (2.1) satisfying Assumption 1.

In the case $d = 2$ and with the standard Euclidian metric, let $\tau \times \sigma \in P_2$ be a block satisfying the admissibility condition (2.5). In the following, we use the notation $x_\tau = (x_{1\tau}, x_{2\tau})$, $x_\sigma = (x_{1\sigma}, x_{2\sigma})$.

We assume that $X(\tau)$ is a rectangle with the boundary $\partial X(\tau) = \cup_{i=1}^4 \Gamma_\tau^i$ and $\partial X(\sigma) = \cup_{i=1}^4 \Gamma_\sigma^i$ with $|\Gamma_\sigma^i| = |\Gamma_\tau^i| = 2\delta$, see Fig. 1. Suppose that the edges Γ_σ^3 and Γ_τ^1 are parallel to the x_2-axis and satisfy $dist(\Gamma_\sigma^3, \Gamma_\tau^1) = 2\delta$. Construct a kernel expansion on the subset $\Gamma_\sigma^3 \times \Gamma_\tau^1 \subset \partial X(\sigma) \times \partial X(\tau)$.

Due to assumptions from above, the coordinates $x_{1\sigma}$ for $x_\sigma \in \Gamma_\sigma^3$ and $x_{1\tau}$ for $x_\tau \in \Gamma_\tau^1$ are both fixed. Hence, in Euclidean distance, the function of two variables $f(x_{2\sigma}, x_{2\tau}) = s(x_{1\tau}, x_{2\tau}, x_{1\sigma}, x_{2\sigma})$ has the family of regularity ellipses $\mathcal{E}_{H_0}$ in the sense of Assumption 6 and with $H_0 = a + b$, where $a^2 = b^2 + \delta^2$ and the small semiaxis b is bounded by $b < dist(\tau, \sigma)$. Due to (2.5), there holds $dist(\tau, \sigma) \ge \sqrt{d}\delta\eta^{-1}$ implying the upper bound $b < \sqrt{d}\delta\eta^{-1}$ and also $a < \sqrt{1 + d\eta^{-2}}\,\delta$. This yields

$$H_0 < (\sqrt{d}\eta^{-1} + \sqrt{1 + d\eta^{-2}})\delta. \tag{4.3}$$

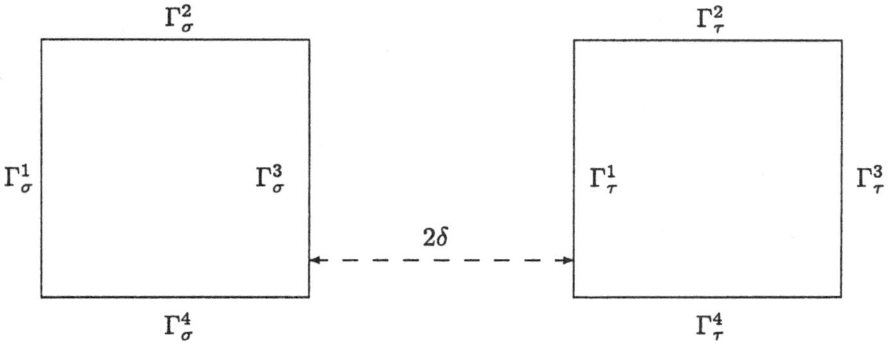

Figure 1: Location of geometrical clusters $X(\tau)$ and $X(\sigma)$.

In particular, for the choice $\eta = \frac{\sqrt{d}}{2}$ with $d = 1, 2, 3$, we obtain

$$H_0 < (2 + \sqrt{5})\delta. \tag{4.4}$$

Applying Proposition 4.3 with the scaling argument from Remark 4.4 leads to the following estimate on the exponent in a convergence rate of the polynomial approximation

$$\left(\frac{H}{\delta}\right)^{-1} \leq \left(\sqrt{d} + \sqrt{d+\eta^2}\right)^{-1} \eta.$$

The constant $M_H(f)$ may be estimated by

$$M_H(f) \leq c \max_{|\xi| \geq \sqrt{d}\delta\eta^{-1} - b} |S(\xi)|.$$

Assume that $|S(\xi)| \leq c|\log|\xi||$. Solving a simple optimization problem,

$$\min_{0<b<\sqrt{d}\delta\eta^{-1}} \frac{H^{-m}}{H-1} M_H(f), \qquad H = b + \sqrt{\delta^2 + b^2}$$

leads to an error estimate of the form

$$||f - \mathbf{I}_m f||_{L^\infty(I_1^2)} \leq c m^q \log^{q_1} m \, (c_1\eta)^m, \qquad q_1 = O(1) \tag{4.5}$$

with $c_1 = \dfrac{1}{\sqrt{d} + \sqrt{d+\eta^2}}$, uniformly with respect to n and $\eta \leq 1$. The bound (4.5) implies the uniform L^∞-estimate

$$||f - \mathbf{I}_m f||_{L^\infty} \leq c h^\alpha = c 2^{-\alpha p}$$

with the degree of polynomials $m = O(p)$ for all $\alpha = O(1)$. Let $\{T_i(x_{2\tau})\}$ and $\{T_j(x_{2\sigma})\}$ be the bases of the corresponding univariate polynomial sets P_N defined on Γ_τ^1 and Γ_σ^3, respectively (say, Chebyshev polynomials). The number of terms k in the product–polynomial interpolant

$$\mathbf{I}_m f := \sum_{i=0}^{m} \sum_{j=0}^{m} a_{ij} T_i(x_{2\tau}) T_j(x_{2\sigma}),$$

where coefficients a_{ij} are the linear combinations of the values $f(\xi_\alpha)$ in the tensor product set of nodes ξ_α, is then estimated by $k = m^2 + 1 = O(p^2)$. Let $|\Gamma_\tau^1| \leq |\Gamma_\sigma^3|$ for definiteness. Then combining all terms having the factor $T_i(x_{2\tau})$, $i = 0, 1, \ldots, m$, we obtain the desired expansion of the order $k = O(p)$,

$$\text{(4.6)} \quad [\mathbf{I}_m f](x_\tau, x_\sigma) := \sum_{i=0}^{m} T_i(x_{2\tau}) \cdot G_i(x_{2\sigma}), \qquad G_i(x_{2\sigma}) = \sum_{j=0}^{m} a_{ij} T_j(x_{2\sigma}).$$

A similar construction for $d = 3$ leads to expansions of the order $O(p^{d-1})$, using Proposition 4.3. The result is based on the polynomial approximations for multivariate analytic functions on a rectangular product piece $\Gamma_\tau^i \times \Gamma_\sigma^j \in \mathbb{R}^4$ of $\partial X(\tau) \times \partial X(\sigma)$, $i, j = 1, ..., 6$.

Remark 4.5. Note that the corresponding Taylor interpolant with respect to the Chebyshev centre of Γ_τ^1 may be also applied for the construction of our wire–basket expansion on the product domain $\Gamma_\tau^i \times \Gamma_\sigma^j$. It has the same order $k = O(p^{d-1})$ but involves a bigger constant $c_1 \simeq 1.0$ in (4.5) (cf. [11]). This is consistent with Chebyshev's classical result that the best polynomial approximation of an analytic function is by far more accurate than the Taylor interpolant. Moreover, (4.6) is based only on the pointwise evaluation of the kernel at the interpolation points.

Remark 4.6. The global degenerate expansion of the order $O(p^{d-1})$ on $\partial X(\tau) \times \partial X(\sigma)$ is constructed in two steps: First, we obtain an expansion on $\Gamma_\tau^1 \times \partial X(\sigma)$ by composing (agglomerating) the expansions for $\Gamma_\sigma^j \times \Gamma_\tau^1$, $j = 1, ..., 4$ based on fixed polynomial basis $\{T_\ell\}_{\ell=0}^m$ on Γ_τ^1, and then by assembling the corresponding representations constructed for each Γ_τ^i, $i = 1, ..., 4$, separately, as above. This approach is suited for the FE approximations of the elliptic extension operator in (3.17).

To digest the performance of our method we proceed as follows. Instead of one global Rk_1-matrix $\mathbf{S} := \sum_{\alpha=1}^{k_1} \mathbf{a}_{h,\alpha} \otimes \mathbf{c}_{h,\alpha}^T$ corresponding to (3.7), we consider the blockwise Rk-approximation of the form

$$\widehat{\mathbf{S}} := \begin{pmatrix} C & B & A & TBT \\ D & F & TBT & G \\ H & TDT & C & TDT \\ TDT & G & TBT & F \end{pmatrix} =: \{\widehat{S}_{ij}\}_{i,j=1}^4,$$

with respect to the degrees of freedom located on the product pieces $\Gamma^i_\tau \times \Gamma^j_\sigma$, $i,j = 1,\ldots,4$. Here T denotes the generic permutation matrix of the size $dim\,\Gamma^i_{h,\tau}$. We introduce the *wire-basket rank* $R_{wb} = R_{wb}(\widehat{\mathbf{S}})$ and the *reduced wire-basket rank* $R_{rwb} = R_{rwb}(\widehat{\mathbf{S}})$ of the matrix-block $\widehat{\mathbf{S}}$ by

$$R_{wb}(\widehat{\mathbf{S}}) := \frac{1}{4}\sum_{i,j=1}^{4} R(\widehat{S}_{ij}), \tag{4.7}$$

and

$$R_{rwb}(\widehat{\mathbf{S}}) := \frac{1}{4}\left(R(A) + R(B) + R(C) + R(D) + R(F) + R(G) + R(H)\right), \tag{4.8}$$

respectively, where $R(A)$ is the rank of A. The value R_{wb} characterises the complexity of matrix-vector multiplication by the block $\widetilde{\mathbf{S}}$, while R_{rwb} specifies the memory requirements. The numerical results estimating R_{rwb} for the harmonic kernel in $2D$ will be presented at the end of this section.

Remark 4.7. Another alternative construction of the wire–basket expansions is designed for using of exact $\mathcal{L}$–harmonic extensions instead of $\mathbf{E}_{\tau,h}$. For this purpose, we apply generalized harmonic polynomials in the tensor–product domain, see also [21] concerning "the operator adapted spectral element methods". For example, in the case of Laplace equation, we apply the system of trigonometric harmonic polynomials $\{Re\, e^{kz}, Im\, e^{kz} | k \in \mathbb{N}_0\}$ in rectangle. Then approximation of the kernel by trigonometric polynomials on the edges Γ^i_τ of a computational cluster allows the exact harmonic extensions into the interior by above defined harmonic polynomials. The system $\{Re\, z^k, Im\, z^k | k \in \mathbb{N}_0\}$ is also dense in the set of harmonic functions and may be applied for approximation within admissible clusters. The details will be discussed in a forthcoming paper.

It is worth to note that in the particular case of *harmonic kernel* $s(x,y) = \frac{1}{4\pi|x-y|}$ for $d = 3$, we obtain an expansion with smaller number of terms compared with the familiar *multipole expansion* of the optimal order $k = O(p^2)$. In fact, let x and y have spherical coordinates (r,θ,ϕ) and (ρ,α,β), respectively. Define spherical harmonics $Y^\mu_\nu(\theta,\phi)$, $\nu = 0,1,2,\ldots$ and $\mu = -\nu,\ldots,\nu$, by

$$Y^\mu_\nu(\theta,\phi) := \sqrt{\frac{2\nu+1}{4\pi}\frac{(\nu-|\mu|)!}{(\nu+|\mu|)!}}\,P^{|\mu|}_\nu(\cos\theta)e^{i\mu\phi},$$

where P^μ_ν are the associated Legendre functions,

$$P^\mu_\nu(x) := (-1)^\mu(1-x^2)^{\mu/2}\frac{d^\mu}{dx^\mu}P_\nu(x),$$

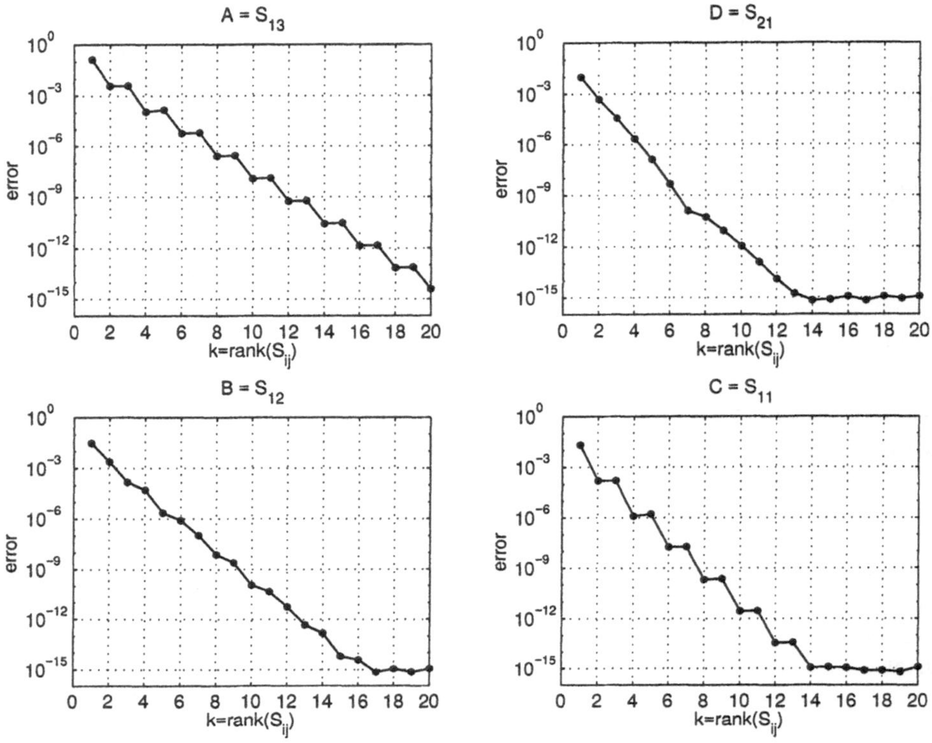

Figure 2: Approximation of blocks A, B, C and D in $\widehat{\mathbf{S}}$.

and $P_\nu(x)$ is the Legendre polynomial of degree ν. It was shown in [6, 4] that the multipole expansion of the form

$$\frac{1}{|x-y|} = \frac{1}{r}\sum_{\nu=0}^{m}\sum_{\mu=-\nu}^{\nu}\left(\frac{\rho}{r}\right)^{\nu} Y_\mu^\nu(\alpha,\beta)Y_\mu^{-\nu}(\theta,\phi) + R_m \tag{4.9}$$

provides the error estimate

$$|R_m| \le \frac{c}{|x-y|}\frac{\gamma^{m+1}}{(1-\gamma)^2}, \qquad \gamma = \frac{1}{2}\frac{diam\tau}{dist(\sigma,\tau_*)} < 1, \tag{4.10}$$

where τ_* is the Chebyshev centre of τ. With the choice $\eta = \frac{\sqrt{3}}{2}$ in the admissibility condition, we obtain $\gamma = \frac{1}{\sqrt{3}} \sim 0.58$, while for $d = 3$, there holds $c_1\eta = (2+\sqrt{5})^{-1} \sim 0.24$, by virtue of (4.4). Therefore, a bound like (4.5) provides better

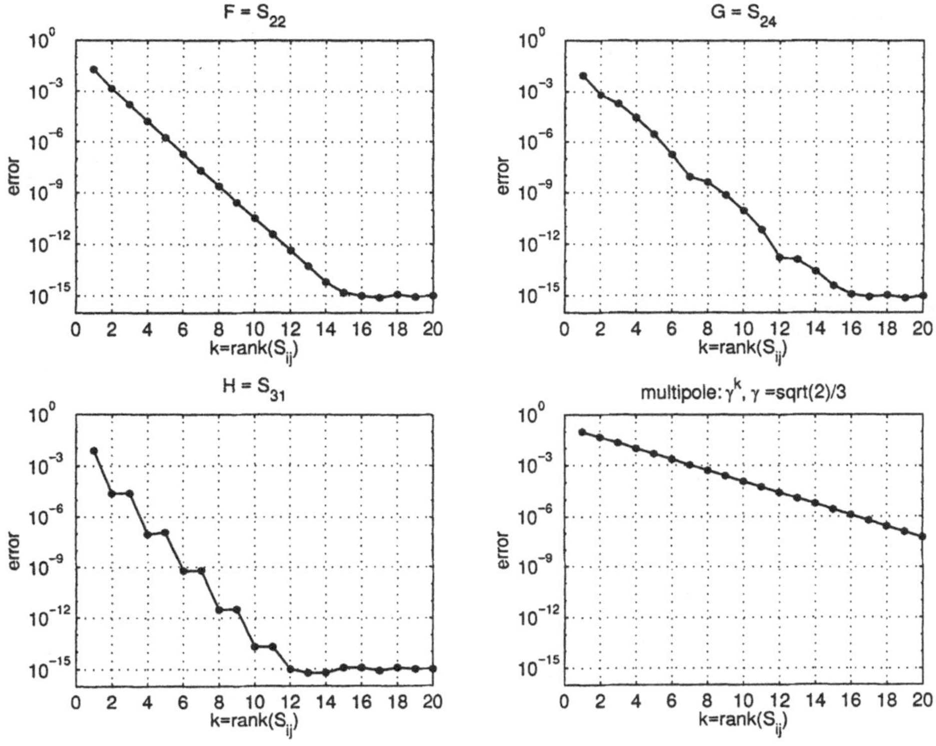

Figure 3: Approximation of blocks F, G and H in $\widehat{\mathbf{S}}$ vs. the multipole expansion.

asymptotic convergence rate than (4.10). At the same time, both the wire–basket and multipole expansions have faster convergence than the Taylor interpolant.

For the $2D$ harmonic potential $s(x,y) = \frac{1}{2\pi}\log|x-y|$, the convergence rate for the multipole expansion with $\eta = \frac{\sqrt{2}}{2}$ is estimated by $\gamma = \frac{\sqrt{2}}{3} \sim 0.47$, while the wire-basket expansion again yields $c_1\eta \sim 0.24$. Applying the degenerate polynomial approximation of kernels on the product domains $\Gamma^i_\tau \times \Gamma^j_\sigma$ by interpolation at the Chebyshev-Gauss-Lobatto points, we obtain the following results for R_{wb} and R_{rwb}, depending on ε, see Fig. 2 and 3. Here we present the maximal approximation error of all blocks in $\widehat{\mathbf{S}}$ versus the degree of the interpolation polynomials. The last picture in Fig. 3 presents the accuracy provided by the 2D multipole expansion corresponding to the exponent $\gamma = \frac{\sqrt{2}}{3}$. Fig. 4 presents the corresponding rank $R_{wb}(\widehat{\mathbf{S}})$ and $R_{rwb}(\widehat{\mathbf{S}})$ defined by (4.7) and (4.8), respectively, depending on the approximation accuracy achieved. Here *Rmp* corresponds to the multipole-like expansion, while *RToepl* stands for the factor $c\log n$ characterising the linear-logarithmic complexity of matrix-vector multiplication by a Toeplitz matrix associated with any pair of parallel edges from $\Gamma_\tau \times \Gamma_\sigma$, see e.g. [26]. This indicates the approximation properties of the wire-basket method.

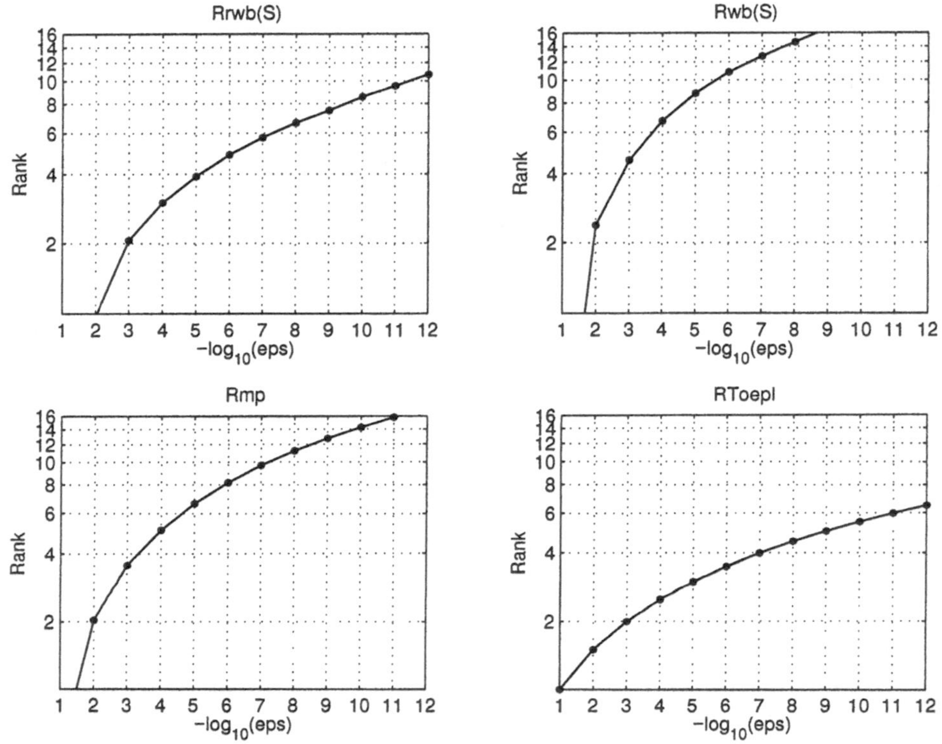

Figure 4: Computational rank for different approximations of $\widehat{\mathbf{S}}$.

Remark 4.8. Sparse approximations may be also applied to the more general class of kernels with bounded mixed derivatives, see (4.11). If the Galerkin ansatz space is defined on the same sparse grid as for the approximation of the kernel function, we obtain an algorithm of complexity $O(n)$ for Case (**B**). In this way the $\mathcal{H}$–matrix technique using sparse grids requires the weakened smoothness condition for $x \neq y$,

$$\left| \frac{\partial^{|\alpha|+|\beta|} s(x,y)}{\partial x_1^{\alpha_1} \ldots \partial x_d^{\alpha_d} \partial y_1^{\beta_1} \ldots \partial y_d^{\beta_d}} \right| \leq c\, |x-y|^{1-3d-2r} \quad \forall\, |\alpha|_\infty, |\beta|_\infty \leq 2, \tag{4.11}$$

for the kernel function $s(x, y)$, where $2r$ is the order of the integral operator (see Assumption 2).

5. Application in BEM

5.1. General Remarks (Case (A))

Consider the case of polygonal/polyhedral boundary $\Gamma = \partial\Omega$, $\Omega \in \mathbb{R}^d$, $d = 2, 3$. To apply the wire-basket expansions in BEM, we assume that for each pair of admissible clusters $(\tau, \sigma) \in P_2^\ell$ there exist two parallelepipeds $\widehat{X}(\tau), \widehat{X}(\sigma) \in \mathbb{R}^d$ which contain $X(\tau)$ and $X(\sigma)$, respectively, and are admissible considered as spatial domains. Specifically, they satisfy the geometrical admissibility condition, similar to (2.5),

$$min\{diam(\widehat{X}(\sigma)), diam(\widehat{X}(\tau))\} \leq 2\eta\, dist(\widehat{X}(\sigma), \widehat{X}(\tau)). \tag{5.1}$$

This condition specifies the choice of $\widehat{X}(\tau)$ and $\widehat{X}(\sigma)$. Then the desired expansion for $s(x, y)$, $(x, y) \in X(\tau) \times X(\sigma)$ is the restriction of the separable $\mathcal{L}$–harmonic extension,

$$s_{\tau,\sigma} := \sum_{\alpha} r_{X(\tau)} E_{\widehat{X}(\tau)} a_\alpha(x) \cdot r_{X(\sigma)} E_{\widehat{X}(\sigma)} c_\alpha(y), \tag{5.2}$$

onto $X(\tau) \times X(\sigma)$, where a_α and c_α provide the kernel expansion on the product boundary of spatial domain $\partial\widehat{X}(\tau) \times \partial\widehat{X}(\sigma)$, see (3.8), and $E_{\widehat{X}(\tau)}$, $E_{\widehat{X}(\sigma)}$ denote the elliptic extension operators into $\widehat{X}(\tau)$ and $\widehat{X}(\sigma)$, respectively. The complexity of (5.2) may depend on the geometry of the pieces $X(\tau)$ and $X(\sigma)$.

If a planar piece $X(\tau)$ (resp. $X(\sigma)$) is parallel to some facet of $\partial\widehat{X}(\tau)$ (resp. $\partial\widehat{X}(\sigma)$) then the evaluation of restrictions in (5.2) needs $O(n_\Gamma \log^2 n_\Gamma)$ arithmetical operations. Here n_Γ is the number of degrees of freedom on $X(\tau)$. For arbitrary plane section of $\widehat{X}(\tau)$ (resp. $\widehat{X}(\sigma)$) this expense is estimated by $O(n_\Gamma \log^3 n_\Gamma)$, see [17, 18] for more details. In the case of curvilinear patches $X(\tau), X(\sigma)$, the linear–logarithmic complexity may be achieved by multilevel expansions which will be discussed separately. However, regardless of the extension procedure, we arrive at expansions of the order $O(p^{d-1})$ in rather general BEM applications.

5.2. Anisotropic Laplacian

We consider the example of kernel expansion in FEM/BEM for $d = 2$ corresponding to the anisotropic Laplace operator. We will be especially interested in the robust and accurate approximations for the singularly perturbed equation.

For the most common operators of the form (2.1), the fundamental solution is known in the explicit form (see [9],[5] and references therein).

Consider the *anisotropic* elliptic operator with

(5.3) $$\mathbf{L} = diag\{a_1, \ldots, a_d\}, \quad a_i = \varepsilon_i^2 > 0, \mathbf{b} = 0, \; c_0 = 0;$$

$$S(x) := \begin{cases} \frac{1}{2\pi\sqrt{det\,\mathbf{L}}} \log \frac{1}{|x|_{\mathbf{L}}} & \text{for } d = 2 \\ \frac{1}{4\pi\sqrt{det\,\mathbf{L}}} \frac{1}{|x|_{\mathbf{L}}} & \text{for } d = 3 \end{cases}$$

in the case $d = 2$, $a_1 = \varepsilon^2$, $a_2 = 1$, $0 < \varepsilon \leq 1$. Let $\tau \times \sigma \in P_2^\ell$ be an admissible (rectangular) block satisfying condition (2.6). The suitable norm involved in the definitions of $diam := diam_{\mathbf{L}}$ and $dist$ is now the *anisotropic norm* $||x|| := ||x||_L := \langle x, \mathbf{L}^{-1}x\rangle^{1/2}$. The coefficient dependent separation scheme is based on the criteria

$$\#\tau' \simeq \#\tau'', \qquad diam\tau' \simeq diam\tau''$$

for all sons $\tau', \tau'' \in \tau$ of the parent cluster τ. Fig. 1 shows those clusters from the resulting tree $T_1(I)$ which correspond to level $\ell = 2$ dependent on the singular perturbation parameter ε. For small enough ε, we have the decomposition into stripes like in the semi-coarsening variant of the multi-grid method. The arising $\mathcal{H}$-matrix becomes close to the block-diagonal one with the blocks corresponding to the vertical grid-lines.

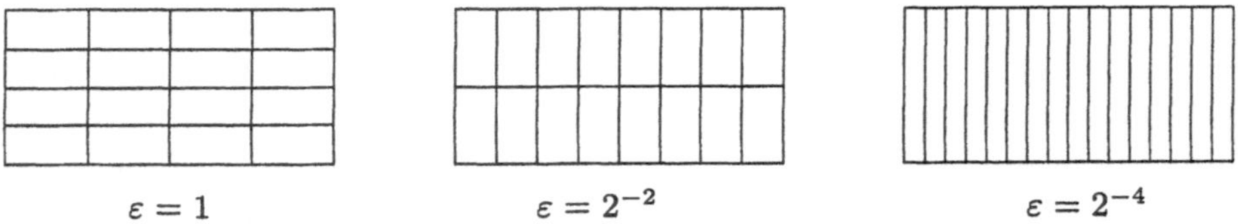

Figure 5: Anisotropy–dependent separation strategies

The error and complexity analysis is identical to the case of the Laplace operator after applying the coordinate transform $y' = y$, $x' = \frac{1}{\varepsilon}x$. Therefore, we skip the details.

References

[1] J.H. Bramble and J.T. King: *A robust finite element method for nonhomogeneous Dirichlet problems in domains with curved boundaries.* Math. Comp. 63(207), 1994, 1-17.

[2] W. Dahmen, S. Prößdorf and R. Schneider: *Wavelet approximation methods for pseudodifferential equations II: Matrix compression and fast solution*, Adv. Comput. Math. 1 (1993) 259-335.

[3] W. Dahmen, S. Prößdorf and R. Schneider, *Wavelet approximation methods for pseudodifferential equations I: Stability and convergence*, Math. Z. **215** (1994) 583-620.

[4] K. Giebermann: *A New Version of Panel Clustering for the Boundary Element Method.* Preprint University of Bonn, January 1999.

[5] M. Gnewuch and S. Sauter: *Boundary Integral Equations for Second Order Elliptic BVPs*, Preprin MPI, 56/1999, Leipzig, 1999.

[6] L. Greengard and V. Rokhlin: *A new version of the fast multipole method for the Laplace equation in three dimensions.* Acta Numerica, 1997, 229-269.

[7] W. Hackbusch: *The panel clustering algorithm.* MAFELAP 1990 (J. R. Whiteman, ed.). Academic Press, London, 1990, 339–348.

[8] W. Hackbusch: *Integral Equations. Theory and Numerical Treatment.* ISNM 128. Birkhäuser, Basel, 1995.

[9] W. Hackbusch: *Elliptic Differential Equations: Theory and Numerical Treatment.* Springer-Verlag, Berlin, 1992.

[10] W. Hackbusch: *A Sparse Matrix Arithmetic based on $\mathcal{H}$-Matrices. Part I: Introduction to $\mathcal{H}$-Matrices.* Computing **62** (1999), 89-108.

[11] W. Hackbusch and B. N. Khoromskij: *A Sparse $\mathcal{H}$-Matrix Arithmetic. Part II. Application to Multi-Dimensional Problems.* Computing **64** (2000), 1, 21-47.

[12] W. Hackbusch and B. N. Khoromskij: *A sparse $\mathcal{H}$-matrix arithmetic: General complexity estimates.* Preprint No. 66, MPI Leipzig, 1999; In: Numerical Analysis in the 20th Century, vol. 6: Ordinary Differential and Integral Equations (J. Pryce, G. Van den Berghe, C.T.H. Baker, G. Monegato, eds.), J. Comp. Appl. Math., Elsevier, 2000.

[13] W. Hackbusch and B. N. Khoromskij: *$\mathcal{H}$-Matrix Approximation on Graded Meshes.* Preprint No. 54, MPI Leipzig, 1999; Proc. of MAFELAP 1999, (J.R. Whiteman, ed.). Academic Press, London, to appear.

[14] W. Hackbusch and B. N. Khoromskij: *Towards $\mathcal{H}$-Matrix Approximation of Linear Complexity.* Preprint No. 15, MPI Leipzig, 2000.

[15] W. Hackbusch, B. N. Khoromskij and S. Sauter: *On $\mathcal{H}^2$-matrices.* Preprint MPI, Leipzig, No. 50, 1999; In: Lectures on Applied Mathematics (H.-J. Bungartz, R. Hoppe, C. Zenger, eds.), Springer-Verlag, 2000, 9-30.

[16] W. Hackbusch and Z. P. Nowak: *On the fast matrix multiplication in the boundary element method by panel clustering.* Numer. Math. **54** (1989) 463–491.

[17] B.N. Khoromskij: *Lectures on Multilevel Schur–Complement Methods for Elliptic Differential Equations.* Preprint 98/4, ICA3, University of Stuttgart, 1998.

[18] B.N. Khoromskij and S. Prößdorf: *Multilevel preconditioning on the refined interface and optimal boundary solvers for the Laplace equation.* Advances in Comp. Math. 4 (1995), 331-355.

[19] B. N. Khoromskij and S. Prößdorf: *Fast computations with harmonic Poincaré-Steklov operators on nested refined meshes.* Advances in Comp. Math., 8 (1998), 111-135.

[20] C. Lage and C. Schwab: *Wavelet Galerkin algorithms for Boundary Integral Equations*, Preprint ETH Zürich, October, 1997.

[21] J.M. Melenk: *Operator adapted spectral element methods I: Harmonic and generalized harmonic polynomials* . Research Report 97-07, ETH Zürich, 1997.

[22] T. von Petersdorff and C. Schwab: *Wavelet approximations for first kind boundary integral equations*, Numer. Math. **74** (1996) 479-516.

[23] E. Tadmor: *The exponential accuracy of Fourier and Chebyshev differencing methods.* SIAM J. Numer. Anal., v. 23, 1986, 1-23.

[24] V. M. Tikhomirov: *Approximation Theory.* In Analysis II, Encyclopaedia of Mathematical Sciences, v. 14, (R.V. Gamkrelidze, ed.) Springer–Verlag, 1990.

[25] E. E. Tyrtyshnikov: *Mosaic-skeleton approximations.* Calcolo **33** (1996) 47–57.

[26] E. P. Zhidkov and B. N. Khoromskij. *Boundary integral equations on special surfaces and their applications.* Sov. J. Numerical Anal. Math. Modelling, North Holland, Antwerpen, **2**, 6, 1988, 463-488.

Max-Planck-Institut für Mathematik in den Naturwissenschaften,
Inselstr. 22-26, D-04103 Leipzig, Germany

1991 Mathematics Subject Classification: Primary 65F05, 65F30; Secondary 65F50

Submitted: 15.5.2000

Operator Theory:
Advances and Applications, Vol. 121

Wavelet Galerkin Schemes for 2D-BEM

HELMUT HARBRECHT AND REINHOLD SCHNEIDER

This paper is concerned with the implementation of the wavelet Galerkin scheme for the Laplacian in two dimensions. We utilize biorthogonal wavelets constructed by A. Cohen, I. Daubechies and J.-C. Feauveau in [3] for the discretization leading to quasi-sparse system matrices which can be compressed without loss of accuracy. We develop algorithms for the computation of the compressed system matrices whose complexity is optimal, i.e., the complexity for assembling the system matrices in the wavelet basis is $\mathcal{O}(N_J)$, where N_J denotes the number of unknowns.

1. Introduction

Various problems in science and engineering can be formulated by integral equations. One of the most prominent type of those integral equations are boundary integral equations. Usually, these equations are solved numerically by the boundary element method (BEM). The boundary element method has been considered as an appropriate tool to solve certain boundary value problems. For example, the boundary element method is a favourable approach for the treatment of exterior boundary value problems. Nevertheless, traditional discretizations of integral equations suffer from a major disadvantage. The corresponding system matrices are densely populated. Therefore, the complexity for solving such equations is at least $\mathcal{O}(N_J^2)$, where N_J denotes the number of equations. This fact restricts the maximal size of the linear equations seriously.

Modern methods for the fast solution of BEM reduce the complexity to a suboptimal rate, i.e., $\mathcal{O}(N_J \log^\alpha N_J)$, or even an optimal rate, i.e., $\mathcal{O}(N_J)$. Prominent examples for such methods are the *fast multipole method* [19], the *panel clustering* [22] or *hierarchical matrices* [21, 36]. As introduced by [1] wavelet bases offer another tool for the fast solution of integral equations. In fact, a Galerkin discretization based on wavelet bases results in numerically sparse matrices, i.e., many matrix entries are negligible and can be treated as zero. Discarding these nonrelevant matrix entries is called matrix compression. Therefore, the full matrix is replaced by a sparse matrix. Of course, the compression procedure induces a perturbation of the original Galerkin discretization. Consequently, the resulting solution differs from the solution of the uncompressed scheme. In [1] this error has been estimated in L_2. It has been shown that, for any $\varepsilon > 0$, a sparse matrix exists such that the compression introduces an error $\leq \varepsilon$.

The article [1] has initiated the investigation of wavelet methods for integral equations, pseudodifferential equations and boundary integral equations [8, 9, 10]. These papers written by Siegfried Prößdorf and coauthors considered also operators of nonzero order by an appropriate preconditioning. Based on norm equivalences [7, 17, 27] it has been shown that for strongly elliptic operators after a diagonal preconditioning the wavelet Galerkin matrices are well conditioned. A new strategy has been introduced to retain the convergence behaviour of the corresponding Galerkin scheme without compromising the complexity of the compression. These early results have been improved by several authors [11, 12, 35] and [30, 31, 33, 34]. Concerning boundary integral equations a strong effort has been spent on the construction of appropriate wavelet bases on surfaces [6, 13, 14, 30, 35]. Furthermore, the efficient computation of the relevant matrix coefficients turned out to be an important task for the successful application of the wavelet Galerkin method [12, 32, 33, 35].

The purpose of the present paper is to describe a fully discrete wavelet Galerkin scheme for boundary integral equations in 2D since this has been omitted in the previous papers. Although the three dimensional boundary value problems are of higher interest, the development and practical realization of 2D wavelet Galerkin methods is of importance by its own. Two dimensional or axial symmetric boundary value problems play an important role in practical applications. Wavelets offer a highly accurate tool to solve these equations. This fact is mainly retained in the presence of piecewise smooth boundaries. In particular, the combination of the finite element method with the boundary element method, or equivalently, the exact use of artificial boundary conditions, is a practically and highly interesting approach for which wavelets seems to be advantageous [24, 25]. For all kind of those problems, the two dimensional case is an excellent object for numerical studies and experiments. One reason is that, in connection with wavelet methods, on curves more scales can be realized. Therefore, one can achieve a relatively high accuracy of the discretization. This allows the study of the asymptotic behaviour of the solution together with the complexity of the algorithm. Furthermore, the implementation is much easier than it is in the three dimensional case. For this reason, testing or modifying algorithms and ideas can be realized quickly. Further developments and improvements of higher dimensional methods will benefit from the experiences in two dimensions. Of course, there are particularities in the two dimensional case which have no counterpart in three dimensions. A proper realization has also to exploit the special properties of the two dimensional case. One purpose of the present paper is to focus on these particular properties, whereas for the theoretical foundations we refer to [5, 11, 12, 24, 31, 35].

The present paper is organized as follows. As typical examples for boundary integral equations, we consider in section 2. the indirect formulations for the Dirichlet problem. Then, only a single function appears on the right hand side of the integral equation. We employ all kind of integral operators resulting from second order boundary value problems and derive Fredholm integral equations of the first kind and of the second kind. For the sake of completeness we mention also Neumann problems. In this respect, we indicate the treatment of the hypersingular operator. It is worth to mention that the present approach can also be applied to direct formulations, see e.g. [24, 25] for more details. For the sake of brevity, in the present paper we treat only the Laplacian explicitly. For further literature on boundary integral equations we refer for example to [4, 20, 28]. The present variational formulations are chosen such that only globally smooth kernels must be handled by numerical integration while the singular parts can be computed analytically. This trick simplifies and accelerates the matrix generation.

In section 3. we define the multiresolution analysis and the wavelet bases on curves through a regular parametrization. Since the coarse scale bases are defined on rather coarse grids, we need a global representation of the boundary curve. ¿From this perspective it becomes obligatory that a wavelet Galerkin scheme has to be a fully discrete one [12, 32, 35], i.e., the computation of the relevant matrix entries requires numerical integration. There are two approximating steps: The first one is the matrix compression, the other one results from numerical integration. The proposed matrix compression strategy has been developed in [8, 12, 31, 35]. In order to avoid logarithmic terms in complexity a second compression step, introduced in [35], is applied to the matrix entries corresponding to wavelets with overlapping supports. Moreover, we briefly recall the wavelet preconditioning.

We explain in section 4. concretely how to establish the matrix pattern of the compressed matrix and how to compute the required matrix entries. It turns out that a naive calculation of the required entries leads to a repeated computation of integrals. To avoid this multiple computation an improved strategy is proposed. We mention that the described algorithms are also practicable in higher dimensions.

Section 5. is devoted to the analysis of the numerical integration. Similarly to the matrix compression, the perturbation resulting from the numerical integration is studied in [12, 32, 35]. It turns out that one has to achieve a certain accuracy $\varepsilon_{j,j'}$ which depends strongly on the corresponding scales (j, j'). In order to achieve the desired accuracies, we utilize exponentially convergent quadratures rules [12, 32, 35], e.g. Gauß-Legendre formulas. For this reason we require (piecewise) analytic curves and parametric representations. The present quadrature algorithm differs from those proposed in [12, 32, 35] since global analyticity of the underlying kernels is exploited. Moreover, we describe a strategy improving the stability and accuracy of the quadrature while the efficiency is not compromised. We want to remark, that this strategy can also be used in 3D, but the present analysis of the convergence and the complexity is valid only for two dimensional problems.

At the end of the paper, in section 6., we present various numerical experiments. These experiments confirm the theoretical results quite well. The accuracy of the Galerkin scheme has never been deteriorated by the matrix compression. In fact, in many cases the solution of the compressed scheme is even slightly better than the solution of the uncompressed scheme. Though it has not been considered in the present paper, the fast wavelet Galerkin scheme can be applied to nonsmooth boundaries as well.

Finally, let us remark, that we have left open several problems. We have not fully exploited the operator splitting $A = A_{\text{sing}} + A_\infty$ into the operator A_{sing} with a singular kernel and the operator A_∞ with an analytical kernel. In fact, A_∞ can be compressed in a more appropriate way, cf. [29]. Whereas, this strategy fails for nonsmooth boundaries, e.g. on boundaries with corners. Perhaps, the treatment of nonsmooth boundaries requires a more careful analysis combined with some minor modifications. The development will be deferred to a forthcoming paper. However, this would be performed in conjunction with an adaptive strategy [2]. Throughout the present paper we only consider asymptotical behaviour. There are situations where this might be far from realistic computations. For instance, if highly oscillating solutions occur, like for the Helmholtz equation with a large Helmholtz number, or homogenization phenomena, the realization of asymptotic convergence and high accuracy is far beyond hardware facilities.

Incorporating the adaptive approximation of the solution or the multiscale approximation of the geometry fit completely into the multiscale concept [2, 16]. ¿From this perspective the fast wavelet method offers a highly powerful tool for the numerical solution of two dimensional problems. However, for three dimensional problems the situation is still quite different. Due to geometric and topological subtleties, the realization of the present concept of wavelet Galerkin schemes is much more difficult in 3D. A strong effort has been spent into this direction during the past five years. Nevertheless, at the present stage of development, the methods based on a hp-approximation of the potentials, like the fast multipole method or the panel clustering, are more flexible and robust to handle complex three dimensional geometries.

Throughout this paper $a \lesssim b$ expresses that a can be bounded by a constant multiple of b uniformly in any parameter on which a and b may depend. Likewise $a \sim b$ means that $a \lesssim b$ and $a \gtrsim b$.

2. The boundary element method

2.1. Dirichlet problems

Let $\Omega^- \in \mathbb{R}^2$ be a bounded and simply connected domain with smooth boundary $\Gamma := \partial\Omega^-$. We set $\Omega^+ = \mathbb{R} \setminus \Omega^-$ and choose $\Omega = \Omega^+$ or $\Omega = \Omega^-$. We denote by $L^2(\Gamma)$ the function space of all square integrable functions on Γ with respect to the canonical inner product

$$(u, v)_{L^2(\Gamma)} = \int_\Gamma u(x)v(x)ds_x \tag{2.1}$$

and by $H^q(\Gamma)$ ($q \in \mathbb{R}$) the corresponding Sobolev spaces. Moreover, $L^2(\Omega)$ indicates the function space of all square integrable functions on Ω with respect to the inner product

$$(u, v)_{L^2(\Omega)} = \int_\Omega u(x)v(x)dx.$$

For $q \in \mathbb{R}$ the space $H^q(\Omega)$ denotes the corresponding Sobolev space.

For a given $f \in H^{1/2}(\Gamma)$ we consider a Dirichlet problem, i.e., we seek $u \in H^1(\Omega)$ such that

$$\begin{aligned} -\Delta u &= 0 \qquad \text{in } \Omega, \\ u &= f \qquad \text{on } \Gamma. \end{aligned} \tag{2.2}$$

For $\Omega = \Omega^-$ the problem is called interior Dirichlet problem. On the other hand, for $\Omega = \Omega^+$ we obtain an exterior Dirichlet problem. In the latter case one additionally demands that

$$u(x) = \mathcal{O}(1) \quad \text{as } |x| \to \infty$$

uniformly for all directions $x/|x|$.

1. Fredholm's integral equation of the first kind for Dirichlet problems: For solving an interior or exterior Dirichlet problem (2.2) by a Fredholm's integral equation of the first kind we introduce the *single layer operator* $\mathcal{V}$

$$(\mathcal{V}\rho)(x) := \int_\Gamma E(x,y)\rho(y)ds_y, \quad x \in \Gamma, \tag{2.3}$$

where the fundamental solution $E(x, y)$ is given by

$$E(x,y) = -\frac{1}{2\pi}\log|x-y|. \tag{2.4}$$

Then, one finds the boundary integral equation

$$\mathcal{V}\rho = f \quad \text{on } \Gamma \tag{2.5}$$

for the unknown density ρ. Knowing ρ the solution u of the Dirichlet problem (2.2) is given by

$$u(x) = \int_\Gamma E(x,y)\rho(y)ds_y, \quad x \in \Omega.$$

In the context of the boundary integral equation (2.5) the single layer operator $\mathcal{V} : H^{-1/2}(\Gamma) \to H^{1/2}(\Gamma)$ defines an operator of order -1, which is symmetric and positive definite if $\operatorname{diam}\Omega^- < 1$, cf. [26].

2. Fredholm's integral equation of the second kind for Dirichlet problems: For solving the Dirichlet problem (2.2) by a Fredholm's integral equation of the second kind we define the *double layer operator*

$$(\mathcal{K}\rho)(x) := \int_\Gamma \frac{\partial}{\partial n_y}E(x,y)\rho(y)ds_y, \quad x \in \Gamma, \tag{2.6}$$

with $E(x, y)$ from (2.4). Note that, here and in the sequel, n_y denotes the unit normal at $y \in \Gamma$ which is oriented to the outside of Ω^-. One finds the equation

$$(\mathcal{K} \pm \tfrac{1}{2}I)\rho = f \quad \text{on } \Gamma, \tag{2.7}$$

where one chooses "+" for an exterior and "−" for an interior problem. In both cases the solution u is represented by

$$u(x) = \int_\Gamma \frac{\partial}{\partial n_y} E(x,y)\rho(y)ds_y, \quad x \in \Omega.$$

The operator on the left hand side of (2.7) defines an operator of order zero, i.e., $\mathcal{K} \pm \frac{1}{2}I : L^2(\Gamma) \to L^2(\Gamma)$. Let us remark that sometimes it is convenient to consider $\mathcal{K} \pm \frac{1}{2}I : H^{1/2}(\Gamma) \to H^{1/2}(\Gamma)$, see e.g. [24]. The equation for the interior problem is uniquely solvable while the equation for the exterior problem has no unique solution since $-\frac{1}{2}$ is an eigenvalue of $\mathcal{K}$. Hence, to ensure uniqueness of the solution ρ in the case of the exterior problem we have to suppose that $\int_\Gamma f(x)ds_x = 0$ and $\int_\Gamma \rho(x)ds_x = 0$.

It is well known [20, 28], that these formulations yield a unique solution $u \in H^1(\Omega)$ of the Dirichlet problem (2.2). Since the solution u is not obtained directly by the boundary integral equations (2.5) and (2.7), respectively, these formulations are called *indirect methods*.

2.2. Variational formulations

The smooth boundary Γ can be parametrized by a 1-periodic function $\gamma : [0,1] \to \Gamma$ such that

$$\alpha(t) := |\gamma'(t)| > 0 \tag{2.8}$$

for all $t \in [0,1]$. Indeed, we suppose that γ is analytic in $[0,1]$. In addition to the spaces $L^2(\Gamma)$ and $H^q(\Gamma)$ we introduce 1-periodic spaces $L^2(0,1)$ and $H^q(0,1)$, respectively. Precisely, let $L^2(0,1)$ be the space of all 1-periodic square integrable functions. Its inner product is denoted by

$$(v,w)_{L^2(0,1)} := \int_0^1 v(t)w(t)dt. \tag{2.9}$$

Then, for any real number q the 1-periodic Sobolev space $H^q(0,1)$ is defined as the closure with respect to the norm

$$\|v\|^2_{H^q(0,1)} = \sum_{n\in\mathbb{Z}} (1+|n|)^{2q}|\hat{v}(n)|^2$$

of the space of all 1-periodic C^∞-functions. Here, $\hat{v}(n)$ indicate the Fourier coefficients

$$\hat{v}(n) = \int_0^1 e^{-2\pi ins}v(s)ds, \quad n \in \mathbb{Z}.$$

Invoking the definition (2.8) the comparison of (2.1) and (2.9) implies

$$(v,w)_{L^2(\Gamma)} = \big(v\circ\gamma, (w\circ\gamma)\alpha\big)_{L^2(0,1)} \tag{2.10}$$

for all $v \in H^q(\Gamma)$ and $w \in H^{-q}(\Gamma)$.

1. Variational formulation for the equation of the first kind: The variational formulation of (2.5) in the $L^2(\Gamma)$ inner product (2.1) is given by

$$\text{seek } \rho \in H^{-1/2}(\Gamma): \quad (\mathcal{V}\rho, \delta)_{L^2(\Gamma)} = (f, \delta)_{L^2(\Gamma)} \quad \forall\, \delta \in H^{-1/2}(\Gamma).$$

We define the integral operator

$$V : H^{-1/2}(0,1) \to H^{1/2}(0,1), \quad (V\mu)(s) := \int_0^1 E\big(\gamma(s), \gamma(t)\big)\mu(t)dt. \tag{2.11}$$

Using (2.10) we find for $\rho, \delta \in H^{-1/2}(\Gamma)$ the identity $(\mathcal{V}\rho, \delta)_{L^2(\Gamma)} = \big(V\big((\rho \circ \gamma)\alpha\big), (\delta \circ \gamma)\alpha\big)_{L^2(0,1)}$. Thereby, V inherits symmetry and positive definiteness from the operator $\mathcal{V}$. Setting $\mu := (\rho \circ \gamma)\alpha \in H^{-1/2}(0,1)$ and $\nu := (\delta \circ \gamma)\alpha \in H^{-1/2}(0,1)$ leads to the variational formulation in $L^2(0,1)$

$$\begin{aligned} &\text{seek } \mu \in H^{-1/2}(0,1): \\ &\quad (V\mu, \nu)_{L^2(0,1)} = (f \circ \gamma, \nu)_{L^2(0,1)} \quad \forall\, \nu \in H^{-1/2}(0,1). \end{aligned} \tag{2.12}$$

Knowing the density μ the solution u is obtained from

$$u(x) = \int_0^1 E\big(x, \gamma(t)\big)\mu(t)dt, \qquad x \in \Omega. \tag{2.13}$$

2. Variational formulation for the equation of the second kind: The variational formulation of (2.7) in $L^2(\Gamma)$ is given by

$$\text{seek } \rho \in L^2(\Gamma): \quad (\mathcal{K}\rho, \delta)_{L^2(\Gamma)} \pm \frac{1}{2}(\rho, \delta)_{L^2(\Gamma)} = (f, \delta)_{L^2(\Gamma)} \quad \forall\, \delta \in L^2(\Gamma).$$

For $t \in [0,1]$ it is convenient to abbreviate here and in the sequel $n_t := n_{\gamma(t)}$. We introduce

$$K : L^2(0,1) \to L^2(0,1), \quad (K\mu)(s) := \int_0^1 \frac{\partial}{\partial n_t} E\big(\gamma(s), \gamma(t)\big)\mu(t)\alpha(t)dt, \tag{2.14}$$

with $E(x,y)$ from (2.4). Comparing the definitions of $\mathcal{K}$ (2.6) and K (2.14) one finds for $\rho, \delta \in L^2(\Gamma)$ the identity $(\mathcal{K}\rho, \delta)_{L^2(\Gamma)} = \big(K(\rho \circ \gamma), (\delta \circ \gamma)\alpha\big)_{L^2(0,1)}$. The substitutions $\mu := \rho \circ \gamma \in L^2(0,1)$ and $\nu := (\delta \circ \gamma)\alpha \in L^2(0,1)$ yield the formulation

$$\begin{aligned} &\text{seek } \mu \in L^2(0,1): \\ &\quad (K\mu, \nu)_{L^2(0,1)} \pm \frac{1}{2}(\mu, \nu)_{L^2(0,1)} = (f \circ \gamma, \nu)_{L^2(0,1)} \quad \forall\, \nu \in L^2(0,1). \end{aligned} \tag{2.15}$$

For the exterior problem we suppose $(f \circ \gamma, \alpha)_{L^2(0,1)} = 0$. Further, the density μ has to satisfy the side condition

$$(\mu, \alpha)_{L^2(0,1)} = 0. \tag{2.16}$$

The solution u is represented by the potential evaluation

$$u(x) = \int_0^1 \frac{\partial}{\partial n_t} E\big(x, \gamma(t)\big)\mu(t)\alpha(t)dt, \quad x \in \Omega. \tag{2.17}$$

2.3. Discretization

For a given $j \in \mathbb{N}$, $j \geq j_0$, choose $N_j = 2^j$ and $\Delta_j := \{0, 1, \ldots, N_j - 1\}$. We subdivide the (periodic) interval $[0,1]$ by $t_k^{(j)} = k/N_j$, $k \in \Delta_j$, to obtain an equidistant partition with step width $h_j := 1/N_j = 2^{-j}$. For the sake of simplicity we identify here and in the sequel the points $t_{k+lN_j}^{(j)}$, $l \in \mathbb{Z}$, with the points $t_k^{(j)}$. For the discretization we employ L^2-normalized piecewise constant functions

$$\phi_{j,k}^{(1)} := 2^{j/2} \chi_{[t_k^{(j)}, t_{k+1}^{(j)}]}, \quad k \in \Delta_j, \tag{2.18}$$

or piecewise linear functions

$$\phi_{j,k}^{(2)}(x) := 2^{3j/2} \begin{cases} x - t_{k-1}^{(j)}, & \text{for } x \in [t_{k-1}^{(j)}, t_k^{(j)}], \\ t_{k+1}^{(j)} - x, & \text{for } x \in [t_k^{(j)}, t_{k+1}^{(j)}], \\ 0, & \text{elsewhere,} \end{cases} \quad k \in \Delta_j. \tag{2.19}$$

The spaces $V_j := \operatorname{span}\{\phi_{j,k}^{(d)} : k \in \Delta_j\}$, $j \geq j_0$, $d = 1, 2$, form a nested sequence of subspaces in $L^2(0,1)$

$$V_{j_0} \subset V_{j_0+1} \subset \ldots, \qquad \overline{\bigcup_{j \geq j_0} V_j} = L^2(0,1), \qquad \bigcap_{j \geq j_0} V_j = V_{j_0}.$$

For the Galerkin scheme we replace the energy spaces $H^{-1/2}(0,1)$ and $L^2(0,1)$ in the variational formulations (2.12) and (2.15) by the finite dimensional spaces V_j, respectively. Then, for the equation of the first kind (2.12), the Ansatz $\mu = \sum \mu_k \phi_{j,k}^{(d)}$ together with

$$\begin{aligned} [\mathbf{V}_\phi]_{k,k'} &= \big(V\phi_{j,k'}^{(d)}, \phi_{j,k}^{(d)}\big)_{L^2(0,1)}, \\ [\boldsymbol{\mu}_\phi]_k = \mu_k, &\quad [\mathbf{f}_\phi]_k = \big(f \circ \gamma, \phi_{j,k}^{(d)}\big)_{L^2(0,1)}, \end{aligned} \tag{2.20}$$

leads to the linear system of equations

$$\mathbf{V}_\phi \boldsymbol{\mu}_\phi = \mathbf{f}_\phi. \tag{2.21}$$

Likewise, with $\boldsymbol{\mu}_\phi$, $\mathbf{f}_\phi$ as in (2.20) and

$$[\mathbf{K}_\phi]_{k,k'} = \big(K\phi_{j,k'}^{(d)}, \phi_{j,k}^{(d)}\big)_{L^2(0,1)}, \qquad [\mathbf{B}_\phi]_{k,k'} = \big(\phi_{j,k'}^{(d)}, \phi_{j,k}^{(d)}\big)_{L^2(0,1)}. \tag{2.22}$$

we obtain for the equation of the second kind (2.15) the linear system of equations

$$(\mathbf{K}_\phi \pm \tfrac{1}{2}\mathbf{B}_\phi)\boldsymbol{\mu}_\phi = \mathbf{f}_\phi. \tag{2.23}$$

For an exterior Dirichlet problem the system matrix $\mathbf{K}_\phi + \frac{1}{2}\mathbf{B}_\phi$ is singular. Hence, we discard in this case one of the unknowns in (2.23) which yields a reduced but regular system matrix with a unique solution $\tilde{\boldsymbol{\mu}}$. Without loss of generality we may assume that the skipped unknown is the last one. Then, the vector $\left[\begin{smallmatrix}\tilde{\mu}\\0\end{smallmatrix}\right]$ solves the original singular linear system of equations. To satisfy the side condition (2.16) one computes the vector $[\mathbf{a}]_k := \left(\phi_{j,k}^{(d)}, \alpha\right)_{L^2(0,1)}$ and sets $\boldsymbol{\mu}_\phi := \left[\begin{smallmatrix}\tilde{\mu}\\0\end{smallmatrix}\right] - \left(\mathbf{a}^T\left[\begin{smallmatrix}\tilde{\mu}\\0\end{smallmatrix}\right]\right)\mathbf{a}$.

The most expensive work for solving a given Dirichlet problem is the computation of the system matrices $\mathbf{V}_\phi$ (2.20) and $\mathbf{K}_\phi$ (2.22). For assembling these matrices one has to evaluate for all $k, k' \in \triangle_j$ the integrals

$$\alpha^{(d)}_{(j,k),(j,k')} := \int_0^1 \int_0^1 k(s,t)\phi^{(d)}_{j,k'}(t)\phi^{(d)}_{j,k}(s)dtds \tag{2.24}$$

with some kernel function $k(s,t)$. More precisely, for piecewise constant functions, that is $d = 1$, one has to compute for all $k, k' \in \triangle_j$ the values

$$\alpha^{(1)}_{(j,k),(j,k')} = h_j \int_0^1 \int_0^1 k\left(t_k^{(j)} + h_j s, t_{k'}^{(j)} + h_j t\right)dtds. \tag{2.25}$$

Since the piecewise linear functions ($d = 2$) are not smooth in their support we compute in this case for all $k, k' \in \triangle_j$ the required integrals on the domain $[t_k^{(j)} + t_{k+1}^{(j)}] \times [t_{k'}^{(j)} + t_{k'+1}^{(j)}]$

$$\alpha^{(2,i)}_{(j,k),(j,k')} := h_j \int_0^1 \int_0^1 k\left(t_k^{(j)} + h_j s, t_{k'}^{(j)} + h_j t\right) p_i(s,t)\, dtds, \tag{2.26}$$

where $p_i(s,t)$, $i = 1,2,3,4$, are the polynomials

$$\begin{aligned} p_1(s,t) &:= st, & p_3(s,t) &:= (1-s)t, \\ p_2(s,t) &:= s(1-t), & p_4(s,t) &:= (1-s)(1-t). \end{aligned} \tag{2.27}$$

Then, we add them to the corresponding matrix entries according to

$$\alpha^{(2)}_{(j,k),(j,k')} = \alpha^{(2,1)}_{(j,k-1),(j,k'-1)} + \alpha^{(2,2)}_{(j,k-1),(j,k')} + \alpha^{(2,3)}_{(j,k),(j,k'-1)} + \alpha^{(2,4)}_{(j,k),(j,k')}. \tag{2.28}$$

A full quadrature scheme for the calculation of the integrals (2.25) and (2.26) by employing Gauß-Legendre quadrature rules is given in section 5..

2.4. Solving Neumann problems

For a given $g \in H^{-1/2}(\Gamma)$ with

$$\int_\Gamma g(x)ds_x = 0, \tag{2.29}$$

we consider a Neumann problem on the domain Ω, that is, we seek $u \in H^1(\Omega)$ such that

$$\begin{aligned} -\Delta u &= 0 && \text{in } \Omega, \\ \frac{\partial u}{\partial n} &= g && \text{on } \Gamma. \end{aligned} \tag{2.30}$$

Analogously to the Dirichlet problem, depending on the choice of Ω the problem indicates an interior Neumann problem and an exterior Neumann problem, respectively. For an exterior Neumann problem it is additionally required that

$$u(x) = \mathcal{O}(1) \quad \text{as } |x| \to \infty$$

uniformly for all directions $x/|x|$.

1. Fredholm's integral equation of the first kind for Neumann problems: The *hypersingular operator* $\mathcal{W}$ is given by

$$(\mathcal{W}\rho)(x) := -\frac{\partial}{\partial n_x} \int_\Gamma \frac{\partial}{\partial n_y} E(x,y)\rho(y) ds_y, \quad x \in \Gamma, \tag{2.31}$$

and defines an operator of order +1, i.e., $\mathcal{W} : H^{1/2}(\Gamma) \to H^{-1/2}(\Gamma)$. We seek the density ρ satisfying the Fredholm integral equation of the first kind

$$\mathcal{W}\rho = g \quad \text{on } \Gamma. \tag{2.32}$$

Since $\mathcal{W}$ is symmetric and positive semidefinite, cf. [20, 28], one restricts ρ by $\int_\Gamma \rho(x) ds_x = 0$. The variational formulation of (2.32) reads

$$\text{seek } \rho \in H^{1/2}(\Gamma): \quad (\mathcal{W}\rho, \delta)_{L^2(\Gamma)} = (g, \delta)_{L^2(\Gamma)} \quad \forall\, \delta \in H^{1/2}(\Gamma).$$

With $W : H^{1/2}(0,1) \to H^{-1/2}(0,1)$,

$$(W\mu)(s) := -\frac{\partial}{\partial n_s} \int_0^1 \frac{\partial}{\partial n_t} E\big(\gamma(s), \gamma(t)\big) \mu(t) \alpha(s) \alpha(t) dt, \tag{2.33}$$

it follows for all $\rho, \delta \in H^{1/2}(\Gamma)$ the equation $(\mathcal{W}\rho, \delta)_{L^2(\Gamma)} = \big(W(\rho \circ \gamma), \delta \circ \gamma\big)_{L^2(0,1)}$. Hence, replacing $\rho \circ \gamma \in H^{1/2}(0,1)$ and $\delta \circ \gamma \in H^{1/2}(0,1)$ by μ and ν, respectively, gives the variational formulation

$$\text{seek } \mu \in H^{1/2}(0,1): \quad (W\mu, \nu)_{L^2(0,1)} = (g \circ \gamma, \nu\alpha)_{L^2(0,1)} \quad \forall\, \nu \in H^{1/2}(0,1).$$

Herein, the restriction on the density takes the form (2.16). Since the energy space is $H^{1/2}(0,1)$ we have to discretize the variational formulation by piecewise linear functions, that is $d = 2$, which yields the Galerkin system

$$\mathbf{W}_\phi \boldsymbol{\mu}_\phi = \mathbf{g}_\phi \tag{2.34}$$

with $\boldsymbol{\mu}_\phi$ as in (2.20) and

$$[\mathbf{W}_\phi]_{k,k'} = \big(W\phi_{j,k'}^{(2)}, \phi_{j,k}^{(2)}\big)_{L^2(0,1)}, \qquad [\mathbf{g}_\phi]_k = \big(g \circ \gamma, \phi_{j,k}^{(2)}\alpha\big)_{L^2(0,1)}.$$

Since the piecewise linear functions lie in $H^1(0,1)$ we may employ the identity $(W\mu,\nu)_{L^2(0,1)} = (V\mu',\nu')_{L^2(0,1)}$, which holds for all $\mu,\nu \in H^1(0,1)$, see [18, 24] for details. Hence, observing that the derivatives of the Ansatz functions are piecewise constant functions, we find for the system matrix $\mathbf{W}_\phi$ the equation

$$\mathbf{W}_\phi = \mathbf{H}\mathbf{V}_\phi\mathbf{H}^T, \tag{2.35}$$

where $\mathbf{H} \in \mathbb{R}^{N_j \times N_j}$ is a band matrix defined by

$$\mathbf{H} := \frac{1}{h_j}\begin{bmatrix} -1 & & & & 1 \\ 1 & -1 & & & \\ & 1 & \ddots & & \\ & & \ddots & -1 & \\ & & & 1 & -1 \end{bmatrix},$$

and $\mathbf{V}_\phi$ given by (2.20) corresponds to the single layer operator discretized via piecewise constant functions ($d=1$). Since $\mathbf{W}_\phi$ inherits the symmetry and positive semidefiniteness from $\mathcal{W}$ the linear system of equations (2.34) is singular. To solve this system under the given side condition, one proceeds exactly as for the equation of the second kind for exterior Dirichlet problems. The density $\mu = \sum \mu_k \phi_{j,k}^{(2)}$ approximates the solution u of the Neumann problem (modulo some constant in the case of an interior problem) by the potential evaluation (2.17).

2. Fredholm's integral equation of the second kind for Neumann problems: We introduce the adjoint $\mathcal{K}^\star$ of the double layer operator

$$(\mathcal{K}^\star u)(x) := \int_\Gamma \frac{\partial}{\partial n_x} E(x,y)v(y)ds_y, \quad x \in \Gamma, \tag{2.36}$$

which indicates an operator of order zero, i.e., $\mathcal{K}^\star : L^2(\Gamma) \to L^2(\Gamma)$. Then, for the Neumann problem (2.30) we formulate the Fredholm integral equation of the second kind

$$\left(\pm \tfrac{1}{2}I - \mathcal{K}^\star\right)\rho = g \quad \text{on } \Gamma, \tag{2.37}$$

where we have to choose "+" for an exterior and "−" for an interior problem. Since $\left(-\tfrac{1}{2}I - \mathcal{K}^\star\right) = -\left(\tfrac{1}{2}I + \mathcal{K}\right)^\star$ this equation is not uniquely solvable in the case of the interior problem. Hence, again the side condition $\int_\Gamma \rho(x)ds_x = 0$ is required for the unknown density ρ. We employ $(\mathcal{K}^\star\rho,\delta)_{L^2(\Gamma)} = (\rho,\mathcal{K}\delta)_{L^2(\Gamma)}$ and substitute $\mu = (\rho\circ\gamma)\alpha$, $\nu = \delta\circ\gamma$ to find the variational formulation

seek $\mu \in L^2(0,1)$:

$$\pm\frac{1}{2}(\mu,\nu)_{L^2(0,1)} - (\mu,K\nu)_{L^2(0,1)} = (g\circ\gamma,\nu\alpha)_{L^2(0,1)} \quad \forall\, \nu \in L^2(0,1).$$

Herein, the required side condition for the interior problem reads $(\mu,1)_{L^2(0,1)} = 0$. With $\boldsymbol{\mu}_\phi$ from (2.20), $\mathbf{K}_\phi$, $\mathbf{B}_\phi$ from (2.22) and $\mathbf{g}_\phi$ similarly as above the Galerkin system is given by

$$\left(\pm\tfrac{1}{2}\mathbf{B}_\phi - \mathbf{K}_\phi^\star\right)\boldsymbol{\mu}_\phi = \mathbf{g}_\phi. \tag{2.38}$$

Note, the system corresponding to the exterior Neumann problem is uniquely solvable while for the interior problem it is singular. To solve the singular system one proceeds likewise to the equation of the second kind for the Dirichlet problem. Note that, this time, we get $[\mathbf{a}]_k := (\phi_{j,k}^{(d)}, 1)_{L^2(0,1)} = 1/\sqrt{h_j}$. The density $\mu = \sum \mu_k \phi_{j,k}^{(d)}$ defines an approximated solution to the Neumann problem (modulo some constant in the case of an interior problem) by (2.13).

3. Wavelet approximation for BEM

3.1. Motivation

We have introduced in subsection 2.3. an equidistant partition of the interval $[0,1]$. On this partition the unknown density is discretized via (periodic) piecewise constant and linear functions, respectively. Instead of using this *single-scale basis* we want to apply wavelets with vanishing moments (more precisely: biorthogonal wavelet bases) yielding numerically sparse system matrices.

The outline is as follows. We first introduce biorthogonal wavelet bases on $\mathbb{R}$. Next, we obtain the wavelet bases on the interval $[0,1]$ by periodization. According to [35] we briefly recall in the third subsection the compression strategy of matrices arising from BEM. Since the system matrices resulting from boundary integral operators of order $\neq 0$ are ill conditioned, we give a simple diagonal preconditioner based on the wavelet expansion in the last subsection.

3.2. Biorthogonal multiresolution on $\mathbb{R}$

On $\mathbb{R}$ the piecewise polynomial splines of degree $d-1$ can be defined as follows. Denoting by $[x_0, \ldots, x_d]f$ the d-th order *divided difference* at the points $x_0, \ldots, x_d \in \mathbb{R}$ (see e.g. [15]) the (centered) cardinal B-spline of order d is given by

$$\phi^{(d)}(x) = d[0,1,\ldots,d]\Big(\cdot - x - \Big\lfloor \frac{d}{2} \Big\rfloor\Big)_+^{d-1}.$$

where $x_+^l := (\max\{0,x\})^l$ and $\lfloor x \rfloor$ ($\lceil x \rceil$) is the largest (smallest) integer less (greater) than or equal to x. This *scaling function* is normalized

$$\|\phi^{(d)}\|_{L^1(\mathbb{R})} = 1,$$

compactly supported

$$\operatorname{supp} \phi^{(d)} = \Big[-\Big\lfloor \frac{d}{2} \Big\rfloor, \Big\lceil \frac{d}{2} \Big\rceil\Big] \tag{3.1}$$

and refinable

$$\phi^{(d)}(x) = \frac{1}{\sqrt{2}} \sum_{k\in\mathbb{Z}} a_k \phi^{(d)}(2x-k) \tag{3.2}$$

with *mask coefficients*

$$a_k = \begin{cases} 2^{1-d}\binom{d}{k}, & -\lfloor \frac{d}{2} \rfloor \le k \le \lceil \frac{d}{2} \rceil, \\ 0, & \text{elsewhere.} \end{cases} \tag{3.3}$$

Introducing for $j, k \in \mathbb{Z}$ the translates and dilates of the scaling function $\phi_{j,k}^{(d)} := 2^{j/2}\phi^{(d)}(2^j \cdot -k)$, the sets $\Phi_j^{(d)} := \{\phi_{j,k}^{(d)} : k \in \mathbb{Z}\}$ generate a sequence of spaces $V_j := \overline{\operatorname{span} \Phi_j^{(d)}}$, which is nested

$$\ldots \subset V_j \subset V_{j+1} \subset \ldots$$

and dense in $L^2(\mathbb{R})$

$$\overline{\bigcup_{j\in\mathbb{Z}} V_j} = L^2(\mathbb{R}), \qquad \bigcap_{j\in\mathbb{Z}} V_j = \{0\}.$$

The spaces V_j are exact of order d, i.e., denoting the space of all polynomials of degree $< d$ by $\Pi_d(\mathbb{R})$, there holds

$$\Pi_d(\mathbb{R}) \subset V_j.$$

Furthermore, $\Phi_j^{(d)}$ forms a Riesz basis in V_j

$$\left\| \Phi_j^{(d)} \mathbf{c} \right\|_{L^2(\mathbb{R})} \sim \|\mathbf{c}\|_{l^2(\mathbb{Z})} \quad \forall\, \mathbf{c} \in l^2(\mathbb{Z}).$$

Due to [3] there exists for every integer $\tilde{d} \ge d$ with $\tilde{d} + d$ even a *dual scaling function* $\tilde{\phi}^{(d,\tilde{d})} \in L^2(\mathbb{R})$ which is biorthogonal to the first scaling function

$$\left(\phi^{(d)}, \tilde{\phi}^{(d,\tilde{d})}(\cdot - k)\right)_{L^2(\mathbb{R})} = \delta_{0,k}, \quad k \in \mathbb{Z}.$$

Moreover, similarly to the primal scaling function, this function is normalized, compactly supported

$$\operatorname{supp} \tilde{\phi}^{d,\tilde{d}} = \left[-\left\lfloor \frac{d}{2} \right\rfloor + 1 - \tilde{d}, \left\lceil \frac{d}{2} \right\rceil - 1 + \tilde{d} \right] \tag{3.4}$$

and refinable

$$\tilde{\phi}^{(d,\tilde{d})}(x) = \frac{1}{\sqrt{2}} \sum_{k\in\mathbb{Z}} \tilde{a}_k \tilde{\phi}^{(d,\tilde{d})}(2x - k). \tag{3.5}$$

The mask coefficients in (3.5) can be derived by the *z-notation*: The coefficient $\tilde{a}_k$ of the sequence $\sum_k \tilde{a}_k z^k = p(z)q(z)$,

$$p(z) = 2^{1-\tilde{d}} \sum_{k=0}^{\tilde{d}} \binom{\tilde{d}}{k} z^{k - \lfloor \frac{\tilde{d}}{2} \rfloor},$$

$$q(z) = \sum_{k=0}^{\frac{d+\tilde{d}}{2}-1} 2^{-k} \binom{\frac{d+\tilde{d}}{2} - 1 + k}{k} \sum_{l=0}^{2k} \binom{2k}{l} (-z)^{l-k},$$

coincides with the mask coefficient $\tilde{a}_k$, see [3] for details. Exactly like the primal side, the translates and dilates of the dual scaling function $\widetilde{\phi}_{j,k}^{(d,\tilde{d})} := 2^{j/2}\widetilde{\phi}^{(d,\tilde{d})}(2^j \cdot -k)$, $j,k \in \mathbb{Z}$, generate collections of Riesz bases $\widetilde{\Phi}_j^{(d,\tilde{d})} := \{\widetilde{\phi}_{j,k}^{(d,\tilde{d})} : k \in \mathbb{Z}\}$ in the spaces $\widetilde{V}_j := \overline{\operatorname{span} \widetilde{\Phi}_j^{(d,\tilde{d})}}$ which are nested, dense in $L^2(\mathbb{R})$ and exact of order $\tilde{d}$.

According to [3] a dual pair of *wavelets* $\psi^{(d,\tilde{d})}$, $\widetilde{\psi}^{(d,\tilde{d})} \in L^2(\mathbb{R})$ satisfying

$$\big(\psi^{(d,\tilde{d})}, \widetilde{\psi}^{(d,\tilde{d})}(\cdot - k)\big)_{L^2(\mathbb{R})} = \delta_{0,k}, \quad k \in \mathbb{Z},$$

is defined by

$$\begin{aligned} \psi^{(d,\tilde{d})}(x) &:= \frac{1}{\sqrt{2}} \sum_{k\in\mathbb{Z}} b_k \phi^{(d)}(2x-k), \\ \widetilde{\psi}^{(d,\tilde{d})}(x) &:= \frac{1}{\sqrt{2}} \sum_{k\in\mathbb{Z}} \tilde{b}_k \widetilde{\phi}^{(d,\tilde{d})}(2x-k), \end{aligned} \tag{3.6}$$

where the mask coefficients b_k and $\tilde{b}_k$ are given by

$$b_k = (-1)^k \tilde{a}_{1-k}, \qquad \tilde{b}_k = (-1)^k a_{1-k}, \qquad k \in \mathbb{Z}, \tag{3.7}$$

with a_k from (3.3) and $\tilde{a}_k$ from (3.5). As a consequence of the finite masks and the compact supports of the scaling functions both wavelets are compactly supported

$$\operatorname{supp} \psi^{(d,\tilde{d})} = \operatorname{supp} \widetilde{\psi}^{(d,\tilde{d})} = \Big[1 - \frac{d+\tilde{d}}{2}, \frac{d+\tilde{d}}{2}\Big]. \tag{3.8}$$

Setting analogously to the scaling functions

$$\psi_{j,k}^{(d,\tilde{d})} := 2^{j/2}\psi^{(d,\tilde{d})}(2^j \cdot -k), \qquad \widetilde{\psi}_{j,k}^{(d,\tilde{d})} := 2^{j/2}\widetilde{\psi}^{(d,\tilde{d})}(2^j \cdot -k),$$

the sets

$$\Psi_j^{(d,\tilde{d})} := \{\psi_{j,k}^{(d,\tilde{d})} : k \in \mathbb{Z}\}, \qquad \widetilde{\Psi}_j^{(d,\tilde{d})} := \{\widetilde{\psi}_{j,k}^{(d,\tilde{d})} : k \in \mathbb{Z}\},$$

generate complement spaces $W_j := \overline{\operatorname{span} \Psi_j^{(d,\tilde{d})}}$ and $\widetilde{W}_j := \overline{\operatorname{span} \widetilde{\Psi}_j^{(d,\tilde{d})}}$ with

$$V_j \oplus W_j = V_{j+1}, \qquad \widetilde{V}_j \oplus \widetilde{W}_j = \widetilde{V}_{j+1},$$

where $\oplus$ denotes the direct sum. Thus, recursively one obtains

$$\overline{\bigoplus_{j\in\mathbb{Z}} W_j} = \overline{\bigoplus_{j\in\mathbb{Z}} \widetilde{W}_j} = L^2(\mathbb{R}).$$

Since biorthogonality implies $W_j \perp \widetilde{V}_j$ the primal wavelets have vanishing moments of order $\tilde{d}$, i.e.,

$$\big((\cdot)^\alpha, \psi_{j,k}^{(d,\tilde{d})}\big)_{L^2(\mathbb{R})} = 0, \quad 0 \le \alpha < \tilde{d}.$$

Furthermore, the wavelets

$$\Psi^{(d,\tilde{d})} := \bigcup_{j\in\mathbb{Z}} \Psi_j^{(d,\tilde{d})}, \qquad \widetilde{\Psi}^{(d,\tilde{d})} := \bigcup_{j\in\mathbb{Z}} \widetilde{\Psi}_j^{(d,\tilde{d})},$$

form Riesz bases in $L^2(\mathbb{R})$

$$\|\mathbf{c}\|^2_{l^2(\mathbb{Z}\times\mathbb{Z})} \sim \left\|\Psi^{(d,\tilde{d})}\mathbf{c}\right\|^2_{L^2(\mathbb{R})} \sim \left\|\widetilde{\Psi}^{(d,\tilde{d})}\mathbf{c}\right\|^2_{L^2(\mathbb{R})} \quad \forall\, \mathbf{c} \in l^2(\mathbb{Z}\times\mathbb{Z}). \tag{3.9}$$

3.3. Periodization

The above setting is clearly not suitable for the treatment of equations which are defined on bounded domains. In the sequel we will utilize a periodic version of a multiscale resolution. It essentially retains all the structural and computational advantages of the stationary and shift-invariant case considered in the previous subsection.

To this end, the simple trick is to replace the meaning of $u_{j,k} := 2^{j/2}u(2^j \cdot -k)$, $k \in \mathbb{Z}$, for compactly supported $u \in L^2(\mathbb{R})$ by its periodized counterpart

$$u_{j,k} := 2^{j/2} \sum_{n\in\mathbb{Z}} u(2^j(\cdot + n) - k).$$

In this way, given any dual pair $\phi^{(d)}$ and $\widetilde{\phi}^{(d,\tilde{d})}$ on $\mathbb{R}$ of compactly supported scaling functions, and setting $\Delta_j := \mathbb{Z} \setminus 2^j\mathbb{Z}$, the corresponding sets

$$\Phi_j^{(d)} := \{\phi_{j,k} : k \in \Delta_j\}, \qquad \Psi_j^{(d,\tilde{d})} := \{\psi_{j,k}^{(d,\tilde{d})} : k \in \Delta_j\}, \qquad j \geq j_0,$$

and likewise $\widetilde{\Phi}_j^{(d,\tilde{d})}$ and $\widetilde{\Psi}_j^{(d,\tilde{d})}$, have finite cardinality 2^j and consist of functions which are 1-periodic. Note that this definition preserves the biorthogonality relations. One easily checks that the scaling functions are biorthogonal

$$\left(\Phi_j^{(d)}, \widetilde{\Phi}_j^{(d,\tilde{d})}\right)_{L^2(0,1)} = \mathbf{I},$$

where $(\cdot,\cdot)$ denotes the inner product on $L^2(0,1)$ defined by (2.9). Moreover, the wavelet bases

$$\Psi^{(d,\tilde{d})} := \Phi_{j_0}^{(d)} \bigcup_{j\geq j_0} \Psi_j^{(d,\tilde{d})}, \qquad \widetilde{\Psi}^{(d,\tilde{d})} := \widetilde{\Phi}_{j_0}^{(d,\tilde{d})} \bigcup_{j\geq j_0} \widetilde{\Psi}_j^{(d,\tilde{d})}$$

are biorthogonal, i.e.,

$$\left(\Psi^{(d,\tilde{d})}, \widetilde{\Psi}^{(d,\tilde{d})}\right)_{L^2(0,1)} = \mathbf{I}.$$

For the sake of simplicity in representation, we will indicate in the sequel the scaling functions on the coarsest level of the wavelet bases by $\Psi_{j_0-1}^{(d,\tilde{d})} := \Phi_{j_0}^{(d)}$ and

$\widetilde{\Psi}_{j_0-1}^{(d,\tilde{d})} := \widetilde{\Phi}_{j_0}^{(d,\tilde{d})}$. Moreover, we set $\triangle_{j_0-1} := \triangle_{j_0}$. Clearly, the coarsest level j_0 has to be chosen large enough to ensure that the diameter of the supports of the scaling functions and the wavelets is smaller that 1. Comparing (3.1), (3.4) and (3.8) implies

$$j_0 \geq \log_2 \left[\operatorname{diam} \left(\operatorname{supp} \widetilde{\phi}^{(d,\tilde{d})} \right) \right] = \log_2 (d + 2\tilde{d} - 2).$$

3.4. Matrix compression

Discretizing an integral operator $A : H^q(0,1) \to H^{-q}(0,1)$ by biorthogonal wavelet bases leads to quasi-sparse matrices. In a first compression step all matrix entries, for which the distances of the supports (on the given boundary Γ) of the corresponding Ansatz and test functions are bigger than a level depending cut-off parameter $\mathcal{B}_{j,j'}$, are set to zero. In the second compression step some of those matrix entries are set to zero, for which the corresponding Ansatz and test functions have overlapping supports.

More precisely, for a given $J > j_0$ we discretize the integral operator A by the wavelet basis $\Psi_J^{(d,\tilde{d})} := \bigcup_{j=j_0-1}^{J-1} \Psi_j^{(d,\tilde{d})}$ instead by the single-scale basis $\Phi_J^{(d)}$. We introduce the abbreviations

$$\begin{aligned} \Theta_{j,k}^{(d,\tilde{d})} &:= \left\{ x = \gamma(s) \in \mathbb{R}^2 : s \in \operatorname{supp} \psi_{j,k}^{(d,\tilde{d})} \right\}, \\ \Xi_{j,k}^{(d,\tilde{d})} &:= \left\{ x = \gamma(s) \in \mathbb{R}^2 : \nexists\, \varepsilon > 0 \mid \psi_{j,k}^{(d,\tilde{d})} \in C^\infty(s - \varepsilon, s + \varepsilon) \right\}. \end{aligned}$$

Note that $\Theta_{j,k}^{(d,\tilde{d})}$ denotes the support of $\psi_{j,k}^{(d,\tilde{d})} \circ \gamma^{-1}$ while $\Xi_{j,k}^{(d,\tilde{d})}$ denotes the so-called *singular support* of $\psi_{j,k}^{(d,\tilde{d})} \circ \gamma^{-1}$, i.e., those points where $\psi_{j,k}^{(d,\tilde{d})} \circ \gamma^{-1}$ is not smooth. The compressed system matrix $\mathbf{A}_\psi$ corresponding to A is given by

$$[\mathbf{A}_\psi]_{(j,k),(j',k')} := \begin{cases} 0, & \operatorname{dist}\left(\Theta_{j,k}^{(d,\tilde{d})}, \Theta_{j',k'}^{(d,\tilde{d})}\right) > \mathcal{B}_{j,j'},\ j,j' \geq j_0, \\ 0, & \operatorname{dist}\left(\Xi_{j,k}^{(d,\tilde{d})}, \Theta_{j',k'}^{(d,\tilde{d})}\right) > \mathcal{B}'_{j,j'},\ j' > j, \\ 0, & \operatorname{dist}\left(\Theta_{j,k}^{(d,\tilde{d})}, \Xi_{j',k'}^{(d,\tilde{d})}\right) > \mathcal{B}'_{j,j'},\ j > j', \\ \left(A\psi_{j',k'}^{(d,\tilde{d})}, \psi_{j,k}^{(d,\tilde{d})}\right)_{L^2(0,1)}, & \text{otherwise.} \end{cases} \tag{3.10}$$

Herein, choosing

$$a, a' > 1, \qquad d < \delta, \delta' < \tilde{d} + 2q, \tag{3.11}$$

the cut-off parameters $\mathcal{B}_{j,j'}$ and $\mathcal{B}'_{j,j'}$ are set as follows

$$\begin{aligned} \mathcal{B}_{j,j'} &= a \max\left\{ 2^{-\min\{j,j'\}}, 2^{\frac{2J(\delta-q)-(j+j')(\delta+\tilde{d})}{2(\tilde{d}+q)}} \right\}, \\ \mathcal{B}'_{j,j'} &= a' \max\left\{ 2^{-\max\{j,j'\}}, 2^{\frac{2J(\delta'-q)-(j+j')\delta'-\max\{j,j'\}\tilde{d}}{\tilde{d}+2q}} \right\}. \end{aligned} \tag{3.12}$$

It is shown in [35] that this compression strategy reduces the number of nonzero entries to $\mathcal{O}(N_J)$ without any loss of accuracy and stability of the underlying Galerkin scheme. The resulting structure of the compressed matrix is figuratively called *finger structure*, cf. figure 1.

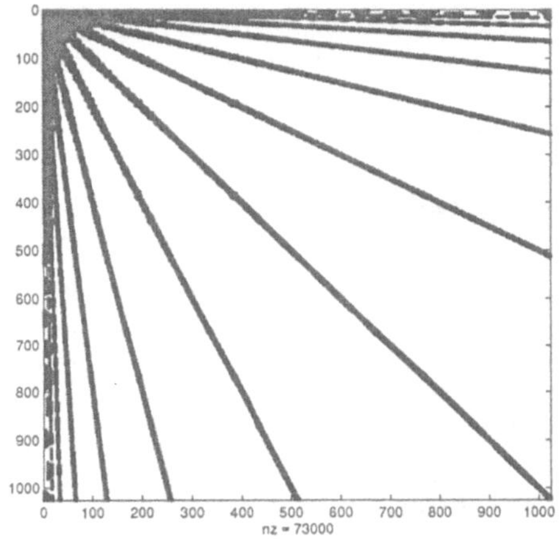

Figure 1: The structure of the compressed system matrix.

3.5. Wavelet preconditioning

The single layer operator and the hypersingular operator are operators of order different from zero. Hence, the system matrix $\mathbf{A}_\psi$ is ill conditioned. According to [5, 35], for example, the wavelet approach offers an simple diagonal preconditioner.

Let us define the number $\overline{q} := \sup\left\{q \in \mathbb{R} : \widetilde{\psi}^{(d,\tilde{d})} \in H^q(0,1)\right\}$ which characterizes the regularity of the dual wavelet basis. Moreover, we introduce the diagonal matrix $\mathbf{D}^r$ by

$$\left[\mathbf{D}^r\right]_{(j,k),(j',k')} = 2^{rj}\delta_{j,j'}\delta_{k,k'}, \quad k \in \triangle_j, \quad k' \in \triangle_{j'}, \quad j_0 - 1 \le j, j' < J. \tag{3.13}$$

Then, if $A : H^q(0,1) \to H^{-q}(0,1)$ denotes an integral operator of order $2q$ with $\overline{>} -q$, the corresponding system matrix $\mathbf{A}_\psi$ is spectrally equivalent to $\mathbf{D}^{2q}$. Therefore, the system matrix $\mathbf{D}^{-q}\mathbf{A}_\psi\mathbf{D}^{-q}$ is well conditioned, i.e.,

$$\mathrm{cond}_{l^2}(\mathbf{D}^{-q}\mathbf{A}_\psi\mathbf{D}^{-q}) \sim 1.$$

Note that the coefficients on the main diagonal of $\mathbf{A}_\psi$ satisfy

$$\left(A\psi_{j,k}^{(d,\tilde{d})}, \psi_{j,k}^{(d,\tilde{d})}\right)_{L^2(0,1)} \sim 2^{2qj}, \quad k \in \triangle_j, \quad j_0 - 1 \le j < J.$$

Therefore, the above preconditioning can be replaced by a diagonal scaling. In fact, the diagonal scaling improves and simplifies the wavelet preconditioning.

4. The discrete wavelet Galerkin scheme

4.1. Changing bases

Generally written, in the single-scale basis we obtain the Galerkin system

$$(\mathbf{A}_\phi + \lambda \mathbf{B}_\phi)\boldsymbol{\mu}_\phi = \mathbf{f}_\phi \tag{4.1}$$

with the unknown density vector $\boldsymbol{\mu}_\phi$. Hereby, $\mathbf{A}_\phi$ is the system matrix corresponding to a boundary integral operator $A : H^q(0,1) \to H^{-q}(0,1)$, $\mathbf{B}_\phi$ is the mass matrix belonging to the identity operator, $\mathbf{f}_\phi$ is the vector of the right hand side and λ is some real constant. We denote by $\mathbf{T}$ the transposed fast wavelet transform, i.e., the basis transform for the dual bases given by

$$\sum_{k\in\triangle_J} a_k \widetilde{\phi}_{k,J}^{(d,\tilde{d})} = \sum_{j=j_0-1}^{J-1} \sum_{k\in\triangle_j} b_{j,k} \widetilde{\psi}_{j,k}^{(d,\tilde{d})}, \qquad [b_{j,k}]_{k\in\triangle_j, j_0-1\le j<J} = \mathbf{T}[a_k]_{k\in\triangle_J},$$

see [3] for details. Then, the change from the single-scale basis into the biorthogonal wavelet basis corresponds to the system

$$\mathbf{T}(\mathbf{A}_\phi + \lambda\mathbf{B}_\phi)\mathbf{T}^\star \boldsymbol{\mu}_\psi = \mathbf{T}\mathbf{f}_\phi, \quad \boldsymbol{\mu}_\phi = \mathbf{T}^\star \boldsymbol{\mu}_\psi.$$

Clearly, assembling $\mathbf{B}_\phi$ and $\mathbf{f}_\phi$ has a complexity $\mathcal{O}(N_J)$. Moreover, it is well known that the application of $\mathbf{T}$ and $\mathbf{T}^\star$ to a vector requires only $\mathcal{O}(N_J)$ operations, cf. [3]. Hence, utilizing iterative methods for the solution of the linear system of equations we may concentrate in the sequel on the efficient computation of the compressed matrix $\mathbf{A}_\psi \approx \mathbf{T}\mathbf{A}_\phi\mathbf{T}^\star$ solving the system

$$(\mathbf{A}_\psi + \lambda\mathbf{T}\mathbf{B}_\phi\mathbf{T}^\star)\boldsymbol{\mu}_\psi = \mathbf{T}\mathbf{f}_\phi, \quad \boldsymbol{\mu}_\phi = \mathbf{T}^\star \boldsymbol{\mu}_\psi. \tag{4.2}$$

Remark 4.1. The equation of the second kind for the exterior Dirichlet problem (2.23), the equation of the second kind for the interior Neumann problem (2.38) and the equation of the first kind for the Neumann problem (2.34), lead to a singular linear system (4.2). Therefore, we discarded one unknown of the single-scale basis solving a reduced system. Analogously in the multiscale basis we omit one of the scaling functions on the coarsest grid. Without loss of generality we discard the first scaling function $\phi_{j_0,0}^{(d)} = \psi_{j_0-1,0}^{(d,\tilde{d})}$. Then, the application of $\mathbf{T}\mathbf{B}_\phi\mathbf{T}^\star$ to the reduced vector $\tilde{\boldsymbol{\mu}}$ in the iterative solver is performed as follows: We compute the vector $\left[\begin{smallmatrix} a \\ \mathbf{b} \end{smallmatrix}\right] := \mathbf{T}\mathbf{B}_\phi\mathbf{T}^\star\left[\begin{smallmatrix} 0 \\ \tilde{\boldsymbol{\mu}} \end{smallmatrix}\right]$ and continue the iteration with $\mathbf{b}$. At the end we get the solution $\boldsymbol{\mu}_\phi = \mathbf{T}^\star\left[\begin{smallmatrix} 0 \\ \tilde{\boldsymbol{\mu}} \end{smallmatrix}\right]$, for which the postulated side conditions are computed as seen in section 2..

4.2. Computing distances between wavelets

To set up the compression pattern according to (3.10), we have to compute the distances $\operatorname{dist}\big(\Theta_{j,k}^{(d,\tilde{d})},\Theta_{j',k'}^{(d,\tilde{d})}\big)$ and $\operatorname{dist}\big(\Xi_{j,k}^{(d,\tilde{d})},\Theta_{j',k'}^{(d,\tilde{d})}\big)$. The first distance is required if $j,j'\geq j_0$ while the second distance is required if $j'>j\geq j_0-1$.

The crucial idea for the evaluation of the distance $\operatorname{dist}\big(\Theta_{j,k}^{(d,\tilde{d})},\Theta_{j',k'}^{(d,\tilde{d})}\big)$ is to find disks $B(m_{j,k},r_{j,k}):=\{x\in\mathbb{R}^2:|x-m_{j,k}|\leq r_{j,k}\}$ with $\Theta_{j,k}^{(d,\tilde{d})}\subset B(m_{j,k},r_{j,k})$. Then, consequently, one has

$$\operatorname{dist}\big(\Theta_{j,k}^{(d,\tilde{d})},\Theta_{j',k'}^{(d,\tilde{d})}\big)\leq\max\{0,|m_{j,k}-m_{j',k'}|-r_{j,k}-r_{j',k'}\}.$$

We start on the finest level $J-1$ to compute these disks. For this, note that we obtain for all $k\in\Delta_j$, $j\geq j_0$, from (3.8)

$$\operatorname{supp}\psi_{j,k}^{(d,\tilde{d})}=2^{-j}\Big[k+1-\frac{d+\tilde{d}}{2},k+\frac{d+\tilde{d}}{2}\Big]=:2^{-j}[k+\underline{l},k+\overline{l}]. \tag{4.3}$$

Invoking (4.3) we calculate

$$\underline{x}_k:=\gamma\big(2^{1-J}(k+\underline{l})\big),\qquad \overline{x}_k:=\gamma\big(2^{1-J}(k+\overline{l})\big),\qquad k\in\Delta_{J-1}.$$

Then, assuming that J is large enough, the disk determined by $m_{J-1,k}:=(\underline{x}_k+\overline{x}_k)/2$ and $r_{J-1,k}:=|\underline{x}_k-\overline{x}_k|/2$ satisfies $\Theta_{J-1,k}^{(d,\tilde{d})}\subset B(m_{J-1,k},r_{J-1,k})$. Next, from (4.3) we deduce

$$\operatorname{supp}\psi_{j,k}^{(d,\tilde{d})}=\operatorname{supp}\psi_{j+1,2k+\underline{l}}^{(d,\tilde{d})}\cup\operatorname{supp}\psi_{j+1,2k+\overline{l}}^{(d,\tilde{d})}$$

for all $k\in\Delta_j$, $j\geq j_0$. Hence, computing recursively $B(m_{j,k},r_{j,k})$ with

$$B(m_{j,k},r_{j,k})\supset B(m_{j+1,2k+\underline{l}},r_{j+1,2k+\underline{l}})\cup B(m_{j+1,2k+\overline{l}},r_{j+1,2k+\overline{l}})$$

we obtain for all $k\in\Delta_j$, $J-1>j\geq j_0$, the disks on the coarser levels. Note, the complexity for determining all the disks $B(m_{j,k},r_{j,k})$ is $\mathcal{O}(N_J)+\mathcal{O}(N_J/2)+\mathcal{O}(N_J/4)+\ldots+\mathcal{O}(1)=\mathcal{O}(N_J)$, i.e., the complexity is linear. Moreover, with these disks we also can determine the distance $\operatorname{dist}\big(\Xi_{j,k}^{(d,\tilde{d})},\Theta_{j',k'}^{(d,\tilde{d})}\big)$ in the following way. For $j\geq j_0$ the singular support $\Xi_{j,k}^{(d,\tilde{d})}$ consists only of points $\gamma(s)$, where $s\in[0,1]$ is a point of the grid on the level $j+1$. Consequently, observing (4.3), we define

$$S_{j,k}:=\{\gamma(2^{-(j+1)}l):l\in\Delta_{j+1}\text{ and }2(k+\underline{l})\leq l\leq 2(k+\overline{l})\}$$

to obtain a set of $2(d+\tilde{d})-1$ points with $\Xi_{j,k}^{(d,\tilde{d})}\subset S_{j,k}$. Furthermore, (3.1) implies that the singular support of the scaling functions $\Phi_{j_0}^{(d)}=\Psi_{j_0-1}^{(d,\tilde{d})}$ consists of the $d+1$ points contained in

$$S_{j_0-1,k}:=\{\gamma(2^{-j_0}l):l\in\Delta_{j_0}\text{ and }k-\lfloor d/2\rfloor\leq l\leq k+\lceil d/2\rceil\}.$$

By these sets we obtain

$$\operatorname{dist}\left(\Xi_{j,k}^{(d,\tilde{d})}, \Theta_{j',k'}^{(d,\tilde{d})}\right) \leq \max\{0, \min_{x \in S_{j,k}} |x - m_{j',k'}| - r_{j',k'}\}.$$

4.3. Setting up the compression pattern

The compression strategy introduced in subsection 3.4. leads to a sparse system matrix $\mathbf{A}_\psi$. We first collect some important properties concerning the structure of this matrix.

- As one easily checks by (3.12), the cut-off parameters $\mathcal{B}_{j,j'}$ and $\mathcal{B}'_{j,j'}$, respectively, are symmetric with respect to the levels j and j', i.e.,

  $$\mathcal{B}_{j,j'} = \mathcal{B}_{j',j}, \qquad \mathcal{B}'_{j,j'} = \mathcal{B}'_{j',j}.$$

  Therefore, due to the symmetry of the distance function, one deduces that the structure of the compressed system matrix is symmetric.

- Naturally, the wavelet basis has a father-son relation with respect to the supports. More precisely, there holds

  $$\Theta_{j+1,2k}^{(d,\tilde{d})}, \Theta_{j+1,2k+1}^{(d,\tilde{d})} \subset \Theta_{j,k}^{(d,\tilde{d})}, \quad k \in \Delta_j, \quad j \geq j_0, \tag{4.4}$$

  cf. (4.3). In addition, (3.12) implies

  $$\mathcal{B}_{j+1,j'+1} \leq \mathcal{B}_{j+1,j'} \leq \mathcal{B}_{j,j'}, \qquad j, j' \geq j_0.$$

  Hence, combining the father-son relation (4.4) with the latter relation, it follows for $j, j' \geq j_0$

  $$\begin{aligned} &\operatorname{dist}\left(\Theta_{j,k}^{(d,\tilde{d})}, \Theta_{j',k'}^{(d,\tilde{d})}\right) > \mathcal{B}_{j,j'} \\ &\qquad \Longrightarrow \quad \begin{cases} \operatorname{dist}\left(\Theta_{j+1,2k}^{(d,\tilde{d})}, \Theta_{j',k'}^{(d,\tilde{d})}\right) > \mathcal{B}_{j+1,j'}, \\ \operatorname{dist}\left(\Theta_{j+1,2k+1}^{(d,\tilde{d})}, \Theta_{j',k'}^{(d,\tilde{d})}\right) > \mathcal{B}_{j+1,j'}, \end{cases} \end{aligned} \tag{4.5}$$

  and

  $$\begin{aligned} &\operatorname{dist}\left(\Theta_{j,k}^{(d,\tilde{d})}, \Theta_{j',k'}^{(d,\tilde{d})}\right) > \mathcal{B}_{j,j'} \\ &\qquad \Longrightarrow \quad \begin{cases} \operatorname{dist}\left(\Theta_{j+1,2k}^{(d,\tilde{d})}, \Theta_{j'+1,2k'}^{(d,\tilde{d})}\right) > \mathcal{B}_{j+1,j'+1}, \\ \operatorname{dist}\left(\Theta_{j+1,2k}^{(d,\tilde{d})}, \Theta_{j'+1,2k'+1}^{(d,\tilde{d})}\right) > \mathcal{B}_{j+1,j'+1}, \\ \operatorname{dist}\left(\Theta_{j+1,2k+1}^{(d,\tilde{d})}, \Theta_{j'+1,2k'}^{(d,\tilde{d})}\right) > \mathcal{B}_{j+1,j'+1}, \\ \operatorname{dist}\left(\Theta_{j+1,2k+1}^{(d,\tilde{d})}, \Theta_{j'+1,2k'+1}^{(d,\tilde{d})}\right) > \mathcal{B}_{j+1,j'+1}. \end{cases} \end{aligned} \tag{4.6}$$

- On the other hand, for the cut-off parameter $\mathcal{B}'_{j,j'}$ we find by (3.12)

$$\mathcal{B}'_{j+1,j'} \leq \mathcal{B}_{j,j'}, \qquad j > j' \geq j_0 - 1.$$

Invoking (4.4) this inequality yields

$$\operatorname{dist}\left(\Theta_{j,k}^{(d,\tilde{d})}, \Xi_{j',k'}^{(d,\tilde{d})}\right) > \mathcal{B}'_{j,j'}$$

$$\Longrightarrow \quad \begin{cases} \operatorname{dist}\left(\Theta_{j+1,2k}^{(d,\tilde{d})}, \Xi_{j',k'}^{(d,\tilde{d})}\right) > \mathcal{B}'_{j+1,j'}, \\ \operatorname{dist}\left(\Theta_{j+1,2k+1}^{(d,\tilde{d})}, \Xi_{j',k'}^{(d,\tilde{d})}\right) > \mathcal{B}'_{j+1,j'}, \end{cases} \tag{4.7}$$

for $j > j' \geq j_0 - 1$.

Performing the compression we are interested in the index sets $\mathcal{I}_{j,j'}$, $j_0 - 1 \leq j, j' < J$, consisting of all pairs (k, k') of indices corresponding to matrix entries which have to be evaluated. Combining (4.5), (4.6) and (4.7) with the symmetric compression structure, one may formulate the algorithm 1, given in pseudo code language. Obviously, the complexity of this algorithm can be estimated by the number of nonzero elements of the compressed system matrix $\mathbf{A}_\psi$, that is by $\mathcal{O}(N_J)$.

4.4. Computing the system matrix

For the computation of the system matrix $\mathbf{A}_\psi$ one has to calculate for $j_0 - 1 \leq j, j' < J$ all coefficients $\left(A\psi_{j',k'}^{(d,\tilde{d})}, \psi_{j,k}^{(d,\tilde{d})}\right)_{L^2(0,1)}$, for which the pair (k, k') is found in the set $\mathcal{I}_{j,j'}$. For the sake of simplicity we assume that $j, j' \geq j_0$, i.e., we omit the scaling functions on the coarsest level $j_0 - 1$. The described strategy can also used in the case $j = j_0 - 1$ or $j' = j_0 - 1$ but the formulas must be modified slightly. We abbreviate in the sequel analogously to (2.24)

$$\alpha_{(j,k),(j',k')}^{(d)} := \left(A\phi_{j',k'}^{(d)}, \phi_{j,k}^{(d)}\right)_{L^2(0,1)}. \tag{4.8}$$

Then, utilizing the refinement relation (3.6) each matrix entry splits into the finite sum

$$\left(A\psi_{j',k'}^{(d,\tilde{d})}, \psi_{j,k}^{(d,\tilde{d})}\right)_{L^2(0,1)} = \sum_l \sum_{l'} b_l b_{l'} \alpha_{(j+1,2k+l),(j'+1,2k'+l')}^{(d)} \tag{4.9}$$

with b_k from (3.7). Since the intersection of the supports of different wavelets might not be empty, a naive calculation of (4.9) leads to repeated computation of the values $\alpha_{(j,k),(j',k')}^{(d)}$. Hence, we develop a strategy to avoid multiple calculation.

In the case of piecewise constant functions one finds analogously to (2.25)

$$\alpha_{(j,k),(j',k')}^{(1)} = \sqrt{h_j h_{j'}} \int_0^1 \int_0^1 k\left(t_k^{(j)} + h_j s, t_{k'}^{(j')} + h_{j'} t\right) dt ds, \tag{4.10}$$

Algorithm 1 The algorithm for setting up the compression pattern. The result of the function **compute**$(\psi_{j,k}^{(d,\tilde{d})}, \psi_{j',k'}^{(d,\tilde{d})})$ is supposed to be **true**, if the matrix entry $(A\psi_{j',k'}^{(d,\tilde{d})}, \psi_{j,k}^{(d,\tilde{d})})_{L^2(0,1)}$ has to be computed according to (3.10). Else it is **false**.

initialisation: $\mathcal{I}_{j_0-1,j_0-1} := \mathcal{I}_{j_0-1,j_0} := \mathcal{I}_{j_0,j_0-1} := \mathcal{I}_{j_0,j_0} := \{\triangle_{j_0} \times \triangle_{j_0}\}$
for $j := j_0 + 1$ **to** $J - 1$ **do begin**
  for $j' := j_0 - 1$ **to** $j - 1$ **do begin**
    $\mathcal{I}_{j,j'} := \{\}$ C: compute $\mathcal{I}_{j,j'}$ from $\mathcal{I}_{j-1,j'}$ (4.5), (4.7)
    for all $(k, k') \in \mathcal{I}_{j-1,j'}$ **do begin**
      if (**compute**$(\psi_{j,2k}^{(d,\tilde{d})}, \psi_{j',k'}^{(d,\tilde{d})})$ = **true**) $\mathcal{I}_{j,j'} := \mathcal{I}_{j,j'} \cup \{(2k, k')\}$
      if (**compute**$(\psi_{j,2k+1}^{(d,\tilde{d})}, \psi_{j',k'}^{(d,\tilde{d})})$ = **true**) $\mathcal{I}_{j,j'} := \mathcal{I}_{j,j'} \cup \{(2k+1, k')\}$
    end
    $\mathcal{I}_{j',j} := \mathcal{I}_{j,j'}$ C: according to symmetry
  end
  $\mathcal{I}_{j,j} := \{\}$ C: compute $\mathcal{I}_{j,j}$ from $\mathcal{I}_{j-1,j-1}$ (4.6)
  for all $(k, k') \in \mathcal{I}_{j-1,j-1}$ **do begin**
    if (**compute**$(\psi_{j,2k}^{(d,\tilde{d})}, \psi_{j,2k'}^{(d,\tilde{d})})$ = **true**) $\mathcal{I}_{j,j} := \mathcal{I}_{j,j} \cup \{(2k, 2k')\}$
    if (**compute**$(\psi_{j,2k+1}^{(d,\tilde{d})}, \psi_{j,2k'}^{(d,\tilde{d})})$ = **true**) $\mathcal{I}_{j,j} := \mathcal{I}_{j,j} \cup \{(2k+1, 2k')\}$
    if (**compute**$(\psi_{j,2k}^{(d,\tilde{d})}, \psi_{j,2k'+1}^{(d,\tilde{d})})$ = **true**) $\mathcal{I}_{j,j} := \mathcal{I}_{j,j} \cup \{(2k, 2k'+1)\}$
    if (**compute**$(\psi_{j,2k+1}^{(d,\tilde{d})}, \psi_{j,2k'+1}^{(d,\tilde{d})})$ = **true**) $\mathcal{I}_{j,j} := \mathcal{I}_{j,j} \cup \{(2k+1, 2k'+1)\}$
  end
end

where $k(s,t)$ is the given kernel function. We introduce the sets $\mathcal{Q}_{j,j'}$, $j_0 < j, j' \le J$, where every calculated triple $((k,k'), \alpha_{(j,k),(j',k')}^{(1)})$ is stored. Then, if a special value $\alpha_{(j,k),(j',k')}^{(1)}$ is required, one first checks if the corresponding triple $((k,k'), \alpha_{(j,k),(j',k')}^{(1)})$ is contained in $\mathcal{Q}_{j,j'}$, else one computes and stores it.

For piecewise linear function we perform a related strategy. We define analogously to (2.26)

$$\alpha_{(j,k),(j',k')}^{(2,i)} := \sqrt{h_j h_{j'}} \int_0^1 \int_0^1 k\big(t_k^{(j)} + h_j s, t_{k'}^{(j')} + h_{j'} t\big)\, p_i(s,t)\, dt ds, \tag{4.11}$$

with $p_i(s,t)$, $i=1,2,3,4$, given by (2.27). Then, similar to (2.28), one finds

$$\alpha^{(2)}_{(j,k),(j',k')} = \alpha^{(2,1)}_{(j,k-1),(j',k'-1)} + \alpha^{(2,2)}_{(j,k-1),(j',k')} + \alpha^{(2,3)}_{(j,k),(j',k'-1)} + \alpha^{(2,4)}_{(j,k),(j',k')}. \quad (4.12)$$

Inserting this equation into (4.9) yields

$$\begin{aligned}\big(A\psi^{(2,\tilde d)}_{j',k'}, \psi^{(2,\tilde d)}_{j,k}\big)_{L^2(0,1)} = \sum_l \sum_{l'} b_l b_{l'} \Big[&\alpha^{(2,1)}_{(j+1,2k+l-1),(j'+1,2k'+l'-1)} \\ &+ \alpha^{(2,2)}_{(j+1,2k+l-1),(j'+1,2k'+l')} + \alpha^{(2,3)}_{(j+1,2k+l),(j'+1,2k'+l'-1)} \\ &+ \alpha^{(2,4)}_{(j+1,2k+l),(j'+1,2k'+l')}\Big].\end{aligned}$$

Resorting the sums leads to

$$\begin{aligned}\big(A\psi^{(2,\tilde d)}_{j',k'}, \psi^{(2,\tilde d)}_{j,k}\big)_{L^2(0,1)} = \sum_l \sum_{l'} \Big[&b_{l+1} b_{l'+1} \alpha^{(2,1)}_{(j+1,2k+l),(j'+1,2k'+l')} \\ &+ b_{l+1} b_{l'} \alpha^{(2,2)}_{(j+1,2k+l),(j'+1,2k'+l')} + b_l b_{l'+1} \alpha^{(2,3)}_{(j+1,2k+l),(j'+1,2k'+l')} \\ &+ b_l b_{l'} \alpha^{(2,4)}_{(j+1,2k+l),(j'+1,2k'+l')}\Big]. \end{aligned} \quad (4.13)$$

Hence, to avoid multiple calculations we define sets $\mathcal{Q}_{j,j'}$, $j_0 < j, j' \le J$, where we store the required tuples $\big((k,k'), \alpha^{(2,1)}_{(j,k),(j',k')}, \alpha^{(2,2)}_{(j,k),(j',k')}, \alpha^{(2,3)}_{(j,k),(j',k')}, \alpha^{(2,4)}_{(j,k),(j',k')}\big)$. Then, the strategy is similar to above, i.e., before calculating values we search for them in the corresponding set $\mathcal{Q}_{j,j'}$. If these values are not contained in this set, we calculate and store them.

Remark 4.2. As we have seen in subsection 2.4., in the single-scale basis the system matrix corresponding to the hypersingular can be derived from the system matrix corresponding to the single layer operator. A similar approach exists for the wavelet bases: According to [3] we conclude $\big(\psi^{(d,\tilde d)}\big)' = \psi^{(d-1,\tilde d+1)}$ which implies

$$\big(W\psi^{(2,\tilde d)}_{j',k'}, \psi^{(2,\tilde d)}_{j,k}\big)_{L^2(0,1)} = \big(V\psi^{(1,\tilde d+1)}_{j',k'}, \psi^{(1,\tilde d+1)}_{j,k}\big)_{L^2(0,1)}.$$

In other words, one computes an entry of the system matrix of the hypersingular operator with respect to piecewise linear wavelets $\Psi^{(2,\tilde d)}$ like an entry of the system matrix of the single layer operator with respect to piecewise constant wavelets $\Psi^{(1,\tilde d+1)}$.

Since we have to compute $\mathcal{O}(N_J)$ entries in $\mathbf{A}_\psi$, the required memory for storing the sets $\mathcal{Q}_{j,j'}$, $j_0 < j, j' \le J$, is $\mathcal{O}(N_J)$. Moreover, note that, if one has evaluated all pairs $(k,k') \in \mathcal{I}_{j,j'}$, one may delete the set $\mathcal{Q}_{j+1,j'+1}$ to reduce the required main memory.

5. Numerical quadrature

5.1. Error estimations on the reference domain

As we have seen in the last sections it is sufficient to develop quadrature schemes on $\square := [0,1]^2$ for the single layer and the double layer kernel, cf. (2.25), (2.26), (4.10) and (4.11). For the quadrature we utilize tensor product Gauß-Legendre rules on $\square$, for which we state first a general error estimation. For this, let us set

$$I^{[0,1]}f := \int_0^1 f(s)ds.$$

The g-point Gauß-Legendre formula on $[0,1]$

$$Q_g^{[0,1]}f := \sum_{i=1}^{g} \omega_{g,i} f(\xi_{g,i})$$

applied to $f \in C^{2g}(0,1)$ can be estimated as

$$R_g^{[0,1]}f := \left|I^{[0,1]}f - Q_g^{[0,1]}f\right| \lesssim \frac{2^{-4g}}{(2g)!} \max_{s\in[0,1]} \left|f^{(2g)}(s)\right|, \tag{5.1}$$

cf. [23]. Further, we define for a given $f \in L^2(\square)$

$$I^{\square}f := \left(I^{[0,1]} \times I^{[0,1]}\right)f = \int_{\square} f(s,t)d(s,t)$$

and the product Gauß-Legendre quadrature formula

$$Q_{g,g'}^{\square}(f) := \left(Q_g^{[0,1]} \times Q_{g'}^{[0,1]}\right)f = \sum_{i=1}^{g}\sum_{i'=1}^{g'} \omega_{g,i}\omega_{g',i'} f\left(\xi_{g,i}, \xi_{g',i'}\right).$$

Since

$$\begin{aligned}
R_{g,g'}^{\square}f &:= I^{\square}f - Q_{g,g'}^{\square}f \\
&= \left[\left(I^{[0,1]} \times I^{[0,1]}\right) - \left(Q_g^{[0,1]} \times Q_{g'}^{[0,1]}\right)\right]f \\
&= \left[\left(I^{[0,1]} \times I^{[0,1]}\right) - \left(I^{[0,1]} \times Q_{g'}^{[0,1]}\right) + \left(I^{[0,1]} \times Q_{g'}^{[0,1]}\right) - \left(Q_g^{[0,1]} \times Q_{g'}^{[0,1]}\right)\right]f \\
&= \left[I^{[0,1]} \times \left(I^{[0,1]} - Q_{g'}^{[0,1]}\right)\right]f + \left[\left(I^{[0,1]} - Q_g^{[0,1]}\right) \times Q_{g'}^{[0,1]}\right]f
\end{aligned}$$

we obtain the estimate

$$\left|R_{g,g'}^{\square}f\right| \le I^{[0,1]} \max_{t\in[0,1]} \left|R_{g'}^{[0,1]}f(\cdot,t)\right| + Q_{g'}^{[0,1]} \max_{s\in[0,1]} \left|R_g^{[0,1]}f(s,\cdot)\right|,$$

Hence, if $f \in C^{2g}(0,1) \times C^{2g'}(0,1)$, we estimate according to (5.1) the quadrature error $R_{g,g'}^{\square}f$ by

$$R_{g,g'}^{\square}f \lesssim \frac{2^{-4g}}{(2g)!} \max_{(s,t)\in\square} \left|\frac{\partial^{2g} f(s,t)}{\partial s^{2g}}\right| + \frac{2^{-4g'}}{(2g')!} \max_{(s,t)\in\square} \left|\frac{\partial^{2g'} f(s,t)}{\partial t^{2g'}}\right|. \tag{5.2}$$

5.2. Quadrature of the double layer operator

We abbreviate

$$k_K(s,t) := \frac{\partial}{\partial n_t} E\big(\gamma(s),\gamma(t)\big)\alpha(t) = \frac{\alpha(t)}{2\pi} \cdot \frac{n_t\big(\gamma(s)-\gamma(t)\big)}{|\gamma(s)-\gamma(t)|^2}. \tag{5.3}$$

Since the parametrization γ is smooth, we may utilize a Taylor expansion

$$\gamma(s) = \gamma(t) + \gamma'(t)(s-t) + \frac{1}{2}\gamma''(t)(s-t)^2 + \mathcal{O}(|s-t|^3).$$

Observing that $n_t\gamma'(t) = 0$ this implies

$$k_K(s,t) = \frac{1}{2\pi}\begin{cases} \frac{n_t(\gamma(s)-\gamma(t))}{|\gamma(s)-\gamma(t)|^2}\alpha(t), & s \neq t, \\ \frac{n_t\gamma''(t)}{2\alpha(t)}, & s = t. \end{cases}$$

Moreover, employing the Taylor expansion of γ, it is easy to verify that the kernel (5.3) is analytic in $\square$ if γ is analytic in $[0,1]$. Hence, there exists a constant $r > 0$ such that

$$\frac{\partial^\alpha k_K(s,t)}{\partial s^\alpha} \lesssim \frac{\alpha!}{r^\alpha}, \qquad \frac{\partial^\alpha k_K(s,t)}{\partial t^\alpha} \lesssim \frac{\alpha!}{r^\alpha}, \tag{5.4}$$

uniformly for $\alpha \in \mathbb{N}$. According to (2.25) and (4.10), we have to calculate for piecewise constant functions

$$\alpha^{(1)}_{(j,k),(j',k')} = \sqrt{h_j h_{j'}} \int_0^1 \int_0^1 k_K\big(t_k^{(j)} + h_j s, t_{k'}^{(j')} + h_{j'} t\big) dt ds.$$

Setting $f(s,t) := \sqrt{h_j h_{j'}} k_K\big(t_k^{(j)} + h_j s, t_{k'}^{(j')} + h_{j'} t\big)$ the estimate (5.4) implies

$$\frac{\partial^\alpha f(s,t)}{\partial s^\alpha} \lesssim \sqrt{h_j h_{j'}} \frac{h_j^\alpha}{r^\alpha}\alpha!, \qquad \frac{\partial^\alpha f(s,t)}{\partial t^\alpha} \lesssim \sqrt{h_j h_{j'}} \frac{h_{j'}^\alpha}{r^\alpha}\alpha!.$$

Combining this with (5.2) yields

$$R^{\square}_{g,g'} f \lesssim \sqrt{h_j h_{j'}} \big(2^{-4g} h_j^{2g} r^{-2g} + 2^{-4g'} h_{j'}^{2g'} r^{-2g'}\big). \tag{5.5}$$

Observing $h_j = 2^{-j}$ and $h_{j'} = 2^{-j'}$, a quadrature error $\lesssim \varepsilon_{j,j'}$ is ensured by the choice

$$g_{j,j'} = \left\lceil -\frac{\log_2 \varepsilon_{j,j'} + \frac{j+j'}{2}}{2(j+2+\log_2 r)} \right\rceil, \qquad g'_{j,j'} = \left\lceil -\frac{\log_2 \varepsilon_{j,j'} + \frac{j+j'}{2}}{2(j'+2+\log_2 r)} \right\rceil, \tag{5.6}$$

For piecewise linear functions we are interested in the degree of the quadrature required for computing

$$\alpha^{(2,i)}_{(j,k),(j',k')} = \sqrt{h_j h_{j'}} \int_0^1 \int_0^1 k_K\big(t_k^{(j)} + h_j s, t_{k'}^{(j')} + h_{j'} t\big) p_i(s,t) dt ds$$

with a given precision $\varepsilon_{j,j'}$. Herein, $p_i(s,t)$ are polynomials of degree one defined by (2.27). One easily checks

$$|p_i(s,t)| \leq 1, \qquad \left|\frac{\partial p_i(s,t)}{\partial s}\right| \leq 1, \qquad \left|\frac{\partial p_i(s,t)}{\partial t}\right| \leq 1, \qquad i = 1,2,3,4. \tag{5.7}$$

Substituting $f(s,t) := \sqrt{h_j h_{j'}} k_K\big(t_k^{(j)} + h_j s, t_{k'}^{(j')} + h_{j'} t\big) p_i(s,t)$ yields by (5.4) and (5.7) the estimate

$$\frac{\partial^\alpha f(s,t)}{\partial s^\alpha} \lesssim \sqrt{h_j h_{j'}} \frac{h_j^{\alpha-1}}{r^{\alpha-1}}(\alpha-1)!, \qquad \frac{\partial^\alpha f(s,t)}{\partial t^\alpha} \lesssim \sqrt{h_j h_{j'}} \frac{h_{j'}^{\alpha-1}}{r^{\alpha-1}}(\alpha-1)!.$$

Consequently, from (5.2) and the latter estimate we deduce

$$R_{g,g'}^{\square} f \lesssim \sqrt{h_j h_{j'}} \big(2^{-4g} h_j^{2g-1} r^{1-2g} + 2^{-4g'} h_{j'}^{2g'-1} r^{1-2g'}\big). \tag{5.8}$$

Therefore, for getting $R_{g,g'}^{\square} f \lesssim \varepsilon_{j,j'}$ we set

$$g_{j,j'} = \left\lceil -\frac{\log_2 \varepsilon_{j,j'} + \frac{j'-j}{2}}{2(j+2+\log_2 r)} \right\rceil, \qquad g'_{j,j'} = \left\lceil -\frac{\log_2 \varepsilon_{j,j'} + \frac{j-j'}{2}}{2(j'+2+\log_2 r)} \right\rceil. \tag{5.9}$$

5.3. Quadrature of the single layer operator

It is sufficient to develop a quadrature scheme for evaluating double integrals of the kind

$$\beta_{(j,k),(j',k')} := \int_0^1 \int_0^1 k_V\big(t_k^{(j)} + h_j s, t_{k'}^{(j')} + h_{j'} t\big) p(s,t)\,dt\,ds,$$

where $k_V(s,t)$ denotes the weakly singular kernel function

$$k_V(s,t) := E\big(\gamma(s),\gamma(t)\big) = -\frac{1}{2\pi}\log|\gamma(s)-\gamma(t)|$$

and $p(s,t)$ indicates a polynomial in s and t of degree ≤ 1. The goal is to split the kernel function

$$k_V(s,t) = k_V^{(1)}(s,t) + k_V^{(2)}(s,t)$$

into a smooth kernel $k_V^{(1)}(s,t)$ and a weakly singular, but analytically integrable, kernel $k_V^{(2)}(s,t)$. Then, the second integral on the right hand side of

$$\begin{aligned}\beta_{(j,k),(j',k')} = &\int_0^1 \int_0^1 k_V^{(1)}\big(t_k^{(j)} + h_j s, t_{k'}^{(j')} + h_{j'} t\big) p(s,t)\,dt\,ds \\ &+ \int_0^1 \int_0^1 k_V^{(2)}\big(t_k^{(j)} + h_j s, t_{k'}^{(j')} + h_{j'} t\big) p(s,t)\,dt\,ds\end{aligned} \tag{5.10}$$

can be computed exactly. In the sequel we assume that $j, j' \geq 2$. Consequently, $h_j = \operatorname{diam}[t_k^{(j)}, t_{k+1}^{(j)}] \leq \frac{1}{4}$ and $h_{j'} = \operatorname{diam}[t_{k'}^{(j')}, t_{k'+1}^{(j')}] \leq \frac{1}{4}$. This is no restriction since we may subdivide one or both integrals to ensure this supposition. Denoting by m, m' the midpoints of the integration intervals,

$$m := \frac{1}{2}\left(t_k^{(j)} + t_{k+1}^{(j)}\right), \qquad m' := \frac{1}{2}\left(t_{k'}^{(j')} + t_{k'+1}^{(j')}\right),$$

there occur three cases:

1. Case: $|m - m'| \leq \frac{1}{2}$

   We split the kernel by

$$\begin{aligned} k_V(s,t) &= -\frac{1}{4\pi} \log|\gamma(s) - \gamma(t)|^2 \\ &= -\frac{1}{4\pi} \log \frac{|\gamma(s) - \gamma(t)|^2}{(s-t)^2} - \frac{1}{4\pi} \log(s-t)^2 =: k_V^{(1)}(s,t) + k_V^{(2)}(s,t). \end{aligned}$$

   Due to the smoothness of γ we find

$$k_V^{(1)}(s,t) = -\frac{1}{4\pi} \begin{cases} \log \frac{|\gamma(s)-\gamma(t)|^2}{(s-t)^2}, & s \neq t, \\ \log\left(\alpha^2(t)\right), & s = t, \end{cases}$$

   hence, the kernel $k_V^{(1)}(s,t)$ is analytic in the domain $Q := \{(s,t) \in \Box : |s-t| < 1\}$. From $|m - m'| < \frac{1}{2}$ and $h_j, h_{j'} \leq \frac{1}{4}$ we deduce that

$$|s - t| < \frac{3}{4} \qquad \forall\ (s,t) \in [t_k^{(j)}, t_{k+1}^{(j)}] \times [t_{k'}^{(j')}, t_{k'+1}^{(j')}],$$

   i.e., the distance to the singularity point (s,t) with $|s-t| = 1$ is $\geq \frac{1}{4}$.

2. Case: $1 + m - m' < \frac{1}{2}$

   We choose now

$$k_V^{(1)}(s,t) := -\frac{1}{4\pi} \log \frac{|\gamma(s) - \gamma(t)|^2}{(1+s-t)^2}, \qquad k_V^{(2)}(s,t) := -\frac{1}{4\pi} \log(1+s-t)^2.$$

   It follows

$$k_V^{(1)}(s,t) = -\frac{1}{4\pi} \begin{cases} \log \frac{|\gamma(s)-\gamma(t)|^2}{(1+s-t)^2}, & (s,t) \neq (0,1), \\ \log\left(\alpha^2(t)\right), & (s,t) = (0,1), \end{cases}$$

   which implies that the kernel $k_V^{(1)}(s,t)$ is analytic in the domain $Q := \{(s,t) \in \Box \setminus (0,1)\}$. From $1 + m - m' < \frac{1}{2}$ and $h_j, h_{j'} \leq \frac{1}{4}$, we deduce

$$|1 + s - t| < \frac{3}{4} \quad \forall\ (s,t) \in [t_k^{(j)}, t_{k+1}^{(j)}] \times [t_{k'}^{(j')}, t_{k'+1}^{(j')}],$$

   i.e., the distance to the singularity $(s,t) = (0,1)$ is $\geq \frac{1}{4}$.

3. Case: $1 + m' - m < \frac{1}{2}$
   Similarly to the second case we set

$$k_V^{(1)}(s,t) := -\frac{1}{4\pi}\log\frac{|\gamma(s)-\gamma(t)|^2}{(1+t-s)^2}, \qquad k_V^{(2)}(s,t) := -\frac{1}{4\pi}\log(1+t-s)^2.$$

   For the kernel $k_V^{(1)}(s,t)$ one finds

$$k_V^{(1)}(s,t) = -\frac{1}{4\pi}\begin{cases}\log\frac{|\gamma(s)-\gamma(t)|^2}{(1+t-s)^2}, & (s,t) \neq (1,0),\\ \log\big(\alpha^2(t)\big), & (s,t) = (1,0).\end{cases}$$

   The kernel $k_V^{(1)}(s,t)$ is analytic in the domain $Q := \{(s,t) \in \square \setminus (1,0)\}$. Furthermore, analogously to the second case, the distance to the singularity $(s,t) = (1,0)$ is $\geq \frac{1}{4}$.

In all three cases the second term on the left hand side of (5.10) can be derived by the following primitives

$$\int\int \log(s-t)^2 dtds = \frac{1}{2}(s-t)^2\big(3 - \log(s-t)^2\big),$$
$$\int\int \log(s-t)^2\, s\, dtds = \frac{1}{18}(s-t)^2\big(16s + 11t - (6s+3t)\log(s-t)^2\big),$$
$$\int\int \log(s-t)^2\, t\, dtds = \frac{1}{18}(s-t)^2\big(16t + 11s - (6t+3s)\log(s-t)^2\big),$$
$$\int\int \log(s-t)^2\, s\, t\, dtds = -\frac{1}{16}(s+t)^2\big((s+t)^2 + 2(s-t)^2\log(s-t)^2\big).$$

We utilize numerical quadrature for the first term on the left hand side of (5.10). As we have seen, in all cases the distance to the weak singularity of $k_V^{(1)}(s,t)$ is $\geq \frac{1}{4}$ independently of (j,k) and (j',k'). Since the kernel $k_V^{(1)}(s,t)$ is analytic except in the singularity, we deduce that there exists a constant $0 < r \leq 1/4$ such that

$$\frac{\partial^\alpha k_V^{(1)}(s,t)}{\partial s^\alpha} \lesssim \frac{\alpha!}{r^\alpha}, \qquad \frac{\partial^\alpha k_V^{(1)}(s,t)}{\partial t^\alpha} \lesssim \frac{\alpha!}{r^\alpha}, \tag{5.11}$$

uniformly for $\alpha \in \mathbb{N}$. Since (5.11) agrees with (5.4) we may adopt the argumentation of the previous subsection. Therefore, to ensure a quadrature error $\lesssim \varepsilon_{j,j'}$ we have to choose the degrees $g_{j,j'}$ and $g'_{j,j'}$ from (5.6) if $d = 1$ and from (5.9) if $d = 2$.

5.4. The recycling scheme

Using the analysis of the wavelet Galerkin scheme as a guideline, it turns out that the required accuracy for the computation of the integrals $\alpha^{(1)}_{(j,k),(j',k')}$ and

$\alpha^{(2,i)}_{(j,k),(j',k')}$, respectively, depends strongly on the involved scales (j,j'), cf. [12, 32, 35]. In particular, the highest accuracy is required on the coarsest levels, that is $(j,j')=(j_0,j_0)$, while for the integrals on the finest levels, that is $(j,j')=(J,J)$, the required accuracy is low. More precisely, it is sufficient to choose

$$\varepsilon_{j,j'} \lesssim 2^{-2J(d-q)}2^{(j+j')d},$$

cf. [5, 12, 32]. According to the formulas (5.6) and (5.9) one has to be very careful if the parameter r is small. In this case, one has to avoid denominators close to zero. This requires that $h_j, h_{j'} < r$. In order to satisfy this condition, one has to subdivide the large domains of integration. On the other hand, on finer scales we already have to compute such integrals over small domains. Therefore, we propose another approach. We compute integrals on the fine scales with the accuracy required on the coarse scales and reuse these results also for the computation of the coarse scale coefficients. This reduces the number of functions significantly.

In particular, by the formulas (4.9) and (4.13) the computation of an matrix entry $\left(A\psi^{(d,\tilde{d})}_{j',k'},\psi^{(d,\tilde{d})}_{j,k}\right)_{L^2(0,1)}$ of the compressed system matrix $\mathbf{A}_\psi$ reduces to the computation of the integrals $\alpha^{(1)}_{(j+1,2k+l),(j'+1,2k'+l')}$ and $\alpha^{(2,i)}_{(j+1,2k+l),(j'+1,2k'+l')}$, respectively. The crucial idea of the *recycling scheme* is to invoke the refinement relation (3.2) of the scaling functions. Then, for $d=1$ one finds for the integrals $\alpha^{(1)}_{(j,k),(j',k')}$ the relations

$$\alpha^{(1)}_{(j,k),(j',k')} = \frac{1}{\sqrt{2}}\left(\alpha^{(1)}_{(j+1,2k),(j',k')} + \alpha^{(1)}_{(j+1,2k+1),(j',k')}\right),$$
$$\alpha^{(1)}_{(j,k),(j',k')} = \frac{1}{\sqrt{2}}\left(\alpha^{(1)}_{(j,k),(j'+1,2k')} + \alpha^{(1)}_{(j,k),(j'+1,2k'+1)}\right).$$

Since for $d=2$ we compute the integrals over the smooth parts of the scaling functions, we additionally invoke (4.11) which gives for the integrals $\alpha^{(2,i)}_{(j,k),(j',k')}$ the relations

$$\alpha^{(2,1)}_{(j,k),(j',k')} = \frac{1}{2\sqrt{2}}\left(\alpha^{(2,1)}_{(j+1,2k),(j',k')} + 2\alpha^{(2,1)}_{(j+1,2k+1),(j',k')} + \alpha^{(2,3)}_{(j+1,2k+1),(j',k')}\right),$$
$$\alpha^{(2,2)}_{(j,k),(j',k')} = \frac{1}{2\sqrt{2}}\left(\alpha^{(2,2)}_{(j+1,2k),(j',k')} + 2\alpha^{(2,2)}_{(j+1,2k+1),(j',k')} + \alpha^{(2,4)}_{(j+1,2k+1),(j',k')}\right),$$
$$\alpha^{(2,3)}_{(j,k),(j',k')} = \frac{1}{2\sqrt{2}}\left(\alpha^{(2,1)}_{(j+1,2k),(j',k')} + 2\alpha^{(2,3)}_{(j+1,2k),(j',k')} + \alpha^{(2,3)}_{(j+1,2k+1),(j',k')}\right),$$
$$\alpha^{(2,4)}_{(j,k),(j',k')} = \frac{1}{2\sqrt{2}}\left(\alpha^{(2,4)}_{(j+1,2k),(j',k')} + 2\alpha^{(2,4)}_{(j+1,2k),(j',k')} + \alpha^{(2,2)}_{(j+1,2k+1),(j',k')}\right),$$

and

$$\begin{aligned}
\alpha^{(2,1)}_{(j,k),(j',k')} &= \frac{1}{2\sqrt{2}}\Big(\alpha^{(2,1)}_{(j,k),(j'+1,2k')} + 2\alpha^{(2,1)}_{(j,k),(j'+1,2k'+1)} + \alpha^{(2,2)}_{(j,k),(j'+1,2k'+1)}\Big),\\
\alpha^{(2,2)}_{(j,k),(j',k')} &= \frac{1}{2\sqrt{2}}\Big(\alpha^{(2,1)}_{(j,k),(j'+1,2k')} + 2\alpha^{(2,2)}_{(j,k),(j'+1,2k')} \quad + \alpha^{(2,2)}_{(j,k),(j'+1,2k'+1)}\Big),\\
\alpha^{(2,3)}_{(j,k),(j',k')} &= \frac{1}{2\sqrt{2}}\Big(\alpha^{(2,3)}_{(j,k),(j'+1,2k')} + 2\alpha^{(2,3)}_{(j,k),(j'+1,2k'+1)} + \alpha^{(2,4)}_{(j,k),(j'+1,2k'+1)}\Big),\\
\alpha^{(2,4)}_{(j,k),(j',k')} &= \frac{1}{2\sqrt{2}}\Big(\alpha^{(2,3)}_{(j,k),(j'+1,2k')} + 2\alpha^{(2,4)}_{(j,k),(j'+1,2k')} \quad + \alpha^{(2,4)}_{(j,k),(j'+1,2k'+1)}\Big).
\end{aligned}$$

Hence, exploiting these *recycling formulas* one may use already computed values for the computation of $\alpha^{(1)}_{(j,k),(j',k')}$ and $\alpha^{(2,i)}_{(j,k),(j',k')}$ instead of applying a direct quadrature. Starting the computation of the required values $\big(A\psi^{(d,\tilde{d})}_{j',k'}, \psi^{(d,\tilde{d})}_{j,k}\big)_{L^2(0,1)}$ of the compressed matrix $\mathbf{A}_\psi$ on the finest levels, i.e., calculating first the entries for which $(j,j') = (J-1, J-1)$, a systematically exploitation of the recycling formulas leads to a natural subdivision of the coarser scales. We mention that the structure of the compressed matrix implies that almost all computed $\alpha^{(1)}_{(j,k),(j',k')}$ can be used for the computation of the required $\alpha^{(1)}_{(j-1,k),(j',k')}$ and $\alpha^{(1)}_{(j,k),(j'-1,k')}$, and likewise the $\alpha^{(2,i)}_{(j,k),(j',k')}$. Moreover, this strategy reduces the number of those values, for which a direct quadrature is required.

By using the above recycling techniques the computed integrals on the finer scales are part of the integrals on the coarser scales. Assuming the worst case, $\alpha^{(1)}_{(j_0,k),(j_0,k')}$ and $\alpha^{(2,i)}_{(j_0,k),(j_0,k')}$ consist of $2^{(j-j_0)+(j'-j_0)}$ values on the scale (j,j'). On the other hand, we have to take into consideration the L^2-normalization of the scaling functions. It turns out that we attain a quadrature error $\lesssim \varepsilon_{j_0,j_0}$ on the coarsest level (j_0,j_0) for both, piecewise constant and piecewise linear functions, by choosing

$$g_{j,j'} = \left\lceil -\frac{\log_2 \varepsilon_{j_0,j_0}}{2(j+2+\log_2 r)} \right\rceil, \qquad g'_{j,j'} = \left\lceil -\frac{\log_2 \varepsilon_{j_0,j_0}}{2(j'+2+\log_2 r)} \right\rceil, \tag{5.12}$$

Herein, $\varepsilon_{j_0,j_0} \lesssim 2^{-2J(d-q)}$ is required to attain the optimal order of convergence $2(d-q)$ of the Galerkin scheme. As one easily checks, for $\varepsilon_{j_0,j_0} = 2^{-2J(d-q)}$ the quadrature degree on the finest scale tends to the (fixed) degree $g_{J,J} = g'_{J,J} = \lceil d-q \rceil$ if $J \to \infty$. We mention that this degree of quadrature is also necessary to obtain the optimal order of convergence in the traditional Galerkin scheme.

The main fact to reduce the number of the nonzero coefficients of $\mathbf{A}_\psi$ from $\mathcal{O}(N_J \log^3 N_J)$ to an optimal rate $\mathcal{O}(N_J)$ relies on the second compression. This compression is active if

$$\operatorname{dist}\big(\Theta^{(d,\tilde{d})}_{j,k}, \Theta^{(d,\tilde{d})}_{j',k'}\big) \lesssim 2^{-\min\{j,j'\}}.$$

As shown in [35, p. 177] the block matrix $[\mathbf{A}_\psi]_{(j,\triangle_j),(j',\triangle_{j'})}$, $j > j'$, consists in this case of $\mathcal{O}\big(\dim(W_j)^\alpha\big)$ instead of $\mathcal{O}\big(\dim(W_j)\big)$ entries. Here, $\alpha = \alpha(j,j')$ denotes some paramter with $0 < \alpha < 1$. Choosing the degrees of quadrature as in (5.12), each matrix element can be computed with $\mathcal{O}\big(J^2/((j+c)(j'+c))\big)$ function calls. From an estimate of the form

$$2^{-\alpha j} j^2 \lesssim 2^{-\alpha' j}$$

with some $0 < \alpha' < \alpha$, the argumentation in [35] can be repeated to prove that the over-all number of function calls is $\mathcal{O}(2^J) = \mathcal{O}(N_J)$. Since the detailed proof of this result is rather technical, we refer to [12] for further details. Let us remark, that the proof in [12] is also valid in the three dimensional case where singular kernel functions appear. Indeed, the present case is much easier since only smooth kernels are involved. Employing the recycling technique gives a further reduction of the complexity. However, the asymptotic behaviour cannot exceed the optimal complexity rate.

6. Numerical results

6.1. The model problem

For testing the algorithms we introduce a numerical example for which the solution is known analytically. We choose Γ given by

$$\gamma : [0,1] \to \Gamma, \qquad \gamma(t) = \tfrac{1}{20}\big(4 + \cos(10\pi t) + \cos(2\pi t)\big)\begin{bmatrix}\cos(2\pi t)\\ \sin(2\pi t)\end{bmatrix},$$

and Ω as the interior domain Ω^-, see figure 2 for this constellation. One easily checks that the function

$$U(x) = \frac{x_1 - 2x_2 + 0.2}{(x_1 - 0.2)^2 + (x_2 - 0.2)^2} \tag{6.13}$$

is harmonic in $\mathbb{R}^2 \setminus \{[\begin{smallmatrix}0.2\\0.2\end{smallmatrix}]\}$. Hence, the interior Dirichlet problem

$$\begin{aligned} -\Delta u &= 0 && \text{in } \Omega,\\ u &= U\big|_\Gamma && \text{on } \Gamma, \end{aligned} \tag{6.14}$$

has the unique solution $u = U$. The function U is plotted in figure 2. Since U is harmonic in $\overline{\Omega}$, it follows that

$$\int_\Gamma \frac{\partial U(x)}{\partial n_x} ds_x = \int_\Gamma n_x^T \nabla U(x) ds_x = 0.$$

Therefore, the interior Neumann problem

$$\begin{aligned} -\Delta u &= 0 && \text{in } \Omega,\\ u &= \tfrac{\partial U}{\partial n} && \text{on } \Gamma, \end{aligned} \tag{6.15}$$

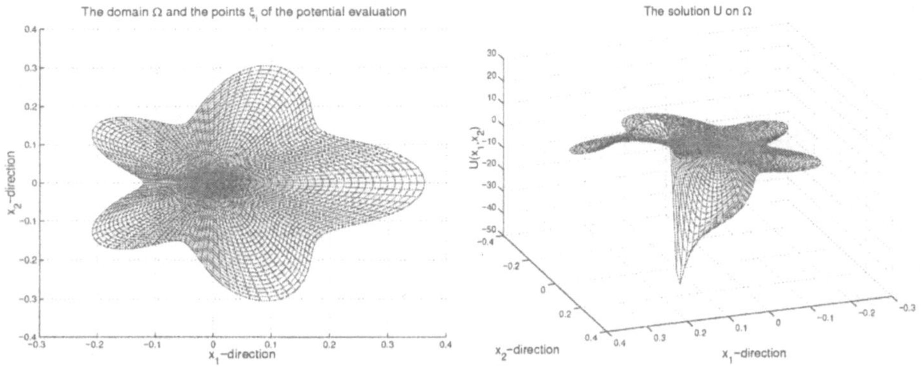

Figure 2: The test domain Ω with the points ξ_i of the potential evaluation (left) and the function $U(x) = \frac{x_1-2x_2+0.2}{(x_1-0.2)^2+(x_2-0.2)^2}$ on Ω (right).

has the unique solution $u = U + c$, where c denotes some real constant.

In the sequel we solve both, the Dirichlet and the Neumann problem, by the boundary integral equations described in this paper. We compute the approximated density vector $\boldsymbol{\mu}_\phi$ corresponding to the single-scale basis by the traditionally single-scale Galerkin scheme (4.1) and by the wavelet Galerkin scheme (4.2). For assembling the Galerkin matrix with respect to the single-scale basis, we choose the fixed degree of quadrature $\lceil d-q \rceil$. In the wavelet Galerkin scheme the degree of quadrature is computed by (5.12) with $\varepsilon_{j_0,j_0} = 2^{-2J(d-q)}$ and $r = 1/4$. We determine the numerical solution u_ϕ and u_ψ, respectively, by the potential evaluations (2.13) and (2.17) in fixed points ξ_i lying in Ω, cf. figure 2. For the Dirichlet problem (6.14) we measure the absolute errors $\max_i |(u-u_\phi)(\xi_i)|$ and $\max_i |(u-u_\psi)(\xi_i)|$, respectively. Since for the Neumann problem (6.15) the solution is given modulo a constant, we calculate in this case the absolute errors $\min_{c\in\mathbb{R}} \max_i |(u-u_\phi+c)(\xi_i)|$ and $\min_{c\in\mathbb{R}}\{\max_i |(u-u_\psi+c)(\xi_i)|\}$, respectively.

6.2. The choice of the compression parameters

First we study the influence of the parameters a, a', δ and δ' to the compression and accuracy of the wavelet Galerkin scheme. For this, we solve the interior Dirichlet problem (6.14) by the equation of the first kind (2.5) and piecewise linear wavelets $\Psi_J^{(2,4)}$. Since the single layer operator $\mathcal{V}$ is an operator of order -1, condition (3.11) implies that $a, a' > 1$ and $2 < \delta, \delta' < 3$. For the sake of simplicity we set $a = a'$ and $\delta = \delta'$ and choose a fixed $N_J := 2048$. We measure the density of $\mathbf{V}_\psi$ and the absolute error of u_ψ for the parameters a and δ lying in the range $1 \le a \le 2$ and $2 \le \delta \le 3$. Herein, the density is the ratio of the number of nonzero elements divided by N_J^2. The obtained results are shown in figure 3. From the plot on the left hand side one deduces that the absolute error of the numerical solution u_ψ does not drop under a fixed value which is independent of a and δ. Obviously, this value is the discretization error. On the other hand, if one chooses $a+\delta$ too small the error of compression is larger than the discretization error. Hence, the error of the numerical solution u_ψ increases. The plot on the right hand side shows the density of the system matrix $\mathbf{V}_\psi$ while changing the parameters. One observes that the density seems to grow linearly with $a+\delta$. Summarizing it turns out, that $a+\delta$ has to be sufficiently large to ensure that the compression error is not higher than the discretization error. But $a+\delta$ should be as small as possible to obtain the best compression rate.

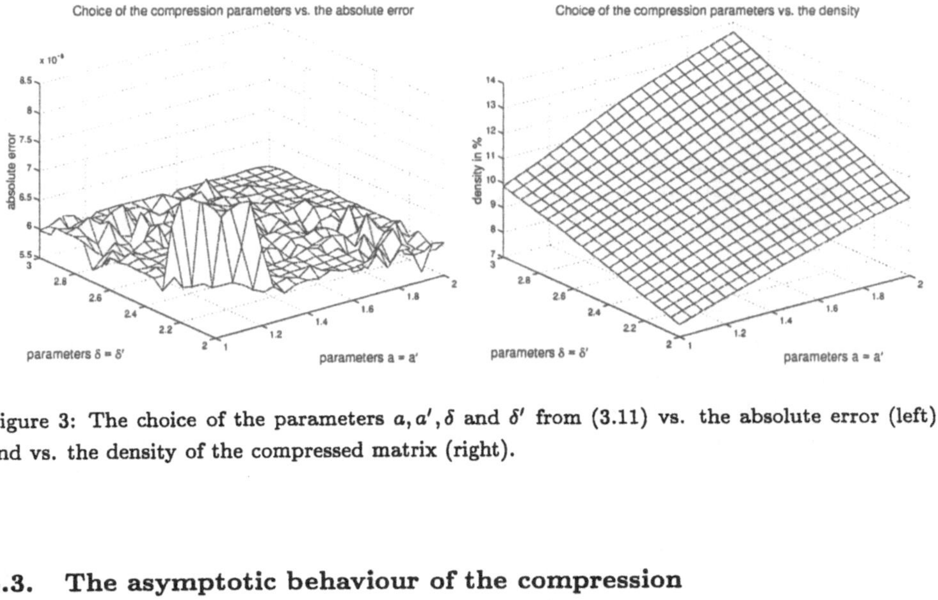

Figure 3: The choice of the parameters a, a', δ and δ' from (3.11) vs. the absolute error (left) and vs. the density of the compressed matrix (right).

6.3. The asymptotic behaviour of the compression

Next, we count the number of nonzero elements in the compressed system matrices while increasing J. To satisfy (3.11) we have to choose wavelets with a sufficient numer of vanishing moments $\tilde{d}$. Since, however, the supports of the wavelets increase proportionally with the number of vanishing moments (which reduces in our experience the compression rates), one has to apply wavelets with minimal number of vanishing moments and minimal supports, respectively. Hence, we choose $\Psi_J^{(1,3)}$ or $\Psi_J^{(2,4)}$ for the discretization of the single and double layer operator and $\Psi_J^{(2,2)}$ for the discretization of the hypersingular operator, respectively. For the computation of the compression we choose the parameters from (3.11) for the double and single layer operator as $a = a' = \delta = \delta' = 1.5$ if $d = 1$ and $a = a' = 1.25$, $\delta = \delta' = 2.5$ if $d = 2$. For the hypersingular operator we choose

$a = a' = 1.5$ and $\delta = \delta' = 2.5$. As the numerical results in the next subsections confirm these settings are sufficient to attain approximatively the accuracy of the traditional single-scale Galerkin scheme.

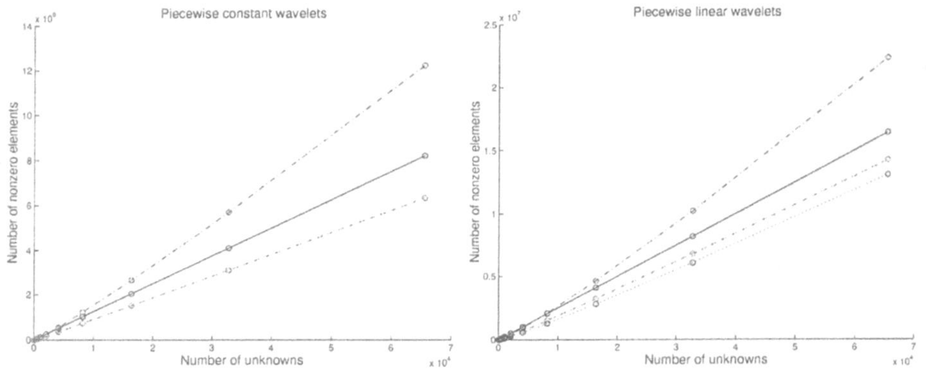

Figure 4: The number of nonzero elements of the compressed system matrices: the dashed line belongs to $\mathbf{V}_\psi$, the dashdotted line belongs to $\mathbf{K}_\psi$, the dotted line belongs to $\mathbf{W}_\psi$ and the solid line indicates linear behaviour ($125N_J$ on the left hand side and $250N_J$ on the right hand side).

The diagrams in figure 4 show the number of nonzero elements of the compressed system matrices corresponding to the double layer operator $\mathcal{K}$, the single layer operator $\mathcal{V}$ and the hypersingular operator $\mathcal{W}$. On the left hand side one finds the number of nonzero elements of the system matrices $\mathbf{K}_\psi$ and $\mathbf{V}_\psi$ computed with respect to the piecewise constant wavelets $\Psi_J^{(1,3)}$. On the right hand side one sees the number of nonzero elements of the system matrices $\mathbf{K}_\psi$, $\mathbf{V}_\psi$ and $\mathbf{W}_\psi$ computed with respect to the piecewise linear wavelets $\Psi_J^{(2,4)}$ and $\Psi_J^{(2,2)}$, respectively. Both diagrams confirm that we obtain asymptotically only $\mathcal{O}(N_J)$ nonzero matrix elements, whereby the constant for the piecewise linear wavelets is larger than for the piecewise constant wavelets.

6.4. The wavelet preconditioner

Since the single layer operator $\mathcal{V}$ and the hypersingular operator $\mathcal{W}$ are operators of order different from zero, the resulting Galerkin matrices $\mathbf{V}_\phi$, $\mathbf{V}_\psi$, $\mathbf{W}_\phi$ and $\mathbf{W}_\psi$ are ill conditioned. More precisely, without any preconditioning the condition numbers in l^2 behave like 2^J, where J indicates the level of discretization. According to subsection 3.5. the wavelet approach provides a simple diagonal preconditioner.

In table 1 we compare the l_2-condition of the matrices $\mathbf{V}_\phi$ and $\mathbf{W}_\phi$ and the preconditioned matrices $\mathbf{V}_\psi$ and $\mathbf{W}_\psi$. The given numbers result from a diagonal scaling of the computed system matrices, which additionally improves the wavelet

Unknowns N_J		32	64	128	256	512	1024	2048
single layer operator	$\Phi_J^{(1)}$	54.37	107.3	219.8	447.2	900.2	1800	3588
	$\Psi_J^{(1,3)}$	8.887	8.967	8.992	9.005	9.010	9.013	9.014
	$\Phi_J^{(2)}$	131.2	268.8	552.8	1137	2308	4640	9291
	$\Psi_J^{(2,4)}$	58.87	65.69	72.71	77.30	81.17	84.01	86.29
hypersingular operator	$\Phi_J^{(2)}$	57.03	130.1	290.6	641.9	1409	3070	6645
	$\Psi_J^{(2,2)}$	4.889	4.946	4.950	4.950	4.951	4.951	4.951

Table 1: The condition numbers of the diagonally scaled system matrices $\mathbf{V}_\phi$, $\mathbf{V}_\psi$, $\mathbf{W}_\phi$ and $\mathbf{W}_\psi$.

preconditioner (3.13). As one figures out of the given results, the behaviour of the condition numbers of the single-scale matrices is $\sim 2^J$ while the condition numbers of the system matrices with respect to the (preconditioned) wavelet bases seem to be bounded. Note, since the matrices $\mathbf{W}_\phi$ and $\mathbf{W}_\psi$, respectively, are only positive semidefinite, we discard one unknown to compute the condition number.

6.5. Numerical results for the Dirichlet problem

We consider the numerical solution of the given Dirichlet problem (6.14). In the tables 2 and 3 we list the over-all computing time and the accuracy obtained by the traditional (single-scale) Galerkin scheme and the compressed wavelet Galerkin scheme. Precisely, table 2 shows the values attained by piecewise constant functions for the equation of the first kind (2.21) while in table 3 we give the results for piecewise linear functions for the equation of the second kind (2.23). We determine for the single-scale scheme and the multiscale scheme the absolute errors of the numerical solutions u_ϕ and u_ψ as described above. The columns titled by "contr." (contraction) give the ratio of the absolute error obtained on the previous level divided by the present absolute error. Since the optimal convergence rate is given by $2(d-r)$, these numbers should be close to $2^{2(d-q)}$. Hence, the optimal ratios 8 in table 2 and 16 in table 3 are achieved approximately. The data listed in brackets are determined via extrapolation, since these problems do not fit into the available main memory of the computer. As one figures out of the tables 2 and 3, the accuracy is not deteriorated by the compression strategy. The break even point of the compressed wavelet Galerkin scheme is about 256 unknowns. Moreover, as the columns titled by "density" (in %) confirm, the compression yields an enormous save of memory.

6.6. Numerical results for the Neumann problem

For the Neumann problem (6.15) we tabulate similarly to the Dirichlet problem the over-all computing time and the accuracy attained by the traditional Galerkin

N_J	single-scale scheme: $d = 1$			multiscale scheme: $(d, \tilde{d}) = (1, 3)$			
	abs. error	contr.	time	abs. error	contr.	time	density
128	8.8e-01	—	0.17	8.7e-01	—	0.21	48.4
256	8.4e-02	11	0.71	8.4e-02	10	0.56	30.5
512	8.7e-03	9.6	2.91	8.7e-03	9.6	1.51	18.0
1024	1.0e-03	8.3	13.2	1.0e-03	8.3	3.75	10.5
2048	1.3e-04	8.1	54.7	1.3e-04	8.2	9.58	5.94
4096	1.6e-05	8.1	225	1.6e-05	8.1	21.4	3.31
8192	(2.0e-06)	(8.0)	(902)	2.0e-06	8.0	50.3	1.82
16384	(2.5e-07)	(8.0)	(3609)	2.5e-07	7.8	116	0.99
32768	(3.1e-08)	(8.0)	(14438)	3.3e-08	7.8	267	0.53
65536	(3.9e-09)	(8.0)	(57751)	4.2e-09	7.8	597	0.28
131072	(4.9e-10)	(8.0)	(231000)	5.4e-10	7.7	1439	0.15

Table 2: Over-all computing times (in seconds) and accuracy of the single-scale and the multiscale scheme obtained for the single layer operator and piecewise constant functions.

N_J	single-scale scheme: $d = 2$			multiscale scheme: $(d, \tilde{d}) = (2, 4)$			
	abs. error	contr.	time	abs. error	contr.	time	density
128	6.0e-01	—	0.13	6.0e-01	—	0.22	62.9
256	2.3e-02	26.5	0.51	2.3e-02	26.5	0.62	39.5
512	1.1e-03	20.3	2.18	1.1e-03	20.3	1.62	23.7
1024	4.0e-05	28.0	10.0	4.0e-05	28.0	4.18	13.6
2048	2.4e-06	16.3	40.6	2.4e-06	16.3	10.5	7.61
4096	1.5e-07	16.1	166	1.5e-07	16.1	23.7	4.18
8192	(9.5e-09)	(16.0)	(663)	9.5e-09	16.0	55.6	2.25
16384	(5.9e-10)	(16.0)	(2653)	5.9e-10	16.0	125	1.20
32768	(3.7e-11)	(16.0)	(10612)	3.7e-11	16.1	268	0.63
65536	(2.3e-12)	(16.0)	(42447)	2.3e-12	16.2	700	0.33

Table 3: Over-all computing times (in seconds) and accuracy of the single-scale and the multiscale scheme obtained for the double layer operator and piecewise linear functions.

scheme and the compressed wavelet Galerkin scheme. Table 4 gives the results for the equation of the second kind (2.38) obtained with piecewise constant functions and table 5 shows the results for the equation of the first kind (2.34) measured for piecewise linear functions. Likewise to the Dirichlet problem the optimal convergence rate is $2(d-q)$. Therefore, the optimal contraction is $2^{2(d-q)}$. This is 4 in table 4 and 8 in table 5. The data listed in brackets signify extrapolated results.

N_J	single-scale scheme: $d = 1$			multiscale scheme: $(d, \tilde{d}) = (1, 3)$			
	abs. error	contr.	time	abs. error	contr.	time	density
128	1.35	—	0.02	1.35	—	0.09	42.2
256	2.0e-01	6.7	0.12	2.0e-01	6.6	0.19	24.9
512	6.1e-02	3.3	0.65	6.1e-02	3.3	0.50	13.9
1024	1.5e-02	4.0	4.03	1.5e-02	4.0	1.09	7.54
2048	3.8e-03	4.0	16.4	3.8e-03	4.0	2.69	4.02
4096	9.5e-04	4.0	66.5	9.5e-04	4.0	6.03	2.10
8192	(2.4e-04)	(4.0)	(266)	2.4e-04	4.0	13.4	1.09
16384	(5.9e-05)	(4.0)	(1064)	5.9e-05	4.0	28.3	0.56
32768	(1.5e-05)	(4.0)	(4255)	1.5e-05	4.0	61.6	0.29
65536	(3.7e-06)	(4.0)	(17019)	3.7e-06	4.0	137	0.15
131072	(9.3e-07)	(4.0)	(68076)	9.4e-07	4.0	326	0.08
262144	(2.3e-07)	(4.0)	(272304)	2.4e-07	4.0	856	0.04

Table 4: Over-all computing times (in seconds) and accuracy of the single-scale and the multiscale scheme obtained for the double layer operator and piecewise constant functions.

N_J	single-scale scheme: $d = 2$			multiscale scheme: $(d, \tilde{d}) = (2, 2)$			
	abs. error	contr.	time	abs. error	contr.	time	density
128	9.6e-01	—	0.18	1.09	—	0.21	48.4
256	7.1e-02	14	0.73	1.0e-01	11	0.56	30.5
512	8.3e-03	8.6	3.10	7.7e-03	13	1.52	18.2
1024	1.0e-03	8.1	17.5	8.1e-04	9.5	3.79	10.7
2048	1.3e-04	8.0	81.8	1.0e-04	8.0	9.68	6.11
4096	1.6e-05	8.0	399	1.3e-05	8.0	21.9	3.44
8192	(2.0e-06)	(8.0)	(1595)	1.6e-06	8.0	51.8	1.91
16384	(2.5e-07)	(8.0)	(6378)	2.0e-07	7.9	121	1.04
32768	(3.1e-08)	(8.0)	(25513)	2.6e-08	7.8	281	0.57
65536	(3.9e-09)	(8.0)	(102050)	3.3e-09	7.8	641	0.31
131072	(4.9e-10)	(8.0)	(408210)	4.3e-10	7.7	1544	0.16

Table 5: Over-all computing times (in seconds) and accuracy of the single-scale and the multiscale scheme obtained for the hypersingular operator and piecewise linear functions.

References

[1] G. Beylkin, R. Coifman, and V. Rokhlin. The fast wavelet transform and numerical algorithms. *Comm. Pure and Appl. Math.*, 44:141–183, 1991.

[2] A. Cohen, W. Dahmen, and R. DeVore. Adaptive wavelet methods for elliptic operator equations – convergence rates. *IGPM-Report, RWTH Aachen*, 1998. to appear in Math. Comp.

[3] A. Cohen, I. Daubechies, and J.-C. Feauveau. Biorthogonal bases of compactly supported wavelets. *Pure Appl. Math.*, 45:485–560, 1992.

[4] M. Costabel. Principles of boundary element methods. *Comp. Phys. Reports*, 6:243–274, 1987.

[5] W. Dahmen. Wavelet and multiscale methods for operator equations. *Acta Numerica*, 6:55–228, 1997.

[6] W. Dahmen, B. Kleemann, S. Prößdorf, and R. Schneider. A multiscale method for the double layer potential equation on a polyhedron. In H.P. Dikshit and C.A. Micchelli, editors, *Advances in Computational Mathematics*, pages 15–57, World Scientific Publ., Singapore, 1994.

[7] W. Dahmen and A. Kunoth. Multilevel preconditioning. *Numer. Math.*, 63:315–344, 1992.

[8] W. Dahmen, S. Prößdorf, and R. Schneider. Multiscale methods for pseudodifferential equations. In L.L. Schumaker and G. Webb, editors, *Wavelet Analysis and its Applications*, volume 3, pages 191–235, 1993.

[9] W. Dahmen, S. Prößdorf, and R. Schneider. Wavelet approximation methods for periodic pseudodifferential equations. Part II – Fast solution and matrix compression. *Advances in Computational Mathematics*, 1:259–335, 1993.

[10] W. Dahmen, S. Prößdorf, and R. Schneider. Wavelet approximation methods for periodic pseudodifferential equations. Part I – Convergence analysis. *Mathematische Zeitschrift*, 215:583–620, 1994.

[11] W. Dahmen, S. Prößdorf, and R. Schneider. Multiscale methods for pseudodifferential equations on smooth manifolds. In C.K. Chui, L. Montefusco, and L. Puccio, editors, *Proceedings of the International Conference on Wavelets: Theory, Algorithms, and Applications*, pages 385–424, 1995.

[12] W. Dahmen and R. Schneider. Wavelets on manifolds II: Application to boundary element methods and pseudodifferential equations. Manuscript.

[13] W. Dahmen and R. Schneider. Composite wavelet basis for operator equations. *Math. Comp.*, 68:1533–1567, 1999.

[14] W. Dahmen and R. Schneider. Wavelets on manifolds I. Construction and domain decomposition. *Math. Anal.*, 31:184–230, 1999.

[15] C. de Boor. *A practical guide to splines.* Springer, 1978.

[16] R. DeVore. Nonlinear approximation. *Acta Numerica*, 7:51–150, 1998.

[17] M. Frazier and B. Jawerth. A discrete transform and decompositions of distribution spaces. *J. Functional Anal.*, 93:34–170, 1990.

[18] G. Gatica and G. Hsiao. *Boundary-field equation methods for a class of nonlinear problems*, volume 331 of *Pitman research notes in mathematics series*. Longman, Essex, 1995.

[19] L. Greengard and V. Rokhlin. A fast algorithm for particle simulation. *J. Comput. Phys.*, 73:325–348, 1987.

[20] W. Hackbusch. *Integralgleichungen*. B.G. Teubner, Stuttgart, 1989.

[21] W. Hackbusch. A sparse matrix arithmetic based on $\mathcal{H}$-matrices. Part I: Introduction to $\mathcal{H}$-matrices. *Computing*, 64:89–108, 1999.

[22] W. Hackbusch and Z.P. Nowak. On the fast matrix multiplication in the boundary element method by panel clustering. *Numer. Math.*, 54:463–491, 1989.

[23] G. Hämmerlin and K.-H. Hoffmann. *Numerische Mathematik*. Springer-Verlag, Berlin-Heidelberg, 1992. 3. Auflage.

[24] H. Harbrecht, F. Paiva, C. Pérez, and R. Schneider. Biorthogonal wavelet approximation for the coupling of FEM-BEM. *Preprint SFB 393/99-32, TU Chemnitz*, 1999. submitted to Numer. Math.

[25] H. Harbrecht, F. Paiva, C. Pérez, and R. Schneider. Wavelet preconditioning for the coupling of FEM-BEM. *Preprint SFB 393/00-07, TU Chemnitz*, 2000. submitted to J. Numer. Lin. Alg. with Appl.

[26] G. Hsiao and W. Wendland. A finite element method for some equations of first kind. *J. Math. Anal. Appl.*, 58:449–481, 1977.

[27] S. Jaffard. Wavelet methods for fast resolution of elliptic equations. *SIAM J. Numer. Anal.*, 29:965–986, 1992.

[28] R. Kress. *Linear Integral Equations*. Springer-Verlag, Berlin-Heidelberg, 1989.

[29] S.V. Pereverzev. Hyperbolic cross and the complexity of the approximate solution of Fredholm integral equations of the second kind with differentiable kernels. *Sib. Math. J.*, 32(1):85–92, 1991. English, Russian original, translation from Sib. Mat. Zh. 32, 1(185):107–115, 1991.

[30] T. von Petersdorff, R. Schneider, and C. Schwab. Multiwavelets for second kind integral equations. *SIAM J. Num. Anal.*, 34:2212–2227, 1997.

[31] T. von Petersdorff and C. Schwab. Wavelet approximation for first kind integral equations on polygons. *Numer. Anal.*, 74:479–516, 1996.

[32] T. von Petersdorff and C. Schwab. Fully discretized multiscale Galerkin BEM. In W. Dahmen, A. Kurdila, and P. Oswald, editors, *Multiscale wavelet methods for PDEs*, pages 287–346, Academic Press, San Diego, 1997.

[33] A. Rathsfeld. On a hierarchical three point basis of piecewise linear functions over smooth boundaries. in this volume.

[34] A. Rathsfeld. A wavelet algorithm for the boundary element solution of a geodetic boundary value problem. *Comput. Methods Appl. Mech. Engrg.*, 157:267–287, 1998.

[35] R. Schneider. *Multiskalen- und Wavelet-Matrixkompression: Analysisbasierte Methoden zur Lösung großer vollbesetzter Gleichungssysteme.* B.G. Teubner, Stuttgart, 1998.

[36] V.M. Tyrtyshnikov. Mosaic sceleton approximation. *Calcolo*, 33:47–57, 1996.

Helmut Harbrecht and Reinhold Schneider
Fakultät für Mathematik
Technische Universität Chemnitz
09107 Chemnitz, Germany

1991 Mathematics Subject Classification: 65N38, 65R20, 65T60, 65F35, 65D32

Submitted: 11.4.2000

Operator Theory:
Advances and Applications, Vol. 121

On the stability of collocation methods for Cauchy singular integral equations on an interval

PETER JUNGHANNS, GIUSEPPE MASTROIANNI

The application of a Banach algebra technique to the investigation of the stability of a collocation method with respect to the Chebyshev nodes of second kind is discussed. The paper is mainly devoted to the proof of the fact that the operator sequence related to the considered collocation method belongs to a certain Banach algebra. Some first conclusions concerning necessary stability conditions are derived.

1. Introduction

During the last few years a Banach algebra approach was developed to study the stability of projection, especially collocation methods for Cauchy singular integral equations on an interval (see [3, 4]).

For given piecewise continuous functions $a, b : [-1,1] \longrightarrow \mathbb{C}$ consider the singular integral equation

$$a(x)u(x) + \frac{b(x)}{\pi \mathrm{i}} \int_{-1}^{1} \frac{u(y)}{y-x}\, dy = f(x)\,, \quad x \in (-1,1)\,,$$

which we write in short form as operator equation

$$Au := (aI + bS)u = f\,. \tag{1.1}$$

Here $aI : \mathbf{L}^2_\sigma \longrightarrow \mathbf{L}^2_\sigma$ and $S : \mathbf{L}^2_\sigma \longrightarrow \mathbf{L}^2_\sigma$ denote the multiplication operator $(au)(x) = a(x)u(x)$ and the Cauchy singular integral operator

$$(Su)(x) = \frac{1}{\pi \mathrm{i}} \int_{-1}^{1} \frac{u(y)}{y-x}\, dy\,,$$

respectively. The Hilbert space $\mathbf{L}^2_\sigma$ is defined as the space of all functions $u : (-1,1) \longrightarrow \mathbb{C}$, which are square integrable w.r.t. the weight $\sigma(x) = v^{\alpha,\beta}(x) := (1-x)^\alpha(1+x)^\beta$, equipped with the inner product

$$\langle u, v\rangle_\sigma = \int_{-1}^{1} u(x)\overline{v(x)}\sigma(x)\, dx$$

and the norm $\|u\|_\sigma = \sqrt{\langle u,u\rangle_\sigma}$. In all that follows we assume $-1 < \alpha, \beta < 1$, which guarantees that the singular integral operator $S : \mathbf{L}^2_\sigma \longrightarrow \mathbf{L}^2_\sigma$ is continuous, i.e. $S \in \mathcal{L}(\mathbf{L}^2_\sigma)$ (see [1, Theorem I.4.1]). By T_n and U_n, $n = 0,1,2,\dots$, we denote the normalized Chebyshev polynomials

$$T_0 = \sqrt{\frac{1}{\pi}}, \quad T_n(\cos s) = \sqrt{\frac{2}{\pi}}\cos ns, \quad n = 1,2,\dots,$$

and

$$U_n(\cos s) = \sqrt{\frac{2}{\pi}}\frac{\sin(n+1)s}{\sin s}, \quad n = 1,2,\dots,$$

of first and second kind, respectively. Moreover, let $\varphi(x) = \sqrt{1-x^2}$ be the Chebyshev weight of second kind, $x^\varphi_{jn} = \cos\frac{j\pi}{n+1}$, $j = 1,\dots,n$, the respective nodes and

$$\widetilde{u}_n(x) = w_{\sigma^{-1}}(x)U_n(x), \quad n = 0,1,2,\dots,$$

with $w_{\sigma^{-1}} = \sqrt{\sigma^{-1}\varphi} = v^{\frac{1}{4}-\frac{\alpha}{2},\frac{1}{4}-\frac{\beta}{2}}$. Then, one aim of the present paper is to study the stability of the collocation method

$$a(x^\varphi_{jn})u_n(x^\varphi_{jn}) + \frac{b(x^\varphi_{jn})}{\pi \mathrm{i}}\int_{-1}^{1}\frac{u_n(y)}{y - x^\varphi_{jn}}\,dy = f(x^\varphi_{jn}), \quad j = 1,\dots,n,$$

to solve (1.1) approximately, where

$$u_n(x) = \sum_{k=0}^{n-1}\xi_{kn}\widetilde{u}_k(x)$$

plays the role of the approximate solution. We write this collocation method in the form

$$A_n u_n = M_n f, \tag{1.2}$$

where $A_n = M_n A P_n$ and $P_n : \mathbf{L}^2_\sigma \longrightarrow \mathbf{L}^2_\sigma$ denotes the Fourier projection

$$P_n u = \sum_{k=0}^{n-1}\langle u, \widetilde{u}_k\rangle_\sigma \widetilde{u}_k$$

w.r.t. the orthonormal system $\{\widetilde{u}_n\}_{n=0}^\infty$ in $\mathbf{L}^2_\sigma$. M_n is the weighted interpolation operator $M_n = w_{\sigma^{-1}}L^\varphi_n(w_{\sigma^{-1}})^{-1}I$, where L^φ_n denotes the usual Lagrange interpolation operator w.r.t. the nodes x^φ_{jn}, $j = 1,\dots,n$. By the stability of the sequence $\{A_n\}$ we mean that the operators $A_n : \operatorname{im} P_n \longrightarrow \operatorname{im} P_n$ are invertible for all sufficiently large n and that their inverses are uniformly bounded, i.e.

$$\sup\left\{\left\|A_n^{-1}P_n\right\|_{\mathbf{L}^2_\sigma\to\mathbf{L}^2_\sigma} : n \ge n_0\right\} < \infty.$$

Introduce the operators $W_n : \mathbf{L}^2_\sigma \longrightarrow \mathbf{L}^2_\sigma$ with $W_n u = \sum_{k=0}^{n-1} \langle u, \widetilde{u}_{n-1-k}\rangle_\sigma \widetilde{u}_k$. By $\mathcal{F}$ we denote the set of all sequences $\{A_n\} = \{A_n\}_{n=1}^\infty$ of linear operators $A_n : \operatorname{im} P_n \longrightarrow \operatorname{im} P_n$, for which $A_n P_n$, $A_n^* P_n$, $\widetilde{A}_n := W_n A_n W_n$, and $\widetilde{A}_n^* P_n$ are strongly convergent. $\mathcal{F}$ is equipped with elementwise algebraic operations and the supremum norm. Moreover, let $\mathcal{J} \subset \mathcal{F}$ denote the set

$$\mathcal{J} = \left\{ \{P_n K_1 P_n + W_n K_2 W_n + C_n\} : K_1, K_2 \in \mathcal{K}(\mathbf{L}^2_\sigma) \,,\ \|C_n\| \longrightarrow 0 \right\} ,$$

where $\mathcal{K}(\mathbf{L}^2_\sigma)$ denotes the ideal of all compact linear operators on the Hilbert space $\mathbf{L}^2_\sigma$. The following lemma is fundamental for our investigations.

Lemma 1.1.([6, Prop. 3], [5, Theorem 7.7]) *The set $\mathcal{F}$ is a C^*-algebra, and $\mathcal{J}$ is a closed two-sided ideal in $\mathcal{F}$. A sequence $\{A_n\} \in \mathcal{F}$ is stable if and only if the strong limits $A = s - \lim A_n P_n$ and $\widetilde{A} = s - \lim \widetilde{A}_n$ are invertible in the space $\mathcal{L}(\mathbf{L}^2_\sigma)$ of all linear and bounded operators in $\mathbf{L}^2_\sigma$ and if the coset $\{A_n\} + \mathcal{J}$ is invertible in $\mathcal{F}/\mathcal{J}$.*

In [4] the special case $\alpha = \beta = -\dfrac{1}{2}$, i.e.

$$\sigma(x) = \frac{1}{\sqrt{1-x^2}} \,, \qquad w_{\sigma^{-1}} = \varphi(x) = \sqrt{1-x^2} \,,$$

is investigated, while the general case $-1 < \alpha, \beta < 1$ was considered in [3]. But, in the last reference it was only possible to show that $\{A_n\} = \{M_n A P_n\}$ belongs to the algebra $\mathcal{F}$ if $b(\pm 1) = 0$. In particular, for the proof of the strong convergence of $\{\widetilde{A}_n P_n\}$ and $\{\widetilde{A}_n^* P_n\}$ the last condition was assumed (see [3, Prop. 3.4]). In Section 2. we give a proof without this condition, and Section 3. concerns with stability conditions for the collocation method (1.2).

2. The strong convergence of $\{\widetilde{A}_n P_n\}$ and $\{\widetilde{A}_n^* P_n\}$

At first we summarize some known results which we need for the following considerations.

Lemma 2.1.([3, Lemma 3.1]) *If f is locally Riemann integrable on $(-1,1)$ and if, for some $\varepsilon > 0$,*

$$|f(x)| \le \operatorname{const} (1-x)^{-\frac{1+\alpha}{2}+\varepsilon} (1+x)^{-\frac{1+\beta}{2}+\varepsilon} \,, \quad x \in (-1,1) \,, \tag{2.1}$$

then $\lim_{n\to\infty} \|M_n f - f\|_\sigma = 0$.

For $\varepsilon > 0$, let $\mathbf{R}_{\sigma,\varepsilon}$ denote the set of all functions $f : (-1,1) \longrightarrow \mathbb{C}$, which are locally bounded and Riemann integrable and which satisfy condition (2.1).

Equipped with the norm

$$\|f\|_{\mathbf{R}_{\sigma,\varepsilon}} = \sup\left\{|f(x)|(1-x)^{\frac{1+\alpha}{2}-\varepsilon}(1+x)^{\frac{1+\beta}{2}-\varepsilon} : x \in (-1,1)\right\}$$

$\mathbf{R}_{\sigma,\varepsilon}$ becomes a Banach space. Since the operators W_n converge weakly to zero, from Lemma 2.1 we get immediately

Corollary 2.2. *If the linear operator* $K : \mathbf{L}^2_\sigma \longrightarrow \mathbf{R}_{\sigma,\varepsilon}$ *is compact then*

$$W_n M_n K W_n \longrightarrow 0 \quad \text{strongly in} \quad \mathbf{L}^2_\sigma \,.$$

By $\mathbf{PC} = \mathbf{PC}[-1,1]$ we denote the algebra of all piecewise continuous functions on $[-1,1]$. We agree that a function $f : [-1,1] \to \mathbb{C}$ belongs to $\mathbf{PC}$ if and only if it possesses one-sided limits $f(x \pm 0)$ with $f(x-0) = f(x)$ at each $x \in (-1,1]$ and is continuous at -1.

Lemma 2.3.([3, Prop.s 3.1-3.3, Lemmata 3.3.,3.4]) *If* $a, b \in \mathbf{PC}$ *and* $\alpha, \beta \in (-1,1) \setminus \{\frac{1}{2}\}$ *then* $A_n P_n \longrightarrow A$ *and* $A_n^* P_n \longrightarrow A^*$ *strongly in* $\mathbf{L}^2_\sigma$.

One can easily check that

$$W_n M_n a W_n = M_n a P_n \tag{2.2}$$

for each function $a : (-1,1) \longrightarrow \mathbb{C}$. Indeed, using the Gaussian rule w.r.t. the Chebyshev weight of second kind, we have, for $k = 0, \ldots, n-1$,

$$\begin{aligned}
\langle M_n f, \widetilde{u}_k \rangle_\sigma &= \langle L_n^\varphi \left(w_{\sigma^{-1}}\right)^{-1} f, U_k \rangle_\varphi \\
&= \frac{\pi}{n+1} \sum_{j=1}^n \left[\varphi(x_{jn}^\varphi)\right]^2 \left[w_{\sigma^{-1}}(x_{jn}^\varphi)\right]^{-1} f(x_{jn}^\varphi) U_k(x_{jn}^\varphi) \\
&= \frac{\pi}{n+1} \sum_{j=1}^n \varphi(x_{jn}^\varphi) \left[w_{\sigma^{-1}}(x_{jn}^\varphi)\right]^{-1} f(x_{jn}^\varphi) \sin\frac{(k+1)j\pi}{n+1} \,.
\end{aligned}$$

This gives, for $m = 0, \ldots, n-1$,

$$\begin{aligned}
W_n M_n a W_n \widetilde{u}_m &= W_n M_n a \widetilde{u}_{n-1-m} \\
&= \sum_{k=0}^{n-1} \frac{\pi}{n+1} \sum_{j=1}^n a(x_{jn}^\varphi) \sin\frac{(n-m)j\pi}{n+1} \sin\frac{(n-k)j\pi}{n+1} \widetilde{u}_k \\
&= \sum_{k=0}^{n-1} \frac{\pi}{n+1} \sum_{j=1}^n a(x_{jn}^\varphi) \sin\frac{(m+1)j\pi}{n+1} \sin\frac{(k+1)j\pi}{n+1} \widetilde{u}_k \\
&= M_n a P_n \widetilde{u}_m \,.
\end{aligned}$$

To prove the strong convergence of $W_nM_nSW_n$ we use the representation

$$S = w_\sigma^{-1} S w_\sigma I + K_\sigma \,, \tag{2.3}$$

where $K_\sigma = (S w_\sigma^{-1} I - w_\sigma^{-1} S) w_\sigma I$, i.e.

$$(K_\sigma u)(x) = \frac{1}{\pi \mathrm{i}} \int_{-1}^{1} \left[1 - \frac{v^{\frac{\alpha}{2}+\frac{1}{4},\frac{\beta}{2}+\frac{1}{4}}(y)}{v^{\frac{\alpha}{2}+\frac{1}{4},\frac{\beta}{2}+\frac{1}{4}}(x)} \right] \frac{u(y)}{y-x}\, dy \,.$$

First we observe that, for $n > m+1$,

$$\begin{aligned}
W_n M_n w_\sigma^{-1} S w_\sigma W_n \widetilde{u}_m &= \mathrm{i} W_n M_n w_\sigma^{-1} T_{n-m} \\
&= \frac{\mathrm{i}}{2} W_n M_n w_\sigma^{-1} (U_{n-m} - U_{n-m-2}) \\
&= \frac{\mathrm{i}}{2} W_n M_n \varphi^{-1} (\widetilde{u}_{n-m} - \widetilde{u}_{n-m-2}) \\
&= \frac{\mathrm{i}}{2} W_n M_n \varphi^{-1} W_n (\widetilde{u}_{m-1} - \widetilde{u}_{m+1}) \\
&= \frac{\mathrm{i}}{2} M_n w_\sigma^{-1} (U_{m-1} - U_{m+1}) \\
&= -\mathrm{i} M_n w_\sigma^{-1} T_{m+1} \\
&\longrightarrow -\mathrm{i} w_\sigma^{-1} T_{m+1} = -w_\sigma^{-1} S w_\sigma \widetilde{u}_m \quad \text{in} \quad \mathbf{L}^2_\sigma \,,
\end{aligned}$$

taking into account the well-known relations

$$S \varphi U_n = \mathrm{i} T_{n+1} \,, \quad n = 0, 1, 2, \ldots \,, \tag{2.4}$$

and

$$T_{n+1} = \frac{1}{2} (U_{n+1} - U_{n-1}) \,, \quad n = 0, 1, 2, \ldots \,, \quad U_{-1} \equiv 0 \,, \tag{2.5}$$

as well as Relation (2.2), Lemma 2.1, and

$$w_\sigma^{-1}(x) = (1-x)^{-\frac{1+\alpha}{2}+\frac{1}{4}} (1+x)^{-\frac{1+\beta}{2}+\frac{1}{4}} \,.$$

By $\mathbf{X}^p_\sigma$ we denote the Banach space $\left\{ f : f w_\sigma \varphi^{-\frac{1}{p}} \in \mathbf{L}^p(-1,1) \right\}$, $p \geq 1$, equipped with the norm $\|f\|_{\mathbf{X}^p_\sigma} = \left\| f w_\sigma \varphi^{-\frac{1}{p}} \right\|_{\mathbf{L}^p(-1,1)}$. The proof of the following lemma will be given in Section 4.

Lemma 2.4. *For sufficiently large* $p > 2$ *and sufficiently small* $\varepsilon > 0$, *the operator* $K_\sigma : \mathbf{X}^p_\sigma \longrightarrow \mathbf{R}_{\sigma,\varepsilon}$ *is compact.*

For the following considerations we need some further notation. Let $\mathbb{T} = \{t \in \mathbb{C} : |t| = 1\}$ be the unit circle of the complex plane and $\mathbf{L}^p(\mathbb{T})$, $p \geq 1$, the respective Lebesgue space of the functions the p-th power of which is summable. The norm in $\mathbf{L}^p(\mathbb{T})$ is defined by

$$\|f\|_{\mathbf{L}^p(\mathbb{T})} = \left(\frac{1}{2\pi} \int_{-\pi}^{\pi} |f(e^{is})|^p \, ds \right)^{\frac{1}{p}} .$$

Moreover, by $S_\mathbb{T}$, $P_\mathbb{T}$, and $Q_\mathbb{T}$ we refer to the Cauchy singular integral operator on the unit circle

$$(S_\mathbb{T} u)(t) = \frac{1}{\pi \mathrm{i}} \int_\mathbb{T} \frac{u(\tau)}{\tau - t} \, d\tau$$

and the projections $P_\mathbb{T} = \frac{1}{2}(I + S_\mathbb{T})$, $Q_\mathbb{T} = \frac{1}{2}(I - S_\mathbb{T})$, respectively. Let

$$\langle f, g \rangle_\mathbb{T} = \frac{1}{2\pi} \int_{-\pi}^{\pi} f(e^{is}) \overline{g(e^{is})} \, ds ,$$

and let $P_n^\mathbb{T} : \mathbf{L}^p(\mathbb{T}) \longrightarrow \mathbf{L}^p(\mathbb{T})$ and $W_n^\mathbb{T} : \mathbf{L}^p(\mathbb{T}) \longrightarrow \mathbf{L}^p(\mathbb{T})$ be defined by

$$P_n^\mathbb{T} f = \sum_{k=-n}^{n} \langle f, e_k \rangle_\mathbb{T} e_k \quad \text{and} \quad W_n^\mathbb{T} f = \sum_{k=-n}^{-1} \langle f, e_{-n-k-1} \rangle_\mathbb{T} e_k + \sum_{k=0}^{n} \langle f, e_{n-k} \rangle_\mathbb{T} e_k ,$$

respectively, where $e_k(t) = t^k$. It is well known that the linear operator $S_\mathbb{T} : \mathbf{L}^p(\mathbb{T}) \longrightarrow \mathbf{L}^p(\mathbb{T})$ is bounded for $1 < p < \infty$. Since the sequence $P_n^\mathbb{T} : \mathbf{L}^p(\mathbb{T}) \longrightarrow \mathbf{L}^p(\mathbb{T})$, $n = 0, 1, 2, \dots$, is uniformly bounded and since

$$W_n^\mathbb{T} = (V_\mathbb{T}^{-n-1} W_\mathbb{T} Q_\mathbb{T} + V_\mathbb{T}^n W_\mathbb{T} P_\mathbb{T}) P_n^\mathbb{T} ,$$

where $(W_\mathbb{T} f)(t) = f(\overline{t})$ and $(V_\mathbb{T} f)(t) = t f(t)$, the sequence $W_n^\mathbb{T} : \mathbf{L}^p(\mathbb{T}) \longrightarrow \mathbf{L}^p(\mathbb{T})$ is also uniformly bounded, which additionally implies that $W_n^\mathbb{T}$ converges weakly to zero in $\mathbf{L}^p(\mathbb{T})$. Finally, define the operator $F : \mathbf{L}^2_\sigma \longrightarrow \mathbf{L}^2(\mathbb{T})$ by

$$Fu = \frac{1}{\sqrt{2}\mathrm{i}} \sum_{n=0}^{\infty} \langle u, \widetilde{u}_n \rangle_\sigma (e_{n+1} - e_{-n-1}) .$$

It is easily seen that, for $0 < s < \pi$,

$$\begin{aligned} (Fu)(e^{\pm is}) &= \pm\sqrt{2} \sum_{n=0}^{\infty} \langle u, \widetilde{u}_n \rangle_\sigma \sin(n+1)s \\ &= \pm\sqrt{\pi} w_\sigma(\cos s) \sum_{n=0}^{\infty} \langle u, \widetilde{u}_n \rangle_\sigma \widetilde{u}_n(\cos s) \\ &= \pm\sqrt{\pi} w_\sigma(\cos s) u(\cos s) , \end{aligned}$$

and, for $u \in \mathbf{X}_\sigma^p$,

$$\begin{aligned}
\|Fu\|_{\mathbf{L}^p(\mathbf{T})}^p &= \frac{1}{2\pi}\int_{-\pi}^{\pi} \left|(Fu)(e^{is})\right|^p ds = \pi^{\frac{p}{2}-1}\int_0^\pi |u(\cos s)w_\sigma(\cos s)|^p\, ds \\
&= \pi^{\frac{p}{2}-1}\int_{-1}^1 \left|u(x)v^{\frac{\alpha}{2}+\frac{1}{4}-\frac{1}{2p},\frac{\beta}{2}+\frac{1}{4}-\frac{1}{2p}}(x)\right|^p dx = \pi^{\frac{p}{2}-1}\|u\|_{\mathbf{X}_\sigma^p}^p .
\end{aligned} \tag{2.6}$$

For $u \in \mathbf{X}_\sigma^p$ define $\alpha_m(u) = \langle u, \widetilde{u}_m\rangle_\sigma$. Because of

$$\begin{aligned}
|\alpha_m(u)| &\le \|u\|_{\mathbf{X}_\sigma^p}\left(\int_{-1}^1 v^{(\frac{\alpha}{2}-\frac{1}{4}+\frac{1}{2p})q,(\frac{\beta}{2}-\frac{1}{4}+\frac{1}{2p})q}(x)|\widetilde{u}_m(x)|^q\,dx\right)^{\frac{1}{q}} \\
&= \|u\|_{\mathbf{X}_\sigma^p}\left(\int_{-1}^1 v^{\frac{q}{2p},\frac{q}{2p}}(x)|U_m(x)|^q\,dx\right)^{\frac{1}{q}}
\end{aligned}$$

the functional α_m belongs to $(\mathbf{X}_\sigma^p)^*$. Moreover, the linear span of $\{\alpha_m : m = 0,1,2,\ldots\}$ is dense in $(\mathbf{X}_\sigma^p)^*$. If $p \ge 2$ then $\mathbf{X}_\sigma^p \subset \mathbf{L}_\sigma^2$. Thus, for $u \in \mathbf{X}_\sigma^p$ and $n > m \ge 0$,

$$\alpha_m(W_n u) = \langle W_n u, \widetilde{u}_m\rangle_\sigma = \langle u, \widetilde{u}_{n-m-1}\rangle_\sigma \longrightarrow 0 \tag{2.7}$$

as $n \longrightarrow \infty$ for each fixed $m = 0,1,2,\ldots$ Using the relation

$$FW_n u = (V_{\mathbf{T}}W_n^{\mathbf{T}}P_{\mathbf{T}} + W_n^{\mathbf{T}}Q_{\mathbf{T}})Fu\,, \quad u \in \mathbf{X}_\sigma^p,$$

together with (2.6) we find, for $u \in \mathbf{X}_\sigma^p$,

$$\|W_n u\|_{\mathbf{X}_\sigma^p} = \pi^{\frac{1}{p}-\frac{1}{2}}\|FW_n u\|_{\mathbf{L}^p(\mathbf{T})} \le \text{const}\ \|u\|_{\mathbf{X}_\sigma^p}, \tag{2.8}$$

where the constant does not depend on u. Thus, the operators $W_n : \mathbf{X}_\sigma^p \longrightarrow \mathbf{X}_\sigma^p$ are uniformly bounded, which implies

$$W_n \longrightarrow 0 \quad \text{weakly in} \quad \mathbf{X}_\sigma^p,\ p \ge 2, \tag{2.9}$$

taking into acount (2.7). Now we are able to prove the following statement.

Lemma 2.5. *Let $\alpha,\beta \in (-1,1)\setminus\{\frac{1}{2}\}$. Then*

$$W_n M_n S W_n \longrightarrow -w_\sigma^{-1}Sw_\sigma I \quad \text{strongly in} \quad \mathbf{L}_\sigma^2 .$$

Proof. From Lemma 2.3 and $\|W_n\|_{\mathbf{L}_\sigma^2\to\mathbf{L}_\sigma^2} = 1$, the uniform boundedness of the operators $W_n M_n S W_n : \mathbf{L}_\sigma^2 \longrightarrow \mathbf{L}_\sigma^2$, $n = 1,2,\ldots$, follows. We use the representation (2.3). Since we have already proved that $W_n M_n S W_n \widetilde{u}_m \longrightarrow -w_\sigma^{-1}Sw_\sigma\widetilde{u}_m$ in $\mathbf{L}_\sigma^2$ it remains to show that $W_n M_n K_\sigma W_n \widetilde{u}_m$ converges to zero in $\mathbf{L}_\sigma^2$ for each fixed $m = 0, 1, 2, \ldots$ As a consequence of Lemma 2.1 and Lemma 2.4 we have $\lim_{n\to\infty} \|(M_n - I)K_\sigma\|_{\mathbf{X}^p_\sigma \to \mathbf{L}^2_\sigma} = 0$ for some $p > 2$. With (2.8) we get

$$\lim_{n\to\infty} \|W_n(M_n - I)K_\sigma W_n\|_{\mathbf{X}^p_\sigma \to \mathbf{L}^2_\sigma} = 0 .$$

Relation (2.9) and Lemma 2.4 imply, for some $p > 2$,

$$\lim_{n\to\infty} \|W_n K_\sigma W_n u\|_\sigma = 0 \quad \text{for all} \quad u \in \mathbf{X}^p_\sigma .$$

Finally we remark that $\widetilde{u}_m \in \mathbf{X}^p_\sigma$ for all $p \geq 1$. □

Corollary 2.6. *If $\alpha, \beta \in (-1,1) \setminus \{\frac{1}{2}\}$ then, for $a, b \in \mathbf{PC}[-1,1]$, the sequence $\{M_n(aI + bS)P_n\}$ belongs to the C^*-algebra $\mathcal{F}$.*

Proof. Due to Lemma 2.3, Relation (2.2), and Lemma 2.5 it remains to prove that the operators $(W_n M_n S W_n)^* P_n$ converge strongly in $\mathbf{L}^2_\sigma$. For this, firstly we remark that one can prove by direct computation that, for each function $a : (-1,1) \longrightarrow \mathbb{C}$,

$$\langle M_n a P_n u, v\rangle_\sigma = \langle u, M_n \overline{a} P_n v\rangle_\sigma \quad \text{for all} \quad u, v \in \mathbf{L}^2_\sigma . \tag{2.10}$$

Secondly, we already know that the operators $(W_n M_n b S W_n)^* P_n$ converge strongly in $\mathbf{L}^2_\sigma$ if $b \in \mathbf{PC}[-1,1]$ and $b(\pm 1) = 0$ ([3, Prop. 3.4]). Now, let $\delta > 0$. For $u, v \in \mathbf{L}^2_\sigma$, from (2.2) and (2.10) it follows

$$\begin{aligned} \langle W_n M_n S W_n u, v\rangle_\sigma &= \langle W_n M_n v^{-\delta,-\delta} W_n W_n M_n v^{\delta,\delta} S W_n u, v\rangle_\sigma \\ &= \langle M_n v^{-\delta,-\delta} P_n W_n M_n v^{\delta,\delta} S W_n u, v\rangle_\sigma \\ &= \langle u, (W_n M_n v^{\delta,\delta} S W_n)^* M_n v^{-\delta,-\delta} P_n v\rangle_\sigma . \end{aligned}$$

Hence,

$$(W_n M_n S W_n)^* P_n = (W_n M_n v^{\delta,\delta} S W_n)^* M_n v^{-\delta,-\delta} P_n . \tag{2.11}$$

Lemma 2.5 implies the uniform boundedness of the sequence $\{(W_n M_n S W_n)^* P_n\}$, such that we only have tho check that $(W_n M_n S W_n)^* P_n \widetilde{u}_k$ converges in $\mathbf{L}^2_\sigma$ for each fixed $k = 0, 1, 2, \ldots$ But this follows, for sufficiently small $\delta > 0$, from (2.11) and Lemma 2.1. □

3. Stability of the sequence $\{M_n(aI + bS)P_n\}$

For further considerations we need the Fredholm and invertibility conditions for the operator $aI + bS : \mathbf{L}^2_\sigma \longrightarrow \mathbf{L}^2_\sigma$, if $a, b \in \mathbf{PC}$. For this we associate to this operator the function

$$\mathbf{c}_\sigma(x,\mu) := \begin{cases} c(x)(1-\mu)+c(x+0)\mu &, \quad \mu\in[0,1]\,,\ x\in(-1,1)\,, \\ c(1)+[1-c(1)]f_{\pi(1+\alpha)}(\mu) &, \quad \mu\in[0,1]\,,\ x=1\,, \\ 1+[c(-1)-1]f_{\pi(1+\beta)}(\mu) &, \quad \mu\in[0,1]\,,\ x=-1\,, \end{cases}$$

where $c(x)=\dfrac{a(x)+b(x)}{a(x)-b(x)}$, and

$$f_\delta(\mu) = \begin{cases} \dfrac{\sin(\pi-\delta)\mu}{\sin(\pi-\delta)}\,e^{\mathrm{i}(\pi-\delta)(\mu-1)} &, \quad \delta\in(0,2\pi)\setminus\{\pi\}\,, \\ \mu &, \quad \delta=\pi\,. \end{cases}$$

Note that, for $z_1,z_2\in\mathbb{C}$ and $0<\delta<\pi$, the image of $z_1+(z_2-z_1)f_\delta(\mu)$, $\mu\in[0,1]$, describes the circular arc $\ell(z_1,z_2,\delta)$ from z_1 to z_2 which lies on the right of the straight line from z_1 to z_2 and from which this straight line is seen under the angle δ. If $\pi<\delta<2\pi$ then the circular arc $\ell(z_1,z_2,\delta)=\ell(z_2,z_1,2\pi-\delta)$ is described. Thus, if $c(x\pm 0)$ is finite for all $x\in[-1,1]$, the image of $\mathbf{c}_\sigma(x,\mu)$ is a closed curve in the complex plane which possesses a natural orientation, and by $\operatorname{wind}\mathbf{c}_\sigma(x,\mu)$ we denote the winding number of this curve w.r.t. the origin 0.

Lemma 3.1.see [1, Theorem IX.4.1]) *Let $a,b\in\mathbf{PC}$. The operator $A=aI+bS:\mathbf{L}^2_\sigma\longrightarrow\mathbf{L}^2_\sigma$ is a Fredholm operator if and only if $a(x\pm0)-b(x\pm0)\neq0$ for all $x\in[-1,1]$ and $\mathbf{c}_\sigma(x,\mu)\neq0$ for all $(x,\mu)\in[-1,1]\times[0,1]$. In this case A is one-sided invertible and* $\operatorname{ind}A=-\operatorname{wind}\mathbf{c}_\sigma(x,\mu)$.

We remark that the multiplication operator $w_\sigma I:\mathbf{L}^2_\sigma\longrightarrow\mathbf{L}^2_{\varphi^{-1}}$ is an isometrical isomorphism. Hence, the invertibility of $\widetilde{A}=aI-w_\sigma^{-1}Sw_\sigma I:\mathbf{L}^2_\sigma\longrightarrow\mathbf{L}^2_\sigma$ is equivalent to the invertibility of $aI-bS:\mathbf{L}^2_{\varphi^{-1}}\longrightarrow\mathbf{L}^2_{\varphi^{-1}}$. So, by Lemma 1.1 and Lemma 3.1 we get the following Lemma. For $0<\gamma,\delta<2\pi$ let

$$\Pi_{\gamma,\delta} := \{re^{\mathrm{i}\theta}: r>0,\ -\gamma<\theta<\delta\}\,.$$

Lemma 3.2. *If $a,b\in\mathbb{C}$ are constants and $\{M_n(aI+bS)P_n\}$ is stable in $\mathbf{L}^2_\sigma$ then $a\neq b$ and $c=(a+b)/(a-b)\in\Pi_{\pi(1+\beta),\pi(1+\alpha)}\cap\Pi_{\pi/2,\pi/2}$.*

Denote by P_n^o and M_n^o the Fourier projection P_n and the weighted interpolation operator M_n in case of $\alpha=\beta=-\frac12$, respectively. Then, one can easily see that $P_n=J^{-1}P_n^oJ$ and $M_n=J^{-1}M_n^oJ$, where $J=w_\sigma I$. This results in

$$M_n(aI+bS)P_n = J^{-1}M_n^o(aI+bJSJ^{-1})P_n^oJ$$

and the following conclusion.

Corollary 3.3. *The sequence $\{M_n(aI+bS)P_n\}$ is stable in the space $\mathbf{L}^2_\sigma$ if and only if the sequence $\{M_n^o(aI+bw_\sigma S w_\sigma^{-1})P_n^o$ is stable in $\mathbf{L}^2_{\varphi^{-1}}$.*

Finally we remark that in the special case $\alpha=\beta=-\frac{1}{2}$ we have $w_\sigma \equiv 1$ and $w_{\sigma^{-1}}=\varphi$, and one can show that $\|M_n^o S P_n^o\|_{\mathbf{L}^2_{\varphi^{-1}}\to\mathbf{L}^2_{\varphi^{-1}}} \le 1$. Thus, in this case the condition $(a+b)/(a-b)\in \Pi_{\frac{\pi}{2},\frac{\pi}{2}}$, of Lemma 3.2, i.e. $|a|>|b|$, is also sufficient for the stability of the sequence $\{M_n^o(aI+bS)P_n^o\}$ in $\mathbf{L}_{\varphi^{-1}}$. This situation is completely investigated also for variable coefficients a and b in [4].

4. Proof of Lemma 2.4

Let $w^\gamma(x)=x^\gamma$, $x>0$, and let $\mathbf{L}^p_{w^\gamma}$, $p\ge 1$, denote the Banach space $\{f: fw^\gamma \in \mathbf{L}^p(0,1)\}$ equipped with the norm

$$\|fw^\gamma\|_p = \left(\int_0^1 x^{p\gamma}|f(x)|^p\,dx\right)^{\frac{1}{p}}.$$

Define the operator K by

$$(Kf)(x)=\int_0^1\left[1-\frac{w^\gamma(y)}{w^\gamma(x)}\right]\frac{f(y)}{x-y}\,dy.$$

Lemma 4.1. *Let $p>2$, $\gamma\in\left(-\frac{1}{4},-\frac{1}{p}\right)\cup\left(\frac{1}{p},1-\frac{1}{2p}\right)$, and $d>0$. Then, for all $f\in\mathbf{L}^p_{w^{\gamma-\frac{1}{2p}}}$,*

$$|(Kf)(x)|\le c_{p\gamma d}\,x^{\min\left\{0,-\gamma-\frac{1}{2p}\right\}}\left\|fw^{\gamma-\frac{1}{2p}}\right\|_p \quad \text{for all}\quad x\in(0,d], \tag{4.1}$$

and, for $0<\varepsilon<\min\left\{\frac{1}{4}-\frac{1}{2p},\frac{1}{4}+\gamma\right\}$,

$$\sup_{\tau>0}\frac{\Omega_{\varphi_d}(Kf,\tau)_{w^{\gamma+\frac{1}{4}-\varepsilon},\infty}}{\tau^{\frac{1}{2}-\frac{1}{p}-2\varepsilon}}\le c_{p\gamma d}\left\|fw^{\gamma-\frac{1}{2p}}\right\|_p, \tag{4.2}$$

where $\varphi_d(x)=\sqrt{x(d-x)}$, $\Omega_{\varphi_d}(g,\tau)_{w^{\gamma+\frac{1}{4}-\varepsilon},\infty}$ is equal to

$$\sup_{0<h\le\tau}\left\|w^{\gamma+\frac{1}{4}-\varepsilon}\left[f\left(\cdot+\frac{h}{2}\varphi_d\right)-f\left(\cdot-\frac{h}{2}\varphi_d\right)\right]\right\|_{\mathbf{L}^\infty(dh^2,d(1-h^2))},$$

and $c_{p\gamma d}$ denotes a positive constant, which depends only on p, γ, and d and which can assume different values at different places.

Proof. We define $F(y) = w^{\gamma - \frac{1}{2p}}(y) f(y)$. Then

$$\begin{aligned} |(Kf)(x)| &= \left| \int_0^1 \frac{y^{-\gamma} - x^{-\gamma}}{x-y} \, y^{\frac{1}{2p}} F(y) dy \right| \\ &\le \left(\int_0^1 \left| \frac{y^{-\gamma} - x^{-\gamma}}{x-y} \, y^{\frac{1}{2p}} \right|^q dy \right)^{\frac{1}{q}} \|F\|_p := A(x) \|F\|_p \,. \end{aligned}$$

To estimate $A(x)$ we write

$$\begin{aligned} A(x) &= \left(\int_0^1 \left| \frac{y^{-\gamma} - x^{-\gamma}}{x-y} \, y^{\frac{1}{2p}} \right|^q dy \right)^{\frac{1}{q}} \\ &= x^{-\gamma - \frac{1}{2p}} \left(\int_0^{\frac{1}{x}} \left[\left| \frac{t^{-\gamma} - 1}{t-1} \right| t^{\frac{1}{2p}} \right]^q dt \right)^{\frac{1}{q}} =: x^{-\gamma - \frac{1}{2p}} B(x) \,, \end{aligned}$$

by using the change of variable $t = \frac{y}{x}$. In case of $\frac{1}{p} < \gamma < 1 - \frac{1}{2p}$ we have

$$B(x)^q \le \int_0^\infty \left[\left| \frac{t^{-\gamma} - 1}{t-1} \right| t^{\frac{1}{2p}} \right]^q dt < \infty \,,$$

since, in view of $|1 - s^\gamma| \le |1 - s|^\gamma$, $s \ge 0$,

$$\int_0^2 \left[\left| \frac{t^{-\gamma} - 1}{t-1} \right| t^{\frac{1}{2p}} \right]^q dt \le \int_0^2 \left[|t-1|^{\gamma-1} t^{-\gamma + \frac{1}{2p}} \right]^q dt \,,$$

$$\begin{aligned} \int_2^\infty \left[\left| \frac{t^{-\gamma} - 1}{t-1} \right| t^{\frac{1}{2p}} \right]^q dt &\le \int_2^\infty \left[\frac{t^{\frac{1}{2p}}}{t-1} \right]^q dt \\ &= \int_2^\infty \left[\frac{t}{t-1} t^{-1+\frac{1}{2p}} \right]^q dt \le 2^q \int_2^\infty t^{(-1+\frac{1}{2p})q} dt \,, \end{aligned}$$

and $(\gamma - 1)q > -1$, $\left(-\gamma + \frac{1}{2p}\right) q > -1$, as well as $\left(-1 + \frac{1}{2p}\right) q < -1$.

Now, let $-\frac{1}{4} < \gamma < -\frac{1}{p}$. If $x \ge 1$ then

$$B(x) \le \left(\int_0^1 \left[(1-t)^{-\gamma-1} t^{\frac{1}{2p}} \right]^q dt \right)^{\frac{1}{q}} < \infty.$$

In case of $x<1$ it remains to take into account $(\gamma+1)q<1$ and to estimate

$$\begin{aligned}\left(\int_1^{\frac{1}{x}}\left[\frac{t^{-\gamma}-1}{t-1}\,t^{\frac{1}{2p}}\right]^q dt\right)^{\frac{1}{q}} &\le x^{-\frac{1}{2p}}\left(\int_1^{\frac{1}{x}}(t-1)^{-(\gamma+1)q}dt\right)^{\frac{1}{q}}\\ &= x^{-\frac{1}{2p}}\left(\frac{\left(\frac{1}{x}-1\right)^{1-(\gamma+1)q}}{1-(\gamma+1)q}\right)^{\frac{1}{q}}\\ &\le \frac{x^{(\gamma+1)-\frac{1}{q}-\frac{1}{2p}}}{1-(\gamma+1)q}=\frac{x^{\gamma+\frac{1}{2p}}}{1-(\gamma+1)q}\,.\end{aligned}$$

Thus, (4.1) is proved.

To prove (4.2) let us estimate $|(Kf)'(x)|$. At first, assume $\frac{1}{p}<\gamma<1-\frac{1}{2p}$. We represent the operator K as follows

$$(Kf)(x)=x^{-\gamma}\int_0^1\frac{x^\gamma-y^\gamma}{x-y}\,y^{-\gamma+\frac{1}{2p}}F(y)\,dy$$

and use the relation

$$\frac{x^\gamma-y^\gamma}{x-y}=\gamma\int_0^1\frac{dz}{[zx+(1-z)y]^{1-\gamma}}\,.$$

Then

$$\begin{aligned}(Kf)'(x) &= -\gamma\,x^{-\gamma-1}\int_0^1\frac{x^\gamma-y^\gamma}{x-y}\,y^{-\gamma+\frac{1}{2p}}F(y)\,dy\\ &\quad+\gamma(\gamma-1)x^{-\gamma}\int_0^1 y^{-\gamma+\frac{1}{2p}}\int_0^1\frac{z\,dz}{[zx+(1-z)y]^{2-\gamma}}\,F(y)\,dy\,.\end{aligned}$$

Due to the Hölder inequality it follows

$$\begin{aligned}|(Kf)'(x)| &\le x^{-\gamma-1}\left(\int_0^1\left[\left|\frac{x^\gamma-y^\gamma}{x-y}\right|y^{-\gamma+\frac{1}{2p}}\right]^q dy\right)^{\frac{1}{q}}\|F\|_p\\ &\quad+x^{-\gamma}\left(\int_0^1\left|y^{-\gamma+\frac{1}{2p}}\int_0^1\frac{z\,dz}{[zx+(1-z)y]^{2-\gamma}}\right|^q dy\right)^{\frac{1}{q}}\|F\|_p\\ &:= x^{-\gamma-1}\widetilde{A}(x)\|F\|_p+x^{-\gamma}\widetilde{B}(x)\|F\|_p\,.\end{aligned}$$

Using the change of variable $t=\frac{y}{x}$ we find

$$\widetilde{A}(x)\le x^{-\frac{1}{2p}}\left(\int_0^\infty\left[|1-t|^{\gamma-1}t^{-\gamma+\frac{1}{2p}}\right]^q dt\right)^{\frac{1}{q}}\le c_{p\gamma}x^{-\frac{1}{2p}}\,,$$

since $\left(-\gamma+\frac{1}{2p}\right)q>-1$ and $\left(\frac{1}{2p}-1\right)q<-1$. Moreover,

$$\widetilde{B}(x)=x^{-1-\frac{1}{2p}}\left(\int_0^{\frac{1}{x}}\left[t^{-\gamma+\frac{1}{2p}}\int_0^1\frac{z\,dz}{[z+(1-z)t]^{2-\gamma}}\right]^q dt\right)^{\frac{1}{q}}.$$

Since

$$\int_0^2\left[t^{-\gamma+\frac{1}{2p}}\int_0^1\frac{z\,dz}{[z+(1-z)t]^{2-\gamma}}\right]^q dt\le\int_0^2 t^{\left(-\gamma+\frac{1}{2p}\right)q}\left[\int_0^1\frac{dz}{z^{1-\gamma}}\right]^q dt=c_{p\gamma},$$

$$\int_0^1\frac{z\,dz}{[z+(1-z)t]^{2-\gamma}}=\frac{1}{(t-1)^2}\int_1^t\frac{t-y}{y^{2-\gamma}}dy\le c_\gamma t^{-1},\quad t>2,$$

and, consequently,

$$\int_2^\infty\left[t^{-\gamma+\frac{1}{2p}}\int_0^1\frac{z\,dz}{[z+(1-z)t]^{2-\gamma}}\right]^q dt\le c_{p\gamma}\int_2^\infty t^{\left(\frac{1}{2p}-1-\gamma\right)q}dt<\infty,$$

we get

$$|(Kf)'(x)|\le c_{p\gamma}x^{-\gamma-1-\frac{1}{2p}}\|F\|_p,\quad x>0. \tag{4.3}$$

Now, let $-\frac{1}{4}<\gamma<-\frac{1}{p}$. Since

$$(Kf)(x)=\int_0^1\frac{y^{-\gamma}-x^{-\gamma}}{x-y}\,y^{\frac{1}{2p}}F(y)\,dy=\gamma\int_0^1 y^{\frac{1}{2p}}\int_0^1\frac{dz}{[zx+(1-z)y]^{1+\gamma}}F(y)\,dy,$$

one has

$$(Kf)'(x)=-\gamma(1+\gamma)\int_0^1 y^{\frac{1}{2p}}\int_0^1\frac{z\,dz}{[zx+(1-z)y]^{2+\gamma}}F(y)\,dy,$$

such that

$$|(Kf)'(x)|\le\left(\int_0^1\left[y^{\frac{1}{2p}}\int_0^1\frac{z\,dz}{[zx+(1-z)y]^{2+\gamma}}\right]^q dy\right)^{\frac{1}{q}}\|F\|_p:=\widehat{A}(x)\|F\|_p.$$

Making the change of variable $t=\frac{y}{x}$ we obtain

$$\widehat{A}(x)=x^{-\gamma-1-\frac{1}{2p}}\left(\int_0^{\frac{1}{x}}\left[t^{\frac{1}{2p}}\int_0^1\frac{z\,dz}{[z+(1-z)t]^{2+\gamma}}\right]^q dt\right)^{\frac{1}{q}}:=x^{-\gamma-1-\frac{1}{2p}}\widehat{B}(x).$$

Obviously,

$$\left(\int_0^2\left[t^{\frac{1}{2p}}\int_0^1\frac{z\,dz}{[z+(1-z)t]^{2+\gamma}}\right]^q dt\right)^{\frac{1}{q}}=c_{p\gamma}<\infty.$$

Since, for $t > 2$,

$$\int_0^1 \frac{z\,dz}{[z+(1-z)t]^{2+\gamma}} = \frac{1}{(t-1)^2}\int_1^t \frac{t-y}{y^{2+\gamma}}\,dy \le c_\gamma t^{-1}$$

we have

$$\int_2^\infty \left[t^{\frac{1}{2p}}\int_0^1 \frac{z\,dz}{[z+(1-z)t]^{2+\gamma}}\right]^q dt \le c_{p\gamma}\int_2^\infty t^{\left(-1+\frac{1}{2p}\right)q}dt < \infty .$$

Hence, (4.3) holds also true in the present case. For $0 < \varepsilon < \min\left\{\frac{1}{4}-\frac{1}{2p}, \frac{1}{4}+\gamma\right\}$ and $dh^2/2 \le x \le d(1-h^2)/2$, it follows

$$\begin{aligned}\left|(Kf)'(x)\sqrt{x(d-x)}x^{\gamma+\frac{1}{4}-\varepsilon}\right| &\le c_{p\gamma}x^{-\gamma-\frac{1}{2p}-1}x^{\gamma+\frac{1}{4}-\varepsilon}\sqrt{x(d-x)}\|fw^{\gamma-\frac{1}{2p}}\|_p \\ &\le c_{p\gamma d}x^{-\frac{1}{2p}-\frac{1}{4}-\varepsilon}\|fw^{\gamma-\frac{1}{2p}}\|_p .\end{aligned}$$

Taking into account the relation

$$\Omega_{\varphi_d}(Kf,\tau)_{w^{\gamma+\frac{1}{4}-\varepsilon},\infty} \le c\sup_{h\le\tau} h \max_{dh^2/2\le x\le d(1-h^2)/2}\left|(Kf)'(x)\sqrt{x(d-x)}x^{\gamma+\frac{1}{4}-\varepsilon}\right|,$$

the estimate (4.2) is proved. □

For $\rho \ge 0$ and $d > 0$, let $\mathbf{C}_{\rho,(d)}$ denote the Banach space of all continuous functions $f : (0,d] \longrightarrow \mathbb{C}$ such that $w^\rho f : [0,d] \longrightarrow \mathbb{C}$ is continuous. The norm in $\mathbf{C}_{\rho,(d)}$ is defined by $\|f\|_{\infty,\rho,(d)} = \|w^\rho f\|_{\infty,(d)} := \sup\{|x^\rho f(x)| : x \in (0,d]\}$.

Corollary 4.2. *If the conditions* $p > 2$, $\gamma \in \left(-\frac{1}{4}, -\frac{1}{p}\right) \cup \left(\frac{1}{p}, 1-\frac{1}{2p}\right)$, *and* $0 < \varepsilon \le \min\left\{\frac{1}{4}-\frac{1}{2p}, \frac{1}{4}+\gamma\right\}$, *as well as* $d > 0$ *are fulfilled, then the operator* $K : \mathbf{L}^p_{w^{\gamma-\frac{1}{2p}}} \longrightarrow \mathbf{C}_{\gamma+\frac{1}{4}-\varepsilon,(d)}$ *is continuous.*

Proof. From (4.1) we get immediately $\left|w^{\gamma+\frac{1}{4}-\varepsilon}(x)(Kf)(x)\right| \le c_{p\gamma d}\left\|fw^{\gamma-\frac{1}{2p}}\right\|_p$ for all $x \in (0,d]$. □

Corollary 4.3. *Let the conditions of Corollary 4.2 be fulfilled. Then the operator* $K : \mathbf{L}^p_{w^{\gamma-\frac{1}{2p}}} \longrightarrow \mathbf{C}_{\gamma+\frac{1}{4}-\varepsilon,(d)}$ *is compact.*

Proof. Let $\mathcal{S} = \left\{f \in L^p_{w^{\gamma-\frac{1}{2p}}} : \|fw^{\gamma-\frac{1}{2p}}\|_p \le 1\right\}$ and $K(\mathcal{S}) = \{Kf : f \in \mathcal{S}\}$. By $\mathcal{P}_m$ we refer to the set of polynomials of degree less or equal to m . Since (see [2, Theorem 8.2.1])

$$\inf_{P\in\mathcal{P}_m}\left\|(Kf-P)w^{\gamma+\frac{1}{4}-\varepsilon}\right\|_{\infty,(d)} \le \text{const}\int_0^{\frac{1}{m}}\frac{\Omega_{\varphi_d}(Kf,\tau)_{w^{\gamma+\frac{1}{4}-\varepsilon},\infty}}{\tau}\,d\tau$$

$$\leq \text{const} \sup_{\tau>0} \frac{\Omega_{\varphi_d}(Kf,\tau)_{w^{\gamma+\frac{1}{4}-\varepsilon},\infty}}{\tau^{\frac{1}{2}-\frac{1}{p}-2\varepsilon}} \int_0^{\frac{1}{m}} t^{\frac{1}{2}-\frac{1}{p}-2\varepsilon-1}\,dt$$

we conclude from (4.2), for $f \in \mathbf{L}_{w^{\gamma-\frac{1}{2p}}}$,

$$\inf_{P\in\mathcal{P}_m} \left\|(Kf-P)w^{\gamma+\frac{1}{4}-\varepsilon}\right\|_{\infty,(d)} \leq \frac{c_{p\gamma d}}{m^{\frac{1}{2}-\frac{1}{p}-2\varepsilon}} \left\|fw^{\gamma-\frac{1}{2p}}\right\|_p ,$$

and, consequently,

$$\sup_{f\in\mathcal{S}} \inf_{P\in\mathcal{P}_m} \left\|(Kf-P)w^{\gamma+\frac{1}{4}-\varepsilon}\right\|_{\infty,(d)} \leq \frac{c_{p\gamma d}}{m^{\frac{1}{2}-\frac{1}{p}-2\varepsilon}} ,$$

where the constant $c_{p\gamma d}$ does not depend on f and m. Now, let $\delta > 0$ be arbitrarily chosen. By the last estimate there exists an m_0, such that for each $f \in \mathcal{S}$ there is a $g_f \in \mathcal{P}_{m_0}$ with $\|Kf - g_f\|_{\infty,\gamma+\frac{1}{4}-\varepsilon,(d)} < \frac{\delta}{2}$. Hence

$$\|g_f\|_{\infty,\gamma+\frac{1}{4}-\varepsilon,(d)} < \frac{\delta}{2} + \|K\|_{\mathbf{L}^p_{w^{\gamma-\frac{1}{2p}}} \to \mathbf{C}_{\gamma+\frac{1}{4}-\varepsilon,(d)}} , \quad f \in \mathcal{S},$$

which implies the compactness of $\mathcal{T} = \{g_f : f \in \mathcal{S}\} \subset \mathbf{C}_{\gamma+\frac{1}{4}-\varepsilon,(d)}$. Consequently, there exists a finite $\frac{\delta}{2}$-net (w.r.t. the $\mathbf{C}_{\gamma+\frac{1}{4}-\varepsilon,(d)}$-norm) for $\mathcal{T}$, which is a δ-net for $K(\mathcal{S})$ in $\mathbf{C}_{\gamma+\frac{1}{4}-\varepsilon,(d)}$. □

For $\gamma, \delta \geq 0$, let $\mathbf{C}_{\gamma,\delta}$ denote the Banach space of all continuous functions $f : (-1,1) \longrightarrow \mathbb{C}$, for which $v^{\gamma,\delta} f : [-1,1] \longrightarrow \mathbb{C}$ is continuous. Moreover, by $\mathbf{L}^p_{v^{\alpha,\beta}}$ we refer to the Banach space of all functions f such that $v^{\alpha,\beta} f$ belongs to $\mathbf{L}^p(-1,1)$. The norms in $\mathbf{C}_{\gamma,\delta}$ and $\mathbf{L}^p_{v^{\alpha,\beta}}$ are defined by

$$\|f\|_{\gamma,\delta,\infty} = \left\|v^{\gamma,\delta} f\right\|_\infty \quad \text{and} \quad \|f\|_{\mathbf{L}^p_{v^{\alpha,\beta}}} = \left\|v^{\alpha,\beta} f\right\|_{\mathbf{L}^p(-1,1)} ,$$

respectively. Define the operator T by

$$(Tu)(x) = \int_{-1}^1 \left[1 - \frac{v^{\gamma,\delta}(y)}{v^{\gamma,\delta}(x)}\right] \frac{u(y)}{y-x}\,dy , \quad x \in (-1,1).$$

Corollary 4.4. *If the conditions $p > 2$, $\gamma, \delta \in \left(-\frac{1}{4}, -\frac{1}{p}\right) \cup \left(\frac{1}{p}, 1-\frac{1}{2p}\right)$, and $0 < \varepsilon < \min\left\{\frac{1}{4} - \frac{1}{2p}, \frac{1}{4}+\gamma, \frac{1}{4}+\delta\right\}$ are fulfilled, then the operator*

$$T : \mathbf{L}^p_{v^{\gamma-\frac{1}{2p},\delta-\frac{1}{2p}}} \longrightarrow \mathbf{C}_{\gamma+\frac{1}{4}-\varepsilon,\delta+\frac{1}{4}-\varepsilon}$$

is compact.

Proof. Of course, it is sufficient to prove the compactness of the operator

$$T_1 : \mathbf{L}^p_{v^{\gamma-\frac{1}{2p},\delta-\frac{1}{2p}}} \longrightarrow \mathbf{C}_{\gamma+\frac{1}{4}-\varepsilon,\delta+\frac{1}{4}-\varepsilon} ,$$

where

$$\begin{aligned}(T_1u)(x) &= \int_0^1 \left[1 - \frac{v^{\gamma,\delta}(y)}{v^{\gamma,\delta}(x)}\right] \frac{u(y)}{y-x}\,dy \\ &= \int_0^1 \left[1 - \frac{v^{\gamma,0}(y)}{v^{\gamma,0}(x)}\right] \frac{u(y)}{y-x}\,dy \\ &\quad + \frac{1}{v^{\gamma,0}(x)} \int_0^1 \left[1 - \frac{v^{0,\delta}(y)}{v^{0,\delta}(x)}\right] \frac{v^{\gamma,0}(y)\,u(y)}{y-x}\,dy \\ &=: (\widetilde{T}_1u)(x) + \left(v^{-\gamma,0}\widehat{T}_1 v^{\gamma,0}u\right)(x)\,.\end{aligned}$$

In view of Corollary 4.3 the operator $\widetilde{T}_1 : \mathbf{L}^p_{v^{\gamma-\frac{1}{2p},\delta-\frac{1}{2p}}} \longrightarrow \mathbf{C}_{\gamma+\frac{1}{4}-\varepsilon,0}$ is compact. In case of $\delta < 0$ the kernel $\left[1 - \frac{v^{0,\delta}(y)}{v^{0,\delta}(x)}\right] \frac{1}{y-x}$ of the operator $\widehat{T}_1$ is continuous for $(x,y) \in [-1,1] \times [0,1]$, which implies the compactness of $\widehat{T}_1 : \mathbf{L}^p_{v^{-\frac{1}{2p},\delta-\frac{1}{2p}}} \longrightarrow \mathbf{C}_{0,0}$. In case of $\delta > 0$ the operator $v^{0,\delta}\widehat{T}_1 : \mathbf{L}^p_{v^{-\frac{1}{2p},\delta-\frac{1}{2p}}} \longrightarrow \mathbf{C}_{0,0}$ is compact, which also leads to the compactness of

$$v^{-\gamma,0}\widehat{T}_1 v^{\gamma,0} : \mathbf{L}^p_{v^{\gamma-\frac{1}{2p},\delta-\frac{1}{2p}}} \longrightarrow \mathbf{C}_{\gamma+\frac{1}{4}-\varepsilon,\delta+\frac{1}{4}-\varepsilon}\,.$$

The corollary is proved. □

Proof of Lemma 2.4. Since $\mathbf{X}^p_\sigma = \mathbf{L}^p_{v^{\frac{\alpha}{2}+\frac{1}{4}-\frac{1}{2p},\frac{\beta}{2}+\frac{1}{4}-\frac{1}{2p}}}$ we can apply Corollary 4.4 for $\gamma = \frac{\alpha}{2} + \frac{1}{4}$ and $\delta = \frac{\beta}{2} + \frac{1}{4}$ to get the compactness of

$$K_\sigma : \mathbf{X}^p_\sigma \longrightarrow \mathbf{C}_{\frac{\alpha+1}{2}-\varepsilon,\frac{\beta+1}{2}-\varepsilon}\,.$$

It remains to use the continuous embedding $\mathbf{C}_{\frac{\alpha+1}{2}-\varepsilon,\frac{\beta+1}{2}-\varepsilon} \subset \mathbf{R}_{\sigma,\varepsilon}$.

References

[1] I. Gohberg, N. Krupnik, *One-dimensional Linear Singular Integral Equations*, Birkhäuser Verlag, Basel, Boston, Berlin, 1992.

[2] Z. Ditzian, V. Totik, *Moduli of Smoothness*, Springer Verlag, New York, Berlin, Heidelberg, London, Paris, Tokyo, 1987.

[3] P. Junghanns, U. Weber, Banach algebra techniques for Cauchy singular integral equations on an interval, in: *Boundary Element Technology XII*, ed.s J.I. Frankel, C.A. Brebbia, and M.A.H. Aliabadi, Computational Mechanics Publications, Southampton, Boston, 1997, pp. 419-428.

[4] P. Junghanns, U. Weber, Local theory of projection methods for Cauchy singular integral equations on an interval, in: *Boundary Integral Methods: Numerical and Mathematical Aspects*, ed. M. Golberg, Computational Mechanics Publications, Series: Computational Engineering, Volume1, Boston, Southampton, 1998, pp. 217-256.

[5] S. Prössdorf, B. Silbermann, *Numerical Analysis for Integral and Related Operator Equations*, Birkhäuser Verlag, Basel, Boston, Berlin, 1991.

[6] B. Silbermann, Lokale Theorie des Reduktionsverfahrens für Toeplitzoperatoren, *Math. Nachr.*, 104 (1981), 137-146.

Department of Mathematics
Technical University Chemnitz
09107 Chemnitz
Germany

Dipartimento di Matematica
Universitá degli Studi della Basilicata
Via Nazario Sauro 85
85100 Potenza
Italy

1991 Mathematics Subject Classification: Primary 45G05; Secondary 65R20; 45E05; 45L05; 45L10

Submitted: 20.4.2000

Operator Theory:
Advances and Applications, Vol. 121

On a new approach for solving Dirac equations with some potentials and Maxwell's system in inhomogeneous media

VLADISLAV V. KRAVCHENKO

A new approach for studying Dirac equations with potentials and Maxwell's system is proposed. It is based on the possibility of the reformulation of these equations in terms of complex quaternions. The corresponding quaternionic equivalents are reduced to the Moisil-Theodoresco operator with some additional multiplicative term. For some cases, for example, when this term depends on one variable, we obtain solutions in explicit form using some specially constructed zero divisors.

1. Introduction

Applications of quaternions to the Dirac and Maxwell equations have more than a century of history, starting from the work of J.C. Maxwell himself. The algebra of quaternions always attracted the attention of physicists as a powerful tool for describing physical phenomena in spaces of three or four dimensions (see an interesting historical review in [17, Section 2.g.]. Nevertheless an overwhelming majority of works related with quaternions in physics do not go beyond the purely algebraic applications of this important algebra, in spite of the significant development of quaternionic analysis in the last decades (see, e.g., [14, 15, 27] and the bibliographies there).

The central object of this theory is the Moisil-Theodoresco equation which plays a role analogous to that of the Cauchy-Riemann conditions in complex analysis. Dozens of works are devoted to this quaternionic equation as well as to its generalizations. The Moisil-Theodoresco operator D (see its definition in Section 3) permits a factorization of the Laplacian $D^2 = -\Delta$, and the major part of generalizations correspond to some operators which also give such a factorization in more dimensions. Nevertheless in some works non-harmonic generalizations of the Moisil-Theodoresco operator are studied (we give a brief history in Section 3) which allow many more possible applications in physics. In the present work we consider the operator $D + \theta(x)I$, where $\theta(x)$ is a complex valued function and I is the identity operator. We propose a method for the solution of the corresponding homogeneous equation. The method is based on the construction of some special complex quaternionic zero divisors. We show that in some cases, e.g., when θ depends only on one variable, using our approach it is possible to obtain exact solutions in explicit form. This is done in Section 4. Section 5 and Section 6 are devoted to the applications of these results to the Dirac equation and to the Maxwell system respectively.

Although many papers have been dedicated to applications of quaternionic algebra to the Dirac spinors, Dirac operator, etc., until recently a suitable direct relation between the massive Dirac equation in its traditional form (written with the aid of Dirac matrices) and some quaternionic differential equation which could be considered as a quaternionic equivalent of the Dirac equation had not been found. A simple matrix transform which satisfies this natural requirement was proposed for the first time in [23]. It allows us to rewrite the Dirac equation in a quaternionic form and to give back the obtained quaternionic solutions in a traditional form habitual for physicists. In the present article we introduce this transform in Subsection 5.1 and use it to obtain quaternionic reformulations of Dirac's operator with different potentials. In Subsections 5.2-5.4 we use the results of Section 4 to obtain solutions of Dirac's equations with pseudoscalar, scalar and electric potentials.

Maxwell's system represents a pair of equations for vectors of an electromagnetic field (here we consider time-harmonic fields). It is a common misconception that the only possible way to separate the equations, obtaining instead of a system, two independent vectorial differential equations, consists in reducing the Maxwell equations to a second order wave equation. In [22] (the reader can find this material also in [27]) there was proposed a quaternionic diagonalization of Maxwell's system which allows us to reduce it to a pair of independent first order quaternionic differential equations. This procedure, as the reduction to the wave equation, works perfectly well in the case of homogeneous and isotropic media. For inhomogeneous media some more assumptions are needed in both cases. In the present work we propose such a diagonalization in the case of slowly changing media. As an illustration of its application we obtain a class of solutions for stratified media (the material parameters depend on one spatial coordinate) using once more the results of Section 4.

2. Complex quaternions

We shall denote by $\mathbb{H}(\mathbb{R})$ and by $\mathbb{H}(\mathbb{C})$ the sets of real and complex quaternions (=biquaternions) respectively (the letter $\mathbb{H}$ is chosen traditionally in honor of the inventor of quaternions, W.R. Hamilton). Each quaternion a is represented in the form $a = \sum_{k=0}^{3} a_k i_k$ where $\{a_k\} \subset \mathbb{R}$ for real quaternions and $\{a_k\} \subset \mathbb{C}$ for complex quaternions, i_0 is the unit and $\{i_k | \quad k = 1,2,3\}$ are the quaternionic imaginary units, that is, the standard basis elements possessing the following properties:

$$i_0^2 = i_0 = -i_k^2; \; i_0 i_k = i_k i_0 = i_k, \quad k = 1,2,3;$$

(2.1) $$i_1 i_2 = -i_2 i_1 = i_3;\ i_2 i_3 = -i_3 i_2 = i_1;\ i_3 i_1 = -i_1 i_3 = i_2.$$

We denote the imaginary unit in $\mathbb{C}$ by i as usual. By definition

(2.2) $$i \cdot i_k = i_k \cdot i, \quad k = \overline{0,3}.$$

Mostly we will consider the set $\mathbb{H}(\mathbb{C})$ of complex quaternions which by the multiplication laws (2.1), (2.2) and by the natural component-wise operation of addition, is a complex non-commutative, associative algebra with zero divisors.

The set of complex quaternions is isomorphic as a real vector space to the set of octonions (Cayley numbers). The difference between the sets lies on the algebraic level. By the definition of octonions the additional imaginary unit i anticommutes with i_k, $k = 1, 2, 3$ and, as a consequence, the algebra of octonions does not enjoy the property of associativity, in contrast with the algebra of complex quaternions (see, e.g., [33, 20, 11]).

Quaternions defined above as linear (real or complex) combinations of the units i_k have many representations in other algebras, useful for applications. First of all if $a \in \mathbb{H}(\mathbb{C})$ then the vector representation $a = a_0 + \vec{a}$, where $\vec{a} := \sum_{k=1}^{3} a_k i_k$, will often be useful. a_0 will be called the scalar part and $\vec{a}$ the vector part of the complex quaternion a: $a_0 =: \mathrm{Sc}\,(a)$, $\vec{a} =: \mathrm{Vec}\,(a)$.

Complex quaternions of the form $a = \vec{a}$ will be called purely vectorial. We identify them with vectors from $\mathbb{C}^3$.

In vector terms, the multiplication of two arbitrary complex quaternions a and b can be rewritten as follows:

$$a \cdot b = a_0 b_0 - < \vec{a}, \vec{b} > + \left[\vec{a} \times \vec{b}\right] + a_0 \vec{b} + b_0 \vec{a},$$

where

$$< \vec{a}, \vec{b} > := \sum_{k=1}^{3} a_k b_k \in \mathbb{C},$$

$$[\vec{a} \times \vec{b}] := \begin{vmatrix} i_1 & i_2 & i_3 \\ a_1 & a_2 & a_3 \\ b_1 & b_2 & b_3 \end{vmatrix} \in \mathbb{C}^3$$

will be called respectively the scalar product and the vector product of complex three-dimensional vectors. We will denote the operator of multiplication by a complex quaternion a from the right-hand side as M^a: $M^a b = b \cdot a$.

There are several representations of quaternions in matrix form. Let $b \in \mathbb{H}(\mathbb{C})$, and introduce the following matrices:

$$B_l(b) := \begin{pmatrix} b_0 & -b_1 & -b_2 & -b_3 \\ b_1 & b_0 & -b_3 & b_2 \\ b_2 & b_3 & b_0 & -b_1 \\ b_3 & -b_2 & b_1 & b_0 \end{pmatrix}, \tag{2.3}$$

$$B_r(b) := \begin{pmatrix} b_0 & -b_1 & -b_2 & -b_3 \\ b_1 & b_0 & b_3 & -b_2 \\ b_2 & -b_3 & b_0 & b_1 \\ b_3 & b_2 & -b_1 & b_0 \end{pmatrix}. \tag{2.4}$$

The matrix subalgebra $\mathcal{B}_l(\mathbb{C}) = \{B_l(b) \quad |b \in \mathbb{H}(\mathbb{C})\}$ of $\mathbb{C}^{4\times 4}$ and $\mathbb{H}(\mathbb{C})$ are isomorphic as complex algebras. The same holds for $\mathcal{B}_r(\mathbb{C}) := \{B_r(b)\,|b \in \mathbb{H}(\mathbb{C})\}$ and $\mathbb{H}(\mathbb{C})$.

We will be in need of two different conjugations in $\mathbb{H}(\mathbb{C})$. For $a \in \mathbb{H}(\mathbb{C})$, the usual complex conjugation is defined by

$$Z_{\mathbb{C}}(a) := a^* = \operatorname{Re} a - i \operatorname{Im} a = \sum_{k=0}^{3} \operatorname{Re}(a_k) i_k - i \sum_{k=0}^{3} \operatorname{Im}(a_k) i_k,$$

and quaternionic conjugation by

$$Z_{\mathbb{H}}(a) := \overline{a} = a_0 - \vec{a}.$$

If $\{a, b\} \subset \mathbb{H}(\mathbb{C})$ then

$$\overline{a \cdot b} = \overline{b} \cdot \overline{a}, \tag{2.5}$$

and if $a = \operatorname{Re} a + i \operatorname{Im} a$, then

$$a \cdot \overline{a} = \overline{a} \cdot a = \sum_{k=0}^{3} a_k^2 = |\operatorname{Re} a|^2 - |\operatorname{Im} a|^2 + 2i < \operatorname{Re} a, \operatorname{Im} a >_{\mathbb{R}^4} \in \mathbb{C}, \tag{2.6}$$

where $|\operatorname{Re} a|$, $|\operatorname{Im} a|$ stand for the usual modulus of a real quaternion (or the Euclidean norm of a four-dimensional vector); $< \operatorname{Re} a, \operatorname{Im} a >_{\mathbb{R}^4}$ the Euclidean scalar product of two four-dimensional vectors. (2.6) shows that the product $a\overline{a}$ can be zero even when a is not, and hence, $\mathbb{H}(\mathbb{C})$ has zero divisors.

Let us denote by $\mathfrak{S}$ the set of zero divisors from $\mathbb{H}(\mathbb{C})$; that is, $\mathfrak{S} := \{a \in \mathbb{H}(\mathbb{C}) \mid a \neq 0; \exists b \neq 0 : ab = 0\}$. Denote by $G\mathbb{H}(\mathbb{C})$ the subset of invertible elements from $\mathbb{H}(\mathbb{C})$. If $a \notin \mathfrak{S} \cup \{0\}$ then

$$a^{-1} := \overline{a}/(a\overline{a}) \tag{2.7}$$

is the inverse of the complex quaternion a. Obviously, $G\mathbb{H}(\mathbb{C}) = \mathbb{H}(\mathbb{C})\backslash(\mathfrak{S} \cup \{0\})$. The following lemma gives some equivalent definitions of the biquaternionic zero divisors.

Lemma 2.1. *([27]) Let $0 \neq a \in \mathbb{H}(\mathbb{C})$. The following conditions are equivalent:*

1. $a \in \mathfrak{S}$.
2. $a\bar{a} = 0$.
3. $a_0^2 = \vec{a}^2$.
4. $a^2 = 2a_0 a = 2\vec{a}a$.

3. Quaternionic differential operators

We shall consider $\mathbb{H}(\mathbb{C})-$valued functions defined in a domain $G \subset \mathbb{R}^3$. The set of all such functions whose components are continuously differentiable in G we will denote by $C^1(G;\mathbb{H}(\mathbb{C}))$. On this set the well known Moisil-Theodoresco operator is defined by the expression

$$D := \sum_{k=1}^{3} i_k \partial_k,$$

where $\partial_k := \frac{\partial}{\partial x_k}$. The operator D was introduced in [29], [30] and studied in a great number of works (see, e.g., [9], [14], [15], [27]). Identifying the quaternionic imaginary units i_1, i_2 and i_3 with the unitary base vectors in $\mathbb{R}^3$ one can rewrite the expression Df, where $f \in C^1(G;\mathbb{H}(\mathbb{C}))$, in the following form

$$Df = -\operatorname{div} \vec{f} + \operatorname{grad} f_0 + \operatorname{rot} \vec{f} \tag{3.1}$$

which would be inconsistent in vector calculus. In the quaternionic context it simply means that the scalar part of the complex quaternion Df is equal to $-\operatorname{div} \vec{f}$ and the vector part is equal to $\operatorname{grad} f_0 + \operatorname{rot} \vec{f}$:

$$\operatorname{Sc}(Df) = -\operatorname{div} \vec{f}, \quad \operatorname{Vec}(Df) = \operatorname{grad} f_0 + \operatorname{rot} \vec{f}.$$

The condition $f \in \ker D$ is equivalent to the Moisil-Theodoresco system

$$\begin{cases} \operatorname{div} \vec{f} = 0, \\ \operatorname{grad} f_0 + \operatorname{rot} \vec{f} = 0, \end{cases} \tag{3.2}$$

which has been studied in a non-quaternionic form, e.g., in [7], [12] and in many other works. The system (3.2) is considered by many as the most natural and simple spatial generalization of the Cauchy-Riemann conditions, but the most interesting applications require the consideration of the operator D plus some additional multiplicative term, physically representing the wave number or the potential, depending on the model under consideration. In the first work in this direction [31] the operator $D + M^{\vec{\alpha}}$ was studied where $\vec{\alpha}$ is a purely vectorial constant quaternion. In that article a technique related to the matrix representation of quaternions (2.3), (2.4) was used, not the quaternions themselves. In [13] the operator $D + \alpha_0 I$ was considered, where $\alpha_0 \in \mathbb{R}$, and I is the identity operator. In both works there were obtained analogues of the Cauchy integral formula, Plemelj-Sokhotski formulas and of some other well known results from complex analysis. The next natural step was the theory of the operators $D + \alpha I$ and $D + M^\alpha$, where α is a constant complex quaternion (see [27]). The first work [36], where the Moisil-Theodoresco operator was studied with an additional functional multiplicative term was dedicated to the following case

$$D - \frac{\operatorname{grad} \eta(x)}{\eta(x)} I + \alpha_0 I, \tag{3.3}$$

where α_0 is a real constant and η is a real differentiable function different from zero. Such a special multiplier as $(\operatorname{grad} \eta)/\eta$ permits an important factorization

$$D - \frac{\operatorname{grad} \eta(x)}{\eta(x)} I = \eta(x) D \eta^{-1}(x) I$$

which practically solves all problems for the operator (3.3), reducing them to the corresponding problems for the operator $D + \alpha_0 I$. It is clear that the situation is similar in the case of the operator

$$D - \frac{\operatorname{grad} \eta(x)}{\eta(x)} I + M^\alpha,$$

where α is a constant complex quaternion (see [25]).

In [24] the operator $D + f(x)I$, where f is an $\mathbb{H}(\mathbb{C})$-valued function, was studied and its relationship with the operator $D + \theta(x)I$, where θ is a scalar complex function, was investigated. We will concentrate on this case.

4. Solutions of the Moisil-Theodoresco equation with potential

4.1. The operator $D + \theta(x)I$

In what follows let θ be a complex valued scalar function defined in $G \subset \mathbb{R}^3$. We consider the following equation

$$(D + \theta(x))u(x) = 0 \quad \text{in } G. \tag{4.1}$$

Suppose that the function φ is some solution of the eikonal equation

(4.2) $$(\nabla\varphi)^2 = \theta^2 \quad \text{in } G.$$

Then let us notice that the $\mathbb{H}(\mathbb{C})$-valued functions $Q^{\pm} := \theta \pm \nabla\varphi$ are zero divisors in G. Let $\eta = e^{\varphi}$. Then $\nabla\varphi = \nabla\eta/\eta$, and the operator $D + \theta I$ can be rewritten in the following manner

$$\begin{aligned} D + \theta I &= D + (\theta + \nabla\varphi - \nabla\varphi)I \\ &= D + (\theta + \nabla\varphi - \frac{\nabla\eta}{\eta})I \\ &= \eta(D + Q^{+})\eta^{-1}I. \end{aligned}$$

Consequently, (4.1) is reduced to the following equation

(4.3) $$(D + Q^{+}(x))v(x) = 0 \quad \text{in } G,$$

where $v = u/\eta$. Note that (4.3) is equivalent to (4.1).

Let us look for the solution of (4.3) in the form

(4.4) $$v = Q^{-}s,$$

where s is an $\mathbb{H}(\mathbb{C})$-valued function. Substituting (4.4) into (4.3) we obtain an equation for s:

(4.5) $$DQ^{-}s = 0.$$

Thus, we should describe the set $\ker D \cap \mathrm{im}\,(Q^{-}I)$, which would be a simple task if Q^{-} were not a zero divisor. Let us consider the following important cases when it is possible to do so.

4.1.1. θ is a function of one variable

Assume that θ depends only on one variable: $\theta = \theta(x_1)$. Then, e.g., the functions

$$\varphi_1 = i\Theta, \quad \varphi_2 = -i\Theta$$

are solutions of (4.2). Here Θ is an antiderivative of θ. Note that any function of the form

(4.6) $$\varphi = \pm i\Theta + \zeta,$$

where ζ is an arbitrary analytic or antianalytic function of complex variable $z = x_2 + ix_3$, will be also a solution of (4.2). Here the trick consists in the fact that (e.g., for an analytic function ζ):

$$\nabla\zeta = \frac{\partial\zeta}{\partial z}(i_2 + ii_3).$$

The complex quaternion $i_2 + ii_3$ is a purely vectorial zero divisor which being squared gives us zero.

Consider first the function φ_1. We have

$$\eta_1(x_1) = e^{\varphi_1(x_1)} = e^{i\Theta(x_1)}, \quad Q_1^{\pm}(x_1) = \theta(x_1) \pm ii_1\theta(x_1) = \theta(x_1)(1 \pm ii_1).$$

The function v is related to u by $v = e^{-\varphi_1}u$, and we suppose that it has the form

$$v(x) = \theta(x_1)(1 - ii_1)s(x).$$

For the function s we obtain the following equation

$$D(\theta(x_1)(1 - ii_1)s(x)) = 0. \tag{4.7}$$

Denote $f = \theta \cdot s$ and use the following quaternionic representation of f:

$$f = F_1 + F_2 i_2,$$

where $F_1 = f_0 + f_1 i_1$, $F_2 = f_2 + f_3 i_1$. Note that F_1 and F_2 commute with $(1 - ii_1)$, and $(1 - ii_1)i_2 = i_2(1 + ii_1)$. Then (4.7) can be rewritten as follows

$$D(F_1(1 - ii_1) + F_2 i_2(1 + ii_1)) = 0. \tag{4.8}$$

Note that $DM^{(1 \pm ii_1)} = M^{(1 \pm ii_1)}D$. Multiplying (4.8) from the right-hand side first by $(1 - ii_1)$ and then by $(1 + ii_1)$ we obtain that (4.8) is equivalent to the system

$$D(F_1)(1 - ii_1) = 0, \tag{4.9}$$

$$D(F_2 i_2)(1 + ii_1) = 0.$$

The last equation can be rewritten in the form

$$D(F_2)(1 - ii_1) = 0. \tag{4.10}$$

Thus, F_1 and F_2 must satisfy the same equation. Consider the equation (4.9). Its solution, obviously, has the form

$$F_1(x) = H_1(x) + S_1(x)(1 + ii_1), \tag{4.11}$$

where S_1 is an arbitrary two-component function, and $H_1 = h_0 + h_1 i_1$ satisfies the equation

$$DH_1 = 0. \tag{4.12}$$

Note that the last term in (4.11) does not contribute in the final solution of (4.1) because of multiplication by Q^{-} (4.4).

In order to solve (4.12) we rewrite it in explicit form $(i_1\partial_1 + i_2\partial_2 + i_3\partial_3)(h_0 + h_1 i_1) = 0$ and obtain that it is equivalent to the following system

$$\partial_1 h_0 = \partial_1 h_1 = 0, \tag{4.13}$$

$$\partial_2 h_0 + \partial_3 h_1 = 0, \tag{4.14}$$

$$\partial_3 h_0 - \partial_2 h_1 = 0. \tag{4.15}$$

From (4.13)-(4.15) we have that H_1 is independent of the variable x_1 and is analytic in the usual complex sense with respect to the complex variable $z = x_3 + i_1 x_2$ as (4.14) and (4.15) represent the corresponding Cauchy-Riemann conditions. More precisely, both Re H_1 = Re $h_0 + i_1$Re h_1, and Im H_1 = Im $h_0 + i_1$Im h_1 are analytic with respect to z.

In a similar way, $F_2 = H_2(x_2, x_3)$ is an analytic function with respect to z. Thus,

$$s(x) = \frac{1}{\theta(x_1)}(H_1(x_2, x_3) + H_2(x_2, x_3)i_2),$$

and the function

$$\begin{aligned} \hat{u}_1(x) &= e^{\varphi_1(x_1)}\theta(x_1)(1 - ii_1)s(x) \\ &= e^{i\Theta(x_1)}(H_1(x_2, x_3)(1 - ii_1) + H_2(x_2, x_3)i_2(1 + ii_1)) \end{aligned}$$

is a solution of (4.1). Moreover, due to the right $\mathbb{H}(\mathbb{C})$-linearity of (4.1), the following function is also its solution

$$u_1(x) = e^{i\Theta(x_1)}(H_1(x_2, x_3)(1 - ii_1)A_1 + H_2(x_2, x_3)i_2(1 + ii_1)A_2), \tag{4.16}$$

where A_1 and A_2 are arbitrary constant complex quaternions.

Taking the function φ_2 as a solution of the eikonal equation (4.2) and repeating all the procedure described above we arrive at another solution of (4.1):

$$u_2(x) = e^{-i\Theta(x_1)}(G_1(x_2, x_3)(1 + ii_1)B_1 + G_2(x_2, x_3)i_2(1 - ii_1)B_2), \tag{4.17}$$

where G_1 and G_2, similarly to H_1 and H_2, are analytic functions with respect to z, and B_1, B_2 are arbitrary constant complex quaternions. Thus, we obtain the following proposition.

Proposition 4.1. *Let $\Theta(x_1)$ be an antiderivative of the function $\theta(x_1)$; H_1, H_2, G_1 and G_2 satisfy the Cauchy-Riemann conditions (4.14), (4.15), and A_1, A_2, B_1 and B_2 be arbitrary constant complex quaternions. Then the functions (4.16) and (4.17) are solutions of the equation $(D + \theta(x_1))u(x) = 0$.*

4.1.2. θ is an analytic function

Let us return to the equation (4.5). We are interested in describing all complex quaternionic solutions of the Moisil-Theodoresco equation

$$Df = 0 \tag{4.18}$$

with the following additional condition

$$f \cdot \overline{f} = 0. \tag{4.19}$$

In [21] the equation

$$\partial_t f + Df + f \cdot \overline{f} = 0 \tag{4.20}$$

was studied and its solutions obtained (in [6] the inhomogeneous equation $Df + f \cdot \overline{f} = g$ was considered in relation with a factorization of the Schroedinger operator). Equation (4.20) is obtained [16, p. 99] from the general self-duality equation taking the Jackiw-Nohl-Rebbi-t' Hooft ansatz for the gauge potential. The main result of [21] reads as follows. For any $u \in \ker(\partial_t + D)(\Omega)$, $u_0 \neq 0$, the function

$$f = -\frac{1}{2u_0}(Du + D_r u)$$

is a solution of (4.20). Here Ω is some domain in $\mathbb{R}^4$ and D_r denotes the right Moisil-Theodoresco operator $D_r u = \sum_{k=1}^{3} \partial_k u i_k$. Note that

$$D_r u = -\text{div}\ \overrightarrow{u} + \text{grad}\ u_0 - \text{rot}\ \overrightarrow{u}$$

(compare with (3.1)). For the purposes of the present work it is sufficient to limit ourselves with the case $\partial_t f \equiv 0$. Then $Du = 0$ and from (3.2) we obtain that $D_r u = 2\text{grad}\ u_0 = -2\text{rot}\ \overrightarrow{u}$. Thus, considering the first of these equalities we arrive at the following

Remark 4.2. Let $u_0 \in \ker \Delta(G)$, $G \subset \mathbb{R}^3$. Then $f = -\nabla u_0 / u_0$ satisfies the equation

$$Df + f \cdot \overline{f} = 0.$$

Then as a class of solutions of (4.18) and (4.19) we can consider $f = \nabla u_0$, where $u_0 \in \ker \Delta(G)$ is such that $\langle \nabla u_0, \nabla u_0 \rangle = 0$. For instance, u_0 can be an arbitrary analytic function of the complex variable $z_1 = x_1 + ix_2$ (and independent of x_3). In this case

$$\nabla u_0 = \frac{\partial u_0}{\partial z_1}(i_1 + ii_2)$$

and $f = \nabla u_0$ satisfies (4.18) and (4.19).

Let us consider the following question. Under which conditions can the expression $Q^- s$ from (4.5) be equal to such a purely vectorial f? In other words, when do we have the equality

$$Q^- = \frac{\partial u_0}{\partial z_1}(i_1 + ii_2)s^{-1} \ \ ? \tag{4.21}$$

Denote $\nu = s^{-1}$ and remember that $Q^- = \theta - \nabla\varphi$. Then (4.21) is equivalent to the system

$$\theta = -\frac{\partial u_0}{\partial z_1}(\nu_1 + i\nu_2), \tag{4.22}$$

$$\partial_1\varphi = -\frac{\partial u_0}{\partial z_1}(\nu_0 + i\nu_3), \tag{4.23}$$

$$\partial_2\varphi = -i\frac{\partial u_0}{\partial z_1}(\nu_0 + i\nu_3), \tag{4.24}$$

$$\partial_3\varphi = i\frac{\partial u_0}{\partial z_1}(\nu_1 + i\nu_2). \tag{4.25}$$

From (4.22) and (4.25) we obtain that

$$\varphi = -i\Theta_3 + \zeta, \tag{4.26}$$

where Θ_3 is an antiderivative of θ with respect to x_3 and ζ is an arbitrary function of x_1 and x_2. Then from (4.23) and (4.24) we obtain the following equality

$$i(\partial_1 + i\partial_2)\Theta_3 = (\partial_1 + i\partial_2)\zeta. \tag{4.27}$$

Taking into account that ζ must not depend on x_3 we see that (4.27) is possible only if

$$(\partial_1 + i\partial_2)\theta = \frac{\partial\theta}{\partial\overline{z}_1} = 0$$

and

$$\frac{\partial\zeta}{\partial\overline{z}_1} = 0.$$

Thus, θ and ζ are analytic functions with respect to $z_1 = x_1 + ix_2$. Then for $\partial_1\varphi$ and $\partial_2\varphi$ we obtain

$$\partial_1\varphi = -i\partial_2\varphi = -i\frac{\partial\Theta_3}{\partial z_1} + \frac{\partial\zeta}{\partial z_1} = \partial_{z_1}\zeta(1 - i\frac{\partial_{z_1}\Theta_3}{\partial_{z_1}\zeta}).$$

Consequently, comparing the last equality with (4.22)-(4.25) we can see that $\zeta = u_0$. Then the following election of the functions ν_k, $k = \overline{0,3}$ is possible

$$\nu_0 = -1, \qquad \nu_1 = -\frac{\theta}{\partial_{z_1}\zeta}, \qquad \nu_2 = 0, \qquad \nu_3 = \frac{\partial_{z_1}\Theta_3}{\partial_{z_1}\zeta}.$$

Then

$$s = \nu^{-1} = \frac{\partial_{z_1}\zeta}{(\partial_{z_1}\zeta)^2 + \theta^2 + (\partial_{z_1}\Theta_3)^2}(-\partial_{z_1}\zeta + \theta i_1 - \partial_{z_1}\Theta_3 i_3).$$

For Q^- we have the expression

$$Q^- = \theta - \nabla\varphi = \theta + (i\frac{\partial\Theta_3}{\partial z_1} - \frac{\partial\zeta}{\partial z_1})(i_1 + ii_2) + i\theta i_3.$$

It is easy to see that

$$Q^- \cdot s = \frac{\partial\zeta}{\partial z_1}(i_1 + ii_2) = \nabla\zeta.$$

Thus, the function v (4.4) is found, and to obtain a solution $\hat{u}$ of (4.1) we have to consider the product

$$\hat{u} = \eta \cdot v = e^{\varphi} \cdot v = e^{-i\Theta_3+\zeta} \cdot \frac{\partial \zeta}{\partial z_1} \cdot (i_1 + ii_2).$$

Finally, the expression $e^{\zeta}\frac{\partial \zeta}{\partial z_1}$, where ζ is an arbitrary analytic function can be substituted by an arbitrary analytic function h, where the relation between them is the following $\zeta(z_1) = \ln H(z_1)$, where H is an antiderivative of h with respect to z_1. We obtain $\hat{u}$ in the form

$$\hat{u} = e^{-i\Theta_3(z_1,x_3)} \cdot h(z_1) \cdot (i_1 + ii_2).$$

Due to the right-linearity of the operator $D + \theta(x)I$ we can multiply our solutions by an arbitrary constant complex quaternion from the right-hand side without loosing the property that it is a solution of (4.1). We summarize the results of this subsection in the following

Proposition 4.3. *Let θ be an analytic function with respect to the complex variable $z_1 = x_1 + ix_2$ and Θ_3 be an antiderivative of θ with respect to x_3. Let h be an arbitrary analytic function of z_1, independent of x_3, and A be an arbitrary constant complex quaternion. Then the function*

$$u(z_1, x_3) = e^{-i\Theta_3(z_1,x_3)} \cdot h(z_1) \cdot (i_1 + ii_2)A \tag{4.28}$$

is a solution of (4.1).

Remark 4.4. Note that the same can be done for an antianalytic function θ. In this case the solution is

$$u(z_1, x_3) = e^{i\Theta_3(z_1,x_3)} \cdot h^*(z_1) \cdot (i_1 - ii_2)A, \tag{4.29}$$

where h^* is an arbitrary antianalytic function with respect to z_1.

Remark 4.5. Note that θ can be independent of x_1 and x_2. Then we return to the case considered in 4.1.1. The solution (4.17) is obtained from (4.28) and (4.16) from (4.29).

4.2. The operator $D + \theta(x)I + M^{\alpha}$

In this section we show that the solution of the equation

$$(D + \theta(x)I + M^{\alpha})w(x) = 0 \quad \text{in } G \tag{4.30}$$

is reduced to the solution of (4.7). Here α is an arbitrary constant complex quaternion, w is an $\mathbb{H}(\mathbb{C})$-valued function and θ as above is a scalar complex function.

We can assume that α is a purely vectorial complex quaternion because otherwise its scalar part can be added to the function θ. Thus, we consider the operator

$$R_\alpha := D + \theta(x)I + M^{\vec{\alpha}}.$$

Let us start with the case $\vec{\alpha} \notin \mathfrak{S}$. Denote by γ any complex square root from $\vec{\alpha}^2$: $\gamma^2 = \vec{\alpha}^2$. Let us introduce the following pair of mutually complementary projection operators

$$P^\pm := \frac{1}{2\gamma} M^{(\gamma \pm \vec{\alpha})} \tag{4.31}$$

acting on the set of $\mathbb{H}(\mathbb{C})$-valued functions. An important algebraic property of the operators P^+ and P^- is the possibility of transformation of the vector $\vec{\alpha}$ to the scalars $\pm\gamma$ and vice versa:

$$P^\pm M^{\vec{\alpha}} = \frac{1}{2\gamma} M^{(\gamma\vec{\alpha} \pm \vec{\alpha}^2)} = \frac{1}{2\gamma} M^{(\gamma\vec{\alpha} \pm \gamma^2)} = \pm\gamma P^\pm.$$

Consequently, the operator R_α can be rewritten in the following form

$$\begin{aligned} R_\alpha &= P^+(D + (\theta(x) + \gamma)I) + P^-(D + (\theta(x) - \gamma)I) \\ &= (D + (\theta(x) + \gamma)I)P^+ + (D + (\theta(x) - \gamma)I)P^-. \end{aligned}$$

This equality together with the obvious commutativity of $P^\pm$ and R_α gives us the following statement.

Proposition 4.6. *Let $\vec{\alpha}$ be an arbitrary constant purely vectorial complex quaternion and $\vec{\alpha} \notin \mathfrak{S}$. Let G be some domain in $\mathbb{R}^3$ and θ be an arbitrary complex-valued function given in G. Then*

$$\ker R_\alpha(G) = P^+ \ker(D + (\theta(x) + \gamma)I)(G) \oplus P^- \ker(D + (\theta(x) - \gamma)I)(G),$$

where $\gamma^2 = \vec{\alpha}^2$ and $P^\pm$ are defined by (4.31).

In other words $w \in \ker R_\alpha$ iff $w = P^+w^+ + P^-w^-$, where w^+ and w^- are solutions of the following equations respectively

$$(D + \theta(x) + \gamma)w^+(x) = 0,$$

$$(D + \theta(x) - \gamma)w^-(x) = 0.$$

Thus, we reduced equation (4.30) to (4.1) in the case $\vec{\alpha} \notin \mathfrak{S}$.

Now let us consider the case $\vec{\alpha} \in \mathfrak{S}$. Then the general solution of the equation

$$R_\alpha w = 0 \tag{4.32}$$

has the form

$$w = u - v\vec{\alpha},$$

where u is an arbitrary solution of (4.1) and v satisfies the equation

$$(D+\theta(x))v = u.$$

For the purposes of this article it will be sufficient to consider the following class of solutions of (4.32)

$$w = u\vec{\alpha},$$

where u is the solution of (4.7).

Using Proposition 4.1 we obtain solutions for equation (4.32) in the case when $\theta = \theta(x_1)$.

Proposition 4.7. *Let $\Theta(x_1)$ be an antiderivative of the function $\theta(x_1)$; $H_1^{\pm}$, $H_2^{\pm}$, $G_1^{\pm}$ and $G_2^{\pm}$ satisfy the Cauchy-Riemann conditions (4.14), (4.15) and $A_1^{\pm}$, $A_2^{\pm}$, $B_1^{\pm}$ and $B_2^{\pm}$ be arbitrary constant complex quaternions. Then*

1) if $\vec{\alpha} \notin \mathfrak{S}$, then the function

(4.33)

$$\begin{aligned} w &= \frac{1}{2\gamma}(e^{i(\Theta(x_1)+\gamma x_1)}(H_1^+(x_2,x_3)(1-ii_1)A_1^+ + H_2^+(x_2,x_3)i_2(1+ii_1)A_2^+) \\ &+ e^{-i(\Theta(x_1)+\gamma x_1)}(G_1^+(x_2,x_3)(1+ii_1)B_1^+ + G_2^+(x_2,x_3)i_2(1-ii_1)B_2^+))(\gamma+\vec{\alpha}) \\ &+ \frac{1}{2\gamma}(e^{i(\Theta(x_1)-\gamma x_1)}(H_1^-(x_2,x_3)(1-ii_1)A_1^- + H_2^-(x_2,x_3)i_2(1+ii_1)A_2^-) \\ &+ e^{-i(\Theta(x_1)-\gamma x_1)}(G_1^-(x_2,x_3)(1+ii_1)B_1^- + G_2^-(x_2,x_3)i_2(1-ii_1)B_2^-))(\gamma-\vec{\alpha}) \end{aligned}$$

is a solution of (4.32),

2) if $\vec{\alpha} \in \mathfrak{S}$, then the function

$$\begin{aligned} (4.34)\quad w &= (e^{i\Theta(x_1)}(H_1^+(x_2,x_3)(1-ii_1)A_1^+ + H_2^+(x_2,x_3)i_2(1+ii_1)A_2^+) \\ &+ e^{-i\Theta(x_1)}(G_1^+(x_2,x_3)(1+ii_1)B_1^+ + G_2^+(x_2,x_3)i_2(1-ii_1)B_2^+))\vec{\alpha} \end{aligned}$$

is a solution of (4.32).

5. Solutions of the classical Dirac equation with potentials

5.1. The Dirac operator in quaternionic form

From the very appearance of Dirac's equation in quantum mechanics many researchers considered that the form of writing it was unsatisfactory. The matrix algebra used for this purpose has dimension 16, and there does not exist any physically meaningful reason for using an algebra of 16 dimensions if the number of equations under consideration is four. A. Sommerfeld posed the following problem: to rewrite the Dirac equation in a form in which the rank of the algebra of the matrices involved coincides with the number of components of the wavefunction, and apparently he made a first attempt to solve it (see [34]). In [19] it was shown that the Dirac equation for a massless field can be represented as an analyticity condition for a function of a quaternionic variable. An important contribution was made in [5], where the symmetrical analysis of the Dirac equation for a free particle with spin 1/2 and non-zero rest mass was essentially simplified after using its quaternionic reformulation (see also [4]).

Let us stress that we do not pretend to mention here even the most important publications dedicated to the quaternionic approach to the Dirac equation. In spite of the fact that up until now this approach is not generally accepted, they are too numerous. We mention only some of the works directly related with the complex quaternionic reformulation of the Dirac equation which is used in the present article and was introduced in [23] (see also [27, Section 12]). This reformulation is properly a simple matrix transformation which allows us to rewrite the classical Dirac operator (given in its "traditional" form using γ-matrices) applied to $\mathbb{C}^4$-vectors as a quaternionic operator applied to $\mathbb{H}(\mathbb{C})$-valued functions. We will not discuss here a quite lengthy procedure which led to the matrix transformation and which was based in part on the Gohberg-Krupnik matrix identity (see, e.g., [28, p. 88]) referring the reader to [23] or [27, Section 12]. We will only use it here in order to apply the results of Section 4 to the Dirac operator with different potentials.

Define $\tilde{f} := f(t, x_1, x_2, -x_3)$. The domain $\tilde{\Omega}$ is assumed to be obtained from the domain $\Omega \subset \mathbb{R}^4$ by the reflection $x_3 \to -x_3$. The above mentioned transformation we denote as $\mathcal{A}$ and define it in the following way. A function $f : \Omega \subset \mathbb{R}^4 \to \mathbb{C}^4$ is transformed into a function $F : \tilde{\Omega} \subset \mathbb{R}^4 \to \mathbb{H}(\mathbb{C})$ by the rule

$$F = \mathcal{A}[f] := \frac{1}{2}\left(-(\tilde{f}_1 - \tilde{f}_2)i_0 + i(\tilde{f}_0 - \tilde{f}_3)i_1 - (\tilde{f}_0 + \tilde{f}_3)i_2 + i(\tilde{f}_1 + \tilde{f}_2)i_3\right).$$

The inverse transformation $\mathcal{A}^{-1}$ is defined as follows

$$\mathcal{A}^{-1}[F] = (-i\tilde{F}_1 - \tilde{F}_2, -\tilde{F}_0 - i\tilde{F}_3, \tilde{F}_0 - i\tilde{F}_3, i\tilde{F}_1 - \tilde{F}_2).$$

Let us present the introduced transformations in a more explicit matrix form which relates the components of a $\mathbb{C}^4$-valued function f with the components of an $\mathbb{H}(\mathbb{C})$-valued function F:

$$F = \mathcal{A}[f] = \frac{1}{2}\begin{pmatrix} 0 & -1 & 1 & 0 \\ i & 0 & 0 & -i \\ -1 & 0 & 0 & -1 \\ 0 & i & i & 0 \end{pmatrix}\begin{pmatrix} \tilde{f}_0 \\ \tilde{f}_1 \\ \tilde{f}_2 \\ \tilde{f}_3 \end{pmatrix}$$

and

$$f = \mathcal{A}^{-1}[F] = \begin{pmatrix} 0 & -i & -1 & 0 \\ -1 & 0 & 0 & -i \\ 1 & 0 & 0 & -i \\ 0 & i & -1 & 0 \end{pmatrix}\begin{pmatrix} \tilde{F}_0 \\ \tilde{F}_1 \\ \tilde{F}_2 \\ \tilde{F}_3 \end{pmatrix}.$$

We will consider the Dirac γ-matrices in the standard [8, 37] Dirac-Pauli form

$$\gamma_0 := \begin{pmatrix} 1 & 0 & 0 & 0 \\ 0 & 1 & 0 & 0 \\ 0 & 0 & -1 & 0 \\ 0 & 0 & 0 & -1 \end{pmatrix}, \qquad \gamma_1 := \begin{pmatrix} 0 & 0 & 0 & -1 \\ 0 & 0 & -1 & 0 \\ 0 & 1 & 0 & 0 \\ 1 & 0 & 0 & 0 \end{pmatrix},$$

$$\gamma_2 := \begin{pmatrix} 0 & 0 & 0 & i \\ 0 & 0 & -i & 0 \\ 0 & -i & 0 & 0 \\ i & 0 & 0 & 0 \end{pmatrix}, \qquad \gamma_3 := \begin{pmatrix} 0 & 0 & -1 & 0 \\ 0 & 0 & 0 & 1 \\ 1 & 0 & 0 & 0 \\ 0 & -1 & 0 & 0 \end{pmatrix};$$

$$\gamma_5 := i\gamma_0\gamma_1\gamma_2\gamma_3 = \begin{pmatrix} 0 & 0 & -1 & 0 \\ 0 & 0 & 0 & -1 \\ -1 & 0 & 0 & 0 \\ 0 & -1 & 0 & 0 \end{pmatrix}.$$

The following algebraic properties of transforms $\mathcal{A}$ and $\mathcal{A}^{-1}$shown in [25] will be needed:

1. $\mathcal{A}\gamma_1\gamma_2\gamma_3\gamma_1\mathcal{A}^{-1}[F] = i_1 F$;
2. $\mathcal{A}\gamma_1\gamma_2\gamma_3\gamma_2\mathcal{A}^{-1}[F] = i_2 F$;
3. $\mathcal{A}\gamma_1\gamma_2\gamma_3\gamma_3\mathcal{A}^{-1}[F] = -i_3 F$;
4. $\mathcal{A}\gamma_1\gamma_2\gamma_3\gamma_0\mathcal{A}^{-1}[F] = F i_1$;
5. $\mathcal{A}\gamma_1\gamma_2\gamma_3\mathcal{A}^{-1}[F] = -iFi_2$.

Let us consider first the Dirac operator without potential

$$\mathbb{D}[\Phi] := \left(\gamma_0\partial_t - \sum_{k=1}^{3}\gamma_k\partial_k + im\right)[\Phi],$$

where $\partial_t := \frac{\partial}{\partial t}$, $m \in \mathbb{R}$, Φ is a $\mathbb{C}^4$-valued function, and let us introduce the following quaternionic operator

$$R := P_1^+(i\partial_t + D) + P_1^-(-i\partial_t + D) - mM^{i_2},$$

where

$$P_k^{\pm} := \frac{1}{2}M^{(1\pm ii_k)}.$$

Using the algebraic properties of $\mathcal{A}$ and $\mathcal{A}^{-1}$ we obtain the following equality [23]:

$$R = -\mathcal{A}\gamma_1\gamma_2\gamma_3\mathbb{D}\mathcal{A}^{-1}. \tag{5.35}$$

Thus, the operator R is the Dirac operator for a free massive particle of spin 1/2 in a quaternionic form.

Now let us see what is the result of transformation of the most frequently studied potentials. The Dirac operator with pseudoscalar potential has the form

$$\mathbb{D}^{ps} := \mathbb{D} + \gamma_0\gamma_5 g I,$$

where g is a scalar function. It is easy to see that

$$R + \vartheta I = -\mathcal{A}\gamma_1\gamma_2\gamma_3\mathbb{D}^{ps}\mathcal{A}^{-1}, \tag{5.36}$$

where $\vartheta = -i\tilde{g}$. We denote

$$R^{ps} := R + \vartheta I.$$

In the same way we obtain quaternionic reformulations similar to (5.36) for the Dirac operator a) with scalar potential $\mathbb{D}^{sc} := \mathbb{D} + gI$, b) with electric potential $\mathbb{D}^{el} := \mathbb{D} + i\gamma_0 gI$, c) with magnetic potential $\mathbb{D}^{m} := \mathbb{D} + \sum_{k=1}^{3}\gamma_k A_k I$. Namely,

$$R^{sc} := R + M^{i\tilde{g}i_2},$$

$$R^{el} := R + M^{-i\tilde{g}i_1},$$

$$R^{m} := R - \vec{B} I,$$

where the purely vectorial quaternionic function $\vec{B}$ has the form $\vec{B} = \tilde{A}_1 i_1 + \tilde{A}_2 i_2 - \tilde{A}_3 i_3$. In all mentioned cases we have equalities similar to (5.35), where the operators R and $\mathbb{D}$ have the corresponding superscripts, e.g.,

$$R^{el} = -\mathcal{A}\gamma_1\gamma_2\gamma_3\mathbb{D}^{el}\mathcal{A}^{-1}.$$

In this work we shall construct a class of time-harmonic solutions of the Dirac equations with one-component static potentials (pseudoscalar, scalar and electric). The same approach can also be used for obtaining results in other situations.

A time-harmonic solution Φ has the form $\Phi(t,x) = q(x)e^{i\omega t}$, where $\omega \in \mathbb{R}$ and q is a $\mathbb{C}^4$-valued function depending only on x.

5.2. Pseudoscalar potential

In the case of pseudoscalar potential we have the following equation for q:

$$\mathbb{D}^{ps}_{\omega,m} q(x) = 0, \tag{5.37}$$

where $\mathbb{D}^{ps}_{\omega,m} := \mathbb{D}_{\omega,m} + \gamma_0\gamma_5 g(x) I$, and $\mathbb{D}_{\omega,m} := i\omega\gamma_0 - \sum_{k=1}^{3}\gamma_k\partial_k + imI$. The corresponding quaternionic reformulation of $\mathbb{D}^{ps}_{\omega,m}$ will be the operator

$$R^{ps}_{\alpha} := D + \vartheta(x)I + M^{\alpha},$$

where as before $\vartheta = -i\tilde{g}$, and $\alpha := -(i\omega i_1 + mi_2)$. Thus, the Dirac operator with pseudoscalar potential for time-harmonic spinor fields is reduced to the quaternionic operator studied in Subsection 4.2.

Now we use the solutions presented in Proposition 4.7 for obtaining solutions of (5.37). As in Subsection 4.2 we must consider separately the cases when $\alpha \notin \mathfrak{S}$ and $\alpha \in \mathfrak{S}$. For $\alpha := -(i\omega i_1 + mi_2)$ this means $\omega^2 \neq m^2$ or $\omega^2 = m^2$ respectively. We start with the first case. In order to obtain solutions of (5.37) we have to apply the transform $\mathcal{A}^{-1}$ to (4.33). For the sake of simplicity we will consider the case when only one of the constant quaternions in (4.33) is different from zero, namely, $A_1^+ \neq 0$. All other cases can be considered in a similar way. Let us omit the superscript "+" and subscript "1" in H_1^+ and A_1^+. We now apply $\mathcal{A}^{-1}$ to the function

$$w = \frac{1}{2\gamma} e^{i(\Theta(x_1)+\gamma x_1)} H(x_2, x_3)(1 - ii_1)A(\gamma - (i\omega i_1 + mi_2)).$$

First, w should be rewritten in the component-wise form. It can be seen that

$$H(x_2, x_3)(1 - ii_1)A(\gamma - (i\omega i_1 + mi_2)) = bh(x_2, x_3)(1 - ii_1)(1 - \frac{m}{\gamma + \omega} i_2),$$

where

$$h = h_0 + ih_1 \tag{5.38}$$

and

$$b = (\gamma + \omega)(a_0 + ia_1) + m(a_2 + ia_3).$$

Thus,

$$w = \frac{b}{2\gamma} e^{i(\Theta(x_1)+\gamma x_1)} h(x_2, x_3)(1 - ii_1 - \frac{m}{\gamma + \omega} i_2 + \frac{im}{\gamma + \omega} i_3).$$

Applying $\mathcal{A}^{-1}$ to w and taking into account that $\theta = -ig$ we obtain the following Dirac spinor

$$\mathcal{A}^{-1} w = \frac{b}{2\gamma(\gamma + \omega)} e^{G(x_1)+\gamma x_1} \tilde{h}(x_2, x_3) \begin{pmatrix} m - (\gamma + \omega) \\ m - (\gamma + \omega) \\ m + \gamma + \omega \\ m + \gamma + \omega \end{pmatrix}, \tag{5.39}$$

where G is an antiderivative of g, and the components of $\tilde{h}(x_2, x_3) = h(x_2, -x_3)$ satisfy the following Cauchy-Riemann conditions with respect to the complex variable $x_2 + ix_3$

$$\partial_2 \tilde{h}_0 - \partial_3 \tilde{h}_1 = 0, \tag{5.40}$$

$$\partial_3 \tilde{h}_0 + \partial_2 \tilde{h}_1 = 0. \tag{5.41}$$

The function (5.39) is a solution of (5.37).

Remark 5.8. This example shows us that an arbitrary constant complex quaternion A in the quaternionic solution degenerates to an arbitrary complex constant b after multiplication by zero divisors and application of the transform $\mathcal{A}^{-1}$. Thus in what follows in order to simplify our calculations we omit the unnecessary constant factors.

Now let us study the second case $\omega^2 = m^2$ which corresponds to the second part of Proposition 4.7. As in the first case, we will consider only one particular solution, namely

$$w = -e^{i\Theta(x_1)}H(x_2, x_3)(1 - ii_1)(i\omega i_1 + mi_2).$$

In a component-wise form we have

$$w = e^{i\Theta(x_1)}h(x_2, x_3)(\omega(1 - ii_1) - m(i_2 - ii_3)),$$

where h is defined by (5.38). Applying $\mathcal{A}^{-1}$ to w we obtain a solution of (5.37):

$$q = \mathcal{A}^{-1}w = e^{i\Theta(x_1)}\tilde{h}(x_2, x_3)\begin{pmatrix} -\omega + m \\ -\omega + m \\ \omega + m \\ \omega + m \end{pmatrix}.$$

5.3. Scalar potential

As was shown in Subsection 5.1, the Dirac operator with scalar potential $\mathbb{D}^{sc}$ is equivalent to the quaternionic operator

$$R^{sc} = P_1^+(i\partial_t + D) + P_1^-(-i\partial_t + D) + M^{(i\tilde{g}(x)-m)i_2}.$$

The equation for time-harmonic spinors

$$\mathbb{D}^{sc}_{\omega,m}q(x) = 0, \tag{5.42}$$

where $\mathbb{D}^{sc}_{\omega,m} := \mathbb{D}_{\omega,m} + g(x)I$, is equivalent to the following quaternionic equation

$$(D + M^{(-i\omega i_1 + (i\tilde{g}(x)-m)i_2)})u(x) = 0, \tag{5.43}$$

which in general is not reduced to the equation $R_\alpha w = 0$ studied in Subsection 4.2. Such a reduction is possible under the additional condition $\omega = 0$. Here we limit ourselves to this case (for $\omega \neq 0$ another approach was proposed in [26] which permits to obtain a class of solutions of (5.43) for some special potentials $g(x)$). Thus, we consider the quaternionic equation

$$R^{sc}_m u(x) = 0, \tag{5.44}$$

where $R^{sc}_m := D + M^{(i\tilde{g}(x)-m)i_2}$. This operator can be rewritten as follows

$$R^{sc}_m = P_2^+(D + (\tilde{g}(x) + im)I) + P_2^-(D - (\tilde{g}(x) + im)I).$$

Remember that $P_2^{\pm} = 1/2M^{(1 \pm i i_2)}$. The operators $P_2^{\pm}$ commute with the operators in parentheses in the above equation. Consequently, we obtain the following

Proposition 5.9. *Let G be some domain in $\mathbb{R}^3$, $\tilde{g}$ be an arbitrary complex valued function defined in G and m be a constant. Then*

$$\ker R_m^{sc}(G) = P_2^+ \ker(D + (\tilde{g}(x) + im)I)(G) \oplus P_2^- \ker(D - (\tilde{g}(x) + im)I)(G).$$

This equality means that any solution u of (5.44) can be represented in the form

$$u = P_2^+ f^+ + P_2^- f^-, \tag{5.45}$$

where $f^+ \in \ker(D + (\tilde{g}(x) + im)I)$ and $f^- \in \ker(D - (\tilde{g}(x) + im)I)$. In this way equation (5.44) is reduced to (4.1).

Let g depend only on x_1. Using Proposition 4.1 we obtain the functions f^+ and f^- in the following form

(5.46)

$$\begin{aligned} f^+(x) &= e^{i\int (g(x_1)+im)dx_1} (H_1^+(x_2, x_3)(1 - ii_1) + H_2^+(x_2, x_3) i_2 (1 + ii_1)) \\ &+ e^{-i\int (g(x_1)+im)dx_1} (G_1^+(x_2, x_3)(1 + ii_1) + G_2^+(x_2, x_3) i_2 (1 - ii_1)), \end{aligned}$$

(5.47)

$$\begin{aligned} f^-(x) &= e^{-i\int (g(x_1)+im)dx_1} (H_1^-(x_2, x_3)(1 - ii_1) + H_2^-(x_2, x_3) i_2 (1 + ii_1)) \\ &+ e^{i\int (g(x_1)+im)dx_1} (G_1^-(x_2, x_3)(1 + ii_1) + G_2^-(x_2, x_3) i_2 (1 - ii_1)). \end{aligned}$$

Here we did not take into account the arbitrary constant complex quaternions according to Remark 5.8. The solution u of (5.44) is now obtained as in (5.45). Finally, the transformation $\mathcal{A}^{-1}$ has to be applied to u. As in Subsection 5.2 we will simplify our considerations limiting ourselves to the case

$$H_1^- = H_2^{\pm} = G_1^{\pm} = G_2^{\pm} = 0.$$

The corresponding solution u of (5.44) has the form

$$u = 2P_2^+ P_1^- e^{i\int (g(x_1)+im)dx_1} H(x_2, x_3),$$

where we omitted the superscript "+" and subscript "1" in H_1^+. The function H as before is required to satisfy the Cauchy-Riemann conditions (4.14), (4.15). In component-wise form we have

$$u = \frac{1}{2} e^{i\int (g(x_1)+im)dx_1} h(x_2, x_3)(1 - ii_1 + ii_2 + i_3), \tag{5.48}$$

where h is defined by (5.38). Now application of $\mathcal{A}^{-1}$ to (5.48) gives us a solution q of the Dirac equation with scalar potential $\mathbb{D}^{sc}_{0,m}q = 0$ in the following form

$$q = \mathcal{A}^{-1}u = -\frac{1}{2}e^{i\int(g(x_1)+im)dx_1}\tilde{h}(x_2,x_3)\begin{pmatrix} 1+i \\ 1+i \\ -1+i \\ -1+i \end{pmatrix},$$

where the function $\tilde{h}$ is an arbitrary solution of (5.40), (5.41).

5.4. Electric potential

Let us consider the Dirac operator with electric potential $\mathbb{D}^{el}$. Its quaternionic equivalent is the operator R^{el} (Subsection 5.1). For time-harmonic spinor fields we have the following equation

$$\mathbb{D}^{el}_{\omega,m}q(x) = 0, \tag{5.49}$$

where $\mathbb{D}^{el}_{\omega,m} := \mathbb{D}_{\omega,m} + i\gamma_0 g(x)I$. This equation is equivalent to the quaternionic equation

$$(D + M^{(-i(\tilde{g}(x)+\omega)i_1 - mi_2)})u(x) = 0. \tag{5.50}$$

The relation between (5.49) and (5.50) is determined by the transformation $\mathcal{A}$: $u = \mathcal{A}q$.

Equation (5.50) cannot in general be reduced to the equations studied in Section 4. Here we will consider a case in which such reduction is possible, namely, $m = 0$. Thus, we consider the quaternionic equation

$$R^{el}_{\omega}u(x) = 0, \tag{5.51}$$

where $R^{el}_{\omega} := D + M^{-i(\tilde{g}(x)+\omega)i_1}$. By analogy with Subsection 5.3 we obtain the following

Proposition 5.10. *Let G be some domain in $\mathbb{R}^3$, $\tilde{g}$ be an arbitrary complex valued function defined in G and ω be a constant. Then*

$$\ker R^{el}_{\omega}(G) = P_1^- \ker(D + (\tilde{g}(x)+\omega)I)(G) \oplus P_1^+ \ker(D - (\tilde{g}(x)+\omega)I)(G).$$

In other words the solution of (5.51) is represented as a sum

$$u = P_1^- f^+ + P_1^+ f^-,$$

where $f^{\pm} \in \ker(D \pm (\tilde{g}(x)+\omega)I)$.

Then in the case $g = g(x_1)$ we immediately arrive at the functions f^+ and f^- in the form (5.46), (5.47), where the expression im must be substituted by ω.

Considering only the "first part" of the solution, as in Subsection 5.3, we obtain a solution of (5.51) in the form

$$u = 2P_1^- e^{i\int(g(x_1)+\omega)dx_1} H(x_2,x_3) = e^{i\int(g(x_1)+\omega)dx_1} h(x_2,x_3)(1 - ii_1)$$

(taking into account that $(P_1^-)^2 = P_1^-$).

Application of $\mathcal{A}^{-1}$ to u gives us a solution q of the Dirac equation with electric potential $\mathbb{D}^{el}_{\omega,0} q = 0$ in the following form

$$q = \mathcal{A}^{-1}u = -e^{i\int(g(x_1)+\omega)dx_1} \tilde{h}(x_2,x_3) \begin{pmatrix} 1 \\ 1 \\ -1 \\ -1 \end{pmatrix},$$

where $\tilde{h} = \tilde{h}_0 + i\tilde{h}_1$ is an arbitrary solution of (5.40), (5.41).

6. Quaternionic diagonalization of Maxwell's equations and wave propagation in stratified media

In this section we give a relation between Maxwell's equations for inhomogeneous media and the quaternionic differential operator studied in Subsection 4.1. The solutions obtained for the case $\theta = \theta(x_1)$ will be used for constructing solutions of Maxwell's equations in stratified media.

In order to explain the principal idea of the diagonalization mentioned in the title of this section, let us consider first the Maxwell equations for isotropic and homogeneous media (in what follows we consider the case of absence of currents and sources).

$$\operatorname{rot} \vec{E}(x) = -i\omega\mu\vec{H}(x), \tag{6.52}$$

$$\operatorname{rot} \vec{H}(x) = i\omega\varepsilon\vec{E}(x). \tag{6.53}$$

The complex vectors $\vec{E}$ and $\vec{H}$ represent complex amplitudes of the electromagnetic field. ε is the complex permittivity of the medium and μ is its complex permeability. ω is the frequency which in general can be also a complex number (circular frequency). The magnitude $\theta = \omega\sqrt{\epsilon\mu}$ is called wave number. The vectors $\vec{E}$ and $\vec{H}$ can be considered as purely vectorial complex quaternions. Then taking into account that from (6.52) and (6.53) one obtains that div $\vec{E}$ = div $\vec{H} = 0$, we rewrite the Maxwell equations in quaternionic form:

$$D\vec{E}(x) = -i\omega\mu\vec{H}(x),$$

$$D\vec{H}(x) = i\omega\varepsilon\vec{E}(x).$$

In [22] (the reader can find this also in [27, Section 9]) the following simple procedure was proposed. Let us denote

$$\vec{\phi} = i\omega\varepsilon\vec{E} + \theta\vec{H}, \tag{6.54}$$

$$\vec{\psi} = -i\omega\varepsilon\vec{E} + \theta\vec{H}. \tag{6.55}$$

Then it is easy to see that $\vec{\phi}$ and $\vec{\psi}$ satisfy the equations

$$(D-\theta)\vec{\phi} = 0, \qquad (D+\theta)\vec{\psi} = 0. \tag{6.56}$$

Due to this simple relation between Maxwell's system and quaternionic equations some new integral representations for different electromagnetic quantities were obtained (see, e.g., [27, Sections 10 and 11]).

Note that this simple procedure is equivalent to the following diagonalization of the Maxwell equations

$$\begin{pmatrix} \text{rot} & i\omega\mu \\ -i\omega\varepsilon & \text{rot} \end{pmatrix}\begin{pmatrix} \vec{E} \\ \vec{H} \end{pmatrix} = \begin{pmatrix} D & i\omega\mu \\ -i\omega\varepsilon & D \end{pmatrix}\begin{pmatrix} \vec{E} \\ \vec{H} \end{pmatrix}$$

$$= B_\theta^{-1}\begin{pmatrix} D-\theta & 0 \\ 0 & D+\theta \end{pmatrix} B_\theta \begin{pmatrix} \vec{E} \\ \vec{H} \end{pmatrix},$$

where

$$B_\theta = \begin{pmatrix} i\omega\varepsilon & \theta \\ -i\omega\varepsilon & \theta \end{pmatrix} \quad \text{and} \quad B_\theta^{-1} = \frac{1}{2}\begin{pmatrix} \frac{1}{i\omega\varepsilon} & -\frac{1}{i\omega\varepsilon} \\ \frac{1}{\theta} & \frac{1}{\theta} \end{pmatrix}.$$

When the medium is not homogeneous, the situation is, of course, much more complicated, but we will apply in principle the same idea. Our considerations will be simpler if we study the case of a medium with constant permeability μ. The case of variable permeability requires only some more quite obvious calculations. Thus, $\varepsilon = \varepsilon(x)$ and $\mu = \text{Const}$. Then from the Maxwell equations (6.52) and (6.53) we have

$$\text{div}\,\vec{H}(x) = 0,$$

and

$$\text{div}\,\vec{E}(x) = -\langle \frac{\text{grad}\,\varepsilon(x)}{\varepsilon(x)}, \vec{E}(x)\rangle.$$

We introduce two auxiliary vectors

$$\vec{\mathcal{E}}(x) = \frac{1}{i\omega\varepsilon(x)}\vec{E}(x) \quad \text{and} \quad \vec{\mathcal{H}}(x) = \frac{1}{\omega}\sqrt{\frac{\mu}{\varepsilon^3(x)}}\vec{H}(x). \tag{6.57}$$

Then we define the analogues of the purely vectorial complex quaternions (6.54) and (6.55) in the following way

$$\vec{\Phi} = \vec{\mathcal{E}} + \vec{\mathcal{H}} \quad \text{and} \quad \vec{\Psi} = -\vec{\mathcal{E}} + \vec{\mathcal{H}}. \tag{6.58}$$

Application of D to $\vec{\Phi}$ gives us the equality

$$(6.59)\quad D\vec{\Phi} = -\frac{\operatorname{grad}\varepsilon}{i\omega\varepsilon^2}\cdot\vec{E} + \frac{1}{i\omega\varepsilon}D\vec{E} - \frac{3\sqrt{\mu}}{2\omega}\frac{\operatorname{grad}\varepsilon}{\varepsilon^{5/2}}\cdot\vec{H} + \frac{1}{\omega}\sqrt{\frac{\mu}{\varepsilon^3}}D\vec{H}.$$

Note that

$$D\vec{E} = -\operatorname{div}\vec{E} + \operatorname{rot}\vec{E} = \langle\frac{\operatorname{grad}\varepsilon}{\varepsilon}, \vec{E}\rangle - i\omega\mu\vec{H},$$

and

$$D\vec{H} = i\omega\varepsilon\vec{E}.$$

Consequently, from (6.59) we obtain

$$\begin{aligned} D\vec{\Phi} &= -\frac{\operatorname{grad}\varepsilon}{i\omega\varepsilon^2}\cdot\vec{E} + \langle\frac{\operatorname{grad}\varepsilon}{i\omega\varepsilon^2}, \vec{E}\rangle - \frac{\mu}{\varepsilon}\vec{H} - \frac{3\sqrt{\mu}}{2\omega}\frac{\operatorname{grad}\varepsilon}{\varepsilon^{5/2}}\cdot\vec{H} + i\sqrt{\frac{\mu}{\varepsilon}}\vec{E} \\ &= -\theta(\frac{1}{i\omega\varepsilon}\vec{E} + \frac{1}{\omega}\sqrt{\frac{\mu}{\varepsilon^3}}\vec{H} - \frac{i\operatorname{grad}\varepsilon}{\omega^2\mu^{1/2}\varepsilon^{5/2}}\cdot\vec{E} \\ &+ \frac{i}{\omega^2\mu^{1/2}}\langle\frac{\operatorname{grad}\varepsilon}{\varepsilon^{5/2}}, \vec{E}\rangle + \frac{3}{2\omega^2}\frac{\operatorname{grad}\varepsilon}{\varepsilon^3}\cdot\vec{H}), \end{aligned}$$

where $\theta = \omega\sqrt{\varepsilon\mu}$.

Finally, introducing the notation

$$\vec{\mathcal{g}}(x) = \frac{\operatorname{grad}\varepsilon(x)}{\omega\sqrt{\mu\varepsilon^3(x)}},$$

we obtain the following equation for $\vec{\Phi}$

$$(6.60)\quad (D+\theta(x))\vec{\Phi}(x) = -\theta(x)(\vec{\mathcal{g}}(x)\cdot\vec{\mathcal{E}}(x) - \langle\vec{\mathcal{g}}(x), \vec{\mathcal{E}}(x)\rangle + \frac{3}{2}\vec{\mathcal{g}}(x)\cdot\vec{\mathcal{H}}(x)).$$

In a similar way we also obtain an equation for $\vec{\Psi}$:

$$(6.61)\quad (D-\theta(x))\vec{\Psi}(x) = \theta(x)(\vec{\mathcal{g}}(x)\cdot\vec{\mathcal{E}}(x) - \langle\vec{\mathcal{g}}(x), \vec{\mathcal{E}}(x)\rangle - \frac{3}{2}\vec{\mathcal{g}}(x)\cdot\vec{\mathcal{H}}(x)).$$

Remember that the relation between the absolute value v of the speed of electromagnetic propagation in the medium and the electromagnetic parameters is

$$v = \frac{1}{\sqrt{\varepsilon\mu}}.$$

In what follows we consider a slowly changing medium, which means that the dimensionless expression $|\operatorname{grad} v|/\omega$ is much less than one (see, e.g., [1]). In other words the medium is called slowly changing if

$$\left|\frac{\operatorname{grad}\varepsilon}{2\omega\sqrt{\mu\varepsilon^3}}\right| \ll 1.$$

This inequality can be rewritten as follows

$$|\overrightarrow{g}| \ll 2.$$

Consequently, for a slowly changing medium the expressions on the right-hand sides of (6.60) and (6.61) are negligible and for $\overrightarrow{\Phi}$ and $\overrightarrow{\Psi}$ we have the homogeneous equations

$$(D + \theta(x))\overrightarrow{\Phi}(x) = 0, \tag{6.62}$$

$$(D - \theta(x))\overrightarrow{\Psi}(x) = 0. \tag{6.63}$$

Thus, once more we arrive at the consideration of the kernel of the operator studied in Subsection 4.1, but with an additional condition. The scalar part of the solution must be zero. Let us see how this restriction appears in the case when θ depends only on x_1. From Proposition 4.1 we obtain the following expressions for $\overrightarrow{\Phi}$ and $\overrightarrow{\Psi}$:

$$\begin{aligned} \overrightarrow{\Phi}(x) &= e^{i\Theta(x_1)}(H_1^+(x_2,x_3)(1-ii_1)A_1^+ + H_2^+(x_2,x_3)i_2(1+ii_1)A_2^+) \\ &+ e^{-i\Theta(x_1)}(G_1^+(x_2,x_3)(1+ii_1)B_1^+ + G_2^+(x_2,x_3)i_2(1-ii_1)B_2^+), \end{aligned} \tag{6.64}$$

$$\begin{aligned} \overrightarrow{\Psi}(x) &= e^{-i\Theta(x_1)}(H_1^-(x_2,x_3)(1-ii_1)A_1^- + H_2^-(x_2,x_3)i_2(1+ii_1)A_2^-) \\ &+ e^{i\Theta(x_1)}(G_1^-(x_2,x_3)(1+ii_1)B_1^- + G_2^-(x_2,x_3)i_2(1-ii_1)B_2^-), \end{aligned} \tag{6.65}$$

where the functions $H_1^\pm$, $H_2^\pm$, $G_1^\pm$and $G_2^\pm$ satisfy the Cauchy-Riemann conditions (4.14), (4.15), Θ is an antiderivative of θ, and $A_1^\pm$, $A_2^\pm$, $B_1^\pm$and $B_2^\pm$ are arbitrary constant complex quaternions which must be chosen in such a way that the scalar parts of the expressions on the right-hand sides of (6.64) and (6.65) be zero.

Let us consider the case $A_2^\pm = B_2^\pm = 0$. Then omitting the subindex "1" we obtain $\overrightarrow{\Phi}$ and $\overrightarrow{\Psi}$ in the form

$$\overrightarrow{\Phi}(x) = e^{i\Theta(x_1)}H^+(x_2,x_3)(1-ii_1)A^+ + e^{-i\Theta(x_1)}G^+(x_2,x_3)(1+ii_1)B^+, \tag{6.66}$$

$$\overrightarrow{\Psi}(x) = e^{-i\Theta(x_1)}H^-(x_2,x_3)(1-ii_1)A^- + e^{i\Theta(x_1)}G^-(x_2,x_3)(1+ii_1)B^-, \tag{6.67}$$

where $H^\pm = h_0^\pm + h_1^\pm i_1$ and $G^\pm = g_0^\pm + g_1^\pm i_1$, and the functions $h_0^\pm$, $h_1^\pm$ and $g_0^\pm$, $g_1^\pm$ satisfy (4.14), (4.15). Note that $H^\pm \cdot (1-ii_1) = h^\pm \cdot (1-ii_1)$, where $h^\pm = h_0^\pm + ih_1^\pm$ and $G^\pm \cdot (1+ii_1) = g^\pm \cdot (1+ii_1)$, where $g^\pm = g_0^\pm - ig_1^\pm$.

The scalar parts of the expressions on the right-hand sides are zero iff

$$\mathrm{Sc}((1-ii_1)A^\pm) = \mathrm{Sc}((1+ii_1)B^\pm) = 0$$

which is equivalent to the following conditions

$$A_0^\pm = -iA_1^\pm \qquad \text{and} \qquad B_0^\pm = iB_1^\pm,$$

where $A^\pm = \sum_{k=0}^3 i_k A_k^\pm$ and $B^\pm = \sum_{k=0}^3 i_k B_k^\pm$.

Under these conditions we obtain

$$(1 - ii_1)A^{\pm} = a^{\pm}(i_2 - ii_3),$$

where $a^{\pm} = A_2^{\pm} + iA_3^{\pm}$ and

$$(1 + ii_1)B^{\pm} = b^{\pm}(i_2 + ii_3),$$

where $b^{\pm} = B_2^{\pm} - iB_3^{\pm}$. Thus, the functions (6.66) and (6.67) take the form

(6.68) $\vec{\Phi}(x) = a^+ e^{i\Theta(x_1)} h^+(x_2, x_3)(i_2 - ii_3) + b^+ e^{-i\Theta(x_1)} g^+(x_2, x_3)(i_2 + ii_3),$

(6.69) $\vec{\Psi}(x) = a^- e^{-i\Theta(x_1)} h^-(x_2, x_3)(i_2 - ii_3) + b^- e^{i\Theta(x_1)} g^-(x_2, x_3)(i_2 + ii_3).$

We proved the following

Proposition 6.11. *Let Θ be an antiderivative of θ; functions $h^+ = h_0^+ + ih_1^+$ and $h^- = h_0^- + ih_1^-$ satisfy (4.14), (4.15); functions $g^+ = g_0^+ - ig_1^+$ and $g^- = g_0^- - ig_1^-$ satisfy (5.40), (5.41); $a^{\pm}$ and $b^{\pm}$ be arbitrary complex numbers. Then functions (6.68) and (6.69) are solutions of (6.62) and (6.63) respectively.*

Now, using this result and the relations (6.57) and (6.58), it is easy to obtain representations for the vectors of the electromagnetic field:

$$\begin{aligned}\vec{E}(x) &= \frac{i\omega\varepsilon(x_1)}{2}(\vec{\Phi}(x) - \vec{\Psi}(x)) \\ &= \frac{i\omega\varepsilon(x_1)}{4}((a^+ e^{i\Theta(x_1)} h^+(x_2, x_3) - a^- e^{-i\Theta(x_1)} h^-(x_2, x_3))(i_2 - ii_3) \\ &+ (b^+ e^{-i\Theta(x_1)} g^+(x_2, x_3) - b^- e^{i\Theta(x_1)} g^-(x_2, x_3))(i_2 + ii_3)),\end{aligned}$$

$$\begin{aligned}\vec{H}(x) &= \frac{\omega\varepsilon^{3/2}(x)}{2\mu^{1/2}}(\vec{\Phi}(x) + \vec{\Psi}(x)) \\ &= \frac{\omega\varepsilon^{3/2}(x)}{4\mu^{1/2}}((a^+ e^{i\Theta(x_1)} h^+(x_2, x_3) + a^- e^{-i\Theta(x_1)} h^-(x_2, x_3))(i_2 - ii_3) \\ &+ (b^+ e^{-i\Theta(x_1)} g^+(x_2, x_3) + b^- e^{i\Theta(x_1)} g^-(x_2, x_3))(i_2 + ii_3)).\end{aligned}$$

The obtained solutions represent a transverse electromagnetic wave and can be applied, for example to the study of the atmospheric propagation of ultrashort radio waves.

Remark 6.12. The results of 4.1.2 also allow us to obtain representations for the vectors of an electromagnetic field in the case when the permittivity is analytic with respect to a complex variable $z = x_k \pm ix_l$ and depends arbitrarily on x_p, where k, l and p are pairwise different and take their values from the set $\{1, 2, 3\}$.

Acknowledgements

This work was partially supported by CONACYT Project 32424-E, Mexico.

References

[1] V. M. Babich and V. S. Buldyrev, *Short-Wavelength Diffraction Theory. Asymptotic Methods.* Springer (1991).

[2] V. G. Bagrov and D. M. Gitman, *Exact solutions of relativistic wave equations.* Kluwer Acad. Publ. (1990).

[3] W. E. Baylis and G. Jones, *Relativistic dynamics of charges in external fields: the Pauli algebra approach.* J. Phys. A, 22, (1) (1989) 17–29.

[4] A. V. Berezin, Yu. A. Kurochkin and E. A. Tolkachev, *Quaternions in relativistic physics.* Nauka y Tekhnika, Minsk, Bielorussia (1989) (in Russian).

[5] A. V. Berezin, E. A. Tolkachev and F. I. Fedorov, *Lorentz transformations and equations for spinor quaternions.* Doklady Akademii Nauk BSSR, 24, (4) (1980) 308–310 (in Russian).

[6] S. Bernstein and K. Gürlebeck, *On a higher dimensional Miura transform.* Complex Variables, 38 (1999) 307–319.

[7] A. V. Bitsadze, *Boundary value problems for second-order elliptic equations.* North-Holland, Amsterdam and Interscience, N.Y. (1968).

[8] N.N. Bogoliubov and D. V. Shirkov, *Quantum fields.* Fizmatlit, Moscow (1993) (in Russian).

[9] F. Brackx, R. Delanghe and F. Sommen, *Clifford analysis.* Pitman Res. Notes in Math., v. 76, London (1982).

[10] G. Casanova, *L'algebre vectorielle.* Presses Universitaires de France (1976).

[11] G. Dixon *Division algebras: octonions, quaternions, complex numbers, and the algebraic design of physics.* Kluwer Acad Publ., Dordrecht (1994).

[12] A. D. Dzhuraev, *Singular integral equation method.* Nauka, Moscow (1987) (in Russian); Engl. transl. Longman Sci. Tech., Harlow and Wiley, N.Y. (1992).

[13] K. Gürlebeck, *Hypercomplex factorization of the Helmholtz equation.* Zeitschrift für Analysis und ihre Anwendungen, 5, (2) (1986) 125–131.

[14] K. Gürlebeck and W. Sprößig, *Q*uaternionic analysis and elliptic boundary value problems. Akademie-Verlag, Berlin (1989).

[15] K. Gürlebeck and W. Sprössig, *Quaternionic and Clifford Calculus for Physicists and Engineers.* John Wiley & Sons (1997).

[16] F. Gürsey and H. C. Tze, *Complex and quaternionic analyticity in chiral and gauge theories. Part 1.* Ann. Phys., 128 (1980) 29–130.

[17] F. Gürsey and H. C. Tze, *On the role of division, Jordan and related algebras in particle physics.* World Scientific, Singapore (1996).

[18] D. Hestenes, *Real Dirac theory.* Advances in Applied Clifford Algebras, 7(S) (1997) 97–144.

[19] D. D. Ivanenko and K. V. Nikolski, Z. Phys., 63 (1930) 129–137.

[20] I. L. Kantor and A. S. Solodovnikov, *Hypercomplex numbers.* Springer (1989).

[21] V. G. Kravchenko and V. V. Kravchenko, *On some nonlinear equations generated by Fueter type operators.* Zeitschrift für Analysis und ihre Anwendungen, 13 (4) (1994) 599–602.

[22] V. V. Kravchenko, *On the relation between holomorphic biquaternionic functions and time-harmonic electromagnetic fields.* Deposited in UkrINTEI, No. 2073-Uk-92, 1992 (in Russian).

[23] V. V. Kravchenko, *On a biquaternionic bag model.* Zeitschrift für Analysis und ihre Anwendungen, 14 (1) (1995) 3–14.

[24] V. V. Kravchenko, *On the Dirac operator with an electromagnetic potential.* Zeitschrift für Analysis und ihre Anwendungen, 17 (3) (1998) 549–556.

[25] V. V. Kravchenko, H. R. Malonek and G. Santana, *Biquaternionic integral representations for massive Dirac spinors in a magnetic field and generalized biquaternionic differentiability.* Mathematical Methods in the Applied Sciences, 19 (1996) 1415–1431.

[26] V. V. Kravchenko and M. Ramírez, *New exact solutions of the massive Dirac equation with electric or scalar potential.* To appear in Mathematical Methods in the Applied Sciences (2000).

[27] V. V. Kravchenko and M. V. Shapiro, *Integral representations for spatial models of mathematical physics.* Pitman Res. Notes in Math. Series, v. 351, Addison-Wesley Longman Ltd., Harlow (1996).

[28] N. Ya. Krupnik, *Banach algebras with symbol and singular integral operators.* Shtiinca, Kishinev (1984) (in Russian).

[29] G. Moisil, *Sur les quaternions monogenes.* Bull. Sci. Math. Paris, 55 (2) (1931) 169–194.

[30] G. Moisil and N. Theodoresco, *Functions holomorphes dans l'espace.* Mathematica (Cluj), 5 (1931) 142–159.

[31] E. I. Obolashvili, *Three-dimensional generalized holomorphic vectors.* Differential Equations, 11 (1) (1975) 82–87.

[32] M. Reed and B. Simon, *Methods of modern mathematical physics IV, Analysis of operators.* Academic Press (1978).

[33] I. R. Shafarevich, *Basic notions of algebra.* Sovremennye problemi matematiki. Fundamentalnye napravleniya, v. 11, VINITI, Moscow (1985) (in Russian); Engl. transl. Encyclopedia of Mathematical Sciences, v. 11, Algebra I, Springer-Verlag, Berlin (1990).

[34] A. Sommerfeld, *Atom structure and spectra.* v. 2, Moscow (1956) (in Russian).

[35] V. Soucek, *Complex–quaternionic analysis applied to spin 1/2 massless fields.* Complex Variables, Theory and Applications, 1 (4) (1983) 327–346.

[36] W. Sprößig, *On the treatment of non-linear boundary value problems of a disturbed Dirac equation by hypercomplex methods.* Complex Variables, 23 (1993) 123–130.

[37] B. Thaller, *The Dirac equation.* Springer–Verlag (1992).

[38] J. Vrbik, *Dirac equation and Clifford algebra.* J. Math. Phys., 35 (1994) 2309–2314.

Departamento de Telecomunicaciones
Escuela Superior de Ingeniería Mecánica y Eléctrica
Instituto Politécnico Nacional
C.P. 07738, D.F., MEXICO

1991 Mathematics Subject Classification: Primary 30G35; Secondary 78A25, 81Q05

Submitted: 3.5.2000

Operator Theory:
Advances and Applications, Vol. 121

Revisiting a quadrature method for CSIE with a weakly singular perturbation kernel

CONCETTA LAURITA, GIUSEPPE MASTROIANNI

In Memory of Siegfried Prößdorf

A quadrature method for Cauchy singular integral equations with constant coefficients is revisited. Estimates in mean weighted norm for the error are given. To construct a polynomial approximation of the solution a linear system has to be solved. It is proved that the system is well conditioned.

1. Introduction

We consider the Cauchy singular integral equation with constant coefficients

$$
\begin{aligned}
aw^{\alpha,-\alpha}(x)v(x) \quad &+ \quad \frac{b}{\pi}\int_{-1}^{1}\frac{w^{\alpha,-\alpha}(t)v(t)}{t-x}dt \\
&+ \quad \int_{-1}^{1}k(x,t)w^{\alpha,-\alpha}(t)v(t)dt = f(x).
\end{aligned} \tag{1.1}
$$

Here a and b are given real constants such that $\sqrt{a^2+b^2}=1$, $b\neq 0$, $w^{\alpha,-\alpha}(x) = (1-x)^{\alpha}(1+x)^{-\alpha}$ with $0<\alpha<1$ defined as follows

$$
a+ib=e^{i\pi\alpha_0}, \quad \alpha := M-\alpha_0, \quad -\alpha = N+\alpha_0, \tag{1.2}
$$

with M and N integers such that $M+N=0$. Moreover f and k are known functions on $[-1,1]$ and $[-1,1]^2$, respectively, with k smooth or weakly singular kernel function of the form

$$
k(x,t) = \frac{h(x,t)-h(x,x)}{t-x}. \tag{1.3}
$$

Finally, v denotes the unknown solution.
By defining the following operators

$$
Dv(x) := aw^{\alpha,-\alpha}(x)v(x) + \frac{b}{\pi}\int_{-1}^{1}\frac{w^{\alpha,-\alpha}(t)v(t)}{t-x}dt, \tag{1.4}
$$

$$
Kv(x) := \int_{-1}^{1}k(x,t)w^{\alpha,-\alpha}(t)v(t)dt, \tag{1.5}
$$

equation (1.1) can be rewritten in operator form as

$$(D+K)v = f. \tag{1.6}$$

In [18] the authors consider the more generale situation in which the Jacobi weight appearing in (1.1) is

$$w^{\alpha,\beta}(x) = (1-x)^\alpha(1+x)^\beta \quad \text{with } \alpha+\beta=0 \quad \text{or} \quad \alpha+\beta=1.$$

A quadrature method for solving the Cauchy singular integral equation is described and its mean weighted convergence is studied. The procedure consists in using Gauss-type quadrature rules to approximate $Dv(x)$ and $Kv(x)$ and collocate at suitable points. In particular by taking the zeros t_j of the orthogonal polynomials $\{p_m\}$ with respect to the weight $w^{\alpha,-\alpha}$ as quadrature nodes and the zeros x_i of the orthogonal polynomials $\{q_m\}$ with respect to $w^{-\alpha,\alpha}$ as collocation knots, one have to solve the following system

$$\sum_{j=1}^{m} \lambda_{mj}(w)\left[\frac{b}{\pi(t_j-x_i)} + k(x_i,t_j)\right]\xi_j = f(x_i), \quad i=1,2,\dots,m. \tag{1.7}$$

This system makes sense since the distribution of the chosen zeros of p_m and q_m is of arccos type. By solving system (1.7) one computes the coefficients of the Lagrange interpolation polynomial

$$v_m(x) = \sum_{j=1}^{m} \frac{p_m(x)}{(x-t_j)p_m'(t_j)}\xi_j,$$

which approximates the solution v of (1.1). A strategy of this type is employed in [2], [13], [14], [15] (see, also, the references given by the authors).
In [18] the authors study the convergence of v_m to the solution v in the weighted space L^2_w and obtain the following results.
"If $k \in C^{r+\lambda}\left([-1,1]^2\right)$ and $f \in C^{r+\lambda}\left([-1,1]\right)$, the quadrature method converges with the rate $O(m^{-r-\lambda})$; if the kernel k is of the form (1.3) with $h \in C^{r+\lambda}\left([-1,1]^2\right)$ the rate of convergence is $O(m^{-r-\lambda}\log m)$."
In this paper we revisit [18] in several aspects and give some improvements in the case $\beta = -\alpha$. The error estimates given in [18] when the kernel k is a smooth function are here obtained making weaker hypotheses on k and f. In the case in which the kernel is weakly singular, under weaker assumptions, we improve the rate of convergence. In fact, in error estimate (3.10) the factor $\log m$ doesn't appear.
Moreover the condition number of system (1.7) is not considered in [18]. Making some little changes in (1.7), we introduce in (3.4) a linear system having the following property: the condition number of the matrix of the system is independent of its dimension m. More precisely we prove that the lim sup of the condition numbers is bounded by the condition number of the original operator. Therefore the system will well conditioned if integral equation (1.1) is well conditioned.

2. Some notations and definitions

Let w be a Jacobi weight function

$$w(x) \equiv w^{\gamma,\delta}(x) = (1-x)^{\gamma}(1+x)^{\delta}, \quad x \in (-1,1), \quad \gamma,\delta > -1$$

and for $X \subseteq [-1,1]$, let $L^2_w(X)$ denote the set of all functions v such that

$$\|v\|_{L^2_w(X)} := \left(\int_X |v(x)|^2 w(x)dx \right)^{\frac{1}{2}} < \infty.$$

If $X = [-1,1]$ we use the following notations

$$L^2_w \equiv L^2_w([-1,1]), \quad \|v\|_{L^2_w([-1,1])} \equiv \|v\|_{w,2}.$$

We want to define, now, a class of subspaces of L^2_w. At first let us introduce the i-th weighted φ-modulus of smoothness of a function $v \in L^2_w$ [5]

$$\Omega^i_{\varphi}(v,t)_{w,2} := \sup_{0<h\leq t} \|\Delta^i_{h\varphi} v\|_{L^2_w(I_{hi})}, \tag{2.1}$$

where $\varphi(x) = \sqrt{1-x^2}$, $\Delta^i_{h\varphi}v(x) = \sum_{j=0}^{i}(-1)^j \begin{pmatrix} i \\ j \end{pmatrix} v\left(x + \left(\frac{i}{2} - j\right) h\varphi(x)\right)$, $i \in \mathbb{N}$, $I_{hi} = [-1+4h^2i^2, 1-4h^2i^2]$ and the error of best approximation by algebraic polynomials of v in L^2_w

$$E_m(v)_{w,2} := \inf_{P\in\mathbb{P}_m} \|v-P\|_{w,2}$$

where $\mathbb{P}_m$ denotes the set of all algebraic polynomials of degree at most m. The following facts, which we shall use in the paper, are well known

$$E_m(v)_{w,2} \leq C \int_0^{\frac{1}{m}} \frac{\Omega^i_{\varphi}(v,t)_{w,2}}{t} dt, \tag{2.2}$$

$$\Omega^i_{\varphi}(v,t)_{w,2} \leq Ct^i \sum_{0\leq j\leq \frac{1}{t}} (1+j)^{i-1} E_j(v)_{w,2} \tag{2.3}$$

where C is a positive constant independent of v, m and t. From (2.2) and (2.3) the following equivalence follows [6]

$$\|v\|_{w,2} + \sup_{t>0} \frac{\Omega^i_{\varphi}(v,t)_{w,2}}{t^r} \sim \|v\|_{w,2} + \sup_{m>0} (m+1)^r E_m(v)_{w,2}, \tag{2.4}$$

where r is a positive real number and the symbol $A \sim B$ means that there exists a constant $C > 0$ such that $C^{-1}A \leq B \leq CA$.

For a real number $r > 0$ and an integer $i > r$, by means of the φ-modulus given by (2.1), we define the following norm

$$\|v\|_{Z_r(w)} := \|v\|_{w,2} + \sup_{t>0} \frac{\Omega_\varphi^i(v,t)_{w,2}}{t^r}, \tag{2.5}$$

and the weighted Hölder-Zygmund-type space $Z_r(w)$ as the set

$$Z_r(w) = \left\{v \in L_w^2 : \|v\|_{Z_r(w)} < \infty\right\}, \tag{2.6}$$

endowed with the previous norm. For $r = 0$ we set $Z_0(w) \equiv L_w^2$.
By equivalence (2.4) the norm on $Z_r(w)$ can be also characterized by means of the error of best polynomial approximation.
In the following we shall denote by w the Jacobi weight function $w^{\alpha,-\alpha}$ appearing in the integral equation (1.1) and by $\{p_m\}$ and $\{q_m\}$ the systems of polynomials with positive leading coefficients orthonormal with respect to the weight w and w^{-1}, respectively. Furthermore, let $t_1, t_2, \ldots, t_m$ be the zeros of the polynomial p_m and $x_1, x_2, \ldots, x_m$ the zeros of q_m. Denote by $l_{mj}(w)$, $j = 1, 2, \ldots, m$ the fundamental polynomials with respect to the nodes $t_1, t_2, \ldots, t_m$ and by $\lambda_{mj}(w)$, $\lambda_{mj}(w^{-1})$, $j = 1, 2, \ldots, m$ the Christoffel numbers related to the weight function w and w^{-1}, respectively.
The dominant operator D satisfies the well known property (see [22], [27])

$$Dp_m(x) = (-1)^M q_m(x), \quad m = 0, 1, \ldots, \tag{2.7}$$

where M is the integer appearing in (1.2).

3. Main results

We shall describe some numerical methods to approximate the solution of integral equation (1.1), revisiting a quadrature one proposed in [18]. The choice of a suitable procedure depends on the smoothness properties of the kernel $k(x,t)$ of the operator K.
We shall consider two cases. In the first one let us assume that the kernel function $k(x,t)$ satisfies the conditions

$$\sup_{|x|<1} \|k(x,\cdot)\|_{Z_r(w)} < \infty \text{ and } \sup_{|t|<1} \|k(\cdot,t)\|_{Z_r(w^{-1})} < \infty, \quad \text{for some } r > \frac{1}{2}. \tag{3.1}$$

For instance, if the kernel is the function $k(x,t) = |x-t|^\mu$, for some $\mu > -\frac{1}{2}$, one has $r = \mu + \frac{1}{2}$. If it is given by $k(x,t) = \log|x-t|$ then $r = \frac{1}{2}$ (see [5]).
We have

Theorem 3.1. *Under hypotheses (3.1), assume that* $\ker(D+K) = \{0\}$ *in* L_w^2. *Then* $D + K : Z_s(w) \longrightarrow Z_s(w^{-1})$ *is an invertible linear bounded operator for all* $0 \le s < r$.

The numerical method consists in approximating the unknown solution v of (1.1) by means of the interpolating polynomial of degree $m-1$

$$v_m = \sum_{j=1}^{m} \xi_j l_{mj}(w), \tag{3.2}$$

whose coefficients $\xi_j,\ j=1,2,\dots,m$ are given by

$$\xi_j = \lambda_{mj}(w)^{-\frac{1}{2}}\eta_j, \quad j=1,2,\dots,m, \tag{3.3}$$

with $\eta_j,\ j=1,2,\dots,m$ equal to the solutions of the following linear system

$$\sqrt{\lambda_{mi}(w^{-1})}\sum_{j=1}^{m}\sqrt{\lambda_{mj}(w)}\left[\frac{b}{\pi(t_j-x_i)}+k(x_i,t_j)\right]\eta_j = \sqrt{\lambda_{mi}(w^{-1})}f(x_i), \tag{3.4}$$

with $i=1,2,\dots,m$.
In [18] the authors suggest to solve the system (1.7) to compute the coefficients ξ_j appearing in (3.2). Anyway it's not sure that system (1.7) in the unknowns $\xi_j,\ j=1,2,\dots,m$, is well conditioned. We shall prove that system (3.4), in the unknowns $\eta_j,\ j=1,2,\dots,m$, equivalent to (1.7), is a well conditioned one.
Let us recall the definition of the condition number of an invertible operator A.

$$\operatorname{cond}(A) = \|A\|\cdot\|A^{-1}\|.$$

In the following statement we will consider D+K as a map acting between the spaces L^2_w and $L^2_{w^{-1}}$ and, for the sake of simplicity, we shall omit the subscripts in the operator norms. Moreover we shall assume the spectral norm, induced by the euclidean norm on $\mathbb{R}^n$.
The main result given in this paper is the following one.

Theorem 3.2. *Assume that k satisfies hypotheses (3.1) and $f \in Z_r(w^{-1})$. If the problem (1.6) has a unique solution $v \in L^2_w$, then the system of equations (3.4) is uniquely solvable for all sufficiently large m, the solution v belongs to $Z_r(w)$ and*

$$\|v-v_m\|_{w,2} \le \frac{C}{m^r}\|v\|_{Z_r(w)}, \tag{3.5}$$

where the polynomial v_m, given by (3.2), is the Lagrange interpolation polynomial corresponding to the solution of (3.4) and C is a constant independent of m, v and f. Moreover, if A_m denotes the matrix of the coefficients of (3.4), one has

$$\limsup_m \operatorname{cond}(A_m) \le \operatorname{cond}(D+K). \tag{3.6}$$

Comparing estimate (3.5) with the error estimate of Theorem 2.1 in [18], we can note that, under weaker hypotheses on the kernel k ($k \in Z_r(w),\ r > \frac{1}{2}$ instead of

$k \in C^{r+\lambda}([-1,1]^2))$ and on the right-hand side function f, the error converges to zero with the same rate. Moreover, the problem which we solve, as (3.6) shows, is well conditioned.

Let us make a further remark. The described quadrature-type method, is equivalent to the discrete collocation method described in [16]. This numerical procedure consists in solving a linear system whose unknowns are the coefficients a_j, $j = 0, 1, \ldots, m-1$ of v_m in the expansion $v_m = \sum_{j=0}^{m-1} a_j p_j$. The system is well conditioned if equation (1.1) is well conditioned. The equivalence between this system and (3.4) can be easily proved taking into account the following relation

$$l_{mj}(w) = \lambda_{mj}(w) \sum_{k=0}^{m-1} p_k(t_j) p_k$$

and the consequent one

$$a_j = \sum_{k=1}^{m} \lambda_{mk}(w) p_j(t_k) \xi_k.$$

Observe that, in both systems, a Gauss-type quadrature rule is used to approximate $Kv_m(x_i)$. This procedure is not suitable for the case in which the kernel $k(x,t)$ of K is a weakly singular function of the form

$$k(x,t) = |x-t|^\mu, \quad -1 < \mu < 0 \quad \text{or} \quad k(x,t) = \log|x-t|. \tag{3.7}$$

In this situation, to construct a polynomial approximation of the solution v of (1.1), we use a numerical procedure given in [17].

At first let us state the following result concerning the solvability of (1.6) (see [17]).

Theorem 3.3. *Under hypotheses (3.7), assume that* $\ker(D+K) = \{0\}$ *in* L^2_w. *Then* $D + K : Z_s(w) \longrightarrow Z_s(w^{-1})$ *is an invertible linear bounded operator for all* $0 \le s < 1 + \mu$ *if* $k(x,t) = |x-t|^\mu$ *and for all* $s \ge 0$ *if* $k(x,t) = \log|x-t|$.

The numerical method consists in constructing a polynomial approximation of the solution v of (1.6) of the kind

$$u_m = \sum_{j=0}^{m-1} a_j p_j, \tag{3.8}$$

where a_j, $j = 0, 1, \ldots, m-1$ are unknown constants.

In order to evaluate the coefficients $a_o, a_1, \ldots, a_{m-1}$ we solve the following linear system

$$\sqrt{\lambda_{mi}(w^{-1})} \sum_{j=0}^{m-1} a_j \left[(-1)^{|M|} q_j(x_i) + M_j(w, x_i)\right] = \sqrt{\lambda_{mi}(w^{-1})} f(x_i), \tag{3.9}$$

with $i = 1, 2, \ldots, m$.
In (3.9) M is the integer defining α in (1.2) and $M_j(w, x_i) = Kp_j(x_i)$. Let us observe that the quantities $M_j(w, x_i)$ are explicitly computable by recurrence relations ([21]), then the elements of the matrix of the system can be evaluated with good precision. The following results hold ([17]).

Theorem 3.4. *Assume that k satisfies hypotheses (3.7) with $\mu > -\frac{1}{2}$ and $f \in Z_s(w^{-1})$, $s > \frac{1}{2}$. If the problem (1.6) has a unique solution $v \in L^2_w$, then the system of equations (3.9) is uniquely solvable for all sufficiently large m, the solution v belongs to $Z_r(w)$, with $r = \min(1+\mu, s)$ if $k(x,t) = |x-t|^\mu$ and $r = s$ if $k(x,t) = \log|x-t|$, and*

$$\|v - u_m\|_{w,2} \leq \frac{C}{m^r}\|v\|_{Z_r(w)}, \tag{3.10}$$

where u_m is the polynomial given by (3.8) corresponding to the solution of (3.9) and C is a constant independent of m, v and f. Moreover, if B_m denotes the matrix of the coefficients of (3.9), one has

$$\limsup_m \operatorname{cond}(B_m) \leq \operatorname{cond}(D + K). \tag{3.11}$$

Let us observe that we can represent the kernel $k(x,t) = |x-t|^\mu$, $-1 < \mu < 0$ as

$$k(x,t) = \frac{h(x,t) - h(x,x)}{t - x}$$

with $h(x,t) = (t-x)|t-x|^\mu$. By applying Theorem 2.2 in [18] we obtain that the error is $O\left(\frac{\log m}{m^{1+\mu}}\right)$. According to (3.10) we have that the error of the prescribed method converges with the rate $O\left(\frac{1}{m^{1+\mu}}\right)$ as $m \to \infty$.
A generalization of the described methods can be obtained by replacing in (3.4) and (3.9) the Christoffel numbers $\lambda_{mi}(w^{-1})$ by $\lambda_m(w^{-1}, y_i)$, where

$$\lambda_m(w^{-1}, x) = \left[\sum_{k=0}^{m-1} q_k^2(x)\right]^{-1}$$

is the m-th Christoffel function related to w^{-1} and y_i, $i = 1, 2, \ldots, m$, are the zeros of the polynomial of degree m orthonormal with respect to a weight function u satisfying the conditions

$$\sqrt{wu\varphi^{\pm 1}} \in L^2, \quad \varphi(x) = \sqrt{1-x^2}. \tag{3.12}$$

The interested reader can consult [17]. Observe that when $u = w^{-1}$ the previous condition is fulfilled and $\lambda_m(w^{-1}, y_i) = \lambda_m(w^{-1}, x_i) = \lambda_{mi}(w^{-1})$.
Let us, also, note that the Chebyshev weight function of the first kind, $\frac{1}{\sqrt{1-x^2}}$, is an other possible choice for u, since it satisfies (3.12). Finally, we want to give a

punctual estimate of the error. The following result is interesting from a numerical point of view.

Theorem 3.5. *Assume the hypotheses of Theorem 3.2 or of Theorem 3.4. If* $w^{\gamma,\delta}(t) = (1-t)^{\gamma}(1+t)^{\delta}$ *with* $\gamma \geq \frac{\alpha}{2} + \frac{1}{4}$ *and* $\delta \geq \frac{-\alpha}{2} + \frac{1}{4}$, *then the following estimate holds*

$$\max_{|t|\leq 1} |v(t) - g_m(t)| w^{\gamma,\delta}(t) \leq \frac{C}{m^{r-\frac{1}{2}}} \|v\|_{Z_r(w)},$$

with $g_m = v_m$ *or* $g_m = u_m$, *respectively.*

4. Proofs of the main results

The proof of Theorem 3.1 is an immediate consequence of the following facts. The dominant operator D, defined by (1.4), is linear, bounded and invertible as a map acting from $Z_r(w)$ into $Z_r(w^{-1})$, for any $r \geq 0$ [17]. Moreover, for the integral operator K, given in (1.5), one has

Proposition 4.1. *Under hypotheses (3.1), the operator* $K : L^2_w \longrightarrow Z_s(w^{-1})$ *is linear and bounded for all* $0 \leq s \leq r$. *Moreover, it is compact for all* $s < r$.

Proof. The proof of the following estimate

$$\|Kv\|_{w^{-1},2} \leq C\|v\|_{w,2}, \tag{4.1}$$

where C is a positive constant independent of v, is trivial. In fact, by applying Schwarz' inequality, one has

$$\begin{aligned} \|Kv\|_{w^{-1},2} &\leq \|v\|_{w,2} \left(\int_{-1}^{1} \int_{-1}^{1} |k(x,t)|^2 w(t) dt w^{-1}(x) dx \right)^{\frac{1}{2}} \\ &\leq C\|v\|_{w,2}. \end{aligned}$$

Moreover, it holds

$$\Omega^i_{\varphi}(Kv,\delta)_{w^{-1},2} \leq C\delta^r \|v\|_{w,2}, \tag{4.2}$$

where the constant C is, also in this case, independent of v. Let us prove (4.2). At first recall the definition of the φ-modulus:

$$\Omega^i_{\varphi}(Kv,\delta)_{w^{-1},2} = \sup_{0<h\leq t} \|\Delta^i_{h\varphi}(Kv)\|_{L^2_{w^{-1}}(I_{hi})}.$$

One has

$$|\Delta^i_{h\varphi}(Kv)(x)|^2 = \left| \Delta^i_{h\varphi} \left(\int_{-1}^{1} k(\cdot,t) v(t) w(t) dt \right)(x) \right|^2$$

$$= \left| \int_{-1}^{1} \Delta_{h\varphi}^{i}\left(k(\cdot,t)\right)(x)v(t)w(t)dt \right|^2$$
$$\leq \left[\int_{-1}^{1} \left|\Delta_{h\varphi}^{i}\left(k(\cdot,t)\right)(x)\right|^2 w(t)dt \right] \|v\|_{w,2}^2$$

from which it follows

$$\|\Delta_{h\varphi}^{i}(Kv)\|_{L^2_{w^{-1}}(I_{hi})} \leq$$
$$\leq \left(\int_{-1+2h^2i^2}^{1-2h^2i^2} \left[\int_{-1}^{1} \left|\Delta_{h\varphi}^{i}\left(k(\cdot,t)\right)(x)\right|^2 w(t)dt \right] w^{-1}(x)dx \right)^{\frac{1}{2}} \|v\|_{w,2}$$
$$= \left(\int_{-1}^{1} \int_{-1+2h^2i^2}^{1-2h^2i^2} \left|\Delta_{h\varphi}^{i}\left(k(\cdot,t)\right)(x)\right|^2 w^{-1}(x)dxw(t)dt \right)^{\frac{1}{2}} \|v\|_{w,2}$$
$$= \left(\int_{-1}^{1} \left\|\Delta_{h\varphi}^{i}\left(k(\cdot,t)\right)\right\|_{L^2_{w^{-1}}(I_{hi})} w(t)dt \right)^{\frac{1}{2}} \|v\|_{w,2}.$$

Taking the supremum for $0 < h \leq \delta$ one has

$$\begin{aligned}
\Omega_{\varphi}^{i}(Kv,\delta)_{w^{-1},2} &\leq \sup_{|t|<1} \Omega_{\varphi}^{i}(k(\cdot,t),\delta)_{w^{-1},2} \left(\int_{-1}^{1} w(t)dt \right)^{\frac{1}{2}} \|v\|_{w,2} \\
&\leq \delta^r \sup_{|t|<1} \sup_{\rho>0} \frac{\Omega_{\varphi}^{i}(k(\cdot,t),\rho)_{w^{-1},2}}{\rho^r} \left(\int_{-1}^{1} w(t)dt \right)^{\frac{1}{2}} \|v\|_{w,2} \\
&\leq C\delta^r \|v\|_{w,2},
\end{aligned}$$

according to relations (3.1).

Then estimate (4.2) is proved and the boundedness of $K : L^2_w \longrightarrow Z_s(w^{-1})$ holds for all $0 \leq s \leq r$.

We have to prove, now, its compactness. At first let us consider $s = 0$, i.e. $K : L^2_w \longrightarrow L^2_{w^{-1}}$. To this end we show that

$$\lim_{m} \sup_{v \in S} E_m(Kv)_{w^{-1},2} = 0, \tag{4.3}$$

where we have set

$$S = \left\{ v \in L^2_w : \|v\|_{w,2} \leq 1 \right\}.$$

Condition (4.3) assures that K is a compact operator since it sends bounded sets into compact ones (see, for instance, [28]).

So, it remains to prove (4.3). Observe that, by (2.2) and (4.2), we can write

$$\begin{aligned}
E_m(Kv)_{w^{-1},2} &\leq C \int_0^{\frac{1}{m}} \frac{\Omega_{\varphi}^{i}(Kv,y)_{w^{-1},2}}{y} dy \\
&\leq C \left(\int_0^{\frac{1}{m}} y^{r-1} dy \right) \|v\|_{w,2} \leq \frac{C}{m^r} \|v\|_{w,2}.
\end{aligned}$$

For a function $v \in S$ we have

$$E_m(Kv)_{w^{-1},2} \leq \frac{C}{m^r} \longrightarrow 0, \quad m \longrightarrow \infty.$$

The compactness of $K : L^2_w \longrightarrow Z_s(w^{-1})$, for $s < r$ immediately follows if one takes into account the estimate (3.15) in [20] for the case $p = 2$ and $q = \infty$. □

To prove Theorem 3.2 we need some preliminary arguments.
Denote by $L^{(1)}_m$ the Lagrange interpolation operator related to the zeros t_j, $j = 1, 2, \ldots, m$ of the orthonormal polynomial p_m. At first let us define the following operator

$$K_m v(x) = \int_{-1}^{1} L^{(1)}_{m,t}\left(k(x,t)\right) v(t) w(t) dt. \tag{4.4}$$

The subscript t of $L^{(1)}_{m,t}$ means that the interpolation is done with respect to the variable t.
The operator K_m satisfies the following properties.

Proposition 4.2. *Under hypotheses (3.1), the operator $K_m : L^2_w \longrightarrow Z_s(w^{-1})$ is linear and bounded for all $0 \leq s \leq r$.*

Proof. At first we want to prove the estimate

$$\|K_m v\|_{w^{-1},2} \leq C \|v\|_{w,2}, \tag{4.5}$$

with C independent of m and v. One has

$$\begin{aligned} \|K_m v\|_{w^{-1},2} &= \left(\int_{-1}^{1} \left| \int_{-1}^{1} L^{(1)}_{m,t}\left(k(x,t)\right) v(t) w(t) dt \right|^2 w^{-1}(x) dx \right)^{\frac{1}{2}} \\ &\leq \left(\int_{-1}^{1} \int_{-1}^{1} \left| L^{(1)}_{m,t}\left(k(x,t)\right) \right|^2 w(t) dt w^{-1}(x) dx \right)^{\frac{1}{2}} \|v\|_{w,2} \\ &\leq \left(\int_{-1}^{1} \left\| L^{(1)}_{m,t}\left(k(x,\cdot)\right) \right\|_{Z_r(w)} w^{-1}(x) dx \right)^{\frac{1}{2}} \|v\|_{w,2}. \end{aligned}$$

Since the Lagrange interpolation operator is bounded on spaces of type (2.6) (see [20]) we can write

$$\begin{aligned} \|K_m v\|_{w^{-1},2} &\leq C \left(\int_{-1}^{1} \|k(x,\cdot)\|_{Z_r(w)} w^{-1}(x) dx \right)^{\frac{1}{2}} \|v\|_{w,2} \\ &\leq C \left(\int_{-1}^{1} w^{-1}(x) dx \right)^{\frac{1}{2}} \|v\|_{w,2} \\ &\leq C \|v\|_{w,2}. \end{aligned}$$

Let us show, now, that also the inequality

$$(4.6) \qquad \Omega_\varphi^i(K_m v, \delta)_{w^{-1},2} \le C\delta^r \|v\|_{w,2},$$

holds for a suitable positive constant C independent of m, δ and v.
To this end let us estimate the norm $\|\Delta_{h\varphi}^i(K_m v)\|_{L^2_{w^{-1}}(I_{hi})}$:

$$\begin{aligned}
&\|\Delta_{h\varphi}^i(K_m v)\|_{L^2_{w^{-1}}(I_{hi})} = \\
&= \left[\int_{-1+2h^2i^2}^{1-2h^2i^2} \left|\Delta_{h\varphi}^i\left(\int_{-1}^1 L_{m,t}^{(1)}(k(\cdot,t))\,v(t)w(t)dt\right)(x)\right|^2 w^{-1}(x)dx\right]^{\frac{1}{2}} \\
&= \left[\int_{-1+2h^2i^2}^{1-2h^2i^2} \left|\Delta_{h\varphi}^i\left(\int_{-1}^1 \sum_{j=1}^m k(\cdot,t_j)l_{mj}(w,t)v(t)w(t)dt\right)(x)\right|^2 w^{-1}(x)dx\right]^{\frac{1}{2}} \\
&\le \left[\int_{-1+2h^2i^2}^{1-2h^2i^2} \left|\sum_{j=1}^m \int_{-1}^1 \left|\Delta_{h\varphi}^i(k(\cdot,t_j))(x)l_{mj}(w,t)v(t)\right| w(t)dt\right|^2 w^{-1}(x)dx\right]^{\frac{1}{2}} \\
&= \left[\int_{-1+2h^2i^2}^{1-2h^2i^2} \left(\sum_{j=1}^m \left|\Delta_{h\varphi}^i(k(\cdot,t_j))(x)\right|\sqrt{\lambda_{mj}(w)}\right)^2 w^{-1}(x)dx\right]^{\frac{1}{2}} \|v\|_{w,2} \\
&\le \left[\int_{-1+2h^2i^2}^{1-2h^2i^2} \left(\sum_{j=1}^m \left|\Delta_{h\varphi}^i(k(\cdot,t_j))(x)\right|^2\right)\left(\sum_{j=1}^m \lambda_{mj}(w)\right) w^{-1}(x)dx\right]^{\frac{1}{2}} \|v\|_{w,2} \\
&\le \left[\sum_{j=1}^m \|\Delta_{h\varphi}^i(k(\cdot,t_j))\|^2_{L^2_{w^{-1}}(I_{hi})}\right]^{\frac{1}{2}} \left(\int_{-1}^1 w(t)dt\right)^{\frac{1}{2}} \|v\|_{w,2}.
\end{aligned}$$

Taking the supremum for $0 < h \le \delta$ one has

$$\begin{aligned}
\Omega_\varphi^i(K_m v, \delta)_{w^{-1},2} &\le \sup_{|t|<1} \Omega_\varphi^i(k(\cdot,t),\delta)_{w^{-1},2}\sqrt{m}\left(\int_{-1}^1 w(t)dt\right)^{\frac{1}{2}} \|v\|_{w,2} \\
\sqrt{m} &\le \delta^r \sup_{|t|<1}\sup_{\rho>0} \frac{\Omega_\varphi^i(k(\cdot,t),\rho)_{w^{-1},2}}{\rho^r}\sqrt{m}\left(\int_{-1}^1 w(t)dt\right)^{\frac{1}{2}} \|v\|_{w,2} \\
&\le C\delta^r\sqrt{m}\|v\|_{w,2},
\end{aligned}$$

and the proof is complete. □

In the following, we will denote by $L_m^{(2)}$ the Lagrange interpolation operator with respect to the zeros of the orthonormal polynomial q_m: $x_i, i = 1, 2, \ldots, m$, .

Proposition 4.3. *If conditions (3.1) are fulfilled, the estimate*

$$\left\|\left(K - L_m^{(2)} K_m\right) v\right\|_{w^{-1},2} \leq \frac{C}{m^r} \|v\|_{w,2}, \tag{4.7}$$

with the constant C independent of m and v, holds.

Proof. At first observe that

$$\begin{aligned}\left\|\left(K - L_m^{(2)} K_m\right) v\right\|_{w^{-1},2} &\leq \left\|\left(K - L_m^{(2)} K\right) v\right\|_{w^{-1},2} \\ &+ \left\|L_m^{(2)} \left(K - K_m\right) v\right\|_{w^{-1},2},\end{aligned} \tag{4.8}$$

then let us estimate the two norms on the right-hand side of (4.8).
By applying the estimate of the error of Lagrange interpolation given in [20] (see Theorem 3.1) to the function Kv, we get

$$\begin{aligned}\left\|\left(K - L_m^{(2)} K\right) v\right\|_{w^{-1},2} &\leq \frac{C}{m^{\frac{1}{2}}} \int_0^{\frac{1}{m}} \frac{\Omega_\varphi^i (Kv, y)_{w^{-1},2}}{y^{\frac{3}{2}}} dy \\ &\leq \frac{C}{m^{\frac{1}{2}}} \left(\int_0^{\frac{1}{m}} y^{r-\frac{3}{2}} dy\right) \|v\|_{w,2} \\ &\leq \frac{C}{m^r} \|v\|_{w,2}\end{aligned}$$

i.e.

$$\left\|\left(K - L_m^{(2)} K\right) v\right\|_{w^{-1},2} \leq \frac{C}{m^r} \|v\|_{w,2}. \tag{4.9}$$

For the norm $\left\|L_m^{(2)} \left(K - K_m\right) v\right\|_{w^{-1},2}$, we can write

$$\begin{aligned}\left\|L_m^{(2)} \left(K - K_m\right) v\right\|_{w^{-1},2} &= \left(\int_{-1}^{1} \left|L_m^{(2)} \left(K - K_m\right) v(x)\right|^2 w^{-1}(x) dx\right)^{\frac{1}{2}} \\ &= \left(\sum_{i=1}^{m} \lambda_{mi}(w^{-1}) \left|(K - K_m) v(x_i)\right|^2\right)^{\frac{1}{2}}.\end{aligned}$$

Since one has

$$\begin{aligned}|(K - K_m) v(x)|^2 &= \left|\int_{-1}^{1} \left[k(x,t) - L_{m,t}^{(1)} (k(x,t))\right] v(t) w(t) dt\right|^2 \\ &\leq \left(\int_{-1}^{1} \left|k(x,t) - L_{m,t}^{(1)} (k(x,t))\right|^2 w(t) dt\right) \|v\|_{w,2}^2 \\ &= \left\|k(x,\cdot) - L_{m,t}^{(1)} k(x,\cdot)\right\|_{w,2}^2 \|v\|_{w,2}^2\end{aligned}$$

and, from hypotheses (3.1) and estimate (3.2) in [20], it follows

$$\begin{aligned}\left\|k(x,\cdot)-L^{(1)}_{m,t}k(x,\cdot)\right\|_2 &\leq \frac{C}{m^{\frac{1}{2}}}\int_0^{\frac{1}{m}}\frac{\Omega^i_\varphi(k(x,\cdot),y)_{w,2}}{y^{\frac{3}{2}}}dy\\ &\leq \frac{C}{m^{\frac{1}{2}}}\left(\int_0^{\frac{1}{m}} y^{r-\frac{3}{2}}dy\right)\sup_{\delta>0}\frac{\Omega^i_\varphi(k(x,\cdot),\delta)_{w,2}}{\delta^r}\\ &\leq \frac{C}{m^r}\sup_{|x|<1}\sup_{\delta>0}\frac{\Omega^i_\varphi(k(x,\cdot),\delta)_{w,2}}{\delta^r}\leq\frac{C}{m^r}\end{aligned}$$

we deduce that

$$\begin{aligned}\left\|L^{(2)}_m(K-K_m)v\right\|_{w^{-1},2} &\leq \frac{C}{m^r}\left(\sum_{i=1}^m\lambda_{mi}(w^{-1})\right)^{\frac{1}{2}}\|v\|_{w,2}\\ &= \frac{C}{m^r}\left(\int_{-1}^1 w^{-1}(x)dx\right)^{\frac{1}{2}}\|v\|_{w,2},\end{aligned}$$

i.e.

$$\left\|L^{(2)}_m(K-K_m)v\right\|_{w^{-1},2}\leq\frac{C}{m^r}\|v\|_{w,2}. \tag{4.10}$$

By combining (4.8), (4.9), (4.10) we get (4.7). □

Now consider the finite dimensional equation

$$\left(D+L^{(2)}_m K_m\right)v_m = L^{(2)}_m f. \tag{4.11}$$

Equation (4.11) can be obtained by projecting (1.6) on $\mathbb{P}_{m-1}$ by means of the operator $L^{(2)}_m$ and taking into account (2.7).
According to the following result, system (3.4) is equivalent to the previous discrete approximating equation.

Proposition 4.4. *Let $v_m=\sum_{j=1}^m\xi_j l_{mj}(w)\in\mathbb{P}_{m-1}$. Then v_m is the solution of (4.11) if and only if its coefficients ξ_j are given by (3.3) and the vector $(\eta_j)^T_{j=1,\dots,m}$ is the solution of (3.4).*

Proof. Let v_m be a solution of (4.11) represented as in (3.2). Using (2.7) and the linearity of the operator $L^{(2)}_m$, equation (4.11) can be rewritten as follows

$$L^{(2)}_m\left((D+K_m)v_m-f\right)=0$$

which is equivalent, in $L^2_{w^{-1}}$, to

$$\left\|L^{(2)}_m\left((D+K_m)v_m-f\right)\right\|_{w^{-1},2}=0. \tag{4.12}$$

The Gaussian rule with respect to the weight w^{-1}, applied to compute the integral in (4.12), is exact, since $L_m^{(2)}$ is a projector on $\mathbb{P}_{m-1}$. Hence (4.12) holds if and only if

$$\sum_{i=1}^{m} \lambda_{mi}(w^{-1}) \left| L_m^{(2)} \left((D+K_m) v_m - f \right)(x_i) \right|^2 = 0$$

i.e.

$$\sqrt{\lambda_{mi}(w^{-1})}\,(D+K_m)\,v_m(x_i) = \sqrt{\lambda_{mi}(w^{-1})} f(x_i), \quad i = 1, 2, \ldots, m. \tag{4.13}$$

Choose now a Gauss-type quadratura formula with the weight w and the nodes t_j. Then

$$Dv_m(x_i) = \frac{b}{\pi} \sum_{j=1}^{m} \lambda_{mj}(w) \frac{v_m(t_j)}{(t_j - x_i)} \tag{4.14}$$

and

$$K_m v_m(x_i) = \sum_{j=1}^{m} \lambda_{mj}(w) k(x_i, t_j) v_m(t_j). \tag{4.15}$$

Setting $\xi_j = v_m(t_j)$, by (4.14) and (3.3) we can conclude that system (4.13) is equivalent to (3.4) and the proof of the proposition is complete. □

In virtue of the equivalence between (3.4) and (4.11), we can finally prove Theorem 3.2, by using Proposition 4.3, too.

Proof. From (4.7), for a well known result, it follows that for sufficiently large m the inverse operators $\left(D + L_m^{(2)} K_m\right)^{-1}$ exist and are uniformly bounded with respect to m. More precisely one has

$$\left\| \left(D + L_m^{(2)} K_m\right)^{-1} \right\| \le \frac{\|(D+K)^{-1}\|}{1 - \|(D+K)^{-1}\| \cdot \left\| K - L_m^{(2)} K_m \right\|}, \tag{4.16}$$

which assures the stability of the method. Then (4.11) has a unique solution v_m which is a polynomial, since we have

$$v_m = D^{-1} \left(L_m^{(2)} (f - K_m v_m) \right)$$

and the inverse operator of D satisfies (see [22], [27])

$$D^{-1} q_m(x) = (-1)^M p_m(x), \quad m = 0, 1, \ldots . \tag{4.17}$$

Then, by Proposition 4.3, we can deduce that also system (3.4) has a unique solution, related to v_m as in (3.2) and (3.3).
To obtain error estimate (3.5), observe

$$v - v_m = \left(D + L_m^{(2)} K_m\right)^{-1} \left[f - L_m^{(2)} f - \left(K - L_m^{(2)} K_m\right) v \right].$$

Moreover, since $f \in Z_r(w^{-1})$, using Lagrange error estimate (3.2) given in [20], we can write

$$\begin{aligned}
\left\| f - L_m^{(2)} f \right\| &\leq \frac{C}{m^{\frac{1}{2}}} \int_0^{\frac{1}{m}} \frac{\Omega_\varphi^i(f,y)_{w^{-1},2}}{y^{\frac{3}{2}}} dy \\
&\leq \frac{C}{m^{\frac{1}{2}}} \left(\int_0^{\frac{1}{m}} y^{r-\frac{3}{2}} dy \right) \sup_{\delta>0} \frac{\Omega_\varphi^i(f,y)_{w^{-1},2}}{\delta^r} \\
&\leq \frac{C}{m^r} \|f\|_{Z_r(w^{-1})}.
\end{aligned}$$

It follows, by using estimate (4.7) and Theorem 3.1

$$\begin{aligned}
\|v - v_m\|_{w,2} &\leq \left\| \left(D + L_m^{(2)} K_m \right)^{-1} \right\| \cdot \left\| f - L_m^{(2)} f - \left(K - L_m^{(2)} K_m \right) v \right\|_{w^{-1},2} \\
&\leq \frac{C}{m^r} \left[\|f\|_{Z_r(w^{-1})} + \|v\|_{w,2} \right] \\
&\leq \frac{C}{m^r} \|v\|_{Z_r(w)}.
\end{aligned}$$

It remains to prove (3.6). At first observe that, since

$$(D + K) - \left(D + L_m^{(2)} K_m \right) = K - L_m^{(2)} K_m,$$

from (4.7), it follows that

$$\left\| D + L_m^{(2)} K_m \right\| \longrightarrow \|D + K\|, \quad \text{as } m \to \infty \tag{4.18}$$

by the continuity of the norm. Furthermore, according to (4.16) we also have

$$\left\| \left(D + L_m^{(2)} K_m \right)^{-1} \right\| \longrightarrow \|(D + K)^{-1}\|, \quad \text{as } m \to \infty. \tag{4.19}$$

Combining (4.18) and (4.19), one has

$$\gamma_m := \frac{\operatorname{cond}\left(D + L_m^{(2)} K_m \right)}{\operatorname{cond}(D + K)} \longrightarrow 1, \quad \text{as } m \to \infty. \tag{4.20}$$

which implies $\gamma_m \leq C$, since $\operatorname{cond}(D + K) \geq 1$.
The next step of the proof is to show that, for sufficiently large m,

$$\operatorname{cond}(A_m) \leq \operatorname{cond}\left(D + L_m^{(2)} K_m \right). \tag{4.21}$$

Let $\underline{\eta}_m = (\eta_{m1}, \eta_{m2}, \ldots, \eta_{mm})^T \in \mathbb{R}^m$ be arbitrary and $\underline{f}_m = A_m \underline{\eta}_m$. Consider the polynomial

$$v_m = \sum_{j=1}^{m} \lambda_{mj}(w)^{-\frac{1}{2}} \eta_{mj} l_{mj}(w), \tag{4.22}$$

and the function

$$f = \left(D + L_m^{(2)} K_m\right) v_m.$$

Then, by (2.7), f is a polynomial belonging to $\mathbb{P}_{m-1}$ and

$$\underline{f}_m = \left[\left(\sqrt{\lambda_{mi}(w^{-1})} f(x_i)\right)_{i=1}^{m}\right]^T.$$

Then, since

$$\begin{aligned} \|v_m\|_{w,2}^2 &= \sum_{k=1}^{m} \lambda_{mi}(w) \left|v_m(t_k)\right|^2 \\ &= \sum_{k=1}^{m} \lambda_{mi}(w) \left|\sum_{j=1}^{m} \lambda_{mj}(w)^{-\frac{1}{2}} \eta_{mj} \delta_{jk}\right|^2 \\ &= \sum_{k=1}^{m} |\eta_{mk}|^2 = \|\underline{\eta}_m\|_2^2, \end{aligned}$$

one has

$$\begin{aligned} \|\underline{f}_m\|_2 &= \|f\|_{w^{-1},2} \\ &\leq \left\|D + L_m^{(2)} K_m\right\| \cdot \|v_m\|_{w,2} \\ &= \left\|D + L_m^{(2)} K_m\right\| \cdot \|\underline{\eta}_m\|_2, \end{aligned}$$

from which, according to the definition of operator norm, it follows

$$\|A_m\| \leq \left\|D + L_m^{(2)} K_m\right\|. \tag{4.23}$$

Consider, now an arbitrary vector $\underline{f}_m = \left[\left(\sqrt{\lambda_{mi}(w^{-1})} f_{mi}\right)_{i=1}^{m}\right]^T \in \mathbb{R}^n$ and the solution $\underline{\eta}_m$ of the system

$$A_m \underline{\eta}_m = \underline{f}_m. \tag{4.24}$$

Take a function $f \in L^2_{w^{-1}}$ such that

$$\left\|L_m^{(2)} f\right\|_{w^{-1},2} = \|\underline{f}_m\|_2.$$

To this end, we can choose f such that $f(x_i) = f_{mi}$, $i = 1, 2, \ldots, m$. Let

$$v_m = \left(D + L_m^{(2)} K_m\right)^{-1} \left(L_m^{(2)} f\right) \in \mathbb{P}_{m-1}.$$

Then v_m can be represented as in (4.22) with $\underline{\eta}_m = (\eta_{m1}, \eta_{m2}, \ldots, \eta_{mm})^T \in \mathbb{R}^m$ equal to the solution of (4.24). Thus, one has

$$\begin{aligned} \|\underline{\eta}_m\|_2 &= \|v_m\|_{w,2} \\ &\leq \left\| \left(D + L_m^{(2)} K_m\right)^{-1} \right\| \cdot \left\| L_m^{(2)} f \right\|_{w^{-1},2} \\ &= \left\| \left(D + L_m^{(2)} K_m\right)^{-1} \right\| \cdot \|\underline{f}_m\|_2, \end{aligned}$$

from which it follows

$$\|(A_m)^{-1}\| \leq \left\| \left(D + L_m^{(2)} K_m\right)^{-1} \right\|. \tag{4.25}$$

Comparing (4.23) and (4.25) we get (4.21) and, by (4.20), the thesis (3.6). □

Let us prove now Theorem 3.5

Proof. Consider the following expansion of a polynomial $v_m \in \mathbb{P}_{m-1}$.

$$v_m(t) = \sum_{j=0}^{m-1} a_j p_j(t)$$

with

$$a_j = \int_{-1}^{1} v_m(t) p_j(t) w(t) dt.$$

By applying Schwarz' inequality we have

$$\begin{aligned} |v_m(t)| &\leq \left(\sum_{j=0}^{m-1} a_j^2\right)^{\frac{1}{2}} \left(\sum_{j=0}^{m-1} p_j^2(t)\right)^{\frac{1}{2}} \\ &= \|v_m\|_{w,2} \sqrt{\lambda_m^{-1}(w;t)}, \end{aligned}$$

where

$$\lambda_m(w,t) = \left[\sum_{j=0}^{m-1} p_j^2(t)\right]^{-1}$$

denotes the m-th Christoffel function related to w. Since ([25])

$$\lambda_m^{-1}(w,t) \sim w(t) \frac{\sqrt{1-t^2}}{m}, \quad |t| \leq 1 - \frac{c}{m^2}$$

one has, for all $|t| \leq 1 - \frac{c}{m^2}$,

$$|v_m(t)| w^{\gamma,\delta}(t) \leq C \|v_m\|_{w,2} \sqrt{m} (1-t)^{\gamma - \frac{\alpha}{2} - \frac{1}{4}} (1+t)^{\delta + \frac{\alpha}{2} - \frac{1}{4}}$$

from which it follows

$$|v_m(t)|w^{\gamma,\delta}(t) \le C\|v_m\|_{w,2}\sqrt{m}, \quad |t| \le 1 - \frac{c}{m^2}. \tag{4.26}$$

Since (3.5) and (3.10) hold, one has

$$(v - v_m)(t)w^{\gamma,\delta}(t) = \sum_{k=0}^{\infty} (v_{2^{k+1}m} - v_{2^k m})(t)w^{\gamma,\delta}(t)$$

almost everywhere in $[-1,1]$. Then by applying (4.26), by Remez inequality we have

$$\begin{aligned} \max_{|t|\le 1} |(v_{2^{k+1}m} - v_{2^k m})(t)| \, w^{\gamma,\delta}(t) &\le \max_{|t|\le 1-\frac{c}{m^2}} |(v_{2^{k+1}m} - v_{2^k m})(t)| \, w^{\gamma,\delta}(t) \\ &\le 2^{\frac{k+1}{2}} \|v_{2^{k+1}m} - v_{2^k m}\|_{w,2} \sqrt{m} \end{aligned}$$

and by (3.5) or (3.10)

$$\max_{|t|\le 1} |(v_{2^{k+1}m} - v_{2^k m})(t)| \, w^{\gamma,\delta}(t) \le 2^{\frac{k+1}{2}} \sqrt{m} \frac{C}{(2^k m)^r} \|v\|_{Z_r(w)}.$$

Therefore, we deduce

$$\sum_{k=0}^{\infty} \max_{|t|\le 1} |(v_{2^{k+1}m} - v_{2^k m})(t)| \, w^{\gamma,\delta}(t) \le \frac{C}{m^{r-\frac{1}{2}}} \|v\|_{Z_r(w)} \sum_{k=0}^{\infty} \frac{1}{2^{(r-\frac{1}{2})k}}.$$

Since, under our assumptions, $r > \frac{1}{2}$, the series converges uniformly in [-1,1]. We deduce that $v(t)w^{\gamma,\delta}(t)$ is a continuous function in $[-1,1]$ and the estimate

$$\max_{|t|\le 1} |v(t) - v_m(t)|w^{\gamma,\delta}(t) \le Cm^{r-\frac{1}{2}} \|v\|_{Z_r(w)}$$

holds. Therefore, the theorem is completely proved. □

Acknowledgements

The authors thank the referees for the accurate reading of the paper and the remarks for its improvement.

References

[1] Belotserkovski S.M., Lifanov I.K., *Chislennye metody dlja singuljarnčh integral'nych urarnenij*, Nauka, Moscow, 1985 (Russian).

[2] Berthold D., Hoppe W., Silbermann B., *A fast algorithm for solving the generalized airfoil equation*, J. Comp. Appl. Math., **43** (1992), 185-219.

[3] Berthold D., Hoppe W., Silbermann B., *The numerical solution of the generalized airfoil equation*, J. of Integr. Eq. and Appl., **4**, No. 3 (1992), 309–336.

[4] Capobianco M.R., Junghanns P., Luther U., Mastroianni G., *Weighted uniform convergence of the quadrature method for Cauchy singular integral equations*, Operator Theory. Advances and Applications **90**, (1996), 153-181.

[5] Ditzian Z., Totik V., *Moduli of smoothness*, SCMG Springer-Verlag, New York Berlin Heidelberg London Paris Tokyo, 1987.

[6] Ditzian Z., Totik V., *Remarks on Besov spaces and best polynomial approximation*, Proceed. of the Amer. Math. Soc., **104** No. 4 (1988), 1059-1066.

[7] Elliott D., *Orthogonal polynomials associated with singular integral equations having a Cauchy kernel*, SIAM J. Math. Anal. **13** (1982), 1041-1052.

[8] Elliott D., *The classical collocation method for singular integral equations*, SIAM J. Numer. Anal. **19** (1982), 816-832.

[9] Elliott D., *A comprehensive approach to the approximate solution of singular integral equations over the arc (-1,1)*, J. of Integr. Eq. and Appl., **2** (1989), 59-94.

[10] Elliott D., *Projection methods for singular integral equations*, J. of Integr. Eq. and Appl., **2** (1989), 95-106.

[11] Junghanns P., *Product integration for the generalized airfoil equation*, in : Beiträge zur Angewandten Analysis und Informatik (ed. E. Schock), Shaker Verlag, Aachen 1994, 171-188.

[12] Junghanns P., Luther U., *Cauchy singular integral equations in spaces of continuous functions and methods for their numerical solution*, J. Comp. Appl. Math., **77**, (1997), 201-237.

[13] Junghanns P., Silbermann B., *Zur Theorie der Näherungsverfahren für singuläre Integralgleichungen auf Intervallen*, Math. Nachr., **103** (1981), 199-244.

[14] Junghanns P., Silbermann B., *The numerical treatment of singular integral equations by means of polynomials approximations*, Preprint, P-Math-35/86, A d W der DDR, Karl-Weierstaß-Institut für Mathematik, Berlin, 1986.

[15] Junghanns P., Silbermann B., *Numerical analysis of the quadrature method for solving linear and nonlinear singular integral equations*, Preprint of the Technical University Chemnitz (1989).

[16] Laurita C.,*Condition numbers for Singular Integral Equations in weighted L^2 spaces*, J. Comp. Appl. Math. **116** (2000), 23-40.

[17] Laurita C., Mastroianni G., Russo M.G., *Revisiting CSIE in L^2: condition numbers and inverse theorems*, to appear on the volume Integral and Integrodifferential equations edited by R. Agarwal and D. O'Regan, Singapore, March (2000), 159-184.

[18] Mastroianni G., Prössdorf S., *A quadrature method for Cauchy integral equations with weakly singular perturbation kernel*, J. of Integr. Eq. and Appl., **4**, No. 2 (1992), 205-228.

[19] Mastroianni G., Prössdorf S., *Some nodes matrices appearing in the numerical analysis for singular integral equations*, BIT **34** (1994), 120-128.

[20] Mastroianni G., Russo M.G., *Lagrange Interpolation in Weighted Besov Spaces*, Constr. Approx. **14** (1998), 1-33.

[21] Mastronardi N., *Recurrence formulas for modified moments*, manuscript.

[22] Mikhlin S.G., Prössdorf S., *Singular Integral Operators*, Akademie-Verlag, Berlin, 1986.

[23] Monegato G., *Numerical resolution of the generalized airfoil equation with Possio kernel*, in Tricomi's ideas and contemporary applied mathematics, special volume of Atti dei Convegni Lincei **147**, (1998), 103-121.

[24] Monegato G., Prössdorf S., *Uniform convergence estimates for a collocation and discrete collocation method for the generalized airfoil equation*, Contributions to Numerical Mathematics (A.G. Agarval, ed.), World Scientific Publishing Company 1993, 285-299 (see also the errata corrige in the Internal Reprint No. 14 (1993) Dip. Mat. Politecnico di Torino).

[25] Nevai P., *Mean convergence of Lagrange interpolation I*, J. Approx. Theory **18** (1976), 363-377.

[26] Prössdorf S., Silbermann B., *Projektionsverfahren und die näherungsweise Lösung singulärer Gleichungen*, Teubner Verlagsges., Leipzig, 1977.

[27] Prössdorf S., Silbermann B., *Numerical Analysis for Integral and related Operator Equations*, Akademie-Verlag, Berlin 1991 and Birkhäuser Verlag, Basel-Boston-Stuttgard 1991.

[28] Timan A.F., *Theory of approximation of functions of a real variable*, Pergamonn Press, Oxford, England, 1963.

Dipartimento di Matematica
Università degli Studi della Basilicata
Via Nazario Sauro 85, 85100 Potenza
Italy

1991 Mathematics Subject Classification: Primary 65R20; Secondary 45E05

Submitted: 13.4.2000

Operator Theory:
Advances and Applications, Vol. 121

Fourier Projections in Weighted $\mathbf{L}^\infty$-Spaces

UWE LUTHER, GIUSEPPE MASTROIANNI

It is well-known, that the norms of the classical Fourier projections S_n in the space of all 2π-periodic L^∞-functions can be estimated by $\|S_n\| \le \text{const}\ \ln(n+1)$. In other words, the $L^\infty(-1,1)$-operator norms of the Fourier projections with respect to the normalized Chebyshev polynomials of first kind are bounded by const $\ln(n+1)$. In this paper we show that this result remains true for Fourier projections with respect to normalized Jacobi polynomials, if we consider them in a weighted L^∞-space, where the weight is a Jacobi weight, which has to fulfil certain conditions. Moreover, we prove that these conditions are necessary, and we also consider the case of pairs of L^∞-spaces with different Jacobi weights. As corollaries we obtain, among others, corresponding results in weighted L^1-spaces and norm estimates of the type $O(\ln^k n)$ for modified Fourier projections in cases where the unmodified Fourier projections can not have a logarithmic norm behaviour.

1. Introduction

Fourier projections and Fourier series are needed in many fields of mathematics, for example for the numerical treatment of operator equations and in the theory of function spaces. For this reason, estimates for operator norms of Fourier projections in given function spaces are of great interest. It is well-known that, under certain assumptions on the underlying scalar product and the orthogonal system, the Fourier projections S_n are uniformly bounded in $\mathbf{L}^p$-spaces with $1 < p < \infty$. But this is not true for $p = \infty$. In the case of classical Fourier expansion of 2π-periodic functions f (i.e. $S_n f$ is the $\mathbf{L}^2_{2\pi}$-orthogonal projection of f onto the space of trigonometric polynomials of degree less than n) it is known that the operator norm of S_n in the space $\mathbf{L}^\infty_{2\pi}$ is less than const $\ln n$ (see e.g. [19], 1.VII.§2). In this paper we will give necessary and sufficient conditions which ensure that the same behaviour holds true for the Fourier projections with respect to algebraic orthogonal polynomials and in weighted $\mathbf{L}^\infty$-spaces on the interval. More precisely, we consider the Fourier projections $S_n^{\alpha,\beta}$ onto $\mathbb{P}_{n-1}$, $\mathbb{P}_m := \{\text{algebraic polynomials of degree} \le m\}$, with respect to the normalized Jacobi-polynomials

$$p_n^{\alpha,\beta}(x) = \gamma_n x^n + \dots\,,\ \gamma_n > 0\,, \quad \int_{-1}^{1} p_n^{\alpha,\beta}(x) p_m^{\alpha,\beta}(x)(1-x)^\alpha(1+x)^\beta\, dx = \delta_{nm}\,,$$

i.e. $S_n^{\alpha,\beta} f$ is defined by

$$S_n^{\alpha,\beta} f = \sum_{j=0}^{n-1} \langle f, p_j^{\alpha,\beta} \rangle_{\alpha,\beta}\, p_j^{\alpha,\beta} \,,$$

where

$$\langle f, g \rangle_{\alpha,\beta} := \int_{-1}^{1} f(x)\,\overline{g(x)}\, v^{\alpha,\beta}(x)\, dx \,, \quad v^{\alpha,\beta}(x) := (1-x)^\alpha (1+x)^\beta \,,$$

and α, β are fixed real numbers with $\alpha, \beta > -1$. In weighted $\mathbf{L}^p$-spaces

$$\mathbf{L}^p_{\rho,\tau} := \{ f : f v^{\rho,\tau} \in \mathbf{L}^p(-1,1) \} \,, \quad \|f\|_{p,\rho,\tau} := \|f v^{\rho,\tau}\|_p \quad (\|g\|_p := \|g\|_{L^p(-1,1)})$$

with $1 < p < \infty$, the following well-known criterion for the uniform boundedness of the norms $\|S_n^{\alpha,\beta}\|_{p,\rho,\tau} := \|S_n^{\alpha,\beta}\|_{L^p_{\rho,\tau} \to L^p_{\rho,\tau}}$ holds true:

$$\sup_{n=1,2,\ldots} \|S_n^{\alpha,\beta}\|_{p,\rho,\tau} < \infty$$

$$\Longleftrightarrow \quad \sup_{n=0,1,\ldots} \|p_n^{\alpha,\beta} v^{\rho,\tau}\|_p < \infty, \quad \sup_{n=0,1,\ldots} \|p_n^{\alpha,\beta} v^{\alpha-\rho,\beta-\tau}\|_{p'} < \infty$$

$$(1.1) \qquad \Longleftrightarrow \quad \max\left\{0, \frac{\alpha}{2} + \frac{1}{4}\right\} < \rho + \frac{1}{p} < \min\left\{\alpha + 1, \frac{\alpha}{2} + \frac{3}{4}\right\},$$

$$\max\left\{0, \frac{\beta}{2} + \frac{1}{4}\right\} < \tau + \frac{1}{p} < \min\left\{\beta + 1, \frac{\beta}{2} + \frac{3}{4}\right\},$$

where p' is defined by $\frac{1}{p} + \frac{1}{p'} = 1$ (cf. [1], Theorem 5.1). Besides the many applications in the Hilbert space setting $\mathbf{L}^p_{\rho,\tau} = \mathbf{L}^2_{\alpha/2,\beta/2}$ ($S_n^{\alpha,\beta}$ is the orthogonal projection in this case), the result (1.1) is, of course, also very useful in case $p \neq 2$, since the uniform boundedness of $S_n^{\alpha,\beta}$ in $\mathbf{L}^p_{\rho,\tau}$ implies that $S_n^{\alpha,\beta} f$ approximates f in $\mathbf{L}^p_{\rho,\tau}$ with the same order as the best possible polynomial approximation. As we said, the result (1.1) is not true for $p = \infty$. But just in (weighted or unweighted) $\mathbf{L}^\infty$-spaces or, more precisely, subspaces of continuous functions, recently there were studied special types of operator equations and numerical methods for them, see e.g. [6], [7], [17], [10], [11] for Cauchy singular integral equations and [5], [13] for Prandtl's integro-differential equation. In all these papers there are studied collocation type methods and it is used that the operator norms of the corresponding interpolation operators in a weighted space of continuous functions increase like $\ln n$ if certain conditions are satisfied (cf. [15], [6], or [16]). To study fast algorithms for these methods, which are based on Fourier projections (see [3], [4], [11]), one also needs logarithmic behaviour for the norms of the Fourier projections in weighted spaces of continuous functions, if one wants to obtain almost optimal convergence rates in the norms of such spaces. Also for studying projection methods instead of collocation methods this behaviour would be of great interest. But until now, such a result was only known for the case $|\alpha| = |\beta| = \frac{1}{2}$ (cf. [5], Lemma 4.1, or [12], Prop.4.60). So we see that, at least for numerical analysis of certain operator equations, a result similar to (1.1) (with logarithmic growth of the norms of $S_n^{\alpha,\beta}$ instead of uniform boundedness) is needed for $p = \infty$. In Sections 2 and 5 we will state and prove such a result, also for pairs of spaces with different weights. In Section 3 we give some Corollaries and in Section 4 we will show that certain modified Fourier projections have norm behaviour $O(\ln^2 n)$ in weighted $\mathbf{L}^\infty$-spaces or weighted spaces of continuous functions, even if the norms of the unmodified Fourier projections increase with a power of n.

2. Main results

Theorem 2.1. *Let $\rho < 1 + \alpha$ and $\tau < 1 + \beta$, such that*

$$\max\left\{0, \frac{\alpha}{2} + \frac{1}{4}\right\} \le \rho \le \frac{\alpha}{2} + \frac{3}{4} \quad \text{and} \quad \max\left\{0, \frac{\beta}{2} + \frac{1}{4}\right\} \le \tau \le \frac{\beta}{2} + \frac{3}{4}. \tag{2.1}$$

Then there exists a constant $c > 0$, independent of n, such that

$$\|S_n^{\alpha,\beta}\|_{\infty,\rho,\tau} \le c \ln(n+1), \quad n \in \mathbb{N} := \{1,2,3,\dots\}.$$

Moreover, the discrete estimates

$$\max_{k=1,\dots,n} \left|(S_n^{\alpha,\beta} f)(x_{nk}^{\alpha,\beta})\right| v^{\rho,\tau}(x_{nk}^{\alpha,\beta}) \le c\,\|f\|_{\infty,\rho,\tau} \ln(n+1), \quad f \in \mathbf{L}_{\rho,\tau}^\infty, \tag{2.2}$$

where $x_{nk}^{\alpha,\beta}$ denote the zeros of $p_n^{\alpha,\beta}$, even hold true if we replace (2.1) *by*

$$\frac{\alpha}{2} - \frac{1}{4} \le \rho \le \frac{\alpha}{2} + \frac{3}{4} \quad \text{and} \quad \frac{\beta}{2} - \frac{1}{4} \le \tau \le \frac{\beta}{2} + \frac{3}{4}. \tag{2.3}$$

Remark 2.2. Later (as a special case of Remark 2.4) we will show that condition (2.1) together with $\rho < 1 + \alpha$ and $\tau < 1 + \beta$ is equivalent to

$$\|p_n^{\alpha,\beta} v^{\alpha-\rho,\beta-\tau}\|_1 \le \text{const} \ln(n+1) \quad \text{and} \quad \sup_{n=0,1,\dots} \|p_n^{\alpha,\beta} v^{\rho,\tau}\|_\infty < \infty.$$

We will prove Theorem 2.1 in Section 5. Now we consider the question whether condition (2.1) is necessary for $\|S_n^{\alpha,\beta}\|_{\infty,\rho,\tau} \le c \ln(n+1)$. The following Theorem gives a positive answer to this question, even if we restrict ourselves to the closed subspace

$$\mathbf{C}_{\rho,\tau} := \{f : (-1,1) \to \mathbb{C} : f v^{\rho,\tau} \in \mathbf{C}[-1,1] \text{ (continuous extension)}\}$$

of $\mathbf{L}^\infty_{\rho,\tau}$. Moreover, we will treat the more general situation of pairs of spaces with possibly different weights.

Theorem 2.3. *Let $\rho, \tau, \gamma, \delta$ be real constants. The following assertions are equivalent.*

(i) $S_n^{\alpha,\beta}$ *maps* $\mathbf{L}^\infty_{\rho,\tau}$ *into* $\mathbf{L}^\infty_{\gamma,\delta}$ *and the operator norm in this pair of spaces is bounded by* $c\ln(n+1)$*, where* c *is independent of* $n \in \mathbb{N}$.

(ii) $S_n^{\alpha,\beta}$ *maps* $\mathbf{C}_{\rho,\tau}$ *into* $\mathbf{C}_{\gamma,\delta}$ *and the operator norm in this pair of spaces is bounded by* $c\ln(n+1)$*, where* c *is independent of* $n \in \mathbb{N}$.

(iii) $v^{\gamma-\rho,\delta-\tau} \in \mathbf{L}^\infty$ *(i.e.* $\gamma \ge \rho$, $\delta \ge \tau$*) and*

$$\sup_{n=0,1,\ldots} \|p_n^{\alpha,\beta} v^{\gamma,\delta}\|_\infty < \infty\,, \quad \|p_n^{\alpha,\beta} v^{\alpha-\rho,\beta-\tau}\|_1 \le c\ln(n+1)\,,$$

where c *is independent of* $n \in \mathbb{N}$.

Remark 2.4. The condition $\|p_n^{\alpha,\beta} v^{\alpha-\rho,\beta-\tau}\|_1 \le c\ln(n+1)$ is equivalent to

$$\rho < 1+\alpha\,,\quad \tau < 1+\beta\,,\quad \rho \le \frac{\alpha}{2}+\frac{3}{4}\,,\quad \tau \le \frac{\beta}{2}+\frac{3}{4}\,. \tag{2.4}$$

If all inequalities in (2.4) hold strictly, then we even have $\sup_n \|p_n^{\alpha,\beta} v^{\alpha-\rho,\beta-\tau}\|_1 < \infty$. If one of the inequalities in (2.4) is not satisfied, then $\|p_n^{\alpha,\beta} v^{\alpha-\rho,\beta-\tau}\|_1 \ge \text{const}\, n^\mu$ with some $\mu > 0$.

The condition $\sup_n \|p_n^{\alpha,\beta} v^{\gamma,\delta}\|_\infty < \infty$ is equivalent to

$$\gamma \ge \max\left\{0,\ \frac{\alpha}{2}+\frac{1}{4}\right\}\,,\quad \delta \ge \max\left\{0,\ \frac{\beta}{2}+\frac{1}{4}\right\}\,. \tag{2.5}$$

If one of the inequalities in (2.5) is not satisfied, then $\|p_n^{\alpha,\beta} v^{\gamma,\delta}\|_\infty \ge \text{const}\, n^\mu$ with some $\mu > 0$.

We will prove this Remark in Section 5. In the present Section we only prove Theorem 2.3, since it can be attributed to Theorem 2.1 and Remark 2.4. Thereby we use the following well-known facts:

$$x_{n,k+1} - x_{nk} \sim \frac{\sqrt{1-x_{nk}^2}}{n}\,, \quad k = 1,\ldots,n-1\,, \tag{2.6}$$

$$\sqrt{1-x_{nk}} \sim \frac{n-k+1}{n}\,,\quad \sqrt{1+x_{nk}} \sim \frac{k}{n}\,,\quad k = 1,\ldots,n \tag{2.7}$$

([20], Theorem 9.22, p.166, or [2], Corollary 3), where $x_{n1} < \ldots < x_{nn}$ are the zeros of $p_n^{\alpha,\beta}$,

$$|p_n^{\alpha,\beta}(x)| \sim n \min_{k=1,\ldots,n} |x - x_{nk}| \left(\sqrt{1-x}+\frac{1}{n}\right)^{-\alpha-\frac{3}{2}} \left(\sqrt{1+x}+\frac{1}{n}\right)^{-\beta-\frac{3}{2}}\,, \quad x \in (-1,1) \tag{2.8}$$

([20], Theorem 9.33, p.171 and Theorem 6.3.28, p.120),

$$\|P_n\|_{p,\eta,\nu} \sim \|P_n v^{\eta,\nu}\|_{L^p[-1+Cn^{-2},1-Cn^{-2}]}\,, \quad P_n \in \mathbb{P}_n\,, \; n > \sqrt{C} \tag{2.9}$$

if C is a positive constant and $1 \le p < \infty$, $\eta,\nu > -\frac{1}{p}$ or $p = \infty$, $\eta,\nu \ge 0$ ([8], Theorem 8.4.8).

All relations hold uniformly in k, n, P_n, x, and the notation $A \sim B$ means that $c_1|B| \le |A| \le c_2|B|$ with positive constants c_1, c_2.

In the sequel, by c we denote a positive constant which can have different values at different places. This also means that, for example in an estimate of the type $A \le cB \le cC$, the constant c in cB can have a value different from that of the constant c in cC.

Proof of Theorem 2.3.

(iii)$\Rightarrow$(i): Set $\rho' = \max\left\{0,\,\rho,\,\frac{\alpha}{2}+\frac{1}{4}\right\}$ and $\tau' = \max\left\{0,\,\tau,\,\frac{\beta}{2}+\frac{1}{4}\right\}$. Then the assumptions of Theorem 2.1 are satisfied with ρ',τ' instead of ρ,τ (see Remark 2.4) and it follows

$$\|S_n^{\alpha,\beta}f\|_{\infty,\gamma,\delta} \le c\,\|S_n^{\alpha,\beta}f\|_{\infty,\rho',\tau'} \le c\,\|f\|_{\infty,\rho',\tau'}\ln(n+1) \le c\,\|f\|_{\infty,\rho,\tau}\ln(n+1)$$

for all $f \in \mathbf{L}^\infty_{\rho,\tau}$ $(\subset \mathbf{L}^\infty_{\rho',\tau'})$.

(i)$\Rightarrow$(ii): This is obvious. (Remark that the polynomial $S_n^{\alpha,\beta}f$ belongs to $\mathbf{L}^\infty_{\gamma,\delta}$ if and only if it belongs to $\mathbf{C}_{\gamma,\delta}$.)

(ii)$\Rightarrow$(iii): $S_n^{\alpha,\beta}f$ is defined for all $f \in \mathbf{C}_{\rho,\tau}$, particularly for $f = v^{-\rho,-\tau}$. This implies that the coefficient $\langle v^{-\rho,-\tau}, p_0^{\alpha,\beta}\rangle_{\alpha,\beta} = c\|v^{\alpha-\rho,\beta-\tau}\|_1$ has to be finite, i. e. $\rho < 1+\alpha$ and $\tau < 1+\beta$. For every $n = 2,3,\dots$ denote by I_{nk}, $k = 1,\dots,n-1$, the closed intervals, which have the same middle points as the intervals $[x_{nk}, x_{n,k+1}]$, but only the half lenghts. We choose continuous functions $g_n : [-1,1] \to [-1,1]$, such that $|g_n(x)| = 1$ for $x \in I_{nk}$ and $\operatorname{sgn} g_n(x) = \operatorname{sgn} p_n^{\alpha,\beta}(x)$ for all $x \in [-1,1]$. Then we have $f_n := g_n v^{-\rho,-\tau} \in \mathbf{C}_{\rho,\tau}$ and $\|f_n\|_{\infty,\rho,\tau} = 1$, so that (ii) implies

$$|\langle f_n, p_n^{\alpha,\beta}\rangle_{\alpha,\beta}|\,\|p_n^{\alpha,\beta}\|_{\infty,\gamma,\delta} = \|(S_{n+1}^{\alpha,\beta} - S_n^{\alpha,\beta})f_n\|_{\infty,\gamma,\delta} \le c\ln n\,, \quad n \ge 2\,. \tag{2.10}$$

From our choice of f_n it follows

$$\begin{aligned} |\langle f_n, p_n^{\alpha,\beta}\rangle_{\alpha,\beta}| &= \int_{-1}^{1} p_n^{\alpha,\beta}(x) g_n(x) v^{\alpha-\rho,\beta-\tau}(x)\,dx \\ &\ge \sum_{k=1}^{n-1} \int_{I_{nk}} |p_n^{\alpha,\beta}(x)| v^{\alpha-\rho,\beta-\tau}(x)\,dx\,. \end{aligned} \tag{2.11}$$

For all $x \in I_{nk}$ and $t \in [x_{nk}, x_{n,k+1}]$ we have

$$|p_n^{\alpha,\beta}(x)| v^{\alpha-\rho,\beta-\tau}(x) \ge cn(x_{n,k+1} - x_{nk}) v^{\frac{\alpha}{2}-\frac{3}{4}-\rho,\frac{\beta}{2}-\frac{3}{4}-\tau}(x)$$

$$\geq cn(x_{n,k+1}-x_{nk})v^{\frac{\alpha}{2}-\frac{3}{4}-\rho,\frac{\beta}{2}-\frac{3}{4}-\tau}(t) \geq c\,|p_n^{\alpha,\beta}(t)|v^{\alpha-\rho,\beta-\tau}(t)\,,$$

where we took (2.8),

$$\begin{aligned} c(1-x_{nk}) &\leq 1-x_{n,k+1} \leq 1-t, 1-x \leq 1-x_{nk}\,, \\ c(1+x_{n,k+1}) &\leq 1+x_{nk} \leq 1+t, 1+x \leq 1+x_{n,k+1} \end{aligned} \tag{2.12}$$

(this follows from (2.7)), i.e. $1 \pm t \sim 1 \pm x$, and

$$\begin{aligned} \sqrt{1-t}+\tfrac{1}{n} &\leq \sqrt{1-t}+c\sqrt{1-x_{nn}} \leq c\sqrt{1-t}\,, \\ \sqrt{1+t}+\tfrac{1}{n} &\leq \sqrt{1+t}+c\sqrt{1+x_{n1}} \leq c\sqrt{1+t} \end{aligned}$$

(see (2.7)) into account. Consequently, with $\xi_{nk} = \frac{1}{2}(x_{nk}+x_{n,k+1})$,

$$\begin{aligned} \int_{I_{nk}} |p_n^{\alpha,\beta}(x)|v^{\alpha-\rho,\beta-\tau}(x)dx &\geq c\,|p_n^{\alpha,\beta}(\xi_{nk})|v^{\alpha-\rho,\beta-\tau}(\xi_{nk}) \int_{I_{nk}} dx \\ &\geq c\,|p_n^{\alpha,\beta}(\xi_{nk})|v^{\alpha-\rho,\beta-\tau}(\xi_{nk}) \int_{x_{nk}}^{x_{n,k+1}} dt \geq c\int_{x_{nk}}^{x_{n,k+1}} |p_n^{\alpha,\beta}(t)|v^{\alpha-\rho,\beta-\tau}(t)dt. \end{aligned} \tag{2.13}$$

So we get, in view of (2.11),

$$\begin{aligned} |\langle f_n, p_n^{\alpha,\beta}\rangle_{\alpha,\beta}| &\geq c\sum_{k=1}^{n-1} \int_{x_{nk}}^{x_{n,k+1}} |p_n^{\alpha,\beta}(t)|v^{\alpha-\rho,\beta-\tau}(t)dt \\ &= c\,\|p_n^{\alpha,\beta}v^{\alpha-\rho,\beta-\tau}\|_{L^1[x_{n1},x_{nn}]}\,. \end{aligned}$$

Together with (2.10) it follows

$$\|p_n^{\alpha,\beta}v^{\alpha-\rho,\beta-\tau}\|_{L^1[x_{n1},x_{nn}]}\|p_n^{\alpha,\beta}v^{\gamma,\delta}\|_\infty \leq c\ln n\,, \quad n=2,3,\ldots\,. \tag{2.14}$$

Taking into account (2.8) we see that

$$|p_n^{\alpha,\beta}| \geq cn(x_{n,k+1}-x_{nk}) \text{ on } I_{nk}\,,$$

which implies $\|p_n^{\alpha,\beta}v^{\gamma,\delta}\|_\infty \geq c$ (see (2.6),(2.7)) and

$$\|p_n^{\alpha,\beta}v^{\alpha-\rho,\beta-\tau}\|_{L^1[x_{n1},x_{nn}]} \geq \sum_{k=1}^{n-1} \int_{I_{nk}} |p_n^{\alpha,\beta}(x)|v^{\alpha-\rho,\beta-\tau}(x)dx$$

$$\geq cn\sum_{k=1}^{n-1}(x_{n,k+1}-x_{nk})^2 v^{\alpha-\rho,\beta-\tau}(x_{nk}) \geq \frac{c}{n}\sum_{k=1}^{n-1} v^{\alpha-\rho+1,\beta-\tau+1}(x_{nk}) \geq c\,, \quad n \geq 2$$

in view of (2.12) and (2.6). So we have proved that even

$$\|p_n^{\alpha,\beta}v^{\alpha-\rho,\beta-\tau}\|_{L^1[x_{n1},x_{nn}]},\ \|p_n^{\alpha,\beta}v^{\gamma,\delta}\|_\infty \leq c\ln n\,, \quad n=2,3,\ldots \tag{2.15}$$

holds true instead of (2.14). From (2.7) it follows the existence of a constant $C > 0$, such that $x_{n1} \le -1 + Cn^{-2}$, $x_{nn} \ge 1 - Cn^{-2}$. Together with (2.9) we obtain

$$
\begin{aligned}
\|p_n^{\alpha,\beta} v^{\alpha-\rho,\beta-\tau}\|_{L^1[x_{n1},x_{nn}]} &\ge \|p_n^{\alpha,\beta} v^{\alpha-\rho,\beta-\tau}\|_{L^1[-1+Cn^{-2},1-Cn^{-2}]} \\
&\ge c\,\|p_n^{\alpha,\beta} v^{\alpha-\rho,\beta-\tau}\|_1
\end{aligned} \tag{2.16}
$$

for $n > \sqrt{C}$. (Remark that (2.9) is applicable since $\alpha - \rho, \beta - \tau > -1$ is already proved.) Consequently,

$$
\|p_n^{\alpha,\beta} v^{\alpha-\rho,\beta-\tau}\|_1,\ \|p_n^{\alpha,\beta} v^{\gamma,\delta}\|_\infty \le c \ln n\,, \quad n = 2, 3, \ldots\,,
$$

and this implies, in view of Remark 2.4,

$$
\sup_{n=0,1,\ldots} \|p_n^{\alpha,\beta} v^{\gamma,\delta}\|_\infty < \infty\,, \quad \|p_n^{\alpha,\beta} v^{\alpha-\rho,\beta-\tau}\|_1 \le c\,\ln(n+1)\,, \quad n \in \mathbb{N}\,.
$$

It remains to prove $\gamma \ge \rho$ and $\delta \ge \tau$. To show $\gamma \ge \rho$ we may assume $\rho > 0$, since the case $\rho \le 0$ is trivial. (Remark that $\|p_0^{\alpha,\beta} v^{\gamma,\delta}\|_\infty \le \sup_{n=0,1,\ldots} \|p_n^{\alpha,\beta} v^{\gamma,\delta}\|_\infty < \infty$ implies $\gamma \ge 0$.) From Remark 2.4 it follows $\sup_n \|p_n^{2\rho-\frac{1}{2},-\frac{1}{2}} v^{\rho,0}\|_\infty < \infty$. If we choose some $k \in \mathbb{N}$, such that $k + \tau \ge 0$, then we obtain

$$
p_n^{2\rho-\frac{1}{2},-\frac{1}{2}} v^{0,k} \in \mathbf{C}_{\rho,\tau} \quad \text{and} \quad \|p_n^{2\rho-\frac{1}{2},-\frac{1}{2}} v^{0,k}\|_{\infty,\rho,\tau} = \|p_n^{2\rho-\frac{1}{2},-\frac{1}{2}} v^{\rho,k+\tau}\|_\infty \le c\,.
$$

Now we conclude from (ii) that

$$
\begin{aligned}
\|p_n^{2\rho-\frac{1}{2},-\frac{1}{2}} v^{\gamma,\delta+k}\|_\infty &= \|p_n^{2\rho-\frac{1}{2},-\frac{1}{2}} v^{0,k}\|_{\infty,\gamma,\delta} = \|S_{n+k+1}^{\alpha,\beta}(p_n^{2\rho-\frac{1}{2},-\frac{1}{2}} v^{0,k})\|_{\infty,\gamma,\delta} \\
&\le c\,\|p_n^{2\rho-\frac{1}{2},-\frac{1}{2}} v^{0,k}\|_{\infty,\rho,\tau} \ln(n+1) \le c \ln(n+1)\,.
\end{aligned} \tag{2.17}
$$

This implies, in view of Remark 2.4, that $\gamma \ge \frac{1}{2}\left(2\rho - \frac{1}{2}\right) + \frac{1}{4} = \rho$. Analogously one can prove $\delta \ge \tau$. □

Remark 2.5. Let $\rho < 1 + \alpha$, $\tau < 1 + \beta$, $\gamma \ge 0$, and $\delta \ge 0$. (These conditions are necessary to ensure that the domain of $S_n^{\alpha,\beta}$ contains $\mathbf{C}_{\rho,\tau}$ and that the image $S_n^{\alpha,\beta}(\mathbf{C}_{\rho,\tau}) = \mathbb{P}_{n-1}$ is contained in $\mathbf{C}_{\gamma,\delta}$; $S_n^{\alpha,\beta}(\mathbf{C}_{\rho,\tau}) = \mathbb{P}_{n-1}$ will be shown in the proof of Lemma 3.1.) The proofs of (2.15) (together with (2.16)) and (2.17) show that the estimates

$$
\begin{aligned}
\|p_n^{\alpha,\beta} v^{\alpha-\rho,\beta-\tau}\|_1,\ \|p_n^{\alpha,\beta} v^{\gamma,\delta}\|_\infty &\le c\left(\|S_n^{\alpha,\beta}\|_{C_{\rho,\tau}\to C_{\gamma,\delta}} + \|S_{n+1}^{\alpha,\beta}\|_{C_{\rho,\tau}\to C_{\gamma,\delta}}\right), \\
\|p_n^{2\rho-\frac{1}{2},-\frac{1}{2}} v^{\gamma,\delta+k}\|_\infty &\le c\,\|S_{n+k+1}^{\alpha,\beta}\|_{C_{\rho,\tau}\to C_{\gamma,\delta}} \quad (\text{if } \rho > 0)\,, \\
\|p_n^{-\frac{1}{2},2\tau-\frac{1}{2}} v^{\gamma+k,\delta}\|_\infty &\le c\,\|S_{n+k+1}^{\alpha,\beta}\|_{C_{\rho,\tau}\to C_{\gamma,\delta}} \quad (\text{if } \tau > 0)
\end{aligned}
$$

($k \in \mathbb{N} \cap [-\min\{\rho,\tau\}, \infty)$) hold true without any further conditions on ρ, τ, γ, δ. Together with Remark 2.4 it follows that $\|S_n^{\alpha,\beta}\|_{C_{\rho,\tau}\to C_{\gamma,\delta}} + \|S_{n+1}^{\alpha,\beta}\|_{C_{\rho,\tau}\to C_{\gamma,\delta}} \ge$

const n^μ with some $\mu > 0$ if one of the conditions $\sup_n \|p_n^{\alpha,\beta} v^{\gamma,\delta}\|_\infty < \infty$, $\|p_n^{\alpha,\beta} v^{\alpha-\rho,\beta-\tau}\|_1 \le c \ln(n+1)$, $v^{\gamma-\rho,\delta-\tau} \in \mathbf{L}^\infty$, i.e. one of the inequalities

$$\rho \le \frac{\alpha}{2} + \frac{3}{4}, \quad \tau \le \frac{\beta}{2} + \frac{3}{4}, \quad \gamma \ge \max\left\{\rho, \frac{\alpha}{2} + \frac{1}{4}\right\}, \quad \delta \ge \max\left\{\tau, \frac{\beta}{2} + \frac{1}{4}\right\},$$

is not satisfied.

3. Some Corollaries

For the sake of simplicity we now consider $S_n^{\alpha,\beta}$ in only one space $\mathbf{L}^\infty_{\rho,\tau}$ (not in pairs of spaces). First we show that the criteria for the behaviour $\|S_n^{\alpha,\beta}\| = O(\ln n)$ in weighted L^∞-spaces imply criteria for the same behaviour in weighted L^1-spaces. Indeed, this is a consequence of the following Lemma.

Lemma 3.1. *For all constants $\rho, \tau > -1$ we have*

$$\|S_n^{\alpha,\beta}\|_{1,\rho,\tau} = \|S_n^{\alpha,\beta}\|_{\infty,\alpha-\rho,\beta-\tau}, \quad n = 0, 1, \dots,$$

where we set $\|S_n^{\alpha,\beta}\|_{1,\rho,\tau} := \infty$ if $S_n^{\alpha,\beta}$ is not defined on the whole space $\mathbf{L}^1_{\rho,\tau}$. (Remark that $S_n^{\alpha,\beta} f$ is defined for all $f \in \mathbf{L}^\infty_{\alpha-\rho,\beta-\tau}$ since $\rho, \tau > -1$.)

Proof. $\|S_n^{\alpha,\beta}\|_{\infty,\alpha-\rho,\beta-\tau}$ is finite if and only if $S_n^{\alpha,\beta}(\mathbf{L}^\infty_{\alpha-\rho,\beta-\tau}) = \mathbb{P}_{n-1}$ is a subset of $\mathbf{L}^\infty_{\alpha-\rho,\beta-\tau}$, i.e. $\rho \le \alpha$, $\tau \le \beta$. ($S_n^{\alpha,\beta}(\mathbf{L}^\infty_{\alpha-\rho,\beta-\tau}) = \mathbb{P}_{n-1}$ can be proved as follows: Choose $k \in \mathbb{N} \cup \{0\}$ such that $k \ge \max\{\rho - \alpha, \tau - \beta\}$ and set $f_j := v^{k,k} p_{j+2k}^{\alpha,\beta} / \langle p_{j+2k}^{\alpha,\beta}, v^{k,k} p_j^{\alpha,\beta}\rangle_{\alpha,\beta}$. Then we have $f_j \in \mathbf{L}^\infty_{\alpha-\rho,\beta-\tau}$ and $S_n^{\alpha,\beta} f_j = p_j^{\alpha,\beta} + c_{j,j+1}\, p_{j+1}^{\alpha,\beta} + \dots + c_{j,n-1}\, p_{n-1}^{\alpha,\beta}$ for $j = 0, \dots, n-1$. Starting with $j = n-1$, we conclude $p_j^{\alpha,\beta} \in S_n^{\alpha,\beta}(\mathbf{L}^\infty_{\alpha-\rho,\beta-\tau})$, $j = n-1, n-2, \dots, 0$.) In this case, also $\|S_n^{\alpha,\beta}\|_{1,\rho,\tau}$ is finite, since the Fourier coefficients $\langle f, p_j^{\alpha,\beta}\rangle_{\alpha,\beta}$ are defined for all $f \in \mathbf{L}^1_{\rho,\tau}$ if $\rho \le \alpha$ and $\tau \le \beta$. On the other hand, if e.g. $\rho = \alpha + \varepsilon$ with some $\varepsilon > 0$, then $S_n^{\alpha,\beta} f$ is not defined for $f = v^{\varepsilon-\rho-1,0} \in \mathbf{L}^1_{\rho,\tau}$, since $\langle f, p_0^{\alpha,\beta}\rangle_{\alpha,\beta} = c \int_{-1}^1 v^{-1,\beta} dx = \infty$. Thus, it remains to consider the case $\rho \le \alpha$, $\tau \le \beta$. If we identify functions $f \in \mathbf{L}^\infty[-1,1]$ with functionals $f \in \mathbf{L}^1[-1,1]^*$ by

$$(f, g) := \int_{-1}^1 f(x)\, g(x)\, dx, \quad g \in \mathbf{L}^1[-1,1],$$

then it is well-known that $\mathbf{L}^\infty[-1,1] = \mathbf{L}^1[-1,1]^*$, where the $\mathbf{L}^\infty$-norm of $f \in \mathbf{L}^\infty[-1,1]$ is equal to its operator norm in $\mathbf{L}^1[-1,1]^*$. Consequently, the dual operator A^* of a bounded linear operator $A : \mathbf{L}^1[-1,1] \to \mathbf{L}^1[-1,1]$, which is defined by $(A^* f, g) = (f, Ag)$ for $f \in \mathbf{L}^\infty[-1,1]$ and $g \in \mathbf{L}^1[-1,1]$, can be considered as an operator in $\mathbf{L}^\infty[-1,1]$ and its operator norm in $\mathbf{L}^\infty[-1,1]$ is equal to the operator norm of A in $\mathbf{L}^1[-1,1]$. Now we apply these facts to the weighted Fourier projections

$$A_n := v^{\rho,\tau} S_n^{\alpha,\beta} v^{-\rho,-\tau} I : \mathbf{L}^1[-1,1] \to \mathbf{L}^1[-1,1] .$$

Obviously, the norm $\|A_n\|_1$ of A_n in $\mathbf{L}^1$ is equal to $\|S_n^{\alpha,\beta}\|_{1,\rho,\tau}$ ($< \infty$, since $-1 < \rho \le \alpha$ and $-1 < \tau \le \beta$). Moreover, the operator $\tilde{A}_n := v^{\alpha-\rho,\beta-\tau} S_n^{\alpha,\beta} v^{\rho-\alpha,\tau-\beta} I$, which has the norm $\|\tilde{A}_n\|_\infty = \|S_n^{\alpha,\beta}\|_{\infty,\alpha-\rho,\beta-\tau}$ ($< \infty$) in $\mathbf{L}^\infty$, is equal to the dual operator A_n^*. Indeed, for all $f \in \mathbf{L}^\infty$ and all $g \in \mathbf{L}^1$, we have

$$\begin{aligned}(f, A_n g) &= \langle v^{\rho-\alpha,\tau-\beta} f,\, S_n^{\alpha,\beta} v^{-\rho,-\tau} \overline{g} \rangle_{\alpha,\beta} = \langle S_n^{\alpha,\beta} v^{\rho-\alpha,\tau-\beta} f,\, v^{-\rho,-\tau} \overline{g} \rangle_{\alpha,\beta} \\ &= (v^{\alpha-\rho,\beta-\tau} S_n^{\alpha,\beta} v^{\rho-\alpha,\tau-\beta} f, g) = (\tilde{A}_n f, g) .\end{aligned}$$

Now the assertion follows from $\|A_n^*\|_\infty = \|A_n\|_1$. □

Corollary 3.2. *Let ρ and τ be real constants. The following assertions are equivalent.*

(i) $S_n^{\alpha,\beta}$ *maps* $\mathbf{L}^1_{\rho,\tau}$ *into* $\mathbf{L}^1_{\rho,\tau}$ *and the operator norm in this space is bounded by* $c \ln(n+1)$, *where c is independent of* $n \in \mathbb{N}$.

(ii) $\sup_n \|p_n^{\alpha,\beta} v^{\alpha-\rho,\beta-\tau}\|_\infty < \infty$ *and* $\|p_n^{\alpha,\beta} v^{\rho,\tau}\|_1 \le \text{const} \ln(n+1)$ *for all* $n \in \mathbb{N}$.

(iii) The following inequalities for ρ and τ hold true:

$$\rho > -1 \quad \textit{and} \quad \frac{\alpha}{2} - \frac{3}{4} \le \rho \le \min\left\{\alpha,\, \frac{\alpha}{2} - \frac{1}{4}\right\},$$

$$\tau > -1 \quad \textit{and} \quad \frac{\beta}{2} - \frac{3}{4} \le \tau \le \min\left\{\beta,\, \frac{\beta}{2} - \frac{1}{4}\right\}.$$

Proof. First we remark that ρ and τ must be greater than -1 if (i) is satisfied. Indeed, if $\rho \le -1$ or $\tau \le -1$, then $(S_{n+1}^{\alpha,\beta} - S_n^{\alpha,\beta})f = \langle f, p_n^{\alpha,\beta}\rangle_{\alpha,\beta}\, p_n^{\alpha,\beta}$ does not belong to $\mathbf{L}^1_{\rho,\tau}$ if we choose, for example, $f = v^{-\rho,-\tau}\operatorname{sgn} p_n^{\alpha,\beta} \in \mathbf{L}^1_{\rho,\tau}$. Now the assertions follow from Lemma 3.1 and Theorem 2.3, Remark 2.4 (applied with $\alpha - \rho$, $\beta - \tau$ instead of ρ, τ). □

An immediate consequence of Theorem 2.1 is the following

Corollary 3.3. *Let* $f \in \mathbf{L}^\infty_{\rho,\tau}$ *and set*

$$E_n^{\rho,\tau}(f) = \inf\{\|f - P_n\|_{\infty,\rho,\tau} : P_n \in \mathbb{P}_n\}, \quad n = 0,1,2,\dots .$$

If

$$\rho < 1 + \alpha, \quad \tau < 1 + \beta,$$

$$\max\left\{0,\, \frac{\alpha}{2} + \frac{1}{4}\right\} \le \rho \le \frac{\alpha}{2} + \frac{3}{4}, \quad \max\left\{0,\, \frac{\beta}{2} + \frac{1}{4}\right\} \le \tau \le \frac{\beta}{2} + \frac{3}{4} \tag{3.1}$$

(i.e. $\|p_n^{\alpha,\beta} v^{\alpha-\rho,\beta-\tau}\|_1 \le \text{const} \ln(n+1)$ *and* $\sup_n \|p_n^{\alpha,\beta} v^{\rho,\tau}\|_\infty < \infty$*; cf. Rem. 2.2), then*

$$\|f - S_n^{\alpha,\beta} f\|_{\infty,\rho,\tau} \le c\, E_{n-1}^{\rho,\tau}(f) \ln(n+1)\,, \quad n = 1,2,\dots\,,$$

where c is independent of n and f.

Proof. This follows from $f - S_n^{\alpha,\beta} f = (f - P_{n-1}) + S_n^{\alpha,\beta}(P_{n-1} - f)$, $P_{n-1} \in \mathbb{P}_{n-1}$, and Theorem 2.1. □

Let $\rho, \tau \ge 0$ and $r \in \mathbb{N}$ be fixed, set $\varphi(x) = \sqrt{1-x^2}$, and define the K-functional of a measurable function $f : (-1,1) \to \mathbb{C}$ by

$$K_{r,\varphi}^{\rho,\tau}(f,t) := \inf_{g^{(r-1)} \in AC_{\text{loc}}} \{\|f - g\|_{\infty,\rho,\tau} + t^r \|\varphi^r g^{(r)}\|_{\infty,\rho,\tau}\}\,, \quad t > 0\,,$$

where AC_{loc} denotes the space of all locally absolutely continuous functions on $(-1,1)$ and $g^{(r)}$ is the r-th derivative of g. Then it follows from results of Ditzian and Totik, [8], that the strong Jackson-type inequality

$$E_n^{\rho,\tau}(f) \le c\, K_{r,\varphi}^{\rho,\tau}(f, n^{-1})\,, \quad n \ge r$$

holds true for all $f \in \mathbf{L}_{\text{loc}}^\infty$ (i.e. $f \in \mathbf{L}^\infty$ on every $[a,b] \subset (-1,1)$), where c is independent of f and n (cf. [14], Theorem 2.1). This implies

$$E_n^{\rho,\tau}(f) \le \frac{c}{n^r} \|f^{(r)}\|_{\infty,\rho+\frac{r}{2},\tau+\frac{r}{2}}\,, \quad n \ge r \tag{3.2}$$

if $f^{(r-1)} \in AC_{\text{loc}}$ and, consequently (apply (3.2) with $f - P_n$, $P_n \in \mathbb{P}_n$, instead of f and take the infimum over all P_n),

$$E_n^{\rho,\tau}(f) \le \frac{c}{n^r} E_{n-r}^{\rho+\frac{r}{2},\tau+\frac{r}{2}}(f^{(r)})\,, \quad n \ge r$$

if $f^{(r-1)} \in AC_{\text{loc}}$ and $f^{(r)} \in \mathbf{L}_{\rho+\frac{r}{2},\tau+\frac{r}{2}}^\infty$. (Especially, we have $f \in \mathbf{L}_{\rho,\tau}^\infty$ in this case.) So we can conclude the following result from Corollary 3.3.

Corollary 3.4. *Let* (3.1) *be satisfied. If* $f^{(r-1)} \in AC_{\text{loc}}$ *and* $f^{(r)} \in \mathbf{L}_{\rho+\frac{r}{2},\tau+\frac{r}{2}}^\infty$, *then*

$$\|f - S_n^{\alpha,\beta} f\|_{\infty,\rho,\tau} \le \frac{c}{n^r} E_{n-r-1}^{\rho+\frac{r}{2},\tau+\frac{r}{2}}(f^{(r)}) \ln(n+1)\,, \quad n \ge r+1\,,$$

where c is independent of f and n.

Now we will consider $S_n^{\alpha,\beta}$ on the space

$$\mathbf{B}_{0,1}^\infty(v^{\rho,\tau}) = \{f \in \mathbf{C}_{\rho,\tau} : \|f\|_{B_{0,1}^\infty(v^{\rho,\tau})} = \|f\|_{\infty,\rho,\tau} + \sum_{m=1}^\infty \frac{E_m^{\rho,\tau}(f)}{m} < \infty\}$$

($\rho, \tau \geq 0$), which can be defined equivalently (in the sense of equivalent norms) by

$$\mathbf{B}^{\infty}_{0,1}(v^{\rho,\tau}) = \{f \in \mathbf{C}_{\rho,\tau} : \|f\|^*_{B^{\infty}_{0,1}(v^{\rho,\tau})} = \|f\|_{\infty,\rho,\tau} + \int_0^1 \frac{K^{\rho,\tau}_{r,\varphi}(f,t)}{t}\, dt < \infty\}$$

(cf. [14], Theorem 2.3). $\mathbf{B}^{\infty}_{0,1}(v^{\rho,\tau})$ appears as a limit case of the well-known Besov spaces $\mathbf{B}^{\infty}_{s,1}(v^{\rho,\tau})$, $s > 0$, which are defined with some $r > s$ (the choice of r has no influence on the topology of the space) by

$$\mathbf{B}^{\infty}_{s,1}(v^{\rho,\tau}) = \{f \in \mathbf{C}_{\rho,\tau} : \|f\|_{B^{\infty}_{s,1}(v^{\rho,\tau})} = \|f\|_{\infty,\rho,\tau} + \int_0^1 \frac{K^{\rho,\tau}_{r,\varphi}(f,t)}{t^{1+s}}\, dt < \infty\}$$

(see [9]), and it can be proved that $\mathbf{B}^{\infty}_{0,1}(v^{\rho,\tau})$ is a Banach space in which the set of all algebraic polynomials is dense ([14], Lemma 4.1). Moreover, one can define $\mathbf{B}^{\infty}_{0,1}(v^{\rho,\tau})$ with the help of the Ditzian-Totik modulus $\omega^r_\varphi(f,t)_{v^{\rho,\tau},\infty}$ of smoothness, which is equivalent to $K^{\rho,\tau}_{r,\varphi}(f,t)$ for $t \leq t_0$; see [8], (8.2.10) for the definition of $\omega^r_\varphi(f,t)_{v^{\rho,\tau},\infty}$ and Theorem 6.1.1 for the equivalence with $K^{\rho,\tau}_{r,\varphi}(f,t)$.

Corollary 3.5. *If (3.1) is satisfied, then the Fourier projections $S^{\alpha,\beta}_n$ are uniformly bounded operators from $\mathbf{B}^{\infty}_{0,1}(v^{\rho,\tau})$ into $\mathbf{C}_{\rho,\tau}$, i.e.*

$$\|S^{\alpha,\beta}_n f\|_{\infty,\rho,\tau} \leq c \left(\|f\|_{\infty,\rho,\tau} + \sum_{m=1}^{\infty} \frac{E^{\rho,\tau}_m(f)}{m} \right), \quad f \in \mathbf{C}_{\rho,\tau}, \quad n \in \mathbb{N},$$

where c is independent of n and f.

Proof. From Corollary 3.3 it follows

$$\|S^{\alpha,\beta}_n f\|_{\infty,\rho,\tau} \leq \|S^{\alpha,\beta}_n f - f\|_{\infty,\rho,\tau} + \|f\|_{\infty,\rho,\tau} \leq \|f\|_{\infty,\rho,\tau} + c\, E^{\rho,\tau}_{n-1}(f) \ln(n+1)\,.$$

For $n = 1$ we have $E^{\rho,\tau}_{n-1}(f)\ln(n+1) = E^{\rho,\tau}_0(f)\ln 2 \leq \|f\|_{\infty,\rho,\tau}\ln 2$ and for $n > 1$ we use that $\{E^{\rho,\tau}_n(f)\}^{\infty}_{n=0}$ is a decreasing sequence, which implies

$$E^{\rho,\tau}_{n-1}(f)\ln(n+1) \leq c\, E^{\rho,\tau}_{n-1}(f) \sum_{m=1}^{n-1} \frac{1}{m} \leq c \sum_{m=1}^{n-1} \frac{E^{\rho,\tau}_m(f)}{m} \leq c \sum_{m=1}^{\infty} \frac{E^{\rho,\tau}_m(f)}{m}\,.$$

□

Remark 3.6. If (3.1) is satisfied, then it follows from Corollary 3.5 and the density of the polynomials in $\mathbf{B}^{\infty}_{0,1}(v^{\rho,\tau})$ that, for all $f \in \mathbf{B}^{\infty}_{0,1}(v^{\rho,\tau})$, $S^{\alpha,\beta}_n f$ converges to f in the norm of the space $\mathbf{C}_{\rho,\tau}$.

Corollaries 3.4 and 3.5 show that

$$\|S^{\alpha,\beta}_n f\|_{\infty,\rho,\tau} \leq c\,\|f\|_{\infty,\rho,\tau}\ln(n+1)\,, \quad f \in \mathbf{C}_{\rho,\tau} \tag{3.3}$$

(i.e. (3.1); see Theorem 2.3 and Remark 2.4) implies

$$(3.4) \qquad \|S_n^{\alpha,\beta} f\|_{\infty,\rho,\tau} \le c \left(\|f\|_{\infty,\rho,\tau} + \sum_{m=1}^{\infty} \frac{E_m^{\rho,\tau}(f)}{m} \right), \quad f \in \mathbf{B}_{0,1}^{\infty}(v^{\rho,\tau})$$

and, for $f \in AC_{\rm loc}$ with $f' \in \mathbf{L}^{\infty}_{\rho+\frac{1}{2},\tau+\frac{1}{2}}$,

$$(3.5) \qquad \|S_n^{\alpha,\beta} f\|_{\infty,\rho,\tau} \le c \left(\|f\|_{\infty,\rho,\tau} + \frac{\ln(n+1)}{n} \|f'\|_{\infty,\rho+\frac{1}{2},\tau+\frac{1}{2}} \right).$$

In some applications one only needs (3.4) or (3.5) instead of (3.3). So it is interesting to ask whether it is possible that (3.4) or (3.5) is satisfied also in cases where (3.3) is not valid. The following Proposition gives a negative answer to this question (at least if $\rho < 1+\alpha$ and $\tau < 1+\beta$).

Proposition 3.7. *Let $0 \le \rho < 1+\alpha$ and $0 \le \tau < 1+\beta$. Then* (3.3), (3.4), *and* (3.5) *are equivalent.*

Proof. We have already proved that (3.3) implies (3.4) and (3.5). Applying (3.4) with $f - P_{n-1}$, where $P_{n-1} \in \mathbb{P}_{n-1}$ such that $\|f - P_{n-1}\|_{\infty,\rho,\tau} = E_{n-1}^{\rho,\tau}(f)$, and taking

$$\begin{aligned} \sum_{m=1}^{\infty} \frac{E_m^{\rho,\tau}(f - P_{n-1})}{m} &\le E_{n-1}^{\rho,\tau}(f) \sum_{m=1}^{n-1} \frac{1}{m} + \sum_{m=n}^{\infty} \frac{E_m^{\rho,\tau}(f)}{m} \\ &\le c\, \frac{\ln(n+1)}{n} \|f'\|_{\infty,\rho+\frac{1}{2},\tau+\frac{1}{2}}, \quad f \in AC_{\rm loc} \end{aligned}$$

(see (3.2)) into account, we even see that (3.5) follows from (3.4). It remains to prove that (3.5) implies (3.3). For this aim, we remark that, in the same way as in the proof of Theorem 2.3, it can be shown

$$(3.6) \quad \|p_n^{\alpha,\beta} v^{\alpha-\rho,\beta-\tau}\|_1,\ \|p_n^{\alpha,\beta} v^{\rho,\tau}\|_\infty \le c \|(S_{n+1}^{\alpha,\beta} - S_n^{\alpha,\beta}) f_n\|_{\infty,\rho,\tau}, \quad n \ge 2,$$

where $f_n = g_n v^{-\rho,-\tau}$ with the continuous and picewise linear spline g_n, defined by the following conditions:

(1) $g_n(x_{nk}) = 0$, $k = 1, \ldots, n$,

(2) $g_n\left(\frac{1}{2}(x_{nk} + x_{n,k+1})\right) = (-1)^k$, $k = 1, \ldots, n-1$,

(3) g_n is linear on $[x_{nk}, \frac{1}{2}(x_{nk} + x_{n,k+1})]$ and $[\frac{1}{2}(x_{nk} + x_{n,k+1}), x_{n,k+1}]$, $k = 1, \ldots, n-1$,

(4) $g_n(x) = 0$ for all $x \in [-1, x_{n1}] \cup [x_{n,n}, 1]$.

Obviously, we have $\|f_n\|_{\infty,\rho,\tau} = 1$, $f_n \in AC_{\mathrm{loc}}$, and

$$f_n' = \left(\rho\, v^{-\rho-1,-\tau} - \tau\, v^{-\rho,-\tau-1}\right) g_n + v^{-\rho,-\tau} g_n' =: A_n + B_n\,.$$

If $x \in [x_{n1}, x_{nn}]$ then $1 \pm x \geq c n^{-2}$ (cf. (2.7)) and so we may estimate $v^{-\rho-1,-\tau}(x)$ and $v^{-\rho,-\tau-1}(x)$ by $c n\, v^{-\rho-\frac{1}{2},-\tau-\frac{1}{2}}(x)$ in this case. For $x \notin [x_{n1}, x_{nn}]$ we have $A_n(x) = 0$. Consequently,

$$\|A_n\|_{\infty,\rho+\frac{1}{2},\tau+\frac{1}{2}} \leq c n\,.$$

On $[x_{nk}, x_{n,k+1}]$, the absolute value of g_n' is equal to $2/(x_{n,k+1} - x_{nk})$. Taking into account (2.6) and (2.12), it follows

$$|g_n'(x)| \leq c\frac{n}{\sqrt{1-x_{nk}^2}} \leq c\frac{n}{\sqrt{1-x^2}}\,, \qquad x \in [x_{nk}, x_{n,k+1}],\ \ k = 0, \ldots, n-1\,.$$

Together with $g_n' = 0$ on $[-1,1] \setminus [x_{n1}, x_{nn}]$ we obtain $\|B_n\|_{\infty,\rho+\frac{1}{2},\tau+\frac{1}{2}} \leq cn$. So we have proved

$$\|f_n\|_{\infty,\rho,\tau} + \frac{\ln(n+1)}{n}\|f_n'\|_{\infty,\rho+\frac{1}{2},\tau+\frac{1}{2}} \leq c\,\ln(n+1)\,.$$

Thus, we conclude from (3.5) and (3.6) that

$$\|p_n^{\alpha,\beta} v^{\alpha-\rho,\beta-\tau}\|_1,\ \|p_n^{\alpha,\beta} v^{\rho,\tau}\|_\infty \leq c\,\ln(n+1)\,, \quad n \geq 2\,,$$

and this implies, in view of Remark 2.4,

$$\sup_{n=0,1,\ldots} \|p_n^{\alpha,\beta} v^{\rho,\tau}\|_\infty < \infty \quad \text{and} \quad \|p_n^{\alpha,\beta} v^{\alpha-\rho,\beta-\tau}\|_1 \leq c\,\ln(n+1)\,,\ \ n \in \mathbb{N},$$

which is equivalent to (3.3) (in view of Theorem 2.3). □

Remark 3.8. Let $0 \leq \rho < 1+\alpha$ and $0 \leq \tau < 1+\beta$. The proof of Proposition 3.7 shows that, for every constant $k \geq 1$, (3.3) is also equivalent to

$$\|S_n^{\alpha,\beta} f\|_{\infty,\rho,\tau} \leq c\left(\|f\|_{\infty,\rho,\tau} + \frac{\|f'\|_{\infty,\rho+\frac{1}{2},\tau+\frac{1}{2}}}{n}\right)\ln^k(n+1)\,, \quad \begin{array}{l} f \in AC_{\mathrm{loc}}\,, \\ f' \in \mathbf{L}^\infty_{\rho+\frac{1}{2},\tau+\frac{1}{2}}\,. \end{array}$$

4. Modified Fourier Projections

In some applications (for example fast algorithms for certain operator equations), in which Fourier projections are used, one needs on one hand side the properties

(4.1) $\langle f - S_n^{\alpha,\beta} f, P_n\rangle_{\alpha,\beta} = 0$ and $S_n^{\alpha,\beta} P_n = P_n$ for $P_n \in \mathbb{P}_{n-1}$ and $f \in \mathbf{L}^2_{\frac{\alpha}{2},\frac{\beta}{2}}$, where $\alpha, \beta > -1$ are given, and on the other hand side a "good" norm behaviour of the Fourier projections $S_n^{\alpha,\beta}$ in a given space $\mathbf{L}_{\rho,\tau}^{\infty}$ or $\mathbf{C}_{\rho,\tau}$. But if the conditions of Theorem 2.1 are not satisfied for these given $\alpha, \beta, \rho, \tau$, then we have no "good" norm behaviour, i.e. the norms increase with a power of n (see Remark 2.5). So it is interesting to ask whether $S_n^{\alpha,\beta}$ can be modified in such a way that properties similar to (4.1) hold true and the norms of the modified operators are bounded by a power of $\ln n$, also if (2.1) is not satisfied. Based on the discrete estimate (2.2) we will give such modifications for the case that the weaker inequalities

$$(4.2)\quad \max\left\{0, \frac{\alpha}{2}-\frac{1}{4}\right\} \le \rho \le \frac{\alpha}{2}+\frac{3}{4} \quad \text{and} \quad \max\left\{0, \frac{\beta}{2}-\frac{1}{4}\right\} \le \tau \le \frac{\beta}{2}+\frac{3}{4}$$

hold true instead of (2.1). For this aim, set

$$(4.3)\qquad r = \begin{cases} 0 & \text{if} \quad \rho \ge \frac{\alpha}{2}+\frac{1}{4} \\ 1 & \text{if} \quad \rho < \frac{\alpha}{2}+\frac{1}{4} \end{cases}, \qquad s = \begin{cases} 0 & \text{if} \quad \tau \ge \frac{\beta}{2}+\frac{1}{4} \\ 1 & \text{if} \quad \tau < \frac{\beta}{2}+\frac{1}{4} \end{cases}$$

and choose, for every $n \in \mathbb{N}$,

$$(4.4)\qquad x_{n0} \in [-1, x_{n1}) \text{ if } s = 1, \quad x_{n,n+1} \in (x_{nn}, 1] \text{ if } r = 1,$$

such that

$$(4.5)\qquad x_{n1} - x_{n0} \sim n^{-2}, \quad x_{n,n+1} - x_{nn} \sim n^{-2}$$

and, additionally,

$$(4.6)\qquad 1 + x_{n0} \sim n^{-2} \text{ if } \tau > 0, \quad 1 - x_{n,n+1} \sim n^{-2} \text{ if } \rho > 0.$$

(In view of (2.7) we see that the conditions (4.5) and (4.6) can easily be fulfiled.)

By $L_{n,r,s}^{\alpha,\beta} f$ we denote the Lagrange interpolation polynomial of degree less than $n+r+s$ interpolating f at the knots $x_{n,1-s}, \ldots, x_{n,n+r}$. It is well-known that, supposed that (4.2) holds true,

$$(4.7)\qquad \|L_{n,r,s}^{\alpha,\beta} f\|_{\infty,\rho,\tau} \le c\, \|f\|_{\infty,\rho,\tau} \ln(n+1), \quad n \in \mathbb{N}, \; f \in \mathbf{C}_{\rho,\tau},$$

where c is independent of n and f (cf. [16] or [6], Theorem 4.1). Using this result, one can easily prove that, if (4.2) is satisfied,

$$(4.8)\qquad \|L_{n,r,s}^{\alpha,\beta} f\|_{\infty,\rho,\tau} \le c \max_{k=1-s,\ldots,n+r} |f v^{\rho,\tau}(x_{nk})| \ln(n+1) \quad \text{for all } f : \{x_{nk}\}_{k=1-s}^{n+r} \to \mathbb{C}.$$

(Use that $L_{n,r,s}^{\alpha,\beta} f = L_{n,r,s}^{\alpha,\beta} \tilde{f}$, where $\tilde{f} = S v^{-\rho,-\tau}$ and S is the picewise linear spline interpolating $f v^{\rho,\tau}$ at the knots $x_{n,1-s}, \ldots, x_{n,n+r}$ and vanishing in $x \in \{-1, 1\} \setminus \{x_{n,1-s}, x_{n,n+r}\}$.)

Besides $L_{n,r,s}^{\alpha,\beta}f$ we will also use the notation $L_{r,s}^n f$ for the Lagrange interpolation polynomial of degree less than $r+s$ interpolating f in x_{n0} if $s=1$ and in $x_{n,n+1}$ if $r=1$, i.e.

$$(L_{r,s}^n f)(x) = \begin{cases} 0 & \text{if } s=r=0 \\ f(x_{n0}) & \text{if } s=1,\ r=0 \\ f(x_{n,n+1}) & \text{if } s=0,\ r=1 \\ f(x_{n0})\dfrac{x_{n,n+1}-x}{x_{n,n+1}-x_{n0}} + f(x_{n,n+1})\dfrac{x-x_{n0}}{x_{n,n+1}-x_{n,0}} & \text{if } s=r=1\,. \end{cases}$$

We will study modified Fourier projections in X, where X is one of the Banach spaces $\mathbf{L}_{\rho,\tau}^\infty$, $\mathbf{C}_{\rho,\tau}$ or

$$\mathbf{B}_{\rho,\tau} = \{\, f : fv^{\rho,\tau} \text{ is defined, measurable and bounded on } (-1,1)\cup\{2^{-\rho},-2^{-\tau}\}\,\}.$$

The norm in $\mathbf{B}_{\rho,\tau}$ is given by

$$\|f\|_{B_{\rho,\tau}} = \sup\{\,|fv^{\rho,\tau}(x)| : x\in(-1,1)\cup\{2^{-\rho},-2^{-\tau}\}\,\}\,.$$

Remark that $\|f\|_{B_{\rho,\tau}}$ can be greater than $\|f\|_{\infty,\rho,\tau}$ for $f\in\mathbf{B}_{\rho,\tau}\setminus\mathbf{C}_{\rho,\tau}$.

Our modifications of the projections $S_n^{\alpha,\beta}$ depend on a given sequence of linear operators I_n, acting from X into a linear space of functions defined on a set $D_n\subset[-1,1]$, which contains the point x_{n0} if $s=1$ and $x_{n,n+1}$ if $r=1$. These operators I_n have to fulfil the following conditions:

$$I_nP = P_{|D_n} \quad \text{for all } P\in\mathbb{P}_n\,, \tag{4.9}$$

$$\left.\begin{aligned} |(v^{\rho,\tau}I_nf)(x_{n0})| &\le c\|f\|_X\ln(n+1) \text{ if } s=1 \\ |(v^{\rho,\tau}I_nf)(x_{n,n+1})| &\le c\|f\|_X\ln(n+1) \text{ if } r=1 \end{aligned}\right\}\quad \text{for all } f\in X\,. \tag{4.10}$$

For example, we may take $I_n = S_{n+1}^{\tilde\alpha,\tilde\beta}$ with $\tilde\alpha\in[2\rho-\frac32,2\rho-\frac12]$, $\tilde\beta\in[2\tau-\frac32,2\tau-\frac12]$ (and, of course, $\tilde\alpha>\rho-1$, $\tilde\beta>\tau-1$) if $X=\mathbf{L}_{\rho,\tau}^\infty$ (see Theorem 2.1) and $I_n=I : X\to X$ if $X=\mathbf{C}_{\rho,\tau}$ or $X=\mathbf{B}_{\rho,\tau}$. Now we are able to state the main result of this section.

Theorem 4.1. *Let $\rho<1+\alpha$, $\tau<1+\beta$ such that (4.2) is fulfiled, let r, s be defined in (4.3) and let $x_{n0}\in[-1,x_{n1})$, $x_{n,n+1}\in(x_{nn},1]$ such that (4.5) and (4.6) are satisfied. Define the modified Fourier projection $S_{n,r,s}^{\alpha,\beta}$ by*

$$S_{n,r,s}^{\alpha,\beta}f = S_n^{\alpha,\beta}f + p_n^{\alpha,\beta}L_{r,s}^n\left(\frac{I_nf - S_n^{\alpha,\beta}f}{p_n^{\alpha,\beta}}\right) \qquad (n\in\mathbb{N})\,.$$

If the linear operators I_n satisfy the conditions (4.9) and (4.10) with one of the spaces $X=\mathbf{L}_{\rho,\tau}^\infty$, $X=\mathbf{B}_{\rho,\tau}$ or $X=\mathbf{C}_{\rho,\tau}$, then $S_{n,r,s}^{\alpha,\beta}$ are linear operators from X into $\mathbb{P}_{n+r+s-1}$ with

$$\|S_{n,r,s}^{\alpha,\beta}\|_{X\to X} \le c\ln^2(n+1)\,,\quad n=1,2,\ldots\,, \tag{4.11}$$

where c is independent of n. Moreover, $S^{\alpha,\beta}_{n,r,s}$ has the following properties:

$$(4.12)\quad \langle f - S^{\alpha,\beta}_{n,r,s}f, P\rangle_{\alpha,\beta} = 0 \quad \textit{for} \quad P \in \mathbb{P}_{n-\max\{r+s,1\}}\,, \quad f \in \mathbf{L}^2_{\frac{\alpha}{2},\frac{\beta}{2}} \cap X$$

and $S^{\alpha,\beta}_{n,r,s}P = P$ for $P \in \mathbb{P}_{n+\min\{r+s-1,0\}}$. If $r + s \le 1$ then $S^{\alpha,\beta}_{n,r,s}$ is a projection from X onto $\mathbb{P}_{n+r+s-1}$.

Proof. Let $f \in X$. In view of (4.8) we have

$$\|S^{\alpha,\beta}_{n,r,s}f\|_{\infty,\rho,\tau} = \|L^{\alpha,\beta}_{n,r,s}S^{\alpha,\beta}_{n,r,s}f\|_{\infty,\rho,\tau} \le c \max_{k=1-s,\dots,n+r} |v^{\rho,\tau}S^{\alpha,\beta}_{n,r,s}f|(x_{nk})\ln(n+1).$$

For $k \in \{1,\dots,n\}$ we may use the estimate (2.2):

$$|v^{\rho,\tau}S^{\alpha,\beta}_{n,r,s}f|(x_{nk}) = |v^{\rho,\tau}S^{\alpha,\beta}_{n}f|(x_{nk}) \le c\|f\|_{\infty,\rho,\tau}\ln(n+1) \le c\|f\|_X\ln(n+1).$$

For $k \notin \{1,\dots,n\}$ we have

$$|v^{\rho,\tau}S^{\alpha,\beta}_{n,r,s}f|(x_{nk}) = |v^{\rho,\tau}I_n f|(x_{nk}) \le c\|f\|_X\ln(n+1)$$

and (4.11) is proved. The relation (4.12) follows from the corresponding relation in (4.1) and the fact that $\langle p^{\alpha,\beta}_n, P\rangle_{\alpha,\beta} = 0$ for all $P \in \mathbb{P}_{n-1}$. It remains to prove $S^{\alpha,\beta}_{n,r,s}P = P$ for all $P \in \mathbb{P}_{n+\min\{r+s-1,0\}}$. For $r = s = 0$ this is clear and for $r + s \ge 1$ one can easily check that $S^{\alpha,\beta}_{n,r,s}p^{\alpha,\beta}_j = p^{\alpha,\beta}_j$, $j = 0,\dots,n$. □

Remark 4.2. Let the assumptions of Theorem 4.1 be satisfied. If $r = s = 1$ then $S^{\alpha,\beta}_{n,r,s}$ is no projection onto $\mathbb{P}_{n+1}$ since, for example, $S^{\alpha,\beta}_{n,r,s}p^{\alpha,\beta}_{n+1} = p^{\alpha,\beta}_n L^n_{r,s}\left(\frac{I_n p^{\alpha,\beta}_{n+1}}{p^{\alpha,\beta}_n}\right) \ne p^{\alpha,\beta}_{n+1}$. If we need a projection in this case, we may define

$$\tilde{S}^{\alpha,\beta}_{n,1,1}f := L^{\alpha,\beta}_{n,1,1}S^{\alpha,\beta}_{n+1,1,1}f = S^{\alpha,\beta}_{n+1}f + L^{\alpha,\beta}_{n,1,1}\left(p^{\alpha,\beta}_{n+1}L^{n+1}_{1,1}\left(\frac{I_{n+1}f - S^{\alpha,\beta}_{n+1}f}{p^{\alpha,\beta}_{n+1}}\right)\right).$$

Here we can take different sequences $\{x_{n0}\}, \{x_{n,n+1}\}$ and $\{\tilde{x}_{n0}\}, \{\tilde{x}_{n,n+1}\}$ of additional knots, satisfying (4.4)–(4.6), for the definitions of $S^{\alpha,\beta}_{n,1,1}$ and $L^{\alpha,\beta}_{n,1,1}$. Since $S^{\alpha,\beta}_{n+1,1,1}$ is exact on $\mathbb{P}_{n+1}$ we get a projection $\tilde{S}^{\alpha,\beta}_{n,1,1}$ from X onto $\mathbb{P}_{n+1}$ in this way. Moreover, in view of (4.11) and (4.7), we have

$$\|\tilde{S}^{\alpha,\beta}_{n,1,1}\|_{X\to X} \le c\ln^3(n+1)\,, \quad n = 1,2,\dots\,.$$

Using the recurrence formula $b_{n+1}p^{\alpha,\beta}_{n+1}(x) = (x - a_n)p^{\alpha,\beta}_n(x) - b_n p^{\alpha,\beta}_{n-1}$ we get the representation

$$L^{\alpha,\beta}_{n,1,1}\left(p^{\alpha,\beta}_{n+1}L^{n+1}_{1,1}\left(\frac{I_{n+1}f - S^{\alpha,\beta}_{n+1}f}{p^{\alpha,\beta}_{n+1}}\right)\right)$$

$$= \frac{p_n^{\alpha,\beta}}{b_{n+1}} L_{1,1}^n \left((x-a_n) L_{1,1}^{n+1} \left(\frac{I_{n+1}f - S_{n+1}^{\alpha,\beta} f}{p_{n+1}^{\alpha,\beta}} \right) \right) - \frac{b_n}{b_{n+1}} p_{n-1}^{\alpha,\beta} L_{1,1}^{n+1} \left(\frac{I_{n+1}f - S_{n+1}^{\alpha,\beta} f}{p_{n+1}^{\alpha,\beta}} \right)$$

($L_{1,1}^{n+1}$ w.r.t. $x_{n+1,0}, x_{n+1,n+2}$ and $L_{1,1}^n$ w.r.t. $\tilde{x}_{n0}, \tilde{x}_{n,n+1}$), which shows that

$$\langle f - \tilde{S}_{n,1,1}^{\alpha,\beta} f, P \rangle_{\alpha,\beta} = 0 \quad \text{for } P \in \mathbb{P}_{n-3} \text{ and } f \in \mathbf{L}^2_{\frac{\alpha}{2},\frac{\beta}{2}} \cap X \,.$$

Explicite formulas for $a_n = a_n(\alpha,\beta)$ and $b_n = b_n(\alpha,\beta)$ can be found in literature.

Example. We take $\rho = \tau = 0$ and $\alpha = \beta = \frac{1}{2}$ ($\Rightarrow r = s = 1$), i.e. we want to modify $S_n^{\frac{1}{2},\frac{1}{2}}$ such that the norms of the modified operators in $\mathbf{L}^\infty[-1,1]$ can be estimated by a power of $\ln(n+1)$. In view of Theorem 2.1 and (2.7), all conditions of Theorem 4.1 (with $X = \mathbf{L}^\infty[-1,1]$) are satisfied for

$$I_n = S_{n+1}^{-\frac{1}{2},-\frac{1}{2}}, \quad x_{n0} = -1, \quad x_{n,n+1} = 1 \,.$$

Let $f = \sum_{j=0}^\infty c_j\, p_j = \sum_{j=0}^\infty d_j\, q_j \in \mathbf{L}^\infty[-1,1]$, where $p_j := p_j^{\frac{1}{2},\frac{1}{2}}$ and $q_j := p_j^{-\frac{1}{2},-\frac{1}{2}}$. (The first series converges in $\mathbf{L}^2_{\frac{1}{4},\frac{1}{4}}$ and the second one in $\mathbf{L}^2_{-\frac{1}{4},-\frac{1}{4}}$.) Then we have

$$S_{n+1}^{-\frac{1}{2},-\frac{1}{2}} f - S_n^{\frac{1}{2},\frac{1}{2}} f = d_n\, q_n + S_n^{\frac{1}{2},\frac{1}{2}} (S_n^{-\frac{1}{2},-\frac{1}{2}} f - f)$$

$$= d_n\, q_n - \sum_{k=0}^{n-1} \left\langle \sum_{j=n}^\infty d_j\, q_j,\, p_k \right\rangle_{\frac{1}{2},\frac{1}{2}} p_k = d_n\, q_n - \sum_{k=0}^{n-1} \sum_{j=n}^\infty d_j \langle q_j,\, p_k \rangle_{\frac{1}{2},\frac{1}{2}}\, p_k \,.$$

Since $\langle q_j,\, p_k \rangle_{\frac{1}{2},\frac{1}{2}} = \langle q_j,\, p_k v^{1,1} \rangle_{-\frac{1}{2},-\frac{1}{2}} = 0$ for $k < j-2$, we get

$$\begin{aligned} S_{n+1}^{-\frac{1}{2},-\frac{1}{2}} f - S_n^{\frac{1}{2},\frac{1}{2}} f &= d_n\, q_n - d_n \langle q_n,\, p_{n-2} v^{1,1} \rangle_{-\frac{1}{2},-\frac{1}{2}}\, p_{n-2} \\ &\quad - \left[d_n \langle q_n,\, p_{n-1} \rangle_{\frac{1}{2},\frac{1}{2}} + d_{n+1} \langle q_{n+1},\, p_{n-1} v^{1,1} \rangle_{-\frac{1}{2},-\frac{1}{2}} \right] p_{n-1} \,. \end{aligned}$$

Using the well-known facts

$$p_l(x) = \sqrt{\frac{2}{\pi}}\, 2^l x^l + \dots, \quad q_m(x) = \sqrt{\frac{2}{\pi}}\, 2^{m-1} x^m + \dots, \quad p_m^{\alpha,\alpha}(x) = (-1)^m p_m^{\alpha,\alpha}(-x)$$

we obtain $\langle q_n,\, p_{n-1} \rangle_{\frac{1}{2},\frac{1}{2}} = 0$ ($q_n p_{n-1}$ is an odd function) and

$$\langle q_m,\, p_{m-2} v^{1,1} \rangle_{-\frac{1}{2},-\frac{1}{2}} = -\sqrt{\frac{2}{\pi}}\, 2^{m-2} \langle q_m,\, x^m \rangle_{-\frac{1}{2},-\frac{1}{2}} = -\frac{1}{2} \langle q_m,\, q_m \rangle_{-\frac{1}{2},-\frac{1}{2}} = -\frac{1}{2} \,.$$

Together with $q_n(x) = x\,p_{n-1}(x) - p_{n-2}(x)$ ([20], Remark 3.1.2, p.9) we get

$$\begin{aligned}(S_{n+1}^{-\frac{1}{2},-\frac{1}{2}} f - S_n^{\frac{1}{2},\frac{1}{2}} f)(x) &= d_n\, q_n(x) + \frac{d_n}{2}\, p_{n-2}(x) + \frac{d_{n+1}}{2}\, p_{n-1}(x) \\ &= \left(d_n\, x + \frac{d_{n+1}}{2}\right) p_{n-1}(x) - \frac{d_n}{2}\, p_{n-2}(x)\,.\end{aligned}$$

Since $p_m(\pm 1) = \sqrt{2/\pi}\,(\pm 1)^m (m+1)$, it follows

$$L_{1,1}^n \left(\frac{S_{n+1}^{-\frac{1}{2},-\frac{1}{2}} f - S_n^{\frac{1}{2},\frac{1}{2}} f}{p_n} \right) = \frac{d_n}{2} + \frac{n\, d_{n+1}}{2(n+1)}\, x\,.$$

Thus, the modified Fourier projections $S_{n,1,1}^{\frac{1}{2},\frac{1}{2}}$ are given by

$$(S_{n,1,1}^{\frac{1}{2},\frac{1}{2}} f)(x) = (S_n^{\frac{1}{2},\frac{1}{2}} f)(x) + \frac{p_n(x)}{2} \left(d_n + \frac{n\, d_{n+1}}{n+1}\, x \right).$$

We have $\|S_{n,1,1}^{\frac{1}{2},\frac{1}{2}}\|_{L^\infty \to L^\infty} \le c \ln^2(n+1)$, but $S_{n,1,1}^{\frac{1}{2},\frac{1}{2}}$ is no projection.

A projection $\tilde{S}_{n,1,1}^{\frac{1}{2},\frac{1}{2}} : L^\infty[-1,1] \to \mathbb{P}_{n+1}$ with $\|\tilde{S}_{n,1,1}^{\frac{1}{2},\frac{1}{2}}\|_{L^\infty \to L^\infty} \le c \ln^3(n+1)$ is given by

$$\begin{aligned}(\tilde{S}_{n,1,1}^{\frac{1}{2},\frac{1}{2}} f)(x) &= (S_{n+1}^{\frac{1}{2},\frac{1}{2}} f)(x) + p_n(x) \left(\frac{(n+1)\, d_{n+2}}{n+2} + d_{n+1}\, x \right) \\ &\quad - \frac{p_{n-1}(x)}{2} \left(d_{n+1} + \frac{(n+1)\, d_{n-2}}{n+2}\, x \right)\end{aligned}$$

(see Remark 4.2 with $\tilde{x}_{n0} = -1$, $\tilde{x}_{n,n+1} = 1$ and use $a_n(\frac{1}{2}, \frac{1}{2}) = 0$, $b_n(\frac{1}{2}, \frac{1}{2}) = \frac{1}{2}$).

5. Proofs of Theorem 2.1 and Remark 2.4

Proof of Remark 2.4, first part.

First we prove the equivalence of $\|p_n^{\alpha,\beta} v^{\alpha-\rho,\beta-\tau}\|_1 \le c \ln(n+1)$ and

(5.1) $$\rho < 1+\alpha\,, \quad \tau < 1+\beta\,, \quad \rho \le \frac{\alpha}{2} + \frac{3}{4}\,, \quad \tau \le \frac{\beta}{2} + \frac{3}{4}\,.$$

If (5.1) is satisfied then, in view of Badkov's Theorem

(5.2) $$|p_n^{\alpha,\beta}(x)| \le c \left(\sqrt{1-x} + \frac{1}{n} \right)^{-\alpha-\frac{1}{2}} \left(\sqrt{1+x} + \frac{1}{n} \right)^{-\beta-\frac{1}{2}}, \qquad x \in [-1,1]\,,\ n \in \mathbb{N}$$

([1], Theorem 1.1; see also [20], Lemma 9.29, p. 170, or (2.6)–(2.8)),

$$\|p_n^{\alpha,\beta}v^{\alpha-\rho,\beta-\tau}\|_{L^1[-1+n^{-2},1-n^{-2}]} \le c\|v^{\frac{\alpha}{2}-\frac{1}{4}-\rho,\frac{\beta}{2}-\frac{1}{4}-\tau}\|_{L^1[-1+n^{-2},1-n^{-2}]}$$

$$= c\int_{-1+n^{-2}}^{1-n^{-2}} v^{\frac{\alpha}{2}-\frac{1}{4}-\rho,\frac{\beta}{2}-\frac{1}{4}-\tau}(x)dx \le c\int_{-1+n^{-2}}^{0}\frac{dx}{1+x} + c\int_0^{1-n^{-2}}\frac{dx}{1-x} = c\ln n$$

for $n \in \mathbb{N}$. If all inequalities in (5.1) hold strictly, then we get the estimate $\|p_n^{\alpha,\beta}v^{\alpha-\rho,\beta-\tau}\|_{L^1[-1+n^{-2},1-n^{-2}]} \le c\|v^{\frac{\alpha}{2}-\frac{1}{4}-\rho,\frac{\beta}{2}-\frac{1}{4}-\tau}\|_1 \le c$. Taking into account (2.9), it follows

$$\|p_n^{\alpha,\beta}v^{\alpha-\rho,\beta-\tau}\|_1 \le c\ln(n+1)\,, \quad n = 1,2,\dots\,, \tag{5.3}$$

even without $\ln(n+1)$ if all inequalities in (5.1) hold strictly. Now, let (5.3) be satisfied. Then it follows $\rho < 1+\alpha$, $\tau < 1+\beta$, since otherwise $\|p_n^{\alpha,\beta}v^{\alpha-\rho,\beta-\tau}\|_1 = \infty$ for all n. Moreover, if $C > 0$ is a constant, such that $-1+Cn^{-2} \le \frac{1}{2}(x_{n1}-1)$ (cf. (2.7)), then

$$|p_n^{\alpha,\beta}(x)| \sim n(1+x_{n1})\left(\sqrt{1+x}+\frac{1}{n}\right)^{-\beta-\frac{3}{2}} \sim n^{\beta+\frac{1}{2}}, \; x \in [-1,-1+Cn^{-2}] \tag{5.4}$$

(see (2.8) and (2.7)), which implies

$$\|p_n^{\alpha,\beta}v^{\alpha-\rho,\beta-\tau}\|_1 \ge \int_{-1}^{-1+Cn^{-2}} |p_n^{\alpha,\beta}(x)|v^{\alpha-\rho,\beta-\tau}(x)dx$$

$$\ge cn^{\beta+\frac{1}{2}}\int_{-1}^{-1+Cn^{-2}}(1+x)^{\beta-\tau}dx = cn^{\beta+\frac{1}{2}}n^{-2(\beta-\tau+1)} = cn^{2\tau-\beta-\frac{3}{2}}\,.$$

Consequently, our assumed estimate (5.3) can only be satisfied if $2\tau-\beta-\frac{3}{2} \le 0$, i.e. $\tau \le \frac{\beta}{2}+\frac{3}{4}$. Otherwise we have $\|p_n^{\alpha,\beta}v^{\alpha-\rho,\beta-\tau}\|_1 \ge cn^\mu$ with $\mu = 2\tau-\beta-\frac{3}{2} > 0$. Analogously, one can prove that (5.3) implies $\rho \le \frac{\alpha}{2}+\frac{3}{4}$, since otherwise $\|p_n^{\alpha,\beta}v^{\alpha-\rho,\beta-\tau}\|_1 \ge cn^\mu$ with $\mu = 2\rho-\alpha-\frac{3}{2} > 0$. □

Proof of Remark 2.4, second part.
Now we prove the equivalence of $\sup_n \|p_n^{\alpha,\beta}v^{\gamma,\delta}\|_\infty < \infty$ and

$$\gamma \ge \max\left\{0, \frac{\alpha}{2}+\frac{1}{4}\right\}, \quad \delta \ge \max\left\{0, \frac{\beta}{2}+\frac{1}{4}\right\}\,. \tag{5.5}$$

If (5.5) is satisfied then, in view of (5.2),

$$\|p_n^{\alpha,\beta}v^{\gamma,\delta}\|_{L^\infty[-1+n^{-2},1-n^{-2}]} \le c\|v^{\gamma-\frac{\alpha}{2}-\frac{1}{4},\delta-\frac{\beta}{2}-\frac{1}{4}}\|_{L^\infty[-1+n^{-2},1-n^{-2}]} \le c\,, \quad n \in \mathbb{N}.$$

Taking into account (2.9), it follows

$$\|p_n^{\alpha,\beta}v^{\gamma,\delta}\|_\infty \le c\,, \quad n = 0,1,2,\dots\,. \tag{5.6}$$

If (5.6) is satisfied then $\gamma,\delta \ge 0$, since otherwise $\|p_n^{\alpha,\beta}v^{\gamma,\delta}\|_\infty = \infty$ for all n. In view of (5.4), there is a constant $C > 0$, such that

$$|p_n^{\alpha,\beta}(-1+Cn^{-2})|\,v^{\gamma,\delta}(-1+Cn^{-2}) \sim n^{\beta+\frac{1}{2}-2\delta}\,, \quad n \in \mathbb{N}\,.$$

So we see that (5.6) implies $\beta + \frac{1}{2} - 2\delta \le 0$, i.e. $\delta \ge \frac{\beta}{2} + \frac{1}{4}$, since otherwise $\|p_n^{\alpha,\beta}v^{\gamma,\delta}\|_\infty \ge cn^\mu$ with $\mu = \beta + \frac{1}{2} - 2\delta > 0$. Analogously, we can prove that (5.6) implies $\gamma \ge \frac{\alpha}{2} + \frac{1}{4}$ and that $\|p_n^{\alpha,\beta}v^{\gamma,\delta}\|_\infty \ge cn^\mu$ with $\mu = \alpha + \frac{1}{2} - 2\gamma > 0$ if $\gamma < \frac{\alpha}{2} + \frac{1}{4}$. □

To prove Theorem 2.1 we need some auxiliary material.

Lemma 5.1. *Let $\eta,\nu \ge 0$. Then*

$$\|P_n\|_\infty \le c n^{2\max\{\eta,\nu\}}\|P_n v^{\eta,\nu}\|_\infty \quad \textit{for all } P_n \in \mathbb{P}_n \textit{ and } n \in \mathbb{N}, \tag{5.7}$$

where c is a constant independent of P_n and n.

Proof. This follows from (2.9) (applied with $p = \infty$). □

Lemma 5.2. **(cf. [21] or [18])** *There are numbers c_n, d_n ($n \in \mathbb{N}$) with $c_n, d_n \sim 1$, such that*

$$\begin{aligned}\sum_{j=0}^{n-1} p_j^{\alpha,\beta}(t)p_j^{\alpha,\beta}(x) &= c_n\, p_n^{\alpha,\beta}(t)p_n^{\alpha,\beta}(x) + \\ & d_n \frac{(1-t^2)p_{n-1}^{\alpha+1,\beta+1}(t)p_n^{\alpha,\beta}(x) - (1-x^2)p_{n-1}^{\alpha+1,\beta+1}(x)p_n^{\alpha,\beta}(t)}{x-t}.\end{aligned} \tag{5.8}$$

Lemma 5.3. *Let $0 \le \eta,\nu \le 1$ and denote by $\chi_{m,x}$ ($x \in (-1,1)$) the characteristic function of the interval*

$$I_m(x) = \begin{cases} \left(-1, 1 - C\frac{1-x}{m}\right) & \textit{if} \quad \eta = 1,\, \nu < 1 \\ \left(-1 + C\frac{1+x}{m}, 1\right) & \textit{if} \quad \eta < 1,\, \nu = 1 \\ \left(-1 + C\frac{1+x}{m}, 1 - C\frac{1-x}{m}\right) & \textit{if} \quad \eta = \nu = 1 \\ (-1,1) & \textit{if} \quad \eta,\nu < 1\,, \end{cases}$$

where $C \in (0,1]$ is fixed. Then there exists a constant c, independent of m and x, such that

$$\left(\int_{-1}^{x-\frac{1+x}{m}} + \int_{x+\frac{1-x}{m}}^{1}\right) \frac{v^{-\eta,-\nu}(t)}{|x-t|}\,\chi_{m,x}(t)\,dt \le c\,v^{-\eta,-\nu}(x)\ln m \tag{5.9}$$

for all $x \in (-1,1)$ *and all* $m \in \mathbb{N}$.

Proof. For $m = 1$ the assertion is trivial. So we may assume $m \geq 2$. It is well-known (see e.g. [6], (2.7) and (2.8)) that

$$\int_{-1}^{x-\frac{1+x}{m}} \frac{v^{-\eta,-\nu}(t)}{|x-t|}\,dt \leq c\,v^{-\eta,-\nu}(x)\ln m \quad \text{if } \nu < 1$$

(use $(1-t)^{-\eta} \leq (1-x)^{-\eta}$ and the substitution $1+t = (1+x)y$) and

$$\int_{x+\frac{1-x}{m}}^{1} \frac{v^{-\eta,-\nu}(t)}{|x-t|}\,dt \leq c\,v^{-\eta,-\nu}(x)\ln m \quad \text{if } \eta < 1$$

(use $(1+t)^{-\nu} \leq (1+x)^{-\nu}$ and the substitution $1-t = (1-x)y$). So it remains to prove that

$$\int_{-1+C\frac{1+x}{m}}^{x-\frac{1+x}{m}} \frac{v^{-\eta,-\nu}(t)}{|x-t|}\,dt \leq c\,v^{-\eta,-\nu}(x)\ln m \quad \text{if } \nu = 1\,, \tag{5.10}$$

$$\int_{x+\frac{1-x}{m}}^{1-C\frac{1-x}{m}} \frac{v^{-\eta,-\nu}(t)}{|x-t|}\,dt \leq c\,v^{-\eta,-\nu}(x)\ln m \quad \text{if } \eta = 1\,. \tag{5.11}$$

If $\nu = 1$ then the left hand side of (5.10) can be estimated by

$$(1-x)^{-\eta}\int_{-1+C\frac{1+x}{m}}^{x-\frac{1+x}{m}} \frac{dt}{(1+t)(x-t)} = \frac{(1-x)^{-\eta}}{1+x}\int_{\frac{C}{m}}^{1-\frac{1}{m}} \frac{dy}{y(1-y)}$$

$$\leq c\,\frac{(1-x)^{-\eta}}{1+x}\left(\int_{\frac{C}{m}}^{\frac{C}{2}} \frac{dy}{y} + \int_{\frac{C}{2}}^{1-\frac{1}{m}} \frac{dy}{1-y}\right) \leq c\,v^{-\eta,-\nu}(x)\ln m\,,$$

where we used the substitution $1+t = (1+x)y$. Analogously one can prove (5.11). □

Lemma 5.4. *Let* $\rho < 1+\alpha$, $\tau < 1+\beta$ *and*

$$\frac{\alpha}{2}+\frac{1}{4} \leq \rho \leq \frac{\alpha}{2}+\frac{3}{4}\,, \quad \frac{\beta}{2}+\frac{1}{4} \leq \tau \leq \frac{\beta}{2}+\frac{3}{4}\,. \tag{5.12}$$

Then, for every fixed $C > 0$, *there exists a constant* c, *independent of* n *and* x, *such that*

$$\int_{-1}^{1} |k_n(x,t)|\,v^{\alpha-\rho,\beta-\tau}(t)\,dt \leq c\,v^{-\rho,-\tau}(x)\ln(n+1)$$

for all $n \in \mathbb{N}$ *and all* $x \in [-1+Cn^{-2}, 1-Cn^{-2}]$, *where*

$$k_n(x,t) = \frac{(1-t^2)p_{n-1}^{\alpha+1,\beta+1}(t)p_n^{\alpha,\beta}(x) - (1-x^2)p_{n-1}^{\alpha+1,\beta+1}(x)p_n^{\alpha,\beta}(t)}{x-t}\,.$$

If $x \in \{x_{n1}, \dots, x_{nn}\}$ then it is sufficient to suppose

$$\frac{\alpha}{2} - \frac{1}{4} \le \rho \le \frac{\alpha}{2} + \frac{3}{4}, \quad \frac{\beta}{2} - \frac{1}{4} \le \tau \le \frac{\beta}{2} + \frac{3}{4} \tag{5.13}$$

instead of (5.12).

Proof. Let ψ_n be the characteristic function of the interval $(-1+\frac{1}{2}n^{-2}, 1-\frac{1}{2}n^{-2})$. In view of (2.9) (applied with $p=1$) we have

$$\int_{-1}^{1} |k_n(x,t)|\, v^{\alpha-\rho,\beta-\tau}(t)\, dt \le c \int_{-1}^{1} |k_n(x,t)|\, \psi_n(t) v^{\alpha-\rho,\beta-\tau}(t)\, dt$$

for all $n \in \mathbb{N}$ and all $x \in (-1,1)$. We split the integral on the right hand side of this estimate into two terms

$$\left(\int_{-1}^{x-\frac{1+x}{m}} + \int_{x+\frac{1-x}{m}}^{1} \right) |k_n(x,t)|\, \psi_n(t) v^{\alpha-\rho,\beta-\tau}(t)\, dt\,, \tag{5.14}$$

$$\int_{x-\frac{1+x}{m}}^{x+\frac{1-x}{m}} |k_n(x,t)|\, \psi_n(t) v^{\alpha-\rho,\beta-\tau}(t)\, dt\,, \tag{5.15}$$

where $m = n^k$ with a fixed $k \in \mathbb{N}$, such that

$$k \ge 2 + \max\{0, 2\alpha+1, 2\beta+1\} + 2\max\{0, -\alpha, -\beta\}\,.$$

Using $|k_n(x,t)| \le c\,\|(v^{1,1}p_{n-1}^{\alpha+1,\beta+1})'\|_\infty \|p_n^{\alpha,\beta}\|_\infty + c\,\|v^{1,1}p_{n-1}^{\alpha+1,\beta+1}\|_\infty \|(p_n^{\alpha,\beta})'\|_\infty$, Markov's inequality $\|P_n'\|_\infty \le n^2 \|P_n\|_\infty$ ($P_n \in \mathbb{P}_n$) and (5.7), we get

$$|k_n(x,t)| \le$$
$$c\,n^{2+\max\{0,2\alpha+1,2\beta+1\}}\, \|p_{n-1}^{\alpha+1,\beta+1} v^{\frac{\alpha}{2}+\frac{5}{4},\frac{\beta}{2}+\frac{5}{4}}\|_\infty\, \|p_n^{\alpha,\beta} v^{\max\{0,\frac{\alpha}{2}+\frac{1}{4}\},\max\{0,\frac{\beta}{2}+\frac{1}{4}\}}\|_\infty.$$

The norms on the right hand side of this inequality are uniformly bounded (see Remark 2.4). Thus, $|k_n(x,t)| \le c\,n^{2+\max\{0,2\alpha+1,2\beta+1\}}$. For $t \in [x - \frac{1+x}{m}, x + \frac{1-x}{m}]$, $x \in [-1+Cn^{-2}, 1-Cn^{-2}]$, and $n \ge 2$ we have $\frac{1}{2}(1-x) \le 1-t \le 1-x+\frac{1+x}{m} \le 1-x+2n^{-2} \le c(1-x)$ and, analogously, $1+t \sim 1+x$. Consequently, if $n \ge 2$, we may estimate (5.15) by

$$c\,v^{\alpha-\rho,\beta-\tau}(x) \int_{x-\frac{1+x}{m}}^{x+\frac{1-x}{m}} |k_n(x,t)|\, dt \le c\,v^{\alpha-\rho,\beta-\tau}(x) \frac{n^{2+\max\{0,2\alpha+1,2\beta+1\}}}{m} \le$$

$$c\,v^{-\rho,-\tau}(x) \frac{n^{2+\max\{0,2\alpha+1,2\beta+1\}+2\max\{0,-\alpha,-\beta\}}}{m} \le c\,v^{-\rho,-\tau}(x) \ \text{ for } |x| \le 1 - \frac{C}{n^2}.$$

If $n = 1$ then it is obvious that, for $|x| \le 1 - C$, (5.15) can be estimated by $cv^{-\rho,-\tau}(x)$ too. In view of (5.2) we have

$$|k_n(x,t)|$$

$$\le c\,\frac{v^{-\frac{\alpha}{2}+\frac{1}{4},-\frac{\beta}{2}+\frac{1}{4}}(t)v^{-\frac{\alpha}{2}-\frac{1}{4},-\frac{\beta}{2}-\frac{1}{4}}(x)+v^{-\frac{\alpha}{2}+\frac{1}{4},-\frac{\beta}{2}+\frac{1}{4}}(x)v^{-\frac{\alpha}{2}-\frac{1}{4},-\frac{\beta}{2}-\frac{1}{4}}(t)}{|x-t|} \tag{5.16}$$

$$\text{for } |t| \le 1 - \frac{1}{2n^2} \text{ and } |x| \le 1 - \frac{C}{n^2}\,.$$

If $x \in \{x_{n1}, \ldots, x_{nn}\}$ then this is even true without the first addend, since the corresponding addend in the definition of $k_n(x,t)$ vanishes and $1 \pm x_{nj} \ge cn^{-2}$ (cf. (2.7)). Obviously, for all $x \in (-1,1)$, the interval $(-1+\frac{1}{2}n^{-2}, 1-\frac{1}{2}n^{-2})$ is contained in $(-1+\frac{1+x}{4m}, 1-\frac{1-x}{4m})$. So we may use Lemma 5.3 and (5.16) to estimate (5.14) by

$$c\left(\int_{-1}^{x-\frac{1+x}{m}} + \int_{x+\frac{1-x}{m}}^{1}\right) \frac{v^{\frac{\alpha}{2}+\frac{1}{4}-\rho,\frac{\beta}{2}+\frac{1}{4}-\tau}(t)v^{-\frac{\alpha}{2}-\frac{1}{4},-\frac{\beta}{2}-\frac{1}{4}}(x)}{|x-t|}\chi_{m,x}(t)\,dt$$

$$+c\left(\int_{-1}^{x-\frac{1+x}{m}} + \int_{x+\frac{1-x}{m}}^{1}\right) \frac{v^{-\frac{\alpha}{2}+\frac{1}{4},-\frac{\beta}{2}+\frac{1}{4}}(x)v^{\frac{\alpha}{2}-\frac{1}{4}-\rho,\frac{\beta}{2}-\frac{1}{4}-\tau}(t)}{|x-t|}\chi_{m,x}(t)\,dt$$

$$\le cv^{-\rho,-\tau}(x)\ln n \quad \text{for } |x| \le 1 - Cn^{-2}\,.$$

If $x \in \{x_{n1}, \ldots, x_{nn}\}$ then we only have to estimate the second integral, which is even possible if (5.13) holds true instead of (5.12). □

Proof of Theorem 2.1.
Let $f \in \mathbf{L}^\infty_{\rho,\tau}$. In view of (5.8) we have the representation

$$\begin{aligned}(S_n^{\alpha,\beta}f)(x) &= \int_{-1}^{1}\left(\sum_{j=0}^{n-1} p_j^{\alpha,\beta}(t)p_j^{\alpha,\beta}(x)\right) f(t)v^{\alpha,\beta}(t)dt \\ &= c_n\int_{-1}^{1} p_n^{\alpha,\beta}(t)p_n^{\alpha,\beta}(x)f(t)v^{\alpha,\beta}(t)dt + d_n\int_{-1}^{1} k_n(x,t)f(t)v^{\alpha,\beta}(t)dt \\ &=: A_n(f,x) + B_n(f,x)\end{aligned} \tag{5.17}$$

with $k_n(x,t)$ from Lemma 5.4 and $c_n, d_n \sim 1$. The conditions for ρ and τ in Theorem 2.1 are equivalent to

$$\|p_n^{\alpha,\beta}v^{\alpha-\rho,\beta-\tau}\|_1 \le c\ln(n+1) \quad \text{and} \quad \|p_n^{\alpha,\beta}v^{\rho,\tau}\|_\infty \le c$$

(cf. Remark 2.4). Thus, it follows

$$
\begin{aligned}
(5.18) \qquad |A_n(f,x)| &\le c\,\|f\|_{\infty,\rho,\tau}\,|p_n^{\alpha,\beta}(x)| \int_{-1}^{1} |p_n^{\alpha,\beta}(t)|\, v^{\alpha-\rho,\beta-\tau}(t)\,dt \\
&\le c\,\|f\|_{\infty,\rho,\tau}\, v^{-\rho,-\tau}(x)\ln(n+1)\,.
\end{aligned}
$$

With the help of Lemma 5.4 we may estimate

$$
(5.19) \qquad |B_n(f,x)| \le c\,\|f\|_{\infty,\rho,\tau}\, v^{-\rho,-\tau}(x)\ln(n+1) \quad \text{for } |x| \le 1-n^{-2}\,.
$$

(5.17)$-$(5.19) together with

$$
\|S_n^{\alpha,\beta} f\|_{\infty,\rho,\tau} \le c \max_{|x|\le 1-n^{-2}} |v^{\rho,\tau}(x)(S_n^{\alpha,\beta} f)(x)|
$$

((2.9) with $p=\infty$) yield

$$
\|S_n^{\alpha,\beta} f\|_{\infty,\rho,\tau} \le c\|f\|_{\infty,\rho,\tau}\ln(n+1)\,, \qquad n=1,2,\ldots\,.
$$

Further, we remark that $A_n(f,x)=0$, $|B_n(f,x)| \le c\,\|f\|_{\infty,\rho,\tau}\, v^{-\rho,-\tau}(x)\ln(n+1)$ for $x \in \{x_{n1},\ldots,x_{nn}\}$ even if (2.3) is fulfiled instead of (2.1) (see Lemma 5.4). □

References

[1] V. M. Badkov, Convergence in the mean and almost everywhere of Fourier series in polynomials orthogonal on an interval, *Math. USSR-Sb.*, **24** (1974), 223-256.

[2] D. Berthold, Lagrangian interpolation operators in Hölder spaces, in *Seminar Analysis, Operator Equations and Numerical Analysis 1989/90*, Karl-Weierstrass-Institute of Mathematics, Akad. Wiss. DDR (now: Institut für Angewandte Anal. und Stochast.), Berlin, 1989, pp. 1-19.

[3] D. Berthold, W. Hoppe, B. Silbermann, A fast algorithm for solving the generalized airfoil equation, *J. Comp. Appl. Math.*, **43** (1992), 185-219.

[4] M. R. Capobianco, G. Criscuolo, P. Junghanns, A fast algorithm for Prandtl's integro-differential equation, *J. Comp. Appl. Math.*, **77** (1997), 103-128.

[5] M. R. Capobianco, G. Criscuolo, P. Junghanns, U. Luther, Uniform convergence of the collocation method for Prandtl's integro-differential equation, to appear in *J. Austral. Math. Soc.*, Ser. B **41** (2000).

[6] M. R. Capobianco, P. Junghanns, U. Luther, G. Mastroianni, Weighted uniform convergence of the quadrature method for cauchy singular integral equations, *Singular Integral Operators and Related Topics*, ed. by A. Böttcher and I. Gohberg, *Operator Theory Advances and Applications*, Vol. **90**, Birkhäuser Verlag, 1996, pp. 153-181.

[7] M. R. Capobianco, M. G. Russo, Uniform convergence estimates for a collocation method for the Cauchy singular integral equation, *J. Integral Equations Appl.*, Vol. **9**, Num. 1 (1997), 21-45.

[8] Z. Ditzian, V. Totik, *Moduli of Smoothness*, Springer-Verlag, 1987.

[9] Z. Ditzian, V. Totik, Remarks on Besov spaces and best polynomial approximation, *Proc. Amer. Math. Soc.*, **104** (1988), 1059-1066.

[10] P. Junghanns, U. Luther, Cauchy singular integral equations in spaces of continuous functions and methods for their numerical solution, *J. Comp. Appl. Math.*, **77** (1997), 201-237.

[11] P. Junghanns, U. Luther, Uniform convergence of a fast algorithm for Cauchy singular integral equations, *Linear Algebra Appl.*, **275-276** (1998), 327-347.

[12] U. Luther, *Generalized Besov Spaces and Cauchy Singular Integral Equations*, PhD thesis, TU Chemnitz, 1998.

[13] U. Luther, Uniform convergence of polynomial approximation methods for Prandtl's integro-differential equation, Preprint 99-11, Dept. Math., Techn. Univ. Chemnitz, 1999.

[14] U. Luther, M. G. Russo, Boundedness of the Hilbert transformation in some weighted Besov type spaces, *Integr. Equ. Oper. Theory*, **36** (2000), 220-240.

[15] G. Mastroianni, Uniform convergence of derivatives of Lagrange interpolation, *J. Comp. Appl. Math.*, **43** (1992), 37-51.

[16] G. Mastroianni, M. G. Russo, Lagrange interpolation in some weighted uniform spaces, *Facta Universitatis (Niš), Ser. Math. Inform.* **12** (1997), 185-201.

[17] G. Monegato, S. Prössdorf, Uniform convergence estimates for a collocation and discrete collocation method for the generalized airfoil equation, in: *Contribution to Numerical Mathematics* (A. G. Agarval, ed.), World Scientific Publishing Company, 1993, 285-299. See also the errata corrige in the Internal Report no. 14, 1993, Dipartimento di Matematica, Politecnico di Torino.

[18] B. Muckenhoupt, Mean convergence of Jacobi series, *Proc. Amer. Math. Soc.*, **23** (1969), 306-310.

[19] I. P. Natanson, *Konstruktive Funktionentheorie*, Akademie-Verlag, Berlin, 1955.

[20] P. Nevai, *Orthogonal Polynomials*, Mem. Amer. Math. Soc. 213, Amer. Math. Soc., Providence, RI, 1979.

[21] H. Pollard, The mean convergence of orthogonal series II, *Trans. Amer. Math. Soc.*, **63** (1948), 355-367.

Technische Universität Chemnitz
Fakultät für Mathematik
D-09107 Chemnitz
Germany

Dipartimento di Matematica
Università della Basilicata
Via N. Sauro 85, Potenza
85100 Italia

1991 Mathematics Subject Classification: Primary 42C10; Secondary 41A10, 33C45

Submitted: 28.3.2000

Operator Theory:
Advances and Applications, Vol. 121

Maximal Banach algebra in spaces of multipliers between Bessel potential spaces

Vladimir Maz'ya, Tatyana Shaposhnikova

It is shown that the maximal Banach algebra $A_p^{m,l}$ imbedded in the space of multipliers between Bessel potential spaces $M(H_p^m(\mathbf{R}^n) \to H_p^l(\mathbf{R}^n))$ is isomorphic to $M(H_p^m(\mathbf{R}^n) \to H_p^l(\mathbf{R}^n)) \cap L_\infty(\mathbf{R}^n)$. A precise description of all imbeddings $A_p^{m,l} \subset A_p^{\mu,\lambda}$ is given.

1. Definitions and notations

The Bessel potential space $H_p^r(\mathbf{R}^n)$, $1 < p < \infty$, $r \geq 0$, is defined as the completion of $C_0^\infty(\mathbf{R}^n)$ in the norm

$$||u; \mathbf{R}^n||_{H_p^r} = ||\Lambda^r u; \mathbf{R}^n||_{L_p},$$

where $\Lambda^r = (-\Delta + 1)^{r/2}$ with Δ denoting the Laplace operator. For integer r this space is isomorphic to the Sobolev space normed by

$$||\nabla_r u; \mathbf{R}^n||_{L_p} + ||u; \mathbf{R}^n||_{L_p}$$

with

$$\nabla_r = \{\partial_{x_1}^{\alpha_1} \dots \partial_{x_n}^{\alpha_n}\}, \quad \alpha_1 + \dots + \alpha_n = r,$$

where ∂_{x_i} is a partial derivative.

Let r be positive and noninteger and let $[r]$ and $\{r\}$ be its integer and fractional parts. Further, let

$$(1.1)\quad (S_r u)(x) = \Big(\int_0^\infty \Big(\int_{|h|<y} |\nabla_{[r]} u(x+h) - \nabla_{[r]} u(x)| dh\Big)^2 \frac{dy}{y^{1+2\{r\}+2n}}\Big)^{1/2}$$

if $\{r\} > 0$ and

$$(S_r u)(x) = |\nabla_r u(x)|$$

if $\{r\} = 0$. According to Strichartz [7], for noninteger r

$$(1.2)\qquad ||u; \mathbf{R}^n||_{H_p^r} \sim ||S_r u; \mathbf{R}^n||_{L_p} + ||u; \mathbf{R}^n||_{L_p},$$

where the equivalence $a \sim b$ means that a/b is bounded and is separated from zero by positive constants.

The Hardy-Littlewood maximal operator over centered balls $\mathcal{M}$ is defined by

$$\mathcal{M}\varphi(x) = \sup_{\rho>0} \fint_{B_\rho(x)} |\varphi(y)|dy,$$

where the integral with the bar means the mean value and $B_\rho(x) = \{y \in \mathbf{R}^n : |y - x| < \rho\}$. We write B_ρ instead of $B_\rho(0)$.

2. Pointwise inequalities involving the Strichartz function

In our recent paper [6] the following inequality was proved and used to describe the maximal subalgebra of the space of multipliers between two Sobolev spaces of integer order

$$|\nabla_k\varphi(x)| \le c\,(\mathcal{M}\varphi(x))^{\frac{j-k}{j}}\,(\mathcal{M}\nabla_j\varphi(x))^{\frac{k}{j}}, \tag{2.1}$$

where k and j are integers, $0 \le k \le j$, and c is a positive constant independent of φ.

Soon after publication of [6] P. Hajłasz kindly draw our attention to the paper by A. Kałamajska [3] dedicated to some integral representation formulas for differentiable functions and pointwise interpolation inequalities on bounded domains. Unlike (2.1), the inequalities in [3] contain the operator $\mathcal{M}$ both in the right- and left-hand sides. Kałamajska proved, in particular, that if

$$\lim_{R\to\infty} R^{-k} \fint_{B_{rR}(aR)} |\varphi(x)|dx = 0, \quad \text{where } a \in \mathbf{R}^n \text{ and } r > 0,$$

then, for any polynomial P of degree less than j,

$$\mathcal{M}\nabla_k\varphi(x) \le c\,(\mathcal{M}(\varphi(x) - P(x)))^{\frac{j-k}{j}}\,(\mathcal{M}\nabla_j\varphi(x))^{\frac{k}{j}}. \tag{2.2}$$

Although weaker than (2.2), inequality (2.1) will be sufficient for our purposes.

In this section we derive two auxiliary inequalities similar to (2.1) involving the function S_r defined by (1.1).

Lemma 1. *Let k and r be integer and noninteger, respectively, with $0 < k < r$. There exists a positive constant $c = c(k, r, n)$ such that*

$$|\nabla_k\varphi(x)| \le c(\mathcal{M}\varphi(x))^{\frac{r-k}{r}}\,(S_r\varphi(x))^{\frac{k}{r}} \tag{2.3}$$

for almost all $x \in \mathbf{R}^n$.

Proof. We start with the following inequality which stems from the Sobolev integral representation formula

$$|\nabla_k\varphi(0)| \le c\Big(\int_{B_1} |\varphi(y)|dy + \int_{B_1} |\nabla_{[r]}\varphi(y)|\frac{dy}{|y|^{n-[r]+k}}\Big) \tag{2.4}$$

(see, for example, [4], Sec. 1.1.10). Clearly, the right-hand side is majorized by

$$c\Big(\mathcal{M}\varphi(0)+\int_{B_1}|\nabla_{[r]}\varphi(y)-\nabla_{[r]}\varphi(0)|\frac{dy}{|y|^{n-[r]+k}}+|\nabla_{[r]}\varphi(0)|\Big). \tag{2.5}$$

Using the notation

$$\psi(y)=|\nabla_{[r]}\varphi(y)-\nabla_{[r]}\varphi(0)|$$

we find that the second term in (2.5) is equal to

$$(n-[r]-k)\int_0^1 t^{-n+[r]-k-1}\int_{B_t}\psi(z)dzdt+\int_{B_1}\psi(z)dz.$$

This is dominated by

$$c\Big(S_r\varphi(0)+\int_{B_2}(2-|z|)\psi(z)dz\Big).$$

We have

$$\int_{B_2}(2-|z|)\psi(z)dz=\int_0^2\int_{B_t}\psi(z)dzdt$$

$$\leq c\Big(\int_0^2 t^{-2n-1-2\{r\}}\Big(\int_{B_t}\psi(z)dz\Big)^2dt\Big)^{1/2}\leq cS_r\varphi(0). \tag{2.6}$$

Therefore,

$$|\nabla_k\varphi(0)|\leq c\big(\mathcal{M}\varphi(0)+S_r\varphi(0)+|\nabla_{[r]}\varphi(0)|\big). \tag{2.7}$$

In order to estimate the third term in (2.5) we use the identity

$$\begin{aligned}\nabla_{[r]}\varphi(0) &= \int_{B_1}\eta(y)\nabla_{[r]}\varphi(y)dy\\ &+\int_{B_1}\eta(y)\big(\nabla_{[r]}\varphi(0)-\nabla_{[r]}\varphi(y)\big)dy,\end{aligned} \tag{2.8}$$

where $\eta\in C_0^\infty(B_1)$ such that

$$\int_{B_1}\eta(y)dy=1.$$

Clearly, the first term on the right in (2.8) is majorized by $c\,\mathcal{M}\varphi(0)$. The second term is estimated by

$$\int_{B_2}(2-|z|)\psi(z)dz$$

and by (2.6) does not exceed $c\, S_r\varphi(0)$. Therefore,

$$|\nabla_{[r]}\varphi(0)| \le c\big(\mathcal{M}\varphi(0) + S_r\varphi(0)\big). \tag{2.9}$$

Combining this with (2.7) we arrive at

$$|\nabla_k\varphi(0)| \le c\big(\mathcal{M}\varphi(0) + S_r\varphi(0)\big).$$

Hence by dilation $x \to x/\rho$ we obtain

$$|\nabla_k\varphi(0)| \le c\big(\rho^{-k}\mathcal{M}\varphi(0) + \rho^{r-k}S_r\varphi(0)\big).$$

The result follows by minimization of the right-hand side in ρ.

Lemma 2. *Let k and l be integer and noninteger, respectively, $0 < k < l$. There exists a positive constant $c = c(k,l,n)$ such that*

$$S_{l-k}\varphi(x) \le c\|\varphi; \mathbf{R}^n\|_{L_\infty}^{\frac{k}{l}}(S_l\varphi(x))^{\frac{l-k}{l}} \tag{2.10}$$

for almost all $x \in \mathbf{R}^n$.

Proof. Clearly,

$$\int_0^2\Big(\int_{B_1}|(\nabla_{[l]-k}\varphi)(\theta y) - (\nabla_{[l]-k}\varphi)(0)|d\theta\Big)^2\frac{dy}{y^{1+2\{l\}}}$$

$$\le \int_0^2\Big(\int_{B_1}\Big|(\nabla_{[l]-k}\varphi)(\theta y) - \sum_{\{\alpha:|\alpha|\le k-1\}}\frac{y^\alpha\theta^\alpha}{\alpha!}(D^\alpha\nabla_{[l]-k}\varphi)(0)\Big|d\theta\Big)^2\frac{dy}{y^{1+2\{l\}}}$$

$$+\sum_{i=0}^{k-1}|\nabla_{[l]-i}\varphi(0)|^2. \tag{2.11}$$

The difference in the integral over B_1 in the right-hand side is equal to

$$k\int_0^1\sum_{\{\alpha:|\alpha|=k\}}\frac{y^\alpha\theta^\alpha}{\alpha!}\Big((D^\alpha\nabla_{[l]-k}\varphi)(\tau\theta y) - (D^\alpha\nabla_{[l]-k}\varphi)(0)\Big)(1-\tau)^{k-1}d\tau.$$

Hence and by Minkowski's inequality the first term on the right in (2.11) is dominated by

$$c\Big(\int_0^1\Big(\int_0^\infty\Big(\int_{B_1}|(\nabla_{[l]}\varphi)(\tau\theta y) - (\nabla_{[l]}\varphi)(0)|d\theta\Big)^2\frac{dy}{y^{1+2\{l\}}}\Big)^{1/2}d\tau\Big)^2$$

$$= c(S_l\varphi(0))^2. \tag{2.12}$$

By Lemma 1,

$$|\nabla_{[l]-i}\varphi(0)| \le c\|\varphi; \mathbf{R}^n\|_{L_\infty}^{\frac{\{l\}+i}{l}}(S_l\varphi(0))^{\frac{[l]-i}{l}},\ \ i = 0,\dots,[l]. \tag{2.13}$$

This together with (2.11) and (2.12) gives

$$\int_0^2 \Big(\int_{B_1} |(\nabla_{[l]-k}\varphi)(\theta y) - (\nabla_{[l]-k}\varphi)(0)|d\theta\Big)^2 \frac{dy}{y^{1+\{l\}}}$$

$$\leq c(||\varphi;\mathbf{R}^n||_{L_\infty} + S_l\varphi(0))^2. \tag{2.14}$$

Obviously,

$$\int_2^\infty \Big(\int_{B_1} |(\nabla_{[l]-k}\varphi)(\theta y) \quad - \quad (\nabla_{[l]-k}\varphi)(0)|d\theta\Big)^2 \frac{dy}{y^{1+\{l\}}}$$

$$\leq c\Big(|(\nabla_{[l]-k}\varphi)(0)|^2 \quad + \quad \int_2^\infty \Big(\int_{B_y} |\nabla_{[l]-k}\varphi(z)|dz\Big)^2 \frac{dy}{y^{1+2\{l\}+2n}}\Big). \tag{2.15}$$

The second term on the right does not exceed

$$\int_2^\infty \Big(\int_{B_{y+1}} d\xi \int_{B_1(\xi)} |\nabla_{[l]-k}\varphi(z)|dz\Big)^2 \frac{dy}{y^{1+2\{l\}+2n}}. \tag{2.16}$$

Since

$$\int_{B_1(\xi)} |\nabla_{[l]-k}\varphi(z)|dz \leq c\int_{B_1(\xi)} \big(|\nabla_{[l]}\varphi(z)| + |\varphi(z)|\big)dz$$

(see, for example, [4], Lemma 1.1.11), integral (2.16) is dominated by

$$c\Big(\int_2^\infty \Big(\int_{B_{y+1}} d\xi \int_{B_1(\xi)} |\nabla_{[l]}\varphi(z)|dz\Big)^2 \frac{dy}{y^{1+2\{l\}+2n}} + ||\varphi;\mathbf{R}^n||^2_{L_\infty}\Big)$$
$$\leq c\Big(\int_2^\infty \Big(\int_{B_{y+2}} |\nabla_{[l]}\varphi(z)|dz\Big)^2 \frac{dy}{y^{1+2\{l\}+2n}} + ||\varphi;\mathbf{R}^n||^2_{L_\infty}\Big)$$
$$\leq c\big(S_l\varphi(0) + |\nabla_{[l]}\varphi(0)| + ||\varphi;\mathbf{R}^n||_{L_\infty}\big)^2.$$

Hence and by (2.15) and (2.13) we find

$$\int_2^\infty \Big(\int_{B_1} |(\nabla_{[l]-k}\varphi)(\theta y) - (\nabla_{[l]-k}\varphi)(0)|d\theta\Big)^2 \frac{dy}{y^{1+2\{l\}}}$$

$$\leq c(||\varphi;\mathbf{R}^n||_{L_\infty} + S_l\varphi(0))^2.$$

Taking into account (2.14) we obtain

$$S_{l-k}\varphi(x) \leq c\big(||\varphi;\mathbf{R}^n||_{L_\infty} + S_l\varphi(x)\big).$$

The result follows by dilation as in Lemma 1.

3. The Banach algebra $A_p^{m,l}$

Let $M(H_p^m(\mathbf{R}^n) \to H_p^l(\mathbf{R}^n))$, where $m \geq l \geq 0$ and $1 < p < \infty$, denote the space of pointwise multipliers acting from $H_p^m(\mathbf{R}^n)$ to $H_p^l(\mathbf{R}^n)$. We write $MH_p^l(\mathbf{R}^n)$ instead of $M(H_p^l(\mathbf{R}^n) \to H_p^l(\mathbf{R}^n))$.

Let $V(\mathbf{R}^n)$ be a Banach space of functions defined on $\mathbf{R}^n$. We shall use the spaces

$$\begin{aligned} V_{loc}(\mathbf{R}^n) &= \{u : \eta u \in V(\mathbf{R}^n) \quad \text{for all } \eta \in C_0^\infty(\mathbf{R}^n)\}, \\ V_{unif}(\mathbf{R}^n) &= \{u : \sup_{\xi \in \mathbf{R}^n} ||\eta_\xi u; \mathbf{R}^n||_V < \infty\}, \end{aligned}$$

where $\eta_\xi = \eta(x - \xi)$, $\eta \in C_0^\infty(\mathbf{R}^n)$, $\eta = 1$ on B_1. The norm in $V_{unif}(\mathbf{R}^n)$ is introduced by

$$||u; \mathbf{R}^n||_{V_{unif}} = \sup_{\xi \in \mathbf{R}^n} ||\eta_\xi u; \mathbf{R}^n||_V.$$

The following assertion which gives explicit necessary and sufficient conditions for a function to belong to $M(H_p^m(\mathbf{R}^n) \to H_p^l(\mathbf{R}^n))$ can be found in [5]. In its statement we use the capacity $\mathrm{cap}_{p,r}(e)$ of a compact set $e \subset \mathbf{R}^n$ defined as

$$\mathrm{cap}_{p,r}(e) = \inf\{||u; \mathbf{R}^n||_{H_p^r}^p : \ u \in C_0^\infty(\mathbf{R}^n),\ u \geq 1 \text{ on } e\}.$$

Lemma 3. (1) *Let $m > l$ and $mp \leq n$. Then $\Gamma \in M(H_p^m(\mathbf{R}^n) \to H_p^l(\mathbf{R}^n))$ if and only if $\Gamma \in (H_{p,loc}^l \cap L_{1,unif})(\mathbf{R}^n)$ and for all compact sets $e \subset \mathbf{R}^n$ with* $\mathrm{diam}(e) \leq 1$ *there exists a constant c such that*

$$||S_l\Gamma; e||_{L_p}^p \leq c\, \mathrm{cap}_{p,m}(e).$$

The equivalence relation

$$||\Gamma; \mathbf{R}^n||_{M(H_p^m \to H_p^l)} \sim \sup_{\substack{e \subset \mathbf{R}^n \\ \mathrm{diam}(e) \leq 1}} \frac{||S_l\Gamma; e||_{L_p}}{(\mathrm{cap}_{p,m}(e))^{1/p}} + ||\Gamma; \mathbf{R}^n||_{L_{1,unif}}$$

is valid.

(2) *Let $lp \leq n$. Then $\Gamma \in MH_p^l(\mathbf{R}^n)$ if and only if $\Gamma \in (H_{p,loc}^l \cap L_\infty)(\mathbf{R}^n)$ and for all compact sets $e \subset \mathbf{R}^n$ with* $\mathrm{diam}(e) \leq 1$

$$||S_l\Gamma; e||_{L_p}^p \leq c\, \mathrm{cap}_{p,l}(e).$$

There holds the equivalence relation

$$||\Gamma; \mathbf{R}^n||_{MH_p^l} \sim \sup_{\substack{e \subset \mathbf{R}^n \\ \mathrm{diam}(e) \leq 1}} \frac{||S_l\Gamma; e||_{L_p}}{(\mathrm{cap}_{p,l}(e))^{1/p}} + ||\Gamma; \mathbf{R}^n||_{L_\infty}.$$

(3) *Let* $m \geq l$, $mp > n$. *Then* $\Gamma \in M(H_p^m(\mathbf{R}^n) \to H_p^l(\mathbf{R}^n))$ *if and only if* $\Gamma \in H_{p,unif}^l(\mathbf{R}^n)$. *The equivalence relation*

$$||\Gamma; \mathbf{R}^n||_{M(H_p^m \to H_p^l)} \sim ||\Gamma; \mathbf{R}^n||_{H_{p,unif}^l}$$

is valid.

Remark. It is well known (see, for instance [5], Sec. 2.1) that for $0 < \rho < 1$

$$\mathrm{cap}_{p,m}(B_\rho) \sim \begin{cases} \rho^{n-pm} & \text{for } pm < n, \\ (\log(2\rho^{-1}))^{1-p} & \text{for } pm = n \end{cases}$$

and that for any compact set $e \subset \mathbf{R}^n$ with $\mathrm{diam}(e) \leq 1$

$$\mathrm{cap}_{p,m}(e) \geq \begin{cases} c\,(\mathrm{mes}_n e)^{1-pm/n} & \text{for } pm < n, \\ c\,|\log(2^{-n}\mathrm{mes}_n e)|^{1-p} & \text{for } pm = n \end{cases} \tag{3.1}$$

For $pm < n$ this implies the estimates

$$c_1\,||S_l\Gamma; \mathbf{R}^n||_{\mathcal{L}^{p,pm}} \leq \sup_{\substack{e \subset \mathbf{R}^n \\ \mathrm{diam}(e) \leq 1}} \frac{||S_l\Gamma; e||_{L_p}}{(\mathrm{cap}_{p,m}(e))^{1/p}} \leq c_2\,||(S_l\Gamma)^p; \mathbf{R}^n||_{L_{n/pm,\infty}}^{1/p},$$

where $\mathcal{L}^{p,\alpha}$ is a Morrey space ([2], Sec. 3.7) and $L_{n/pm,\infty}$ is a Marcinkiewicz space ([8], Sec. 1.18.6). In other words, the capacitary criteria for $\Gamma \in M(H_p^m(\mathbf{R}^n) \to H_p^l(\mathbf{R}^n))$ in Lemma 3 give separately necessary and sufficient conditions in terms of Morrey and Marcinkiewicz spaces, respectively.

The following imbedding will be of use.

Lemma 4. ([5], Sec. 2.4) *If* $k \in [0, l]$, *then*

$$M(H_p^m(\mathbf{R}^n) \to H_p^l(\mathbf{R}^n)) \subset M(H_p^{m-l+k}(\mathbf{R}^n) \to H_p^k(\mathbf{R}^n))$$

and the inequality

$$||\Gamma; \mathbf{R}^n||_{M(H_p^{m-l+k} \to H_p^k)} \leq c||\Gamma; \mathbf{R}^n||_{M(H_p^m \to H_p^l)}$$

holds.

Here is our first result.

Theorem 1. *The maximal Banach algebra* $A_p^{m,l}$ *imbedded in* $M(H_p^m(\mathbf{R}^n) \to H_p^l(\mathbf{R}^n))$, $m \geq l$, *is isomorphic to the space*

$$M(H_p^m(\mathbf{R}^n) \to H_p^l(\mathbf{R}^n)) \cap L_\infty(\mathbf{R}^n). \tag{3.2}$$

The estimate

$$\begin{aligned} ||\gamma_1\gamma_2; \mathbf{R}^n||_{M(H_p^m \to H_p^l)} \;\leq\;& c\big(||\gamma_1; \mathbf{R}^n||_{L_\infty}||\gamma_2; \mathbf{R}^n||_{M(H_p^m \to H_p^l)} \\ &+ ||\gamma_2; \mathbf{R}^n||_{L_\infty}||\gamma_1; \mathbf{R}^n||_{M(H_p^m \to H_p^l)}\big) \end{aligned}$$

is valid.

Proof. Let $A_p^{m,l}$ be a subset of $M(H_p^m(\mathbf{R}^n) \to H_p^l(\mathbf{R}^n))$ and let $\gamma \in A_p^{m,l}$. For any $N = 1, 2, \ldots$ and $u \in H_p^m(\mathbf{R}^n)$

$$||\gamma^N u||_{L_p}^{1/N} \le ||\gamma^N u||_{H_p^l}^{1/N} \le ||\gamma^N u||_{M(H_p^m \to H_p^l)}^{1/N} ||u||_{H_p^m}^{1/N}$$

$$\le (c||\gamma^N u||_{A_p^{m,l}})^{1/N} ||u||_{H_p^m}^{1/N} \le c^{1/N} ||\gamma||_{A_p^{m,l}} ||u||_{H_p^m}^{1/N}.$$

Here and elsewhere in this proof we omit $\mathbf{R}^n$ in the notations of the norms. Passing to the limit as $N \to \infty$ we find that $\gamma \in L_\infty(\mathbf{R}^n)$ and

$$||\gamma||_{L_\infty} \le ||\gamma||_{A_p^{m,l}}. \tag{3.3}$$

Assume that γ_1 and γ_2 belong to the space (3.2). For any $u \in H_p^m(\mathbf{R}^n)$,

$$||S_l(\gamma_1\gamma_2 u)||_{L_p} = ||S_{\{l\}}\nabla_{[l]}(\gamma_1\gamma_2 u)||_{L_p} \le c \sum_{|\alpha|+|\beta|+|\sigma|=[l]} ||S_{\{l\}}(D^\alpha\gamma_1 D^\beta\gamma_2 D^\sigma u)||_{L_p}$$

$$\le c \sum_{|\alpha|+|\beta|+|\sigma|=[l]} \big(A_{\alpha,\beta,\sigma} + B_{\alpha,\beta,\sigma} + C_{\alpha,\beta,\sigma}\big),$$

where

$$A_{\alpha,\beta,\sigma} = ||D^\alpha\gamma_1 D^\beta\gamma_2 S_{\{l\}} D^\sigma u||_{L_p}, \tag{3.4}$$

$$B_{\alpha,\beta,\sigma} = ||D^\alpha\gamma_1 S_{\{l\}} D^\beta\gamma_2 D^\sigma u||_{L_p}, \tag{3.5}$$

$$C_{\alpha,\beta,\sigma} = \Big(\int_{\mathbf{R}^n} |D^\alpha\gamma_1(x)|^p \Big(\int_0^\infty \Big(\int_{B_1} |D^\beta\gamma_2(x+\theta y)||D^\sigma u(x+\theta y) - D^\sigma u(x)|d\theta\Big)^2 \frac{dy}{y^{1+2\{l\}}}\Big)^{p/2} dx\Big)^{1/p}. \tag{3.6}$$

Applying Lemma 1 we obtain

$$A_{\alpha,\beta,\sigma} \le c||\gamma_1||_{L_\infty}^{1-\frac{|\alpha|}{[l]-|\sigma|}} ||\gamma_2||_{L_\infty}^{1-\frac{|\beta|}{[l]-|\sigma|}}$$

$$\times ||(\mathcal{M}\nabla_{[l]-|\sigma|}\gamma_1)^{\frac{|\alpha|}{[l]-|\sigma|}} (\mathcal{M}\nabla_{[l]-|\sigma|}\gamma_2)^{\frac{|\beta|}{[l]-|\sigma|}} S_{\{l\}+|\sigma|}u||_{L_p}. \tag{3.7}$$

By Hölder's inequality the last norm is dominated by

$$c||(\mathcal{M}\nabla_{[l]-|\sigma|}\gamma_1) S_{\{l\}+|\sigma|}u||_{L_p}^{\frac{|\alpha|}{[l]-|\sigma|}} ||(\mathcal{M}\nabla_{[l]-|\sigma|}\gamma_2) S_{\{l\}+|\sigma|}u||_{L_p}^{\frac{|\beta|}{[l]-|\sigma|}}. \tag{3.8}$$

Since by Minkowski's inequality

$$S_{\{l\}}v \le \Lambda^{|\sigma|+\{l\}-m} S_{\{l\}} \Lambda^{m-|\sigma|-\{l\}} v,$$

it follows that for $\Gamma \in M(H_p^{m-\{l\}-|\sigma|}(\mathbf{R}^n) \to L_p(\mathbf{R}^n))$

$$||\Gamma S_{\{l\}} v||_{L_p} \leq ||\Gamma||_{M(H_p^{m-\{l\}-|\sigma|} \to L_p)} ||\Lambda^{|\sigma|+\{l\}-m} S_{\{l\}} \Lambda^{m-|\sigma|-\{l\}} v||_{H_p^{m-|\sigma|-\{l\}}}$$

$$\leq c||\Gamma||_{M(H_p^{m-|\sigma|-\{l\}} \to L_p)} ||v||_{H_p^{m-|\sigma|}}.$$

Putting here $\Gamma = \mathcal{M}\nabla_{[l]-|\sigma|}\gamma_i$, $i = 1, 2$, and $v = \nabla_{|\sigma|} u$ we majorize (3.8) by

$$c||\mathcal{M}\nabla_{[l]-|\sigma|}\gamma_1||^{\frac{|\alpha|}{[l]-|\sigma|}}_{M(H_p^{m-|\sigma|-\{l\}} \to L_p)} ||u||^{\frac{|\alpha|}{[l]-|\sigma|}}_{H_p^m}$$

$$\times||\mathcal{M}\nabla_{[l]-|\sigma|}\gamma_2||^{\frac{|\beta|}{[l]-|\sigma|}}_{M(H_p^{m-|\sigma|-\{l\}} \to L_p)} ||u||^{\frac{|\beta|}{[l]-|\sigma|}}_{H_p^m}. \tag{3.9}$$

We estimate (3.9) by Verbitsky's theorem (cf. [5], Sec. 2.6)

$$||\mathcal{M}\Gamma||_{M(H_p^r \to L_p)} \leq c||\Gamma||_{M(H_p^r \to L_p)}$$

combined with the inequality

$$||\nabla_j \Gamma||_{M(H_p^{m-l+j} \to L_p)} \leq c||\Gamma||_{M(H_p^m \to H_p^l)} \tag{3.10}$$

which follows from Lemmas 3 and 4. Taking into account the equality $|\alpha| + |\beta| + |\sigma| = [l]$, we obtain that (3.9) does not exceed

$$c\,||\gamma_1||^{\frac{|\alpha|}{|\alpha|+|\beta|}}_{M(H_p^m \to H_p^l)} ||\gamma_2||^{\frac{|\beta|}{|\alpha|+|\beta|}}_{M(H_p^m \to H_p^l)} ||u||_{H_p^m}. \tag{3.11}$$

Hence, by (3.7), and by Hölder's inequality,

$$A_{\alpha,\beta,\sigma} \leq c\big(||\gamma_1||_{L_\infty} ||\gamma_2||_{M(H_p^m \to H_p^l)} + ||\gamma_2||_{L_\infty} ||\gamma_1||_{M(H_p^m \to H_p^l)}\big)||u||_{H_p^m}. \tag{3.12}$$

To estimate $B_{\alpha,\beta,\sigma}$, defined by (3.5), we apply Lemma 1 to the function $D^\alpha\gamma_1$ and Lemma 2 to the function $S_{\{l\}}D^\beta\gamma_2$. Then, by Hölder's inequality,

$$B_{\alpha,\beta,\sigma} \leq c||\gamma_1||^{1-\frac{|\alpha|}{l-|\sigma|}}_{L_\infty} ||\gamma_2||^{1-\frac{|\beta|+\{l\}}{l-|\sigma|}}_{L_\infty} ||(S_{l-|\sigma|}\gamma_1)^{\frac{|\alpha|}{l-|\sigma|}} (S_{l-|\sigma|}\gamma_2)^{\frac{|\beta|+\{l\}}{l-|\sigma|}} D^\sigma u||_{L_p}$$

$$\leq c||\gamma_1||^{1-\frac{|\alpha|}{l-|\sigma|}}_{L_\infty} ||\gamma_2||^{1-\frac{|\beta|+\{l\}}{l-|\sigma|}}_{L_\infty} ||(S_{l-|\sigma|}\gamma_1)D^\sigma u||^{\frac{|\alpha|}{l-|\sigma|}}_{L_p} ||(S_{l-|\sigma|}\gamma_2)D^\sigma u||^{\frac{|\beta|+\{l\}}{l-|\sigma|}}_{L_p}.$$

By Lemmas 3 and 4, for $i = 1, 2$

$$\begin{aligned} ||(S_{l-|\sigma|}\gamma_i)D^\sigma u||_{L_p} &\leq ||S_{l-|\sigma|}\gamma_i||_{M(H_p^{m-|\sigma|} \to L_p)} ||u||_{H_p^m} \\ &\leq c||\gamma_i||_{M(H_p^m \to H_p^l)} ||u||_{H_p^m}. \end{aligned}$$

Hence $B_{\alpha,\beta,\sigma}$ has the same majorant (3.12) as $A_{\alpha,\beta,\sigma}$.

In order to estimate $C_{\alpha,\beta,\sigma}$ defined by (3.6) we use Lemma 1

$$C_{\alpha,\beta,\sigma} \leq c||\gamma_1||_{L_\infty}^{1-\frac{|\alpha|}{[l]-|\sigma|}} ||\gamma_2||_{L_\infty}^{1-\frac{|\beta|}{[l]-|\sigma|}} K_{\alpha,\beta,\sigma},$$

where

$$K_{\alpha,\beta,\sigma} = \Big(\int_{\mathbf{R}^n} (\mathcal{M}\nabla_{[l]-|\sigma|}\gamma_1(x))^{\frac{p|\alpha|}{[l]-|\sigma|}}$$

$$\times \int_0^\infty \Big(\Big(\int_{B_1} (\mathcal{M}\nabla_{[l]-|\sigma|}\gamma_2(x+\theta y))^{\frac{p|\beta|}{[l]-|\sigma|}} |\nabla_{|\sigma|}u(x+\theta y) - \nabla_{|\sigma|}u(x)|d\theta\Big)^2 \frac{dy}{y^{1+2\{l\}}}\Big)^{p/2} dx\Big)^{1/p}$$

By Hölder's inequality $K_{\alpha,\beta,\sigma}$ is dominated by (3.9) which, as was shown above, has the majorant (3.11). Therefore, (3.12) is valid with $A_{\alpha,\beta,\sigma}$ replaced by $C_{\alpha,\beta,\sigma}$. This completes the proof.

Corollary 1. *The maximal Banach algebra $A_p^{m,l}$ in $M(H_p^m(\mathbf{R}^n) \to H_p^l(\mathbf{R}^n))$, $m \geq l$, $1 < p < \infty$, is the set of functions $\gamma \in H_{p,loc}^l(\mathbf{R}^n)$ such that*

$$\sup_{\substack{e\subset R^n \\ \mathrm{diam}(e)\leq 1}} \frac{||S_l\gamma; e||_{L_p}}{(\mathrm{cap}_{p,m}(e))^{1/p}} + ||\gamma; \mathbf{R}^n||_{L_\infty} < \infty. \tag{3.13}$$

In the case $mp > n$ this condition can be simplified as

$$||S_l\gamma; \mathbf{R}^n||_{L_p,unif} + ||\gamma; \mathbf{R}^n||_{L_\infty} < \infty.$$

Proof. The result follows from Theorem 1 and Lemma 3.

In the next theorem we deal with the imbedding $A_p^{m,l} \subset A_p^{\mu,\lambda}$. We fix an arbitrary μ and find the maximum value of λ for which the imbedding holds. Taking into account that for $\mu \geq m$ the best value of λ is l we can restrict ourselves to $\mu < m$.

Theorem 2. *For any $\theta \in (0,1)$ the following imbeddings are valid*

(i) *if $pm \leq n$, then $A_p^{m,l} \subset A_p^{m\theta,l\theta}$,*

(ii) *if $pm\theta > n$, then $A_p^{m,l} \subset A_p^{m\theta,\min\{m\theta,l\}}$,*

(iii) *if $pm\theta = n$, then*

$$A_p^{m,l} \subset A_p^{m\theta,m\theta} \quad \text{for } m\theta < l$$

and

$$A_p^{m,l} \subset A_p^{m\theta,l-\epsilon} \quad \text{for } m\theta \geq l$$

with an arbitrary small $\epsilon > 0$,

(iv) *if $pm\theta < n < pm$, then*

$$A_p^{m,l} \subset A_p^{m\theta,m\theta} \quad \text{for } pl > n$$

and

$$A_p^{m,l} \subset A_p^{m\theta, m\theta lp/n} \quad \text{for } pl \le n.$$

All these imbeddings are best possible.

Proof. (i) Since $\gamma \in A_p^{m,l}$ continuously maps $L_p(\mathbf{R}^n)$ to $L_p(\mathbf{R}^n)$ and $H_p^m(\mathbf{R}^n)$ to $H_p^l(\mathbf{R}^n)$, the imbedding $A_p^{m,l} \subset A_p^{m\theta, l\theta}$ results by complex interpolation (see [9], Sec. 2.4.7).

In the cases (ii)-(iv) we have $pm > n$. Thus, by Corollary 1, $A_p^{m,l} = (H^l_{p,unif} \cap L_\infty)(\mathbf{R}^n)$.

(ii) Since $pm\theta > n$, Corollary 1 implies $A_p^{m\theta,\lambda} = (H^\lambda_{p,unif} \cap L_\infty)(\mathbf{R}^n)$ for $\lambda \le m\theta$. The result follows from the imbedding $H^l_{p,unif}(\mathbf{R}^n) \subset H^{\min\{m\theta,l\}}_{p,unif}(\mathbf{R}^n)$.

(iii) If $pm\theta = n$ and $m\theta < l$, then $pl > n$ and by Lemma 3 $H^l_{p,unif}(\mathbf{R}^n) = MH^l_p(\mathbf{R}^n)$. The last space is imbedded into $MH_p^{m\theta}(\mathbf{R}^n) = A_p^{m\theta,m\theta}$ by complex interpolation between $H_p^l(\mathbf{R}^n)$ and $L_p(\mathbf{R}^n)$. Hence $A_p^{m,l} \subset A_p^{m\theta,m\theta}$.

Let $pm\theta = n$ and $m\theta \ge l$. By Corollary 1, a norm in $A_p^{m\theta,l-\epsilon}$ can be given by

$$\sup_{\substack{e\subset \mathbf{R}^n \\ \text{diam}(e)\le 1}} \frac{\|S_{l-\epsilon}\gamma; e\|_{L_p}}{(\text{cap}_{p,m\theta}(e))^{1/p}} + \|\gamma;\mathbf{R}^n\|_{L_\infty}.$$

Since for $pm\theta = n$ estimate (3.1) gives

$$\text{cap}_{p,m\theta}(e) \ge c\,(\text{mes}_n e)^{\epsilon/n}$$

with an arbitrary $\epsilon > 0$, it follows that

$$\|\gamma;\mathbf{R}^n\|_{A_p^{m\theta,l-\epsilon}} \le c\left(\|\gamma;\mathbf{R}^n\|_{H^{l-\epsilon}_{q,unif}} + \|\gamma;\mathbf{R}^n\|_{L_\infty}\right)$$

with $q = pn/(n-p\epsilon)$. It remains to use the Sobolev imbedding $H^l_{p,unif}(\mathbf{R}^n) \subset H^{l-\epsilon}_{q,unif}(\mathbf{R}^n)$.

(iv) Let $pm\theta < n < pm$ and $pl > n$. Then $A_p^{m,l} = H^l_{p,unif}(\mathbf{R}^n) = MH^l_p(\mathbf{R}^n)$. Since $l > m\theta$ we have $MH_p^l(\mathbf{R}^n) \subset MH_p^{m\theta}(\mathbf{R}^n) = A_p^{m\theta,m\theta}$. The result follows.

Now let $pm\theta < n < pm$ and $lp \le n$. In view of (3.1) we obtain from Corollary 1 that

$$\begin{aligned}\|\gamma;\mathbf{R}^n\|_{A_p^{m\theta,m\theta lp/n}} &\le c\Big(\sup_{\substack{e\subset \mathbf{R}^n \\ \text{diam}(e)\le 1}} \frac{\|S_{m\theta lp/n}\gamma; e\|_{L_p}}{(\text{mes}_n e)^{(n-pm\theta)/np}} + \|\gamma;\mathbf{R}^n\|_{L_\infty}\Big)\\ &\le c\left(\|\gamma;\mathbf{R}^n\|_{H^{m\theta lp/n}_{n/m\theta,unif}} + \|\gamma;\mathbf{R}^n\|_{L_\infty}\right).\end{aligned}$$

The imbedding $A_p^{m,l} \subset A_p^{m\theta,m\theta lp/n}$ follows from the Gagliardo-Nirenberg type inequality (see [1], Lemma 3.4)

$$\|\gamma;\mathbf{R}^n\|_{H^{m\theta lp/n}_{n/m\theta,unif}} \le c\,\|\gamma;\mathbf{R}^n\|^{pm\theta/n}_{H^l_{p,unif}}\|\gamma;\mathbf{R}^n\|^{1-pm\theta/n}_{L_\infty}.$$

We now show that the imbedding $A_p^{m,l} \subset A_p^{m\theta,\lambda}$ with λ given in (i)-(iv) cannot be improved. Let

$$\gamma_\mu(x) = \exp(i|x|^{-\mu}) \tag{3.14}$$

with $\mu > 0$. From the equivalence relations

$$|\nabla_{[l]}\gamma_\mu(x)| \sim |x|^{-[l](\mu+1)}$$

and

$$|\nabla_{[l]}\gamma_\mu(x+h) - \nabla_{[l]}\gamma_\mu(x)| \sim \frac{\min\{|h|, |x|^{1+\mu}\}}{|x|^{([l]+1)(1+\mu)}},$$

where $|x|$ is sufficiently small, it follows that

$$S_l\gamma_\mu(x) \sim |x|^{-l(\mu+1)} \tag{3.15}$$

for $|x| < 1$. Furthermore, $S_l\gamma_\mu(x)$ is bounded for $|x| \geq 1$. Hence and by Lemma 3,

$$\gamma_\mu \in A_p^{m,l} \iff \begin{cases} (\mu+1)l \leq m & \text{if } pm < n, \\ (\mu+1)l < m & \text{if } pm = n, \\ p(\mu+1)l < n & \text{if } pm > n. \end{cases} \tag{3.16}$$

From (3.16) we conclude that in the case $pm \leq n$

$$\gamma \in A_p^{m,l} \iff \gamma \in A_p^{m\theta,l\theta} \quad \text{for } \theta \in (0,1)$$

which shows that the imbedding (i) is precise.

Since the imbedding (ii) is equivalent to

$$(H^l_{p,unif} \cap L_\infty)(\mathbf{R}^n) \subset (H^{\min\{m\theta,l\}}_{p,unif} \cap L_\infty)(\mathbf{R}^n)$$

and, obviously, $\lambda \leq m\theta$ in $A_p^{m,l} \subset A_p^{m\theta,\lambda}$, the imbedding (ii) cannot be improved.

We turn to the imbeddings (iii). The optimality of the first one (corresponding to $m\theta < l$) is obvious. Let $m\theta \geq l$. We show that $A_p^{m,l}$ is not imbedded into $A_p^{m\theta,l}$.

Let $\mu \geq 0$ and $pl(\mu+1) = n$. We introduce the function

$$\Gamma_{\mu,\delta}(x) = \eta(x)\exp(i|x|^{-\mu}(\log|x|^{-1})^{-\delta}),$$

where $\delta > -1$, η is a function in $C_0^\infty(\mathbf{R}^n)$ with support in a small neighbourhood of the origin, equal to 1 near the origin. For $\mu > 0$ direct calculations imply

$$|\nabla_{[l]}\Gamma_{\mu,\delta}(x)| \sim |x|^{-[l](\mu+1)}(\log|x|^{-1})^{-[l]\delta}$$

and

$$|\nabla_{[l]}\Gamma_{\mu,\delta}(x+h) - \nabla_{[l]}\Gamma_{\mu,\delta}(x)| \sim \frac{\min\{|h|, |x|^{1+\mu}(\log|x|^{-1})^{\delta}\}}{|x|^{([l]+1)(1+\mu)}(\log|x|^{-1})^{([l]+1)\delta}},$$

where $|x|$ is sufficiently small. Therefore, for $\mu > 0$,

$$S_l \Gamma_{\mu,\delta}(x) \sim |x|^{-l(\mu+1)} (\log|x|^{-1})^{-l\delta}, \tag{3.17}$$

where $|x|$ is sufficiently small.

Analogously,

$$|\nabla_{[l]} \Gamma_{0,\delta}(x)| \sim |x|^{-[l]} (\log|x|^{-1})^{-[l](\delta+1)}$$

and

$$|\nabla_{[l]} \Gamma_{0,\delta}(x+h) - \nabla_{[l]} \Gamma_{0,\delta}(x)| \sim \frac{\min\{|h|, |x|(\log|x|^{-1})^{\delta+1}\}}{|x|^{[l]+1} (\log|x|^{-1})^{([l]+1)(\delta+1)}}$$

for small $|x|$. Hence,

$$S_l \Gamma_{0,\delta}(x) \sim |x|^{-l} (\log|x|^{-1})^{-l(\delta+1)}. \tag{3.18}$$

Now it is straightforward that $\Gamma_{\mu,\delta} \in A_p^{m,l} = (H^l_{p,unif} \cap L_\infty)(\mathbf{R}^n)$ if and only if $pl\delta > 1$ for $\mu > 0$ and $pl(\delta+1) > 1$ for $\mu = 0$. On the other hand, by Corollary 1 and (3.1),

$$||\Gamma_{\mu,\delta}; \mathbf{R}^n||_{A_p^{m\theta,l}} \geq c\,(\log \rho^{-1})^{(p-1)/p} ||S_l \Gamma_{\mu,\delta}; B_\rho||_{L_p}$$

for small $\rho > 0$. Applying (3.17), (3.18) we obtain

$$||\Gamma_{\mu,\delta}; \mathbf{R}^n||_{A_p^{m\theta,l}} \geq c\,(\log \rho^{-1})^{(p-1)/p + (1-pl\delta)/p} \quad \text{for } \mu > 0$$

and

$$||\Gamma_{0,\delta}; \mathbf{R}^n||_{A_p^{m\theta,l}} \geq c\,(\log \rho^{-1})^{(p-1)/p + (1-pl(\delta+1))/p}.$$

This, obviously, implies that $\Gamma_{\mu,\delta} \in A_p^{m,l}$ and $\Gamma_{\mu,\delta} \notin A_p^{m\theta,l}$ if $1 > l\delta > 1/p$ for $\mu > 0$ and $\Gamma_{0,\delta} \in A_p^{m,l}$ and $\Gamma_{0,\delta} \notin A_p^{m\theta,l}$ if $1 > l(\delta+1) > 1/p$. The result follows.

We pass to (iv). It suffices to consider only the case $pl < n$. Assume that $A_p^{m,l} \subset A_p^{m\theta,\lambda}$ with $\lambda = m\theta pl/n(1-\epsilon)$ for some $\epsilon > 0$. We choose μ to satisfy $pl(\mu+1) > n(1-\epsilon)$. Then the function γ_μ, introduced by (3.14), belongs to $H^l_{p,unif}(\mathbf{R}^n) = A_p^{m,l}$. On the other hand, by Corollary 1 and (3.15), for $\rho < 1$

$$||\gamma_\mu; \mathbf{R}^n||_{A_p^{m\theta,\lambda}} \geq c\,\rho^{m\theta - n/p} ||S_\lambda \gamma_\mu; B_\rho||_{L_p} \geq c\,\rho^{m\theta - \lambda(\mu+1)}.$$

Since $m\theta - \lambda(\mu+1) = m\theta(1 - pl(\mu+1)/n(1-\epsilon)) < 0$, we have $\gamma_\mu \notin A_p^{m\theta,\lambda}$. The proof is complete.

References

[1] D.R. Adams, M. Frazier, BMO and smooth truncation in Sobolev spaces, Studia Math. **89** (1988), pp. 241–260.

[2] D.R. Adams, L.I. Hedberg, Function Spaces and Potential Theory, Springer-Verlag, 1996.

[3] A. Kałamajska, Pointwise multiplicative inequalities and Nirenberg type estimates in weighted Sobolev spaces, Studia Math **108**:3 (1994), pp. 275–290.

[4] V. Maz'ya, Sobolev Spaces, Springer-Verlag, 1985.

[5] V. Maz'ya, T. Shaposhnikova, Theory of Multipliers in Spaces of Differentiable Functions, Pitman, London, 1985.

[6] V. Maz'ya, T. Shaposhnikova, On pointwise interpolation inequalities for derivatives, Mathematica Bohemica **124**:2-3 (1999), pp. 131–148.

[7] R.S. Strichartz, Multipliers on fractional Sobolev spaces, J. Math. and Mech. **16**:9 (1967), pp. 1031–1060.

[8] H. Triebel, Interpolation Theory. Function Spaces. Differential Operators, VEB Deutscher Verlag der Wissenschaften, Berlin, 1978.

[9] H. Triebel, Theory of Function Spaces, Akademische Verlagsgesellschaft Geest und Portig K.-G., Leipzig, Birkhäuser Verlag, Basel-Boston-Stuttgart, 1983.

Department of Mathematics,
Linköping University,
S-58183 Linköping, Sweden

1991 Mathematics Subject Classification: Primary 46E35, 46E25

Submitted: 25.10.1999

Operator Theory:
Advances and Applications, Vol. 121

Mixed Type Direct and Inverse Scattering Problems

DAVID NATROSHVILI, ZURAB TEDIASHVILI

The uniqueness theorem for a three-dimensional mixed type inverse scattering problem is proved provided that the boundary of an unknown scatterer contains disjoint sound-soft and sound-hard submanifolds.

1. Introduction

The inverse scattering problems for scatterers with a sound-soft boundary or a sound-hard boundary are well investigated together with inverse penetrable scattering problems and stability (see [CK1], [CK2], [Is2] and references therein).

Here we apply the ideas of Isakov [Is1] and Kirsch and Kress [KiKr1] to study a mixed type inverse scattering problem. We assume that the boundary of an unknown scatterer contains both – sound-soft and sound-hard disjoint submanifolds (see Section 4).

First we consider the direct mixed type boundary value problem for a generalized Helmholtz equation when on the sound-soft part the Dirichlet condition is given, while on the sound-hard part of the boundary the Neumann condition is prescribed (see (3.2)-(3.6)). We present the existence and regularity results for the solution to the direct mixed boundary value problem. Moreover, we derive explicit representation of the solution by means of potential type integrals.

On the basis of these results we prove the uniqueness theorem for the corresponding mixed type inverse scattering problem. Namely, we show that, if for two scatterers the far-field patterns coincide for a fixed wave number and for all incident directions, then the scatterers and the sound-soft and sound-hard parts of their boundaries coincide as well (see Theorem 5.1).

2. Preliminary material

Let $\Omega^+ \subset \mathbb{R}^3$ be a bounded domain ($\operatorname{diam}\Omega^+ < +\infty$) with a Liapunov type smooth, connected, nonselfintersecting boundary $S = \partial\Omega^+$ and $\overline{\Omega^+} = \Omega^+ \cup S$. We assume that $\Omega^- = \mathbb{R}^3 \backslash \overline{\Omega^+}$ is a connected domain which is filled up by a homogeneous anisotropic medium with a constant density ρ.

Further, let some physical process (the propagation of acoustic waves say) in Ω^- be described by a complex-valued scalar function $w(x)$ being a solution of the homogeneous "wave equation" (generalized Helmholtz equation)

(2.1) $$a(D,\omega)w(x) := a(D)w(x) + \rho\omega^2 w(x) = 0, \quad x \in \Omega^-,$$

where

$$a(D) = a_{pq}D_pD_q, \; a_{pq} = a_{qp}, \; D := (D_1, D_2, D_3), \; D_p = \frac{\partial}{\partial x_p}, \; p,q = 1,2,3,$$

ρ is a positive constant (density), ω is the so-called oscillation parameter. Moreover, a_{pq} are real constants defining a positive definite matrix $\widetilde{a} := [a_{pq}]_{3\times 3}$. Here and in what follows the summation over repeated indices is meant from 1 to 3.

To prove the uniqueness and existence theorems of solutions to the exterior boundary value problems (BVPs) we require some special decay conditions and structural restrictions at infinity which are similar to the Sommerfeld radiation conditions for solutions to the Helmholtz equation (cf. [Vek1], [Ku1] [CK1], [Va1], [JN1], [JN2]).

Denote by Σ_ω the characteristic ellipsoid (cf. [JN1])

(2.2) $$\Sigma_\omega := \{\xi \in \mathbb{R}^3 \; : \; a_{pq}\xi_p\xi_q = \rho\omega^2\}.$$

It is evident that for an arbitrary vector $\eta \in \mathbb{R}^3$ with $|\eta| = 1$ there exists only one point $\xi(\eta) \in \Sigma_\omega$ such that the outward unit normal vector $n(\xi(\eta))$ to Σ_ω at the point $\xi(\eta)$ has the same direction as η, i.e., $n(\xi(\eta)) = \eta$.

It can be easily verified that

(2.3) $$\xi(\eta) = \omega\sqrt{\rho}\,(\widetilde{a}^{-1}\eta\cdot\eta)^{-1/2}\,\widetilde{a}^{-1}\eta,$$

where $\widetilde{a}^{-1}$ is the matrix inverse to $\widetilde{a}$ and the dot denotes the usual scalar product of two vectors.

We say that a function w satisfies the generalized Sommerfeld type radiation conditions ($w \in \mathrm{Som}(\Omega^-)$) if $w \in C^1(\Omega^-)$ and for sufficiently large $|x|$

(2.4) $$w(x) = O(|x|^{-1}), \quad D_k w(x) - i\,\xi_k(\hat{x})w(x) = O(|x|^{-2}), \quad k = 1,2,3,$$

where $\xi(\hat{x}) \in \Sigma_\omega$ corresponds to the vector $\hat{x} = x/|x|$ (cf., e.g., [Va1], [JN1], [JN2]).

Obviously, the conditions (2.4) coincide with the classical Sommerfeld radiation conditions for the Helmholtz equation, i.e., when $a(D)$ is the Laplace operator (see, for example, [Vek1], [CK1]). In the sequel elements of the class $\mathrm{Som}(\Omega^-)$ will also be referred to as radiating functions.

The fundamental solution to the equation (2.1) reads as (cf., [Mi1], [JN1], [JN2])

$$\gamma(x,\omega) = -\frac{\exp\{i\omega\sqrt{\rho}(\widetilde{a}^{-1}x\cdot x)^{1/2}\}}{4\pi|\widetilde{a}|^{1/2}(\widetilde{a}^{-1}x\cdot x)^{1/2}},$$

where $|\widetilde{a}| = \det\widetilde{a}$ and i is the imaginary unit.

For sufficiently large $|x|$ we have the following asymptotic formula

$$(2.5)\ \gamma(x-y,\omega)=c^*\,\frac{\exp\{i(\xi\cdot(x-y))\}}{|x|}+O(|x|^{-2}),\quad c^*=-\frac{|\widetilde{a}\xi|}{4\pi\omega\,(\rho|\widetilde{a}|)^{1/2}},$$

where y varies in a bounded subset of $\mathbb{R}^3$ and $\xi=\xi(\hat{x})\in\Sigma_\omega$ (for details see [JN1]).

Clearly, $\gamma\in\mathrm{Som}(\mathbb{R}^3\setminus\{0\})$.

Let us introduce the single- and double-layer potentials

$$(2.6)\qquad V(g)(x)\ :=\ \int_S \gamma(x-y,\omega)\,g(y)\,dS_y,\quad x\in\mathbb{R}^3\setminus S,$$

$$(2.7)\qquad W(g)(x)\ :=\ \int_S [\partial_{n(y)}\gamma(y-x,\omega)]\,g(y)\,dS_y,\quad x\in\mathbb{R}^3\setminus S,$$

where g is a scalar function and $\partial_{n(x)}$ denotes the co-normal derivative at the point $x\in S$

$$\partial_{n(x)}:=a_{kj}n_k(x)\,D_j.$$

Note that, if w is a solution of the homogeneous equation (2.1), then w is an analytic function of the real variable x in the domain Ω^-. If, in addition, $w\in C^1(\overline{\Omega^-})\cap\mathrm{Som}(\Omega^-)$, then the integral representation formula holds

$$(2.8)\qquad V\left([\partial_n w]^-\right)\ -\ W\left([w(y)]^-\right)=\begin{cases} w(x) & \text{for}\quad x\in\Omega^-,\\ 0 & \text{for}\quad x\in\Omega^+,\end{cases}$$

where n is again the outward unit normal vector to S and the symbols $[\,\cdot\,]^\pm$ denote limits on S from $\Omega^\pm$.

We have the following analogue of the classical Rellich's lemma (for details see [JN2]).

Lemma 2.1. *Let $w\in\mathrm{Som}(\Omega^-)$ be a solution of (2.1) in Ω^- and let*

$$\lim_{R\to+\infty}\ \mathrm{Im}\left\{\int_{\Sigma(O,R)}\overline{w(x)}\,\partial_n w(x)\,d\Sigma(O,R)\right\}=0,$$

where $\Sigma(O,R)$ is the sphere centered at the origin and radius R.

Then $w=0$ in Ω^-.

The basic properties of single- and double-layer potentials and integral operators generated by them are described by the lemmata (see [Mi1], [JN1], [JN2]).

Lemma 2.2. *Let $S\in C^{k+1+\alpha'}$ with integer $k\geq 0$ and $0<\alpha<\alpha'\leq 1$, $0\leq m\leq k$.*

Then

i) *the operators*

$$\begin{aligned} V\ &:\ C^{m+\alpha}(S)\to C^{m+1+\alpha}(\overline{\Omega^+}),\\ &:\ C^{m+\alpha}(S)\to C^{m+1+\alpha}(\overline{\Omega^-})\cap\mathrm{Som}(\Omega^-),\\ W\ &:\ C^{m+\alpha}(S)\to C^{m+\alpha}(\overline{\Omega^+}),\\ &:\ C^{m+\alpha}(S)\to C^{m+\alpha}(\overline{\Omega^-})\cap\mathrm{Som}(\Omega^-),\end{aligned}$$

are bounded;

ii) *for arbitrary $g \in C^{k+\alpha}(S)$ and $z \in S$ the following jump relations hold on S :*

$$(2.9)\qquad [V(g)(z)]^{\pm} = \int_S \gamma(z-y,\omega)\, g(y)\, dS_y =: \mathcal{H}\, g(z), \quad m \geq 0,$$

$$[\partial_{n(z)} V(g)(z)]^{\pm} = \mp 2^{-1} g(z) + \int_S [\partial_{n(z)}\gamma(z-y,\omega)]\, g(y)\, dS_y$$

$$(2.10)\qquad =: [\mp 2^{-1} I + \mathcal{K}^{(1)}]\, g(z), \quad m \geq 0,$$

$$[W(g)(z)]^{\pm} = \pm 2^{-1} g(z) + \int_S [\partial_{n(y)}\gamma(y-z,\omega)]\, g(y)\, dS_y$$

$$(2.11)\qquad =: [\pm 2^{-1} I + \mathcal{K}^{(2)}]\, g(z), \quad m \geq 0,$$

$$(2.12)\qquad [\partial_{n(z)} W(g)(z)]^{+} = [\partial_{n(z)} W(g)(z)]^{-} =: \mathcal{L}\, g(z), \quad m \geq 1,$$

where I stands for the identical operator;

iii) *the operators*

$$\begin{aligned} \mathcal{H} &: C^{m+\alpha}(S) \to C^{m+1+\alpha}(S), \\ \mathcal{K}^{(1)},\ \mathcal{K}^{(2)} &: C^{m+\alpha}(S) \to C^{m+\alpha}(S), \\ \mathcal{L} &: C^{m+1+\alpha}(S) \to C^{m+\alpha}(S), \end{aligned}$$

are bounded; moreover, $\mathcal{H}$ has a weakly singular kernel-function of type $O(|x-y|^{-1})$, while $\mathcal{K}^{(1)}$ and $\mathcal{K}^{(2)}$ have weakly singular kernel-functions of type $O(|x-y|^{-2+\alpha'})$ on S for $k=0$ and $O(|x-y|^{-1})$ for $k \geq 1$, and $\mathcal{L}$ is a singular integro-differential operator;

iv) *the operators $\mathcal{H}$, $\mp 2^{-1} I + \mathcal{K}^{(1)}$, $\pm 2^{-1} I + \mathcal{K}^{(2)}$, and $\mathcal{L}$ are elliptic pseudodifferential operators of order* -1, 0, 0, *and* 1, *respectively, with the index zero;*

In what follows W_p^s, H_p^s, $B_{p,q}^s$ denote the well-known Sobolev-Slobodetskii, Bessel-potential and Bessov spaces. Note that $H_2^s = W_2^s = B_{2,2}^s$, $W_p^t = B_{p,p}^t$ and $H_p^k = W_p^k$ $(p>1)$ for any $s \in \mathbb{R}$, for any positive and non-integer t and for any non-negative integer k (see, e.g., [Tr1]).

For $w \in W^1_{p,loc}(\Omega^-)$ by the trace theorem (see, e.g., [Tr1], Theorem 3.3.3) we have $[w]_S^- \in B_{p,p}^{1-1/p}(S) = W_p^{1-1/p}(S)$, $1 < p < \infty$. If, in addition, $a(D,\omega)w(x) = 0$ in Ω^-, using the corresponding Green formula, we can also define $[\partial_n w]_S^-$ as a functional in $B_{p,p}^{-1/p}(S)$ by the duality relation (cf. [CS1], [DNS1], [DNS2])

$$(2.13)\qquad \langle [\partial_n w]_S^-\, ,\, [v]_S^- \rangle_S := \int_{\Omega^-} [\rho\,\omega^2\, w\, v \; - \; a_{pq} D_p w\, D_q v]\, dx,$$

where $1/p + 1/p' = 1$ and v is an arbitrary element of the space $W^1_{p',comp}(\Omega^-)$ (i.e., $v \in W^1_{p',loc}(\Omega^-)$ and has a compact support); obviously, $[v]_S^- \in B^{1/p}_{p',p'}(S)$.

Clearly,

$$\langle f\,,\,g\rangle_S = \int_S f\,g\,dS$$

for arbitrary smooth functions f and g.

Lemma 2.3. *Let $s \in \mathbb{R}$, $1 < p < \infty$, $1 \le q \le \infty$, and $S \in C^\infty$. The operators V, W, $\mathcal{H}$, $\mathcal{K}^{(1)}$, $\mathcal{K}^{(2)}$, and $\mathcal{L}$ can be extended by continuity to the bounded operators:*

$$\begin{aligned}
V \ &: \ B^s_{p,p}(S) \to H^{s+1+1/p}_p(\Omega^+) \cap C^\infty(\Omega^+),\\
&: \ B^s_{p,p}(S) \to H^{s+1+1/p}_{p,loc}(\Omega^-) \cap C^\infty(\Omega^-) \cap \mathrm{Som}(\Omega^-),\\
&: \ B^s_{p,q}(S) \to B^{s+1+1/p}_{p,q}(\Omega^+) \cap C^\infty(\Omega^+),\\
&: \ B^s_{p,q}(S) \to B^{s+1+1/p}_{p,q,loc}(\Omega^-) \cap C^\infty(\Omega^-) \cap \mathrm{Som}(\Omega^-),\\
W \ &: \ B^s_{p,p}(S) \to H^{s+1/p}_p(\Omega^+) \cap C^\infty(\Omega^+),\\
&: \ B^s_{p,p}(S) \to H^{s+1/p}_{p,loc}(\Omega^-) \cap C^\infty(\Omega^-) \cap \mathrm{Som}(\Omega^-),\\
&: \ B^s_{p,q}(S) \to B^{s+1/p}_{p,q}(\Omega^+) \cap C^\infty(\Omega^+),\\
&: \ B^s_{p,q}(S) \to B^{s+1/p}_{p,q,loc}(\Omega^-) \cap C^\infty(\Omega^-) \cap \mathrm{Som}(\Omega^-),\\
\mathcal{H} \ &: \ H^s_p(S) \to H^{s+1}_p(S) \quad [B^s_{p,q}(S) \to B^{s+1}_{p,q}(S)],\\
\mathcal{K}^{(j)} \ &: \ H^s_p(S) \to H^s_p(S) \qquad [B^s_{p,q}(S) \to B^s_{p,q}(S)], \quad j=1,2,\\
\mathcal{L} \ &: \ H^{s+1}_p(S) \to H^s_p(S) \quad [B^{s+1}_{p,q}(S) \to B^s_{p,q}(S)].
\end{aligned}$$

For these extended operators formulae (2.9)-(2.12) remain valid in the corresponding spaces. Representation (2.8) holds for $w \in W^1_{p,loc}(\Omega^-)$ if $a(D,\omega)w(x) = 0$ in Ω^-.

Lemma 2.4. *Let $g \in B^{1-1/p}_{p,q}(S)$, $1 < p < \infty$, $1 \le q \le \infty$, and $S \in C^\infty$. Moreover, let*

$$w(x) = W(g)(x) - i\,V(g)(x), \quad x \in \Omega^-. \tag{2.14}$$

If w vanishes in Ω^-, then $g = 0$ on S.

Further, we introduce the elliptic operators

$$\mathcal{N} g := \Big[(-2^{-1} I + \mathcal{K}^{(2)}) - i\,\mathcal{H}\Big]\,g, \tag{2.15}$$

$$\mathcal{M} g := \Big[\mathcal{L} - i\,(2^{-1} I + \mathcal{K}^{(1)})\Big]\,g, \tag{2.16}$$

which are generated by the limiting values on S of the potential (2.14) and its co-normal derivative.

Lemma 2.5. *Let $S \in C^{k+1+\alpha'}$ where $k \ge 1$ is a positive integer and $0 < \alpha' \le 1$. Then the operators*

$$\begin{aligned}
\mathcal{N} &: C^{m+1+\alpha}(S) \to C^{m+1+\alpha}(S),\\
\mathcal{M} &: C^{m+1+\alpha}(S) \to C^{m+\alpha}(S), \quad 0 \le m \le k,
\end{aligned}$$

are isomorphisms.

Lemma 2.6. *Let $S \in C^\infty$ and $s \in \mathbb{R}$, $1 < p < \infty$, $1 \le q \le \infty$.*
Then the operators

$$\begin{array}{lll} \mathcal{N} & : & H_p^s(S) \to H_p^s(S) \quad [B_{p,q}^s(S) \to B_{p,q}^s(S)], \\ \mathcal{M} & : & H_p^{s+1}(S) \to H_p^s(S) \quad [B_{p,q}^{s+1}(S) \to B_{p,q}^s(S)], \end{array}$$

are isomorphisms.

In particular, if $S \in C^{2+\alpha'}$ and $1 < p < \infty$, then the operators

$$\mathcal{N} : W_p^{1-1/p}(S) \to W_p^{1-1/p}(S) \quad \text{and} \quad \mathcal{M} : W_p^{1-1/p}(S) \to W_p^{-1/p}(S)$$

are isomorphisms. Moreover,

$$\mathcal{N}^{-1} : W_p^{1-1/p}(S) \to W_p^{1-1/p}(S)$$

is a singular integral operator on S of normal type, while

$$\mathcal{M}^{-1} : W_p^{-1/p}(S) \to W_p^{1-1/p}(S)$$

is a smoothing integral operator on S with the kernel-function of type $O(|x-y|^{-1})$.

The exterior Dirichlet and Neumann type boundary value problems (BVPs) for the equation (2.1) are studied in [JN1], [JN2].

Lemma 2.7. *The exterior Dirichlet BVP*

$$\begin{array}{l} a(D,\omega)w(x) = 0 \ \text{ in } \ \Omega^-, \quad w \in W_{p,loc}^1(\Omega^-) \cap \mathrm{Som}(\Omega^-), \ \ p > 1, \\ [w]^- = \varphi \ \text{ on } \ S, \quad \varphi \in W_p^{1-1/p}(S), \ \ S \in C^{2+\alpha'}, \end{array}$$

is uniquely solvable and the solution is representable in the form

$$w(x) = [W - iV]\left(\mathcal{N}^{-1}\varphi\right)(x), \quad x \in \Omega^-.$$

Lemma 2.8. *The exterior Neumann BVP*

$$\begin{array}{l} a(D,\omega)w(x) = 0 \ \text{ in } \ \Omega^-, \quad w \in W_{p,loc}^1(\Omega^-) \cap \mathrm{Som}(\Omega^-), \ \ p > 1, \\ [\partial_n w]^- = \psi \ \text{ on } \ S, \quad \psi \in W_p^{-1/p}(S), \ \ S \in C^{2+\alpha'}, \end{array}$$

is uniquely solvable and the solution is representable in the form

$$w(x) = [W - iV]\left(\mathcal{M}^{-1}\psi\right)(x), \quad x \in \Omega^-.$$

Remark that, if, in Lemmata 2.7 and 2.8,

$$\varphi \in B_{p,q}^{1-1/p}(S) \quad \text{and} \quad \psi \in B_{p,q}^{-1/p}(S), \quad p > 1, \ \ 1 \le q \le \infty,$$

then the solutions of the above BVPs belong to the appropriate functional space $B^1_{p,q,loc}(\Omega^-) \cap \mathrm{Som}(\Omega^-)$.

Let us now consider the interior Robin BVP

$$\begin{aligned} &a(D,\omega)w(x) = 0 \ \text{ in } \ \Omega^+, \\ &[\partial_n w \ - \ i\,w]^+ = \mu \ \text{ on } \quad S \in C^{2+\alpha'}. \end{aligned} \tag{2.17}$$

If we look for a solution to the BVP (2.17) in the form of a single-layer potential

$$w(x) \ = \ V(g)(x), \quad x \in \Omega^+,$$

we arrive at the Fredholm integral equation on S

$$\mathcal{R}g := \left(-2^{-1}I + \mathcal{K}^{(1)} - i\,\mathcal{H}\right) g = \mu.$$

There hold the following assertions.

Lemma 2.9. *Let*

(2.18) $S \in C^{k+1+\alpha'}$ with integer $k \geq 0,\ 0 < \alpha < \alpha' \leq 1,\ 0 \leq m \leq k,$

(2.19) $s \in \mathbb{R},\quad 1 < p < \infty,\quad 1 \leq q \leq \infty.$

Then the operators

$$\begin{aligned} \mathcal{R} \ &: \ C(S) \to C(S), \\ &: \ C^{m+\alpha}(S) \to C^{m+\alpha}(S), \quad 0 \leq m \leq k, \\ &: \ B^s_{p,q}(S) \to B^s_{p,q}(S), \quad S \in C^\infty, \end{aligned}$$

are isomorphisms.

Lemma 2.10. *Let the conditions* (2.18) *and* (2.19) *be fulfilled and* μ *belong to one of the spaces* $C(S)$, $C^{k+\alpha}(S)$, *and* $B^s_{p,q}(S)$. *Then the interior Robin BVP* (2.17) *is uniquely solvable in the classes* $C^1(\overline{\Omega^+})$, $C^{k+1+\alpha}(\overline{\Omega^+}))$, *and* $B^{s+1+1/p}_{p,q}(\Omega^+)$, *respectively.*

Moreover, an arbitrary solution to the equation (2.1) *in* Ω^+ *is uniquely representable in the form*

$$w(x) = V\left(\mathcal{R}^{-1}\left[\partial_n w - iw\right]^+_S\right)(x), \quad x \in \Omega^+.$$

For $\overline{\Omega^+_0} \in \Omega^+$ *there holds the uniform estimate*

$$|w(x)| \leq c\, ||[\partial_n w - iw]^+;\ X(S)||, \quad x \in \overline{\Omega^+_0},$$

where $X(S)$ *stands for one of the spaces* $C(S)$, $C^{k+\alpha}(S)$, *and* $B^s_{p,q}(S)$, *and* $||\cdot;\ X(S)||$ *denotes the corresponding norm.*

The proofs of these lemmata are quite similar to the corresponding ones in [CK1], [JN1], [JN2], [DNS1], and [DNS2].

From Lemma 2.10 it follows that the plane wave $\exp\{i\,d\cdot x\}$, where $d \in \Sigma_\omega$, can be uniquely represented in the form

$$e^{i\,d\cdot x} = V\left(\mathcal{R}^{-1}\left[(\partial_{n(\zeta)} - i)e^{i\,d\cdot\zeta}\right]_S^+\right)(x), \quad x \in \Omega^+. \tag{2.20}$$

Note, that $\exp\{i\,d\cdot x\}$ with $d \in \Sigma_\omega$ is a non-radiating solution to the homogeneous equation (2.1) in $\mathbb{R}^3$.

Let

$$P(S) := \left\{(\partial_{n(x)} - i)\, e^{i\,d\cdot x},\; x \in S,\; d \in \Sigma_\omega\right\},$$

$$P_{sp}(S) := \left\{\sum_{q=1}^{m} c_q\, p(x; d^{(q)}) \;:\; p(x; d^{(q)}) \in P(S),\; c_q \in \mathbb{C},\; d^{(q)} \in \Sigma_\omega,\; m \in \mathbb{N}\right\},$$

$$P_{sp}(\mathbb{R}^3) := \left\{\sum_{q=1}^{m} c_q\, e^{i\,d^{(q)}\cdot x},\; x \in \mathbb{R}^3,\; c_q \in \mathbb{C},\; d^{(q)} \in \Sigma_\omega,\; m \in \mathbb{N}\right\};$$

here $\mathbb{N}$ and $\mathbb{C}$ are the sets of all natural and complex numbers, respectively.

Lemma 2.11. *The set $P(S)$ is complete in $L_2(S)$.*

Proof. Let $f \in L_2(S)$ and

$$\int_S \left[(\partial_{n(y)} - i)e^{i\,d\cdot y}\right] f(y)\, dS_y = 0 \tag{2.21}$$

for all $d \in \Sigma_\omega$.

Let us consider the function

$$w(x) = (W - i\,V)(f)(x), \quad x \in \mathbb{R}^3 \backslash S,$$

where V and W are potential type operators given by (2.6) and (2.7).

Clearly, we have

$$w(x) = c^* \frac{\exp\{i\,\xi\cdot x\}}{|x|} \int_S \left[(\partial_{n(y)} - i)e^{-i\,\xi\cdot y}\right] f(y)\, dS_y + O(|x|^{-2})$$

as $|x| \to +\infty$, where $\xi \in \Sigma_\omega$ corresponds to x and c^* is defined by (2.5).

By (2.21) we then conclude

$$w(x) = O(|x|^{-2}),$$

which implies $w(x) = 0$ in Ω^- due to Lemma 2.1. Therefore, we obtain

$$[w(x)]^- = \mathcal{N} f = 0 \quad \text{on} \quad S.$$

By ellipticity of the operator $\mathcal{N}$ we have the inclusion $f \in C^{1+\alpha}(S)$, whence by Lemma 2.5 we arrive at the equation $f = 0$ on S. This completes the proof. □

Lemma 2.12. *Let Ω^+ be a bounded domain with C^2 boundary such that Ω^- be connected and let $w \in C^1(\overline{\Omega^+}) \cap C^2(\Omega^+)$ be a solution to the equation (2.1) in Ω^+.*

Then there exists a sequence $v_m \in P_{sp}(\mathbb{R}^3)$ such that $v_m \to w$ and $D^\beta v_m \to D^\beta w$ as $m \to \infty$ uniformly on compact subsets of Ω^+ ($\beta = (\beta_1, \beta_2, \beta_3)$ is an arbitrary multiindex).

Proof. From Lemma 2.11 it follows that there exists in $P_{sp}(S)$ a sequence of type

$$\sum_{q=1}^{m} c_q \, (\partial_{n(x)} - i) \exp\{i \, d^{(q)} \cdot x\}, \quad x \in S,$$

which converges (in the L_2-sense) to the function $[(\partial_{n(x)} - i)w]^+ \in C(S) \subset L_2(S)$.

We set

$$v_m(x) = \sum_{q=1}^{m} c_q \, e^{i \, d^{(q)} \cdot x}, \quad x \in \overline{\Omega^+}.$$

Hence,

$$(\partial_{n(x)} - i)v_m(x) \to [(\partial_{n(x)} - i)w(x)]^+ \ \text{ in } \ L_2(S).$$

By Lemma 2.10 the functions v_m and w can be represented in the form

$$v_m(x) = V(\mathcal{R}^{-1}[(\partial_n - i)v_m]^+)(x), \quad x \in \Omega^+,$$
$$w(x) = V(\mathcal{R}^{-1}[(\partial_n - i)w]^+)(x), \quad x \in \Omega^+.$$

Now, let $\overline{\Omega_0^+} \subset \Omega^+$ and $x \in \overline{\Omega_0^+}$. Denote by δ the distance between $\overline{\Omega_0^+}$ and $S = \partial\Omega^+$. The above representations of v_m and w together with Lemmata 2.9 and 2.10 then imply

$$\begin{aligned} &|D^\beta w(x) - D^\beta v_m(x)| \\ &\le c_1(\delta) \, ||\mathcal{R}^{-1}[(\partial_n - i)v_m]^+ - \mathcal{R}^{-1}[(\partial_n - i)w]^+; \ L_2(S)|| \\ &\le c_2(\delta) \, ||[(\partial_n - i)v_m]^+ - [(\partial_n - i)w]^+; \ L_2(S)|| \to 0 \end{aligned}$$

as $m \to +\infty$ (uniformly in $\overline{\Omega_0^+}$) for arbitrary multiindex β. □

Corollary 2.13. *Let $x_0 \notin \overline{\Omega^+}$. Then there exists a sequence $v_m \in P_{sp}(\mathbb{R}^3)$ such that (for arbitrary multiindex β)*

$$D^\beta v_m(x) \to D^\beta \gamma(x - x_0, \omega)$$

uniformly in $\overline{\Omega^+}$, i.e., for arbitrary $k \in \mathbb{N} \cup \{\not{\kappa}\}$ and $\alpha \in (0; 1)$

$$||v_m(x) - \gamma(x - x_0, \omega) \ : \ C^{k+\alpha}(\overline{\Omega^+})|| \to 0$$

as $m \to \infty$.

3. The direct mixed BVP

Now we are in the position to consider the direct mixed BVP of the wave scattering theory. In this case, a compact connected boundary S of a bounded domain Ω^+ is devided into two disjoint parts S' and S'': $S = S' \cup S'' \cup \ell$ with $\ell = \partial S' = \partial S''$. In what follows we assume that S and ℓ are $C^{k+\alpha}$-smooth with $k \geq 3$, $0 < \alpha \leq 1$. As above $\Omega^- = \mathbb{R}^3 \setminus \overline{\Omega^+}$. In the sequel the Ω^+ will be referred to as a scatterer.

A total wave field in Ω^- is represented as a sum of incident and scattered fields

$$w^{tot}(x) = w^{inc}(x) + w^{sc}(x),$$

where the incident field $w^{inc}(x)$ is taken in the form of a plane wave

$$w^{inc}(x) = w^{inc}(x;d) = e^{i\,d\cdot x}, \quad x \in \mathbb{R}^3, \quad d \in \Sigma_\omega, \tag{3.1}$$

while the scattered field solves the following mixed BVP:

$$a(D,\omega)\,w^{sc}(x) = 0 \quad \text{in} \quad \Omega^-, \tag{3.2}$$

$$w^{sc} \in \mathrm{Som}(\Omega^-), \tag{3.3}$$

$$[w^{sc}(x)]^-_{S'} = -w^{inc}(x) =: f^{inc}(x) \quad \text{on} \quad S', \tag{3.4}$$

$$[\partial_{n(x)} w^{sc}(x)]^-_{S''} = -\partial_{n(x)} w^{inc}(x) =: F^{inc}(x) \quad \text{on} \quad S''. \tag{3.5}$$

The submanifolds S' and S'' are called a sound-soft part and a sound-hard part of S, respectively.

We have to consider the problem (3.2)-(3.5) in the Sobolev space $W^1_{p,loc}(\Omega^-)$ with $p > 1$ since, in general, a mixed BVP does not possess a regular (C^1-smooth say) solution in the closed domain $\overline{\Omega^-}$. Therefore, we assume that

$$w^{sc} \in W^1_{p,loc}(\Omega^-) \cap \mathrm{Som}(\Omega^-). \tag{3.6}$$

We need also the following functional spaces:

$$B^s_{p,q}(S^*) = \{\pi_{S^*} f \;:\; f \in B^s_{p,q}(S)\},$$
$$\widetilde{B}^s_{p,q}(S^*) = \{f \in B^s_{p,q}(S);\ \mathrm{supp}\, f \subset \overline{S^*}\},$$

where S^* is a submanifold of S ($S^* \subset S$) and π_{S^*} is the restriction operator on S^*.

The boundary functions in (3.4) and (3.5) meet the natural conditions

$$f^{inc} = -\pi_{S'} w^{inc} \in B^{1-1/p}_{p,p}(S'), \tag{3.7}$$

$$F^{inc} = -\pi_{S''} \partial_n w^{inc} \in B^{-1/p}_{p,p}(S''). \tag{3.8}$$

The boundary condition (3.4) is considered in the usual trace sense, while (3.5) is to be understood in the functional sense (see (2.13))

$$\langle [\partial_n w^{sc}]^-_S\,,\, [v]^-_S\rangle_S := \int_{\Omega^-} [\rho\,\omega^2\, w^{sc}\, v \;-\; a_{pq} D_p w^{sc}\, D_q v]\, dx = \langle [F^{inc}\,,\, [v]^-_S\rangle_S$$

for arbitrary $v \in W^1_{p',comp}(\Omega^-)$ such that $[v]^-_S \in \widetilde{B}^{1/p}_{p',p'}(S'')$. We look for a solution to the mixed BVP (3.2)-(3.6) in the form

$$w^{sc}(x) = [W - i\,V](\mathcal{N}^{-1}[f_0^{inc} + \widetilde{\varphi}])(x), \quad x \in \Omega^-, \tag{3.9}$$

where V, W, and $\mathcal{N}$ are the operators given by (2.6), (2.7), and (2.15), respectively; f_0^{inc} is some fixed extension of the function f^{inc} from S' onto the whole surface S preserving the space

$$f_0^{inc} \in B^{1-1/p}_{p,p}(S), \quad \pi_{S'} f_0^{inc} = f^{inc},$$

and $\widetilde{\varphi} \in \widetilde{B}^{1-1/p}_{p,p}(S'')$ is an unknown function.

Moreover, we assume

$$\|f_0^{inc}\,;\,B^{1-1/p}_{p,p}(S)\| \le 2\,\|f^{inc}\,;\,B^{1-1/p}_{p,p}(S')\|. \tag{3.10}$$

It is evident that the conditions (3.2), (3.3), (3.4), and (3.6) are automatically satisfied, while the condition (3.5) leads to the pseudodifferential equation

$$\pi_{S''}\,\mathcal{M}\,\mathcal{N}^{-1}[f_0^{inc} + \widetilde{\varphi}] = F^{inc} \quad \text{on} \quad S'', \tag{3.11}$$

where $\mathcal{M}$ is defined by (2.16).

Let us introduce the operators

$$\widetilde{\Phi} := \pi_{S''}\mathcal{M}\,\mathcal{N}^{-1} \;:\; \widetilde{B}^{1-1/p}_{p,p}(S'') \to B^{-1/p}_{p,p}(S''), \tag{3.12}$$

$$\Phi := \pi_{S''}\mathcal{M}\,\mathcal{N}^{-1} \;:\; B^{1-1/p}_{p,p}(S) \to B^{-1/p}_{p,p}(S''). \tag{3.13}$$

Note that the restriction of the operator Φ on $\widetilde{B}^{1-1/p}_{p,p}(S'')$ coincides with the $\widetilde{\Phi}$, i.e.,

$$\Phi\, g = \widetilde{\Phi}\, g \quad \text{for} \quad g \in \widetilde{B}^{1-1/p}_{p,p}(S'').$$

Rewrite (3.11) as follows

$$\widetilde{\Phi}\,\widetilde{\varphi} = F^{inc} - \Phi\, f_0^{inc} \quad \text{on} \quad S''. \tag{3.14}$$

Clearly, $F^{inc} - \Phi\, f_0^{inc} \in B^{-1/p}_{p,p}(S'')$.

Thus, the mixed BVP under consideration is reduced to the pseudodifferential equation (3.14).

Now we go over to the investigation of the uniqueness, existence and regularity questions. We start with the particular case ($p = 2$) and formulate the uniqueness theorem which is a consequence of Lemma 2.1.

Lemma 3.1. *The homogeneous exterior mixed BVP* (3.2)-(3.5) ($f^{inc} = 0$, $F^{inc} = 0$) *has only the trivial solution in the class* $W^1_{2,loc}(\Omega^-) \cap \mathrm{Som}(\Omega^-)$.

Proof. It is verbatim the proof of Lemma 9.6 in [JN3]. □

Applying this lemma together with the general theory of elliptic pseudodifferential equations on manifold with boundary (see [Esk1], [DNS2]) and word for word arguments of the proof of Lemma 16.11 in [JN3] we can prove the following assertions.

Lemma 3.2. *The operator*

$$\widetilde{\Phi} \, : \, \widetilde{B}^{r}_{p,q}(S'') \to B^{r-1}_{p,q}(S'') \tag{3.15}$$

is bounded for arbitrary $p \in (1, +\infty)$, $q \in [1, +\infty]$, $r \in \mathbb{R}$.

The operator (3.15) *is an isomorphism if and only if the inequalities*

$$\frac{1}{p} - \frac{1}{2} < r < \frac{1}{p} + \frac{1}{2}$$

hold.

Lemma 3.3. *Let* $4/3 < p < 4$. *Then the equation* (3.14) *is uniquely solvable and*

$$\widetilde{\varphi} = \widetilde{\Phi}^{-1} \left(F^{inc} - \Phi \, f_0^{inc}\right) \in \widetilde{B}^{1-1/p}_{p,p}(S''),$$

$$\|\widetilde{\varphi};\, \widetilde{B}^{1-1/p}_{p,p}(S'')\| \le c \left\{\|f^{inc};\, B^{1-1/p}_{p,p}(S')\| + \|F^{inc};\, B^{-1/p}_{p,p}(S'')\|\right\}$$

with some positive constant c.

Further, we formulate the existence result.

Theorem 3.4. *Let conditions* (3.7) *and* (3.8) *be fulfilled and* $4/3 < p < 4$.

Then the direct mixed BVP (3.2)-(3.6) *is uniquely solvable and the solution is representable in the form* (3.9) *where* $\widetilde{\varphi}$ *solves the pseudodifferential equation* (3.14). *The solution*

$$w^{sc}(x) = [W - i\,V] \left(\mathcal{N}^{-1} \, (I - \widetilde{\Phi}^{-1}\Phi) \, f_0^{inc} + \mathcal{N}^{-1} \, \widetilde{\Phi}^{-1} F^{inc}\right)(x), \tag{3.16}$$

does not depend on $\pi_{S''} f_0^{inc}$ *and there holds the uniform estimate on* $\overline{\Omega_0^-} \subset \Omega^-$

$$|D_x^{\beta} \, w^{sc}(x)| \le c \left\{\|f^{inc};\, B^{1-1/p}_{p,p}(S')\| + \|F^{inc};\, B^{-1/p}_{p,p}(S'')\|\right\}$$

with arbitrary multiindex β *and some constant* $c = c(\delta)$ *where* $\delta = \mathrm{dist}\{\overline{\Omega_0^-}, \, \partial\Omega^-\}$; *moreover,* $c(\delta) \to 0$ *as* $\delta \to +\infty$, *and* $c(\delta) \to +\infty$ *as* $\delta \to 0$.

Proof. The first part of the theorem easily follows from Lemmata 3.1, 3.2, and 3.3.

The second part can be proved as follows.

Let f_{01}^{inc} and f_{02}^{inc} be two extensions of the f^{inc} from S' onto the whole of S:

$$f_{0j}^{inc} \in B_{p,p}^{1-1/p}(S), \quad \pi_{S'} f_{0j}^{inc} = f^{inc}.$$

Denote by w_j^{sc} the scattered field corresponding to the function f_{0j}^{inc}. Then due to the linearity of the operators involved in (3.16) we get

$$w_1^{sc}(x) - w_2^{sc}(x) = [W - i\,V] \left(\mathcal{N}^{-1}\,(I - \widetilde{\Phi}^{-1}\Phi)\,(f_{01}^{inc} - f_{02}^{inc})\right)(x).$$

Taking into account that $f_{01}^{inc} - f_{02}^{inc} \in \widetilde{B}_{p,p}^{1-1/p}(S'')$ we conclude

$$\widetilde{\Phi}^{-1}\Phi\,(f_{01}^{inc} - f_{02}^{inc}) = \widetilde{\Phi}^{-1}\widetilde{\Phi}\,(f_{01}^{inc} - f_{02}^{inc}) = f_{01}^{inc} - f_{02}^{inc},$$

whence

$$(I - \widetilde{\Phi}^{-1}\Phi)\,(f_{01}^{inc} - f_{02}^{inc}) = 0.$$

Consequently, $w_1^{sc}(x) = w_2^{sc}(x)$ in Ω^-.

Let $\overline{\Omega_0^-} \subset \Omega^-$ and x be an arbitrary point in $\overline{\Omega_0^-}$. By virtue of (3.16) then we have

$$\begin{aligned}
&\left|D_x^\beta\, w^{sc}(x)\right| \\
&= \left|\langle\, D_x^\beta\,[(\partial_{n(y)} - i)\gamma(x-y,\omega)],\, \mathcal{N}^{-1}\,[(I - \widetilde{\Phi}^{-1}\Phi)\, f_0^{inc} + \widetilde{\Phi}^{-1}\, F^{inc}]\,\rangle_S\right| \\
&\le c_1(\delta)\, \Big\{ ||\mathcal{N}^{-1}\,(I - \widetilde{\Phi}^{-1}\Phi)\, f_0^{inc};\ B_{p,p}^{1-1/p}(S)|| \\
&\qquad + ||\mathcal{N}^{-1}\,\widetilde{\Phi}^{-1}\, F^{inc};\ B_{p,p}^{1-1/p}(S)|| \Big\} \\
&\le c_2(\delta)\, \Big\{ ||f_0^{inc};\ B_{p,p}^{1-1/p}(S)|| + ||F^{inc};\ B_{p,p}^{-1/p}(S'')|| \Big\} \\
&\le c_3(\delta)\, \Big\{ ||f^{inc};\ B_{p,p}^{1-1/p}(S')|| + ||F^{inc};\ B_{p,p}^{-1/p}(S'')|| \Big\},
\end{aligned}$$

where the positive constants $c_k(\delta)$ have the properties: $c_k(\delta) = O(\delta^{-2-|\beta|})$ as $\delta \to 0$ and $c_k(\delta) = O(\delta^{-1})$ as $\delta \to +\infty$, $k = 1, 2, 3$. This completes the proof. □

Note that at infinity (i.e., as $|x| \to +\infty$) we have the asymptotic formula

$$w^{sc}(x) = \frac{\exp\{i\,\xi \cdot x\}}{|x|}\, w_\infty^{sc}(\xi) + O(|x|^{-2}),$$

where $\xi \in \Sigma_\omega$ corresponds to $\hat{x} = x/|x|$, and

$$\begin{aligned}
w_\infty^{sc}(\xi) \;&=\; c^* \int_S [(\partial_{n(y)} - i) e^{-i\,\xi\cdot y}]\, \Big(\mathcal{N}^{-1}\,(I \\
&-\; \widetilde{\Phi}^{-1}\Phi)\, f_0^{inc} + \mathcal{N}^{-1}\,\widetilde{\Phi}^{-1}\, F^{inc}\Big)\, dS_y \\
&=\; c^* \langle (\partial_{n(y)} - i) e^{-i\,\xi\cdot y},\, \mathcal{N}^{-1}\,(I - \widetilde{\Phi}^{-1}\Phi)\, f_0^{inc} \rangle_S \\
&+\; c^* \langle (\partial_{n(y)} - i) e^{-i\,\xi\cdot y},\, \mathcal{N}^{-1}\,\widetilde{\Phi}^{-1}\, F^{inc} \rangle_S,
\end{aligned}$$

where c^* is defined by (2.5). Whence

$$
\begin{aligned}
|w^{sc}_{\infty}(\xi)| \le |c^*| \Big\{ & ||(\partial_{n(y)} - i)e^{-i\xi\cdot y};\ B^{-(1-1/p)}_{p',p'}(S)|| \\
& \times ||\mathcal{N}^{-1}\,(I - \widetilde{\Phi}^{-1}\Phi)\, f_0^{inc};\ B^{1-1/p}_{p,p}(S)|| \\
& + ||(\partial_{n(y)} - i)e^{-i\xi\cdot y};\ B^{-(1-1/p)}_{p',p'}(S)|| \\
& \times ||\mathcal{N}^{-1}\,\widetilde{\Phi}^{-1}\, F^{inc};\ B^{1-1/p}_{p,p}(S)|| \Big\} \\
\le c_1 \Big\{ & ||\mathcal{N}^{-1}\,(I - \widetilde{\Phi}^{-1}\Phi)\, f_0^{inc};\ B^{1-1/p}_{p,p}(S)|| \\
& + ||\mathcal{N}^{-1}\,\widetilde{\Phi}^{-1}\, F^{inc};\ B^{1-1/p}_{p,p}(S)|| \Big\} \\
\le c_2 \Big\{ & ||f_0^{inc};\ B^{1-1/p}_{p,p}(S)|| + ||F^{inc};\ B^{-1/p}_{p,p}(S'')|| \Big\} \\
\le c \Big\{ & ||f^{inc};\ B^{1-1/p}_{p,p}(S')|| + ||F^{inc};\ B^{-1/p}_{p,p}(S'')|| \Big\}, \qquad (3.17)
\end{aligned}
$$

where c_1 , c_2, and c are some positive constants; in this case, c does not depend on f^{inc} and F^{inc} (it depends only on S', S'', ω, and the coefficients a_{pq}).

As we have mentioned above a mixed BVP, in general, does not possess a C^1-smooth solution in $\overline{\Omega^-}$ even for C^∞-smooth boundary data. However, applying the arguments of the references [CS1], [DNS2], [Fi2], [JN3] (Chapter 6), we can establish the following regularity result for the solution to the mixed BVP (3.2)-(3.5).

Theorem 3.5. *Let the conditions of Theorem* 3.4 *be fulfilled and let the function* $w^{sc} \in W^1_{p,loc}(\Omega^-) \cap \mathrm{Som}(\Omega^-)$ *be the unique solution to the BVP* (3.2)-(3.5).

In addition to (3.7) *and* (3.8), *let*

$$f^{inc} \in C^\alpha(S'), \quad F^{inc} \in B^{\alpha-1}_{\infty,\infty}(S''),$$

for some $\alpha \in (0;\ 1)$.

Then

$$w^{sc} \in C^\nu(\overline{\Omega^-}) \cap \mathrm{Som}(\Omega^-)$$

with any $\nu \in (0; \alpha_0)$, $\alpha_0 := \min\{\alpha,\ 1/2\}$.

In particular, if

$$f^{inc} \in C^{k+\alpha}(S'), \quad F^{inc} \in C^{k-1+\alpha}(S''), \ \ k \ge 1, \ \ 0 < \alpha < 1,$$

then

$$w^{sc} \in C^{1/2-\varepsilon}(\overline{\Omega^-}) \cap C^{k+\alpha}(\Omega^- \cup \overline{S^*})$$

where $S^* \subset S$ *and* $\overline{S^*} \cap \ell = \emptyset$ *with* $\ell = \partial S' = \partial S''$, *and* ε *is an arbitrary small number.*

Proof. It is verbatim the proof of Theorem 16.5 in [JN3]. □

Remark 3.6. In the direct mixed scattering BVP in question we have

$$f^{inc}(x;d) = -\pi_{S'}\, e^{i\,d\cdot x} \in C^{k+\alpha}(\overline{S'}) \subset B^{1-1/p}_{p,p}(S'), \tag{3.18}$$

$$F^{inc}(x;d) = -\pi_{S''}\, \partial_{n(x)}\, e^{i\,d\cdot x} \in C^{k-1+\alpha}(\overline{S''}) \subset B^{-1/p}_{p,p}(S''), \tag{3.19}$$

and, therefore, due to Theorems 3.4 and 3.5 we get the following representation of the unique solution

$$w^{sc}(x;d) = [W - i\,V]\,(g(\cdot\,;d))\,(x), \tag{3.20}$$

where

$$g(x;d) = \Big[\mathcal{N}^{-1}\,(I - \widetilde{\Phi}^{-1}\Phi)\, f_0^{inc}(\cdot\,;d) + \mathcal{N}^{-1}\,\widetilde{\Phi}^{-1} F^{inc}(\cdot\,;d)\Big]\,(x). \tag{3.21}$$

Here $f^{inc}(\cdot\,;d)$ and $F^{inc}(\cdot\,;d)$ are given by (3.18) and (3.19), and $f_0^{inc}(\cdot\,;d)$ is some fixed extension of $f^{inc}(\cdot\,;d)$ from S' onto the whole of S satisfying the condition (3.10) ($f_0^{inc}(x;d) = e^{i\,d\cdot x}\,\chi(x)$ with $\chi \in C^\infty(\mathbb{R}^3)$ where $\chi(x) = 1$ in some small ε-neighbourhood of S' and χ vanishes outside of a 2ε-neighbourhood of S').

Moreover,

$$w^{sc}(x;d) = w^{sc}_\infty(\xi;d)\,\frac{\exp\{i\,\xi\cdot x\}}{|x|} + O(|x|^{-2}) \quad \text{as} \quad |x| \to +\infty, \tag{3.22}$$

where $\xi \in \Sigma_\omega$ corresponds to x and where $w^{sc}_\infty(\cdot;d)$ is the so-called far-field pattern corresponding to the scatterer Ω^+ and to the direction vector $d \in \Sigma_\omega$:

$$w^{sc}_\infty(\xi;d) = c^* \int_S \big[(\partial_{n(y)} - i)e^{-i\,\xi\cdot y}\big]\, g(y;d)\, dS_y \tag{3.23}$$

with $g(\cdot\,;d)$ and c^* defined by (3.21) and (2.5).

Clearly, $w^{sc}_\infty(\xi;d)$ is a real analytic complex-valued function in $\xi \in \Sigma_\omega$. For arbitrary $\xi \in \Sigma_\omega$ and $d \in \Sigma_\omega$ we have

$$\begin{aligned} |w^{sc}_\infty(\xi;d)| &\le c_1 \Big\{ \|\pi_{S'}\, e^{i\,d\cdot x};\; B^{1-1/p}_{p,p}(S')\| + \|\pi_{S''}\, \partial_{n(x)} e^{i\,d\cdot x};\; B^{-1/p}_{p,p}(S'')\| \Big\} \\ &\le c_2\, \|e^{i\,d\cdot x};\; B^{1-1/p}_{p,p}(S)\| \le c_3\, \|e^{i\,d\cdot x};\; C^1(S)\|. \end{aligned}$$

Remark 3.7. Applying the asymptotic technique developed in the references [Esk1], [Ben1], [ChDu1],we can obtain the following estimate for the solution of the direct mixed scattering BVP in a vicinity of the curve ℓ:

$$|D_k\, w^{sc}(x;d)| \le c\,[r(x)]^{-1/2}, \quad r(x) = \mathrm{dist}\{x,\, \ell\}, \tag{3.24}$$

where c is some positive constant.

4. Formulation of the mixed inverse scattering problem

The mixed inverse scattering problem consists in finding the surface S (i.e., the scatterer Ω^+) together with sound-soft and sound-hard subsurfaces, S' and S'', respectively, if the corresponding far-field pattern $w^{sc}_\infty(\cdot\,;d)$ is known for several or all direction vectors $d \in \Sigma_\omega$. More precise mathematical formulation of the mixed inverse scattering problem reads as follows:

Find a compact connected surface $S \in C^{k+\alpha}$ and its disjoint sound-soft and sound-hard parts – S' and S'' such that $S = \overline{S'} \cup \overline{S''}$ and the total wave field

$$w^{tot}(x;d) = w^{inc}(x;d) + w^{sc}(x;d)$$

with $w^{inc}(x;d)$ given by (3.1) as an incident wave and $w^{sc}(x;d)$ as a corresponding (unknown) radiating scattered wave, meets the following conditions (cf. (3.4) and (3.5))

$$\begin{aligned} &[w^{tot}(x;d)]^-_{S'} = 0 \quad \text{on} \quad S',\\ &[\partial_{n(x)} w^{tot}(x;d)]^-_{S''} = 0 \quad \text{on} \quad S'',\\ &w^{tot}(\cdot\,;d) \in W^1_{p,loc}(\Omega^-), \quad p > 1. \end{aligned}$$

In addition, the far-field pattern $w^{sc}_\infty(\cdot\,;d)$ is known for some (or all) direction vectors $d \in \Sigma_\omega$ (see (3.23))

$$w^{sc}_\infty(\xi;d) = \mathcal{G}(\xi;d), \quad \xi \in \Sigma_\omega,$$

where $\mathcal{G}(\cdot\,;d)$ is a given function of ξ on Σ_ω.

In what follows we assume that the common boundary of the subsurfaces S' and S'' ($\ell = \partial S' = \partial S''$) is also of the class $C^{k+\alpha}$ ($k \geq 3$, $0 < \alpha < 1$).

The bounded domain Ω^+ surrounded by the surface S and the corresponding function $w^{sc}_\infty(\cdot\,;d)$ we shall refer to as a scatterer Ω^+ and a far-field pattern for the scatterer Ω^+, respectively.

5. Uniqueness theorem

Here we prove the following uniqueness

Theorem 5.1. *Let Ω^+_j, $j = 1, 2$, be two scatterers with a sound-soft part S'_j and a sound-hard part S''_j:*

$$S_j = \partial\Omega^+_j = \overline{S'_j} \cup \overline{S''_j}, \quad \ell_j = \partial S'_j = \partial S''_j, \quad S'_j \cap S''_j = \emptyset.$$

Moreover, let for a fixed wave number ω the far-field patterns $w^{sc}_{j,\infty}(\cdot\,;d)$ for the both scatterers coincide for all incident directions $d \in \Sigma_\omega$.

Then $\Omega_1^+ = \Omega_2^+$ *(i.e.,* $S_1 = S_2$*),* $S_1' = S_2'$*, and* $S_1'' = S_2''$*.*

Proof. First we show that $\Omega_1^+ = \Omega_2^+$. Assume the contrary: $\Omega_1^+ \neq \Omega_2^+$. Denote by $w_j^{tot}(x;d)$ and $w_j^{sc}(x;d)$ the total and the scattered fields which correspond to the domains $\Omega_j^- = \mathbb{R}^3 \setminus \overline{\Omega_j^+}$ $(j = 1,2)$, i.e.,

$$w_j^{tot}(x;d) = w_j^{inc}(x;d) + w_j^{sc}(x;d), \quad w_j^{inc}(x;d) = e^{i\,d\cdot x}, \quad x \in \Omega_j^-, \quad d \in \Sigma_\omega,$$

$$a(D,\omega)\, w_j^{sc}(x;d) = 0 \quad \text{in} \quad \Omega_j^-,$$

$$w_j^{sc}(\cdot\,;d) \in W^1_{p,loc}(\Omega_j^-) \cap C^\infty(\Omega_j^-) \cap C^{1/2-\varepsilon}(\overline{\Omega_j^-}) \cap \mathrm{Som}(\Omega_j^-),$$

$$[w_j^{sc}(x;d)]^-_{S_j'} = -\pi_{S_j'} e^{i\,d\cdot x} =: f_j^{inc}(x;d) \quad \text{on} \quad S_j', \tag{5.1}$$

$$[\partial_{n(x)} w_j^{sc}(x;d)]^-_{S_j''} = -\pi_{S_j''} e^{i\,d\cdot x} =: F_j^{inc}(x;d) \quad \text{on} \quad S_j'', \tag{5.2}$$

where ε is an arbitrary small positive number.

Due to formulae (3.20) and (3.21) we have the representation

$$w_j^{sc}(x;d) = [W_j - i\,V_j]\left(\mathcal{N}_j^{-1}\,(I - \widetilde{\Phi}_j^{-1}\Phi_j)\, f_{0j}^{inc}(\cdot\,;d) + \mathcal{N}_j^{-1}\,\widetilde{\Phi}_j^{-1} F_j^{inc}(\cdot\,;d)\right)(x),$$

where the operators W_j, V_j, $\mathcal{N}_j$, $\widetilde{\Phi}_j$, and Φ_j are defined by (2.6), (2.7), (2.15), (3.12), and (3.13) with S_j, S_j', and S_j'' in the place of S, S', and S'', respectively; moreover, here f_{0j}^{inc} is some fixed extension of the function $f_j^{inc}(x;d) = -\pi_{S_j'} e^{i\,d\cdot x}$ from the S_j' onto the whole of S_j, such that

$$\|f_{0j}^{inc}(\cdot\,;d)\ ;\ B^{1-1/p}_{p,p}(S_j)\| \le 2\, \|\pi_{S_j'} e^{i\,d\cdot x}\ ;\ B^{1-1/p}_{p,p}(S_j')\|.$$

Note that $w_j^{sc}(x;d)$ does not depend on $\pi_{S_j''} f_{0j}^{inc}$.

By the assumption of the theorem

$$w^{sc}_{1,\infty}(\xi;d) = w^{sc}_{2,\infty}(\xi;d), \quad \xi \in \Sigma_\omega, \tag{5.3}$$

for all $d \in \Sigma_\omega$. Here $\xi \in \Sigma_\omega$ corresponds to $\hat{x} = x/|x|$ and is given by (2.3) (with $\eta = \hat{x}$).

By formulae (3.23) and (3.21) we have

$$\begin{aligned} w^{sc}_{j,\infty}(\xi;d) &= c^* \int_{S_j} \left[(\partial_{n(y)} - i)e^{-i\,\xi\cdot y}\right] \left(\mathcal{N}_j^{-1}\,(I - \widetilde{\Phi}_j^{-1}\Phi_j)\, f_{0j}^{inc}(\cdot\,;d)\right. \\ &+ \left.\mathcal{N}_j^{-1}\,\widetilde{\Phi}_j^{-1} F_j^{inc}(\cdot\,;d)\right) dS_y. \end{aligned}$$

Further, let $\Omega_{12}^- = \mathbb{R}^3 \setminus \{\overline{\Omega_1^+} \cup \overline{\Omega_2^+}\}$ and $\Omega_{12}^+ = \Omega_1^+ \cup \Omega_2^+$.

The condition (5.3) together with Lemma 2.1 implies that

$$w_1^{sc}(x;d) = w_2^{sc}(x;d) \quad \text{in} \quad \Omega_{12}^-, \tag{5.4}$$

since the difference $w_1^{sc}(x;d) - w_2^{sc}(x;d)$ solves the equation (2.1) in Ω_{12}^- and at infinity decays as $O(|x|^{-2})$.

It is evident that, if in the place of an incident wave $w^{inc}(x;d)$ we take an arbitrary function $v_m(x)$ from the linear span $P_{sp}(\mathbb{R}^3)$ and denote the corresponding scattered waves by $w_{m,j}^{sc}(x)$, $j = 1,2$, then

$$\left.\begin{array}{lll} w_{m,1,\infty}^{sc}(\xi) = w_{m,2,\infty}^{sc}(\xi) & \text{on} & \Sigma_\omega \\ w_{m,1}^{sc}(x) = w_{m,2}^{sc}(x) & \text{in} & \Omega_{12}^- \end{array}\right\} \tag{5.5}$$

where

$$w_{m,j}^{sc}(x) = [W_j - i\,V_j]\left(\mathcal{N}_j^{-1}\,(I - \widetilde{\Phi}_j^{-1}\Phi_j)\,f_{0j,m}^{inc} + \mathcal{N}_j^{-1}\,\widetilde{\Phi}_j^{-1}F_{j,m}^{inc}\right)(x),$$

$$w_{m,j,\infty}^{sc}(\xi) = c^* \int_{S_j} [(\partial_{n(y)} - i)e^{-i\,\xi\cdot y}]\,\left(\mathcal{N}_j^{-1}\,(I - \widetilde{\Phi}_j^{-1}\Phi_j)\,f_{0j,m}^{inc}\right.$$
$$\left. + \mathcal{N}_j^{-1}\,\widetilde{\Phi}_j^{-1}\,F_{j,m}^{inc}\right)\,dS_y;$$

here

$$f_{j,m}^{inc}(x) = -\pi_{S_j'}\,v_m(x), \quad F_{j,m}^{inc}(x) = -\pi_{S_j''}\,\partial_{n(x)}v_m(x),$$

and $f_{0j,m}^{inc}$ is some extension of $f_{j,m}^{inc}$ from S_j' onto the whole of S_j preserving the functional space and satisfying the inequality

$$\|f_{0j,m}^{inc}\,;\,B_{p,p}^{1-1/p}(S_j)\| \le 2\,\|f_{j,m}^{inc}\,;\,B_{p,p}^{1-1/p}(S_j')\|.$$

Next, we consider the following BVP

$$a(D,\omega)\,w_j^{sc}(x;x_0) = 0 \quad \text{in} \quad \Omega_j^-, \quad j = 1,2, \tag{5.6}$$

$$w_j^{sc}(\cdot\,;x_0) \in W_{p,loc}^1(\Omega_j^-) \cap C^\infty(\Omega_j^-) \cap C^{1/2-\varepsilon}(\overline{\Omega_j^-}) \cap \mathrm{Som}(\Omega_j^-), \tag{5.7}$$

$$[w_j^{sc}(x;x_0)]_{S_j'}^- = -\pi_{S_j'}\gamma(x - x_0,\omega) =: f_j^{inc}(x;x_0) \quad \text{on} \quad S_j', \tag{5.8}$$

$$[\partial_{n(x)}w_j^{sc}(x;x_0)]_{S_j''}^- = -\pi_{S_j''}\partial_{n(x)}\gamma(x - x_0,\omega) =: F_j^{inc}(x;x_0) \quad \text{on} \quad S_j'', \tag{5.9}$$

where $x_0 \in \Omega_{12}^-$.

Clearly, this problem is solvable and

$$\begin{aligned} w_j^{sc}(x;x_0) \;&=\; [W_j - i\,V_j]\left(\mathcal{N}_j^{-1}\,(I - \widetilde{\Phi}_j^{-1}\Phi_j)\,f_{0j}^{inc}(\cdot\,;x_0)\right. \\ &\quad+\; \left.\mathcal{N}_j^{-1}\,\widetilde{\Phi}_j^{-1}F_j^{inc}(\cdot\,;x_0)\right)(x), \\ w_{j,\infty}^{sc}(\xi;x_0) \;&=\; c^* \int_{S_j} [(\partial_{n(y)} - i)e^{-i\,\xi\cdot y}]\,\left(\mathcal{N}_j^{-1}\,(I - \widetilde{\Phi}_j^{-1}\Phi_j)\,f_{0j}^{inc}(\cdot\,;x_0)\right. \\ &\quad+\; \left.\mathcal{N}_j^{-1}\,\widetilde{\Phi}_j^{-1}\,F_j^{inc}(\cdot\,;x_0)\right)\,dS_y. \end{aligned} \tag{5.10}$$

In what follows we show that

$$w^{sc}_{1,\infty}(\xi;x_0)=w^{sc}_{2,\infty}(\xi;x_0)\qquad\text{for all}\quad \xi\in\Sigma_\omega, \tag{5.11}$$

and, consequently,

$$w^{sc}_1(x;x_0)=w^{sc}_2(x;x_0),\quad x\in\Omega^-_{12}. \tag{5.12}$$

In fact, since $x_0\in\Omega^-_{12}$, we have: $\operatorname{dist}\{x_0,\ S_1\cup S_2\}\geq\delta>0$. Therefore, there exist bounded domains Ω^* and Ω^*_0 with C^∞-smooth boundaries such that $x_0\notin\Omega^*$, $\overline{\Omega^+_{12}}\subset\Omega^*_0$, and $\overline{\Omega^*_0}\subset\Omega^*$. By Corollary 2.13 then there exists a sequence of plane waves $v_m\in P_{sp}(\mathbb{R}^3)$ such that

$$D^\beta v_m(x)\to D^\beta\gamma(x-x_0,\omega)\quad\text{as}\quad m\to+\infty \tag{5.13}$$

uniformly on $\overline{\Omega^+_{12}}$ (for arbitrary multiindex β).

This implies (see (3.17))

$$\begin{aligned}
&|w^{sc}_{m,j,\infty}(\xi)-w^{sc}_{j,\infty}(\xi;x_0)|\\
&\leq c\,\Big\{||f^{inc}_{j,m}(\cdot)-f^{inc}_j(\cdot\,;x_0);\ B^{1-1/p}_{p,p}(S'_j)||\\
&+||F^{inc}_{j,m}(\cdot)-F^{inc}_j(\cdot\,;x_0);\ B^{-1/p}_{p,p}(S''_j)||\,\Big\}\\
&=c\,\Big\{||\pi_{S'_j}[v_m(\cdot)-\gamma(\cdot-x_0;\omega)];\ B^{1-1/p}_{p,p}(S'_j)||\\
&+||\pi_{S''_j}[\partial_n v_m(\cdot)-\partial_n\gamma(\cdot-x_0;\omega)];\ B^{-1/p}_{p,p}(S''_j)||\,\Big\}\to 0
\end{aligned}$$

as $m\to+\infty$ $(j=1,2)$, due to (5.13).

With the help of equations (5.5) we derive

$$w^{sc}_{1,\infty}(\xi;x_0)=\lim_{m\to\infty}w^{sc}_{m,1,\infty}(\xi)=\lim_{m\to\infty}w^{sc}_{m,2,\infty}(\xi)=w^{sc}_{2,\infty}(\xi;x_0).$$

Whence (5.11) and (5.12) follow.

Now, we show that (5.12) contradicts to the assumption $\Omega^+_1\neq\Omega^+_2$.

In fact, since $\Omega^+_1\neq\Omega^+_2$, there exists a point $x^*\in[\partial\Omega^+_{12}\setminus(\overline{\Omega^+_1}\cap\overline{\Omega^+_2})]\setminus[\ell_1\cup\ell_2]$, such that the closed ball $\overline{B(x^*,2\delta)}$ centered at x^* and radius $2\delta>0$ does not intersect ℓ_1, ℓ_2, and either $\overline{\Omega^+_1}$ or $\overline{\Omega^+_2}$. Without restriction of generality, let us assume that $\overline{B(x^*,2\delta)}\cap(\ell_1\cup\ell_2\cup\overline{\Omega^+_2})=\emptyset$. Evidently, $S^*_1=\partial\Omega^+_1\cap\overline{B(x^*,2\delta)}$ is a subset of either S'_1 or S''_1, and $\operatorname{dist}\{\overline{B(x^*,\delta)},\ \ell_1\cup\overline{\Omega^+_2}\}\geq\delta>0$.

Again without loss of generality, let $S^*_1\subset S''_1$ (the case $S^*_1\subset S'_1$ can be considered quite similarly).

Let us choose a sequence $x^p\in B(x^*,\delta)\cap\Omega^-_{12}$ on the normal line to S_1 at the point $x^*\in S^*_1$ such that $|x^p-x^*|\to 0$ as $p\to+\infty$.

Further, let us take x^p in the place of x_0 in the problem (5.6)-(5.9). Due to the inequality $\operatorname{dist}\{\overline{B(x^*,\delta)}\cap\Omega^-_{12},\ \partial\Omega^-_2\}\geq\delta>0$, we conclude that the function

$w_2^{sc}(\cdot\,, x^p)$ is C^∞-regular in $\overline{B(x^*,\delta)} \subset \Omega_2^-$. Moreover, for arbitrary multiindex β we have

$$\lim_{p\to+\infty} D_x^\beta w_2^{sc}(x;x^p) = D_x^\beta w_2^{sc}(x;x^*)$$

uniformly with respect to x in $\overline{B(x^*,\delta)}$, which follows from the representation formula (5.10) with x^p and x^* in the place of x_0. In fact, we easily derive

$$\begin{aligned}
&|D_x^\beta\, w_2^{sc}(x;x^p) - D_x^\beta\, w_2^{sc}(x;x^*)| \\
&\le c_1(\delta)\ \Big\{||\pi_{S_2'}[\gamma(\cdot - x^p;\omega) - \gamma(\cdot - x^*;\omega)];\ B_{p,p}^{1-1/p}(S_2')|| \\
&+||\pi_{S_2''}[\partial_n\gamma(\cdot - x^p;\omega) - \partial_n\gamma(\cdot - x^*;\omega)];\ B_{p,p}^{-1/p}(S_2'')||\Big\} \\
&\le c_2(\delta)\ |x^p - x^*|.
\end{aligned}$$

In particular, the limit

$$\lim_{p\to+\infty} \partial_{n(x^*)} w_2^{sc}(x^*;x^p) = \partial_{n(x^*)} w_2^{sc}(x^*;x^*) \tag{5.14}$$

exists and is a finite number.

On the other hand, due to the equation (5.12) with x^p in the place of x_0, we get

$$w_1^{sc}(\cdot\,;x^p) = w_2^{sc}(\cdot\,;x^p) \quad \text{in} \quad \Omega_{12}^-.$$

Since the both functions are continuously differentiable in $\overline{B(x^*,\delta)} \cap \overline{\Omega_{12}^-}$ for arbitrary x^p (see Theorem 3.5), we conclude

$$\begin{aligned}
&[\partial_{n(x^*)} w_1^{sc}(x^*;x^p)]^- = \lim_{x\to x^*} a_{kj}\, n_k(x^*)\, D_j\, w_1^{sc}(x;x^p) \\
&= \lim_{x\to x^*} a_{kj}\, n_k(x^*)\, D_j\, w_2^{sc}(x;x^p) = \partial_{n(x^*)} w_2^{sc}(x^*;x^p).
\end{aligned}$$

But due to the boundary condition (5.9) with $j=1$ and with x^p in the place of x_0 we have

$$[\partial_{n(x^*)} w_1^{sc}(x^*;x^p)]^- = -\partial_{n(x^*)}\gamma(x^* - x^p,\omega).$$

Therefore, we arrive at the equation

$$[\partial_{n(x^*)} w_2^{sc}(x^*;x^p)]^- = -\partial_{n(x^*)}\gamma(x^* - x^p,\omega)$$

which contradicts to the equation (5.14), since the right-hand side expression in the previous equation is not bounded when x^p approaches x^*:

$$\partial_{n(x^*)}\gamma(x^* - x^p,\omega) = \frac{1}{4\pi|\widetilde{a}|^{1/2}[\widetilde{a}^{-1}n(x^*)\cdot n(x^*)]^{3/2}} \cdot \frac{1}{|x^* - x^p|^2} + O(1)$$

(we recall that $x^p - x^* = |x^p - x^*|\, n(x^*)$ where $n(x^*)$ is the outward unit normal vector to S_1 at the point $x^* \in S_1$).

This contradiction proves that

$$\Omega_1^+ = \Omega_2^+, \quad \text{i.e.,} \quad S_1 = S_2 =: S.$$

Further, we show that $S_1' = S_2'$ and $S_1'' = S_2''$.

Let us assume that $S_1' \neq S_2'$. Moreover, without loss of generality we provide that $S_1' \cap S_2' \neq S_1'$.

Clearly, in this case, there exists a subsurface $S^* \subset S$ with a smooth boundary ∂S^* such that $S^* \subset S_1' \cap S_2''$. We put (see (5.4))

$$w^{sc}(x;d) := w_1^{sc}(x;d) = w_2^{sc}(x;d) \quad \text{in} \quad \Omega^- := \Omega_1^- = \Omega_2^-,$$

where $d \in \Sigma_\omega$ is some direction vector.

It is easy to see that on S^* the following conditions

$$[w^{sc}(x;d)]_{S^*}^- = [w_1^{sc}(x;d)]_{S^*}^- = -\pi_{S^*} e^{i\,d\cdot x}, \tag{5.15}$$

$$[\partial_n w^{sc}(x;d)]_{S^*}^- = [\partial_n w_2^{sc}(x;d)]_{S^*}^- = -\pi_{S^*} \partial_n e^{i\,d\cdot x}, \tag{5.16}$$

hold due to the boundary conditions (5.1) and (5.2).

Next, we show that (5.15) and (5.16) imply

$$w^{sc}(x;d) = -e^{i\,d\cdot x} \quad \text{in} \quad \Omega^-. \tag{5.17}$$

In fact, let

$$w(x) := w^{sc}(x;d) + e^{i\,d\cdot x}, \quad x \in \Omega^-,$$

and let

$$\Omega_R^- = B(O,R) \cap \Omega^-, \quad \Omega^+ := \Omega_1^+ = \Omega_2^+ \subset B(O,R),$$

where $B(O,R)$ is the ball centered at the origin and radius R. Evidently,

$$[w(x)]_{S^*}^- = 0, \quad [\partial_n w(x)]_{S^*}^- = 0 \quad \text{on} \quad S^*. \tag{5.18}$$

We have the following Green formula in the domain Ω_R^-

$$\begin{aligned} &\int_{\Omega_R^-} \left[a(D,\omega) v^{(1)}\, v^{(2)} - v^{(1)}\, a(D,\omega) v^{(2)} \right] dx \\ &= -\int_S \left[(\partial_n v^{(1)})^-\, (v^{(2)})^- - (v^{(1)})^-\, (\partial_n v^{(2)})^- \right] dS \\ &+ \int_{\Sigma(O,R)} \left[(\partial_n v^{(1)})^+\, (v^{(2)})^+ - (v^{(1)})^+\, (\partial_n v^{(2)})^+ \right] dS, \end{aligned} \tag{5.19}$$

where $\Sigma(O,R) = \partial B(O,R)$ and n is the outward normal vector on the both surfaces S and $\Sigma(O,R)$; moreover, here we assume that $a(D,\omega)v^{(j)} \in L_2(\Omega_R^-)$, $v^{(j)} \in C(\overline{\Omega_R^-}) \cap C^1(\overline{\Omega_R^-} \setminus \ell)$, $\ell = \ell_1 \cup \ell_2$, and in some vicinity of ℓ (cf. Remark 3.7):

$$\left| D_k v^{(j)}(x) \right| \leq c\,[r(x)]^{-1/2}, \quad r(x) = \operatorname{dist}\{x, \ell\}, \quad k = 1,2,3, \; j = 1,2.$$

First, let us put $v^{(2)} = w$ and $v^{(1)} = \gamma(\cdot - x, \omega)$ with $x \in \Omega^+$. These functions solve the equation (2.1) in Ω_R^- and, due to the results of Section 3, satisfy the above conditions. Therefore, (5.19) yields

$$\int_{\Sigma(O,R)} \{\partial_{n(y)}\gamma(y-x,\omega)\,[w(y)]^+ - \gamma(y-x,\omega)\,[\partial_n w(y)]^+\}\, dS_y$$

$$(5.20) \qquad - \int_{S\setminus S^*} \{\partial_{n(y)}\gamma(y-x,\omega)\,[w(y)]^- - \gamma(y-x,\omega)\,[\partial_n w(y)]^-\}\, dS_y = 0$$

for $x \in \Omega^+$, due to the condition (5.18) on S^*.

Applying (5.19) by standard arguments we can derive the integral representation formula for $w(x)$ in Ω_R^-

$$\int_{\Sigma(O,R)} \{\partial_{n(y)}\gamma(y-x,\omega)\,[w(y)]^+ - \gamma(y-x,\omega)\,[\partial_n w(y)]^+\}\, dS_y,$$

$$(5.21) \quad - \int_{S\setminus S^*} \{\partial_{n(y)}\gamma(y-x,\omega)\,[w(y)]^- - \gamma(y-x,\omega)\,[\partial_n w(y)]^-\}\, dS_y = w(x)$$

for $x \in \Omega_R^-$. Denote the function defined by the left-hand side expression in (5.20) and (5.21) by $\widetilde{w}$:

$$\widetilde{w}(x) = \int_{\Sigma(O,R)} \{\partial_{n(y)}\gamma(y-x,\omega)\,[w(y)]^+ - \gamma(y-x,\omega)\,[\partial_n w(y)]^+\}\, dS_y$$
$$- \int_{S\setminus S^*} \{\partial_{n(y)}\gamma(y-x,\omega)\,[w(y)]^- - \gamma(y-x,\omega)\,[\partial_n w(y)]^-\}\, dS_y.$$

It is evident that $\widetilde{w}$ is a real-analytic (complex-valued) function of x in the connected domain

$$B(O,R)\setminus\{S\setminus S^*\} = \Omega_R^- \cup \Omega^+ \cup S^*.$$

Moreover, the restriction of $\widetilde{w}$ on Ω^+ vanishes, i.e., $\widetilde{w}|_{\Omega^+} = 0$ due to (5.20) and, therefore, $\widetilde{w}(x) = 0$ in $\Omega_R^- \cup \Omega^+ \cup S^*$. Since $w = \widetilde{w}|_{\Omega_R^-}$, we conclude $w(x) = 0$ in Ω_R^-. Whence, $w(x) = 0$ in Ω^- follows immediately by analyticity of w in Ω^-, i.e., (5.17) holds. In turn, this leads to the contradiction, since $w^{sc} \in \mathrm{Som}(\Omega^-)$, while $e^{i\,d\cdot x} \notin \mathrm{Som}(\Omega^-)$. Thus $S_1' = S_2'$ and $S_1'' = S_2''$, which completes the proof. □

References

[Ben1] Bennish, J., *Asymptotics for elliptic boundary value problems for systems of pseudodifferential equations*, Journal of Mathematical Analysis and Applications, **179**, No.2 (1993), 417-445.

[CK1] Colton, D. and Kress, R., *Integral Equation Methods in Scattering Theory*, John Wiley, New York, 1983.

[CK2] Colton, D. and Kress, R., *Inverse Acoustic and Electromagnetic Scattering Theory*, Springer-Verlag: Berlin, Heidelberg, New York, 1992.

[CS1] Costabel, M. and Stephan, E.P., *An improved boundary element Galerkin method for three-dimensional crack problems*, Integral Equations and Operator Theory, **10** (1987), 467-504.

[ChDu1] Chkadua, O. and Duduchava, R.,*Asymptotics of functions represented by potentials*, Preprint, Universität Stuttgart, Sonderforschungsbereich 404, Bericht **98/12**, 1998.

[DNS1] Duduchava, R., Natroshvili, D. and Shargorodsky, E., *Boundary value problems of the mathematical theory of cracks*, Proceedings of I.Vekua Institute of Applied Mathematics Tbilisi State University, **39** (1990), 68-84.

[DNS2] Duduchava, R., Natroshvili, D. and Shargorodsky, E., *Basic boundary value problems of thermoelasticity for anisotropic bodies with cuts, I,II*, Georgian Mathematical Journal, **2**, No. 2 (1995), 123-140; **2**, No. 3 (1995), 259-276.

[Esk1] Eskin, G., *Boundary Value Problems for Elliptic Pseudodifferential Equtions*, Transl. of Mathem. Monographs, Amer. Math. Soc., **52**, Providence, Rhode Island, 1981.

[Fi1] Fichera, G., *Sull' esistenza e sul calcolo delle soluzioni dei problemi al cantorno, relativi all' equilibrio di un corpo elastico*, Ann. Scuola Norm. Sup. Pisa, **4, s.3, 1–2**,(1950), 35-99

[Fi2] Fichera, G., *Existence Theorems in Elasticity*. Handb. der Physik, Bd. **6/2**, Springer-Verlag, Heidelberg, 1973.

[Is1] Isakov, V., *On uniqueness in the inverse transmission scattering problem*, Commun. Part. Diff. Eq., **15**(1990), 1565-1587.

[Is2] Isakov, V., *New stability results for soft obstacles in inverse scattering*, Inverse Problems, **9**(1993), 535-543.

[JN1] Jentsch, L. and Natroshvili, D., *Non-local approach in mathematical problems of fluid-structure interaction*, Math. Methods Appl. Sci., **22**, No. 1 (1999), 13-42.

[JN2] Jentsch, L. and Natroshvili, D., *Interaction between thermoelastic and scalar oscillation fields*, Integral Equations and Operator Theory, **28** (1997), 261-288.

[JN3] Jentsch, L. and Natroshvili, D.,*Three-dimensional mathematical problems of thermoelasticity of anisotropic bodies, I, II,* Memoirs on Differential Equations and Mathematical Phisics, **17**, 7-126, **18**, 3-52, 1999.

[KiKr1] Kirsch, A. and Kress, R., *Uniqueness in inverse obstacle scattering*, Inverse Problems **9** (1993), 285-299.

[Ku1] Kupradze, V.D., *Potential methods in the theory of elasticity*, Jerusalem, 1965.

[LiMa1] Lions, J.-L. and Magenes, E., *Problèmes aux limites non homogènes et applications*, Vol. 1, Dunod-Paris, 1968.

[Maz1] Maz'ya, V.G., *Boundary integral equations*, in R.V.Gamkrelidze (Ed.): Encyclopedia of Mathematical Sciences, **27**, Springer-Verlag: Berlin, Heidelberg, (1991), 127-222.

[Mi1] Miranda, C., *Partial Differential Equations of Elliptic Type,* 2-nd ed., Springer-Verlag, Berlin-Heidelberg-New York, 1970.

[Tr1] Triebel, H., *Theory of function spaces*, Birkhäuser Verlag, Basel-Boston-Stuttgart, 1983.

[Va1] Vainberg, B.R., *Asymptotic Methods in Mathematical Physics*, Moscow University Press, Moscow,1982.

[Vek1] Vekua, I.N., *On metaharmonic functions*. Proc. Tbilisi Mathem. Inst. of Acad. Sci. Georgian SSR, **12** (1943), 105-174.

Department of Mathematics
Georgian Technical University
Kostava str. 77
Tbilisi - 75
Georgia

1991 Mathematics Subject Classification: Primary 35P25, 47A40; Secondary 35J05, 35J25,

Submitted: 19.4.2000

Operator Theory:
Advances and Applications, Vol. 121

On an inverse problem in groundwater filtration and its regularization by the conjugate gradient method

Robert Plato

Dedicated to the memory of Professor Siegfried Prößdorf

In this paper we start with the consideration of a parameter estimation problem with noisy data which arises as an inverse problem in groundwater filtration. It turns out that in appropriate Hilbert spaces this problem can be formulated as a linear non-compact ill-posed problem with a model perturbation that can be estimated only at the solution of the problem. In the remaining part of the paper we deal with those problems in general Hilbert spaces and consider the CGNR method, this is, the classical method of conjugate gradients by Hestenes and Stiefel applied to the associated normal equations. Two a posteriori stopping rules are introduced to obtain stable numerical solutions, and convergence results are provided for the corresponding approximations, respectively. Finally, being a main concern of this paper, we present numerical illustrations with the CGNR method applied to a non-compact linear perturbed test problem.

1 A parameter estimation problem

1.1 Formulation of the problem

In the sequel we recall the basic facts on a specific parameter estimation problem with noisy data which arises as an inverse problem in groundwater filtration. For more details see, e.g., G. Vainikko ([19], [21]), G. Vainikko and Kunisch [20], E. Vainikko [18] and Bruckner, Handrock and Langmach [2]. General results on parameter identification problems can be found, e.g., in Alt, Hoffmann and Sprekels [1], Chicone and Gerlach [3], Richter [16] and Sprekels [17].

Let $\Omega \subset \mathbb{R}^N$ with $N \geq 2$ be an open bounded domain with a piecewise smooth boundary $\partial\Omega$, and let $\Gamma \subset \partial\Omega$ be a relatively open subset. In the sequel we consider the inverse problem of determining a function $a \in L^2(\Omega)$ that satisfies the following conditions,

$$-\mathrm{div}\Big(a(x)\nabla u(x)\Big) \;=\; h(x), \qquad x \in \Omega, \tag{1.1}$$

$$\Big[a(x)\nabla u(x)\Big] \cdot \nu(x) \;=\; g(x), \qquad x \in \Gamma, \tag{1.2}$$

where $u \in W^{1,\infty}(\Omega)$, $h \in L^2(\Omega)$ and $g \in L^2(\Gamma)$ are given functions, ν denotes the outer normal with respect to $\partial\Omega$, and derivates are understood in a distributional sense. Moreover, $u \cdot v \in \mathbb{R}$ denotes the inner product of two vectors u, $v \in \mathbb{R}^N$.

A weak formulation associated with the inverse problem for (1.1)–(1.2) is to determine a function $a \in L^2(\Omega)$ that satisfies

$$\int_\Omega a(x)\nabla u(x)\cdot\nabla w(x)\,dx = \int_\Omega h(x)w(x)\,dx + \int_\Gamma g(x)w(x)\,dS(x) \qquad \text{for all } w \in H^1(\Omega,\Gamma), \tag{1.3}$$

where

$$H^1(\Omega,\Gamma) = \Big\{\, w \in H^1(\Omega) \;:\; w(x) = 0 \text{ for } x \in \partial\Omega\backslash\Gamma \,\Big\}.$$

1.2 An abstract setting for the weak formulation (1.3)

In order to present an abstract setting for the weak formulation (1.3), in the sequel the following notation is introduced.

(a) Consider the space G of gradients of functions $w \in H^1(\Omega,\Gamma)$, i.e.,

$$G = \Big\{\, \nabla w \;:\; w \in H^1(\Omega,\Gamma) \,\Big\} \subset \big(L^2(\Omega)\big)^N,$$

and let $Q_G : \big(L^2(\Omega)\big)^N \to \big(L^2(\Omega)\big)^N$ be the orthogonal projection onto the subspace $G \subset \big(L^2(\Omega)\big)^N$.

(b) Let the operator T be defined by

$$Ta = Q_G\big(a\nabla u\big), \qquad a \in L^2(\Omega), \tag{1.4}$$

and then $T : L^2(\Omega) \to G$ is a linear bounded operator which in fact is non-compact and has a non-closed range. Note moreover that the operator T depends on the function u.

(c) Let the function $\psi \in H^1(\Omega,\Gamma)$ be a solution of the following direct problem,

$$\begin{aligned} -\Delta\psi(x) &= h(x), && x \in \Omega, \\ \nabla\psi(x)\cdot\nu(x) &= g(x), && x \in \Gamma; \qquad \psi(x) = 0, \qquad x \in \partial\Omega\backslash\Gamma. \end{aligned}$$

By using the notations from (a)–(c), the weak formulation (1.3) then in fact is equivalent to the following ill-posed equation,

$$Ta = \nabla\psi. \tag{1.5}$$

1.3 Noisy data

We suppose additionally that equation (1.5) has a $L^2(\Omega)$-minimum norm solution a_* that is bounded almost everywhere,

$$a_* \in L^\infty(\Omega) \cap \mathcal{N}(T)^\perp, \qquad Ta_* = \nabla\psi. \tag{1.6}$$

On the other hand it is supposed that $u \in W^{1,\infty}(\Omega)$ is available approximately only, more precisely, a function $u^\delta \in W^{1,\infty}(\Omega)$ and a noise level δ are given so that the following condition is satisfied,

$$\|\nabla u^\delta - \nabla u\|_{(L^2(\Omega))^N} \le \delta. \tag{1.7}$$

This leads to a perturbation of the model: in place of the operator T only a linear bounded operator $T_\delta : L^2(\Omega) \to G$ defined by $T_\delta a = Q_G\big(a\nabla u^\delta\big)$ is available, and from (1.6)–(1.7) in our specific situation we obtain the following estimate for the perturbation at the L^2-minimum norm solution a_* of the original problem:

$$\underbrace{\Big(\int_\Omega |a_*(x)|^2 \big|\nabla u^\delta(x) - \nabla u(x)\big|_2^2 \,dx\Big)^{1/2}}_{= \|T_\delta a_* - T a_*\|_{(L^2(\Omega))^N}} \le \|a_*\|_{L^\infty(\Omega)}\delta, \tag{1.8}$$

where $|\cdot|_2 : \mathbb{R}^N \to \mathbb{R}$ denotes the Euclidian norm. The following property will be also needed in the general presentations in section 2: if in addition to (1.7) the following condition is satisfied for some constant $M > 0$ independent of δ,

$$\sup_{x\in\Omega} \big|\nabla u^\delta(x)\big|_2 \le M,$$

then we moreover have

$$T_\delta^* \nabla w \to T^* \nabla w \qquad \text{as } \delta \to 0 \qquad\qquad \text{for any } \nabla w \in G, \tag{1.9}$$

where $\big(T_\delta^* \nabla w\big)(x) = \nabla u^\delta(x)\cdot\nabla w(x)$ and $\big(T^*\nabla w\big)(x) = \nabla u(x)\cdot\nabla w(x)$ for $\nabla w \in G$.

The two conditions (1.8) and (1.9) are the basic ingredients for the results on conjugate gradient-type methods for linear ill-posed problems presented in section 2. This section is concluded with the following remark on a different noise condition.

Remark 1.1. If the condition (1.7) on the noise is replaced by the following stronger condition,

$$\sup_{x\in\Omega} \big|\nabla u(x) - \nabla u^\delta(x)\big|_2 \le \delta, \tag{1.10}$$

then in place of (1.8) the uniform estimate

$$\big\|T_\delta a - T a\big\|_{(L^2(\Omega))^N} \le \|a\|_{L^2(\Omega)}\delta, \qquad\qquad a \in L^2(\Omega),$$

is satisfied. In this situation one may apply the results in Nemirovskiĭ [12] on conjugate gradient-type methods for linear ill-posed problems with perturbations of the underlying operator which are small with respect to the operator norm. However, condition (1.10) typically is not satisfied in the application presented above since the function ∇u may have jumps in Ω, and the results in [12] thus are not applicable here.

1.4 Outline of the paper

In section 2 we present the basic properties of the CGNR method in a general Hilbert space setting which is applicable to the specific parameter estimation problem considered in section 1, and the regularizing properties of two specific stopping rules are stated. Finally, in section 3 numerical experiments are provided which extend the considerations in [15].

2 The CGNR method

2.1 The general setting

In the sequel we consider general equations of the form

$$Ta = f_*, \tag{2.1}$$

where

$$T \in \mathcal{L}(\mathcal{H}_1, \mathcal{H}_2) \qquad \big(\mathcal{H}_1,\ \mathcal{H}_2 \text{ real Hilbert spaces}\big) \tag{2.2}$$

$$f_* \in \mathcal{R}(T). \tag{2.3}$$

Here, $\mathcal{L}(\mathcal{H}_1, \mathcal{H}_2)$ denotes the space of bounded linear operators from $\mathcal{H}_1$ into $\mathcal{H}_2$, and $\mathcal{R}(T) \subset \mathcal{H}_2$ denotes the range of T which in general is non-closed, and then equation (2.1) is ill-posed. We denote by $a_* \in \mathcal{H}_1$ the minimum norm solution of (2.1), i.e.,

$$a_* \in \mathcal{N}(T)^\perp, \qquad Ta_* = f_*, \tag{2.4}$$

where $\mathcal{N}(T)^\perp \subset \mathcal{H}_1$ denotes the orthogonal complement of the null space of T. In the sequel we assume that for $\delta > 0$, approximations f^δ and T_δ for f_* and T are available, respectively; more precisely,

$$T_\delta \in \mathcal{L}(\mathcal{H}_1, \mathcal{H}_2), \qquad f^\delta \in \mathcal{H}_2, \qquad \big\| T_\delta a_* - f^\delta \big\| \le \gamma_* \delta, \tag{2.5}$$

where $\gamma_* > 0$ denotes a constant. Moreover we suppose that

$$\big\| T_\delta^* z - T^* z \big\| \to 0 \quad \text{as } \delta \to 0 \qquad \text{for any } z \in \mathcal{H}_2, \tag{2.6}$$

where $A^* \in \mathcal{L}(\mathcal{H}_2, \mathcal{H}_1)$ denotes the adjoint of an operator $A \in \mathcal{L}(\mathcal{H}_1, \mathcal{H}_2)$, and the arising norms in the Hilbert spaces $\mathcal{H}_1$ and $\mathcal{H}_2$ are denoted by $\|\cdot\|$, respectively, if not further specified.

Remark 2.2. (1) Note that the parameter estimation problem presented in section 1 fits into the general framework considered in (2.1)–(2.6).

(2) If $Q_\delta \in \mathcal{L}(\mathcal{H}_2, \mathcal{H}_2)$ denote orthogonal projections that satisfy $\|Q_\delta f - f\| \to 0$ as $\delta \to 0$ for each $f \in \mathcal{H}_2$, then (2.5), (2.6) remain valid when T_δ and f^δ are

replaced by $Q_\delta T_\delta$ and $Q_\delta f^\delta$, respectively, with γ_* and δ not being modified. This means that the results presented in this paper cover not only an error in the model or the right-hand side but also a discretization of equation (2.1) by the minimal error method, this is, the Galerkin scheme with respect to the spaces $\mathcal{W}_h = \mathcal{R}(Q_\delta) \subset \mathcal{H}_2$ and $\mathcal{V}_h = T_\delta^*(\mathcal{W}_h) \subset \mathcal{H}_1$.

2.2 Formulation of the CGNR method

In the sequel we suppose that the conditions (2.2)–(2.6) are satisfied and assume that the noise level $\delta > 0$ in condition (2.5) is fixed, and for notational convenience we shall suppress the dependence of δ on some vectors and scalars that arise in the course of iteration to be defined next. First we introduce the Krylov subspaces with respect to $T_\delta^* T_\delta$ and a vector $r \in \mathcal{H}_1$,

$$\mathcal{K}_n\big(T_\delta^* T_\delta, r\big) = \operatorname{span}\Big\{r, T_\delta^* T_\delta r, \ldots, (T_\delta^* T_\delta)^{n-1} r\Big\} \subset \mathcal{H}_1, \qquad n = 0, 1, \ldots .$$

We next present a formal definition of the CGNR method for solving equation (2.1), this is, the classical method of conjugate gradients by Hestenes and Stiefel being applied to the normal equations associated with equation (2.1). We admit noisy data given by (2.5), and for notational convenience $a_0^\delta = 0$ is taken as initial vector.

Definition 2.3. Let the conditions (2.2)–(2.4) be satisfied. The (possibly terminating) sequence $a_n^\delta \in \mathcal{H}_1$, $n = 0, 1, \ldots$, associated with the CGNR method, applied to equation (2.1) with noisy data given by (2.5), is defined as follows,

$$(2.7) \qquad a_n^\delta \in \mathcal{K}_n\big(T_\delta^* T_\delta, T_\delta^* f^\delta\big),$$

$$(2.8) \qquad \big\| T_\delta a_n^\delta - f^\delta \big\| = \min_{a \in \mathcal{K}_n(T_\delta^* T_\delta, T_\delta^* f^\delta)} \big\| T_\delta a - f^\delta \big\|.$$

The sequence terminates, if the residual associated with the normal equations,

$$(2.9) \qquad r_n^\delta = T_\delta^* \big(T_\delta a_n^\delta - f^\delta\big) \in \mathcal{H}_1, \qquad n = 0, 1, \ldots,$$

vanishes for some $n =: n_*$.

The letter "R" in CGNR indicates that in (2.8) the residual associated with (the perturbed version of) the equation $Ta = f_*$ is minimized while the letter "N" in CGNR corresponds to the fact that this minimization is employed over Krylov subspaces generated by (perturbations of) the normal equations $T^*Ta = T^* f_*$. A survey on conjugate gradient type methods can be found, e. g., in Freund, Golub and Nachtigal [6].

The algorithm for the computation of the iterates $a_n^\delta \in \mathcal{H}_1$ is as follows:

Algorithm 2.4. (*CGNR method for equation (2.1) with noisy given data as in (2.5)*)
Let the conditions (2.2)–(2.5) be satisfied. *Step 0: Let* $a_0^\delta = 0$.

For $n = 0, 1, \ldots$:
(1) If $r_n^\delta = 0$ then terminate, $n_* := n$.
(2) Otherwise, proceed with step $n+1$:

$$d_n = \begin{cases} -r_n^\delta + \beta_{n-1} d_{n-1}, & \beta_{n-1} = \dfrac{\|r_n^\delta\|^2}{\|r_{n-1}^\delta\|^2}, \qquad \text{if} \quad n \geq 1, \\ -r_0^\delta, & \qquad\qquad\qquad\qquad \text{if} \quad n = 0, \end{cases}$$

$$a_{n+1}^\delta = a_n^\delta + \alpha_n d_n, \qquad \alpha_n = \frac{\|r_n^\delta\|^2}{\|T_\delta d_n\|^2}.$$

It is well-known that a_n^δ determined by Algorithm 2.4 satisfies (2.7)–(2.8).

2.3 Further properties of the CGNR method

The following representation for a_n^δ exists,

(2.10) $\quad a_n^\delta = q_n(T_\delta^* T_\delta) T_\delta^* f^\delta \qquad$ for some $q_n \in \Pi_{n-1} \qquad (n = 0, 1, \ldots).$

Here, $\Pi_{-1} := \{0\}$, and $\Pi_{n-1} = \{\ q : q$ is a polynomial of degree $\leq n-1\ \}$, $n = 1, 2, \ldots$. For the first stopping rule to be defined next, the real numbers

(2.11) $\qquad S_n^\delta := q_n(0), \qquad n = 0, 1, \ldots,$

will be needed. These numbers S_0^δ, $S_1^\delta, \ldots$, in fact increase, and they can be computed easily by the following three-term recurrence which results from Algorithm 2.4 and the uniqueness of q_n in (2.10),

$$S_0^\delta = 0, \qquad S_1^\delta = \alpha_0,$$

$$S_{n+1}^\delta = \Big(1 + \frac{\alpha_n \beta_{n-1}}{\alpha_{n-1}}\Big) S_n^\delta - \frac{\alpha_n \beta_{n-1}}{\alpha_{n-1}} S_{n-1}^\delta + \alpha_n, \qquad n = 1, 2, \ldots .$$

2.4 A first stopping rule and its regularizing properties

It follows from the ill-posedness of equation (2.1) that the iterates $a_1^\delta, a_2^\delta, \ldots$ typically are semiconvergent, i.e., the error $\|a_n^\delta - a_*\|$ decreases in each iteration step as long as n does not exceed a certain threshold, and the error begins to increase when n passes this threshold. Due to this semiconvergence the iteration should be stopped after an appropriate number of steps. For this purpose we consider two stopping rules which have the desired regularizing properties, respectively, in a sense which is specified in the Theorems 2.7 and 2.9 below.

The basic idea associated with the following stopping rule is to choose n_δ such that $S_{n_\delta+1}^\delta \to \infty$ and $\big(S_{n_\delta}^\delta\big)^{1/2} \delta \to 0$ holds for $\delta \to 0$.

Definition 2.5. Choose real numbers $\ell_\delta > 0$ with

$$\ell_\delta \to \infty, \qquad \ell_\delta^{1/2}\delta \to 0 \qquad \text{as } \delta \to 0.$$

Let the conditions (2.2)–(2.6) be satisfied, and for fixed $\delta > 0$ let $a_n^\delta \in \mathcal{H}_1$, $n = 0, 1, \ldots$, be generated by the CGNR method, and let S_n^δ and r_n^δ, $n = 0, 1, \ldots$, be given by (2.11) and (2.9), respectively. By n_δ we denote the first integer such that one of the following two conditions is satisfied,

$$S_{n_\delta+1}^\delta > \ell_\delta \qquad \text{or} \qquad r_{n_\delta}^\delta = 0. \tag{2.12}$$

Remark 2.6. If the first stopping criterion in (2.12) applies then necessarily $r_{n_\delta}^\delta \neq 0$ holds, since only then $S_{n_\delta+1}^\delta$ is well-defined.

Theorem 2.7. *Let the conditions (2.2)–(2.6) be satisfied, and let $a_n^\delta \in \mathcal{H}_1$, $n = 0, 1, \ldots$, be generated by the CGNR method. The stopping rule given by Definition 2.5 leads to a termination after a finite number of iteration steps n_δ, and*

$$\left\| a_{n_\delta}^\delta - a_* \right\| \to 0 \qquad \text{as } \delta \to 0.$$

A proof of Theorem 2.7 is presented in [15].

2.5 A second stopping rule and its regularizing properties

The next stopping rule is a discrepancy principle and depends on values of the defect,

$$\Delta_n^\delta := \left\| T_\delta a_n^\delta - f^\delta \right\|, \qquad n = 0, 1, \ldots . \tag{2.13}$$

Definition 2.8. Let the conditions (2.2)–(2.6) be satisfied, and fix some constant $b > \gamma_*$. For fixed $\delta > 0$ let $a_n^\delta \in \mathcal{H}_1$, $n = 0, 1, \ldots$, be generated by the CGNR method, and let Δ_n^δ be given by (2.13). Stop the iteration at step $n_\delta := n$, if the following condition is satisfied for the first time,

$$\Delta_{n_\delta}^\delta \leq b\delta.$$

Note that the discrepancy principle considered in Definition 2.8 requires a knowledge of the constant γ_* in condition (2.5). Considering the application in section 1 and in particular the estimate (1.8) this means that an estimate for the L^∞-norm of the solution a_* has to be available.

Theorem 2.9. *Let the conditions (2.2)–(2.6) be satisfied, and let* $a_n^\delta \in \mathcal{H}_1$, $n = 0, 1, \ldots$, *be generated by the CGNR method. The stopping rule given by Definition 2.8 leads to a termination after a finite number of iteration steps* n_δ, *and*

$$
\begin{aligned}
\left\| a_{n_\delta}^\delta - a_* \right\| &\to 0 \qquad \text{as } \delta \to 0, \\
n_\delta \delta &\to 0 \qquad \text{as } \delta \to 0.
\end{aligned} \tag{2.14}
$$

A proof of Theorem 2.9 is presented in [15].

Remark 2.10. A similar discrepancy principle for a stationary iteration method like the Landweber iteration typically yields an asymptotic behaviour $n_\delta^{1/2}\delta \to 0$ as $\delta \to 0$, cf. Vainikko [21], while the complexity in each iteration step is the same as it is for conjugate gradient-type methods. The property (2.14) thus reflects the efficiency of the CGNR method.

2.6 Bibliographical remarks on conjugate gradient-type methods for linear ill-posed problems

In the sequel we refer to related results obtained for conjugate gradient-type methods to solve linear ill-posed problems.

(a) Results for noisy right-hand sides (and precisely given operator) are obtained, e.g., in Eicke, Louis and Plato [4], Hanke [10], Gilyazov [8], and in [13], [14]. For recent monographs containing associated results we refer to Gilyazov [7], Hanke [9], Engl, Hanke and Neubauer [5], and Kirsch [11].

(b) More generally as in the situation (a) with perturbed right-hand sides, Nemirovskiĭ [12] considers conjugate gradient-type methods for linear ill-posed problems where perturbations of the underlying operator are admitted which are small with respect to the operator norm. It is the purpose of the present paper to modify the associated results in [12] to obtain regularizing stopping rules for the situation when only pointwise estimates for the operator perturbations are available, cf. Remark 1.1.

3 Numerical experiments

We next consider some numerical experiments which is the basic purpose of this paper.

3.1 The specific equation

In our test equation we shall consider the operator

$$(Ta)(x) := a(x)\nabla u(x), \quad \text{with} \quad \nabla u(x) := 20 \begin{pmatrix} x_1 + 0.25 \\ x_2 - 0.33 \end{pmatrix}, \quad x = (x_1,\, x_2)^\top \in \mathcal{Q},$$

where the notation

$$\mathcal{Q} \;:=\; [0,1]^2$$

is used. One has $\nabla u \in \left(L^\infty(\mathcal{Q})\right)^2$, and thus $T : L^2(\mathcal{Q}) \to \left(L^2(\mathcal{Q})\right)^2$ is a linear operator which has a non-closed range and is non-compact with respect to this space setting. In the numerical experiments the following specific equation is considered,

$$\begin{aligned} Ta \;&=\; f_*, \qquad\qquad \text{with } f_* \;:=\; Ta_*, \\ a_*(x_1,x_2) :=&\; \frac{\sin 8r}{8r}, \quad x_1,\; x_2 \in [0,1], \quad r = \sqrt{\left(x_1 - 0.5\right)^2 \;+\; \left(x_2 - 0.5\right)^2}, \end{aligned}$$

and the following perturbed right-hand sides and operators are chosen,

$$\begin{aligned} f^\delta \;&=\; f_* + V^\delta, \qquad\qquad T_\delta a = a\nabla u^\delta, \\ &\nabla u^\delta(x_1,x_2) := \nabla u(x_1,x_2) + \begin{pmatrix} W_1^\delta(x_1,x_2)\left(x_1x_2\right)^{-1/4} \\ W_2^\delta(x_1,x_2)\left((1-x_1)(1-x_2)\right)^{-1/4} \end{pmatrix}, \end{aligned}$$

where the functions $V^\delta \in \left(L^\infty(\mathcal{Q})\right)^2$ and $W_1^\delta,\; W_2^\delta \in L^\infty(\mathcal{Q})$ have uniformly distributed random values with $\|V^\delta\|_{(L^\infty(\mathcal{Q}))^2} \le \delta/\sqrt{2}$ and $\|W_j^\delta\|_{L^\infty(\mathcal{Q})} \le \sqrt{2}\delta$, respectively. Note that $\|\nabla u^\delta - \nabla u\|_{(L^2(\mathcal{Q}))^2} \le \delta$ and $a_* \in L^\infty(\mathcal{Q})$, $\|a_*\|_{L^\infty(\mathcal{Q})} = 1$, thus $\|T_\delta a_* - Ta_*\|_{(L^2(\mathcal{Q}))^2} \le \delta$ holds. On the other hand one has $\nabla u^\delta \notin \left(L^\infty(\mathcal{Q})\right)^2$, hence T_δ does not approximate the operator T with respect to the operator norm as $\delta \to 0$, cf. Remark 1.1. We consider the following specific noise levels,

$$\delta \;=\; \|f_*\|_{(L^2(\mathcal{Q}))^2} \cdot \%/100, \qquad \text{with} \quad \% \in \{\, 0.25,\ 0.50,\ 1.0,\ 2.0,\ 4.0,\ 8.0 \,\}.$$

3.2 The numerical experiments obtained with CGNR method

In the numerical experiments the CGNR method is terminated by the stopping rule given by Definition 2.8, with

$$b = 2.2.$$

Note that in our situation we have $\left\|T_\delta a_* - f^\delta\right\|_{(L^2(\mathcal{Q}))^2} \le 2\delta$, thus this specific choice of b is admissible. We are now in a position to present the numerical results, cf. Table 1.

Remark 3.11. The CGNR method in fact was applied to a discretized version of the test equation which is obtained by the minimal error method (cf. Remark 2.2) by using spaces of step functions,

$$\mathcal{W}_h = \left\{U_h : \mathcal{Q} \to \mathbb{R}^2 \,:\, U_h = \text{const. on } \left[t_{k-1}, t_k\right] \times \left[t_{l-1}, t_l\right],\; k,l = 1,\ldots,N\right\},$$

% noise	δ	$\Vert a^{\delta}_{n_\delta} - a_* \Vert_{L^2(Q)}$	n_δ
8.0	0.470	0.079	2
4.0	0.235	0.046	3
2.0	0.118	0.024	5
1.0	0.059	0.019	6
0.50	0.029	0.013	8
0.25	0.015	0.011	10

Table 1: Numerical results for the test equation

where $h = 1/N$ and $t_j = jh$, $j = 0, 1, \ldots, N$, with $N = 64$ in fact. As basis functions for $\mathcal{W}_h$, the following piecewise constant functions $\Psi_1, \Psi_2, \ldots, \Psi_{N^2} \in \mathcal{W}_h$ are taken,

$$\Psi_j = \begin{pmatrix} 1 \\ 0 \end{pmatrix} \chi_{I_{k,l}}, \qquad \Psi_{j+1} = \begin{pmatrix} 0 \\ 1 \end{pmatrix} \chi_{I_{k,l}}, \qquad j = (k-1)2N + 2l - 1,$$
$$k,\ l = 1, \ldots, N,$$

where $\chi_{I_{k,l}}$ denotes the characteristic function with respect to the square $I_{k,l} = [t_{k-1}, t_k] \times [t_{l-1}, t_l]$. Finally, all computations are performed in MATLAB on an IBM RISC/6000.

We conclude our presentations with Figures 1 and 2 showing the approximations $a^{\delta}_{n_\delta}$ that are obtained for noisy data chosen according to section 3.1, with $\% = 8$ and $\% = 0.5$, respectively.

Acknowledgements

The author would like to thank the referees for some useful comments.

References

[1] H. W. Alt, K.-H. Hoffmann, and J. Sprekels. *A numerical procedure to solve certain identification problems.* In Optimal Control of Partial Differential Equations, Conf. Oberwolfach 1982, pp. 11–43, 1984.

[2] G. Bruckner, S. Handrock-Meyer, and H. Langmach. *An inverse problem from 2D-groundwater modelling.* Inverse Problems, 14(4):835–851, 1998.

[3] C. Chicone and J. Gerlach. *Identifiability of distributed parameters.* In H. W. Engl and C. W. Groetsch, editors, Inverse and Ill-Posed Problems, Proc. St. Wolfgang 1986, pp. 513–521, Academic Press, Boston, 1987.

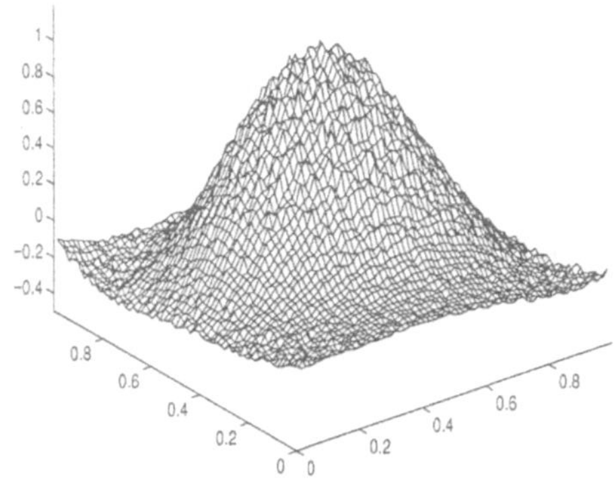

Figure 1: Reconstruction for noise level $\delta = 0.47$

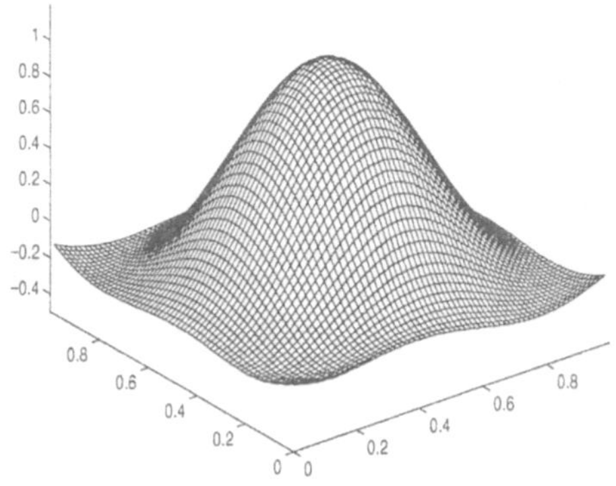

Figure 2: Reconstruction for noise level $\delta = 0.029$

[4] B. Eicke, A. K. Louis, and R. Plato. *The instability of some gradient methods for ill-posed problems.* Numer. Math., 58(1):129–134, 1990.

[5] H. W. Engl, M. Hanke, and A. Neubauer. *Regularization of Inverse Problems.* Kluwer, Dordrecht, 1st edition, 1996.

[6] R. Freund, G. Golub, and N. Nachtigal, *Iterative solution of linear systems.* Acta Numerica, pp. 1–44, 1991.

[7] S. F. Gilyazov. *Approximate solution of ill-posed problems* (in Russian). Publisher Moscow University, Moscow, 1st edition, 1995.

[8] S. F. Gilyazov. *Regularizing conjugate-direction methods.* Comput. Maths. Math. Phys., 35(4):385–394, 1995.

[9] M. Hanke. *Conjugate Gradient Type Methods for Ill-Posed Problems.* Longman House, Harlow, 1995.

[10] M. Hanke. *The minimal error method is a regularization method.* Proc. Am. Math. Soc., 123(11):3487–3497, 1995.

[11] A. Kirsch. *An Introduction to the Mathematical Theory of Inverse Problems.* Springer-Verlag, New York, 1st edition, 1996.

[12] A. S. Nemirovskiĭ. *The regularizing properties of the adjoint gradient method in ill-posed problems.* U.S.S.R. Comput. Math. Math. Phys., 26(2):7–16, 1986.

[13] R. Plato. *Optimal algorithms for linear ill-posed problems yield regularization methods.* Numer. Funct. Anal. Optim., 11(1-2):111–118, 1990.

[14] R. Plato. *The method of conjugate residuals for solving the Galerkin equations associated with symmetric positive semidefinite ill-posed problems.* SIAM J. Numer. Anal., 35(4):1621–1645, 1998.

[15] R. Plato. *The conjugate gradient method for linear ill-posed problems with operator perturbations.* Numer. Algorithms, 20(1):1–22, 1999.

[16] G. Richter. *An inverse problem for the steady state diffusion equation.* SIAM J. Appl. Math., 41(2):210–221, 1981.

[17] J. Sprekels. *Identification of parameters in distributed systems: an overview.* In Methods Oper. Res. 54, pp. 163–176, 1986.

[18] E. Vainikko. *Courant elements for the inverse problem in the filtration coefficient identification.* In M. Krizek et. al., editor, Finite Element Methods. 50 years of the Courant element. Proc. Univ. Jyvaeskylae, Finland, 1993., pp. 451–459, Marcel Dekker, New York, 1994.

[19] G. Vainikko. *Identification of filtration coefficient in an elliptic boundary problem.* In A.N. Tikhonov, editor, Proc. Moscow 1991, pp. 202–213, VSP, Utrecht, Tokyo, 1992.

[20] G. Vainikko and K. Kunisch. *Identifiability of the transmissivity coefficient in an elliptic boundary value problem.* Z. Anal. Anwendungen, 12:327–341, 1993.

[21] G. Vainikko. *On the discretization and regularization of ill-posed problems with noncompact operators.* Numer. Funct. Anal. Optim., 13(3-4):381–396, 1992.

Fachbereich Mathematik, Technische Universität Berlin,
Straße des 17. Juni 135, D–10623 Berlin, Germany

2000 Mathematics Subject Classification: Primary 65J20 65N21; Secondary 47A52 65F22

Submitted: 27.4.2000

Operator Theory:
Advances and Applications, Vol. 121

On a constructive representation of an orthogonal trigonometric Schauder basis for $C_{2\pi}$

JÜRGEN PRESTIN, KATHI K. SELIG

Dedicated to the memory of Professor Dr. Siegfried Prößdorf

A particular class of orthogonal trigonometric Schauder bases for $C_{2\pi}$ is given by periodic wavelet packet functions. These bases are of minimal growth of the polynomial degree. The focus of attention is their construction and the estimation of the Lebesgue constant. The corresponding approximation error is asymptotically optimal.

1. Introduction

The Fourier series of a continuous function does not converge, in general, in the supremum norm. Faber [Fa] proved for $C_{2\pi}$ that any polynomial set $\{t_\mu : \mu \in \mathbb{N}\}$ with $\deg t_\mu \leq \mu/2$ cannot be a basis. So, a long standing question was which minimal degree does one need in order to preserve pointwise convergence? Privalov gave a conclusive answer by two results [Pri1, Pri2]. First, he showed that, for any basis $\{t_\mu : \mu \in \mathbb{N}\}$ in $C_{2\pi}$, there exists an $\varepsilon > 0$ such that for sufficiently large μ, one has $\deg t_\mu \geq (1+\varepsilon)\mu/2$. Second, for any such $\varepsilon > 0$ he verified the existence of a Schauder basis satisfying $\deg t_\mu \leq (1+\varepsilon)\mu/2$.

Lorentz and Sahakian showed in [LS] that the additional condition on the basis to be orthogonal does not affect the growth of the degree. Their proof was based on Meyer wavelets and corresponding wavelet packets on the real line which then were periodized.

We were studying de la Vallée Poussin means and related polynomial wavelets in [PS1, PS2, Se1, Se2] which led us to a similar basis with optimal growth of the degree. Our construction by means of periodic wavelet and wavelet packet spaces allows an asymptotically optimal estimation of the norm of the corresponding partial sum operator.

For our basis $\{t_k : k \in \mathbb{N}\}$, we will consider its partial sum operator

$$S_\mu f \;=\; \sum_{k=1}^{\mu} \langle f\,,\, t_k\rangle\, t_k \tag{1.1}$$

and verify the Schauder basis property

$$\|f - S_\mu f\|_\infty \;\to\; 0 \qquad (\mu \to \infty)$$

for all $f \in C_{2\pi}$. In this paper, we prove it in the form

$$\|f - S_\mu f\|_\infty \;\leq\; (1 + \|S_\mu\|_{C\to C})\, E_{\lfloor \frac{\mu}{2}(1-\varepsilon)\rfloor}(f)\,,$$

where the best approximation of $f \in C_{2\pi}$ by trigonometric polynomials of degree at most n,

$$E_n(f) \;:=\; \inf_{p_n \in \mathbb{T}_n} \|f - p_n\|_\infty\,,$$

tends to zero as $n \to \infty$. Here and in the sequel, $\mathbb{T}_n$ denotes the set of trigonometric polynomials of degree at most n, and we set $\lceil r \rceil := \min\{k \in \mathbb{Z} : \; k \geq r\}$ and $\lfloor r \rfloor := \max\{k \in \mathbb{Z} : \; k \leq r\}$, for $r \in \mathbb{R}$. In fact, we construct the functions t_k depending on $\varepsilon \in (0, 2/3]$ such that

$$\mathbb{T}_{\lfloor \frac{\mu}{2}(1-\varepsilon)\rfloor} \;\subseteq\; \operatorname{span}\{t_k \;:\; k = 1, \ldots, \mu\}\,, \tag{1.2}$$

and

$$\deg t_\mu \;\leq\; \frac{\mu}{2}(1+\varepsilon)\,.$$

Our estimation of the uniform boundedness of the partial sum operator in the sup-norm,

$$\|S_\mu\|_{C\to C} \;=\; \sup_{\|f\|_\infty = 1} \|S_\mu f\|_\infty \;<\; 15 + \frac{4}{\pi^2} \ln \frac{1}{\varepsilon} \tag{1.3}$$

is asymptotically optimal as $\varepsilon \to 0$ (cf. [Pri1]).

In Section 2, we define the orthonormal Schauder basis of optimal degree and state the main theorem. Then, Section 3 shall explain the constructive aspects of our trigonometric wavelet packets. Section 4 is exclusively devoted to the proof of the estimation of the operator norm. We emphasize here that the essential point of our construction to obtain (1.3) is the combination of the ideas of a periodic multiresolution analysis with the explicite determination of frequencies as given in Lemma 3.4.

Let us mention that our polynomial system $\{t_\mu\}$ is also a basis in some separable Hölder spaces. The investigation of approximation in Hölder spaces was initiated by Siegfried Prößdorf in [Prö]. To describe a particular case let, for $0 < \alpha < 1$,

$$C^\alpha = \{f \in C_{2\pi} : \|f\|_{\infty,\alpha} := \|f\|_\infty + \sup_{x \neq y} |x-y|^{-\alpha} |f(x) - f(y)| < \infty\}.$$

With the help of the results in [Pre] and [PP] one can easily prove that our orthonormal system of polynomials $\{t_\mu\}$ is a basis in the subspace

$$\tilde{C}^\alpha = \{f \in C^\alpha : \lim_{y \to x} |x-y|^{-\alpha} |f(x) - f(y)| = 0 \quad \text{for all } x\}.$$

2. Orthogonal bases of optimal degree

We are going to define the Schauder basis polynomials t_μ depending on the number ε to which the degree of the polynomials must be tuned.

Let $\varepsilon > 0$ be given and fix $\lambda \in \mathbb{N}$,

$$\lambda := \begin{cases} 1 & \text{if } \varepsilon \geq \frac{2}{3}, \\ \lceil \log_2 \frac{4}{3\varepsilon} \rceil & \text{if } \varepsilon < \frac{2}{3}, \end{cases}$$

i.e., $4 \leq 3\varepsilon \cdot 2^\lambda < 8$ for $\varepsilon \leq 2/3$. For any $\varepsilon > 2/3$, we get by definition the same λ and therewith the same Schauder basis as for $\varepsilon = 2/3$. Hence, we can restrict ourselves to $\varepsilon \leq 2/3$ in the rest of the paper.

We denote *dimension numbers*

$$N_j := 3 \cdot 2^j \qquad (j \in \mathbb{N}_0),$$

and, for our fixed $\lambda \in \mathbb{N}$,

$$N_j^{\lambda,p} := N_{j-\lambda}(2^\lambda + p) \qquad (j \geq \lambda; \quad p = 0, \ldots, 2^\lambda).$$

Then, for all $\mu \in \mathbb{N}$, $\mu > 2N_\lambda$ there is a *one-to-one correspondence of* μ *and the triple* (j, p, s), with $j \geq \lambda$, $p \in \{0, \ldots, 2^\lambda - 1\}$ and $s \in \{1, \ldots, 2N_{j-\lambda}\}$, by the equality

$$\mu = 2N_{j-\lambda}(2^\lambda + p) + s = 2N_j^{\lambda,p} + s.$$

We define the *shift operator* $T_j : C_{2\pi} \to C_{2\pi}$ *of level* $j \in \mathbb{N}_0$ by

$$T_j f := f\left(\cdot - \frac{\pi}{N_j}\right).$$

We denote the *bit-reversed number of* $p \in \{0, \ldots, 2^\lambda - 1\}$ *with respect to* λ *digits* by

$$b(\lambda, p) := \sum_{\ell=0}^{\lambda-1} \epsilon_{\lambda-\ell-1}(p)\, 2^\ell, \tag{2.1}$$

where $\epsilon_\ell(p)$ are the bits (or binary digits) of p in its binary decomposition

$$p = \sum_{\ell=0}^{\lambda-1} \epsilon_\ell(p)\, 2^\ell, \qquad \text{i.e.,} \quad \epsilon_\ell(p) = \left\lfloor \frac{p}{2^\ell} \right\rfloor \bmod 2.$$

Further, let

$$M_j := \begin{cases} 1 & \text{if } j < \lambda, \\ 2^{j-\lambda} & \text{if } j \geq \lambda, \end{cases} \tag{2.2}$$

and

$$j(p) := \begin{cases} j & \text{for } p = 0, \dots, 2^\lambda - 2, \\ j+1 & \text{for } p = 2^\lambda - 1. \end{cases} \tag{2.3}$$

Definition 2.1. Let $\varepsilon > 0$ be given. The set of polynomials $\{t_\mu : \mu \in \mathbb{N}\}$ is defined as follows. Let

$$\begin{aligned} t_1 &:= 1, \\ t_{2k} &:= \sqrt{2}\cos k\,\cdot \qquad \text{for } k = 1, \dots, N_\lambda, \\ t_{2k+1} &:= \sqrt{2}\sin k\,\cdot \qquad \text{for } k = 1, \dots, N_\lambda - 1, \end{aligned}$$

and for all $j \ge \lambda$, $p = 0, \dots, 2^\lambda - 1$, $s = 1, \dots, 2N_{j-\lambda}$, let

$$t_{2N_j^{\lambda,p}+s} := \frac{1}{\sqrt{N_{j-\lambda}}} T_{j-\lambda}^{s-1} T_{j+1}^{b(\lambda+1,2^\lambda+p)} \left(\sum_{k=-M_j+1}^{M_j-1} \frac{M_j + k}{\sqrt{M_j^2 + k^2}} \cos(N_j^{\lambda,p} + k)\,\cdot \right.$$

$$\left. + \sqrt{2} \sum_{k=N_j^{\lambda,p}+M_j}^{N_j^{\lambda,p+1}-M_{j(p)}} \cos k\,\cdot + \sum_{k=-M_{j(p)}+1}^{M_{j(p)}-1} \frac{M_{j(p)} - k}{\sqrt{M_{j(p)}^2 + k^2}} \cos(N_j^{\lambda,p+1} + k)\,\cdot \right).$$

Note that, for $\mu > 2N_\lambda$, the polynomials t_μ are shifted cosine sums which obviously satisfy

$$t_{2N_j^{\lambda,p}+s} \in \mathbb{T}_{N_j^{\lambda,p+1}+M_{j(p)}-1} \quad \text{and} \quad t_{2N_j^{\lambda,p}+s} \perp \mathbb{T}_{N_j^{\lambda,p}-M_j}.$$

Theorem 2.2. *For given $\varepsilon > 0$, the polynomial system $\{t_\mu : \mu \in \mathbb{N}\}$ from Definition 2.1 is an orthonormal Schauder basis in $C_{2\pi}$ of optimal degree, i.e., for all $\mu, \nu \in \mathbb{N}$, it holds that*

$$\langle t_\mu, t_\nu \rangle = \delta_{\mu,\nu}, \tag{2.4}$$

and

$$\deg t_\mu \le \frac{\mu}{2}(1+\varepsilon). \tag{2.5}$$

In particular, for all $f \in C_{2\pi}$ we have

$$\left\| f - \sum_{k=1}^{\mu} \langle f, t_k \rangle\, t_k \right\|_\infty < \left(16 + \frac{4}{\pi^2} \ln \frac{1}{\varepsilon}\right) E_{\lfloor \frac{\mu}{2}(1-\varepsilon) \rfloor}(f). \tag{2.6}$$

Proof. First, we prove the degree inequality (2.5) which is trivial for $\mu \le 2N_\lambda$. For $\mu > 2N_\lambda$, we deduce for $p = 0, \dots, 2^\lambda - 2$

$$\begin{aligned} \frac{\deg t_{2N_j^{\lambda,p}+s}}{2N_j^{\lambda,p} + s} &= \frac{N_j^{\lambda,p+1} + M_j - 1}{2N_j^{\lambda,p} + s} < \frac{N_j^{\lambda,p} + N_{j-\lambda} + M_j}{2N_j^{\lambda,p}} \\ &= \frac{1}{2}\left(1 + \frac{4 \cdot 2^{j-\lambda}}{3 \cdot 2^{j-\lambda}(2^\lambda + p)}\right) \le \frac{1}{2}\left(1 + \frac{4}{3 \cdot 2^\lambda}\right) \le \frac{1}{2}(1+\varepsilon), \end{aligned}$$

since $\lambda \geq \log_2 \frac{4}{3\varepsilon}$. For $p = 2^\lambda - 1$, we have

$$\frac{\deg t_{2N_j^{\lambda,p}+s}}{2N_j^{\lambda,p}+s} = \frac{N_{j+1}+M_{j+1}-1}{2N_j^{\lambda,2^\lambda-1}+s} < \frac{N_{j+1}+M_{j+1}}{2(N_{j+1}-N_{j-\lambda})}$$
$$= \frac{1}{2}\left(1+\frac{5\cdot 2^{j-\lambda}}{3\cdot 2^{j-\lambda}(2^{\lambda+1}-1)}\right) = \frac{1}{2}\left(1+\frac{5}{3\cdot(2^{\lambda+1}-1)}\right).$$

Since $\lambda \geq 1$,

$$\frac{2^\lambda}{2^{\lambda+1}-1} = \frac{2^{\lambda+1}}{2(2^{\lambda+1}-1)} = \frac{1}{2}\left(1+\frac{1}{2^{\lambda+1}-1}\right) \leq \frac{1}{2}\left(1+\frac{1}{3}\right) = \frac{2}{3} < \frac{4}{5}.$$

Hence, (2.5) follows analogously for $p = 2^\lambda - 1$.

The orthonormality (2.4) can be deduced from the construction of $\{t_\mu : \mu \in \mathbb{N}\}$ which we will explain in Section 3. It could also be proved straightforward.

The estimation in (2.6) is a consequence of the results in (1.2) and (1.3), which are given in Lemma 3.4 and Theorem 4.4, respectively. □

3. Construction methods

3.1. General theory

In this section, let us focus on Hilbert space theory in $L^2_{2\pi}$. Our construction is strongly based upon multiresolution, wavelet and wavelet packet spaces in $L^2_{2\pi}$ being shift-invariant spaces spanned by one function and its translates. The wavelet spaces and the wavelet packet spaces, respectively, ought to be mutually orthogonal. Finally, the translates of the wavelet packets have to be orthonormalized and will be orthonormal translates of another wavelet packet of the same space.

A periodic multiresolution analysis (PMRA) shall be, as usual, a chain of spaces of the following kind.

Definition 3.1. A *periodic multiresolution analysis (PMRA) in* $L^2_{2\pi}$ is a sequence of subspaces $\{V_j\}_{j\in\mathbb{N}_0}$ of $L^2_{2\pi}$ with the properties:

(MR1) For all $j \in \mathbb{N}_0$ there exists a function $\phi_j \in V_j$, such that $\{T_j^s\,\phi_j : s = 0,\ldots,2N_j-1\}$ is a basis for V_j.

(MR2) For all $j \in \mathbb{N}_0$ it holds $V_j \subset V_{j+1}$.

(MR3) It holds $\operatorname{clos}_{L^2_{2\pi}}(\bigcup_{j\in\mathbb{N}_0} V_j) = L^2_{2\pi}$.

While sticking to shift-invariance, we have to give up dilation invariance for the case of periodic functions, simply due to the fact that $\phi \in L^2_{2\pi} \implies \phi(\frac{\cdot}{2}) \in L^2_{4\pi}$. So, in each level j, ϕ_j can be any function that belongs to the superior space

V_{j+1} and has linearly independent translates with respect to the shift operator T_j. Hence, such a *scaling function* ϕ_j has to satisfy a two-scale relation

$$\phi_j = \sum_{s=0}^{N_{j+2}-1} \alpha_{j,s} \, T_{j+1}^s \phi_{j+1} \, .$$

In terms of Fourier coefficients

$$c_k(f) := \langle f, e^{ik\cdot} \rangle := \frac{1}{2\pi} \int_{-\pi}^{\pi} f(x) \, e^{-ikx} \, dx$$

it means

$$(3.1) \qquad c_{k+mN_{j+2}}(\phi_j) = \hat{\alpha}_{j,k} \, c_{k+mN_{j+2}}(\phi_{j+1}) \, ,$$

for all $k = 0, \ldots, N_{j+2}-1$, $m \in \mathbb{Z}$, where $(\alpha_{j,s})_{s=0}^{N_{j+2}-1}$ is a vector of complex numbers, and $(\hat{\alpha}_{j,k})_{k=0}^{N_{j+2}-1}$ is its discrete Fourier transform. The linear independence of the $2N_j$ translates of ϕ_j, i.e. (MR1), is equivalent to

$$(3.2) \qquad \sum_{m \in \mathbb{Z}} |c_{k+mN_{j+1}}(\phi_j)|^2 > 0 \qquad (k = 0, \ldots, N_{j+1} - 1) \, ,$$

and then (MR2) holds iff there exist $(\hat{\alpha}_{j,k})_{k=0}^{N_{j+2}-1}$ such that (3.1) is satisfied for all $j \in \mathbb{N}_0$.

The wavelet spaces contain what is remaining when cutting V_j out of V_{j+1}.

Definition 3.2. Let $\{V_j\}_{j \in \mathbb{N}_0}$ in $L^2_{2\pi}$ be a given PMRA. For all $j \in \mathbb{N}_0$, the *wavelet space* W_j *of level* j is the orthogonal complement of V_j in V_{j+1}, i.e.,

$$W_j = V_{j+1} \ominus V_j \, .$$

From this definition it follows that W_j is invariant with respect to the shift T_j and that there exists a function ψ_j such that $W_j = \operatorname{span}\{T_j^s \psi_j : s = 0, \ldots, 2N_j - 1\}$. Such a function ψ_j shall be called *wavelet*, and again, no dilation property is required.

Given any PMRA and, in particular, coefficient vectors $(\hat{\alpha}_{j,k})_{k=0}^{N_{j+2}-1}$ for which (3.1) holds, $\psi_j \in L^2_{2\pi}$ is a wavelet iff we have

(W) *The Fourier coefficients of* ψ_j *satisfy*

$$c_k(\psi_j) = \frac{\sigma_{j,(k \bmod N_{j+2})} \, \overline{\hat{\alpha}}_{j,((k+N_{j+1}) \bmod N_{j+2})}}{\sum_{\ell \in \mathbb{Z}} |c_{k+\ell N_{j+2}}(\phi_{j+1})|^2} \, c_k(\phi_{j+1}) \qquad (k \in \mathbb{Z})$$

with some numbers $\sigma_{j,k} \in \mathbb{C} \setminus \{0\}$, $k = 0, \ldots, N_{j+2} - 1$, *for which*

$$\sigma_{j,k} \quad = \quad -\sigma_{j,k+N_{j+1}} \qquad (k = 0, \ldots, N_{j+1} - 1).$$

So we have exactly $2N_j$ degrees of freedom for the choice of ψ_j.

For the concept of periodic wavelet packets, we do not simply adopt the decomposition algorithms from the scaling spaces as done in $L^2(\mathbb{R})$ (see e.g. [CMW, CW]), but we persist on splitting any T_j-invariant subspace into two (orthogonal) T_{j-1}-invariant subspaces.

Definition 3.3. For a given wavelet space $W_j = W_j^{0,0}$ of level j, $j \in \mathbb{N}$, and for any number $\lambda \in \mathbb{N}$ with $\lambda \leq j$, the *wavelet packet spaces of level j and depth* λ are the subspaces $\{W_j^{\lambda,p}\}_{p=0}^{2^\lambda - 1}$ of $L^2_{2\pi}$ satisfying the following properties:

(WP1) For all $p = 0, \ldots, 2^\lambda - 1$, there exists a function $\psi_j^{\lambda,p} \in W_j^{\lambda,p}$ such that $\{T^s_{j-\lambda}\, \psi_j^{\lambda,p} :\ s = 0, \ldots, 2N_{j-\lambda} - 1\}$ is a basis for $W_j^{\lambda,p}$.

(WP2) For all $\ell = 0, \ldots, \lambda - 1$ and $p = 0, \ldots, 2^\ell - 1$, we have

$$W_j^{\ell,p} \quad = \quad W_j^{\ell+1,2p} \oplus W_j^{\ell+1,2p+1}.$$

The functions $\psi_j^{\lambda,p}$ are called *wavelet packets*.

The properties (WP1) and (WP2) can be written equivalently in the following form:

(WP) *There exist vectors* $(\hat{\beta}_{j,k}^{\ell,p})_{k=0}^{N_{j-\ell+2}-1} \in \mathbb{C}^{N_{j-\ell+2}}$, *for all* $\ell = 1, \ldots, \lambda$ *and* $p = 0, \ldots, 2^\ell - 1$, *such that*

$$c_k(\psi_j^{\ell,p}) \quad = \quad \hat{\beta}^{\ell,p}_{j,(k \bmod N_{j-\ell+2})}\, c_k(\psi_j^{\ell-1,\lfloor \frac{p}{2} \rfloor}) \qquad (k \in \mathbb{Z}),$$

with

$$\begin{aligned} & \hat{\beta}_{j,k}^{\ell,2\lfloor \frac{p}{2} \rfloor}\, \overline{\hat{\beta}_{j,k}^{\ell,2\lfloor \frac{p}{2} \rfloor+1}} \sum_{m\in\mathbb{Z}} |c_{k+mN_{j-\ell+2}}(\psi_j^{\ell-1,\lfloor \frac{p}{2} \rfloor})|^2 \\ = & -\hat{\beta}_{j,k+2N_{j-\ell}}^{\ell,2\lfloor \frac{p}{2} \rfloor}\, \overline{\hat{\beta}_{j,k+2N_{j-\ell}}^{\ell,2\lfloor \frac{p}{2} \rfloor+1}} \sum_{m\in\mathbb{Z}} |c_{k+2N_{j-\ell}+mN_{j-\ell+2}}(\psi_j^{\ell-1,\lfloor \frac{p}{2} \rfloor})|^2 \end{aligned}$$

and

$$\left|\hat{\beta}_{j,k}^{\ell,p}\right|^2 + \left|\hat{\beta}_{j,k+2N_{j-\ell}}^{\ell,p}\right|^2 \quad > \quad 0,$$

for all $k = 0, \ldots, 2N_{j-\ell} - 1$.

Last but not least we need to know how to orthonormalize the translates of a function without loosing the translation invariance property of the basis. Given $f \in L^2_{2\pi}$ for which the translates $\{T^s_j f : s = 0, \dots, 2N_j - 1\}$ are linearly independent, i.e., (3.2) is satisfied by $f = \phi_j$, we obtain orthonormal translates $\{T^s_j g : s = 0, \dots, 2N_j - 1\}$ by choosing

$$(3.3) \qquad c_{k+2N_j\ell}(g) = c_{k+2N_j\ell}(f) \left(\sum_{m\in\mathbb{Z}} |c_{k+2N_jm}(f)|^2 \right)^{-\frac{1}{2}},$$

for all $k = 0, \dots, 2N_j - 1$ and $\ell \in \mathbb{Z}$.

3.2. Particular polynomial functions

Now, let us define special scaling functions that will lead us to our polynomials in Definition 2.1. A *de la Vallée Poussin mean* is given, for any $N, M \in \mathbb{N}$ with $N \geq M$, by

$$(3.4) \qquad \varphi_N^M(x) = 1 + 2 \sum_{k=1}^{N-M} \cos kx + \sum_{k=-M+1}^{M-1} \frac{M-k}{M} \cos(N+k)x .$$

Its interpolating property $\varphi_N^M(\frac{s\pi}{N}) = 2N\,\delta_{s,0}$, for $s = 0, \dots, 2N-1$, is clear from

$$(3.5) \qquad \varphi_N^M(x) = \begin{cases} \dfrac{\sin Nx \sin Mx}{2M \sin^2 \frac{x}{2}} & \text{if } x \notin 2\pi\mathbb{Z}, \\ 2N & \text{if } x \in 2\pi\mathbb{Z} \end{cases}$$

and yields also the linear independence of its translates $\{\varphi_N^M(\cdot - \frac{s\pi}{N}) : s = 0, \dots, 2N-1\}$.

With the numbers $N_j = 3 \cdot 2^j$, the $\phi_j = \varphi_{N_j}^{M_j}$ are scaling functions of a PMRA iff $M_j + M_{j+1} \leq N_j + 1$ for all $j \in \mathbb{N}_0$ (see [Se3]) which is satisfied by M_j as given in (2.2). Indeed, since

$$c_k(\varphi_{N_j}^{M_j}) = \begin{cases} 1 & \text{if } \quad 0 \leq |k| \leq N_j - M_j, \\ \frac{M_j - (|k| - N_j)}{2M_j} & \text{if } \ N_j - M_j < |k| < N_j + M_j, \\ 0 & \text{if } \ N_j + M_j \leq |k|, \end{cases}$$

condition (3.2) holds, and with

$$\hat{\alpha}_{j,k} = \begin{cases} 1 & \text{if } \quad 0 \leq k \leq N_j - M_j, \\ \frac{M_j - (k - N_j)}{2M_j} & \text{if } \ N_j - M_j < k < N_j + M_j, \\ 0 & \text{if } \ N_j + M_j \leq k \leq 3N_j - M_j, \\ \frac{M_j + (k - 3N_j)}{2M_j} & \text{if } \ 3N_j - M_j < k < 3N_j + M_j, \\ 1 & \text{if } \ 3N_j + M_j \leq k < 4N_j, \end{cases}$$

(3.1) is true, for all $j \in \mathbb{N}_0$. Moreover, one easily verifies

$$\mathbb{T}_{N_j - M_j} \subset V_j \subset \mathbb{T}_{N_j + M_j - 1} .$$

Due to $N_j - M_j \geq 2N_j/3$ and the Weierstrass theorem, (MR3) is valid, too.

Based on the PMRA generated by de la Vallée Poussin means, we further define functions

$$\psi_j = T_{j+1}\left(\varphi_{N_{j+1}}^{M_{j+1}} - \varphi_{N_j}^{M_j}\right) \tag{3.6}$$

for all $j \in \mathbb{N}_0$ and

$$\psi_j^{\lambda,p} = T_{j+1}^{b(\lambda+1,2^\lambda+p)}\left(\varphi_{N_j^{\lambda,p+1}}^{M_{j(p)}} - \varphi_{N_j^{\lambda,p}}^{M_j}\right) \tag{3.7}$$

for all $j \geq \lambda$ and $p = 0, \ldots, 2^\lambda - 1$, where $b(\lambda, p)$ is the bit-reversed number (2.1). Using the above theory, we now show that the functions ψ_j and $\psi_j^{\lambda,p}$ are wavelets and wavelet packets, respectively.

The functions ψ_j in (3.6) are wavelets since they satisfy property (W) if we choose

$$\sigma_{j,(k \bmod N_{j+2})} = e^{-\frac{ik\pi}{N_{j+1}}} \times \begin{cases} \sum\limits_{\ell\in\mathbb{Z}} |c_{k+\ell N_{j+2}}(\varphi_{N_{j+1}}^{M_{j+1}})|^2 & \text{if } \ N_j \leq k < 3N_j, \\ \sum\limits_{\ell\in\mathbb{Z}} |c_{k+N_{j+1}+\ell N_{j+2}}(\varphi_{N_{j+1}}^{M_{j+1}})|^2 & \text{if } -N_j \leq k < N_j. \end{cases}$$

These functions ψ_j have already been used, e.g. in [Pri3]. For the functions $\psi_j^{\lambda,p}$ in (3.7), we have property (WP) in form of

$$c_k(\psi_j^{\lambda,p}) = e^{-\frac{i(p \bmod 2)k\pi}{N_{j-\lambda+1}}} \hat{\beta}^{\lambda,2\lfloor \frac{p}{2} \rfloor}_{j,((k+(p \bmod 2)N_{j-\lambda+1}) \bmod N_{j-\lambda+2})} \, c_k\left(\psi_j^{\lambda-1,\lfloor \frac{p}{2} \rfloor}\right)$$

with the factors

$$\hat{\beta}_{j,k}^{\lambda,2\lfloor \frac{p}{2} \rfloor} = \begin{cases} 1 & \text{if } \quad 0 \leq k \leq N_{j-\lambda} - M_j\,, \\ \frac{M_j-(k-N_{j-\lambda})}{2M_j} & \text{if } \quad N_{j-\lambda} - M_j < k < N_{j-\lambda} + M_j\,, \\ 0 & \text{if } \quad N_{j-\lambda} + M_j \leq k \leq 3N_{j-\lambda} - M_j\,, \\ \frac{M_j+(k-3N_{j-\lambda})}{2M_j} & \text{if } \quad 3N_{j-\lambda} - M_j < k < 3N_{j-\lambda} + M_j\,, \\ 1 & \text{if } \quad 3N_{j-\lambda} - M_j \leq k < 4N_{j-\lambda}\,. \end{cases}$$

If we compute the Fourier coefficients of each $\psi_j^{\lambda,p}$ explicitely, then we see that their absolute value is equal to the Fourier coefficients of $\varphi_{N_j^{\lambda,p+1}}^{M_{j(p)}} - \varphi_{N_j^{\lambda,p}}^{M_j}$, and the modulation that is multiplied in the iteration step from $\psi_j^{\lambda-1,\lfloor \frac{p}{2} \rfloor}$ to $\psi_j^{\lambda,p}$ is exactly $e^{-\frac{i(p \bmod 2)k\pi}{N_{j-\lambda+1}}}$. Since

$$c_k(T_j f) = e^{-\frac{ik\pi}{N_j}} c_k(f)\,,$$

it follows that the total shift of $\varphi^{M_{j(p)}}_{N_j^{\lambda,p+1}} - \varphi^{M_j}_{N_j^{\lambda,p}}$ to gain $\psi_j^{\lambda,p}$ is indeed

$$T_{j-\lambda+1}^{p \bmod 2} \cdots T_j^{\lfloor \frac{p}{2^{\lambda-1}} \rfloor \bmod 2} T_{j+1} = T_j^{b(\lambda,p)} T_{j+1} = T_{j+1}^{b(\lambda+1,2^\lambda+p)} .$$

So, our functions in (3.7) are wavelet packet functions.

Now, we want to orthonormalize $\{T_{j-\lambda}^s \psi_j^{\lambda,p} : s = 0, \ldots, 2N_{j-\lambda}-1\}$. Their Fourier coefficients are changed following the rule in (3.3), i.e., they are divided by the square root of

$$\sum_{m\in\mathbb{Z}} |c_{k+2N_{j-\lambda}m}(\psi_j^{\lambda,p})|^2 = \begin{cases} \frac{M_{j(p)}^2+k^2}{2M_{j(p)}^2} & 0 \le |k| < M_{j(p)} , \\ 1 & M_{j(p)} \le |k| \le N_{j-\lambda} - M_j , \\ \frac{M_j^2+(k-N_{j-\lambda})^2}{2M_j^2} & N_{j-\lambda} - M_j < k < N_{j-\lambda} + M_j , \end{cases}$$

for $k = -N_{j-\lambda}+M_j, \ldots, N_{j-\lambda}+M_j-1 \bmod 2N_{j-\lambda}$. Thus we obtain the functions in Definition 2.1 for $\mu > 2N_\lambda$ via their Fourier coefficients

$$c_k(t_{2N_j^{\lambda,p}+1}) = c_k(\psi_j^{\lambda,p}) \left(\sum_{m\in\mathbb{Z}} |c_{k+2N_{j-\lambda}m}(\psi_j^{\lambda,p})|^2 \right)^{-\frac{1}{2}} \qquad (k \in \mathbb{Z}) .$$

So, for any $j \ge \lambda$, we have by wavelet and wavelet packet decomposition

$$\begin{aligned} V_j &= V_\lambda \oplus \bigoplus_{h=\lambda}^{j-1} W_h \\ &= V_\lambda \oplus \bigoplus_{h=\lambda}^{j-1} \bigoplus_{q=0}^{2^\lambda-1} W_h^{\lambda,q} \qquad (3.8) \\ &= \operatorname{span}\{t_k : k = 1, \ldots, 2N_j\} , \end{aligned}$$

and, for any $p \in \{1, \ldots, 2^\lambda - 1\}$,

$$\bigoplus_{q=0}^{p-1} W_j^{\lambda,q} = \operatorname{span}\{t_k : k = 2N_j+1, \ldots, 2N_j^{\lambda,p}\} . \qquad (3.9)$$

Finally, we determine the trigonometric polynomials which are included in our polynomial spaces. As pointed out in the introduction, the change of basis from wavelet packets to polynomials consisting of one or two frequencies is the main ingredient to obtain optimal Lebesgue constants.

Lemma 3.4. *For given $\varepsilon > 0$, the span of the first μ polynomials t_k from Definition 2.1 contains all trigonometric polynomials of degree less or equal to $(1-\varepsilon)\mu/2$, i.e.,*

$$\mathbb{T}_{\lfloor \frac{\mu}{2}(1-\varepsilon) \rfloor} \subseteq \operatorname{span}\{t_k : k = 1, \ldots, \mu\} . \qquad (3.10)$$

Moreover, for $j \geq \lambda$,

$$(3.11)\qquad \begin{aligned} &\operatorname{span}\{t_k \, : \, k = 1, \dots, 2N_j\} \\ &= \mathbb{T}_{N_j - M_j} \oplus \operatorname{span}\{\theta^{\lambda,2^\lambda-1}_{j-1,N_{j-\lambda-1}+k} \, : \, k = -M_j+1, \dots, M_j - 1\}, \end{aligned}$$

and for $j \geq \lambda$, $p \in \{1, \dots, 2^\lambda - 1\}$,

$$(3.12)\qquad \begin{aligned} &\operatorname{span}\{t_k \, : \, k = 1, \dots, 2N_j^{\lambda,p}\} \\ &= \mathbb{T}_{N_j^{\lambda,p} - M_j} \oplus \operatorname{span}\{\theta^{\lambda,p-1}_{j,N_{j-\lambda}+k} \, : \, k = -M_j+1, \dots, M_j - 1\}, \end{aligned}$$

with

$$\theta^{\lambda,p}_{j,N_{j-\lambda}} = \sqrt{2}\; T_{j+1}^{b(\lambda+1,2^\lambda+p)} \;\cos N_j^{\lambda,p+1}\cdot\,,$$

$$\begin{aligned} \theta^{\lambda,p}_{j,N_{j-\lambda}-k} \quad &= \\ T_{j+1}^{b(\lambda+1,2^\lambda+p)} &\left(\frac{M_{j(p)}+k}{\sqrt{M_{j(p)}^2+k^2}} \cos(N_j^{\lambda,p+1} - k)\cdot + \frac{M_{j(p)}-k}{\sqrt{M_{j(p)}^2+k^2}} \cos(N_j^{\lambda,p+1} + k)\cdot \right), \\ \theta^{\lambda,p}_{j,N_{j-\lambda}+k} \quad &= \\ T_{j+1}^{b(\lambda+1,2^\lambda+p)} &\left(\frac{M_{j(p)}+k}{\sqrt{M_{j(p)}^2+k^2}} \sin(N_j^{\lambda,p+1} - k)\cdot - \frac{M_{j(p)}-k}{\sqrt{M_{j(p)}^2+k^2}} \sin(N_j^{\lambda,p+1} + k)\cdot \right) \end{aligned}$$

for $k = 1, \dots, M_{j(p)} - 1$, where the shift operator in $\theta^{\lambda,2^\lambda-1}_{j,N_{j-\lambda}+k}$ can be omitted.

Proof. Here and in the proof of Lemma 4.3 we need coefficient functions given in the interval $[-\frac{\pi}{3}, 2\pi - \frac{\pi}{3}]$ by

$$(3.13)\qquad g_p(x) \;=\; \begin{cases} \frac{1+\frac{3x}{\pi}}{\sqrt{1+(\frac{3x}{\pi})^2}} & \text{if} \quad -\frac{\pi}{3} \leq x < \frac{\pi}{3}\,, \\ \sqrt{2} & \text{if} \quad \frac{\pi}{3} \leq x < \pi - \frac{\pi}{3}\,, \\ \frac{1-3(\frac{x}{\pi}-1)}{\sqrt{1+(3(\frac{x}{\pi}-1))^2}} & \text{if} \quad \pi - \frac{\pi}{3} \leq x < \pi + \frac{\pi}{3}\,, \\ 0 & \text{if} \quad \pi + \frac{\pi}{3} \leq x \leq 2\pi - \frac{\pi}{3}\,, \end{cases}$$

for $p = 0, \dots, 2^\lambda - 2$, and

$$(3.14)\qquad g_{2^\lambda-1}(x) \;=\; \begin{cases} \frac{1+\frac{3x}{\pi}}{\sqrt{1+(\frac{3x}{\pi})^2}} & \text{if} \quad -\frac{\pi}{3} \leq x < \frac{\pi}{3}\,, \\ \sqrt{2} & \text{if} \quad \frac{\pi}{3} \leq x < \pi - \frac{2\pi}{3}\,, \\ \frac{2-3(\frac{x}{\pi}-1)}{\sqrt{4+(3(\frac{x}{\pi}-1))^2}} & \text{if} \quad \pi - \frac{2\pi}{3} \leq x < \pi + \frac{2\pi}{3}\,, \\ 0 & \text{if} \quad \pi + \frac{2\pi}{3} \leq x \leq 2\pi - \frac{\pi}{3}\,. \end{cases}$$

Hence, we can rewrite our basis functions

$$t_{2N_j^{\lambda,p}+s} \;=\; \frac{1}{\sqrt{N_{j-\lambda}}}\; T_{j-\lambda}^{s-1}\; T_{j+1}^{b(\lambda+1,2^\lambda+p)} \sum_{k=-M_j+1}^{N_{j-\lambda}+M_{j(p)}-1} g_p\left(\tfrac{k\pi}{N}\right) \cos(N_j^{\lambda,p} + k)\cdot\;.$$

Rewriting the shift by $T_{j-\lambda}^{s-1}$ and regrouping the frequencies, we obtain, now for $s = 0, \dots, 2N_{j-\lambda} - 1$,

$$\begin{aligned}
&\sqrt{N_{j-\lambda}}\, T_{j+1}^{-b(\lambda+1,2^\lambda+p)}\, t_{2N_j^{\lambda,p}+s+1} \\
&= \sum_{k=-M_j+1}^{N_{j-\lambda}+M_{j(p)}-1} g_p(\tfrac{k\pi}{N}) \cos(N_{j-\lambda}(2^\lambda+p)+k)\left(\cdot - \tfrac{s\pi}{N_{j-\lambda}}\right) \\
&=(-1)^{ps} \sum_{k=-M_j+1}^{N_{j-\lambda}+M_{j(p)}-1} g_p(\tfrac{k\pi}{N})\Big(\cos \tfrac{ks\pi}{N_{j-\lambda}} \cos(N_j^{\lambda,p}+k)\cdot + \sin \tfrac{ks\pi}{N_{j-\lambda}} \sin(N_j^{\lambda,p}+k)\cdot\Big) \\
&= (-1)^{ps}\Big\{ \sum_{k=1}^{M_j-1} \cos \tfrac{ks\pi}{N_{j-\lambda}} \Big(g_p(\tfrac{k\pi}{N}) \cos(N_j^{\lambda,p}+k)\cdot + g_p(-\tfrac{k\pi}{N}) \cos(N_j^{\lambda,p}-k)\cdot\Big) \\
&\qquad + \sum_{k=1}^{M_j-1} \sin \tfrac{ks\pi}{N_{j-\lambda}} \Big(g_p(\tfrac{k\pi}{N}) \sin(N_j^{\lambda,p}+k)\cdot - g_p(-\tfrac{k\pi}{N}) \sin(N_j^{\lambda,p}-k)\cdot\Big) \\
&+ \sqrt{2} \sum_{k=M_j}^{N_{j-\lambda}-M_{j(p)}} \Big(\cos \tfrac{ks\pi}{N_{j-\lambda}} \cos(N_j^{\lambda,p}+k)\cdot + \sin \tfrac{ks\pi}{N_{j-\lambda}} \sin(N_j^{\lambda,p}+k)\cdot\Big) \\
&+ \sqrt{2} \cos N_j^{\lambda,p}\cdot + \sqrt{2}(-1)^s \cos N_j^{\lambda,p+1}\cdot \\
&+ \sum_{k=1}^{M_{j(p)}-1} \cos \tfrac{(N_{j-\lambda}-k)s\pi}{N_{j-\lambda}} \Big(g_p(\pi - \tfrac{k\pi}{N}) \cos(N_j^{\lambda,p+1}-k)\cdot + g_p(\pi+\tfrac{k\pi}{N}) \cos(N_j^{\lambda,p+1}+k)\cdot\Big) \\
&+ \sum_{k=1}^{M_{j(p)}-1} \sin \tfrac{(N_{j-\lambda}-k)s\pi}{N_{j-\lambda}} \Big(g_p(\pi - \tfrac{k\pi}{N}) \sin(N_j^{\lambda,p+1}-k)\cdot \\
&\qquad\qquad - g_p(\pi + \tfrac{k\pi}{N}) \sin(N_j^{\lambda,p+1}+k)\cdot\Big)\Big\}.
\end{aligned}$$

Here, we observe a change of basis in $W_j^{\lambda,p}$ from $\{t_{2N_j^{\lambda,p}+s} : s = 1, \dots, 2N_{j-\lambda}\}$ to another orthonormal, more frequency-localized basis, $\{\theta_{j,k}^{\lambda,p} : k = 0, \dots, 2N_{j-\lambda} - 1\}$, by means of the cosine and sine transform in form of

$$t_{2N_j^{\lambda,p}+s+1} = \frac{(-1)^{ps}}{\sqrt{N_{j-\lambda}}} \left(\sum_{k=0}^{N_{j-\lambda}} \cos \tfrac{ks\pi}{N_{j-\lambda}}\, \theta_{j,k}^{\lambda,p} + \sum_{k=1}^{N_{j-\lambda}-1} \sin \tfrac{ks\pi}{N_{j-\lambda}}\, \theta_{j,2N_{j-\lambda}-k}^{\lambda,p} \right).$$

Hence, the union of the smallest PMRA space V_λ and the wavelet packet spaces, as given in (3.8) and (3.9), provides us with a basis of the cosine and sine functions $\{t_k : k = 1, \dots, 2N_\lambda\}$ and

$$\{\cos(N_h^{\lambda,q}+k)\cdot,\ \sin(N_h^{\lambda,q}+k)\cdot,: k = M_h, \dots, N_{h-\lambda} - M_{h(q)}\} \subset W_h^{\lambda,q}$$

for all indices h, q which are included. Here, the meaning of $h(q)$ is the same as of $j(p)$ in (2.3). Moreover, in the overlapping ranges of frequencies, we have bases

$$\begin{aligned}
S_h^{\lambda,q} &:= \{\theta_{h,N_{h-\lambda}+k}^{\lambda,q} : k = -M_{h(q)}+1, \dots, M_{h(q)}-1\}, \\
T_h^{\lambda,q} &:= \{\theta_{h,k}^{\lambda,q} : k = 0, \dots, M_h - 1\} \cup \{\theta_{h,2N_{h-\lambda}-k}^{\lambda,q} : k = 1, \dots, M_h - 1\}
\end{aligned}$$

for which $S_h^{\lambda,q-1}, T_h^{\lambda,q} \subset I\!T_{N_h^{\lambda,q}+M_h-1} \ominus I\!T_{N_h^{\lambda,q}-M_h+1}$ and $S_h^{\lambda,q-1} \perp T_h^{\lambda,q}$ hold, where for $q=0$, we have to write $S_{h-1}^{\lambda,2^\lambda-1}$. From the dimensions, we see that, for $h \le \lambda$ and $q = 0, \ldots, 2^\lambda - 1$,

$$\operatorname{span}(S_h^{\lambda,q-1} \cup T_h^{\lambda,q}) = I\!T_{N_h^{\lambda,q}+M_h-1} \ominus I\!T_{N_h^{\lambda,q}-M_h+1}.$$

The shift operator in the definition of $\theta_{j-1,N_{j-\lambda}+k}^{\lambda,2^\lambda-1}$, for $k = -M_j+1, \ldots, M_j-1$, can be omitted because $T_j^{b(\lambda+1,2^\lambda+2^\lambda-1)} = T_j^{2^{\lambda+1}-1}$, and a shift of $\theta_{j-1,N_{j-\lambda}+k}^{\lambda,2^\lambda-1}$ by a multiple of T_j yields a linear combination of $\{\theta_{j-1,N_{j-\lambda}+\ell}^{\lambda,2^\lambda-1} : |\ell| = k\}$.

Inclusion (3.10) is trivial for $\mu \le 2N_\lambda$. For $\mu > 2N_\lambda$, it follows from (3.11) and (3.12) due to

$$N_j^{\lambda,p} - M_j \ \ge\ \frac{\mu}{2}(1-\varepsilon) \ \ge\ \left\lfloor \frac{\mu}{2}(1-\varepsilon) \right\rfloor,$$

which can be shown similar to the proof of the inequality (2.5) for the degree. □

4. Norm of the Fourier sum operator

First, we prove some auxiliary results.

Lemma 4.1. *For any $N, M \in \mathbb{N}$ $N \ge M$, the functions φ_N^M in (3.4) satisfy*

$$\|\varphi_{2N}^{2M}\|_1 = \|\varphi_N^M\|_1$$

and

$$\|\varphi_N^1\|_1 < 1.6 + \frac{4}{\pi^2}\ln N. \tag{4.1}$$

Proof. From (3.5) it follows

$$\varphi_{2N}^{2M}(x) = \frac{\sin 2Nx \sin 2Mx}{4M\sin^2\frac{x}{2}} = \frac{\sin N(2x)\sin M(2x)}{M\sin^2\frac{2x}{2}}\cos^2\frac{x}{2}$$
$$= \varphi_N^M(2x)\,(1+\cos x).$$

Since φ_N^M and φ_{2N}^{2M} are even and 2π-periodic, and $\cos\frac{y}{2} = -\cos\frac{2\pi-y}{2}$, we have

$$\|\varphi_{2N}^{2M}\|_1 = \frac{1}{\pi}\int_0^\pi |\varphi_N^M(2x)|\,(1+\cos x)\,dx = \frac{1}{\pi}\int_0^\pi |\varphi_N^M(y)|\,(1+\cos\tfrac{y}{2})\,dy$$
$$= \|\varphi_N^M\|_1 + \frac{1}{4\pi}\int_0^{2\pi} |\varphi_N^M(y)|\left(\cos\frac{y}{2} - \cos\frac{2\pi-y}{2}\right)dy = \|\varphi_N^M\|_1.$$

Concerning (4.1), we follow [Zy], Chap. II.12, and obtain

$$\begin{aligned}
\|\varphi_N^1\|_1 &= \frac{1}{\pi}\int_0^\pi \left|\frac{\sin Nx \sin x}{2\sin^2 x/2}\right| dx \\
&= \frac{2}{\pi}\int_0^\pi \frac{|\sin Nx|}{x}\,dx - \frac{2}{\pi}\int_0^\pi |\sin Nx| \left(\frac{1}{x} - \frac{1}{2}\cot\frac{x}{2}\right) dx \\
&< \frac{2}{\pi}\sum_{k=0}^{N-1}\int_0^{\pi/N} \frac{|\sin Nx|}{x + k\pi/N}\,dx \\
&< \frac{2}{\pi}\int_0^{\pi/N} \frac{\sin Nx}{x}\,dx + \frac{2N}{\pi^2}\sum_{k=1}^{N-1}\frac{1}{k}\int_0^{\pi/N} \sin Nx\,dx \\
&= \frac{2}{\pi}\,\mathrm{Si}(\pi) + \frac{4}{\pi^2}\sum_{k=1}^{N-1}\frac{1}{k} \;<\; 1.6 + \frac{4}{\pi^2}\ln N\,.
\end{aligned}$$

□

Lemma 4.2. *For any $M \in \mathbb{N}$ it holds that*

$$\frac{1}{2M}\left\|\sum_{s=0}^{2M-1}\left|1 + 2\sum_{k=1}^{M-1}\frac{M^2-k^2}{M^2+k^2}\cos k\left(\cdot - \frac{s\pi}{M}\right)\right|\right\|_\infty \;<\; 2.1\,.$$

Proof. The function to be estimated in the supremum norm is obviously $\frac{\pi}{M}$-periodic, so it suffices to consider its supremum over the interval $[0, \frac{\pi}{M})$. Defining a 2π-periodic absolutely continuous (coefficient) function by

$$g(x) \;=\; \frac{1-(\frac{x}{\pi})^2}{1+(\frac{x}{\pi})^2} \qquad (x \in [-\pi,\pi])\,,$$

we can write

$$\begin{aligned}
&\left\|\sum_{s=0}^{2M-1}\left|1 + 2\sum_{k=1}^{M-1}\tfrac{M^2-k^2}{M^2+k^2}\cos k\left(\cdot - \tfrac{s\pi}{M}\right)\right|\right\|_\infty \\
&= \sup_{x\in[0,\frac{\pi}{M})}\sum_{s=0}^{2M-1}\left|\sum_{k=-M+1}^{M} g\left(\tfrac{k\pi}{M}\right)\cos k\left(x - \tfrac{s\pi}{M}\right)\right| \\
&= \sup_{\xi\in[0,1)}\sum_{s=0}^{2M-1}\left|\sum_{k=-M+1}^{M}\left(f^\xi\left(\tfrac{k\pi}{M}\right)\cos\tfrac{ks\pi}{M} + h^\xi\left(\tfrac{k\pi}{M}\right)\sin\tfrac{ks\pi}{M}\right)\right|,
\end{aligned}$$

where the functions

$$f^\xi(x) \;=\; g(x)\,\cos\xi x\,, \qquad h^\xi(x) \;=\; g(x)\,\sin\xi x \qquad (x \in (-\pi,\pi])$$

are even and odd, respectively, and are extended 2π-periodically. Expanding them in Fourier series,

$$f^\xi(x) \;=\; \tfrac{1}{2}a_0(f^\xi) + \sum_{n=1}^{\infty} a_n(f^\xi)\cos nx\,, \qquad h^\xi(x) \;=\; \sum_{n=1}^{\infty} b_n(h^\xi)\sin nx$$

and using the well-known identities

$$\sum_{k=-M+1}^{M} \cos m\tfrac{k\pi}{M} = 2M\,\delta_{0,m\bmod 2M}\,, \qquad \sum_{k=-M+1}^{M} \sin m\tfrac{k\pi}{M} = 0\,, \quad (m \in \mathbb{Z})\,,$$

we obtain, for arbitrary $\xi \in [0,1)$,

$$\begin{aligned}
&\tfrac{1}{2M}\sum_{s=0}^{2M-1}\left|\sum_{k=-M+1}^{M}\left(f^{\xi}\left(\tfrac{k\pi}{M}\right)\cos\tfrac{ks\pi}{M} + h^{\xi}\left(\tfrac{k\pi}{M}\right)\sin\tfrac{ks\pi}{M}\right)\right| \\
&= \sum_{s=0}^{2M-1}\left|\tfrac{1}{2}\sum_{m=0}^{\infty}\left(a_{2Mm+s}(f^{\xi}) + a_{2M(m+1)-s}(f^{\xi}) + b_{2Mm+s}(h^{\xi}) - b_{2M(m+1)-s}(h^{\xi})\right)\right| \\
&= \sum_{s=0}^{2M-1}\left|\sum_{m=0}^{\infty}\tfrac{1}{2\pi}\int_{-\pi}^{\pi} g(x)\left(\cos(2Mm+s-\xi)x + \cos(2M(m+1)-s+\xi)x\right)dx\right|,
\end{aligned}$$

with $b_0(h^{\xi}) := 0$. These integrals can be estimated by means of the total variation. Since g is a 2π-periodic function having a first derivative in $[-\pi,\pi]$ of finite total variation $V[g']$, it holds that

$$\left|\tfrac{1}{\pi}\int_{-\pi}^{\pi} g(x)\cos rx\,dx\right| \;\le\; \frac{V[g']}{\pi r^2}\,, \qquad (r \in \mathbb{R}\setminus\{0\})\,. \tag{4.2}$$

Note that (4.2) is well-known for $r \in \mathbb{N}$, see e.g. [Fi], Chap. 19. As $g(-\pi) = g(\pi) = 0$, the result can be extended to all $r \in \mathbb{R}\setminus\{0\}$. From the first and second derivative,

$$g'(x) \;=\; -\frac{4x}{\pi^2\left(1+(\frac{x}{\pi})^2\right)^2}\,, \qquad g''(x) \;=\; -\frac{4\left(1-3(\frac{x}{\pi})^2\right)}{\pi^2\left(1+(\frac{x}{\pi})^2\right)^3}\,,$$

we see that g' is monotonously increasing in $[-\pi,-\frac{\pi}{\sqrt{3}}]$ and $[\frac{\pi}{\sqrt{3}},\pi]$ and monotonously decreasing in $[-\frac{\pi}{\sqrt{3}},\frac{\pi}{\sqrt{3}}]$. Further, the extrema are at $-\frac{\pi}{\sqrt{3}}$ and $\frac{\pi}{\sqrt{3}}$ where

$$\sup_x |g'(x)| \;=\; g'\left(-\frac{\pi}{\sqrt{3}}\right) \;=\; -\,g'\left(\frac{\pi}{\sqrt{3}}\right) \;=\; \frac{3\sqrt{3}}{4\pi}\,.$$

Hence, by $g'(-\pi) > 0$ and $g'(\pi) < 0$, we obtain

$$V[g'] \;=\; 4\sup_x |g'(x)| \;=\; \frac{3\sqrt{3}}{\pi}\,.$$

For $-1 \le r \le 1$, instead of (4.2), we estimate the integral by the zeroth Fourier coefficient,

$$a_0(g) \;=\; \frac{2}{\pi}\int_0^{\pi} g(x)\,dx \;=\; 2\int_0^{1}\frac{1-y^2}{1+y^2}\,dy \;=\; \pi - 2\,.$$

So we can summarize

$$\frac{1}{2M}\left\|\sum_{s=0}^{2M-1}\left|1+2\sum_{k=1}^{M-1}\tfrac{M^2-k^2}{M^2+k^2}\cos k\left(\cdot-\tfrac{s\pi}{M}\right)\right|\right\|_\infty$$
$$\leq \quad \pi-2+\frac{V[g']}{\pi}\sum_{k=1}^{\infty}\frac{1}{k^2} \quad = \quad \pi-2+\frac{\sqrt{3}}{2} \quad < \quad 2.1\,.$$

□

By similar techniques we prove the following estimate.

Lemma 4.3. *For the functions t_k given in Definition 2.1, it holds that*

$$\left\|\sum_{s=1}^{2N_{j-\lambda}}\left|t_{2N_j^{\lambda,p}+s}\right|\right\|_\infty \quad < \quad 2.8\sqrt{2N_{j-\lambda}}\,.$$

Proof. By definition of the functions t_k we have

$$\left\|\sum_{s=1}^{2N_{j-\lambda}}\left|t_{2N_j^{\lambda,p}+s}\right|\right\|_\infty \quad = \quad \left\|\sum_{s=1}^{2N_{j-\lambda}}\left|T_{j+1}^{-b(\lambda+1,2^\lambda+p)}\,t_{2N_j^{\lambda,p}+s}\right|\right\|_\infty$$
$$= \quad \sup_{x\in[0,2\pi)}\frac{1}{\sqrt{N_{j-\lambda}}}\sum_{s=0}^{2N_{j-\lambda}-1}\left|\sum_{k=-M_j+1}^{M_j-1}\tfrac{M_j+k}{\sqrt{M_j^2+k^2}}\cos(N_j^{\lambda,p}+k)(x-\tfrac{s\pi}{N_{j-\lambda}})\right.$$
$$+\sqrt{2}\sum_{k=N_j^{\lambda,p}+M_j}^{N_j^{\lambda,p+1}-M_{j(p)}}\cos k(x-\tfrac{s\pi}{N_{j-\lambda}})$$
$$\left.+\sum_{k=-M_{j(p)}+1}^{M_{j(p)}-1}\tfrac{M_{j(p)}-k}{\sqrt{M_{j(p)}^2+k^2}}\cos(N_j^{\lambda,p+1}+k)(x-\tfrac{s\pi}{N_{j-\lambda}})\right|.$$

Hence, we have to estimate the supremum of a $\frac{\pi}{N_{j-\lambda}}$-periodic function which can be done on an interval of length $\frac{\pi}{N_{j-\lambda}}$. Because of $2M_j \leq M_j + M_{j+1} \leq N_{j-\lambda}$, the degree of the cosine terms is less than $N_j^{\lambda,p} + N_{j-\lambda} + M_{j(p)} \leq N_{j-\lambda}(2^\lambda + p) + 2N_{j-\lambda} - M_j$. For simplicity we set $N = N_{j-\lambda}$ and $M = M_j$. So, using the 2π-periodic absolutely continuous coefficient functions g_p defined in (3.13) and (3.14), for all $p = 0, \ldots, 2^\lambda - 1$, we can write analogously to the proof of the previous lemma,

$$\begin{aligned}
&\left\| \sum_{s=0}^{2N_{j-\lambda}-1} \left| t_{2N_j^{\lambda,p}+s} \right| \right\|_\infty \\
&= \sup_{x\in[0,\frac{\pi}{N})} \frac{1}{\sqrt{N}} \sum_{s=0}^{2N-1} \left| \sum_{k=-M+1}^{2N-M} g_p\left(\tfrac{k\pi}{N}\right) \cos(N(2^\lambda+p)+k)(x-\tfrac{s\pi}{N}) \right| \\
&= \sup_{\xi\in[0,1)} \frac{1}{\sqrt{N}} \sum_{s=0}^{2N-1} \left| \sum_{k=-M+1}^{2N-M} g_p\left(\tfrac{k\pi}{N}\right) \cos\left((2^\lambda+p)\pi+\tfrac{k\pi}{N}\right)(\xi-s) \right| \\
&= \sup_{\xi\in[0,1)} \frac{1}{\sqrt{N}} \sum_{s=0}^{2N-1} \left| \sum_{k=-M+1}^{2N-M} \left(f_p^\xi\left(\tfrac{k\pi}{N}\right) \cos s\tfrac{k\pi}{N} + h_p^\xi\left(\tfrac{k\pi}{N}\right) \sin s\tfrac{k\pi}{N} \right) \right|,
\end{aligned}$$

with

$$\begin{aligned}
f_p^\xi(x) &= g_p(x)\ \cos((2^\lambda+p)\xi\pi+\xi x), \\
h_p^\xi(x) &= g_p(x)\ \sin((2^\lambda+p)\xi\pi+\xi x).
\end{aligned}$$

Note that the g_p as defined in (3.13) and (3.14) are given in $[-\frac{\pi}{3}, 2\pi-\frac{\pi}{3}]$ with $g_p(-\frac{\pi}{3}) = g_p(2\pi-\frac{\pi}{3}) = 0$. Extending g_p 2π-periodically, the functions f_p^ξ and h_p^ξ are absolutely continuous functions with first derivative of bounded variation on an interval of length 2π. We expand these functions in Fourier series,

$$\begin{aligned}
f_p^\xi(x) &= \tfrac{1}{2}a_0(f_p^\xi) + \sum_{n=1}^{\infty} a_n(f_p^\xi)\cos nx + \sum_{n=1}^{\infty} b_n(f_p^\xi)\sin nx, \\
h_p^\xi(x) &= \tfrac{1}{2}a_0(h_p^\xi) + \sum_{n=1}^{\infty} a_n(h_p^\xi)\cos nx + \sum_{n=1}^{\infty} b_n(h_p^\xi)\sin nx,
\end{aligned}$$

and use

$$\sum_{k=-M+1}^{2N-M} \cos m\tfrac{k\pi}{N} = 2N\,\delta_{0,m \bmod 2N}, \qquad \sum_{k=-M+1}^{2N-M} \sin m\tfrac{k\pi}{N} = 0 \quad (m\in\mathbb{Z}),$$

to rewrite, for any $\xi \in [0,1)$

$$\begin{aligned}
&\sum_{s=0}^{2N-1} \left| \sum_{k=-M+1}^{2N-M} \left(f_p^\xi\left(\tfrac{k\pi}{N}\right) \cos s\tfrac{k\pi}{N} + h_p^\xi\left(\tfrac{k\pi}{N}\right) \sin s\tfrac{k\pi}{N} \right) \right| \\
&= N \sum_{s=0}^{2N-1} \left| \sum_{m=0}^{\infty} \left(a_{2Nm+s}(f_p^\xi) + a_{2N(m+1)-s}(f_p^\xi) + b_{2Nm+s}(h_p^\xi) - b_{2N(m+1)-s}(h_p^\xi) \right) \right| \\
&= N \sum_{s=0}^{2N-1} \left| \sum_{m=0}^{\infty} \frac{1}{2\pi} \int_{-\pi}^{\pi} g_p(x) \left(\cos(2Nm+s-\xi)x + \cos(2N(m+1)-s+\xi)x \right) dx \right|.
\end{aligned}$$

Again, we estimate the Fourier coefficients by means of the total variation of the first derivative of g_p. Here we consider the closed interval $[-\frac{\pi}{3}, 2\pi-\frac{\pi}{3}]$. For

$p = 0, \ldots, 2^\lambda - 2$, we observe that g'_p is monotonously increasing in $[-\frac{\pi}{3}, \frac{3-\sqrt{17}}{4}\frac{\pi}{3}]$ and $[\pi - \frac{3-\sqrt{17}}{4}\frac{\pi}{3}, \pi + \frac{\pi}{3}]$ and monotonously decreasing in $[\frac{3-\sqrt{17}}{4}\frac{\pi}{3}, \frac{\pi}{3}]$ and in $[\pi - \frac{\pi}{3}, \pi - \frac{3-\sqrt{17}}{4}\frac{\pi}{3}]$. Further, the extrema are at $\frac{3-\sqrt{17}}{4}\frac{\pi}{3}$ and $\pi - \frac{3-\sqrt{17}}{4}\frac{\pi}{3}$ where

$$\sup_x |g'_p(x)| = g'_p\left(\tfrac{3-\sqrt{17}}{4}\tfrac{\pi}{3}\right) = -g'_p\left(\pi - \tfrac{3-\sqrt{17}}{4}\tfrac{\pi}{3}\right) = \tfrac{1+\sqrt{17}}{(42-6\sqrt{17})^{3/2}}\tfrac{48}{\pi}.$$

Since $g'_p(-\frac{\pi}{3}) > 0$ and $g'_p(\pi + \frac{\pi}{3}) < 0$, the total variation of g'_p is

$$V[g'_p] = 4g'_p\left(\tfrac{3-\sqrt{17}}{4}\tfrac{\pi}{3}\right) = \tfrac{1+\sqrt{17}}{(42-6\sqrt{17})^{3/2}}\tfrac{192}{\pi}.$$

For $p = 2^\lambda - 1$, we have analogously the local minimum

$$g'_{2^\lambda-1}\left(\pi - \tfrac{3-\sqrt{17}}{2}\tfrac{\pi}{3}\right) = -\tfrac{1+\sqrt{17}}{(42-6\sqrt{17})^{3/2}}\tfrac{24}{\pi}.$$

Hence,

$$\begin{aligned} V[g'_{2^\lambda-1}] &= 2g'_{2^\lambda-1}\left(\tfrac{3-\sqrt{17}}{4}\tfrac{\pi}{3}\right) - 2g'_{2^\lambda-1}\left(\pi - \tfrac{3-\sqrt{17}}{2}\tfrac{\pi}{3}\right) \\ &= \tfrac{1+\sqrt{17}}{(42-6\sqrt{17})^{3/2}}\tfrac{144}{\pi}. \end{aligned}$$

The zeroth coefficient can be computed explicitely. For $p = 0, \ldots, 2^\lambda - 2$,

$$\begin{aligned} |a_0(g_p)| &= \frac{1}{\pi}\int_{-\frac{\pi}{3}}^{2\pi-\frac{\pi}{3}} g_p(x)\,dx = \sqrt{2}(1 - \frac{2}{3}) + \frac{2}{3}\int_{-1}^{1} \frac{1+y}{\sqrt{1+y^2}}\,dy \\ &= \frac{\sqrt{2}}{3} + \frac{2}{3}\ln\frac{\sqrt{2}+1}{\sqrt{2}-1}. \end{aligned}$$

Analogously,

$$|a_0(g_{2^\lambda-1})| = \frac{1}{\pi}\int_{-\frac{\pi}{3}}^{2\pi-\frac{\pi}{3}} g_{2^\lambda-1}(x)\,dx \frac{1+y}{\sqrt{1+y^2}}\,dy = \ln\frac{\sqrt{2}+1}{\sqrt{2}-1}.$$

Finally, we get the estimates

$$\left\|\sum_{s=0}^{2N_{j-\lambda}-1}\left|t_{2N_j^{\lambda,p}+s}\right|\right\|_\infty \le \sqrt{N_{j-\lambda}}\left(|a_0(g_p)| + \frac{V[g'_p]}{\pi}\sum_{k=1}^{\infty}\frac{1}{k^2}\right)$$

$$\le \begin{cases} \sqrt{2N_{j-\lambda}}\left(\frac{1}{3} + \frac{\sqrt{2}}{3}\ln\frac{\sqrt{2}+1}{\sqrt{2}-1} + \frac{16\sqrt{2}(1+\sqrt{17})}{(42-6\sqrt{17})^{3/2}}\right) & \text{for } p = 0, \ldots, 2^\lambda - 2, \\ \sqrt{2N_{j-\lambda}}\left(\frac{\sqrt{2}}{2}\ln\frac{\sqrt{2}+1}{\sqrt{2}-1} + \frac{12\sqrt{2}(1+\sqrt{17})}{(42-6\sqrt{17})^{3/2}}\right) & \text{for } p = 2^\lambda - 1. \end{cases}$$

A simple calculation proves the Lemma. □

Now, let us estimate the norm of the partial sum operator with respect to our basis of Definition 2.1.

Theorem 4.4. *Let $\varepsilon > 0$ be given. The orthogonal projection operator S_μ in* (1.1) *for the functions $\{t_\mu : \mu \in \mathbb{N}\}$ from Definition* 2.1, *acting as an operator from $C_{2\pi}$ to $C_{2\pi}$, is uniformly bounded for all $\mu \in \mathbb{N}$. For $\varepsilon \leq \frac{2}{3}$, it holds that*

$$\|S_\mu\|_{C\to C} \quad < \quad 15 + \frac{4}{\pi^2}\,\ln\frac{1}{\varepsilon}\,.$$

Proof. With the kernel function

$$K_\mu(x,\xi) := \sum_{k=1}^{\mu} t_k(\xi)\; t_k(x)$$

we have the well-known representation

$$\begin{aligned}
\|S_\mu\|_{C\to C} &= \sup_{\|f\|_\infty=1}\ \sup_{x\in[0,2\pi)} \left|\sum_{k=1}^{\mu}\langle f\,,\,t_k\rangle\; t_k(x)\right| \\
&= \sup_{\|f\|_\infty=1}\ \sup_{x\in[0,2\pi)} \left|\frac{1}{2\pi}\int_0^{2\pi} f(\xi)K_\mu(x,\xi)d\xi\right| \\
&= \sup_{x\in[0,2\pi)} \tfrac{1}{2\pi}\int_0^{2\pi} |K_\mu(x,\xi)|\,d\xi \\
&= \sup_{x\in[0,2\pi)} \|K_\mu(x,\cdot)\|_1\,. \qquad (4.3)
\end{aligned}$$

Now we distinguish 3 cases.

Case I. Let $\mu \leq 2N_\lambda$: For $\mu \in \{1,2\}$, it is trivial that $\|K_\mu(x,\cdot)\|_1 \;=\; 1$. For $k = 2,\ldots,N_\lambda$, one easily obtains with (3.4) and (4.1)

$$\begin{aligned}
\|K_{2k-1}(x,\cdot)\|_1 &= \frac{1}{2\pi}\int_0^{2\pi}\left|1 + 2\sum_{m=1}^{k-1}\cos m(\xi - x)\right| d\xi \\
&= \|\varphi^1_{k-1}(\cdot - x) - \cos k(\cdot - x)\|_1 \\
&\leq \|\varphi^1_{k-1}\|_1 + \frac{2}{\pi} \\
&< 2.3 + \frac{4}{\pi^2}\,\ln k
\end{aligned}$$

and analogously

$$\begin{aligned}
\|K_{2k}(x,\cdot)\|_1 &= \frac{1}{2\pi}\int_0^{2\pi}\left|1 + 2\sum_{m=1}^{k-1}\cos m(\xi - x) + 2\cos k\xi\cos kx\right| d\xi \\
&= \|\varphi^1_{k-1}(\cdot - x) + \cos k(\cdot + x)\|_1 \\
&\leq \|\varphi^1_{k-1}\|_1 + \frac{2}{\pi} \\
&< 2.3 + \frac{4}{\pi^2}\,\ln k.
\end{aligned}$$

Hence, the projection operator is bounded for all indices $\mu \le 2N_\lambda = 3 \cdot 2^{\lambda+1}$. As $\lambda < \log_2 \frac{8}{3\varepsilon}$, we have

$$\begin{aligned}\sup_{x\in[0,2\pi)} \|K_\mu(x,\cdot)\|_1 &\le 2.3 + \frac{4}{\pi^2}\ln\frac{\mu}{2} \le 2.3 + \frac{4}{\pi^2}\ln(3\cdot 2^\lambda)\\ &< 2.3 + \frac{4}{\pi^2}\ln\frac{8}{\varepsilon} < 3.2 + \frac{4}{\pi^2}\ln\frac{1}{\varepsilon}.\end{aligned}$$

Therewith, the assertion is proved for $\mu \le 2N_\lambda$.

Case II. Let $\mu = 2N_j^{\lambda,p}$: We consider the projection operator $S_{2N_j^{\lambda,p}}$. Since the range of the orthogonal projection operator is independent of the choice of its basis, for every real orthonormal basis $\{s_{j,k}^{\lambda,p} : k = 0,\dots, 2N_j^{\lambda,p}-1\}$ of the range of $S_{2N_j^{\lambda,p}}$ it holds that

$$K_{2N_j^{\lambda,p}}(x,\xi) = \sum_{k=0}^{2N_j^{\lambda,p}-1} s_k(\xi)\, s_k(x).$$

By (4.3), we have that

$$\|S_{2N_j^{\lambda,p}}\|_{C\to C} = \sup_{x\in[0,2\pi)} \|K_{2N_j^{\lambda,p}}(x,\cdot)\|_1 = \sup_{x\in[0,2\pi)} \left\| \sum_{k=0}^{2N_j^{\lambda,p}-1} s_k(\cdot)\, s_k(x)\right\|_1.$$

Using Lemma 3.4 and the functions $\theta_{j,k}^{\lambda,p}$ defined therein, we choose $s_0 = 1$, and

$$\begin{aligned} s_k &= \sqrt{2}\,\cos k\cdot && (k = 1,\dots, N_j - M_j),\\ s_{N_j+k} &= \theta^{\lambda,2^\lambda-1}_{j-1,N_j-\lambda-1+k} && (k = -M_j+1,\dots, M_j-1),\\ s_{2N_j-k} &= \sqrt{2}\,\sin k\cdot && (k = 1,\dots, N_j - M_j),\end{aligned}$$

if $p = 0$, or

$$\begin{aligned} s_k &= \sqrt{2}\, T_{j+1}^{b(\lambda+1,2^\lambda+p-1)}(\cos k\cdot) && (k = 1,\dots, N_j^{\lambda,p} - M_j),\\ s_{N_j^{\lambda,p}+k} &= \theta^{\lambda,p-1}_{j,N_j-\lambda+k} && (k = -M_j+1,\dots, M_j-1),\\ s_{2N_j^{\lambda,p}-k} &= \sqrt{2}\, T_{j+1}^{b(\lambda+1,2^\lambda+p-1)}(\sin k\cdot) && (k = 1,\dots, N_j^{\lambda,p} - M_j),\end{aligned}$$

if $p \in \{1,\dots, 2^\lambda - 1\}$. Since all summands in the kernel $K_{2N_j^{\lambda,p}}(x,\xi)$ have the same shift, we can take both of the shift operators $T_{j+1}^{b(\lambda+1,2^\lambda+p-1)}$, acting on ξ and x, respectively, in front of the sum and finally neglect them totally when we integrate over ξ and take the supremum over x. Hence, for $p = 1,\dots, 2^\lambda - 1$, we insert the inverse shift operators and consider suitably

$$\|S_{2N_j^{\lambda,p}}\|_{C\to C} = \sup_{x\in[0,2\pi)} \left\| \sum_{k=0}^{2N_j^{\lambda,p}-1} T_{j+1}^{-b(\lambda+1,2^\lambda+p-1)} s_k(\cdot)\; T_{j+1}^{-b(\lambda+1,2^\lambda+p-1)} s_k(x)\right\|_1.$$

For the sake of simplicity, we set $N = N_j^{\lambda,p}$ and $M = M_j$. Using trigonometric formulas we obtain, for all $p = 0, \ldots, 2^\lambda - 1$,

$$\begin{aligned}\|S_{2N}\|_{C\to C} &= \sup_{x\in[0,2\pi)} \frac{1}{2\pi}\int_0^{2\pi}\Bigg|1 + 2\sum_{k=1}^{N-M}\cos k(\xi - x) + 2\cos N\xi\cos Nx \\ &\quad + \sum_{k=1}^{M-1}\left(\tfrac{(M+k)^2}{M^2+k^2}\cos(N-k)(\xi-x) + \tfrac{(M-k)^2}{M^2+k^2}\cos(N+k)(\xi-x)\right) \\ &\quad + \sum_{k=1}^{M-1}\tfrac{M^2-k^2}{M^2+k^2}\, r_k(x,\xi)\Bigg|\, d\xi\end{aligned}$$

with remainder terms

$$\begin{aligned} r_k(x,\xi) &= \cos(N(\xi + x) - k(\xi - x)) + \cos(N(\xi+x) + k(\xi - x)) \\ &= 2\ \cos N(\xi + x)\ \cos k(\xi - x)\,. \end{aligned}$$

Decomposing $2\cos N\xi\cos Nx = \cos N(\xi - x) + \cos N(\xi + x)$, we can write

$$\begin{aligned}\|S_{2N}\|_{C\to C} &= \sup_{x\in[0,2\pi)} \|Q_N^M(\cdot - x) + \cos N(\cdot + x)\ R_M(\cdot - x)\|_1 \\ &\le \sup_{x\in[0,2\pi)} \|Q_N^M(\cdot - x)\|_1 + \sup_{x\in[0,2\pi)} \|R_M(\cdot - x)\|_1 \\ &= \|Q_N^M\|_1\ + \|R_M\|_1\,,\end{aligned}$$

with

$$Q_N^M(x) \quad := \quad 1 + 2\sum_{k=1}^{N-M}\cos kx + \sum_{k=-M+1}^{M-1}\tfrac{(M-k)^2}{M^2+k^2}\cos(N+k)x$$

and

$$R_M(x) \quad := \quad 1 + 2\sum_{k=1}^{M-1}\tfrac{M^2-k^2}{M^2+k^2}\cos kx\,.$$

For the $L^1_{2\pi}$ norm of a polynomial $p_n \in \mathbb{T}_n$, we use an inequality from [Ti], Chap. 4,

$$\text{(4.4)} \qquad \frac{1}{2\pi}\int_0^{2\pi}|p_n(\xi)|\, d\xi \;\le\; \sup_x \frac{1}{m}\sum_{s=0}^{m-1}\left|p_n\left(x - \tfrac{2\pi s}{m}\right)\right| \qquad (m\in\mathbb{N})\,.$$

Applying this estimate to R_M of degree $M-1$ with $m = 2M$ we obtain from Lemma 4.2,

$$\text{(4.5)} \qquad \|R_M\|_1 \;\le\; \frac{1}{2M}\left\|\sum_{s=0}^{2M-1}\left|R_M(\cdot - \tfrac{s\pi}{M})\right|\right\|_\infty \;\le\; 2.1\,.$$

From the form (3.4) of the de la Vallée Poussin mean we deduce

$$Q_N^M = \varphi_N^M + \varrho_N^M$$

with

$$\begin{aligned}\varrho_N^M(x) &:= \sum_{k=-M+1}^{M-1} \left(\tfrac{(M-k)^2}{M^2+k^2} - \tfrac{M-k}{M}\right)\cos(N+k)x \\ &= \tfrac{2}{M}\sin Nx \sum_{k=1}^{M-1} k\,\tfrac{M^2-k^2}{M^2+k^2}\sin kx \\ &= -\tfrac{1}{M}\sin Nx\, R'_M(x)\,.\end{aligned}$$

By Bernstein's inequality, $\|p'_n\|_1 \le n\|p_n\|_1$, for all $p_n \in \mathbb{T}_n$, we further get

$$\|\varrho_N^M\|_1 \le \frac{1}{M}\|R'_M\|_1 \le \|R_M\|_1 < 2.1\,.$$

The $L^1_{2\pi}$ norm of $\varphi_N^M = \varphi_{N_j^{\lambda,p}}^{M_j}$ comes by iteration of Lemma 4.1,

$$\|\varphi_{N_j^{\lambda,p}}^{M_j}\|_1 = \|\varphi^1_{N_j^{\lambda,p}/M_j}\|_1 = \|\varphi^1_{3(2^\lambda+p)}\|_1 < 1.6+\frac{4}{\pi^2}\ln(3\cdot 2^{\lambda+1}) < 2.8+\frac{4}{\pi^2}\ln\frac{1}{\varepsilon}.$$

Altogether, we obtain

$$\begin{aligned}\|S_{2N_j^{\lambda,p}}\|_{C\to C} &\le \|Q_{N_j^{\lambda,p}}^{M_j}\|_1 + \|R_{M_j}\|_1 \\ &\le \|\varphi_{N_j^{\lambda,p}}^{M_j}\|_1 + 2\|R_{M_j}\|_1 < \frac{4}{\pi^2}\ln\frac{1}{\varepsilon} + 2.8 + 4.2\,. \qquad (4.6)\end{aligned}$$

Case III. Let $\mu > 2N_\lambda$ be arbitrary: For every $\mu > 2N_\lambda$ we have exactly one triple (j,p,s) of numbers $j \ge \lambda$, $p \in \{0,\dots,2^\lambda-1\}$ and $s \in \{1,\dots,2N_{j-\lambda}\}$ such that $\mu = 2N_j^{\lambda,p} + s$. We decompose the orthogonal projection S_μ into the operator $S_{2N_j^{\lambda,p}}$ and a remainder,

$$S_\mu f = S_{2N_j^{\lambda,p}} f + \sum_{k=2N_j^{\lambda,p}+1}^{\mu} \langle f, t_k\rangle\, t_k\,.$$

Estimating the norm of the remainder operator we obtain

$$\begin{aligned}\sup_{\|f\|_\infty=1}\sup_{x\in[0,2\pi)}\left|\sum_{k=2N_j^{\lambda,p}+1}^{\mu}\langle f, t_k\rangle t_k(x)\right| &= \sup_{x\in[0,2\pi)}\frac{1}{2\pi}\int_0^{2\pi}\left|\sum_{k=2N_j^{\lambda,p}+1}^{\mu} t_k(\xi)\, t_k(x)\right|d\xi \\ &\le \sup_{x\in[0,2\pi)}\sum_{k=2N_j^{\lambda,p}+1}^{2N_j^{\lambda,p+1}} |t_k(x)|\,\|t_k\|_1 \\ &= \left\|t_{2N_j^{\lambda,p}+1}\right\|_1 \left\|\sum_{s=1}^{2N_{j-\lambda}}\left|t_{2N_j^{\lambda,p}+s}\right|\right\|_\infty \\ &\le \frac{1}{2N_{j-\lambda}}\left\|\sum_{s=1}^{2N_{j-\lambda}}\left|t_{2N_j^{\lambda,p}+s}\right|\right\|_\infty^2,\end{aligned}$$

where the last inequality is obtained from (4.4) with $m = 2N_{j-\lambda}$. Applying Lemma 4.3 and adding together with (4.6) we finally get, for all $\mu > 2N_\lambda$,

$$\begin{aligned}\|S_\mu\|_{C\to C} &\leq \|S_{2N_j^{\lambda,p}}\|_{C\to C} + \sup_{\|f\|_\infty=1}\sup_{x\in[0,2\pi)}\left|\sum_{k=2N_j^{\lambda,p}+1}^{\mu}\langle f\,,\,t_k\rangle\,t_k(x)\right| \\ &< 7+\frac{4}{\pi^2}\ln\frac{1}{\varepsilon}+(2.8)^2\,,\end{aligned}$$

which concludes the proof. □

Acknowledgements

The second author gratefully acknowledges the support of the "Deutsche Forschungsgemeinschaft" through the graduate program "Angewandte Algorithmische Mathematik", Technische Universität München.

References

[CMW] R. R. Coifman, Y. Meyer, M. V. Wickerhauser, Size properties of wavelet packets, in: *Wavelets and their applications* (M. B. Ruskai, G. Beylkin, R. Coifman, I. Daubechies, S. Mallat, Y. Meyer, L. Raphael, eds.), Jones and Bartlett, Boston, 1992, 453–470.

[CW] R. R. Coifman, M. V. Wickerhauser, Wavelets and adapted waveform analysis. A toolkit for signal processing and numerical analysis, in: *Different perspectives on wavelets* (I. Daubechies, ed.), Amer. Math. Soc., Providence, 1993, 119–153.

[Fa] G. Faber, Über die interpolatorische Darstellung stetiger Funktionen, *Jber. Deutsch. Math. Verein.* **23** (1914), 192–210.

[Fi] G. M. Fichtenholz, *Differential- und Integralrechnung III*, Deutscher Verlag der Wissenschaften, Berlin, 1970.

[KLT] Y. M. Koh, S. L. Lee, H. H. Tan, Periodic orthogonal splines and wavelets, *Appl. Comput. Harmon. Anal.* **2**(3) (1995), 201–218.

[LS] R. A. Lorentz, A. A. Sahakian, Orthogonal trigonometric Schauder bases of optimal degree for $C(0, 2\pi)$, *J. Fourier Anal. Appl.* **1** (1994), 103–112.

[PT2] G. Plonka, M. Tasche, A unified approach to periodic wavelets, in: *Wavelets: theory, algorithms, and applications* (C.K. Chui, L. Montefusco, L. Puccio, eds.), Academic Press, San Diego, 1994, 137–151.

[Pre] J. Prestin, Best approximation in Lipschitz spaces, in: *Coll. Math. Soc. Jan. Bolyai* 49: Alfred Haar Memorial Conf. (J. Szabados, K. Tandori, eds.) Budapest Jan. Bolyai Math. Soc., 1987, and North Holland Publ. Comp., Amsterdam-Oxford-New York, 1987, 753–759.

[PP] J. Prestin, S. Prößdorf, Error estimates in generalized trigonometric Hölder-Zygmund norms, *Z. Anal. Anwendungen* **9**(4) (1990), 343–349.

[PS1] J. Prestin, K. Selig, Interpolatory and orthonormal trigonometric wavelets, in: *Signal and image representation in Combined Spaces* (J. Zeevi, R. Coifman, eds.), Academic Press, San Diego, 1998, 201–255.

[PS2] J. Prestin, K. Selig, On the Gram matrix of translates of de la Vallée Poussin kernels, *Rostock. Math. Kolloq.* **49** (1995), 105–114.

[Pri1] A. A. Privalov, On the growth of degrees of polynomial bases and approximation of trigonometric projectors, *Mat. Zametki* **42**(2) (1987), 207–214; Engl. Transl.: *Math. Notes* **42**(2) (1987), 619–623.

[Pri2] A. A. Privalov, Growth of degrees of polynomial bases, *Mat. Zametki* **48**(4) (1990), 69–78; Engl. Transl.: *Math. Notes* **48**(4) (1990), 1017–1024.

[Pri3] A. A. Privalov, On an orthogonal trigonometric basis, *Mat. Sb.* **182**(3) (1991), 384–394; Engl. Transl.: *Math. USSR Sb.* **72**(2) (1992), 363–372.

[Prö] S. Prößdorf, Zur Konvergenz der Fourierreihen hölderstetiger Funktionen, *Math. Nachr.* **69** (1975), 7–14.

[Se1] K. Selig, Trigonometric wavelets for time–frequency–analysis, in: *Approximation theory, wavelets and applications* (S. P. Singh, ed.), Kluwer Academic Publ., Dordrecht, 1995, 453–464.

[Se2] K. Selig, Trigonometric wavelets and the uncertainty principle, in: *Approximation theory* (M. W. Müller, M. Felten, D. H. Mache, eds.), Akademie Verlag, Berlin, 1995, 293–304.

[Se3] K. Selig, *Periodische Wavelet-Packets und eine gradoptimale Schauderbasis*, Thesis, Univ. Rostock 1997, Shaker Verlag, Aachen, 1998.

[Ti] A. F. Timan, *Theory of approximation of functions of a real variable*, Pergamon Press, Oxford, 1963.

[Zy] A. Zygmund, *Trigonometric series*, Cambridge University Press, Cambridge, 1968.

Institute of Mathematics
Medical University of Lübeck
Wallstr. 40
D-23560 Lübeck, Germany

Center for Mathematical Sciences
TUM - Munich University of Technology
D-80290 Munich
Germany

1991 Mathematics Subject Classification: Primary 42A10; Secondary 42C15

Submitted: 16.5.2000

Operator Theory:
Advances and Applications, Vol. 121

Discretization methods for the Lavrent'ev regularization

SIEGFRIED PRÖSSDORF AND MASAHIRO YAMAMOTO

In this paper we consider an ill-posed operator equation $Lu = f$ in a Banach space X where L is a linear compact operator in X from itself and there exists a constant $C > 0$ such that $\|(\rho + L)^{-1}\| \le \frac{C}{\rho^k}$ for some $k \in \mathcal{N}$ and small $\rho > 0$. Let f_δ be available data and polluted with noise level $\delta > 0$: $\|f - f_\delta\| \le \delta$. For reconstruction of u from f_δ, we adopt a discretization of the Lavrent'ev regularization: solve $u_{\rho,\delta,h}$ in

$$\rho u + P_h L u = P_h f_\delta$$

where $\rho > 0$ is a parameter and $\{P_h\}_{h>0}$ is a family of projections used for the discretization with discretization parameter h. With suitable a-priori choice of ρ and h for δ, we will establish optimal convergence rates of $u_{\rho,\delta,h}$ towards u as $\delta \to 0$. Furthermore we discuss a similar regularization in the case where f_δ is replaced by discretely observated data $f_{\delta,h}$. Finally we show that the Lavrent'ev regularization can suppress the condition numbers of the resulting discretized problems.

1. Introduction

We consider Symm's integral equation

$$\int_\Gamma \log \frac{1}{|x-y|} u(y)dy = f(x), \qquad x \in \Gamma \tag{1.1}$$

where $\Omega \subset \mathcal{R}^2$ is a bounded domain whose boundary Γ is sufficiently smooth. Symm's integral equation arises from solving the Dirichlet boundary value problem for the Laplace equation in Ω (e.g. Hsiao and Wendland [10], Kress [12]). We set

$$(Lv)(x) = \int_\Gamma v(y) \log \frac{1}{|x-y|} dy, \qquad x \in \Gamma. \tag{1.2}$$

Professor Dr. Siegfried Prössdorf passed away on 19 July 1998. Most part of this work had been done during M. Yamamoto's stay at Weierstrass Institut für Angewandte Analysis und Stochastik (Berlin) in September of 1997, which was realized by an invitation of Professor Prössdorf. Professor Prössdorf and Yamamoto had tried to finish but his untimely death made completion by the both impossible, which is most deplorable to Yamamoto. Yamamoto has hesitated to complete this work by himself but he has finally decided to do so because he believes that this work can convey a sign of small portion of great spirits of Professor Prössdorf as a distinguished mathematician and our common friend. Needless to say, if this present version may be less complete in comparison with Professor Prössdorf's other papers, then everything should be owing to M. Yamamoto. Yamamoto wishes that Professor Prössdorf in Heaven could generously accept his modest trial. Completion of this work without Professor Prössdorf has been very sorrowful and painful, but it is a consolation to M. Yamamoto that he has worked with Professor Prössdorf and shared many various beautiful memories.

Here we assume that

$$(1.3) \qquad 1 \;>\; \textit{the diameter of } \Omega$$

It is known that

$$L : L^2(\Gamma) \longrightarrow H^1(\Gamma) \quad \textit{and}$$

$$(1.4) \qquad L^{-1} : H^1(\Gamma) \longrightarrow L^2(\Gamma) \quad \textit{are continuous}$$

(e.g. Hsiao and Wendland [10]). Now we consider the solution to (1.1) with noisy data for f. The regularity property of L^{-1} guarantees that the problem of solving $Lu = f$ for $f \in H^1(\Gamma)$ is well-defined and is stable for errors in $H^1(\Gamma)$: if $\|f - \widetilde{f}\|_{H^1(\Gamma)}$ is small, then $\|u - \widetilde{u}\|_{L^2(\Gamma)}$ is small where $u = L^{-1}f$ and $\widetilde{u} = L^{-1}\widetilde{f}$.

In this paper, however, we assume that we have to measure errors in data with the same norm for solutions and are concerned with the stability in reconstruction scheme. By (1.4) we know that $\|u - \widetilde{u}\|_{L^2(\Gamma)}$ is not necessarily small even though $\|f - \widetilde{f}\|_{L^2(\Gamma)}$ is small. For the stable reconstruction of u from data polluted with $L^2(\Gamma)$-errors, several regularization methods are proposed (e.g. Baumeister [1], Groetsch [7], Hofmann [8], Lavrent'ev, Romanov and Shishatskiĭ[13], Natterer [14], Tikhonov and Arsenin [19]).

Let $f_\delta \in L^2(\Gamma)$ be our available data and be polluted with errors in $L^2(\Gamma)$:

$$(1.5) \qquad \|f - f_\delta\|_{L^2(\Gamma)} \le \delta,$$

where $\delta > 0$ is a noise level corresponding to an a-priori known upper limit of errors. The requirements for the stable reconstruction are

Find $u_\delta \in L^2(\Gamma)$ for f_δ such that

1. $f_\delta \longmapsto u_\delta$ is continuous from $L^2(\Gamma)$ to itself.

2. $\lim_{\delta\downarrow 0} \|u_\delta - u\|_{L^2(\Gamma)} = 0$.

Here we take the Lavrent'ev regularization, on the basis of singular perturbation for reconstructing $u_\delta = u_{\rho,\delta}$, the solution to

$$(1.6) \qquad \rho u_\delta + L u_\delta = f_\delta$$

for a small parameter $\rho > 0$. Suitable choices of ρ guarantee the above-mentioned requirements 1 and 2, and this regularization is called the Lavrent'ev regularization (e.g Lavrent'ev, Romanov and Shishatskiĭ[13]). For this regularization, we can refer also to Gerlach and von Wolfersdorf [3], Gorenflo and Yamamoto [5], [6], Plato [15], [16]. Other than the Lavrent'ev regularization, the Tikhonov regularization is widely used. The advantage of the Lavrent'ev regularization in comparison with the Tikhonov regularization, is that we need not consider the adjoint operator of L. In the Tikhonov regularization, one must treat $(\rho + L^*L)^{-1}$, and in discretized forms, L^*L is not necessarily sparse even though L is discretized to an upper (or a lower) diagonal matrix.

Moreover we can take discretization itself as a regularization method for a stable reconstruction, and we can refer to Bruckner, Prössdorf and Vainikko [2], Hsiao and Prössdorf [9] for the operator equation $Lu = f$.

In this paper, we discuss a discretized version of the Lavrent'ev regularization, and our purpose is to give an optimal a-priori choice of regularizing parameter ρ and discretizing parameter h in order that discretized approximate solutions of $u_{\rho,\delta}$ may converge to the exact solution u.

This paper is composed of six sections. In §2 and §3, we state two theorems giving strategies of a-priori choices of regularization parameters in the cases of noises measured continuously and discretely respectively. In §4 and §5, we prove them. In §6, in a simple case, we show that the Lavrent'ev regularization can suppress the condition numbers of the discretized problems.

2. Result on a-priori choices of parameters in the case where noises are measured continuously

Henceforth H_λ denotes a Banach scale such that the parameter λ varies over a subset of $\mathcal{R}$ and the embedding $H_\lambda \subset H_0$ is compact for $\lambda > 0$. We denote the norm in H_λ by $\|\cdot\|_\lambda$. Let us consider an operator $L : H_0 \longrightarrow H_\beta$ with some $\beta > 0$ with the property:

there exists a constant $C > 0$ such that

$$C^{-1}\|v\|_0 \le \|Lv\|_\beta \le C\|v\|_0 \quad for \quad v \in H_0. \tag{2.1}$$

Remark: The operator L is not necessarily surjective. That is, LH_0 is a proper subset of H_λ in general. We take a Banach scale for showing the flexibility of our scheme, although the examples treated are concerned only with Hilbert scales.

Example 1: We set

$$(Lv)(x) = \int_\Gamma v(y)\log\frac{1}{|x-y|}dy, \qquad x \in \Gamma, \tag{2.2}$$

where $\Omega \subset \mathcal{R}^2$ is a bounded domain with smooth boundary Γ and let the diameter of $\Omega < 1$. Then L satisfies (2.1) with $H_s = H^s(\Gamma)$, $s \in \mathcal{R}$, the Sobolev(-Slobodetski) spaces, and $\beta = 1$ (e.g. Hsiao and Wendland [10]).

Example 2: We set

$$(Lv)(x) = \int_0^x (x-y)^{\beta-1}v(y)dy, \quad x \in (0,1) \tag{2.3}$$

where $\beta > 0$. Then, by Theorem 4.3.2 (p.80 in Gorenflo and Vessella [4]) for example, we can prove that L satisfies (2.1) with a scale of Sobolev spaces.

Let

$$Lu = f, \qquad u \in H_0. \tag{2.4}$$

That is, f is the exact data and $u \in H_0$ is unknown.

For discretization, we introduce a family $\{X_h, P_h\}_{h>0}$ where X_h is a finite dimensional space of H_0 and P_h is a projection from H_0 to X_h. Furthermore we assume that

there exists a constant $C = C(\beta) > 0$ such that

$$\|(1 - P_h)v\|_0 \le Ch^\beta \|v\|_\beta. \qquad for \quad all \quad v \in H_\beta \quad h > 0. \tag{2.5}$$

Henceforth f_δ is considered as our available data which is polluted with errors in H_0, and $\delta > 0$ is a noise level. More precisely, for some $\delta > 0$, let $f_\delta \in H_0$ satisfy

$$\|f_\delta - f\|_0 \le \delta. \tag{2.6}$$

The condition (2.6) means that noises in our data are measured in the space H_0. In §3, we treat the case where noises are measured in a discrete space.

Example 3: Let H_λ, $\lambda \in \mathcal{R}$, denote the Sobolev spaces of 1-periodic functions on the real line with the norm

$$\|v\|_\lambda = \left(|\widehat{v}(0)|^2 + \sum_{m \in Z, m \neq 0} |m|^{2\lambda} |\widehat{v}(m)|^2 \right)^{\frac{1}{2}}$$

where $\widehat{v}(m) = \int_0^1 v(t)e^{-2m\pi it}dt$, $m \in Z$. We introduce the finite dimensional space of trigonometric functions

$$T_n = \{v_n = \sum_{m \in Z_n} c_m e^{2m\pi it}; c_m \in C\},$$

where $Z_n = \{m \in Z; -\frac{n}{2} \le m \le \frac{n}{2}\}$. Here $\dim T_n = n$ if n is odd and $\dim T_n = n + 1$ if n is even. Let P_n be the orthogonal projection:

$$(P_n v)(t) = \sum_{m \in Z_n} \widehat{v}(m) e^{2m\pi it} \in T_n.$$

Then we have

$$\|(1 - P_n)v\|_0 \le \left(\frac{n}{2}\right)^{-\beta} \|v\|_\beta, \quad v \in H^\beta$$

for $\beta \ge 0$ (e.g. Prössdorf and Silbermann [17]). Therefore setting $h = \frac{1}{n}$, $n \in \mathcal{N}$, we see that $\{P_h, X_h\}_{h>0}$ satisfies (2.5). For other examples, we refer to Prössdorf and Silbermann [17].

Moreover we require that the polluted data f_δ satisfies the estimate

$$\|(1 - P_h)f_\delta\|_0 \le M(h^\alpha + \delta) \tag{2.7}$$

with given $M > 0$ and $\alpha > 0$ not depending on f_δ. If $\sup_{h>0} \|P_h\| < \infty$ in addition to (2.5), then (2.7) follows with $\alpha = \beta$. In fact,

$$\begin{aligned}\|(1-P_h)f_\delta\|_0 &= \|(1-P_h)f + (1-P_h)(f_\delta - f)\|_0 \\ &\le \|(1-P_h)f\|_0 + \|1-P_h\|\|f_\delta - f\|_0 \\ &\le Ch^\beta \|Lu\|_\beta + (1 + \sup_{h>0} \|P_h\|)\delta\end{aligned}$$

by (2.1), (2.4), (2.5) and (2.6).

In Example 3, since $\|P_n\| = 1$, the inequality (2.7) holds.

We further introduce an equation

$$\rho u + P_h L u = P_h f_\delta \tag{2.8}$$

as an approximate equation to the original equation (2.4). Here $\rho > 0$ and $h > 0$ are parameters regularizing the ill-posedness in the original problem $Lu = f$.

Henceforth we denote by $\mathcal{L}(H_0)$ the set of all bounded linear operators mapping the Banach space H_0 into itself and by $\|\cdot\|$ the norm in $\mathcal{L}(H_0)$. Henceforth for positive variables x and y, by $x \sim y$ we mean that there exists a constant $C > 0$ independent of x and y such that $C^{-1}x \le y \le Cx$.

Now we are ready to state the first main result giving optimal choices of ρ and h guaranteeing convergence of solutions to (2.8) towards the exact solution u.

Theorem 1: We assume (2.1), (2.4) - (2.7), and moreover we suppose:

(A1) There exist $C > 0$, $\rho_0 > 0$ and $k \in \mathcal{N}$ such that

$$\|(\rho + L)^{-1}\| \le \frac{C}{\rho^k}$$

for $0 < \rho \le \rho_0$.

(A2) There exist constants $C_0 > 0$, $\rho_0 > 0$ and $0 < \gamma \le 1$ such that

$$\|(\rho + L)^{-1}u\|_0 \le C_0 \rho^{\gamma - 1} \qquad \textit{for} \quad \textit{all} \quad 0 < \rho \le \rho_0.$$

We take

$$q \ge \max\left\{\frac{2k+\gamma}{\beta(k+\gamma)}, \frac{1}{\alpha}\right\} \tag{2.9}$$

and we choose ρ and h satisfying

$$\rho \sim \delta^{\frac{1}{k+\gamma}}, \qquad h \sim \delta^q. \tag{2.10}$$

Then for small $\delta > 0$, we have

$$(\rho + P_h L)^{-1} \in \mathcal{L}(H_0) \tag{2.11}$$

and setting

$$u_{\rho,\delta,h} = (\rho + P_h L)^{-1} P_h f_\delta, \tag{2.12}$$

we have an error estimate

$$\|u - u_{\rho,\delta,h}\|_0 = O\left(\delta^{\frac{\gamma}{k+\gamma}}\right) \tag{2.13}$$

as $\delta \longrightarrow 0$.

The order of the convergence in (2.13) is optimal among resulting rates yielded by choices $\rho \sim \delta^p$ and $h \sim \delta^q$ with all possible $p > 0$ and $q > 0$.

We will realize this theorem for the operator L given by (1.2) in relation with Symm's integral equation. Let $H_0 = L^2(\Gamma)$. Then by Hsiao and Wendland [10], we see that there exists a constant $c_0 > 0$ such that

$$(Lv, v)_{L^2(\Gamma)} \ge c_0\|Lv\|^2_{H^{\frac{1}{2}}(\Gamma)} \ge 0, \quad v \in L^2(\Gamma). \tag{2.14}$$

Moreover $(\rho + L)^{-1}$ exists for $\rho > 0$ and is bounded defined on H_0. In fact, since L is compact by (1.4), it is sufficient to prove that $(\rho + L)v = 0$ implies $v = 0$ in terms of the Fredholm alternative. We have $((\rho + L)v, v)_{L^2(\Gamma)} = 0$ and by (2.14), we obtain $0 = \rho\|v\|^2_{L^2(\Gamma)} + (Lv, v)_{L^2(\Gamma)} \ge \rho\|v\|^2_{L^2(\Gamma)}$ which implies that $v = 0$.

Hence L is m-accretive and in particular, (A1) is satisfied with $k = 1$ (e.g Tanabe [18]). If

$$u \in \mathcal{R}(L) = H^1(\Gamma), \tag{2.15}$$

then (A2) is true with $\gamma = 1$. In fact, by (2.15), we have $u = Lv$ for some $v \in L^2(\Gamma)$. Then

$$\begin{aligned} \|(\rho + L)^{-1}u\|_{L^2(\Gamma)} &= \|(\rho + L)^{-1}Lv\|_{L^2(\Gamma)} = \|v - \rho(\rho + L)^{-1}v\|_{L^2(\Gamma)} \\ &\le \|v\|_{L^2(\Gamma)} + \rho\|(\rho + L)^{-1}v\|_{L^2(\Gamma)} \le \|v\|_{L^2(\Gamma)} + C\|v\|_{L^2(\Gamma)} \end{aligned}$$

by (A1). As for more general a-priori information on the solution, we refer to Gorenflo and Yamamoto [6].

3. Result on a-priori choices of parameters in the case where noises are measured in a discrete form

We consider the same Banach scale H_λ and the same operator L as in §2. That is, we assume (2.1) and we are required to construct approximate solutions for the exact solution u to (2.4) on the basis of noisy data. In §2, the noise level δ is given in (2.6), in other words, δ is an upper bound of noises measured by the H_0-norm.

From the practical point of view, we can not measure data continuously in independent variables, namely, by the norm in H_0. For example, in Example 1, it is reasonable that our available data are a set of values of f with errors at discrete points on Γ. If the accuracy in the measurements is improved, then such discrete data can reduce errors in the noisy data better.

Now we formulate discrete measurements by means of a family $\{Y_h, Q_h\}_{h>0}$ where Y_h is a finite dimensional subspace of H_0 and Q_h is a projection from H_0 to Y_h for every $h > 0$.

We assume that our available noisy data $f_{\delta,h}$ satisfies

$$\|f_{\delta,h} - Q_h f\|_0 \leq C_0\delta \tag{3.1}$$

where $C_0 > 0$ depends on f, not on δ and h.

Remark 3.1: The assumption (3.1) implies $\lim_{\delta\downarrow 0} f_{\delta,h} = Q_h f$, and we can interpret (3.1) that $f_{\delta,h}$ approximates a finite dimensional component $Q_h f$ of f for fixed $h > 0$ and we can not assume anything about other components.

Remark 3.2: Let us assume (2.6). That is, we can obtain noisy data f_δ with noise level δ in H_0: $\|f - f_\delta\|_0 \leq \delta$. Then, setting $f_{\delta,h} = Q_h f_\delta$, we can construct discrete measured data meeting (3.1), provided that

$$\|Q_h v\|_0 \leq C_1\|v\|_0, \quad v \in H_0 \tag{3.2}$$

where $C_1 > 0$ is independent of v. In fact, we have

$$\|f_{\delta,h} - Q_h f\|_0 = \|Q_h f_\delta - Q_h f\|_0 \leq C_1\delta.$$

In this sense, our formulation here for noisy data is more general than (2.6).

Now we proceed to construction of u from discrete noisy data $f_{\delta,h}$ satisfying (3.1). To this end, we introduce an approximate equation

$$\rho v + P_h L v = f_{\delta,h}. \tag{3.3}$$

Correspondingly to Theorem 1, we can show

Theorem 2: We assume (2.1), (2.4), (2.5), (3.1). Moreover we suppose (A1), (A2), the same as in Theorem 1, and

(A3)

$$\|(1 - P_h)f\|_0 \leq Ch^\alpha, \quad \|(1 - Q_h)f\|_0 \leq Ch^\alpha \qquad \textit{for} \quad \textit{all} \quad h > 0$$

with given $C > 0$ and some $\alpha > 0$ depending on f, not on h.

We take

$$q > \begin{cases} \max\left\{\frac{k}{\beta(k+\gamma)}, \frac{1}{\alpha}\right\}, & \textit{if} \quad P_h \neq Q_h \\ \frac{k}{\beta(k+\gamma)}, & \textit{if} \quad P_h = Q_h, \quad h > 0 \end{cases} \tag{3.4}$$

and we choose ρ and h satisfying

$$\rho \sim \delta^{\frac{1}{k+\gamma}}, \qquad h \sim \delta^q. \tag{3.5}$$

Then for small $\delta > 0$, we have

$$(\rho + P_h L)^{-1} \in \mathcal{L}(H_0) \tag{3.6}$$

and, setting

(3.7) $$u_{\rho,\delta,h} = (\rho + P_h L)^{-1} f_{\delta,h},$$

we have an error estimate

(3.8) $$\|u - u_{\rho,\delta,h}\|_0 = \left(\delta^{\frac{\gamma}{k+\gamma}}\right)$$

as $\delta \longrightarrow 0$.

Remark 3.3: In Theorem 1, we have to extra assume (2.7) for noisy data f_δ. On the other hand, in Theorem 2, we assume (3.1).

Remark 3.4: In Theorem 2, $\{P_h, X_h\}_{h>0}$ for discretization of regularization needs not to coincide with $\{Q_h, Y_h\}_{h>0}$ which describes the discrete measurements. In particular, if

(3.9) $$X_h \subset Y_h, \qquad h > 0,$$

and the uniform boundedness (3.2) holds, then $\|(1 - Q_h)f\|_0 \le Ch^\alpha$ follows from $\|(1-P_h)f\|_0 \le Ch^\alpha$. In fact, $X_h \subset Y_h$ and P_h, Q_h are projections, so that $Q_h P_h = P_h$, $h > 0$. Therefore we have $1 - Q_h = 1 - P_h + Q_h(P_h - 1) = (1 - Q_h)(1 - P_h)$, and so $\|(1 - Q_h)f\|_0 = \|(1 - Q_h)(1 - P_h)f\|_0 \le (1 + C_1)\|(1 - P_h)f\|_0$ by (3.2).

4. Proof of Theorem 1

We set

(4.1) $$u_{\rho,\delta} = (\rho + L)^{-1} f_\delta.$$

Then, since $Lu = f$, we have

$$\begin{aligned} \|u_{\rho,\delta} - u\|_0 &= \|(\rho + L)^{-1} f_\delta - u\|_0 \\ &= \|(\rho + L)^{-1}(f_\delta - f) + (\rho + L)^{-1} f - u\|_0 \\ &\le \|(\rho + L)^{-1}(f_\delta - f)\|_0 + \|(\rho + L)^{-1} f - u\|_0 \\ &= \|(\rho + L)^{-1}(f_\delta - f)\|_0 + \rho\|(\rho + L)^{-1} u\|_0. \end{aligned} \tag{4.2}$$

By (A1), (A2) and (2.6), we obtain

(4.3) $$\|u_{\rho,\delta} - u\|_0 \le \frac{C}{\rho^k}\delta + \rho^\gamma C_0, \quad 0 < \rho \le \rho_0.$$

Next we estimate $u_{\rho,\delta} - u_{\rho,\delta,h}$. Since $\rho u_{\rho,\delta,h} + P_h L u_{\rho,\delta,h} = P_h f_\delta$, we subtract it from $\rho u_{\rho,\delta} + L u_{\rho,\delta} = f_\delta$ by (4.1), so that

$$\begin{aligned} \rho(u_{\rho,\delta} - u_{\rho,\delta,h}) &= f_\delta - L u_{\rho,\delta} - P_h f_\delta + P_h L u_{\rho,\delta,h} \\ &= (1 - P_h) f_\delta + L(u_{\rho,\delta,h} - u_{\rho,\delta}) + (P_h - 1) L u_{\rho,\delta,h}. \end{aligned}$$

Therefore we obtain

$$(\rho + L)(u_{\rho,\delta} - u_{\rho,\delta,h}) = (1 - P_h) f_\delta + (P_h - 1) L u_{\rho,\delta,h}$$

and by (A1)

$$
\begin{aligned}
& \| \quad u_{\rho,\delta} - u_{\rho,\delta,h} \|_0 \\
(4.4) \qquad & \leq \quad \frac{C}{\rho^k} \left(\|(1-P_h)f_\delta\|_0 + \|(P_h-1)Lu_{\rho,\delta,h}\|_0 \right).
\end{aligned}
$$

We have to estimate $\|(P_h-1)Lu_{\rho,\delta,h}\|_0$. For this, we show

$$
(4.5) \qquad \|(\rho + P_h L)^{-1}\| \leq \frac{1}{\rho^k} \frac{1}{1 - \frac{Ch^\beta}{\rho^k}} \qquad if \quad \frac{Ch^\beta}{\rho^k} < 1.
$$

In fact, we have $\rho + P_h L = (\rho + L)(1 + (\rho + L)^{-1}(P_h - 1)L)$. For $v \in H_0$, by (A1) and (2.1), (2.5), we have

$$
\begin{aligned}
\|(\rho + L)^{-1}(P_h - 1)Lv\|_0 & \leq \frac{C}{\rho^k} \|(P_h - 1)Lv\|_0 \\
& \leq \frac{C}{\rho^k} h^\beta \|Lv\|_\beta \leq \frac{C}{\rho^k} h^\beta \|v\|_0.
\end{aligned}
$$

Therefore by the Neumann series and (A1), we see (4.5).
We apply (4.5) to $u_{\rho,\delta,h} = (\rho + P_h L)^{-1} P_h f_\delta$, so that

$$
(4.6) \qquad \|u_{\rho,\delta,h}\|_0 \leq \frac{1}{\rho^k} \frac{1}{1 - \frac{Ch^\beta}{\rho^k}} \|P_h f_\delta\|_0 \leq \frac{Mh^\alpha + (M+1)\delta + \|f\|_0}{\rho^k - Ch^\beta}
$$

if

$$
(4.7) \qquad \rho^k > Ch^\beta.
$$

Here we have used

$$
\begin{aligned}
& \|P_h f_\delta\|_0 \leq Mh^\alpha + M\delta + \|f_\delta\|_0 \\
& \leq Mh^\alpha + M\delta + \delta + \|f\|_0
\end{aligned}
$$

by (2.7) and (2.6). Hence by (2.1), (2.5) and (4.6), for small $h > 0$ and $\delta > 0$, we obtain

$$
\begin{aligned}
\|(P_h - 1)Lu_{\rho,\delta,h}\|_0 & \leq Ch^\beta \|Lu_{\rho,\delta,h}\|_\beta \leq Ch^\beta \|u_{\rho,\delta,h}\|_0 \\
(4.8) \qquad & \leq \frac{Ch^\beta (Mh^\alpha + (M+1)\delta + \|f\|_0)}{\rho^k - Ch^\beta} \leq \frac{Ch^\beta}{\rho^k - Ch^\beta}.
\end{aligned}
$$

Substitution of (2.7) and (4.8) into (4.4) yields

$$
(4.9) \qquad \|u_{\rho,\delta} - u_{\rho,\delta,h}\|_0 \leq \frac{CMh^\alpha}{\rho^k} + \frac{CM\delta}{\rho^k} + \frac{Ch^\beta}{\rho^k(\rho^k - Ch^\beta)}
$$

with which we combine (4.3) to obtain

$$\begin{aligned}\|u-u_{\rho,\delta,h}\|_0 &\le \|u-u_{\rho,\delta}\|_0+\|u_{\rho,\delta}-u_{\rho,\delta,h}\|_0\\ &\le C\left(\frac{\delta}{\rho^k}+\rho^\gamma+\frac{h^\alpha}{\rho^k}+\frac{h^\beta}{\rho^k(\rho^k-Ch^\beta)}\right)\end{aligned}$$

if $\rho^k > Ch^\beta$. We set

$$\rho \sim \delta^p, \qquad h \sim \delta^q \tag{4.10}$$

with $p > 0$ and $q > 0$. If $pk < q\beta$, then $\rho^k - Ch^\beta \sim \delta^{pk}(1 - C\delta^{q\beta-pk})$, so that $\rho^k > Ch^\beta$ as $\delta \downarrow 0$. Moreover

$$\begin{aligned}\frac{\delta}{\rho^k} &+ \rho^\gamma+\frac{h^\alpha}{\rho^k}+\frac{h^\beta}{\rho^k(\rho^k-Ch^\beta)}\\ &\le C(\delta^{1-pk}+\delta^{p\gamma}+\delta^{q\alpha-pk}+\delta^{q\beta-2pk}).\end{aligned}$$

We have to choose $p > 0$ and $q > 0$ so that $F(p,q) \equiv \min\{1-pk, p\gamma, q\alpha - pk, q\beta-2pk\}$ is maximum and positive. First let us choose $p > 0$ so that $f(p) \equiv \min\{1-pk, p\gamma\}$ is maximum. It is easy to see that $p_0 = \frac{1}{k+\gamma}$ gives the maximum $\frac{\gamma}{k+\gamma}$ of f. Hence

$$\max_{p,q} F(p,q) \le \max_p f(p) = \frac{\gamma}{k+\gamma}.$$

If $q_0 > 0$ satisfies (2.9), then we see that

$$\alpha q_0 - \frac{k}{k+\gamma} \ge \frac{\gamma}{k+\gamma}, \quad \beta q_0 - \frac{2k}{k+\gamma} \ge \frac{\gamma}{k+\gamma}$$

and $F(p_0,q_0) = \frac{\gamma}{k+\gamma}$. Therefore $\max_{p,q} F(p,q) = F(p_0,q_0) = \frac{\gamma}{k+\gamma}$. Thus the proof of Theorem 1 is complete.

5. Proof of Theorem 2

We set $\rho \sim \delta^{\frac{1}{k+\gamma}}$ and $h \sim \delta^q$ with $q > 0$ for small $\delta > 0$. By (3.4), we have $q\beta > \frac{k}{k+\gamma}$, so that (3.5) implies

$$\frac{Ch^\beta}{\rho^k} \sim C\delta^{q\beta}\delta^{-\frac{k}{k+\gamma}} = C\delta^{q\beta-\frac{k}{k+\gamma}}.$$

Therefore we obtain

$$\frac{Ch^\beta}{\rho^k} < 1 \tag{5.1}$$

for small $\delta > 0$. Therefore by (4.5) we have

$$\|(\rho + P_h L)^{-1}\| \le \frac{1}{\rho^k} \frac{1}{1 - \frac{Ch^\beta}{\rho^k}} \le \frac{C}{\rho^k}. \tag{5.2}$$

Here and henceforth $C > 0$ denotes generic constants independent of ρ, δ and h. We set

$$u_{\rho,h} = (\rho + P_h L)^{-1} Q_h f. \tag{5.3}$$

First we estimate:

$$\|u_{\rho,\delta,h} - u_{\rho,h}\|_0 = \|(\rho + P_h L)^{-1}(f_{\delta,h} - Q_h f)\|_0 \le \frac{C\delta}{\rho^k} \tag{5.4}$$

by (3.1) and (5.2).

Next, by $Lu = f$, (A3) and (5.2), we have

$$\begin{aligned} \|u_{\rho,h} - u\|_0 &= \|(\rho + P_h L)^{-1} Q_h f - u\|_0 \\ &= \|(\rho + P_h L)^{-1}(Q_h f - (\rho + P_h L)u)\|_0 \\ &= \|(\rho + P_h L)^{-1}(Q_h - P_h) f - \rho(\rho + P_h L)^{-1} u\|_0 \\ &\le \|(\rho + P_h L)^{-1}\|(\|(Q_h - 1) f\|_0 + \|(1 - P_h) f\|_0) \\ &\quad + \rho\|(\rho + P_h L)^{-1} u\|_0 \\ &\le \frac{Ch^\alpha}{\rho^k} + \rho\|(\rho + P_h L)^{-1} u\|_0. \end{aligned} \tag{5.5}$$

By (2.1), (2.5) and (A1), we see that

$$\|(\rho + L)^{-1}(P_h - 1)L\| \le \frac{Ch^\beta}{\rho^k} < 1$$

for sufficiently small $\delta > 0$. Then by the Neumann series, we have

$$\|(1 + (\rho + L)^{-1}(P_h - 1)L)^{-1}\| \le \frac{1}{1 - \frac{Ch^\beta}{\rho^k}} \le C$$

for small $\delta > 0$. Since $(\rho + P_h L)^{-1} u = (1 + (\rho + L)^{-1}(P_h - 1)L)^{-1}(\rho + L)^{-1} u$, we obtain

$$\rho\|(\rho + P_h L)^{-1} u\|_0 \le C\rho\|(\rho + L)^{-1} u\|_0 \le C\rho^\gamma$$

by (A2). With (5.5) we obtain

$$\|u_{\rho,h} - u\|_0 \le \frac{Ch^\alpha}{\rho^k} + C\rho^\gamma$$

with which we combine (5.4) to see

$$\|u_{\rho,\delta,h} - u\|_0 \le C\left(\frac{h^\alpha}{\rho^k} + \rho^\gamma + \frac{\delta}{\rho^k}\right).$$

Let $\rho \sim \delta^{\frac{1}{k+\gamma}}$ and $h \sim \delta^q$. Then

$$\|u_{\rho,\delta,h} - u\|_0 \leq C(\delta^{\alpha q - \frac{k}{k+\gamma}} + \delta^{\frac{\gamma}{k+\gamma}}). \tag{5.6}$$

Therefore under $q > \max\left\{\frac{k}{\beta(k+\gamma)}, \frac{1}{\alpha}\right\}$, we have $q\alpha \geq 1$ and $\delta^{\alpha q - \frac{k}{k+\gamma}} \leq \delta^{1-\frac{k}{k+\gamma}}$ for $\delta \leq 1$. Hence

$$\|u_{\rho,\delta,h} - u\|_0 \leq C\delta^{\frac{\gamma}{k+\gamma}}$$

for $\rho \sim \delta^{\frac{1}{k+\gamma}}$ and $h \sim \delta^q$.

Finally let $P_h = Q_h$ for all $h > 0$. Then (5.5) can be replaced by

$$\|u_{\rho,h} - u\|_0 \leq \rho\|(\rho + P_h L)^{-1} u\|_0.$$

Therefore (5.6) is reduced to

$$\|u_{\rho,\delta,h} - u\|_0 \leq C\delta^{\frac{\gamma}{k+\gamma}},$$

so that $q > \frac{k}{\beta(k+\gamma)}$ gives the desired rate. Thus the proof of Theorem 2 is complete.

6. Control of condition numbers by the regularization

We can interpret a suitable discretization itself as a regularization for the problem (2.4), which is called self-regularization (e.g. Bruckner, Prössdorf and Vainikko [2], Hsiao and Prössdorf [9]). In this section, in comparison with the self-regularization, we show one advantage of our hybrid method (i.e. combination of the Lavrent'ev regularization and the discretization) as one regularization technique. That is, our discretization method for the Lavrent'ev regularization can control condition numbers better than the self-regularization.

For testing such a control, we take the following simple setting although it is not adjusted to the framework in the previous sections. Let X be a separable Hilbert space over $\mathcal{R}$ and let

$$\begin{aligned} L \ &: \ X \longrightarrow X \quad \textit{be a compact self-adjoint} \\ &\textit{linear operator such that} \quad (Lu, u) \geq 0, \quad u \in X. \end{aligned} \tag{6.1}$$

Henceforth $(\cdot,\cdot)$ denotes the scalar product in X. Moreover we assume that L is in the Schmidt class, that is,

$$|||L||| = \left(\sum_{k=1}^{\infty} \|L\varphi_k\|^2\right)^{\frac{1}{2}} < \infty \tag{6.2}$$

where $\{\varphi_k\}_{k\in\mathcal{N}}$ is an orthonormal basis in X (e.g. Kato [11]). We note that $|||L|||$ is independent of a choice of an orthonormal basis.

As discretization, we adopt the Galerkin method: Let $\{\varphi_k\}_{k\in\mathcal{N}}$ be an orthonormal basis in X and we set

$$L_N = (L\varphi_i, \varphi_j)_{1\le i,j\le N} \tag{6.3}$$

and

$$f_N = (f, \varphi_i)_{1\le i\le N}, \qquad f \in X. \tag{6.4}$$

By (6.1) we note that L_N is a non-negative symmetric matrix. Then in the Galerkin method, we are required to solve u_N in

$$L_N u_N = f_N \tag{6.5}$$

as approximations to the solution u to $Lu = f$. The condition number $\gamma(L_N)$ is defined by

$$\gamma(L_N) = \frac{\textit{the greatest eigenvalue of } L_N}{\textit{the least non}-\textit{negative eigenvalue of } L_N}. \tag{6.6}$$

The condition number is an indicator for the ill-conditioning of the problem. Then there exists a constant $c_0 > 0$ dependent on L but independent of N such that

$$\gamma(L_N) \ge c_0 N^{\frac{1}{2}} \tag{6.7}$$

for large $N \in \mathcal{N}$ (Wing [20]).

The problem (2.4) is ill-posed, so that the discretized problem is ill-conditioned in any case. In the self-regularization, the condition numbers may increase with any rate as N increases. On the other hand, we can show how the Lavrent'ev regularization controls the condition numbers of the resulting matrices in the Galerkin method.

Proposition 1: We consider the discretization of the Lavrent'ev regularization by the Galerkin method

$$\rho u_N + L_N u_N = f_N \tag{6.8}$$

where L_N and f_N are given by (6.3) and (6.4) respectively. Let $\delta > 0$ be a noise level in (2.6), and let us choose the regularization parameter ρ by

$$\rho = C\left(\frac{1}{N}\right)^p$$

with some $p > 0$ and $C > 0$. Then

$$\gamma(\rho + L_N) \le 1 + C^{-1}\|L\|N^p. \tag{6.9}$$

The estimate (6.9) guarantees that the condition numbers may grow up with at most polynomial rate of N. Its exponent p is given according to the a-priori choices of regularization parameters in Theorems 1 and 2.

Associated with the Galerkin method, when in Theorem 1 we can take $\{P_h, X_h\}_{h>0}$ with $h = \frac{1}{N}$ for example, the strategy (2.10) implies

$$\rho \sim \left(\frac{1}{N}\right)^{\frac{1}{q(k+\gamma)}}.$$

Therefore, as long as we adopt the strategy (2.10), in the case of $h = \frac{1}{N}$, we see that

$$p = \frac{1}{q(k+\gamma)}.$$

Proof: Let $0 < \lambda_N \le \lambda_{N-1} \le \cdots \le \lambda_1$ be the eigenvalues of $\rho + L_N$. Then since $\rho + L_N$ is positive and symmetric, we have

$$\begin{aligned}
\lambda_1 &= \max_{\|v\|=1, v\in\mathcal{R}^N}((\rho + L_N)v, v) \\
&= \rho + \max_{v_1^2+\cdots+v_N^2=1} \sum_{i=1}^{N}\sum_{j=1}^{N}(L\varphi_i, \varphi_j)v_iv_j \\
&= \rho + \left(L\left(\sum_{i=1}^{N} v_i\varphi_i\right), \sum_{i=1}^{N} v_i\varphi_i\right) \\
&\le \rho + \|L\|. \qquad (6.10)
\end{aligned}$$

On the other hand, since L_N is non-negative, we have $\lambda_N \ge \rho$. Hence

$$\gamma(\rho + L_N) = \frac{\lambda_1}{\lambda_N} \le \frac{\rho + \|L\|}{\rho},$$

which completes the proof of the proposition.

Acknowledgements

Masahiro Yamamoto would like to express his hearty thanks to Weierstrass Institut für Angewandte Analysis und Stochastik (Berlin, Germany) for the kind hospitality during his stay. This work is partially supported by Sanwa Systems Development Co., Ltd (Tokyo, Japan). He thanks the referees for their very careful reviewing and many precise valuable comments. In particular, for an essential improvement of the condition (2.7), he is indebted. Furthermore he is grateful to Professor Dr. Rudolf Gorenflo (Free University of Berlin) for his kind comments and to Professor Jin Cheng (Fudan University, China) for his assistance in preparing the text-file.

References

[1] Baumeister, J., Stable Solutions of Inverse Problems. Vieweg, Braunschweig, 1987.

[2] Bruckner,G., Prössdorf, S. and Vainikko, G., Error bounds of discretization methods for boundary integral equations with noisy data. Applicable Analysis 63 (1996), 25–37.

[3] Gerlach, W and von Wolfersdorf, L., On approximation computation of the values of the normal derivative of solutions to linear partial differential equations of second order with application to Abel's integral equation. Zeitschrift für Angewandte Mathematik und Mechanik 66 (1986), 31–36.

[4] Gorenflo, R. and Vessella, S., Abel Integral Equations. Lecture Notes in Mathematics, 1461, Springer-Verlag, Berlin, 1991.

[5] Gorenflo, R. and Yamamoto, M., On regularized inversion of Abel integral operators. Proceedings of International Conference on Analysis and Mechanics of Continuous Media (HoChiMinh City, 27-29 December 1995) in honour of D.D. Ang, Publications of The HoChiMinh City Mathematical Society (1995), 162–182.

[6] Gorenflo, R. and Yamamoto, M., A-priori strategy for regularizing parameters in the Lavrent'ev regularization. preprint.

[7] Groetsch, C.W., The Theory of Tikhonov Regularization for Fredholm Equations of the First Kind. Pitman, Boston, 1984.

[8] Hofmann, B., Regularization for Applied Inverse and Ill-posed Problems. Teubner, Leipzig, 1986.

[9] Hsiao, G.C. and Prössdorf, S., On the stability of the spline collocation method for a class of integral equatios of the first kind. Applicable Analysis 30 (1988), 249–261.

[10] Hsiao, G.C. and Wendland, W.L., A finite element method for some integral equations of the first kind. J. Math. Anal. Appl. 58 (1977), 449–481.

[11] Kato, T., Perturbation Theory for Linear Operators. (second edition) Springer-Verlag, Berlin, 1976.

[12] Kress, R., Linear Integral Equations. (second edition) Springer-Verlag, Berlin, 1999.

[13] Lavrent'ev, M.M., Romanov, V.G. and Shishatskiĭ, S.P., Ill-posed Problems of Mathematical Physics and Analysis. American Mathematical Society, Providence, Rhode Island, English translation, 1986.

[14] Natterer, F., Error bounds for Tikhonov regularization in Hilbert scales. Applicable Analysis 18 (1984), 25–37.

[15] Plato, R., The Galerkin scheme for Lavrentiev's m-times iterated method to solve linear accretive Volterra integral equations of the first kind. BIT 37 (1997), 404–423.

[16] Plato, R., Resolvent estimates for Abel integral operators and the regularization of associated first kind integral equations. J. Integral Equations and Appl. 9 (1997), 253–278.

[17] Prössdorf, S. and Silbermann, B., Numerical Analysis for Integral and Related Operator Equations. Akademie-Verlag, Birkhäuser-Verlag, Basel, Berlin, 1991.

[18] Tanabe, H., Equations of Evolution. Pitman, London, 1979.

[19] Tikhonov, A.N. and Arsenin, V.Y., Solutions of Ill-posed Problems. (English translation) John Wiley & Sons, New York, 1977.

[20] Wing, G.M., Condition numbers of matrices arising from the numerical solution of linear integral equations of the first kind. J. Integral Equations 9 (Suppl.) (1985), 191–204.

Department of Mathematical Sciences, the University of Tokyo,
3-8-1 Komaba, Meguro, Tokyo 153-8914, Japan

1991 Mathematics Subject Classification: Primary 65R30; Secondary 65R20

Submitted: 22.5.2000

Operator Theory:
Advances and Applications, Vol. 121

On a Hierarchical Three-Point Basis in the Space of Piecewise Linear Functions over Smooth Surfaces

A. RATHSFELD

Dedicated to my teacher Professor Siegfried Prößdorf

In this paper we consider a smooth boundary surface of a three-dimensional domain and the space of piecewise linear functions defined over a uniform triangular grid. We introduce a wavelet basis which is a variant of the well-known three-point hierarchical basis with a simple modification near the boundary points of the global patches of parametrization. Each wavelet is the linear combination of no more than three finite element functions defined over a grid from a hierarchy of triangulations. For the spaces spanned by this basis, the approximation and inverse properties in a certain range of Sobolev spaces are well known. Consequently, the simple basis is a Riesz basis and can be used to precondition operator equations, e.g. boundary element methods. Since the construction of wavelets with and without zero boundary values is part of the setting, the wavelets can also be used for finite element methods.

1. Introduction

It is well known that wavelet methods can be applied to the numerical solution of partial differential and integral equations (cf. e.g. the overview paper [7]). The wavelets can be used to construct preconditioners, to convert fully populated matrices into sparse matrices, and to design adaptive numerical procedures. The construction of these wavelets over regular grids in the Euclidean space is rather standard (cf. e.g [16]). To get an appropriate basis over arbitrary domains and manifolds, these functions have to be modified and, depending on the desired properties, the construction is quite complicated. Let us mention here the general constructions by Canuto, Cohen, Dahmen, Daubechies, Feaveau, Schneider, Stevenson, Tabacco, Urban [5, 9, 12, 13, 14, 2, 3, 15].

Clearly, the good asymptotic rates of compression for functions or operators are related to the vanishing moment conditions. However, to get cheap preconditioners for finite element methods and to obtain effective compression and quadrature algorithms for boundary element methods, it is important to have basis functions with minimal support and minimal sums in the two-scale relations where the wavelets are expressed as a linear combination of scaling functions. For piecewise linear finite element spaces and for prescribed second order vanishing moments, this leads naturally to the so called three-point hierarchical basis functions which have been introduced by Dahmen, Kleemann, Prößdorf, and Schneider [8] and which have been analyzed by Junkherr, Lorentz, Oswald, Stevenson [18, 21, 19].

A nice construction of three-point hierarchical wavelets over general surfaces is due to Stevenson [21]. Unfortunately, the resulting basis can be shown to be stable in Sobolev spaces of orders greater than one half, only. To get a wider range of Sobolev orders for the Riesz basis property we follow a different direction. Our intention is to take the basis functions defined over the two-dimensional Euclidean space and to define corresponding functions over a two-dimensional manifold by taking compositions with parameter mappings and by simple modifications at the joint lines of different parametrization patches. In contrast to the constructions in the papers [12, 13, 14, 2, 3, 15], our aim is to keep the three-point character and to sacrifice rather the vanishing moments than the smallness of the support. In fact it has turned out that the violation of the vanishing moment property over a lower dimensional submanifold is not essential for applications in boundary element methods (cf. [20, 17]). The basic idea for the glueing of basis functions along the joint boundary of different parametrization patches should be the simple symmetry trick which has been used already for the univariate case in [5]. Unfortunately, it has turned out that this trick does not work over hexagonal grids. Therefore, we change to triangular criss-cross grids like in Figure 1.

For criss-cross grids, we shall define the three-point hierarchical functions and describe the construction of the corresponding wavelets over smooth manifolds in Sect. 2. We shall apply Stevenson's technique in Sect. 3 to show the Riesz stability on the plane and use simple symmetry arguments to get the stability as well as the approximation and inverse properties over the manifold (cf. Theorem 1). In Sect. 4 we present numerical results for the constants in the stability estimates and for an application to a boundary integral method.

2. The Definition of the Hierarchical Basis Functions

2.1. The Manifold

Suppose we are given a closed boundary manifold $\Gamma \subset \mathbb{R}^3$ with finite degree of smoothness. More exactly, we assume that Γ is the union of m_Γ triangular patches Γ_m, i.e.

$$\begin{aligned} \Gamma &= \cup_{m=1}^{m_\Gamma} \Gamma_m, \quad \Gamma_m := \kappa_m(T), \\ T &:= \left\{(s,t) \in \mathbb{R}^2 : 0 \le s \le 1,\ 0 \le t \le \min\{s, 1-s\}\right\}. \end{aligned} \tag{2.1}$$

Here the κ_m denote parametrization mappings from the standard triangle T to the manifold Γ. We assume that the κ_m extend to mappings from a small neighbourhood of $T \subseteq \mathbb{R}^2$ to Γ and that these extensions are d_Γ times continuously differentiable. Here d_Γ is an integer which is assumed to be greater or equal to three. Further we suppose that the intersection of two patches Γ_m and $\Gamma_{m'}$ is either empty or a corner point for both patches or a whole side for Γ_m and $\Gamma_{m'}$.

In the last case we assume that the representations

$$\begin{aligned} \Gamma_m \cap \Gamma_{m'} &= \left\{\kappa_m\Big(c_1 + \lambda(c_2 - c_1)\Big) : 0 \le \lambda \le 1\right\}, \\ \Gamma_m \cap \Gamma_{m'} &= \left\{\kappa_{m'}\Big(c_1' + \lambda(c_2' - c_1')\Big) : 0 \le \lambda \le 1\right\} \end{aligned}$$

satisfy the condition

$$\kappa_m\Big(c_1 + \lambda(c_2 - c_1)\Big) = \kappa_{m'}\Big(c_1' + \lambda(c_2' - c_1')\Big), \quad 0 \le \lambda \le 1. \tag{2.2}$$

In the construction of the wavelet basis the numbering of the patches will play a crucial role since the basis functions will first be defined on Γ_1, then on Γ_2, and so on. To secure stability of the so constructed basis, we even need an assumption connected with the numbering. We suppose that, if the corner P of a patch Γ_m is contained in the union $\cup_{m'=1}^{m-1}\Gamma_{m'}$ of the preceding patches, then at least one of the sides of Γ_m ending at P is contained in $\cup_{m'=1}^{m-1}\Gamma_{m'}$. It is not hard to see that, for a boundary manifold Γ homeomorphic to the sphere and for any fixed triangulation, there always exists a numbering of the triangular patches which fulfills the assumption. However, the numbering assumption seems to be a severe topological restriction. It seems to us that, for boundaries homeomorphic to the torus a construction of similar basis systems is possible only if the triangular patches are combined with rectangular ones and if the piecewise linear functions over the triangular patches are combined with piecewise bilinear functions over the rectangular patches (cf. [20]).

To secure stability of the wavelet construction, we need a final assumption on the parametrizations. For any $m = 2, \ldots, m_\Gamma - 1$, we suppose that, if one of the two "shorter" sides $\kappa_m(\{(s,s) : 0 \le s \le 0.5\})$ and $\kappa_m(\{(s,1-s) : 0.5 \le s \le 1\})$ is contained in $\cup_{m'=1}^{m-1}\Gamma_m$, then the other must also be contained in $\cup_{m'=1}^{m-1}\Gamma_m$. This last assumption can always be satisfied if the parameter mappings κ_m are replaced by a composition of κ_m with a suitable affine automorphism of T.

The Sobolev spaces $H^s(\Gamma)$ over Γ can be defined in the usual way. We define the space $H^s(\Gamma_m)$ over Γ_m as the image of the Sobolev space over T, i.e.

$$H^s(\Gamma_m) := \{f : f \circ \kappa_m \in H^s(T)\}.$$

Consequently, we get

$$H^s(\Gamma) = \left\{(f_m)_{m=1}^{m_\Gamma} \in \bigoplus_{m=1}^{m_\Gamma} H^s(\Gamma_m) : f_m|_{\Gamma_m \cap \Gamma_{m'}} = f_{m'}|_{\Gamma_m \cap \Gamma_{m'}}\right\}, \quad \frac{1}{2} < s < \frac{3}{2},$$

$$H^s(\Gamma) = \bigoplus_{m=1}^{m_\Gamma} H^s(\Gamma_m), \quad -\frac{1}{2} < s < \frac{1}{2}, \tag{2.3}$$

$$\|f\|_{H^s(\Gamma)} \sim \sqrt{\sum_{m=1}^{m_\Gamma} \|f|_{\Gamma_m}\|^2_{H^s(\Gamma_m)}}, \quad f \in H^s(\Gamma), \; -\frac{1}{2} < s < \frac{3}{2}.$$

Finally, we note that the sphere can serve as a simple example for a boundary manifold fulfilling all assumptions. To get the corresponding parametrization mappings, we inscribe a tetrahedron and take the projections from the midpoint mapping the triangular faces of the tetrahedron onto triangular patches of the sphere. Composing these parametrizations with suitable affine mappings, we arrive at a representation (2.1) for the sphere. The numbering of these four parameter patches can be chosen arbitrarily.

2.2. Grid and Collocation Points

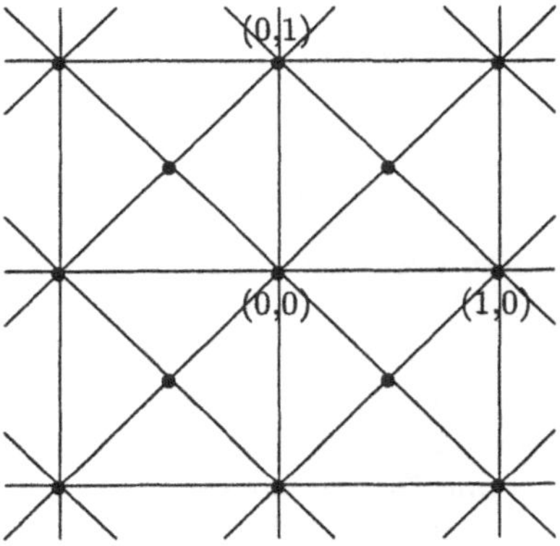

Figure 1: Grid $\triangle_0^{\mathbb{R}^2}$.

Let us introduce a hierarchy of uniform grids over the standard triangle T. For the step sizes 2^{-l}, $l = 0, \ldots, L$, we set

$$\begin{aligned}
\triangle_l^T &:= {}^1\!\triangle_l^T \cup {}^2\!\triangle_l^T, \\
{}^1\!\triangle_l^T &:= \left\{(i2^{-l}, j2^{-l}) : \; 0 \le i \le 2^l, \; 0 \le j \le \min\{2^l - i, i\}\right\}, \\
{}^2\!\triangle_l^T &:= \left\{(2^{-l-1}, 2^{-l-1}) + (i2^{-l}, j2^{-l}) : \; 0 \le i < 2^l, \; 0 \le j < \min\{2^l - i, i+1\}\right\}
\end{aligned}$$

and denote the grid points by $\tau = (s,t) \in \triangle_l^T$. The grid $\triangle_l^T$ is the restriction of the grid (cf. Figure 1)

$$\triangle_l^{\mathbb{R}^2} := \left\{(i2^{-l}, j2^{-l}) : i, j \in \mathbb{Z}^2\right\} \cup \left\{(2^{-l-1}, 2^{-l-1}) + (i2^{-l}, j2^{-l}) : i, j \in \mathbb{Z}^2\right\}$$

to the triangle T. Using the parametrizations, we arrive at a grid hierarchy on Γ.

$$\Delta_l^\Gamma \quad := \quad \{\kappa_m(\tau) : m = 1, \ldots, m_\Gamma, \tau \in \Delta_l^T\}.$$

Clearly, a grid point $P = \kappa_m(\tau)$ may have more than one representation. If P is in the interior of a side of the triangular patch Γ_m which is a common side with $\Gamma_{m'}$, then there are exactly two representations $P = \kappa_m(\tau)$ and $P = \kappa_{m'}(\tau')$. If P is a corner point of a patch, then there exist $k > 2$ representations $P = \kappa_{m_1}(\tau_1) = \kappa_{m_2}(\tau_2) = \ldots = \kappa_{m_k}(\tau_k)$. We introduce ${}^i\Delta_l^\Gamma$ as the set of those $P \in \Delta_l^\Gamma$ whose representation $P = \kappa_m(\tau)$ with the smallest m satisfies $\tau \in {}^i\Delta_l^T$, i.e.,

$${}^i\Delta_l^\Gamma \quad := \quad \cup_{m=1}^{m_\Gamma} \{\kappa_m(\tau) : \tau \in {}^i\Delta_l^T, \ \kappa_m(\tau) \notin \cup_{m'=1}^{m-1} \kappa_{m'}(\Delta_l^T)\},$$

and arrive at $\Delta_l^\Gamma = {}^1\Delta_l^\Gamma \cup {}^2\Delta_l^\Gamma$. The points of Δ_l^Γ will be denoted by upper capital letters like P and Q.

To each grid Δ_l^Γ there corresponds a partition of Γ into triangular pieces. Indeed, let us introduce the sets of centroids

$$\begin{aligned}
\square_0^{\mathbb{R}^2} &:= \left\{\left(\frac{1}{2}, \frac{1}{6}\right) + k, \ \left(\frac{1}{2}, \frac{5}{6}\right) + k, \ \left(\frac{1}{6}, \frac{1}{2}\right) + k, \ \left(\frac{5}{6}, \frac{1}{2}\right) + k : \ k \in \mathbb{Z}^2\right\}, \\
\square_l^{\mathbb{R}^2} &:= \left\{2^{-l}\tau : \ \tau \in \square_0^{\mathbb{R}^2}\right\}, \quad \square_l^T := T \cap \square_l^{\mathbb{R}^2}, \\
\square_l^\Gamma &:= \left\{\kappa_m(\tau) : \ \tau \in \square_l^T, \ m = 1, 2, \ldots, m_\Gamma\right\}.
\end{aligned}$$

For each point $\tau \in \square_l^T$, there exist three uniquely defined neighbour points τ_1, τ_2, and τ_3 such that $\tau_1, \tau_2, \tau_3 \in \Delta_l^T$, that the triangle T_τ spanned by the three corners τ_1, τ_2, and τ_3 is of square measure $2^{-2l}/4$, and that τ is the centroid of T_τ. We arrive at the triangulation $\{T_\tau : \ \tau \in \square_l^T\}$ of T. Note that, for $l' > l$, the centroids in $\square_l^T$ are located at the boundaries of the smaller triangles $T_{\tau'}$ with $\tau' \in \square_{l'}^T$. Hence there is a one to one correspondence between the triangles T_τ over several levels and the centroids in $\cup_{l=0}^L \square_l^T$. Similarly to the triangulation over T, we define the triangulation $\{T_\tau : \ \tau \in \square_l^{\mathbb{R}^2}\}$ of $\mathbb{R}^2$. For Γ and a point $Q = \kappa_m(\tau) \in \square_l^\Gamma$, we set $\Gamma_Q := \{\kappa_m(\sigma) : \ \sigma \in T_\tau\}$ and arrive at the triangulation $\{\Gamma_Q : \ Q \in \square_l^\Gamma\}$. Further, we denote the level l of the points $Q \in \square_l^\Gamma$ by $l(Q)$. Notice that each partition triangle Γ_Q, $Q \in \square_l^\Gamma$, of the generation l splits into four subtriangles of the generation $l+1$. We call Γ_Q the father of the four subtriangles and, for $Q \in \square_l^\Gamma$, $l > 0$, we denote the father of Γ_Q by Γ_{Q^F}.

Beside the grids Δ_l^Γ we introduce the difference grids

$$\nabla_l^\Gamma \quad := \quad \begin{cases} \Delta_0^\Gamma & \text{if } l = -1 \\ \Delta_{l+1}^\Gamma \setminus \Delta_l^\Gamma & \text{if } l = 0, \ldots, L-1, \end{cases}$$

and obtain $\Delta_L^\Gamma = \bigcup_{l=-1}^{L-1} \nabla_l^\Gamma$. For $P \in \Delta_L^\Gamma$, we denote the unique level l for which $P \in \nabla_l^\Gamma$ by $l(P)$. Analogously to ∇_l^Γ, we define the difference grids and the point levels over T and $\mathbb{R}^2$ and get $\Delta_L^T = \bigcup_{l=-1}^{L-1} \nabla_l^T$ as well as $\Delta_L^{\mathbb{R}^2} = \bigcup_{l=-1}^{L-1} \nabla_l^{\mathbb{R}^2}$.

Finally, in accordance to the splitting $\triangle_l^T = {}^1\!\triangle_l^T \cup {}^2\!\triangle_l^T$, we introduce ${}^i\nabla_l^T = \nabla_l^T \cap {}^i\!\triangle_{l+1}^T$ for $i = 1,2$ and get $\nabla_l^T = {}^1\nabla_l^T \cup {}^2\nabla_l^T$ as well as ${}^2\nabla_l^T = {}^2\!\triangle_{l+1}^T$. Similarly, we define ${}^i\nabla_l^{\mathbb{R}^2}$ and ${}^i\nabla_l^\Gamma$.

2.3. The Piecewise Linear Functions

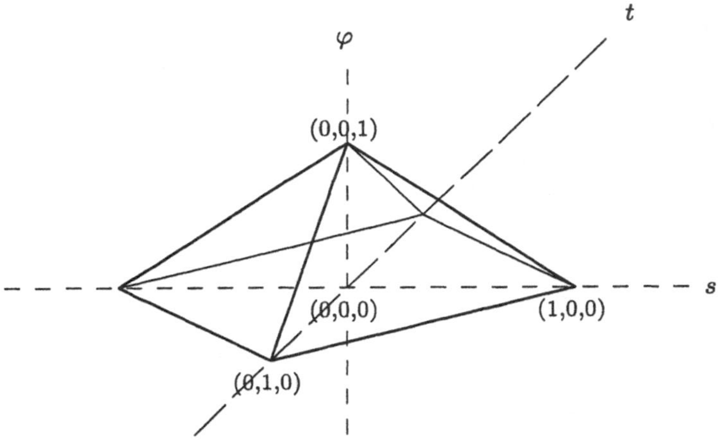

Figure 2: Hat function $(s,t) \mapsto {}^1\varphi(s,t)$.

To prepare the introduction of linear spaces, we first define two-dimensional hat functions for the grid $\triangle_0^{\mathbb{R}^2}$.

$$
\begin{aligned}
{}^1\varphi\,(s,t) &:= \max\left\{0, 1 - \max\{|s-t|, |s+t|\}\right\},\\
{}^2\varphi\,(s,t) &:= \max\left\{0, 1 - 2\max\{|s|, |t|\}\right\}.
\end{aligned}
$$

Clearly, the function ${}^1\varphi$ and the function ${}^2\varphi$ shifted to the point $(0.5, 0.5)$ are piecewise linear functions subordinate to the triangulation $\{T_\tau : \tau \in \square_0^{\mathbb{R}^2}\}$ (cf. the grid in Figure 1, the graph of ${}^1\varphi$ in Figure 2, and the graph of ${}^2\varphi$ shifted to the point $(0.5, 0.5)$ in Figure 3). Note that ${}^2\varphi$ can be obtained from ${}^1\varphi$ by rotation with angle $\pi/4$ and by dilation with factor $\sqrt{2}$, i.e.,

$$
{}^2\varphi(s,t) := {}^1\varphi(s+t, s-t).
$$

Now we get piecewise linear basis functions by dilating and shifting ${}^1\varphi$ and ${}^2\varphi$ to each grid point. More precisely, for each grid point on T, we set

$$
\varphi_\tau^l(\sigma) \;:=\; {}^i\varphi\left(2^l(\sigma-\tau)\right), \quad \tau \in {}^i\!\triangle_l^T.
$$

With the help of the parametrizations we introduce the piecewise linear (with respect to the parametrization) hat functions over Γ. For each grid point $P \in \Delta_l^\Gamma$, we set

$$(2.4)\quad \varphi_P^l(Q) := \begin{cases} \varphi_\tau^l(\sigma) & \text{if there exist } m, \tau, \sigma \text{ s.t. } Q = \kappa_m(\sigma),\ P = \kappa_m(\tau) \\ 0 & \text{else.} \end{cases}$$

Due to the assumptions on the parametrizations (cf. (2.2)) the basis functions are well defined. Note that if $P \in \Delta_l^\Gamma$ is in the interior of the parametrization patch Γ_m, then the support $\operatorname{supp}\varphi_P^l$ of φ_P^l is contained in Γ_m. If $P = \kappa_m(\tau) = \kappa_{m'}(\tau)$ is in the interior of a side, then $\operatorname{supp}\varphi_P^l \subseteq \Gamma_m \cup \Gamma_{m'}$. For corner points $P = \kappa_{m_1}(\tau_1) = \kappa_{m_2}(\tau_2) = \ldots = \kappa_{m_k}(\tau_k)$ of the triangular parametrization patches we get $\operatorname{supp}\varphi_P^l \subseteq \cup_{n=1}^k \Gamma_{m_n}$. We denote the span of the functions φ_P^l, $P \in \Delta_l^\Gamma$ by Lin_l^Γ. Obviously, this is the space of all continuous and piecewise linear functions over the partition $\{\Gamma_Q : Q \in \Box_l^\Gamma\}$ corresponding to the grid Δ_l^Γ, where linearity is understood with respect to the parametrization. The maximal space of all piecewise linear and continuous functions over the finest grid Δ_L^Γ is Lin_L^Γ.

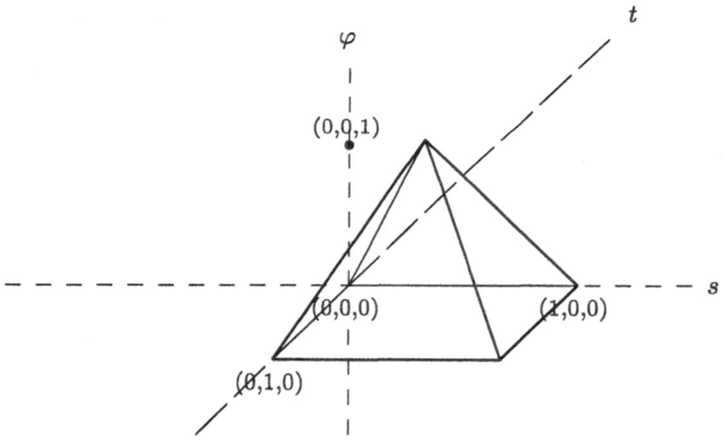

Figure 3: Hat function $(s,t) \mapsto {}^2\varphi(s-0.5, t-0.5)$.

2.4. The Wavelet Basis of the Piecewise Linear Space

Now we introduce a simple wavelet basis for the piecewise linear space. These functions have been considered first for the case of different grids in the plane $\mathbb{R}^2$ (cf. [18, 21, 19]) and are called three-point hierarchical basis functions. More

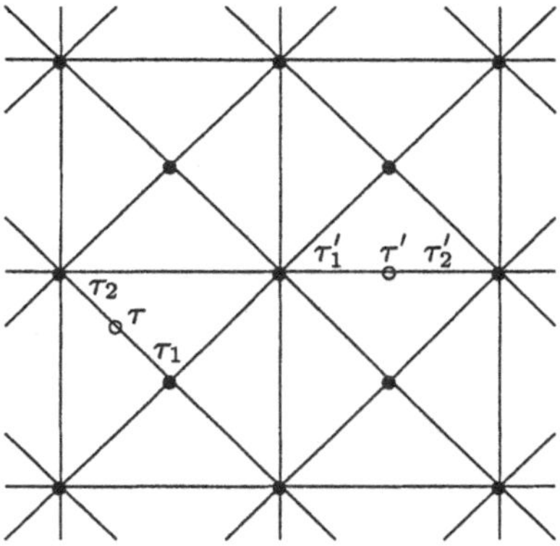

Figure 4: Neighbours τ_1 and τ_2.

precisely, for the plane and for any point $\tau \in \Delta_L^{\mathbb{R}^2}$, we set (cf. Figure 5 for the supports of such functions)

$$
(2.5)\quad \psi_\tau := \begin{cases} \varphi_\tau^0 & \text{if } \tau \in \nabla_{-1}^{\mathbb{R}^2} \\ \varphi_\tau^{l+1} - \frac{1}{2}\left\{\varphi_{\tau_1}^{l+1} + \varphi_{\tau_2}^{l+1}\right\} & \text{if } \tau \in {}^1\nabla_l^{\mathbb{R}^2}, \\ & \quad l = l(\tau) \in \{0, \ldots, L-1\} \\ \varphi_\tau^{l+1} - \frac{1}{4}\left\{\varphi_{\tau_1}^{l+1} + \varphi_{\tau_2}^{l+1}\right\} & \text{if } \tau \in {}^2\nabla_l^{\mathbb{R}^2}, \\ & \quad l = l(\tau) \in \{0, \ldots, L-1\}. \end{cases}
$$

Here τ_1 and τ_2 denote the uniquely defined neighbours of τ on $\Delta_{l+1}^{\mathbb{R}^2}$ (cf. Figure 4). Indeed any difference grid point $\tau \in {}^2\nabla_l^{\mathbb{R}^2} \subset \Delta_{l+1}^{\mathbb{R}^2}$ has exactly two neighbour points τ_1 and τ_2 at minimal distance which belong to $\Delta_l^{\mathbb{R}^2} \subset \Delta_{l+1}^{\mathbb{R}^2}$. Any difference grid point $\tau' \in {}^1\nabla_l^{\mathbb{R}^2} \subset \Delta_{l+1}^{\mathbb{R}^2}$ has exactly two neighbour points τ_1' and τ_2' at minimal distance which belong to ${}^1\Delta_l^{\mathbb{R}^2} \subset \Delta_{l+1}^{\mathbb{R}^2}$. The functions ψ_τ with $\tau \in \nabla_l^{\mathbb{R}^2}$, $l = 0, \ldots, L-1$ have two vanishing moments, i.e. they are orthogonal to all constant and linear functions.

The wavelet functions ψ_τ on the manifold Γ are slight modifications of (2.5). The definition is not very difficult. However, to motivate this definition, we shortly explain the construction:

- We start with the first parametrization patch Γ_1 and the definition of functions ψ_P such that $P \in \Delta_L^\Gamma \cap \Gamma_1$. First we restrict the functions ψ_τ from (2.5) to T. If these restrictions intersect the boundary of T, then we modify them adding restrictions of three-point basis functions $\psi_{\tau'}$ with τ' outside of T. The resulting basis functions $\psi_\tau^{\&}$ are restrictions of functions which are

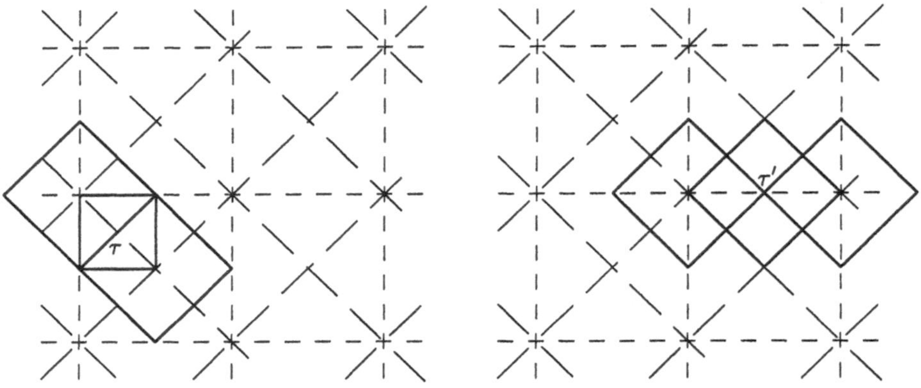

Figure 5: Supports of wavelets ψ_τ and $\psi_{\tau'}$.

symmetric (even) with respect to the boundary of T. For $P = \kappa_1(\tau)$, we take the composition $\psi_P = \psi_\tau^{\&} \circ \kappa_1^{-1}$ to arrive at functions over the parametrization patch Γ_1. To get continuous functions over Γ, we extend the ψ_P with $P \in \nabla_l^\Gamma \cap \Gamma_1$, $l = -1, 0, \ldots, L-1$ from Γ_1 to Γ such that the extensions are piecewise linear on the partition $\{\Gamma_Q : Q \in \square_{l+1}^\Gamma\}$ corresponding to the grid Δ_{l+1}^Γ and vanish at all grid points from $\Delta_{l+1}^\Gamma \setminus \Gamma_1$.

- Next we define the functions ψ_P with $P \in \Delta_L^\Gamma \cap \{\Gamma_2 \setminus \Gamma_1\}$. We start again with the restrictions of (2.5) to T. Since we have already basis functions over the boundary $\Gamma_1 \cap \Gamma_2$, we need basis functions on Γ_2 vanishing over $\Gamma_1 \cap \Gamma_2$, i.e. basis functions on T vanishing on the side S' for which $\kappa_2(S') = \Gamma_2 \cap \Gamma_1$. Therefore, we modify the functions on T such that they are restrictions of functions antisymmetric (odd) with respect to the side S' and symmetric (even) with respect to the sides S of T with $\kappa_2(S) \not\subset \Gamma_1$. Clearly all these functions vanish on S'. We take the composition with κ_2^{-1} to arrive at functions over the parametrization patch Γ_2 which vanish over $\Gamma_2 \cap \Gamma_1$. To get continuous functions, we extend these functions ψ_P with $P \in \nabla_l^\Gamma \cap \{\Gamma_2 \setminus \Gamma_1\}$, $l = -1, 0, \ldots, L-1$ from Γ_2 to Γ such that the extensions are piecewise linear on the partition $\{\Gamma_Q : Q \in \square_{l+1}^\Gamma\}$ corresponding to the grid Δ_{l+1}^Γ and vanish at all grid points from $\Delta_{l+1}^\Gamma \setminus \Gamma_2$.

- Analogously to the previous step, we define the basis functions ψ_P with the point P in the set $\Delta_L^\Gamma \cap \{\Gamma_3 \setminus (\Gamma_1 \cup \Gamma_2)\}$. Having done this, we construct the functions ψ_P with point P in the set $\Delta_L^\Gamma \cap \{\Gamma_4 \setminus (\Gamma_1 \cup \Gamma_2 \cup \Gamma_3)\}$ and so on. Finally, we define basis functions ψ_P with the point P in the set $\Delta_L^\Gamma \cap \{\Gamma_{m_\Gamma} \setminus \cup_{m=1}^{m_\Gamma - 1} \Gamma_m\}$.

For more details and the properties of the basis we refer to Sect. 3. The final definition of the three-point hierarchical wavelet functions over the manifold Γ is

$$(2.6)\quad \psi_P := \begin{cases} \varphi_P^0 & \text{if } P \in \nabla_{-1}^\Gamma \\ \varphi_P^{l+1} - \frac{1}{2}\left\{\varepsilon^{P,P_1}\varphi_{P_1}^{l+1} + \varepsilon^{P,P_2}\varphi_{P_2}^{l+1}\right\} & \text{if } P \in {}^1\nabla_l^\Gamma, \\ & \quad l \in \{0,\dots,L-1\} \\ \varphi_P^{l+1} - \frac{1}{4}\left\{\varepsilon^{P,P_1}\varphi_{P_1}^{l+1} + \varepsilon^{P,P_2}\varphi_{P_2}^{l+1}\right\} & \text{if } P \in {}^2\nabla_l^\Gamma, \\ & \quad l \in \{0,\dots,L-1\}, \end{cases}$$

where P_1 and P_2 are the uniquely defined neighbours on Δ_{l+1}^Γ of $P \in \nabla_l^\Gamma$, i.e. $P_1 = \kappa_m(\tau_1)$ and $P_2 = \kappa_m(\tau_2)$ if $P = \kappa_m(\tau)$ is the representation with the minimal $m \in \{1,\dots,m_\Gamma\}$ and if τ_1, τ_2 are the neighbours of τ. The coefficients $\varepsilon^{P,P'}$ are equal to one in almost all cases. Only if the point $P' = P_1, P_2$ is at the boundary of a parametrization patch, then a value $\varepsilon^{P,P'}$ different from one is needed. More precisely, the coefficients $\varepsilon^{P,P'}$ are given by (cf. Sect. 2.2 for the definition of ${}^i\Delta_L^\Gamma$)

(2.7)

$$\varepsilon^{P,P'} := \begin{cases} 1 & \text{if there is a parametrization patch } \Gamma_m \text{ such that } P \\ & \quad \text{and } P' \text{ belong to the interior of the triangle } \Gamma_m \\ & \text{or there exists a side } \Gamma_m \cap \Gamma_{m'} \text{ of a parametrization} \\ & \quad \text{patch such that } P \text{ and } P' \text{ belong to the interior of} \\ & \quad \text{the side } \Gamma_m \cap \Gamma_{m'} \\ 2 & \text{if there exists a side } \Gamma_m \cap \Gamma_{m'} \text{ of a parametrization} \\ & \quad \text{patch such that } m < m', \text{ that } P \text{ is an interior point} \\ & \quad \text{of } \Gamma_m, \text{ and that } P' \text{ belongs to the interior of the side} \\ & \quad \Gamma_m \cap \Gamma_{m'} \\ & \text{or } P' = \cap_{i=1}^k \Gamma_{m_i} \text{ is a corner of a parametrization patch,} \\ & \quad P' \in {}^2\Delta_0^\Gamma, \text{ the point } P \text{ is an interior point of a side} \\ & \quad \Gamma_{m_1} \cap \Gamma_{m_2}, \text{ and } m_1 < m_i,\ i = 2,\dots,k \\ 4 & \text{if } P' = \cap_{i=1}^k \Gamma_{m_i} \text{ is a corner of a parametrization patch,} \\ & \quad P' \in {}^1\Delta_0^\Gamma, \text{ the point } P \text{ is an interior point of a side} \\ & \quad \Gamma_{m_1} \cap \Gamma_{m_2}, \text{ and } m_1 < m_i,\ i = 2,\dots,k \\ & \text{or } P' = \cap_{i=1}^k \Gamma_{m_i} \text{ is a corner of a parametrization patch,} \\ & \quad P' \in {}^2\Delta_0^\Gamma, \text{ the point } P \text{ is an interior point of the face} \\ & \quad \Gamma_{m_1}, \text{ and } m_1 < m_i,\ i = 2,\dots,k \\ 0 & \text{else.} \end{cases}$$

Clearly, the support of ψ_P is contained in the union of all those Γ_m in which P or at least one of the neighbour points P_1 or P_2 is located. The basis $\{\psi_P : P \in \Delta_L^\Gamma\}$ spans the piecewise linear space Lin_L^Γ since the system is linearly independent (cf. (3.20)). Moreover, it represents a hierarchical basis, i.e.

$$\{\psi_P : P \in \Delta_L^\Gamma\} = \bigcup_{l=-1}^{L-1} \{\psi_P : P \in \nabla_l^\Gamma\},$$

$$Lin_0^\Gamma \subset Lin_1^\Gamma \subset \ldots \subset Lin_L^\Gamma,$$

$$Lin_{l'}^\Gamma = \operatorname{span} \bigcup_{l=-1}^{l'-1} \{\psi_P : P \in \nabla_l^\Gamma\}.$$

The function ψ_P with $P \in \nabla_l^\Gamma$, $l = 0, \ldots, L-1$ and with $\operatorname{supp}\psi_P$ contained in the interior of only one parametrization patch has two vanishing moments, i.e. it is orthogonal to the set of all functions that are constant or linear with respect to the parametrization. Orthogonality means here orthogonality with respect to the L^2 scalar product in the parameter domain.

3. The Properties of the Three-Point Hierarchical Basis

3.1. The Functions in the Plane

The three-point hierarchical basis is well analyzed in the case of a hierarchy of uniform triangulations over the plane (cf. [18, 21, 19]). The triangles of level l in this hierarchy are obtained by splitting the level $l-1$ triangles into four subtriangles. This splitting is realized by connecting the three midpoints of the three sides. Unfortunately, we are not able to prove Riesz stability for the corresponding three-point hierarchical wavelets over triangles and manifolds. The reason is that the grids, where three straight lines meet in each grid point, are not suitable for the symmetric extensions which we present after Lemma 1. Therefore, we define our basis over the triangulations $\{T_\tau : \tau \in \square_l^{\mathbb{R}^2}\}$ (cf. Figure 1). For these partitions, the triangles of level l are obtained from those of level $(l-1)$ by cutting each triangle along the lines connecting one midpoint of a side with the opposite corner and with the two other midpoints. Fortunately, the techniques of proof from e.g. [21] apply also to our situation. To describe the results we need some notation. To avoid ambiguities we write $\psi_\tau^{\mathbb{R}^2}$ for ψ_τ in this section. We define the level $l(\tau)$ of τ by $l(\tau) := l$ if τ in $\nabla_l^{\mathbb{R}^2}$. From now on C stands for a generic constant the value of which varies from instance to instance. For two expressions E_1 and E_2, we write $E_1 \sim E_2$ if there is a constant independent of the parameters involved in E_1 and E_2 such that $E_1/C \le E_2 \le C\, E_1$. We get

Lemma 1 *For $-\alpha_H < s < 1.5$, the basis $\{\psi_\tau^{\mathbb{R}^2} : \tau \in \cup_{L=0}^\infty \Delta_L^{\mathbb{R}^2}\}$ is a Riesz basis, i.e., for any vector of real numbers $(\xi_\tau)_\tau$ we get*

$$\left\| \sum_{\tau\in\cup_{L=0}^\infty \Delta_L^{\mathbb{R}^2}} \xi_\tau \psi_\tau^{\mathbb{R}^2} \right\|_{H^s(\mathbb{R}^2)} \sim \sqrt{\sum_{\tau\in\cup_{L=0}^\infty \Delta_L^{\mathbb{R}^2}} 2^{2l(\tau)(s-1)} |\xi_\tau|^2}. \tag{3.1}$$

The positive real constant α_H is greater or equal to $0.559\ldots$.

Proof. i) In this proof we shall use the technique of Stevenson [21]. The reader is supposed to be familiar with that paper. Following [21] we introduce the quadrature approximation of the L^2-scalar product and the norm

$$\begin{aligned}\langle u,v\rangle_{\triangle_l^{\mathbb{R}^2}} &:= 2^{-2l}\left\{\frac{2}{3}\sum_{\tau\in{}^1\triangle_l^{\mathbb{R}^2}} u(\tau)\overline{v(\tau)}+\frac{1}{3}\sum_{\tau\in{}^2\triangle_l^{\mathbb{R}^2}} u(\tau)\overline{v(\tau)}\right\},\\ \|u\|_{\triangle_l^{\mathbb{R}^2}} &:= \sqrt{\langle u,u\rangle_{\triangle_l^{\mathbb{R}^2}}}.\end{aligned}$$

With respect to this scalar product the basis $\{\varphi_\tau^l : \tau\in\triangle_l^{\mathbb{R}^2}\}$ is orthogonal, it is $\langle\cdot,\cdot\rangle_{\triangle_{l+1}^{\mathbb{R}^2}}$- biorthogonal to the basis $\{\varphi_\tau^{l+1} : \tau\in\triangle_l^{\mathbb{R}^2}\}$, and the wavelet functions can be represented as

$$\psi_\tau^{\mathbb{R}^2} = \varphi_\tau^{l+1} - \sum_{\tau'\in\triangle_l^{\mathbb{R}^2}} \frac{\langle\varphi_\tau^{l+1},\varphi_{\tau'}^l\rangle_{\triangle_{l+1}^{\mathbb{R}^2}}}{\langle\varphi_{\tau'}^{l+1},\varphi_{\tau'}^l\rangle_{\triangle_{l+1}^{\mathbb{R}^2}}}\varphi_{\tau'}^{l+1},\quad \tau\in\nabla_l^{\mathbb{R}^2}.$$

In other words, the wavelets $\psi_\tau^{\mathbb{R}^2}$, $\tau\in\nabla_l^{\mathbb{R}^2}$ are orthogonal to the space $Lin_l^{\mathbb{R}^2} = \operatorname{span}\{\varphi_\tau^l : \tau\in\triangle_l^{\mathbb{R}^2}\}$ with respect to the scalar product $\langle\cdot,\cdot\rangle_{\triangle_{l+1}^{\mathbb{R}^2}}$, i.e., they are prewavelets (semi-orthogonal wavelets) with respect to a non-standard scalar product.

We introduce the mappings $\tilde{m}_l : Lin_l^{\mathbb{R}^2}\longrightarrow Lin_l^{\mathbb{R}^2}$ and $\tilde{Y}_l : Lin_{l+1}^{\mathbb{R}^2}\longrightarrow Lin_l^{\mathbb{R}^2}$ by

$$\begin{aligned}\langle\tilde{m}_l u_l, v_l\rangle_{\triangle_l^{\mathbb{R}^2}} &= \langle u_l,v_l\rangle_{\triangle_{l+1}^{\mathbb{R}^2}},\quad u_l,v_l\in Lin_l^{\mathbb{R}^2},\\ \left\langle\tilde{Y}_l u_{l+1}, v_l\right\rangle_{\triangle_l^{\mathbb{R}^2}} &= \langle u_{l+1},v_l\rangle_{\triangle_{l+1}^{\mathbb{R}^2}},\quad u_{l+1}\in Lin_{l+1}^{\mathbb{R}^2},\ v_l\in Lin_l^{\mathbb{R}^2}.\end{aligned}$$

For a function $v_l\in Lin_l^{\mathbb{R}^2}$, we observe that $\langle v_l,v_l\rangle_{\triangle_l^{\mathbb{R}^2}} - \langle v_l,v_l\rangle_{\triangle_{l+1}^{\mathbb{R}^2}}$ is equivalent to $2^{-2l}\langle\nabla v_l,\nabla v_l\rangle_{L^2}$, i.e., $\|\nabla v_l\|^2\sim\langle 2^{2l}(I-\tilde{m}_l)v_l,v_l\rangle_{\triangle_l^{\mathbb{R}^2}}$. Thus we introduce the norms

$$\|v_l\|_{\triangle_l^{\mathbb{R}^2},s} := \left\|\left[I+2^{2l}(I-\tilde{m}_l)\right]^{s/2} v_l\right\|_{\triangle_l^{\mathbb{R}^2}} \tag{3.2}$$

which are equivalent to $\|v_l\|_{H^s(\mathbb{R}^2)}$. Then it is proved in [21] (cf. the paragraph before [21], Theorem 4.7) that the norm equivalences (3.1) hold for

$$-1+{}^2\log\left\{\frac{1}{2}\sup_{l=0,1,\dots}\left\|\tilde{Y}_l\tilde{m}_{l+1}^{-1}\right\|_{\triangle_l^{\mathbb{R}^2},-2\leftarrow\triangle_{l+1}^{\mathbb{R}^2},-2}\right\} < s < \frac{3}{2}. \tag{3.3}$$

Moreover, a simple modification of the derivation of (3.3) yields even the s-range

$$-1+{}^2\log\left\{\frac{1}{2}\sup_{l=l_0,l_0+1,\dots}\left\|\prod_{j=l+1}^{l+k}\tilde{Y}_j\tilde{m}_{j+1}^{-1}\right\|^{1/k}_{\triangle_{l+1}^{\mathbb{R}^2},-2\leftarrow\triangle_{l+k+1}^{\mathbb{R}^2},-2}\right\} < s < \frac{3}{2}, \tag{3.4}$$

where l_0 and k are arbitrarily fixed positive integers. All what left is to compute the lower bound of the s-range, i.e., to estimate $\|\prod \tilde{Y}_j \tilde{m}_{j+1}^{-1}\|^{1/k}$.

ii) Now we derive the standard representation of $\tilde{Y}_j$ and $\tilde{m}_j$ from the theory of wavelets (cf. [16]). We consider the $\langle\cdot,\cdot\rangle_{\triangle_{l+1}^{\mathbb{R}^2}}$-orthonormalized bases

$$\left\{{}^1\Phi_k^l := \sqrt{\frac{3}{2}}2^l\varphi_{2^{-l}k}^l \,:\, k\in\mathbb{Z}^2\right\}\cup\left\{{}^2\Phi_k^l := \sqrt{3}2^l\varphi_{(2^{-l-1},-2^{-l-1})+2^{-l}k}^l \,:\, k\in\mathbb{Z}^2\right\}$$

of the spaces $Lin_l^{\mathbb{R}^2}$ and represent the mappings $\tilde{m}_l$ and $\tilde{Y}_l$ as matrices with respect to these bases. As mentioned already by Stevenson, we get $\tilde{m}_l = p_l^* p_l$ and $\tilde{Y}_l = p_l^*$, where p_l is the matrix of the embedding operator $Lin_l^{\mathbb{R}^2} \longrightarrow Lin_{l+1}^{\mathbb{R}^2}$. Due to the refinement equations

$$\begin{aligned}
{}^1\Phi_{(0,0)}^0 &= \frac{1}{2}{}^1\Phi_{(0,0)}^1 + \frac{1}{4}\left\{{}^1\Phi_{(0,1)}^1 + {}^1\Phi_{(0,-1)}^1 + {}^1\Phi_{(1,0)}^1 + {}^1\Phi_{(-1,0)}^1\right\} + \\
&\qquad \frac{\sqrt{2}}{8}\left\{{}^2\Phi_{(0,0)}^1 + {}^2\Phi_{(-1,0)}^1 + {}^2\Phi_{(0,1)}^1 + {}^2\Phi_{(-1,1)}^1\right\}, \\
{}^2\Phi_{(0,0)}^0 &= \frac{\sqrt{2}}{2}{}^1\Phi_{(1,-1)}^1 + \frac{1}{4}\left\{{}^2\Phi_{(0,0)}^1 + {}^2\Phi_{(1,0)}^1 + {}^2\Phi_{(0,-1)}^1 + {}^2\Phi_{(1,-1)}^1\right\},
\end{aligned}$$

we get (cf. e.g. [16]), for the function $u_0 = \sum_{k\in\mathbb{Z}^2}[\xi_k^1\,{}^1\Phi_k^0 + \xi_k^2\,{}^2\Phi_k^0]$ embedded as $u_0 = \sum_{k\in\mathbb{Z}^2}[\eta_k^1\,{}^1\Phi_k^1 + \eta_k^2\,{}^2\Phi_k^1]$,

$$\begin{aligned}
\eta_k &= \sum_{k'\in\mathbb{Z}^2} h_{k-2k'}^T \xi_{k'}\,, \quad \eta_k := \begin{pmatrix} \eta_k^1 \\ \eta_k^2 \end{pmatrix}, \\
\xi_k &= \begin{pmatrix} \xi_k^1 \\ \xi_k^2 \end{pmatrix}, \quad h_k^T := \begin{pmatrix} h_k^{1,1} & h_k^{2,1} \\ h_k^{1,2} & h_k^{2,1} \end{pmatrix},
\end{aligned}$$

$$(3.5)\quad h_k^{i,j} := \begin{cases} \frac{1}{2} & \text{if } i=j=1,\ k=(0,0) \\ \frac{\sqrt{2}}{2} & \text{if } i=2,\ j=1,\ k=(1,-1) \\ \frac{1}{4} & \text{if } i=j=1,\ k\in\{(0,1),(0,-1),(1,0),(-1,0)\} \\ & \quad\text{or } i=j=2,\ k\in\{(0,0),(1,0),(0,-1),(1,-1)\} \\ \frac{\sqrt{2}}{8} & \text{if } i=1,\ j=2,\ k\in\{(0,0),(-1,0),(0,1),(-1,1)\}. \end{cases}$$

As usually in the theory of wavelets, we identify the coefficient vectors $(\xi_k)_{k\in\mathbb{Z}^2}$ and $(\eta_k)_{k\in\mathbb{Z}^2}$ with the generator functions $\xi(x,y) := \sum \xi_{(k_1,k_2)}e^{\mathrm{i}2\pi k_1 x}e^{\mathrm{i}2\pi k_2 y}$ and $\eta(x,y) := \sum \eta_{(k_1,k_2)}e^{\mathrm{i}2\pi k_1 x}e^{\mathrm{i}2\pi k_2 y}$, respectively. Then the l^2 spaces of coefficient vectors are isometric to the space of L^2 functions over $\mathbb{R}^2/\mathbb{Z}^2$, and (3.5) is equivalent to the equation $\eta(x,y) = h^T(x,y)\xi(2x,2y)$ with the matrix function

$$\begin{aligned}
h^T(x,y) &:= \sum_{(k_1,k_2)\in\mathbb{Z}^2} h_{(k_1,k_2)}^T e^{\mathrm{i}2\pi k_1 x}e^{\mathrm{i}2\pi k_2 y} \\
&= \begin{pmatrix} \frac{1}{2}\{1+\cos(2\pi x)+\cos(2\pi y)\} & \frac{\sqrt{2}}{2}e^{\mathrm{i}2\pi(x-y)} \\ \frac{\sqrt{2}}{2}e^{\mathrm{i}\pi(y-x)}\cos(\pi x)\cos(\pi y) & e^{\mathrm{i}\pi(x-y)}\cos(\pi x)\cos(\pi y) \end{pmatrix}.
\end{aligned}$$

In other words, the embedding operator $(\xi_k)_k \mapsto (\eta_k)_k = p_l(\xi_k)_k$ corresponds to the multiplication operator $\xi(x,y) \mapsto \eta(x,y) := h^T(x,y)\xi(2x,2y)$. We denote the adjoint matrix function of $(x,y) \mapsto h^T(x,y)$ by $(x,y) \mapsto \bar{h}(x,y)$. The formula $\tilde{m}_l = p_l^* p_l$ and an easy computation reveal that the operator $\tilde{m}_l$ acting in the space of generator functions is simply the operator of multiplication by the invertible matrix function

$$\begin{aligned} \tilde{m}(x,y) &:= \frac{1}{4}\sum_{i,j=0}^{1} \bar{h}\left(\frac{x}{2}+\frac{i}{2},\frac{y}{2}+\frac{j}{2}\right) h^T\left(\frac{x}{2}+\frac{i}{2},\frac{y}{2}+\frac{j}{2}\right) \\ &= \begin{pmatrix} \frac{5}{8}+\frac{1}{8}\{\cos(2\pi x)+\cos(2\pi y)\} & \frac{\sqrt{2}}{8}e^{\mathrm{i}\pi(x-y)}\cos(\pi x)\cos(\pi y) \\ \frac{\sqrt{2}}{8}e^{\mathrm{i}\pi(y-x)}\cos(\pi x)\cos(\pi y) & \frac{3}{4} \end{pmatrix}. \end{aligned}$$

We denote the self adjoint and non-negative matrix $I - \tilde{m}(x,y)$ by $a(x,y)$ and conclude that $\tilde{m}_{l+1}^{-1} p_l$ corresponds to

$$\xi(x,y) \mapsto \tilde{m}^{-1}(x,y)h^T(x,y)\xi(2x,2y).$$

The H^{-2} operator norm $\|\tilde{Y}_l \tilde{m}_{l+1}^{-1}\|$ is equal to the H^2 operator norm $\|\tilde{m}_{l+1}^{-1} p_l\|$, and, due to the norm definition in (3.2), the last is equal to the operator norm of the multiplication operator

(3.6)
$$\xi(x,y) \mapsto \left[I + 2^{2(l+1)}a(x,y)\right] \tilde{m}^{-1}(x,y)h^T(x,y)\left[I + 2^{2l}a(2x,2y)\right]^{-1}\xi(2x,2y)$$

acting in the L^2 space over $\mathbb{R}^2/\mathbb{Z}^2$. Thus, to compute the lower bound in (3.4), we have to estimate the norm of the operators (3.6) depending on l and the norm of their products, respectively.

iii) To estimate the norm of (3.6), we introduce the auxiliary operator Te^ϵ depending on a non-negative parameter ϵ by

(3.7)
$$Te^\epsilon \xi(x,y) \quad := \quad [\epsilon I + 4a(x,y)]\,\tilde{m}^{-1}(x,y)h^T(x,y)\,[\epsilon I + a(2x,2y)]^{-1}\,\xi(2x,2y)$$

and observe that the operator in (3.6) is Te^ϵ for $\epsilon = 2^{-2l}$. This 2^{-2l} can be made small by choosing l_0 large in (3.4). In what follows we shall derive an estimate for Te^0. We shall split Te^ϵ for $\epsilon = 2^{-2l}$ into the sum of three terms, and, using the bound for Te^0, we shall estimate each term separately.

Following the announced program, we observe

$$\begin{aligned} Te^0\xi(x,y) &= Ma(x,y)\xi(2x,2y), \\ Ma(x,y) &:= 4a(x,y)\tilde{m}^{-1}(x,y)h^T(x,y)a(2x,2y)^{-1}. \end{aligned} \tag{3.8}$$

The determinant $\det(a(x,y))$ of $a(x,y)$ has a zero only at $(x,y) = (0,0)$ and $\det(\tilde{m}(x,y))$ does not vanish at all. Moreover, we get $\det(a(x,y)) \sim x^2 + y^2$ for

$(x,y) \longrightarrow (0,0)$. Since $a(x,y)^{-1} = a(x,y)^A/\det(a(x,y))$ with

$$a(x,y)^A := \begin{pmatrix} a_{2,2}(x,y) & -a_{1,2}(x,y) \\ -a_{2,1}(x,y) & a_{1,1}(x,y) \end{pmatrix},$$

we arrive at

$$Ma(x,y) = \frac{4a(x,y)\tilde{m}(x,y)^A h^T(x,y)a(2x,2y)^A}{[\det(a(2x,2y))\det(\tilde{m}(x,y))]}.$$

A lengthy but trivial calculation shows that all entries of the matrix function $a(x,y)\tilde{m}(x,y)^A h^T(x,y)a(2x,2y)^A$ vanish together with their first derivatives at the points $(0,0)$, $(1/2,0)$, $(0,1/2)$, and $(1/2,1/2)$, where $\det(a(2x,2y))$ has its zeros. Hence, $Ma(x,y)$ is bounded over $\mathbb{R}^2/\mathbb{Z}^2$. Using the periodicity of the function $\xi \in L^2(\mathbb{R}^2/\mathbb{Z}^2)$, the norm of operator Te^0 can be estimated as follows.

$$\begin{aligned} &\|Ma(x,y)\xi(2x,2y)\|_{L^2}^2 \\ &= \int_0^1\int_0^1 \Big\langle [Ma^*Ma](x,y)\xi(2x,2y), \xi(2x,2y) \Big\rangle \,dxdy \\ &= \int_0^1\int_0^1 \left\langle \frac{1}{4}\sum_{i=0}^{1}\sum_{j=0}^{1}[Ma^*Ma]\left(\frac{x}{2}+\frac{i}{2},\frac{y}{2}+\frac{i}{2}\right)\xi(x,y),\xi(x,y) \right\rangle dxdy, \end{aligned}$$

$$\|Te^0\| \le \sup_{(x,y)} \left\| \frac{1}{4}\sum_{i,j=0}^{1}[Ma^*Ma]\left(\frac{x}{2}+\frac{i}{2},\frac{y}{2}+\frac{j}{2}\right)\right\|^{1/2}. \tag{3.9}$$

The matrix norm on the right-hand side of the last equation is the l^2 matrix norm, i.e., the operator norm in the two-dimensional Euclidean space. A numerical evaluation of (3.9) yields $\|Te^0\| \le 10.37\dots$.

Next we fix a small positive number δ and introduce the cut off function $\chi(x,y)$ on $\mathbb{R}^2/\mathbb{Z}^2$ which is equal to one for $|x|,|y| \le \delta$ and zero else. Using this function, we split

$$\begin{aligned} Te^\epsilon &= \sum_{i=1}^{3} Te_i^\epsilon, \qquad (3.10)\\ Te_1^\epsilon\xi(x,y) &:= \Big(1-\chi(2x,2y)\Big)\,Te^\epsilon\xi(x,y),\\ Te_2^\epsilon\xi(x,y) &:= \chi(2x,2y)\left[4a(x,y)\tilde{m}^{-1}(x,y)h^T(x,y)a(2x,2y)^{-1}\right]\\ &\qquad a(2x,2y)\left[\epsilon I + a(2x,2y)\right]^{-1}\xi(2x,2y),\\ Te_3^\epsilon\xi(x,y) &:= \chi(2x,2y)\left[\tilde{m}^{-1}(x,y)h^T(x,y)\right]\epsilon I\left[\epsilon I + a(2x,2y)\right]^{-1}\xi(2x,2y). \end{aligned}$$

Since $\chi^2 = \chi$ and since

$$\begin{aligned} Te_i^\epsilon[\chi\xi](x,y) &= \chi(2x,2y)Te_i^\epsilon\xi(x,y),\\ \|\xi(x,y)\|^2 &= \|\chi(2x,2y)\xi(x,y)\|^2 + \|(1-\chi(2x,2y))\xi(x,y)\|^2,\\ \|\xi(x,y)\|^2 &= \|\chi(x,y)\xi(x,y)\|^2 + \|(1-\chi(x,y))\xi(x,y)\|^2, \end{aligned}$$

we conclude

$$\begin{aligned} \|Te^\epsilon\xi\|^2 &= \|\chi(2\cdot,2\cdot)Te^\epsilon\xi\|^2 + \|[1-\chi(2\cdot,2\cdot)]Te^\epsilon\xi\|^2 \\ &= \|Te_1^\epsilon[\chi\xi]\|^2 + \|[Te_2^\epsilon + Te_3^\epsilon][(1-\chi)\xi]\|^2 \\ &\le \max\left\{\|Te_1^\epsilon\|, \|Te_2^\epsilon + Te_3^\epsilon\|\right\}^2 \left\{\|\chi\xi\|^2 + \|(1-\chi)\xi\|^2\right\}, \end{aligned}$$

$$\|Te^\epsilon\| \le \max\left\{\|Te_1^\epsilon\|, \|Te_2^\epsilon\| + \|Te_3^\epsilon\|\right\}. \tag{3.11}$$

The matrices $a(2x, 2y)$ are invertible on the support of $(1 - \chi(2x, 2y))$ and the inverses are uniformly bounded. Hence,

$$\begin{aligned} \|Te_1^\epsilon - Te_1^0\| &\le C\,\epsilon, \\ \|Te_1^\epsilon\| &\le \|Te_1^0\| + C\,\epsilon \le 10.37\ldots + C\,\epsilon. \end{aligned} \tag{3.12}$$

Clearly, the last constant C depends on the δ from the definition of the cut off function χ. On the other hand, the adjoint operator $[Te_2^\epsilon]^*$ is given by

$$[Te_2^\epsilon]^*\xi(x,y) = \chi(x,y)a(x,y)\left[\epsilon I + a(x,y)\right]^{-1}[Te^0]^*\xi(x,y).$$

From this and from the matrix inequality $a(x,y)\left[\epsilon I + a(x,y)\right]^{-1} \le I$ we obtain $\|[Te_2^\epsilon]^*\| \le \|[Te^0]^*\|$ and

$$\|Te_2^\epsilon\| \le \|Te^0\| \le 10.37\ldots\ . \tag{3.13}$$

Now we turn to $\|Te_3^\epsilon\|$. The non-negative self adjoint matrix $a(x,y)$ can be represented as $a(x,y) = \lambda(x,y)q(x,y) + \mu(x,y)o(x,y)$, where $\lambda(x,y)$ and $\mu(x,y)$ are the eigenvalues of $a(x,y)$. The matrices $q(x,y)$ and $o(x,y) = I - q(x,y)$ are the orthogonal projections onto the spaces of eigenvectors. In particular, we get

$$\begin{aligned} a(0,0) &= \begin{pmatrix} \frac{1}{8} & -\frac{\sqrt{2}}{8} \\ -\frac{\sqrt{2}}{8} & \frac{1}{4} \end{pmatrix}, \quad q(0,0) = \begin{pmatrix} \frac{1}{3} & -\frac{\sqrt{2}}{3} \\ -\frac{\sqrt{2}}{3} & \frac{2}{3} \end{pmatrix}, \\ o(0,0) &= \begin{pmatrix} \frac{2}{3} & \frac{\sqrt{2}}{3} \\ \frac{\sqrt{2}}{3} & \frac{1}{3} \end{pmatrix}, \\ \lambda(0,0) &= \frac{3}{8}, \quad \mu(0,0) = 0. \end{aligned}$$

Since $\lambda(x,y)$ is separated from 0 by a positive constant, we get

$$\begin{aligned} \epsilon I\left[\epsilon + a(2x,2y)\right]^{-1} &= \frac{\epsilon}{[\epsilon + \lambda(2x,2y)]}q(2x,2y) + \frac{\epsilon}{[\epsilon + \mu(2x,2y)]}o(2x,2y), \\ \left\|\frac{\epsilon}{[\epsilon + \lambda(2x,2y)]}q(2x,2y)\right\| &\le C\epsilon. \end{aligned}$$

Consequently, we arrive at

$$\begin{aligned} \|Te_3^\varepsilon - Te_4^\varepsilon\| &\leq C\epsilon, \\ Te_4^\varepsilon \xi(x,y) &:= \chi(2x,2y)\left[\tilde{m}^{-1}(x,y)h^T(x,y)\right] \frac{\epsilon}{[\epsilon + \mu(2x,2y)]} o(2x,2y)\xi(2x,2y). \end{aligned}$$

In other words, the norm $\|Te_3^\varepsilon\|$ is less than $C\epsilon$ plus the norm $\|Te_5\|$ of the operator

$$Te_5 : \xi(x,y) \mapsto \chi(2x,2y)\left[\tilde{m}^{-1}h^T\right](x,y)\, o(2x,2y)\xi(2x,2y),$$

and we even get $\|Te_3^\varepsilon\| \leq C\epsilon + C\delta + \|Te_6\|$ with Te_6 defined by

(3.14)

$$\xi(x,y) \mapsto \begin{cases} \left[\tilde{m}^{-1}h^T\right](0,0)\ o(0,0)\xi(2x,2y) & \text{if } |2x| \leq \delta \text{ and } |2y| \leq \delta \\ \left[\tilde{m}^{-1}h^T\right]\left(\frac{1}{2},0\right) o(0,0)\xi(2x,2y) & \text{if } |2x-1| \leq \delta \text{ and } |2y| \leq \delta \\ \left[\tilde{m}^{-1}h^T\right]\left(0,\frac{1}{2}\right) o(0,0)\xi(2x,2y) & \text{if } |2x| \leq \delta \text{ and } |2y-1| \leq \delta \\ \left[\tilde{m}^{-1}h^T\right]\left(\frac{1}{2},\frac{1}{2}\right) o(0,0)\xi(2x,2y) & \text{if } |2x-1| \leq \delta \\ & \text{and } |2y-1| \leq \delta \\ 0 & \text{else}. \end{cases}$$

Since we have

$$\begin{aligned} h^T(0,0)\,o(0,0) &= 2\,o(0,0), & h^T\left(\tfrac{1}{2},0\right) o(0,0) &= 0, \\ h^T\left(0,\tfrac{1}{2}\right) o(0,0) &= 0, & h^T\left(\tfrac{1}{2},\tfrac{1}{2}\right) o(0,0) &= 0, \\ \tilde{m}^{-1}(0,0)\,o(0,0) &= o(0,0), \end{aligned} \tag{3.15}$$

we conclude

$$Te_6 : \xi(x,y) \mapsto \begin{cases} 2\,o(0,0)\xi(2x,2y) & \text{if } |2x| \leq \delta \text{ and } |2y| \leq \delta \\ 0 & \text{else} \end{cases}$$

and $\|Te_3^\varepsilon\| \leq 2 + C\epsilon + C\delta$. This and the estimates (3.11), (3.12), and (3.13), lead us to $\|Te^\epsilon\| \leq 12.37\ldots + C\epsilon + C\delta$. Choosing δ small and choosing ϵ small in comparison to δ, we get $\|Te^\epsilon\| \leq 12.37\ldots$. Using (3.4) with $k=1$ and sufficiently large l_0, the Riesz property (3.1) follows for "$1.62\ldots < s < 1.5$".

iv) To improve the lower bound of the Sobolev range, we apply (3.4) with larger k. Analogously to (3.7) and (3.8), we define

$$\begin{aligned} Te^\epsilon\xi(x,y) &:= \left[\epsilon I + 4^k a(x,y)\right] \prod_{i=0}^{k-1} \left\{ \left[\tilde{m}^{-1}h^T\right](2^i x, 2^i y) \right\} \\ &\quad \times \left[\epsilon I + a(2^k x, 2^k y)\right]^{-1} \xi(2^k x, 2^k y) \\ Ma(x,y) &:= 4^k a(x,y) \prod_{i=0}^{k-1} \left\{ \left[\tilde{m}^{-1}h^T\right](2^i x, 2^i y) \right\} a(2^k x, 2^k y)^{-1}. \end{aligned}$$

For $k = 10$, numerical computations lead us to the estimate (compare (3.9))

$$\sup_{(x,y)} \left\| \frac{1}{4^k} \sum_{j,j'=0}^{2^k-1} [Ma^* Ma] \left(\frac{x}{2^k} + \frac{j}{2^k}, \frac{y}{2^k} + \frac{j'}{2^k} \right) \right\|^{1/2} \leq 20661.3\ldots .$$

Analogously to (3.14), we define Te_6 by

$$(3.16)\ \xi(x,y) \mapsto \begin{cases} \prod_{i=0}^{k-1} \left\{ [\tilde{m}^{-1} h^T] \left(2^i \frac{j}{2^k}, 2^i \frac{j'}{2^k} \right) \right\} o(0,0) \xi(2^k x, 2^k y) \\ \qquad \text{if } |2^k x - j| \leq \delta \text{ and } |2^k y - j'| \leq \delta \\ 0 \qquad \text{else .} \end{cases}$$

In view of (3.15) we conclude

$$Te_6 : \xi(x,y) \mapsto \begin{cases} 2^k o(0,0) \xi(2^k x, 2^k y) & \text{if } |2^k x| \leq \delta \\ & \text{and } |2^k y| \leq \delta \\ 0 & \text{else ,} \end{cases}$$

and the arguments from part iii) of the present proof lead us to the estimate $\|Te^\epsilon\| \leq 20661.3\ldots + 2^{10} + C\epsilon + C\delta$. Choosing small values ϵ and δ, we get $\|Te^\epsilon\| \leq 21685.3\ldots$, and (3.4) implies the Riesz property (3.1) for $-0.559\ldots < s < 1.5$. ∎

3.2. The Functions over Triangles

In the construction of Sect. 2.4 we need basis functions which admit symmetric (even) or antisymmetric (odd) extensions with respect to the boundary of T. To construct such functions, we shall extend the piecewise linear functions on T by symmetry mappings to periodic functions over the plane $\mathbb{R}^2$. More precisely, we shall suppose that a subset of the three sides of T is given through which the functions should possess an even extension. Through the rest of the sides there should exist odd extensions. In accordance to these symmetry properties, we shall define an extension procedure from functions over T to periodic functions over $\mathbb{R}^2$. For the periodic extensions, however, there exists a natural basis. Restricting this basis to the triangle, we shall arrive at our basis over T.

In view of the assumptions in Sect. 2.1, the two shorter sides $\{(s,s) : 0 \leq s \leq 0.5\}$ and $\{(s,1-s) : 0.5 \leq s \leq 1\}$ simultaneously belong to the fixed subset of sides or not. For the sake of definiteness, we suppose the only side with odd extension is the lower side $\{(s,0) : 0 \leq s \leq 1\}$. To prepare the definition of the extension, we introduce the points (cf. Figure 6)

$$\begin{array}{llllll} P := (0,0), & U := (1,0), & Z := (0.5,0.5), \\ W := (0,1), & X := (1,1), & Q := (0,-1), \\ R := (1,-1), & S := (2,-1), & Y := (2,1), \\ & V := (2,0). & \end{array}$$

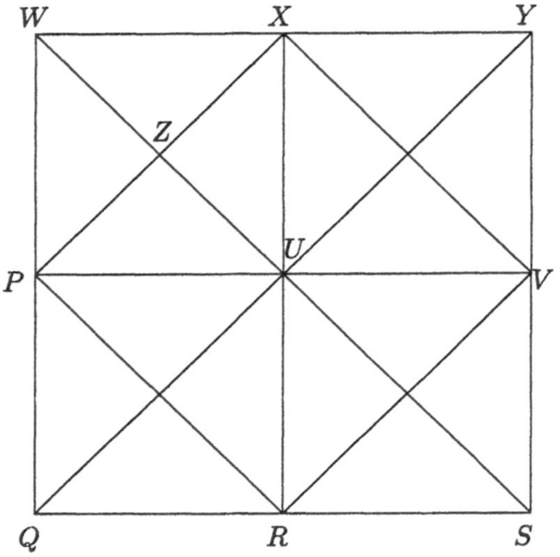

Figure 6: Torus $\mathbb{T}$.

Clearly, a piecewise linear function u_L on T admits a continuous extension through the boundary if and only if u_L vanishes on the side of odd extension. If a function u_L vanishing on $\{(s,0) : 0 \le s \le 1\}$ is given, then we can extend u_L to triangle PZW by symmetry with respect to the line through P and Z, i.e. $v_L(s,t) := u_L(t,s)$. The extended function on triangle PUW will be denoted by v_L. We can extend v_L to triangle WUX as a function symmetric with respect to the line through W and U by $v_L(s,t) := v_L(1-t, 1-s)$. Similarly, we extend v_L to the square $QRUP$ as a function antisymmetric with respect to the line through T and U by $v_L(s,t) := -v_L(s,-t)$. Again we extend v_L to the rectangle $RSYX$ as a function antisymmetric with respect to the line through R and X by $v_L(s,t) := -v_L(2-s,t)$. In other words, the function u_L is extended to a continuous piecewise linear function v_L on the square $QSYW$. This function extends to a function which is 2-periodic with respect to both variables, and we denote the periodic extension v_L by u_L^{ext}.

Let us consider the periodic functions more carefully. Periodicity of a piecewise linear function w_L means that w_L satisfies

$$w_L(s,t) \quad = \quad w_L(s+2k, t+2k'), \quad (k,k') \in \mathbb{Z}^2.$$

The periodic functions are functions defined on the torus, i.e., on the quotient space

$$\mathbb{T} \quad := \quad \mathbb{R}^2 / \left\{(2k, 2k') : (k,k') \in \mathbb{Z}^2\right\}.$$

We denote the space of periodic linear functions by $Lin_L^{\mathbb{T}}$. To get periodic basis functions, we take periodizations ψ_τ^{per} of $\psi_\tau^{\mathbb{R}^2}$ defined by

$$\psi_\tau^{per}(s,t) \quad := \quad \sum_{(k,k')\in\mathbb{Z}^2} \psi_\tau^{\mathbb{R}^2}(s+2k, t+2k') \quad = \quad \sum_{(k,k')\in\mathbb{Z}^2} \psi_{\tau+(2k,2k')}(s,t).$$

If we define the grid $\triangle_L^{\mathbb{T}}$ by

$$\triangle_L^{\mathbb{T}} := \left\{ (s,t) \in \triangle_L^{\mathbb{R}^2} : 0 \leq s,t < 2 \right\},$$

then $\{\psi_\tau^{per} : \tau \in \triangle_L^{\mathbb{T}}\}$ is a finite system of basis functions of $Lin_L^{\mathbb{T}}$. It is well known that Lemma 1 remains true for periodic functions and for the Sobolev spaces over the torus, i.e., for $-0.559\ldots < s < 1.5$ and for all vectors of coefficients ξ_τ,

$$(3.17) \qquad \left\| \sum_{\tau\in\cup_{L=0}^\infty \triangle_L^{\mathbb{T}}} \xi_\tau \psi_\tau^{per} \right\|_{H^s(\mathbb{T})} \quad \sim \quad \sqrt{\sum_{\tau\in\cup_{L=0}^\infty \triangle_L^{\mathbb{T}}} 2^{2l(\tau)(s-1)} |\xi_\tau|^2}.$$

On the other hand, the extension $v_L = u_L^{ext}$ of a linear function u_L on triangle T belongs to the subspace

$$Lin_L^{Sym} := \left\{ w_L \in Lin_L^{\mathbb{T}} : [w_L|_T]^{ext} = w_L \right\}$$

which is determined by the properties of symmetry included in the extension procedure $[w_L|_T] \mapsto [w_L|_T]^{ext}$. For a point $\tau \in \triangle_L^{\mathbb{T}} \setminus \{(s,0) : 0 \leq s \leq 1\}$, we denote by $\tau_1^\tau, \ldots, \tau_k^\tau$ those points of $\triangle_L^{\mathbb{T}}$ for which the function value $[u_L]^{ext}(\tau_i^\tau)$ is set to $\pm u_L(\tau)$ in the extension procedure $u_L \mapsto [u_L]^{ext}$. We define $\lambda_i \in \{1,-1\}$ by $[u_L]^{ext}(\tau_i^\tau) = \lambda_i u_L(\tau)$. Clearly, the points τ_i^τ are obtained by the symmetric reflections mapping the triangle T to the subtriangles of the quadrangle $QSYW$. The number of these points is $k = 16$ if τ is an interior point of T, $k = 8$ if τ is on a side of T, and $k = 4$ if τ is the corner Z. Now a function w_L belongs to Lin_L^{Sym}, if and only if, $w_L(\tau) = \lambda_i w_L(\tau_i^\tau)$, $i = 1, \ldots, k$. Obviously, the set of functions $\sum_{i=1}^k \lambda_i [\varphi_{\tau_i^\tau}^L]^{per}$ with $\tau \in \triangle_L^{\mathbb{T}} \setminus \{(s,0) : 0 \leq s \leq 1\}$ forms a basis of Lin_L^{Sym} and the cardinality of $\triangle_L^T \setminus \{(s,0) : 0 \leq s \leq 1\}$ is the dimension of Lin_L^{Sym}. Another basis is formed by $\sum_{i=1}^k \lambda_i \psi_{\tau_i^\tau}^{per}$ with $\tau \in \triangle_L^T \setminus \{(s,0) : 0 \leq s \leq 1\}$. Indeed this system of functions has the right cardinality, all its elements belong to the space Lin_L^{Sym}, and they are linearly independent since the functions ψ_τ^{per}, $\tau \in \triangle_L^T \setminus \{(s,0) : 0 \leq s \leq 1\}$ are linearly independent. We introduce the functions

$$\psi_\tau^{ext} \quad := \quad \sum_{i=1}^k \lambda_i \psi_{\tau_i^\tau}^{per}, \quad \psi_\tau^T := \psi_\tau^{ext}|_T, \quad \tau \in \triangle_L^T \setminus \{(s,0) : 0 \leq s \leq 1\}$$

and obtain $w_L = \sum_{\tau\in\Delta_L^T\setminus\{(s,0):\,0\le s\le 1\}} \xi_\tau \psi_\tau^{ext}$. Applying this to the extension $w_L = [u_L]^{ext}$ of a function u_L on T, we arrive at $u_L = \sum_{\tau\in\Delta_L^T\setminus\{(s,0):\,0\le s\le 1\}} \xi_\tau \psi_\tau^T$. It turns out that $\{\psi_\tau^T : \tau \in \Delta_L^T\}$ is a basis of the space of piecewise linear functions over T vanishing over the side $\{(s,0) : 0 \le s \le 1\}$. Using $\|u_L\|_{H^s(T)} \sim \|[u_L]^{ext}\|_{H^s(\mathbb{T})}$, the Riesz property (3.17) implies

$$\left\| \sum_{\tau\in\cup_{L=0}^\infty \Delta_L^T\setminus\{(s,0):\,0\le s\le 1\}} \xi_\tau \psi_\tau^T \right\|_{H^s(T)} \sim \sqrt{\sum_{\tau\in\cup_{L=0}^\infty \Delta_L^T\setminus\{(s,0):\,0\le s\le 1\}} 2^{2l(\tau)(s-1)}|\xi_\tau|^2} \tag{3.18}$$

for $-0.559\ldots < s < 1.5$. We note that, for $\tau \in \Delta_L^T \setminus \{(s,0) : 0 \le s \le 1\}$,

(3.19)

$$\psi_\tau^T := \begin{cases} \varphi_\tau^0|_T & \text{if } \tau \in \Delta_L^T \cap \nabla_{-1}^T \\ \varphi_\tau^{l+1}|_T - \frac{1}{2}\left\{\varepsilon^{\tau,\tau_1}\varphi_{\tau_1}^{l+1}|_T + \varepsilon^{\tau,\tau_2}\varphi_{\tau_2}^{l+1}|_T\right\} & \text{if } \tau \in \Delta_L^T \cap {}^1\nabla_l^T, \\ & \quad l = 0,\ldots,L-1 \\ \varphi_\tau^{l+1}|_T - \frac{1}{4}\left\{\varepsilon^{\tau,\tau_1}\varphi_{\tau_1}^{l+1}|_T + \varepsilon^{\tau,\tau_2}\varphi_{\tau_2}^{l+1}|_T\right\} & \text{if } \tau \in \Delta_L^T \cap {}^2\nabla_l^T, \\ & \quad l = 0,\ldots,L-1, \end{cases}$$

$$\varepsilon^{\tau,\tau'} := \begin{cases} 1 \ \text{if} & \tau \text{ and } \tau' \text{ belong to the interior of } T \\ & \text{or there exists a side of } T \text{ such that } \tau \text{ and } \tau' \\ & \text{belong to the interior of this side} \\ 2 \ \text{if} & \tau \text{ is an interior point of } T \text{ and } \tau' \text{ belongs to} \\ & \text{a side of } T \\ & \text{or } \tau' = Z \text{ and } \tau \text{ is on a side of } T \\ 4 \ \text{if} & \tau' = Z \text{ and } \tau \text{ is an interior point of } T \\ 0 \ \text{else.} & \end{cases}$$

With ψ_τ^T we have constructed a three-point wavelet basis for the space of linear functions on T vanishing on $\{(s,0) : 0 \le s \le 1\}$. Completely analogously, we can construct a basis for the linear functions on T vanishing on three, two or no sides. These functions are the basis ingredients for the wavelet basis on the manifold.

3.3. The Functions over Manifolds

As indicated in Sect. 2.4, the three-point hierarchical basis of (2.6) is constructed as follows: We start with functions ψ_P such that $P \in \Delta_L^\Gamma \cap \Gamma_1$. We just take the basis $\{\psi_\tau^T\}$ on T with no zero condition for boundary sides. For $P = \kappa_1(\tau)$, we take the composition $\psi_P = \psi_\tau^T \circ \kappa_1^{-1}$ to get functions over the parametrization patch Γ_1. To get continuous piecewise linear functions, we extend these functions ψ_P with $P \in \nabla_l^\Gamma \cap \Gamma_1 \subset \Delta_{l+1}^\Gamma \cap \Gamma_1$ from Γ_1 to Γ such that the extension is piecewise linear on the partition $\{\Gamma_Q : Q \in \square_{l+1}^\Gamma\}$ corresponding to the grid Δ_{l+1}^Γ

and vanishes at all grid points from $\Delta^\Gamma_{l+1} \setminus \Gamma_1$. This simply means that, if $\psi^T_\tau = \varphi^{l+1}_\tau - \frac{1}{2}\{\varepsilon^{\tau,\tau_1}\varphi^{l+1}_{\tau_1} + \varepsilon^{\tau,\tau_2}\varphi^{l+1}_{\tau_2}\}$ resp. $\psi^T_\tau = \varphi^{l+1}_\tau - \frac{1}{4}\{\varepsilon^{\tau,\tau_1}\varphi^{l+1}_{\tau_1} + \varepsilon^{\tau,\tau_2}\varphi^{l+1}_{\tau_2}\}$, then $\psi_P = \varphi^{l+1}_P - \frac{1}{2}\{\varepsilon^{\tau,\tau_1}\varphi^{l+1}_{P_1} + \varepsilon^{\tau,\tau_2}\varphi^{l+1}_{P_2}\}$ resp. $\psi_P = \varphi^{l+1}_P - \frac{1}{4}\{\varepsilon^{\tau,\tau_1}\varphi^{l+1}_{P_1} + \varepsilon^{\tau,\tau_2}\varphi^{l+1}_{P_2}\}$, where φ^{l+1}_P and $\varphi^{l+1}_{P_i}$ are the continuous hat functions introduced in Sect. 2.3.

Next we define the functions ψ_P for $P \in \Delta^\Gamma_L \cap \Gamma_2 \setminus \Gamma_1$. The patch Γ_2 has one or no common side with Γ_1. We take the basis $\{\psi^T_\tau\}$ on T which vanishes on those sides (one ore maybe no side) which are mapped by κ_2 into a side common with Γ_1. Again we take the composition with κ_2^{-1} to get functions over the parametrization patch Γ_2 which vanish over $\Gamma_2 \cap \Gamma_1$. To get continuous piecewise linear functions, we extend these functions ψ_P with $P \in \nabla^\Gamma_l \cap \Gamma_2 \setminus \Gamma_1 \subset \Delta^\Gamma_{l+1} \cap \Gamma_2$ from Γ_2 to Γ such that the extension is piecewise linear on the partition corresponding to the grid Δ^Γ_{l+1} and vanishes at all grid points from $\Delta^\Gamma_{l+1} \setminus \Gamma_2$. In other words, if $\psi^T_\tau = \varphi^{l+1}_\tau - \frac{1}{2}\{\varepsilon^{\tau,\tau_1}\varphi^{l+1}_{\tau_1} + \varepsilon^{\tau,\tau_2}\varphi^{l+1}_{\tau_2}\}$ resp. $\psi^T_\tau = \varphi^{l+1}_\tau - \frac{1}{4}\{\varepsilon^{\tau,\tau_1}\varphi^{l+1}_{\tau_1} + \varepsilon^{\tau,\tau_2}\varphi^{l+1}_{\tau_2}\}$, then $\psi_P = \varphi^{l+1}_P - \frac{1}{2}\{\varepsilon^{\tau,\tau_1}\varphi^{l+1}_{P_1} + \varepsilon^{\tau,\tau_2}\varphi^{l+1}_{P_2}\}$ resp. $\psi_P = \varphi^{l+1}_P - \frac{1}{4}\{\varepsilon^{\tau,\tau_1}\varphi^{l+1}_{P_1} + \varepsilon^{\tau,\tau_2}\varphi^{l+1}_{P_2}\}$, where φ^{l+1}_P and $\varphi^{l+1}_{P_i}$ are the continuous hat functions introduced in Sect. 2.3.

Analogously to the previous step, we define the functions ψ_P for $P \in \Delta^\Gamma_L \cap \Gamma_3 \setminus \{\Gamma_1 \cup \Gamma_2\}$ which vanish over $[\cup^2_{m=1}\Gamma_m] \cap \Gamma_3$. Then we construct the functions ψ_P with $P \in \Delta^\Gamma_L \cap \Gamma_4 \setminus \{\Gamma_1 \cup \Gamma_2 \cup \Gamma_3\}$ vanishing over $[\cup^3_{m=1}\Gamma_m] \cap \Gamma_4$ and so on. Finally, we define ψ_P with $P \in \Delta^\Gamma_L \cap \Gamma_{m_\Gamma} \setminus \cup^{m_\Gamma - 1}_{m=1}\Gamma_m$ vanishing over the boundary of Γ_{m_Γ}. We arrive at the basis of (2.6). If the level $l(P)$ of P is defined by $l(P) := l$ for P in ∇^Γ_l, then we get

Theorem 1 *i) For $-0.5 < s < 1.5$, the basis $\{\psi_P : P \in \cup^\infty_{L=0}\Delta^\Gamma_L\}$ is a Riesz basis, i.e., for any vector of real numbers $(\xi_P)_P$, we get*

$$\left\| \sum_{P \in \Delta^\Gamma_L} \xi_P \psi_P \right\|_{H^s(\Gamma)} \sim \sqrt{\sum_{P \in \Delta^\Gamma_L} 2^{2l(P)(s-1)} |\xi_P|^2}. \tag{3.20}$$

ii) For the Sobolev space orders $s \le t \le 2$, $s < 1.5$, the functions from Lin^Γ_L fulfill the approximation property (Jackson type theorem)

$$\inf_{u_L \in Lin^\Gamma_L} \|u - u_L\|_{H^s(\Gamma)} \le C 2^{-L(t-s)} \|u\|_{H^t(\Gamma)}. \tag{3.21}$$

iii) For the interpolation projection R_L defined by $R_L f := \sum_{P \in \Delta^\Gamma_L} f(P)\varphi^L_P$, for $u \in H^t(\Gamma)$, and for the Sobolev space orders $0 \le s \le t \le 2$, $s < 1.5$, $t > 1$, we get

$$\|u - R_L u\|_{H^s(\Gamma)} \le C 2^{-L(t-s)} \|u\|_{\oplus^{m_\Gamma}_{m=1} H^t(\Gamma_m)}. \tag{3.22}$$

iv) For the $L^2(\Gamma)$ orthogonal projection P_L and for the Sobolev space orders $-2 \le s \le t \le 2$, $s < 1.5$, $t > -1.5$, we get

$$\|u - P_L u\|_{H^s(\Gamma)} \le C 2^{-L(t-s)} \|u\|_{H^t(\Gamma)}. \tag{3.23}$$

v) For the Sobolev space orders $s \le t < 1.5$, the functions u_L from Lin_L^Γ fulfill the inverse property (Bernstein inequality)

$$\|u_L\|_{H^t(\Gamma)} \le C2^{L(t-s)}\|u_L\|_{H^s(\Gamma)}. \tag{3.24}$$

Proof. The assertions ii) - v) are well known. It remains to proof the Riesz property. Let $-0.5 < s < 1.5$ and $f = \sum_{P\in\Delta_L^\Gamma} \xi_P\psi_P$. Since $\psi_P = \psi_\tau \circ \kappa_1^{-1}$ for any $P = \kappa_1(\tau) \in \Gamma_1 \cap \Delta_L^\Gamma$ and since all the ψ_P with $P \notin \Gamma_1$ vanish over Γ_1, the corresponding estimate over Γ_1 analogous to (3.19) implies

$$\left\| \sum_{P\in\Delta_L^\Gamma\cap\Gamma_1} \xi_P\psi_P \right\|_{H^s(\Gamma_1)} \sim \sqrt{\sum_{P\in\Delta_L^\Gamma\cap\Gamma_1} 2^{2l(P)(s-1)}|\xi_P|^2}. \tag{3.25}$$

Now we set $f_2^+ := \sum_{P\in\Delta_L^\Gamma\cap\Gamma_2\setminus\Gamma_1} \xi_P\psi_P$ and $f_2^- := (f - f_2^+)|_{\Gamma_2}$. Clearly, the second function f_2^- is $\sum_{P\in\Delta_L^\Gamma\cap\Gamma_1} \xi_P\psi_P|_{\Gamma_2}$, and we observe that, for each restriction $\psi_P|_{\Gamma_2}$, $P \in \Delta_L^\Gamma\cap\Gamma_1$, the function $\psi_P \circ \kappa_2$ is equal to a restriction to T of a wavelet $\psi_\tau^{\mathbb{R}^2}$ or at least to the linear combinations of three restrictions to T of wavelets $\psi_\tau^{\mathbb{R}^2}$. First suppose all $\psi_P \circ \kappa_2|_T$ with $P \in \Delta_L^\Gamma\cap\Gamma_1$ are restrictions of wavelets $\psi_\tau^{\mathbb{R}^2}$. Then the upper estimate of the Riesz properties (3.1) applied to the $\psi_P \circ \kappa_2|_T$ and the lower estimate (3.25) yield (cf. also (2.3))

$$\begin{aligned} \|f\|_{H^s(\Gamma_2)} &\le \|f_2^+\|_{H^s(\Gamma_2)} + \|f_2^-\|_{H^s(\Gamma_2)} \\ &\le \|f_2^+\|_{H^s(\Gamma_2)} + C\sqrt{\sum_{P\in\Delta_L^\Gamma\cap\Gamma_1} 2^{2l(P)(s-1)}|\xi_P|^2} \\ &\le \|f_2^+\|_{H^s(\Gamma_2)} + C\,\|f\|_{H^s(\Gamma_1)}, \qquad (3.26) \\ \|f\|_{H^s(\Gamma_2)} &\ge \|f_2^+\|_{H^s(\Gamma_2)} - C\,\|f\|_{H^s(\Gamma_1)}. \qquad (3.27) \end{aligned}$$

The case that not all $\psi_P \circ \kappa_2|_T$ are restrictions of wavelets $\psi_\tau^{\mathbb{R}^2}$ occurs only if $P = \kappa_2(\tau)$ is at the boundary of $\kappa_2(T)$, if $\psi_P = \varphi_P^{l+1} - \frac{1}{4}\{\varphi_{P_1}^{l+1} + \varphi_{P_2}^{l+1}\}$ resp. $\psi_P = \varphi_P^{l+1} - \frac{1}{2}\{\varphi_{P_1}^{l+1} + \varphi_{P_2}^{l+1}\}$, and if the corresponding wavelet on $\mathbb{R}^2$ is $\psi_\tau^{\mathbb{R}^2} = \varphi_\tau^{l+1} - \frac{1}{2}\{\varphi_{\tau_1}^{l+1} + \varphi_{\tau_2}^{l+1}\}$ resp. $\psi_\tau^{\mathbb{R}^2} = \varphi_\tau^{l+1} - \frac{1}{4}\{\varphi_{\tau_1}^{l+1} + \varphi_{\tau_2}^{l+1}\}$. The functions $\varphi_{\tau_i}^{l+1}|_T$, however, are restrictions to T of wavelets $\psi_{\tau_i'}^{\mathbb{R}^2}$ with $\tau_i' \in \mathbb{R}^2 \setminus T$ and $\tau_i' \in \nabla_l^{\mathbb{R}^2}$. Moreover these $\varphi_{\tau_i}^{l+1}|_T$ coincide with the restrictions of $\psi_{P_i} \circ \kappa_2|_T$ for certain $P_i \in \Delta_L^\Gamma \cap \Gamma_1$. Hence, for an upper bound of $\|f_2^-\|^2$, we get the sum of terms $2^{2l(P)(s-1)}|\xi_P|^2$ and $2^{2l(P_i)(s-1)}|\xi_{P_i} \pm \frac{1}{4}\xi_P|^2$, and the estimates (3.26) and (3.27) remain valid. From these and (3.25) we get that

$$\|f\|_{H^s(\Gamma_1)} + \|f\|_{H^s(\Gamma_2)} \sim \sqrt{\sum_{P\in\Delta_L^\Gamma\cap\Gamma_1} 2^{2l(P)(s-1)}|\xi_P|^2} + \|f_2^+\|_{H^s(\Gamma_2)},$$

and the estimate over Γ_2 analogous to (3.19) leads to

$$\|f\|_{H^s(\Gamma_1)} + \|f\|_{H^s(\Gamma_2)} \sim \sqrt{\sum_{P\in\Delta_L^\Gamma\cap[\Gamma_1\cup\Gamma_2]} 2^{2l(P)(s-1)}|\xi_P|^2}.$$

Repeating the last arguments with Γ_1 replaced by $\Gamma_1 \cup \Gamma_2$ and Γ_2 replaced by Γ_3, we arrive at

$$\|f\|_{H^s(\Gamma_1)} + \|f\|_{H^s(\Gamma_2)} + \|f\|_{H^s(\Gamma_3)} \sim \sqrt{\sum_{P\in\Delta_L^\Gamma\cap[\Gamma_1\cup\Gamma_2\cup\Gamma_3]} 2^{2l(P)(s-1)}|\xi_P|^2}.$$

Further applications of the arguments lead finally to

$$\sum_{m=1}^{m_\Gamma} \|f\|_{H^s(\Gamma_m)} \sim \sqrt{\sum_{P\in\Delta_L^\Gamma\cap\bigcup_{m=1}^{m_\Gamma}\Gamma_m} 2^{2l(P)(s-1)}|\xi_P|^2} = \sqrt{\sum_{P\in\Delta_L^\Gamma} 2^{2l(P)(s-1)}|\xi_P|^2}$$

■

The Riesz property implies the existence of a projection Q_L, which is defined by

$$u = \sum_{P\in\bigcup_{l=0}^{\infty}\Delta_l^\Gamma} \xi_P\psi_P \mapsto Q_L u := \sum_{P\in\Delta_L^\Gamma} \xi_P\psi_P$$

and which is bounded in H^s, $-0.5 < s < 1.5$. For the wavelet coefficients of smooth functions, we obtain the following decay estimate.

Lemma 2 *Suppose the continuous function u belongs to $\oplus_{m=1}^{m_\Gamma} H^s(\Gamma_m)$ for an s with $-0.5 < s \leq 2$ and suppose $\sum_{P\in\Delta_L^\Gamma}\xi_P\psi_P$ is the representation of either the interpolation $R_L u$ or the orthogonal projection $P_L u$ or the projection $Q_L u$. Then*

(3.28)

$$\sqrt{\sum_{P\in\Delta_L^\Gamma} 2^{2l(P)(s-1)}|\xi_P|^2} \leq C\,\|u\|_{\oplus_{m=1}^{m_\Gamma}H^s(\Gamma_m)} \cdot \begin{cases} 1 & \text{if } -0.5 < s < 1.5 \\ \sqrt{L} & \text{if } 1.5 \leq s \leq 2. \end{cases}$$

Proof. The case $-0.5 < s < 1.5$ follows immediately from the Riesz property (3.20), and it remains to consider $1.5 \leq s \leq 2$. First we suppose that $\sum \xi_P\psi_P$ is the projection $Q_L u$. The Riesz property and the approximation property of Lemma 1, iii), which remains valid for R_L replaced by the uniformly bounded Q_L (cf. (3.20)), imply

$$\begin{aligned}
\sqrt{\sum_{P\in\nabla_{l-1}^\Gamma} 2^{-2(l-1)}|\xi_P|^2} &\sim \|Q_l u - Q_{l-1}u\|_{L^2} \\
&\leq \|Q_l u - u\|_{L^2} + \|u - Q_{l-1}u\|_{L^2} \\
&\leq C\,2^{-ls}\|u\|_{\oplus_{m=1}^{m_\Gamma}H^s(\Gamma_m)}, \\
\sqrt{\sum_{P\in\nabla_{l-1}^\Gamma} 2^{2(l-1)(s-1)}|\xi_P|^2} &\leq C\,\|u\|_{\oplus_{m=1}^{m_\Gamma}H^s(\Gamma_m)}.
\end{aligned}$$

L	0	1	2	3	4	5
c_0	0.86	0.92	0.78	0.85	0.92	0.95
C_0	1.32	2.17	3.50	3.97	4.27	4.54

Table 1: Constants in the norm equivalence for L^2.

L	0	1	2	3	4	5
c_1	0.33	0.20	0.16	0.16	0.15	0.14
C_1	1.96	2.90	3.48	3.63	3.77	4.03

Table 2: Constants in the norm equivalence for H^1.

Passing to the squares and summing up over $l = -1, \ldots, L-1$, we get the upper bound $C\,L\,\|u\|^2$. Taking square roots we obtain the assertion for $1.5 \le s \le 2$.

Now we denote the coefficients of $R_L u$ by $\tilde{\xi}_P$ in order to distinguish them from those of $Q_L u$. From the assertion with $Q_L u$ and from Lemma 1 ii) and iii) we get

$$\begin{aligned}
\sqrt{\sum_{P\in\triangle_L^\Gamma} 2^{-2l(P)}|\xi_P - \tilde{\xi}_P|^2} &\sim \|Q_L u - R_L u\|_{L^2} \\
&\le \|Q_L u - u\|_{L^2} + \|u - R_L u\|_{L^2} \\
&\le C\,2^{-Ls}\|u\|_{\oplus_{m=1}^{m_\Gamma} H^s(\Gamma_m)}, \\
\sqrt{\sum_{P\in\triangle_L^\Gamma} 2^{2l(P)(s-1)}|\xi_P - \tilde{\xi}_P|^2} &\le C\,\|u\|_{\oplus_{m=1}^{m_\Gamma} H^s(\Gamma_m)}.
\end{aligned}$$

This together with the estimate (3.28) for the coefficients ξ_P of $Q_L u$ implies (3.28) for the coefficients $\tilde{\xi}_P$ of $R_L u$. Similarly we can prove the corresponding assertion for the orthogonal projection. ∎

4. Numerical test

To check our hierarchical three-point basis, we have performed two different tests. First we have computed the constants in the norm equivalence (3.20) for $s = 0$ and $s = 1$ over the sphere. More precisely, we have inscribed a regular tetrahedron into the sphere and have chosen the parametrization mappings as compositions of affine mappings from the standard triangle T to the sides of the tetrahedron with the stereographical projection of the sides to the sphere. We have computed L^2 scalar products of piecewise linear functions from Lin_L^Γ by applying a simple ten point quadrature rule over the finest partition corresponding to grid $\triangle_L^\Gamma$. This

L	0	1	2	3	4	5
c_0	1.22	1.23	1.36	1.49	1.59	1.64
C_0	1.00	1.97	2.68	3.01	3.25	3.40

Table 3: Constants in the norm equivalence for L^2 with improved scaling.

L	0	1	2	3	4	5
c_1	1.53	1.52	1.41	1.39	1.40	1.40
C_1	1.90	2.44	2.57	2.62	2.63	2.67

Table 4: Constants in the norm equivalence for H^1 with improved scaling.

way we have computed the Gramian $(\langle\psi_j^l, \psi_{j'}^{l'}\rangle)_{(l',j'),(l,j)}$ and the Euclidean matrix norms of the Gramian and of its inverse. Obviously, these norms are the squares of the constants c_0 and C_0 in

(4.1)
$$\frac{1}{C_s}\sqrt{\sum_P 2^{2(s-1)l(P)}\,|\xi_P|^2} \le \left\| c_s \sum_P \xi_P \psi_P \right\|_{H^s(\Gamma)} \le C_s\sqrt{\sum_P 2^{2(s-1)l(P)}\,|\xi_P|^2}.$$

To get C_1 and c_1, we have computed an approximate H^1 scalar product replacing the surface gradient by the gradient with respect to the parametrization. Using this we proceeded analogously to the L^2 case. The results are contained in Tables 1 and 2. The constants $C_0 = 4.54$ and $C_1 = 4.03$ seem to be quite satisfactory. However, we have to remark that, in the worst case, these constants lead to an additional factor of $[C_s]^2[C_t]^2$ in the estimate of the condition number if we use the wavelet basis to precondition an operator mapping H^s into H^t. The constants can be improved replacing the scaling factors $2^{2(s-1)l(P)}$ for ψ_P by the reciprocal diagonal entries of the Gramians. With this modification we get the results in the Tables 3 and 4, i.e., $C_0 = 3.40$ and $C_1 = 2.67$.

In a second test we have used our basis over the sphere Γ to solve a boundary integral equation. In particular, we have solved the double layer equation

$$x(R) + \frac{1}{2\pi}\int_\Gamma \frac{n_S \cdot (R-S)}{|R-S|^3} x(S)\mathrm{d}_S\Gamma = 2y(R), \quad R \in \Gamma \tag{4.2}$$

numerically by a wavelet algorithm for the piecewise linear collocation. The three-point hierarchical basis has been used as the wavelet basis in the trial space. The number of degrees of freedom *DOF* is the cardinality of Δ_L^Γ. In the test space of Dirac delta functionals we have chosen linear combinations of three Dirac deltas.

L	DOF	LE	DE	LO	DO	IT	RA
3	130	2.7^{-2}	2.4^{-3}	1.97	2.23	22	1.59
4	514	5.5^{-3}	4.0^{-4}	2.29	2.59	23	3.68
5	2 050	1.4^{-3}	9.6^{-5}	2.00	2.04	23	10.74
6	8 194	3.5^{-4}	4.1^{-5}	1.97	1.23	24	35.09
7	32 770		2.9^{-5}		0.53	24	120.67

Table 5: Computation with higher rates of compression.

For the full details of the wavelet algorithm we refer the reader to [17] (cf. also the fundamental papers [1, 10, 11]). The data of our numerical test is collected in Tables 5 and 6.

The solution x of (4.2) is the layer function of the double layer representation for the Dirichlet problem solving Laplace's equation in the interior of the sphere. In our computations we have chosen the Dirichlet data $y(R) := 1/\sqrt{R-(2,0,0)}$. Thus the solution of the Dirichlet problem is given by the same formula, and, substituting the collocation approximation into the double layer representation, we can determine the error of the approximate solution to Laplace's equation. The supremum of such errors over a regular $8 \times 8 \times 8$ grid in the cube $[-0.35, 0.35] \times [-0.35, 0.35] \times [-0.35, 0.35]$ is denoted by DE. The exact values of the layer function are unknown of course. Thus, to get an error estimate for the layer, we have compared the layer function with the approximate solution on the next finer grid. The L^2 norm error estimate obtained by the weighted l^2 norm of the wavelet coefficients (cf. (4.1)) is denoted by LE. By LO and DO we denote estimates of the convergence order for LE and DE, respectively. In other words $LE \sim h^{LO}$ and $DE \sim h^{DO}$ with $h = 2^{-L}$.

We have solved the linear systems arising in the collocation method by GMRes up to an error of 10^{-7}. If the wavelet basis is stable as shown in Theorem 1, i), then the condition number of the linear system should be uniformly bounded with respect to L. So a bounded number IT of GMRes iterations is to be expected. Indeed, Tables 5 and 6 confirm that the numbers of iterations are bounded.

The reason to consider a wavelet method for the piecewise linear collocation is to replace the fully populated stiffness matrix by a sparse approximation. In fact, the majority of the entries are very small due to the vanishing moment properties and due to the small supports of the trial functions and test functionals. So we can neglect a large amount of entries such that the additional compression error is not essential. The compression ratio RA is the number of all entries divided by the number of entries remaining after the compression step. The results in Tables 5 and 6 prove that our three-point hierarchical basis is a good candidate for a wavelet algorithm.

L	*DOF*	*LE*	*DE*	*LO*	*DO*	*IT*	*RA*
3	130	2.7^{-2}	1.5^{-3}	1.97	2.64	22	1.43
4	514	5.5^{-3}	5.2^{-4}	2.29	1.59	23	3.15
5	2 050	1.4^{-3}	1.3^{-4}	2.00	2.02	23	8.78
6	8 194	3.5^{-4}	3.3^{-5}	1.97	1.94	24	27.65
7	32 770		6.4^{-6}		1.92	24	92.79

Table 6: Computation with lower rates of compression.

Acknowledgements

The author has been supported by a grant of Deutsche Forschungsgemeinschaft under grant numbers Pr 336/5-1 and Pr 336/5-2.

References

[1] Beylkin, G., Coifman, R., & Rokhlin, V., Fast wavelet transforms and numerical algorithms I, *Comm. Pure Appl. Math.*, **44**, pp. 141–183, 1991.

[2] Canuto, C., Tabacco, A., & Urban, K., The wavelet element method Part I: Construction and Analysis, *Appl. Comp. Harm. Anal.*, **6**, pp. 1–52, 1999.

[3] Canuto, C., Tabacco, A., & Urban, K., The wavelet element method Part II: Realization and additional features in 2D and 3D, to appear in *Appl. Comp. Harm. Anal.*.

[4] Canuto, C., Tabacco, A., & Urban, K., Numerical solution of elliptic problems by the wavelet element method, in: *Proceedings of the 2nd ENUMATH 97 conference*, H.G. Bock et al. (eds.), World Scientific, Singapore, pp. 17–37, 1998.

[5] Cohen, A., Daubechies, I., & Feauveau, J.-C., Biorthogonal bases of compactly supported wavelets, *Comm. Pure and Appl. Math.*, **45**, pp. 485–560, 1992.

[6] Dahmen, W., Stability of multiscale transformations, *Journal of Fourier Analysis and Applications*, **2**, pp. 341–361, 1996.

[7] Dahmen, W., Wavelet and multiscale methods for operator equations, *Acta Numerica*, **6**, pp. 55–228, Cambridge University Press, 1997.

[8] Dahmen, W., Kleemann, B., Prößdorf, S. & Schneider, R., A multiscale method for the double layer potential equation on a polyhedron, in: Dikshit, H. P. et al. (ed.), *Proceedings of the conference on advances in computational mathematics*, held at New Delhi, India, January 5-9, 1993. Singapore: World Scientific. Ser. Approx. Decompos. 4, pp. 15-57, 1994.

[9] Dahmen, W., Kunoth, A., & Urban, K. Biorthogonal spline-wavelets on the interval - stability and moment conditions, *Appl. Comp. Harm. Anal.*. **6**, No.2, pp. 132–196, (1999).

[10] Dahmen, W., Prößdorf, S. & Schneider, R., Wavelet approximation methods for pseudo-differential equations I: Stability and convergence, *Math. Zeitschr.*, **215**, pp. 583–620, 1994.

[11] Dahmen, W., Prößdorf, S. & Schneider, R., Wavelet approximation methods for pseudo-differential equations II: Matrix compression and fast solution, *Advances in Comp. Math.*, **1**, pp. 259-335, 1993.

[12] Dahmen, W. & Schneider, R., Composite wavelet bases for operator equations,*Math. Comp.*, **68**, pp. 1533–1567, 1999.

[13] Dahmen, W. & Schneider, R., Wavelets on manifolds I: Construction and Domain Decomposition, to appear in *SIAM J. Math. Anal.*.

[14] Dahmen, W. & Schneider, R., Wavelets with complementary boundary conditions - Function spaces on the cube, *Results in Mathematics*, **34**, pp. 255–293, 1998.

[15] Dahmen, W. & Stevenson, R., Element-by element construction of wavelets satisfying stability and moment conditions, to appear in *SIAM J. Numer. Anal.*, **37**, pp. 319–325, 1999.

[16] Daubechies, I., *Ten lectures on wavelets*, CBMS Lecture Notes **61**, SIAM, Philadelphia, 1992.

[17] Ehrich, S. & Rathsfeld, A., Piecewise Linear Wavelet Collocation, Approximation of the Boundary Manifold and Quadrature, to appear.

[18] Junkherr, J., *Effiziente Lösung von Gleichungssystemen, die aus der Diskretisierung von schwach singulären Integralgleichungen 1. Art herrühren*, Dissertation, Christian-Albrechts-Universität, Kiel, 1994.

[19] Lorentz, R. & Oswald, P., Constructing economical Riesz bases for Sobolev spaces, *GMD-Bericht* **993**, GMD, Sankt Augustin, 1996.

[20] Rathsfeld, A., A wavelet algorithm for the solution of a singular integral equation over a smooth two-dimensional manifold, *J. Integr. Equ. Appl.*, **10**, pp. 445–501, 1998.

[21] Stevenson, R., Stable three-point wavelet bases on general meshes, *Numer. Math.*, **80**, pp. 131–158, 1998.

A. Rathsfeld
Weierstraß-Institut für Angewandte Analysis und Stochastik
Mohrenstr. 39, D-10117 Berlin, Germany

1991 Mathematics Subject Classification: Primary 41A15; Secondary 65N60, 65N38

Submitted: 2.5.2000

Operator Theory:
Advances and Applications, Vol. 121

Algebras of approximation sequences: Fractality

STEFFEN ROCH

Dedicated to the memory of Siegfried Prössdorf

This paper deals with a fundamental property of an approximation sequence which is responsible for the uniformity of certain limiting processes: its fractality. Roughly speaking, a sequence is fractal if the knowledge of any of its infinite subsequences allows to reconstruct the whole sequence up to a sequence tending to zero in the norm. Typical features of a fractal sequence (A_n) are the existence of the limit of the condition numbers of the matrices A_n and the existence of the limit in the sense of the Hausdorff metric of the spectra, the pseudospectra, and the numerical ranges of the A_n. It will be moreover shown that every approximation sequence possesses a fractal subsequence.

1. Introduction

Fractality is a property of an approximation method which makes certain limiting processes uniform. To motivate and illustrate this notion we start with a simple and transparent example.

Stable approximation methods and condition numbers. Let H be a separable complex Hilbert space, $L(H)$ the C^*-algebra of the linear and bounded operators on H, and (P_n) a sequence of orthogonal projections on H which converges strongly (= pointwisely) to the identity operator I on H:

$$\text{s-lim}\, P_n = I \quad \Leftrightarrow \quad P_n x \to x \quad \text{for all } x \in H.$$

Assume for a moment that $\dim \operatorname{Im} P_n = n$. Then $\operatorname{Im} P_n$ and $L(\operatorname{Im} P_n)$ can be identified with the linear space $\mathbb{C}^n$ and with the algebra $\mathbb{C}^{n \times n} = L(\mathbb{C}^n)$, respectively.

Let $A \in L(H)$. An *approximation method for* A is a sequence (A_n) of matrices $A_n \in \mathbb{C}^{n \times n}$ such that $A_n P_n \to A$ and $A_n^* P_n \to A^*$ strongly as $n \to \infty$. As a consequence of the Banach-Steinhaus theorem, every approximation method is bounded in the sense that

$$\|(A_n)\| := \sup \|A_n\| < \infty. \tag{1.1}$$

The method (A_n) *converges*, or *is applicable to* A, if the equations

$$A_n x^{(n)} = P_n y \tag{1.2}$$

possess unique solutions $x^{(n)} \in \operatorname{Im} P_n$ for all sufficiently large n and all right hand sides $y \in H$, and if these solutions converge in the norm of H to a solution of the equation

$$Ax = y. \tag{1.3}$$

By the Banach-Steinhaus theorem again, the method (A_n) for A is applicable if and only if the sequence (A_n) is *stable* in the sense that the matrices A_n are invertible for all sufficiently large n and that

$$\sup \|A_n^{-1}\| < \infty. \tag{1.4}$$

If the sequence (A_n) is stable, then the *condition numbers* $\operatorname{cond} A_n := \|A_n\| \, \|A_n^{-1}\|$ are well-defined for n large enough. Without further information, all one can say in the moment is that the sequence $(\operatorname{cond} A_n)$ is *bounded* (due to (1.1) and (1.4)).

Fredholm integral equations of second kind. An obvious example of an approximation method where even *convergence* of the sequence $(\operatorname{cond} A_n)$ can be verified, is the Galerkin method for a Fredholm integral equation of the second kind:

$$Ax = (I - K)x = y, \quad y \in H, \tag{1.5}$$

where K is a compact operator on H. For the Galerkin method, choose an orthonormal basis $(e_i)_{i\geq 1}$ of H, specify P_n as the orthogonal projection from H onto the span of $\{e_1, \ldots, e_n\}$, and replace the equation (1.5) by the sequence of the approximation equations

$$A_n x^{(n)} := (I - P_n K P_n) x^{(n)} = P_n y, \quad n \geq 1. \tag{1.6}$$

Since

$$\|(I - P_n K P_n) - (I - K)\| \to 0 \quad \text{as} \quad n \to \infty, \tag{1.7}$$

a standard Neumann series argument yields that the sequence $(I - P_n K P_n)$ is stable if and only if the operator $I - K$ is invertible, and that

$$\|(I - P_n K P_n)^{-1} - (I - K)^{-1}\| \to 0 \quad \text{as} \quad n \to \infty \tag{1.8}$$

in this case. From (1.7) and (1.8) one immediately gets

Proposition 1.1. *If $K \in L(H)$ is compact and $I - K$ is invertible, then the sequence $(\operatorname{cond}(I - P_n K P_n))_{n\geq 1}$ of the condition numbers of the Galerkin method is convergent, and its limit is $\|I - K\| \, \|(I - K)^{-1}\|$, the condition number of $I - K$.*

Thanks to the norm convergence (1.7), projection methods for Fredholm integral equations of second kind play an exceptional role. A similar norm convergence cannot hold if one wishes to approximate non-compact operators by $n \times n$ matrices. In these instances, one has to deal with strong (= pointwise) convergence, and the above arguments to prove Proposition 1.1 do not apply. Nevertheless, one *can* verify the convergence of the condition numbers also for many strongly convergent approximation methods, see [2, 16]. To illustrate this, and to motivate the subsequent introduction of the notion of a fractal algebra, we proceed with an alternative proof of Proposition 1.1.

Alternative approach. This approach requires two ingredients. The first one is a formula for the limes superior of the norms $\|A_n\|$ which is valid for an arbitrary bounded approximation sequence (A_n). For its formulation, let $\mathcal{F}$ stand for the C^*-algebra of all bounded sequences (A_n) of operators $A_n \in L(H)$, provided with elementwise operations and the supremum norm, and let $\mathcal{G}$ refer to the closed ideal of $\mathcal{F}$ consisting of all sequences (A_n) with $\lim \|A_n\| = 0$. The quotient algebra $\mathcal{F}/\mathcal{G}$ plays an outstanding role for several problems in numerical analysis; for example the sequence $(A_n) \in \mathcal{F}$ is stable if and only if its coset $(A_n) + \mathcal{G}$ is invertible in $\mathcal{F}/\mathcal{G}$. The following identity, due to Böttcher [2], relates the norms $\|A_n\|$ to the norm of the coset $(A_n) + \mathcal{G}$.

Proposition 1.2. *For every sequence* $(A_n) \in \mathcal{F}$,

$$\limsup \|A_n\| = \|(A_n) + \mathcal{G}\|_{\mathcal{F}/\mathcal{G}}.$$

The second ingredient is a characteristic property of the Galerkin method for Fredholm integral equations which, of course, is a consequence of the norm convergence (1.7), but which also belongs to many other (not necessarily norm-convergent) approximation sequences.

Let $\mathcal{A}$ stand for the C^*-subalgebra of $\mathcal{F}$ consisting of all sequences (A_n) which converge in the norm. For example, the sequence $(I - P_nKP_n)$ of the Galerkin method is an element of $\mathcal{A}$. Further let $(A_n) \in \mathcal{A}$, let η be an infinite subsequence of $\mathbb{N}$, and suppose that we only know the subsequence $(A_{\eta(n)})$ of (A_n). Then we know $\lim A_{\eta(n)}$, hence we know $\lim A_n$, and consequently we know the coset $(A_n) + \mathcal{G}$. Thus, knowledge of an arbitrary subsequence of (A_n) allows us to reconstruct the complete sequence (A_n) up to a sequence tending to zero in the norm. Precisely this property means that the mapping

$$\pi_\eta : (A_{\eta(n)}) \mapsto (A_n) + \mathcal{G}$$

is a continuous *-homomorphism with norm 1.

Now we can prove Proposition 1.1 without (directly) invoking the norm convergence. What we have to show is that the limes superior in Proposition 1.2 is actually a limes if the sequence (A_n) belongs to $\mathcal{A}$. Assume it is not. Then there exists an infinite subsequence η of $\mathbb{N}$ such that

$$\sup \|A_{\eta(n)}\| < \limsup \|A_n\|.$$

Since $\|\pi_\eta\| = 1$ and by Proposition 1.2, we further have

$$\limsup \|A_n\| = \|(A_n) + \mathcal{G}\| = \|\pi_\eta(A_{\eta(n)})\| \leq \|(A_{\eta(n)})\| = \sup \|A_{\eta(n)}\|,$$

which is a contradiction.

Contents of this paper. In the following section, the notions of fractal homomorphisms, fractal algebras and fractal approximation sequences are defined, and some of their properties are discussed. In Section 3, these notions are applied to certain spectral approximation processes which will yield the convergence in the Hausdorff sense of the sets of the eigenvalues, the sets of the pseudo-eigenvalues, and of the numerical ranges of A_n for several approximation sequences (A_n). The concluding section is devoted to some special features of self-adjoint fractal sequences, and to the question whether every approximation sequence possesses a fractal subsequence.

The notion of a fractal homomorphism has been introduced in [16]. In that paper, algebras of approximation sequences were studied for which there exists a sufficient number of fractal homomorphisms of a special kind (related with so-called lifting theorems). For these algebras, analogues of Theorem 2.11, Corollary 2.12 and Propositions 3.7 and 3.11 (see below) were derived. The (at the first glance strange) notion of fractality has been chosen to emphasize the *self-similarity* effect of fractal sequences (see Definition 2.1).

2. Fractal sequences

Definitions. It is not important in what follows that the elements of the sequences under consideration are matrices or operators. So we will use slighly generalized definitions of $\mathcal{F}$ and $\mathcal{G}$. Given unital C^*-algebras $\mathcal{C}_n$ $(n = 0, 1, 2, ...)$ with identity elements e_n, let $\mathcal{F}$ stand for the set of all sequences $(c_0, c_1, c_2, ...)$ with $c_n \in \mathcal{C}_n$ which are bounded in the sense that

$$\|(c_n)\|_{\mathcal{F}} := \sup \|c_n\|_{\mathcal{C}_n} < \infty,$$

and write $\mathcal{G}$ for the set of all sequences $(c_0, c_1, c_2, ...)$ in $\mathcal{F}$ with $\|c_n\| \to 0$ as $n \to \infty$. Provided with elementwise operations, an elementwise involution and the supremum norm, $\mathcal{F}$ becomes a unital C^*-algebra and $\mathcal{G}$ a closed ideal of $\mathcal{F}$. Observe that, thanks to this extended definition, the results of this paper become applicable also to not fully discretized approximation sequences such as $(P_t A P_t)_{t>0}$ where A is an operator on $L^2(\mathbb{R})$ and P_t is the orthogonal projection from $L^2(\mathbb{R})$ onto its closed subspace $L^2([-t, t])$.

For every strongly monotonically increasing subsequence η of $\mathbb{N}$, define $\mathcal{F}_\eta$ and $\mathcal{G}_\eta$ analogously as the C^*-algebra consisting of all bounded sequences and the ideal of all zero sequences $(c_{\eta(0)}, c_{\eta(1)}, c_{\eta(2)}, ...)$ with $c_{\eta(n)} \in \mathcal{C}_{\eta(n)}$, and let R_η stand for the restriction mapping

$$R_\eta : \mathcal{F} \to \mathcal{F}_\eta, \quad (a_n) \mapsto (a_{\eta(n)}).$$

The mapping R_η is a *-homomorphism from $\mathcal{F}$ onto $\mathcal{F}_\eta$ which maps $\mathcal{G}$ onto $\mathcal{G}_\eta$. Further, given a C^*-subalgebra $\mathcal{A}$ of $\mathcal{F}$, let $\mathcal{A}_\eta$ refer to the image of $\mathcal{A}$ under R_η. Clearly, $\mathcal{A}_\eta$ is a C^*-subalgebra of $\mathcal{F}_\eta$.

Definition 2.1. Let $\mathcal{A}$ be a C^*-subalgebra of $\mathcal{F}$.
(*a*) A *-homomorphism $W : \mathcal{A} \to \mathcal{B}$ of $\mathcal{A}$ into a C^*-algebra $\mathcal{B}$ is *fractal* if, for every strongly monotonically increasing sequence η, there is a *-homomorphism $W_\eta : \mathcal{A}_\eta \to \mathcal{B}$ such that $W = W_\eta R_\eta$.
(*b*) The algebra $\mathcal{A}$ is *fractal*, if the canonical homomorphism $\pi : \mathcal{A} \to \mathcal{A}/(\mathcal{A} \cap \mathcal{G})$ is fractal.
(*c*) A sequence $(a_n) \in \mathcal{F}$ is *fractal* if the smallest C^*-subalgebra of $\mathcal{F}$ which contains (a_n) is fractal.

Thus, given a subsequence $(a_{\eta(n)})$ of a sequence (a_n) which belongs to a fractal algebra $\mathcal{A}$, it is possible to reconstruct the original sequence (a_n) from its subsequence modulo sequences in $\mathcal{A} \cap \mathcal{G}$. This assumption is quite natural for sequences arising from discretization procedures. For example, the algebra $\mathcal{A}$ examined in the introduction is fractal, and $W : \mathcal{A} \to L(H)$, $(A_n) \mapsto \lim A_n$ is a fractal homomorphism.

Fractal homomorphisms. We proceed with two lemmas on fractal homomorphisms.

Lemma 2.2. *Let $\mathcal{A}$ be a C^*-subalgebra of $\mathcal{F}$, and let $\mathcal{B}$ be a C^*-algebra. Then $\mathcal{A} \cap \mathcal{G}$ lies in the kernel of every fractal *-homomorphism $W : \mathcal{A} \to \mathcal{B}$.*

Proof. Let $(g_n) \in \mathcal{A} \cap \mathcal{G}$. Given $\varepsilon > 0$ there is an n_0 such that $\|g_n\| \le \varepsilon$ for $n \ge n_0$. Consider the sequence $\eta(n) := n + n_0$. Since W is fractal,

$$W(g_n) = (W_\eta R_\eta)(g_n) = W_\eta(g_{\eta(n)}) = W_\eta(g_{n+n_0}).$$

Consequently,

$$\|W(g_n)\| = \|W_\eta(g_{n+n_0})\| \le \|W_\eta\| \; \|(g_{n+n_0})\|_{\mathcal{F}_\eta} \le \varepsilon$$

(observe that $\|W_\eta\| \le 1$). Letting ε go to zero we get the assertion. □

Thus, if $W : \mathcal{A} \to \mathcal{B}$ is a fractal *-homomorphism, then $W(a_n)$ does only depend on the coset $\pi(a_n) = (a_n) + \mathcal{A} \cap \mathcal{G}$.

Lemma 2.3. *Let $\mathcal{A}$ be a fractal C^*-subalgebra of $\mathcal{F}$ and $\mathcal{B}$ be a C^*-algebra. A *-homomorphism $W : \mathcal{A} \to \mathcal{B}$ is fractal if and only if $\mathcal{A} \cap \mathcal{G} \subseteq \operatorname{Ker} W$.*

Proof. If W is fractal, then $\mathcal{A} \cap \mathcal{G} \subseteq \operatorname{Ker} W$ by Lemma 2.2. Let, conversely, $\mathcal{A} \cap \mathcal{G} \subseteq \ker W$, and let η be a monotonically increasing sequence. If W^π denotes the quotient mapping

$$W^\pi : \mathcal{A}/(\mathcal{A} \cap \mathcal{G}) \to \mathcal{B}, \quad \pi(a_n) \mapsto W(a_n), \tag{2.1}$$

then, evidently, $W = W^\pi \pi$. Further, since π is fractal by assumption, there is a *-homomorphism π_η such that $\pi = \pi_\eta R_\eta$. Thus,

$$W = W^\pi \pi = W^\pi(\pi_\eta R_\eta) = (W^\pi \pi_\eta) R_\eta$$

which shows fractality of W. □

Criteria for fractality of algebras. The following two theorems provide equivalent characterizations of fractal subalgebras of $\mathcal{F}$.

Theorem 2.4. *A C^*-subalgebra $\mathcal{A}$ of $\mathcal{F}$ is fractal if and only if the following implication holds for every element $(a_n) \in \mathcal{A}$ and every strongly monotonically increasing sequence η:*

$$R_\eta(a_n) \in \mathcal{G}_\eta \;\Rightarrow\; (a_n) \in \mathcal{A} \cap \mathcal{G}.$$

Proof. Let $\mathcal{A}$ be fractal, and assume there is a strongly monotonically increasing sequence η as well as an element $(a_n) \in \mathcal{A}$ with $R_\eta(a_n) \in \mathcal{G}_\eta$ for which (a_n) is not in $\mathcal{A} \cap \mathcal{G}$, i.e.

$$\|\pi(a_n)\| = \|(a_n) + \mathcal{A} \cap \mathcal{G}\| =: C > 0.$$

Since $R_\eta(a_n) \in \mathcal{G}_\eta$, there is an n_0 such that $\|a_{\eta(n)}\| \le C/2$ for $n \ge n_0$. Define a sequence μ by $\mu(n) := \eta(n + n_0)$. Then $\|R_\mu(a_n)\| \le C/2$. On the other hand, the fractality of $\mathcal{A}$ implies that

$$C \le \|\pi_\mu R_\mu(a_n)\| \le \|\pi_\mu\| \, \|R_\mu(a_n)\| = \|R_\mu(a_n)\|. \tag{2.2}$$

for every strongly monotonically increasing sequence μ, which is a contradiction. Hence, (a_n) belongs to $\mathcal{A} \cap \mathcal{G}$.

For the reverse direction, let $(a_n), (b_n) \in \mathcal{A}$ be sequences with $R_\eta(a_n) = R_\eta(b_n)$. Then, by hypotheses, $(a_n) - (b_n)$ is in $\mathcal{A} \cap \mathcal{G}$, and so it is correct to define a *-homomorphism π_η via

$$\pi_\eta : \mathcal{A}_\eta \to \mathcal{A}/(\mathcal{A} \cap \mathcal{G}), \quad R_\eta(a_n) \mapsto (a_n) + \mathcal{A} \cap \mathcal{G}.$$

Evidently, $\pi_\eta R_\eta = \pi$ whence follows that π is a fractal homomorphism and $\mathcal{A}$ is a fractal algebra. □

Corollary 2.5. *(a) Every C^*-subalgebra of a fractal algebra $\mathcal{A} \subseteq \mathcal{F}$ is fractal.*
(b) A C^*-subalgebra $\mathcal{A}$ of $\mathcal{F}$ is fractal if and only if each of its elements is fractal.

Proof. (*a*) Let $\mathcal{B}$ be a C^*-subalgebra of $\mathcal{A}$, and let $(b_n) \in \mathcal{B}$ be an element with $R_\eta(b_n) \in \mathcal{G}_\eta$. Since $\mathcal{A}$ is fractal, Theorem 2.4 implies that (b_n) belongs to $\mathcal{A} \cap \mathcal{G}$. Then, clearly, (b_n) lies in $\mathcal{B} \cap \mathcal{G}$, and invoking Theorem 2.4 once more one gets the fractality of $\mathcal{B}$.
(*b*) If $\mathcal{A}$ is a fractal C^*-subalgebra of $\mathcal{F}$, then each of its elements is fractal due to assertion (*a*). For the reverse assertion we claim that, whenever $\mathcal{A}$ is not fractal, then it contains a non-fractal sequence. Indeed, if $\mathcal{A}$ fails to be fractal, then, by Theorem 2.4, there is an element $(a_n) \in \mathcal{A}$ as well as a sequence η such that $R_\eta(a_n) \in \mathcal{G}_\eta$, i.e. $\|a_{\eta(n)}\| \to 0$ as $n \to \infty$, but $(a_n) \notin \mathcal{A} \cap \mathcal{G}$, i.e. $\|a_n\| \not\to 0$ as $n \to \infty$. Applying Theorem 2.4 once again this shows that the C^*-subalgebra

of $\mathcal{A}$ which is generated by this sequence (a_n) cannot be fractal, i. a. (a_n) is a non-fractal sequence. □

Corollary 2.6. *If $\mathcal{A}$ is a fractal C^*-subalgebra of $\mathcal{F}$, then $\mathcal{A}_\eta \cap \mathcal{G}_\eta = (\mathcal{A} \cap \mathcal{G})_\eta$.*

The following criterion proves to be useful for verifying the fractality of several concrete algebras of approximation methods.

Theorem 2.7. *Let $\mathcal{A}$ be a unital C^*-subalgebra of $\mathcal{F}$. The algebra $\mathcal{A}$ is fractal if and only if there exists a family $\{W_t\}_{t\in T}$ of unital and fractal *-homomorphisms W_t from $\mathcal{A}$ into unital C^*-algebras $\mathcal{B}_t$ such that the following equivalence holds for every sequence $(a_n) \in \mathcal{A}$: The coset $(a_n) + \mathcal{A} \cap \mathcal{G}$ is invertible in $\mathcal{A}/(\mathcal{A} \cap \mathcal{G})$ if and only if $W_t(a_n)$ is invertible in $\mathcal{B}_t$ for every $t \in T$.*

Proof. If $\mathcal{A}$ is fractal, then the canonical homomorphism $\pi : \mathcal{A} \to \mathcal{A}/(\mathcal{A} \cap \mathcal{G})$ yields the 'family' $\{\pi\}$ which obviously has the desired properties.

Let, conversely, $\{W_t\}_{t\in T}$ be a family of unital and fractal *-homomorphisms which is subject to the conditions of the theorem. Given a strongly monotonically increasing sequence η, define the operator T_η by

$$T_\eta : \mathcal{A}/\mathcal{A} \cap \mathcal{G} \to \mathcal{A}_\eta/(\mathcal{A} \cap \mathcal{G})_\eta, \quad (a_n) + \mathcal{A} \cap \mathcal{G} \mapsto R_\eta(a_n) + (\mathcal{A} \cap \mathcal{G})_\eta.$$

One easily checks that this definition is correct, and that T_η is a *-homomorphism from $\mathcal{A}/\mathcal{A} \cap \mathcal{G}$ onto $\mathcal{A}_\eta/(\mathcal{A} \cap \mathcal{G})_\eta$. We claim that T_η is an isomorphism.

For this goal, it is sufficient to show that T_η is a symbol mapping in the sense that, if $(a_n) \in \mathcal{A}$, and if $T_\eta((a_n) + \mathcal{A} \cap \mathcal{G}) = R_\eta(a_n) + (\mathcal{A} \cap \mathcal{G})_\eta$ is invertible, then $(a_n) + \mathcal{A} \cap \mathcal{G}$ is invertible.

If the coset $R_\eta(a_n) + (\mathcal{A} \cap \mathcal{G})_\eta$ is invertible, then there are a sequence $(b_n) \in \mathcal{A}$ as well as sequences $(g_n), (h_n) \in \mathcal{A} \cap \mathcal{G}$ such that

$$(2.3)\; R_\eta(a_n)R_\eta(b_n) = R_\eta(e_n) + R_\eta(g_n), \quad R_\eta(b_n)R_\eta(a_n) = R_\eta(e_n) + R_\eta(h_n).$$

The homomorphisms W_t are fractal by assumption, i.e. there are homomorphisms $W_{t,\eta}$ such that $W_t = W_{t,\eta}R_\eta$. Applying $W_{t,\eta}$ to the equalities (2.3) we get

$$W_t(a_n)W_t(b_n) = W_t(e_n) + W_t(g_n), \quad W_t(b_n)W_t(a_n) = W_t(e_n) + W_t(h_n).$$

Since $W_t(e_n)$ is the identity element of $\mathcal{B}_t$ and $W_t(g_n) = W_t(h_n) = 0$ by Lemma 2.2, all elements $W_t(a_n)$ are invertible in $\mathcal{B}_t$. Thus, the coset $(a_n) + \mathcal{A} \cap \mathcal{G}$ is invertible, and T_η is indeed a symbol mapping and, thus, a *-isomorphism between $\mathcal{A}/\mathcal{A} \cap \mathcal{G}$ and $\mathcal{A}_\eta/(\mathcal{A} \cap \mathcal{G})_\eta$.

Let finally Π_η denote the canonical homomorphism from $\mathcal{A}_\eta$ onto $\mathcal{A}_\eta/(\mathcal{A} \cap \mathcal{G})_\eta$. Then, evidently, $T_\eta^{-1}\Pi_\eta R_\eta = \pi$, i.e. π is fractal (with $\pi_\eta = T_\eta^{-1}\Pi_\eta$). □

Example: Spline projection methods for singular integral equations. To mention at least one concrete class of fractal approximation sequences we consider spline projection methods for singular integrals on the real line. All details, as well as more general results (pertaining arbitrary spline spaces, singular integrals on arbitrary Lyapunov curves etc.) can be found in [7]. Here we restrict our attention to a situation without technical complications.

Let S_n stand for the smallest closed subspace of $L^2(\mathbb{R})$ which contains, for every k, the functions $\varphi_{kn}(t) := \chi_{[0,1)}(nt-k)$ where $\chi_{[0,1)}$ is the characteristic function of the interval $[0, 1)$. Every spline space S_n is isometric to the Hilbert space $l^2(\mathbb{Z})$, with the isometry given by

$$E_n : l^2(\mathbb{Z}) \to S_n, \quad (x_k)_{k\in\mathbb{Z}} \mapsto n^{1/2} \sum_{k\in\mathbb{Z}} x_k \varphi_{kn}.$$

The inverse of $E_n : l^2(\mathbb{Z}) \to S_n$ will be denoted by E_{-n}. Further write L_n for the orthogonal projection from $L^2(\mathbb{R})$ onto S_n; these projections converge strongly to the identity operator as n tends to infinity.

On $l^2(\mathbb{Z})$ we introduce the following operators. Given a piecewise continuous function a on the unit circle $\mathbb{T}$ with kth Fourier coefficient a_k, let $T^0(a) \in L(l^2(\mathbb{Z}))$ stand for the Laurent operator with matrix representation $(a_{i-j})_{i,j\in\mathbb{Z}}$. The projection operator $(x_k) \mapsto (\ldots, 0, 0, x_0, x_1, \ldots)$ will be denoted by P. Finally, given $n \in \mathbb{Z}$ and $t \in \mathbb{T}$, define operators V_n and Y_t by

$$V_n : l^2(\mathbb{Z}) \to l^2(\mathbb{Z}),\ (x_k) \mapsto (x_{k-n}), \quad Y_t : l^2(\mathbb{Z}) \to l^2(\mathbb{Z}),\ (x_k) \mapsto (t^{-k} x_k).$$

Now let the C^*-algebra $\mathcal{F}$ of all bounded approximation sequences be defined as above with the algebras $\mathcal{C}_n$ specified as $L(S_n)$, and let $\mathcal{A}$ refer to the smallest closed subalgebra of $\mathcal{F}$ which contains all sequences of the form

$$(E_n V_{\{sn\}} P V_{-\{sn\}} E_{-n})_{n\geq 1} \quad \text{and} \quad (E_n T^0(a) E_{-n})_{n\geq 1}$$

with a running through the piecewise continuous functions on $\mathbb{T}$, with $s \in \mathbb{R}$, and with $\{sn\}$ referring to the smallest integer which is not smaller that sn. This definition of $\mathcal{A}$ seems to be artificial at the first glance, but one can show that this algebra indeed contains a bulk of interesting and concrete approximation methods. For example, the sequence $(L_n(aI+bS)L_n)$ of the Galerkin method for the singular integral operator

$$((aI + bS)u)(t) := a(t)u(t) + \frac{b(t)}{\pi i} \int_{-\infty}^{\infty} \frac{u(s)}{s-t}\, ds, \quad t \in \mathbb{R}$$

belongs to $\mathcal{A}$ if a and b are piecewise continuous functions on the real line with jumps only at integer points and at infinity.

Proposition 2.8. *Let $(A_n) \in \mathcal{A}$.*
(a) For every $t \in \mathbb{T}$, the strong limit

$$W^t(A_n) := \text{s-lim}\, E_n Y_{t^{-1}} E_{-n} A_n E_n Y_t E_{-n} L_n$$

exists, and the mapping W^t is a *-homomorphism from $\mathcal{A}$ into $L(L^2(\mathbb{R}))$.
(b) For every $s \in \mathbb{R}$, the strong limit

$$W_s(A_n) := \text{s-lim}\, V_{-\{sn\}} E_{-n} A_n E_n V_{\{sn\}}$$

exists, and the mapping W_s is a *-homomorphism from $\mathcal{A}$ into $L(l^2(\mathbb{Z}))$.
(c) The cosets of the operators $E_{-n}A_nE_n$ modulo the ideal $K(l^2(\mathbb{Z}))$ of the compact operators on $l^2(\mathbb{Z})$ are independent of n. If $W_\infty(A_n)$ denotes one of these cosets, then W_∞ is a *-homomorphism from $\mathcal{A}$ into $L(l^2(\mathbb{Z}))/K(l^2(\mathbb{Z}))$.

The stability criterion for sequences in $\mathcal{A}$ reads as follows (Theorem 3.3 in [7]):

Theorem 2.9. *A sequence $(A_n) \in \mathcal{A}$ is stable if and only if the operators resp. cosets $W^t(A_n)$ and $W_s(A_n)$ are invertible for every $t \in \mathbb{T}$ and $s \in \mathbb{R} \cup \{\infty\}$.*

It is evident that the homomorphisms W^t and W_s are fractal for every $t \in \mathbb{T}$ and $s \in \mathbb{R} \cup \{\infty\}$. Hence, Theorem 2.9 implies in combination with Theorem 2.7:

Corollary 2.10. *The algebra $\mathcal{A}$ is fractal.*

Fractal algebras and convergence of norms. Here we turn back once again to the problem whether the limit $\lim \|a_n\|$ exists for a sequence $(a_n) \in \mathcal{F}$. It will be pointed out that fractality of (a_n) is indeed enough to guarantee the existence of this limit.

Theorem 2.11. *Let $\mathcal{A}$ be a fractal C^*-subalgebra of $\mathcal{F}$.*
(a) If $(a_n) \in \mathcal{A}$ and η is a strongly monotonically increasing sequence, then

$$\|(a_n) + \mathcal{G}\|_{\mathcal{F}/\mathcal{G}} = \|(a_{\eta(n)}) + \mathcal{G}_\eta\|_{\mathcal{F}_\eta/\mathcal{G}_\eta}.$$

(b) If $(a_n) \in \mathcal{A}$, then the limit $\lim \|a_n\|$ exists and is equal to $\|(a_n) + \mathcal{G}\|$.

Proof. (a) The third isomorphy theorem for C^*-algebras states that, if $\mathcal{B}$ is a C^*-subalgebra of $\mathcal{F}$ and $(b_n) \in \mathcal{B}$, then

$$\|(b_n) + \mathcal{G}\|_{\mathcal{F}/\mathcal{G}} = \|(b_n) + \mathcal{G}\|_{(\mathcal{B}+\mathcal{G})/\mathcal{G}} = \|(b_n) + \mathcal{B} \cap \mathcal{G}\|_{\mathcal{B}/(\mathcal{B}\cap\mathcal{G})}. \tag{2.4}$$

Let now $(a_n) \in \mathcal{A}$ and $(g_n) \in \mathcal{A} \cap \mathcal{G}$. Then

$$\begin{aligned}
\|(a_n) + \mathcal{G}\|_{\mathcal{F}/\mathcal{G}} &= \|(a_n) + \mathcal{A} \cap \mathcal{G}\|_{\mathcal{A}/(\mathcal{A}\cap\mathcal{G})} && \text{(by (2.4))} \\
&= \|\pi(a_n + g_n)\|_{\mathcal{A}/(\mathcal{A}\cap\mathcal{G})} && \text{(definition of } \pi) \\
&= \|\pi_\eta R_\eta(a_n + g_n)\|_{\mathcal{A}/(\mathcal{A}\cap\mathcal{G})} && \text{(fractality of } \mathcal{A}) \\
&\le \|(a_{\eta(n)} + g_{\eta(n)})_{n=0}^\infty\|_{\mathcal{A}_\eta} && \text{(since } \|\pi_\eta\| \le 1).
\end{aligned}$$

Taking the infimum over all $(g_n) \in \mathcal{A} \cap \mathcal{G}$, and invoking Corollary 2.6, we obtain

$$\begin{aligned}
\|(a_n) + \mathcal{G}\|_{\mathcal{F}/\mathcal{G}} &\le \|(a_{\eta(n)})_{n=0}^\infty + (\mathcal{A} \cap \mathcal{G})_\eta\|_{\mathcal{A}_\eta/(\mathcal{A}\cap\mathcal{G})_\eta} \\
&= \|(a_{\eta(n)})_{n=0}^\infty + \mathcal{A}_\eta \cap \mathcal{G}_\eta\|_{\mathcal{A}_\eta/(\mathcal{A}_\eta\cap\mathcal{G}_\eta)},
\end{aligned}$$

and a further application of (2.4) gives

$$\|(a_n) + \mathcal{G}\|_{\mathcal{F}/\mathcal{G}} \le \|(a_{\eta(n)})_{n=0}^{\infty} + \mathcal{G}_\eta\|_{\mathcal{F}_\eta/\mathcal{G}_\eta} . \tag{2.5}$$

On the other hand, the lim sup-formula for the norm in $\mathcal{F}/\mathcal{G}$ (i.e. Proposition 1.2, which also holds in the present, slightly more general situation) yields

$$\|(a_{\eta(n)}) + \mathcal{G}_\eta\|_{\mathcal{F}_\eta/\mathcal{G}_\eta} = \limsup \|a_{\eta(n)}\| \le \limsup \|a_n\| = \|(a_n) + \mathcal{G}\|_{\mathcal{F}/\mathcal{G}}$$

which together with (2.5) proves the assertion.
(*b*) Let η be a strongly monotonically increasing sequence such that the limit $\lim \|a_{\eta(n)}\|$ exists and is equal to $\liminf \|a_n\|$. By part (*a*) of this theorem, and by the lim sup-formula again,

$$\begin{aligned}\limsup \|a_n\| &= \|(a_n) + \mathcal{G}\|_{\mathcal{F}/\mathcal{G}} = \|(a_{\eta(n)}) + \mathcal{G}_\eta\|_{\mathcal{F}_\eta/\mathcal{G}_\eta} \\ &= \limsup \|a_{\eta(n)}\| = \lim \|a_{\eta(n)}\| = \liminf \|a_n\|\end{aligned}$$

which proves assertion (*b*). □

Corollary 2.12. *Let $\mathcal{A}$ be a fractal C^*-subalgebra of $\mathcal{F}$, and let $(a_n) \in \mathcal{A}$ be a stable sequence. Then the limit* $\lim \operatorname{cond} a_n := \lim \|a_n\| \, \|a_n^{-1}\|$ *exists and is equal to* $\operatorname{cond}((a_n) + \mathcal{G}) := \|(a_n) + \mathcal{G}\| \, \|((a_n) + \mathcal{G})^{-1}\|$.

As a particular consequence we obtain that the norms $\|A_n\|$ as well as (in case of a stable sequence) the condition numbers $\operatorname{cond} A_n$ converge for every sequence (A_n) in the algebra $\mathcal{A}$ of the spline projection methods. Moreover, the limit $\lim \|A_n\|$ is equal to $\sup\{\|W^t(A_n)\|, \|W_s(A_n)\|\}$ where the supremum is taken over all $t \in \mathbb{T}$ and $s \in \mathbb{R} \cup \{\infty\}$. An analogous result holds for the limit of the condition numbers.

It is obvious that the decisive point in this approach is a precise knowledge on the stability of sequences in $\mathcal{A}$: it is both important to have a criterion for the stability of an arbitrary sequence in $\mathcal{A}$, and that this criterion has the 'right form', i.e. via a fractal mappings. Appropriate stability criteria are available for many algebras of concrete approximation methods (see, e.g. [7] and the references therein).

3. Examples and applications

The goal of this section is to illustrate how fractality affects spectral approximation problems, i.e. the problem whether the spectrum of a given operator A can be determined approximately by computing the spectrum of certain approximations A_n of A. Spectral approximation problems have been studied from several points of view; see e.g. the monographs [5] and [10]. It is well known that this spectral approximation is, roughly speaking, possible if A and the A_n are self-adjoint or normal, whereas it can fail drastically in case A is a non-self-adjoint operator. Thus, one considers at least two alternative approaches to the approximation of spectra, namely via pseudospectra and via numerical ranges. Both pseudospectra and numerical ranges can be viewed as approximants of the usual spectra, and both exhibit a much better asymptotic behaviour than the latter. In all these cases, we will observe that the approximation processes under consideration become more uniform in the presence of fractality. As a rule, we will obtain convergence with respect to the Hausdorff metric instead of pointwise convergence.

Set sequences. We start with recalling some elementary facts on set sequences. A general reference is [8], Section 28. A *set sequence* is a sequence of subsets of the complex plane $\mathbb{C}$. If (A_n) is a sequence of operators, then the mapping $n \mapsto \sigma(A_n)$ which assigns with $n \in \mathbb{N}$ the spectrum of A_n is a set sequence in this sense.

Definition 3.1. (a) Let $(M_n)_{n=1}^{\infty}$ be a set sequence. The *partial limiting set* or *limes superior* $\limsup M_n$ (resp. the *uniform limiting set* or *limes inferior* $\liminf M_n$) of the sequence (M_n) consists of all points $m \in \mathbb{C}$ which are a partial limit (resp. the limit) of a sequence (m_n) of points $m_n \in M_n$.

Observe that the partial limiting set $\limsup M_n$ is non-empty if infinitely many of the M_n are non-empty and if $\cup_n M_n$ is bounded, whereas the uniform limiting set can be empty even under these restrictions as the trivial example $M_n = \{(-1)^n\}$ shows. Both the partial and the uniform limiting set are closed.

Let $\mathbb{C}^C$ denote the set of all non-empty and compact subsets of the complex plane. A criterion for the coincidence of the partial and uniform limiting sets of a set sequence taking values in $\mathbb{C}^C$ can be given in terms of the convergence of that sequence with respect to the Hausdorff metric on $\mathbb{C}^C$ where the distance of two sets $A, B \subseteq \mathbb{C}^C$ is defined as

$$h(A, B) := \max\{\max_{a\in A} \operatorname{dist}(a, B), \max_{b\in B} \operatorname{dist}(b, A)\}$$

with $\operatorname{dist}(a, B) = \min_{b\in B} |a - b|$. We denote the limit of a sequence (M_n) with respect to this metric by $\lim M_n$.

Proposition 3.2. *Let (M_n) be a sequence taking values in $\mathbb{C}^C$. Then* $\limsup M_n$ *and* $\liminf M_n$ *coincide if and only if the sequence (M_n) is convergent in sense of the Hausdorff metric. If one of these conditions is satisfied, then*

$$\limsup M_n = \liminf M_n = \lim M_n.$$

In Section 4, we will need the following compactness result saying that bounded subsets of $\mathbb{C}^C$ are relatively compact with respect to h.

Theorem 3.3. *Every bounded set sequence (M_n) in $\mathbb{C}^C$ possesses a subsequence which converges with respect to the Hausdorff metric.*

Spectra and their limiting sets. Again, let $(\mathcal{C}_n)$ be a family of C^*-algebras with identity elements e_n and define the algebra $\mathcal{F}$ and its ideal $\mathcal{G}$ as above. The spectrum of an element $a_n \in \mathcal{C}_n$ will be denoted by $\sigma(a_n)$. Our first goal is the asymptotic behaviour of the spectra $\sigma(a_n)$ for bounded sequences $(a_n) \in \mathcal{F}$.

Let $(a_n) \in \mathcal{F}$. It turns out that the partial limiting set $\limsup \sigma(a_n)$ is related with a new notion of stability which might be called 'spectral' stability (in contrast to the conventional 'normal' stability defined in the introduction). A sequence (a_n) is *spectrally stable* if its entries a_n are invertible for all sufficiently large n, and if the spectral radii $\rho(a_n^{-1})$ of their inverses are uniformly bounded (whereas the conventional stability requires uniform boundedness of the norms $\|a_n^{-1}\|$). Clearly, every conventionally stable sequence is also spectrally stable. The simple proof of the following result can be found in [15].

Proposition 3.4. *Let $(a_n) \in \mathcal{F}$. Then $s \in \mathbb{C}$ belongs to the partial limiting set* $\limsup \sigma(a_n)$ *if and only if the sequence $(a_n - se_n)$ is not spectrally stable.*

Thus, the determination of the partial limiting set $\limsup \sigma(a_n)$ requires to investigate the spectral stability of the sequences $(a_n - se_n)$ which can be is a rather involved problem in general. These difficulties disappear for normal sequences (i.e. for sequences which commute with their adjoint) in which case the conventional stability and the spectral stability coincide. So one has:

Corollary 3.5. *Let $(a_n) \in \mathcal{F}$ be a sequence of normal elements. Then*

$$\limsup \sigma(a_n) = \sigma_{\mathcal{F}/\mathcal{G}}((a_n) + \mathcal{G}).$$

The fractality of the sequence (a_n) again guarantees that the limes superior in the latter spectral identity (which plays a similar role as the limsup identity for norms in Proposition 1.2) is actually a Hausdorff limit.

Theorem 3.6. *Let $\mathcal{A}$ be a fractal C^*-subalgebra of $\mathcal{F}$ which contains the identity.*
(a) A sequence $(a_n) \in \mathcal{A}$ is conventionally stable if and only if it possesses a conventionally stable subsequence.
(b) If $(a_n) \in \mathcal{A}$ is normal, then $\limsup \sigma(a_n) = \liminf \sigma(a_n) = \lim \sigma(a_n)$.

Proof. (*a*) Let the notations be as in Section 2, and let η be a strongly monotonically increasing sequence such that the sequence $(a_{\eta(n)}) = R_\eta(a_n)$ is conventionally stable. Clearly, η can be chosen in such a way that all $a_{\eta(n)}$ are invertible. By the inverse closedness of $\mathcal{A}_\eta$ in $\mathcal{F}_\eta$, there is a sequence $(b_n) \in \mathcal{A}$ such that

$$R_\eta(a_n)\, R_\eta(b_n) = R_\eta(b_n)\, R_\eta(a_n) = R_\eta(e_n). \tag{3.1}$$

The canonical homomorphism $\pi : \mathcal{A} \to \mathcal{A}/(\mathcal{A} \cap \mathcal{G})$ is fractal by hypothesis, i.e. $\pi = \pi_\eta R_\eta$ with certain homomorphism π_η. Applying π_η to (3.1) one gets the invertibility of $\pi(a_n) = (a_n) + \mathcal{A} \cap \mathcal{G}$ in $\mathcal{A}/(\mathcal{A} \cap \mathcal{G})$, whence the stability of (a_n) follows.

(*b*) Assume, there is a $t \in \limsup \sigma(a_n) \setminus \liminf \sigma(a_n)$. Then one can find a $\delta > 0$ as well as a strongly monotonically increasing sequence η such that $\operatorname{dist}(t, \sigma(a_{\eta(n)})) \geq$

δ for all n. The normality of the $a_{\eta(n)}$ guarantees that $(a_{\eta(n)} - te_{\eta(n)})$ is a conventionally stable sequence (with the norms of the inverses being bounded above by $1/\delta$). Then, by part (a), the sequence $(a_n - te_n)$ itself is conventionally stable. Hence, $t \notin \sigma((a_n) + \mathcal{G}) = \limsup \sigma(a_n)$ by Corollary 3.5 which is a contradiction. □

As an example, consider once more the spline Galerkin method for singular integral operators, i.e. specify $\mathcal{F}$, $\mathcal{G}$ and $\mathcal{A}$ as in the previous section. By Corollary 2.10, $\mathcal{A}$ is a fractal algebra.

Proposition 3.7. *Let $(A_n) \in \mathcal{A}$ be a normal sequence. Then*

$$\limsup \sigma(A_n) = \liminf \sigma(A_n) = \bigcup_{s \in \mathbb{R} \cup \{\infty\}} \sigma(W_s(A_n)) \cup \bigcup_{t \in \mathbb{T}} \sigma(W^t(A_n)).$$

The proof follows immediately from the previous theorem, Theorem 2.9 and Corollary 3.5. This result applies, e.g., to the sequence (A_n), $A_n = L_n(aI + bS + K)L_n$ where a and b are real-valued functions with a being piecewise continuous and continuous on all non-integer points and b continuous on $\mathbb{R} \cup \{\infty\}$, and where the compact operator K is chosen such that the operator $aI + bS + K$ becomes self-adjoint.

Pseudospectra and their limiting sets. A computer working with finite accuracy cannot distinguish between a non-invertible matrix and an invertible matrix the inverse of which has a very large norm. This suggests the following definition of ε- invertibility which reflects finite accuracy. It is a remarkable observation mainly due to Landau, Reichel and Trefethen [9, 11, 17] that the related ε-pseudospectra also exhibit a much better convergence behaviour than the common spectra. Beautiful plots of limiting sets of pseudospectra can be found in [11, 17, 3].

Definition 3.8. Let $\mathcal{B}$ be a C^*- algebra with identity e and let $\varepsilon > 0$. An element $a \in \mathcal{B}$ is ε-invertible if it is invertible and if $\|a^{-1}\| < 1/\varepsilon$. The ε-pseudospectrum $\sigma^{(\varepsilon)}(a)$ of a consists of all $\lambda \in \mathbb{C}$ for which $a - \lambda e$ is not ε-invertible.

Let the C^*-algebras $\mathcal{C}_n$, $\mathcal{F}$ and $\mathcal{G}$ as above. The following result identifies the partial limiting set $\limsup \sigma^{(\varepsilon)}(a_n)$ of the ε-pseudospectra of a sequence $(a_n) \in \mathcal{F}$ with the ε-pseudospectra of the coset $(a_n) + \mathcal{G}$ in $\mathcal{F}/\mathcal{G}$. In contrast to the case $\varepsilon = 0$ (Corollary 3.5) this result holds for arbitrary sequences in $\mathcal{F}$.

Theorem 3.9. *Let $(a_n) \in \mathcal{F}$ and $\varepsilon > 0$. Then*

$$\limsup \sigma^{(\varepsilon)}_{\mathcal{C}_n}(a_n) = \sigma^{(\varepsilon)}_{\mathcal{F}/\mathcal{G}}((a_n) + \mathcal{G}).$$

The proof of Theorem 3.9 follows essentially by adapting Böttcher's proof for a special class of sequences in $\mathcal{F}$ which is related with the finite section method for

Toeplitz operators [2]. A generalization of Theorem 3.9 to the case of pseudospectra of operator polynomials is derived in [14].

In the presence of fractality of the sequence (a_n), one will expect equality between the partial and the uniform limiting sets of the pseudospectra of the a_n. Indeed, this equality holds as we will see now. For, we agree upon calling a sequence $(a_n) \in \mathcal{F}$ *ε-stable* if the coset $(a_n) + \mathcal{G}$ is ε-invertible in $\mathcal{F}/\mathcal{G}$.

Theorem 3.10. *Let $\mathcal{A}$ be a fractal C^*-subalgebra of $\mathcal{F}$ which contains the identity of $\mathcal{F}$, and let $\varepsilon > 0$.*
(a) A sequence $(a_n) \in \mathcal{A}$ is ε-stable if and only if it possesses an ε-stable subsequence.
(b) If $(a_n) \in \mathcal{A}$, then $\limsup \sigma^{(\varepsilon)}(a_n) = \liminf \sigma^{(\varepsilon)}(a_n) = \lim \sigma^{(\varepsilon)}(a_n)$.

Proof. (a) Suppose η is a strongly monotonically increasing sequence such that $(a_{\eta(n)}) = R_\eta(a_n)$ is an ε-stable subsequence of (a_n). Then the sequence (a_n) is stable by Theorem 3.6(a), and Theorem 2.11 implies that

$$1/\varepsilon > \|((a_{\eta(n)}) + \mathcal{G}_\eta)^{-1}\|_{\mathcal{F}/\mathcal{G}} = \|((a_n) + \mathcal{G})^{-1}\|_{\mathcal{F}/\mathcal{G}},$$

i.e. (a_n) is ε-stable.
(b) Assume t lies in $\limsup \sigma^{(\varepsilon)}(a_n) \setminus \liminf \sigma^{(\varepsilon)}(a_n)$. Then there is a subsequence η of $\mathbb{N}$ such that $t \notin \limsup \sigma^{(\varepsilon)}(a_{\eta(n)})$. By Theorem 3.9, the sequence $(a_{\eta(n)} - te_{\eta(n)})$ is ε-stable, whence via assertion (a) the ε-stability of the complete sequence $(a_n - te_n)$ follows. Again by Theorem 3.9, this implies $t \notin \limsup \sigma^{(\varepsilon)}(a_n)$ which contradicts the assumption. □

A consequence is the following result on spectral approximation of singular integral operators by a spline Galerkin method which, in contrast to Proposition 3.7, holds for arbitrary sequences $(L_n(aI + bS + K)L_n) \in \mathcal{A}$.

Proposition 3.11. *Let $\mathcal{A}$ be the algebra of spline approximation methods for singular integral operators as introduced in Section 2, and let $(A_n) \in \mathcal{A}$ and $\varepsilon > 0$. Then*

$$\limsup \sigma^{(\varepsilon)}(A_n) = \liminf \sigma^{(\varepsilon)}(A_n) = \bigcup_{s \in \mathbb{R} \cup \{\infty\}} \sigma^{(\varepsilon)}(W_s(A_n)) \cup \bigcup_{t \in \mathbb{T}} \sigma^{(\varepsilon)}(W^t(A_n)).$$

The proof of the latter identity follows from the fact that the function, which is defined on the disjoint union of $\mathbb{R} \cup \{\infty\}$ with $\mathbb{T}$ and which takes the value $\|W_s(a_n)\|$ at $s \in \mathbb{R} \cup \{\infty\}$ and the value $\|W^t(a_n)\|$ at $t \in \mathbb{T}$, attains its maximum (see [16]).

Numerical ranges and their limiting sets The (several kinds of) numerical ranges provide further examples of (upper) spectral approximants which share with the ε-pseudospectra their good asymptotic behaviour for all approximation sequences.

Let H be a (complex) Hilbert space with inner product $\langle \cdot , \cdot \rangle$, and let A be a linear bounded operator on H. The *spatial numerical range* $SN_H(A)$ of A is the set

$$SN_H(A) := \{\langle Ax, x\rangle : \ x \in H, \ \|x\| = 1\}.$$

In what follows we want to consider the asymptotic behaviour of the spatial numerical ranges $SN_{\mathbb{C}^n}(A_n)$ of a sequence (A_n) of approximation operators, and our goal is to identify the limiting set $\lim SN_{\mathbb{C}^n}(A_n)$ with a subset of $\mathbb{C}$ which depends on the coset $(A_n)+\mathcal{G}$ in $\mathcal{F}/\mathcal{G}$ only. So we need an analogue of the spatial numerical range for elements of a C^*-algebra. If $\mathcal{B}$ is a C^*-algebra with identity element e, then the *algebraic numerical range* $AN_{\mathcal{B}}(b)$ of $b \in \mathcal{B}$ is the set

$$AN_{\mathcal{B}}(b) := \{\varphi(b) : \varphi \in \mathcal{B}^* \ \text{ with } \ \|\varphi\| = \varphi(e) = 1\}.$$

For a detailed account on numerical ranges see, e.g., [4] and [6]. Here we only mention that $AN_{\mathcal{B}}(b)$ is a convex and compact subset of the complex plane, that $AN_{\mathcal{B}}(c) = AN_{\mathcal{C}}(c)$ for every C^*-subalgebra $\mathcal{C}$ of $\mathcal{B}$ and for every element $c \in \mathcal{C}$ (inverse closedness), and that

$$AN_{\mathcal{B}/\mathcal{J}}(b + \mathcal{J}) = \cap_{j\in\mathcal{J}} AN_{\mathcal{B}}(b + j) \tag{3.2}$$

for every closed ideal $\mathcal{J}$ of $\mathcal{B}$ and every element $b \in \mathcal{B}$. Further, for every bounded linear operator A on a Hilbert space H,

$$AN_{L(H)}(A) = \operatorname{clos} SN_H(A). \tag{3.3}$$

Again we start with a general result relating the partial limiting set of the numerical ranges of a sequence (a_n) to the numerical range of the coset of (a_n) modulo zero sequences.

Theorem 3.12. *For every sequence $(a_n) \in \mathcal{F}$ one has*

$$\operatorname{conv} \limsup_{n\to\infty} AN_{\mathcal{C}_n}(a_n) = AN_{\mathcal{F}/\mathcal{G}}((a_n) + \mathcal{G}).$$

Observe that the *partial* limiting set of a sequence of convex sets needs not to be convex again which explains the conv operator on the left hand side of this equality. For the proof of Theorem 3.12 we need two auxiliary results: one describing a convexity property of sequences of convex sets, and one characterizing the algebraic numerical range of elements of $\mathcal{F}$.

Lemma 3.13. *Let (M_k) be a monotonically decreasing (i.e. $M_k \supseteq M_{k+1}$ for all k) sequence of compact subsets of $\mathbb{C}$. Then* $\operatorname{conv} \cap_k M_k = \cap_k \operatorname{conv} M_k$.

The elementary proof is omitted.

Theorem 3.14. *For every sequence $(a_n) \in \mathcal{F}$ one has*

$$AN_{\mathcal{F}}((a_n)) \ = \ \operatorname{conv}\operatorname{clos} \cup_k AN_{\mathcal{C}_k}(a_k).$$

Proof. Let $m \in AN_{\mathcal{C}_k}(a_k)$ for some fixed k, and let ϕ be a state of $\mathcal{C}_k$ with $\phi(a_k) = m$. Then

$$\psi_\phi : \mathcal{F} \to \mathbb{C}, \quad (b_n)_{n=1}^\infty \mapsto \phi(b_k)$$

is a state of $\mathcal{F}$, and $\psi_\phi((a_n)) = \phi(a_k) = m$. Thus, $m \in AN_{\mathcal{F}}((a_n))$, which implies that $\cup_k AN_{\mathcal{C}_k}(a_k) \subseteq AN_{\mathcal{F}}((a_n))$. Since $AN_{\mathcal{F}}((a_n))$ is a closed convex set we arrive at the inclusion conv clos $\cup_k AN_{\mathcal{C}_k}(a_k) \subseteq AN_{\mathcal{F}}((a_n))$.

For the reverse inclusion, think of $\mathcal{C}_k$ as a C^*-algebra of linear bounded operators on some Hilbert space H_k, and let $\oplus H_k$ refer to the orthogonal sum of these Hilbert spaces. Evidently, the elements of $\mathcal{F}$ can be identified with diagonal operators acting on $\oplus H_k$ in an obvious manner. For simplicity, we denote the diagonal operator which corresponds to the sequence $(a_k) \in \mathcal{F}$ by (a_k) again.

¿From the inverse closedness with respect to algebraic numerical ranges and from (3.3) we infer that

$$AN_{\mathcal{F}}((a_n)) = AN_{L(\oplus H_k)}((a_n)) = \operatorname{clos} SN_{\oplus H_k}((a_n)). \tag{3.4}$$

Thus, given $m \in AN_{\mathcal{F}}((a_n))$ and $\varepsilon > 0$, there is a vector $(x_k) \in \oplus H_k$ with norm 1 such that

$$|m - \langle (a_k)(x_k), (x_k) \rangle_{\oplus H_k}| = |m - \sum_{k=1}^{\infty} \langle a_k x_k, x_k \rangle_{H_k}| < \varepsilon.$$

Let $\mathbb{M} \subseteq \mathbb{N}$ denote the set of all k with $x_k \neq 0$, choose arbitrary elements $y_k \in H_k$ with $\|y_k\| = 1$ for $k \notin \mathbb{M}$, and set

$$x'_k := \begin{cases} x_k / \|x_k\| & \text{if} \quad k \in \mathbb{M} \\ y_k & \text{if} \quad k \notin \mathbb{M}. \end{cases}$$

Then

$$\sum_{k=1}^{\infty} \langle a_k x_k, x_k \rangle = \sum_{k \in \mathbb{M}} \|x_k\|^2 \left\langle a_k \frac{x_k}{\|x_k\|}, \frac{x_k}{\|x_k\|} \right\rangle = \sum_{k=1}^{\infty} \|x_k\|^2 \langle a_k x'_k, x'_k \rangle$$

and, consequently,

$$\left| m - \sum_{k=1}^{\infty} \|x_k\|^2 \langle a_k x'_k, x'_k \rangle_{H_k} \right| < \varepsilon.$$

Because of $\|x_k\|^2 \geq 0$ and $\sum_{k=1}^{\infty} \|x_k\|^2 = \|(x_k)\|^2_{\oplus H_k} = 1$, this shows that m can be approximated by convex linear combinations of points $\langle a_k x'_k, x'_k \rangle \in \cup_k SN_{H_k}(a_k)$ as closely as desired. Hence, and by (3.3),

$$m \in \operatorname{clos}\operatorname{conv} \cup_k SN_{H_k}(a_k) \subseteq \operatorname{clos}\operatorname{conv} \cup_k AN_{L(H_k)}(a_k),$$

and employing inverse closedness once more we find

$$m \in \operatorname{clos}\operatorname{conv} \cup_k AN_{\mathcal{C}_k}(a_k). \tag{3.5}$$

Since $\cup_k AN_{\mathcal{C}_k}(a_k)$ is bounded (the radius of this set is not greater than $\|(a_k)\|_{\mathcal{F}}$), and since $\operatorname{clos}\operatorname{conv} M = \operatorname{conv}\operatorname{clos} M$ for every bounded subset M of the complex plane, (3.5) is just the assertion. □

We are now in position to prove Theorem 3.12.

Proof. Recall from (3.2) that

$$AN_{\mathcal{F}/\mathcal{G}}((a_n)+\mathcal{G}) = \cap_{(g_n)\in\mathcal{G}} AN_{\mathcal{F}}((a_n)+(g_n))$$

whence, in combination with Theorem 3.14,

$$AN_{\mathcal{F}/\mathcal{G}}((a_n)+\mathcal{G}) = \cap_{(g_n)\in\mathcal{G}} \operatorname{conv}\operatorname{clos}\cup_n AN_{\mathcal{C}_n}(a_n+g_n). \tag{3.6}$$

For every $k \in \mathbb{N}$, choose an $m_k \in AN_{\mathcal{C}_n}(a_n)$ and define a sequence $(G_n^{(k)}) \in \mathcal{G}$ by

$$g_n^{(k)} = \begin{cases} -a_n + m_k e_n & \text{if} \quad n \le k-1 \\ 0 & \text{if} \quad n \ge k. \end{cases}$$

Then

$$AN_{\mathcal{C}_n}(a_n + g_n^{(k)}) = \begin{cases} \{m_k\} \subseteq AN_{\mathcal{C}_n}(a_n) & \text{if} \quad n \le k-1 \\ AN_{\mathcal{C}_n}(a_n) & \text{if} \quad n \ge k, \end{cases}$$

whence

$$AN_{\mathcal{F}/\mathcal{G}}((a_n)+\mathcal{G}) \subseteq \cap_{k=1}^{\infty} \operatorname{conv}\operatorname{clos}\cup_{n\ge k} AN_{\mathcal{C}_n}(a_n).$$

Applying Lemma 3.13 to the sets $M_k := \operatorname{clos}\cup_{n\ge k} AN_{\mathcal{C}_n}(a_n)$ gives

$$AN_{\mathcal{F}/\mathcal{G}}((a_n)+\mathcal{G}) \subseteq \operatorname{conv} \cap_{k=1}^{\infty} \operatorname{clos}\cup_{n\ge k} AN_{\mathcal{C}_n}(a_n),$$

and it is not hard to check that the set on the right hand side coincides with $\operatorname{conv}\limsup AN_{\mathcal{C}_n}(a_n)$ which proves one half of the theorem.

For the second half, let $m \in \limsup AN_{\mathcal{C}_n}(a_n)$. Then there exist a subsequence $(n_k) \subseteq \mathbb{N}$ tending to infinity as $k \to \infty$ as well as points $m_k \in AN_{\mathcal{C}_{n_k}}(a_{n_k})$ which converge to m as $k \to \infty$. Choose states ϕ_k of $\mathcal{C}_{n_k}$ with $\phi_k(a_{n_k}) = m_k$, and let $\mathcal{L}$ stand for the linear subspace of $\mathcal{F}$ consisting of all sequences

$$(b_n) := \alpha(a_n) + \beta(e_n) + (g_n)$$

where e_n is again the identity of $\mathcal{C}_n$, $\alpha, \beta \in \mathbb{C}$, and (g_n) runs through the ideal $\mathcal{G}$. If $(b_n) \in \mathcal{L}$, then the limit $\lim_{k\to\infty} \phi_k(b_{n_k})$ exists. Indeed, it is sufficient to check the existence of this limit for the generating sequences of $\mathcal{L}$, for which one has

$$\phi_k(a_{n_k}) = m_k \to m, \ \phi_k(e_{n_k}) = 1 \to 1 \text{ and } |\phi_k(g_{n_k})| \le \|\phi_k\| \cdot \|g_{n_k}\| = \|g_{n_k}\| \to 0.$$

Thus, we get a linear functional ϕ on $\mathcal{L}$ defined by

$$\phi((b_n)) = \lim_{k\to\infty} \phi_k(b_{n_k})$$

which maps the sequences $(a_n), (e_n)$ and (g_n) into $m, 1$ and 0, respectively, and which is continuous with norm 1:

$$|\phi((b_n))| = |\lim \phi_k(b_{n_k})| \leq \sup_k |\phi_k(b_{n_k})| \leq \sup_k \|b_{n_k}\| \leq \|(b_n)\|_{\mathcal{F}}.$$

Using the Hahn-Banach theorem, one can extend ϕ to a linear functional with norm 1 on all of $\mathcal{F}$. We denote this extension by ϕ again. Since $\|\phi\| = \phi((e_n)) = 1$, this functional is a state on $\mathcal{F}$. Further, the ideal $\mathcal{G}$ lies in the kernel of ϕ by construction, so one can define a further functional ψ acting on $\mathcal{F}/\mathcal{G}$ by

$$\psi : \mathcal{F}/\mathcal{G} \to \mathbb{C}, \quad (c_n) + \mathcal{G} \mapsto \phi((c_n)).$$

Clearly, $\psi((e_n) + \mathcal{G}) = 1$ and $\psi((a_n) + \mathcal{G}) = m$, and ψ is continuous with norm 1. Thus, ψ is a state of $\mathcal{F}/\mathcal{G}$ which implies that $m \in AN_{\mathcal{F}/\mathcal{G}}((a_n) + \mathcal{G})$ and, consequently,

$$\limsup AN_{\mathcal{C}_n}(a_n) \subseteq AN_{\mathcal{F}/\mathcal{G}}((a_n) + \mathcal{G}).$$

Finally, since algebraic numerical ranges are convex,

$$\text{conv} \limsup AN_{\mathcal{C}_n}(a_n) \subseteq AN_{\mathcal{F}/\mathcal{G}}((a_n) + \mathcal{G})$$

which verifies the second half of the theorem. □

Our final goal in this section is consequences of the fractality of the sequence (a_n) for the asymptotic behaviour of the partial and uniform limiting sets of the numerical ranges of a_n.

Theorem 3.15. *Let $\mathcal{A}$ be a fractal C^*-subalgebra of $\mathcal{F}$ which contains the identity. Then, for every $(a_n) \in \mathcal{A}$,*

$$\limsup AN_{\mathcal{C}_n}(a_n) = \liminf AN_{\mathcal{C}_n}(a_n).$$

Proof. The inclusion $\liminf AN_{\mathcal{C}_n}(a_n) \subseteq \limsup AN_{\mathcal{C}_n}(a_n)$ is obvious. For the reverse inclusion recall that the numerical ranges $AN_{\mathcal{C}_n}(a_n)$ are convex, and that the *uniform* limiting set of convex sets is convex again. Thus, the inclusion $\limsup AN_{\mathcal{C}_n}(a_n) \subseteq \liminf AN_{\mathcal{C}_n}(a_n)$ holds if and only if

$$\text{conv} \limsup AN_{\mathcal{C}_n}(a_n) \subseteq \liminf AN_{\mathcal{C}_n}(a_n).$$

So, by Theorem 3.12, what one has to prove is that

$$AN_{\mathcal{F}/\mathcal{G}}((a_n) + \mathcal{G}) \subseteq \liminf AN_{\mathcal{C}_n}(a_n) \quad \text{for every } (a_n) \in \mathcal{A}.$$

¿From the inverse closedness with respect to algebraic numerical ranges we conclude that $AN_{\mathcal{F}/\mathcal{G}}((a_n) + \mathcal{G}) = AN_{(\mathcal{A}+\mathcal{G})/\mathcal{G}}((a_n) + \mathcal{G})$, and the third isomorphy theorem implies that

$$AN_{\mathcal{F}/\mathcal{G}}((a_n) + \mathcal{G}) = AN_{\mathcal{A}/(\mathcal{A}\cap\mathcal{G})}((a_n) + \mathcal{A} \cap \mathcal{G}).$$

Let $m \in AN_{\mathcal{A}/(\mathcal{A}\cap\mathcal{G})}((a_n) + \mathcal{A}\cap\mathcal{G})$, and let ϕ be a state of $\mathcal{A}/(\mathcal{A}\cap\mathcal{G})$ such that $m = \phi((a_n) + \mathcal{A}\cap\mathcal{G}) = \phi(\pi(a_n))$ where we use the notations of Section 2. Since π is fractal, $m = \phi(\pi_\eta R_\eta(a_n))$ for every monotonically increasing sequence η. It is obvious that $\phi \circ \pi_\eta$ is a state on $\mathcal{A}_\eta = R_\eta \mathcal{A}$. Hence,

$$m \in AN_{\mathcal{A}_\eta}(R_\eta(a_n)) = AN_{\mathcal{F}_\eta}(R_\eta(a_n)),$$

whence, by Theorem 3.14,

$$m \in \operatorname{conv} \operatorname{clos} \cup_n AN_{\mathcal{C}_{\eta(n)}}(a_{\eta(n)}) \quad \text{for every } \eta. \tag{3.7}$$

Now assume there exists an $m \in AN_{\mathcal{F}/\mathcal{G}}((a_n) + \mathcal{G})$ which does not belong to the uniform limiting set of the $AN_{\mathcal{C}_n}(a_n)$. Then there is a strongly monotonically increasing sequence $\eta^* \subseteq \mathbb{N}$ such that $\operatorname{dist}(m, AN_{\mathcal{C}_{\eta^*(n)}}(a_{\eta^*(n)})) \geq d > 0$ for all n.

Due to the compactness and convexity of $AN_{\mathcal{C}_{\eta^*(n)}}(a_{\eta^*(n)})$, there exist points $m_{\eta^*(n)}$ in $AN_{\mathcal{C}_{\eta^*(n)}}(a_{\eta^*(n)})$ such that

$$|m - m_{\eta^*(n)}| = \operatorname{dist}(m, AN_{\mathcal{C}_{\eta^*(n)}}(a_{\eta^*(n)})),$$

and these points $m_{\eta^*(n)}$ are unique for every n. It is further clear that all numbers $m_{\eta^*(n)}$ lie in the disk around the origin with radius $r := \sup \|a_n\| = \|(a_n)\|_{\mathcal{F}}$. Hence, there is at least one cluster point m^* of the sequence $(m_{\eta^*(n)})_{n\in\mathbb{N}}$. Given $\varepsilon > 0$, consider those $\eta^*(n)$ for which the $m_{\eta^*(n)}$ belong to the ε-neighbourhood U of m^*. These $\eta^*(n)$ single out an subsequence η_ε of η^*. A little thought reveals that then

$$\operatorname{dist}(m, \operatorname{conv} \operatorname{clos} \cup_{\eta_\varepsilon(n)} AN_{\mathcal{C}_{\eta_\varepsilon(n)}}(a_{\eta_\varepsilon(n)})) \geq d/2 \tag{3.8}$$

if only ε is small enough. Since (3.7) holds for every sequence (in particular for the sequence η_ε), (3.8) contradicts (3.7). □

Uniform limiting sets of convex sets are convex again. Hence, Theorems 3.12 and 3.15 yield

Corollary 3.16. *Let $\mathcal{A}$ be a fractal C^*-subalgebra of $\mathcal{F}$ which contains the identity. Then, for every sequence $(a_n) \in \mathcal{A}$,*

$$\limsup AN_{\mathcal{C}_n}(a_n) = \liminf AN_{\mathcal{C}_n}(a_n) = AN_{\mathcal{F}/\mathcal{G}}((a_n) + \mathcal{G}).$$

To illlustrate these results, we consider once again spline Galerkin methods.

Proposition 3.17. *Let $\mathcal{A}$ be the algebra of spline approximation methods for singular integral operators and $(A_n) \in \mathcal{A}$. Then*

$$\begin{aligned}
\limsup AN_{\mathbb{C}^{n\times n}}(A_n) &= \limsup SN_{\mathbb{C}^n}(A_n) \\
&= \liminf AN_{\mathbb{C}^{n\times n}}(A_n) = \liminf SN_{\mathbb{C}^n}(A_n) \\
&= AN_{\mathcal{F}/\mathcal{G}}((A_n) + \mathcal{G}) \\
&= \operatorname{conv}\left(\bigcup_{s\in\mathbb{R}\cup\{\infty\}} AN_{L(l^2)}(W_s(A_n)) \cup \bigcup_{t\in\mathbb{T}} AN_{L(l^2)}(W^t(A)) \right).
\end{aligned}$$

Proof. The identities in the first lines are consequences of the preceding corollary and of (3.3). The last identity can be checked by repeating the arguments of the proof of Theorem 3.14, where an analogous result for sequences instead of pairs of operators is derived, and by taking into account the convexity of algebraic numerical ranges. □

4. On non-fractal sequences

The main goal of this concluding section is to prove that *every* approximation sequence possesses a fractal subsequence. For, we need some results on fractality of self-adjoint sequences.

Fractality of self-adjoint sequences. If (A_n) is a self-adjoint fractal approximation sequence then, as we know from Theorem 3.6(b),

$$\limsup \sigma(A_n) = \liminf \sigma(A_n). \tag{4.1}$$

Now we will see that, conversely, (4.1) is the only obstruction for a self-adjoint bounded sequence to be fractal.

Theorem 4.1. *Let $\mathcal{F}$ be as above and let $(a_n) \in \mathcal{F}$ be a self-adjoint sequence. Then (a_n) is fractal if and only if the equality (4.1) holds.*

Proof. The 'only if'-part is Theorem 3.6(b). For the reverse conclusion suppose that (4.1) holds. Let $\mathcal{A}$ denote the smallest closed subalgebra of $\mathcal{F}$ which contains the sequence (a_n) and the identity sequence (e_n) of $\mathcal{F}$. Further, given a monotonically increasing sequence η, write $\mathcal{A}_\eta$ for the algebra $R_\eta \mathcal{A}$. We claim that

$$\mathcal{A}/(\mathcal{A} \cap \mathcal{G}) \quad \text{is isomorphic to} \quad \mathcal{A}_\eta/(\mathcal{A}_\eta \cap \mathcal{G}_\eta) \tag{4.2}$$

with the isomorphism given by

$$(b_n) + (\mathcal{A} \cap \mathcal{G}) \mapsto (b_{\eta(n)}) + (\mathcal{A}_\eta \cap \mathcal{G}_\eta). \tag{4.3}$$

To get the claim recall that $\mathcal{A}/(\mathcal{A}\cap\mathcal{G})$ and $\mathcal{A}_\eta/(\mathcal{A}_\eta\cap\mathcal{G}_\eta)$ are isomorphic to $(\mathcal{A}+\mathcal{G})/\mathcal{G}$ and $(\mathcal{A}_\eta + \mathcal{G}_\eta)/\mathcal{G}_\eta$, respectively. The latter algebras are singly generated by their elements $(a_n)+\mathcal{G}$ and $(a_{\eta(n)})+\mathcal{G}_\eta$, and the spectra of these cosets are $\limsup \sigma(a_n)$ and $\limsup \sigma(a_{\eta(n)})$ due to Corollary 3.5, respectively. The assumption (4.1) guarantees that these spectra coincide; hence, by the Gelfand-Naimark theorem for singly generated C^*-algebras, the isomorphy (4.2) follows.

Let now π stand for the canonical homomorphism from $\mathcal{A}$ onto $\mathcal{A}/(\mathcal{A}\cap\mathcal{G})$ and let η be a monotonically increasing sequence. Then, evidently, $\pi = \pi_\eta R_\eta = \varphi_\eta \psi_\eta R_\eta$

where ψ_η is the canonical homomorphism from $\mathcal{A}_\eta$ onto $\mathcal{A}_\eta/(\mathcal{A}_\eta \cap \mathcal{G}_\eta)$ and where φ_η if the inverse of the isomorphism (4.3). Hence, π is fractal. □

As a first application of the previous theorem we derive a fractality result for the sequence of the finite sections of a self-adjoint operator. Observe in this connection that the spectrum of a self-adjoint operator A for which the finite section method (P_nAP_n) is fractal can be as complicated as possible: Given an arbitrary compact subset K of the real line, choose a dense subsequence $(k_i)_{i=1}^\infty$ in K, and consider the operator $A = \operatorname{diag}(k_1, k_2, k_3, \ldots)$. Then

$$\limsup \sigma(P_nAP_n) = \liminf \sigma(P_nAP_n) = \lim \sigma(P_nAP_n) = K.$$

Theorem 4.2. *If $A \in L(H)$ is a self-adjoint operator with connected spectrum, then the sequence (P_nAP_n) is fractal.*

Proof. Let $\lambda \in \mathbb{R}$. If the operator $A - \lambda I$ is invertible, then it is either positively or negatively definite. Clearly, definiteness implies the stability of the sequence $(P_n(A - \lambda I)P_n)$. Conversely, if the sequence $(P_n(A - \lambda I)P_n)$ is stable, then the operator $A - \lambda I$ is invertible due to Polski's theorem ([7], Theorem 1.2). Thus,

$$\sigma(A) = \sigma_{\mathcal{F}/\mathcal{G}}((P_nAP_n) + \mathcal{G}). \tag{4.4}$$

Further we know from Corollary 3.5 that

$$\sigma_{\mathcal{F}/\mathcal{G}}((P_nAP_n) + \mathcal{G}) = \limsup \sigma(P_nAP_n). \tag{4.5}$$

Finally, we claim that

$$\sigma(A) \subseteq \liminf \sigma(P_nAP_n). \tag{4.6}$$

Once (4.6) is verified, (4.4) - (4.6) give $\limsup \sigma(P_nAP_n) = \liminf \sigma(P_nAP_n)$ which is equivalent to the fractality of (P_nAP_n) by Theorem 4.1.

It remains to check (4.6) where we follow arguments from [1]. Suppose λ is a real number which is not in $\liminf \sigma(P_nAP_n)$. We claim that then $A - \lambda I$ is invertible.

Let H_n denote the range of P_n. Since $\lambda \in \mathbb{R} \setminus \liminf \sigma(P_nAP_n)$, there is an $\varepsilon > 0$ as well as an infinite subsequence η of N such that

$$\sigma(P_nAP_n) \cap (\lambda - \varepsilon, \lambda + \varepsilon) = \emptyset \quad \text{for all } n \in \eta.$$

Thus, the distance of $\sigma(P_nAP_n)$ to λ is at least ε, which shows that the operators $P_nAP_n - \lambda I|_{H_n}$ are invertible and that their inverses are uniformly bounded:

$$\sup_{n \in \eta} \|(P_nAP_n - \lambda I|_{H_n})^{-1}\| \leq 1/\varepsilon. \tag{4.7}$$

The strong convergence of $(P_nAP_n - \lambda I)$ to $A - \lambda I$ together with the uniform boundedness (4.7) readily implies the strong convergence of the sequence of the inverses $((P_nAP_n - \lambda I)^{-1})$. Write B for the strong limit of this sequence. Letting n go to infinity in

$$(P_nAP_n - \lambda I|_{H_n})^{-1}(P_nAP_n - \lambda I|_{H_n}) = (I|_{H_n})$$

yields $B(A - \lambda I) = I$. Analogously, $(A - \lambda I)B = I$ whence the invertibility of $A - \lambda I$. □

Existence of fractal subsequences. There are simple examples such as

$$A = \operatorname{diag}\left(\begin{pmatrix} 0 & 1 \\ 1 & 0 \end{pmatrix}, \begin{pmatrix} 0 & 1 \\ 1 & 0 \end{pmatrix}, \ldots\right) \in L(l^2) \tag{4.8}$$

which show that the finite section method for self-adjoint operators is not necessarily fractal: For A as in (4.8) one has

$$\sigma(P_{2n}AP_{2n}) = \{-1, 1\} \neq \sigma(P_{2n+1}AP_{2n+1}) = \{-1, 1, 0\}$$

for all n. In the case at hand, it turns out that the (non-fractal) sequence (P_nAP_n) possesses a fractal subsequence (formed by the matrices of even order). So one might ask whether *every* (self-adjoint or not) approximation sequence possesses a fractal subsequence. For a long time I conjectured that the answer is *no* and, motivated by Theorem 4.2, I moreover conjectured that if (A_n) is a *completely non-fractal sequence* (i.e. if no infinite subsequence of (A_n) is fractal), then $\sigma(A)$ is a set of Cantor type. Then T. Ehrhardt draw my attention to the connection between the coincidence of the partial and the uniform limiting set on the one hand and the convergence with respect to the Hausdorff metric on the other hand, as well as to Theorem 3.3, which in combination with Theorem 4.1 gave a surprisingly simple proof of the fact that the converse of my conjecture is true.

Theorem 4.3. *Let $\mathcal{F}$ be as above. Every self-adjoint sequence $(a_n) \in \mathcal{F}$ possesses a fractal subsequence.*

Proof. Consider the sets $M_n := \sigma(a_n)$. By Theorem 3.3, there exists a subsequence $(M_{\eta(n)})_{n\geq 1}$ of (M_n) which converges with respect to the Hausdorff metric. Hence,

$$\limsup_{n\to\infty} (M_{\eta(n)}) = \liminf_{n\to\infty} (M_{\eta(n)}),$$

and the sequence $(a_{\eta(n)})_{n\geq 1}$ is fractal due to Theorem 4.1. □

Moreover, based on this result, it is now possible to verify the existence of a fractal subsequence for *every* (not necessarily self-adjoint) sequence. Actually, a little bit more can be shown.

Theorem 4.4. *Let $\mathcal{F}$ be as above, and let $\mathcal{A}$ be a separable C^*-subalgebra of $\mathcal{F}$. Then there exists a sequence $\eta \subseteq \mathbb{N}$ such that the algebra $\mathcal{A}_\eta = R_\eta \mathcal{A} \subseteq \mathcal{F}_\eta$ is fractal.*

Since every finitely generated C^*-algebra is separable, this result immediately implies:

Theorem 4.5. *Every sequence in $\mathcal{F}$ possesses a fractal subsequence.*

One cannot expect that Theorem 4.4 holds for arbitrary C^*-subalgebras of $\mathcal{F}$; for example it is certainly not valid for l^∞. On the other hand, there exist non-separable but fractal algebras; the algebra $\mathcal{A}$ related with spline Galerkin methods for singular integral operators can serve as an example.

The proof of Theorem 4.4 will make use of the following equivalent characterization of fractal algebras which is a simple consequence of the third isomorphy theorem.

Lemma 4.6. *A C^*-algebra $\mathcal{A} \subseteq \mathcal{F}$ is fractal if and only if the restriction of the canonical homomorphism $\pi : \mathcal{F} \to \mathcal{F}/\mathcal{G}$ onto $\mathcal{A}$ is fractal.*

Here is the proof of Theorem 4.4.

Proof. Let $\mathcal{A}$ be a separable C^*-subalgebra of $\mathcal{F}$ with a countable dense subset $((a_n^{(k)})_{n\geq 1})_{k\geq 1}$. Denote by $b_n^{(2k)}$ and $b_n^{(2k+1)}$ the real and the imaginary part of $a_n^{(k)}$, respectively, and write $\mathcal{B} \subseteq \mathcal{A}$ for the set of all sequences $(b_n^{(k)})_{n\geq 1}$ with $k \geq 1$ and $\mathcal{D} \subseteq \mathcal{A}$ for the set of all difference sequences $(b_n^{(k)})_{n\geq 1} - (b_n^{(l)})_{n\geq 1}$ with $k, l \geq 1$. The set $\mathcal{B} \cup \mathcal{D}$ is countable, and each of its elements is self-adjoint.

Let $((d_n^{(k)})_{n\geq 1})_{k\geq 1}$ be any numeration of the elements of $\mathcal{B} \cup \mathcal{D}$. By Theorem 4.3, every sequence $(d_n^{(k)})_{n\geq 1}$ possesses a fractal subsequence. By employing a standard diagonalization process, we will construct a sequence $\eta \in \mathbb{N}$ such that the sequence $(d_{\eta(n)}^{(k)})_{n\geq 1}$ is fractal for every $k \geq 1$.

Let $\eta_1 \subseteq \mathbb{N}$ be a sequence such that $(d_{\eta_1(n)}^{(1)})_{n\geq 1}$ is a fractal sequence and, for every $k \geq 2$, choose a subsequence η_k of η_{k-1} such that $(d_{\eta_k(n)}^{(k)})_{n\geq 1}$ is a fractal sequence. Then define η by $\eta(n) := \eta_n(n)$. The sequence η coincides (with possible exception of at most finitely many entries) with a subsequence of η_k for every k. Hence, every sequence in

$$\mathcal{D}_\eta := R_\eta \mathcal{D} = \{(d_{\eta(n)}^{(k)})_{n\geq 1},\ k \geq 1\}$$

is fractal.

We claim that the algebra $\mathcal{A}_\eta := R_\eta \mathcal{A}$ is fractal. What we have to verify is that, given a subsequence μ of η, there is a homomorphism $\hat{\pi}_\mu$ such that

$$\hat{\pi}|_{\mathcal{A}_\eta} = \hat{\pi}_\mu R_\mu|_{\mathcal{A}_\eta}$$

where $\hat{\pi}$ is the canonical homomorphism from $\mathcal{F}_\eta$ onto $\mathcal{F}_\eta/\mathcal{G}_\eta$ (Lemma 4.6). Observe that the set of all sequences

$$(a_{\eta(n)}^{(k)})_{n\geq 1} = (b_{\eta(n)}^{(2k)})_{n\geq 1} + i\,(b_{\eta(n)}^{(2k+1)})_{n\geq 1}$$

with $k = 1, 2, \ldots$ is dense in $\mathcal{A}_\eta$. So, without loss of generality, we can assume $\eta = \mathbb{N}$ in what follows. That is, we will have to deal with the following situation:

$\mathcal{A}$ is a C^*-subalgebra of $\mathcal{F}$ with countable dense subset $((a_n^{(k)})_{n\geq 1})_{k\geq 1}$ such that each of the self-adjoint sequences $(b_n^{(k)})_{n\geq 1}$ and $(b_n^{(k)})_{n\geq 1} - (b_n^{(l)})_{n\geq 1}$ with $k, l \geq 1$ is fractal. We have to show that $\mathcal{A}$ is fractal, i.e. given a subsequence μ of $\mathbb{N}$, we have to define a homomorphism π_μ such that $\pi|_\mathcal{A} = \pi_\mu R_\mu|_\mathcal{A}$, π referring to the canonical homomorphism from $\mathcal{F}$ onto $\mathcal{F}/\mathcal{G}$.

Write $\mathcal{B}$ and $\mathcal{D}$ in place of $\mathcal{B}_\eta$ and $\mathcal{D}_\eta$. Let μ be a subsequence of $\mathbb{N}$. We start with defining the mapping π_μ on the set of the self-adjoint elements of $\mathcal{A}_\mu$. So let $(a_n) \in \mathcal{A}$ and assume $(a_{\mu(n)})_{n\geq 1}$ is self-adjoint.

Claim 1. *There is a sequence $((c_n^{(k)})_{n\geq 1})_{k\geq 1} \subseteq \mathcal{B}$ such that*

$$\|(a_{\mu(n)})_{n\geq 1} - (c_{\mu(n)}^{(k)})_{n\geq 1}\|_{\mathcal{F}_\mu} \to 0 \quad as \quad k \to \infty. \tag{4.9}$$

Indeed, write a_n as $\operatorname{Re} a_n + i\operatorname{Im} a_n$. Since $((a_n^{(k)})_{n\geq 1})_{k\geq 1}$ is a dense subset of $\mathcal{A}$, we can approximate the sequence $(\operatorname{Re} a_n)_{n\geq 1}$ as closely as desired by sequences of the form $(\operatorname{Re} a_n^{(k)})_{n\geq 1} = (b_n^{(2k)})_{n\geq 1}$. Then, clearly, the sequence $(\operatorname{Re} a_{\mu(n)})_{n\geq 1}$ can be approximated as closely as desired by sequences of the form $(b_{\mu(n)}^{(2k)})_{n\geq 1} \in \mathcal{B}_\mu$, and since $\operatorname{Re} a_{\mu(n)} = a_{\mu(n)}$ by hypothesis, this gives the claim.

Now, given a self-adjoint sequence $(a_{\mu(n)}) \in \mathcal{A}_\mu$, choose and fix a sequence $((c_{\mu(n)}^{(k)})_{n\geq 1})_{k\geq 1} \subseteq \mathcal{B}_\mu$ with property (4.9) and, for every k, choose a sequence $(\hat{c}_n^{(k)}) \in \mathcal{B}$ with $R_\mu(\hat{c}_n^{(k)}) = (\hat{c}_{\mu(n)}^{(k)})$.

Let $\mathcal{A}^{(k)}$ refer to the smallest C^*-subalgebra of $\mathcal{A}$ which contains the sequence $(\hat{c}_n^{(k)})_{n\geq 1}$. The algebras $\mathcal{A}^{(k)}$ are fractal by construction, hence (Lemma 4.6), there are homomorphisms $\pi_\mu^{(k)}$ such that

$$\pi|_{\mathcal{A}^{(k)}} = \pi_\mu^{(k)} R_\mu|_{\mathcal{A}^{(k)}} \quad \text{for every } k.$$

In particular,

$$\pi_\mu^{(k)}(\hat{c}_{\mu(n)}^{(k)}) = (\hat{c}_n^{(k)}) + \mathcal{G}.$$

Moreover, the coset $(\hat{c}_n^{(k)}) + \mathcal{G}$ turns out to be independent of the choice of the 'representative' $(\hat{c}_n^{(k)})$ of the sequence $(\hat{c}_{\mu(n)}^{(k)})$:

Claim 2. *Let (c_n), $(d_n) \in \mathcal{B}$ be sequences with $c_{\mu(n)} = d_{\mu(n)}$ for every n. Then $(c_n) + \mathcal{G} = (d_n) + \mathcal{G}$.*

Indeed, the sequence $(c_n - d_n)$ belongs to $\mathcal{D}$ and is, thus, fractal by construction. By Theorem 2.11 (*b*), the limit $\lim \|c_n - d_n\|$ exists and is equal to $\|(c_n - d_n) + \mathcal{G}\|$. On the other hand, this limit is zero since infinitely many of the differences $c_n - d_n$ are zero by assumption. Hence, $(c_n - d_n) \in \mathcal{G}$, proving our claim.

Thus, knowing only the subsequences $(c^{(k)}_{\mu(n)})_{n\geq1}$, one can rediscover the cosets $(c^{(k)}_n)+\mathcal{G}$ uniquely.

Claim 3. *As $k\to\infty$, the cosets $(c^{(k)}_n)+\mathcal{G}$ converge in $\mathcal{F}/\mathcal{G}$.*

This follows from

$$\begin{aligned}\|(c^{(k)}_n)+\mathcal{G}-(c^{(l)}_n)+\mathcal{G}\|_{\mathcal{F}/\mathcal{G}} &= \|(c^{(k)}_{\mu(n)})+\mathcal{G}_\mu-(c^{(l)}_{\mu(n)})+\mathcal{G}_\mu\|_{\mathcal{F}_\mu/\mathcal{G}_\mu}\\ &\leq \|(c^{(k)}_{\mu(n)})-(c^{(l)}_{\mu(n)})\|_{\mathcal{F}_\mu},\end{aligned} \tag{4.10}$$

where the equality is a consequence of the fractality of the sequence $(c^{(k)}_n-c^{(l)}_n)\in\mathcal{D}$, and from the convergence of the sequences $(c^{(k)}_{\mu(n)})_{n\geq1}$ to $(a_{\mu(n)})_{n\geq1}$. Hence $((c^{(k)}_n)+\mathcal{G})_{k\geq1}$ is a Cauchy sequence and thus convergent, which verifies Claim 3.

Let $(c_n)+\mathcal{G}$ denote the limit of the sequence $((c^{(k)}_n)+\mathcal{G})_{k\geq1}$.

Claim 4. *The coset $(c_n)+\mathcal{G}$ does not depend on the choice of the sequence $((c^{(k)}_{\mu(n)})_{n\geq1})_{k\geq1}$ which approximates $(a_{\mu(n)})$.*

Indeed, consider besides $((c^{(k)}_{\mu(n)})_{n\geq1})_{k\geq1}$ another sequence $((d^{(k)}_{\mu(n)})_{n\geq1})_{k\geq1}\subseteq\mathcal{B}_\mu$ which also converges to $(a_{\mu(n)})$ and which generates (in the same way as the sequence $((c^{(k)}_{\mu(n)}))$ does) a coset $(d_n)+\mathcal{G}$. Then we have as in (4.10)

$$\begin{aligned}\|(c_n)+\mathcal{G}-(d_n)+\mathcal{G}\|_{\mathcal{F}/\mathcal{G}} &= \lim_{k\to\infty}\|(c^{(k)}_n)+\mathcal{G}-(d^{(k)}_n)+\mathcal{G}\|_{\mathcal{F}/\mathcal{G}}\\ &\leq \limsup_{k\to\infty}\|(c^{(k)}_{\mu(n)})-(d^{(k)}_{\mu(n)})\|_{\mathcal{F}_\mu}\end{aligned} \tag{4.11}$$

(recall that the sequence $(c^{(k)}_n-d^{(k)}_n)$ is fractal by construction). Since both sequences $((c^{(k)}_{\mu(n)})_{n\geq1})_{k\geq1}$ and $((d^{(k)}_{\mu(n)})_{n\geq1})_{k\geq1}$ have the same limit as $k\to\infty$, the right hand side of (4.11) tends to zero which gives the claim.

Thus, every self-adjoint sequence $(a_{\mu(n)})\in\mathcal{A}_\mu$ corresponds to a uniquely determined coset $(c_n)+\mathcal{G}$ which we denote by $\pi_\mu(a_{\mu(n)})$.

Claim 5. *For every self-adjoint sequence $(a_n)\in\mathcal{A}$, and every sequence μ,*

$$\pi_\mu(a_{\mu(n)})_{n\geq1}=(a_n)+\mathcal{G}. \tag{4.12}$$

Indeed, possibly there are several sequences $((c^{(k)}_{\mu(n)})_{n\geq1})_{k\geq1}$ which converge to $(a_{\mu(n)})$. But among these sequences there is, by assumption, at least one such that $(c^{(k)}_n)\to(a_n)$ in $\mathcal{F}$. For this special sequence, one has

$$(c^{(k)}_n)+\mathcal{G}\to(a_n)+\mathcal{G}\quad\text{in }\mathcal{F}/\mathcal{G}.$$

The limit $\lim_{k\to\infty}((c^{(k)}_n)+\mathcal{G})$ is, as we have seen in Claim 4, independent of the choice of $(c^{(k)}_n)$. Hence,

$$\pi_\mu(a_{\mu(n)})=(a_n)+\mathcal{G},$$

which settles the construction of π_μ on the set of the self-adjoint sequences of $\mathcal{A}_\mu$.

If $(a_{\mu(n)})_{n\geq 1}$ is an arbitrary (not necessarily self-adjoint) sequence in $\mathcal{A}_\mu$, then define

$$\pi_\mu(a_{\mu(n)}) := \pi_\mu(\mathrm{Re}\, a_{\mu(n)}) + i\, \pi_\mu(\mathrm{Im}\, a_{\mu(n)}).$$

Due to (4.12),

$$\pi_\mu(a_{\mu(n)}) = (\mathrm{Re}\, a_n) + \mathcal{G} + i\,((\mathrm{Im}\, a_n) + \mathcal{G}) = (a_n) + \mathcal{G},$$

whence $\pi_\mu R_\mu|_{\mathcal{A}} = \pi|_{\mathcal{A}}$ as desired. □

References

[1] W. Arveson, *C^*-algebras and numerical linear algebra.* J. Funct. Anal. **122** (1994), 333–360.

[2] A. Böttcher, *Pseudospectra and singular values of large convolution operators.* J. Integral Equations Appl. **6** (1994), 3, 267–301.

[3] A. Böttcher, B. Silbermann, *Introduction to Large Truncated Toeplitz Matrices.* Springer-Verlag, New York, Berlin, Heidelberg 1999.

[4] F. F. Bonsall, J. Duncan, *Numerical Ranges of Operators on Normed Spaces and of Elements of Normed Algebras.* London Math. Soc. Lecture Note Series **2**, Cambridge University Press, Cambridge 1971.

[5] F. Chatelin, *Spectral Approximation of Linear Operators.* Academic Press, New York 1983.

[6] K. E. Gustafson, D. K. M. Rao, *Numerical Range. The Field of Values of Linear Operators and Matrices.* Springer Verlag, New York 1996.

[7] R. Hagen, S. Roch, B. Silbermann, *Spectral Theory of Approximation Methods for Convolution Equations.* Birkhäuser Verlag, Basel, Boston, Berlin 1995.

[8] F. Hausdorff, *Set Theory.* Chelsea Publ. Comp., New York 1957.

[9] H. Landau, *The notion of approximate eigenvalues applied to an integral equation of laser theory.* Q. Appl. Math., April 1977, 165 – 171.

[10] B. V. Limaye, *Spectral Perturbation and Approximation with Numerical Experiments.* Proc. of the Centre for Math. Analysis Vol. 13, Australian National University 1986.

[11] L. Reichel, L. N. Trefethen, *Eigenvalues and pseudo-eigenvalues of Toeplitz matrices.* Linear Algebra Appl. **162**(1992), 153–185.

[12] S. Roch, *Numerical ranges of large Toeplitz matrices.* Linear Algebra Appl. **282** (1998), 185–198.

[13] S. Roch, *Spectral approximation of Wiener-Hopf operators with almost periodic generating function. Submitted to:* Numer. Funct. Anal. Optimization.

[14] S. Roch, *Pseudospectra of operator polynomials. Submitted to:* Proceedings of the IWOTA 1998, Groningen.

[15] S. Roch, B. Silbermann, *Limiting sets of eigenvalues and singular values of Toeplitz matrices.* Asymptotic Anal. **8** (1994), 293–309.

[16] S. Roch, B. Silbermann, *C^*-algebra techniques in numerical analysis.* J. Oper. Theory **35** (1996), 2, 241–280.

[17] L. N. Trefethen, *Pseudospectra of matrices.* – *In:* D. F. Griffiths, G. A. Watson (Eds.), Numerical Analysis 1991, Longman 1992, 234–266.

Technische Universität Darmstadt
Fachbereich Mathematik
Schlossgartenstrasse 7
D 64289 Darmstadt

1991 Mathematics Subject Classification: Primary 65R20; Secondary 47A10, 47A12

Submitted: 4.4.2000

Operator Theory:
Advances and Applications, Vol. 121

Fast solvers of generalized airfoil equation of index 1

GENNADI VAINIKKO

The generalized airfoil equation is periodized with the help of cosine transformation. For the periodic problem, a fast solver is constructed on the basis of a fully discrete version of trigonometric collocation method with product integration, and conjugate gradient iterations to solve the discretized problem.

1. Introduction

In this paper we propose some fast algorithms to solve the generalized airfoil equation

$$(1.1)\int_{-1}^{1}\Big(\frac{1}{\pi}\frac{1}{x-y}+b_1(x,y)\log|x-y|+b_2(x,y)\Big)v(y)dy=g(x)\,,\quad -1<x<1$$

with smooth kernel functions b_1 and b_2. We periodize the problem using the cosine transformation and after that discretize it using a fully discrete version of trigonometric collocations combined with product integration and some dimension reduction. Here we follow some ideas of [31–33]. To solve the discretized problem, we follow the idea of [29] and apply the conjugate gradients iteration method. In this way we obtain an order optimal approximate solution in the scale of periodic Sobolev norms:

$$(1.2)\qquad \|u_N^k-u\|_\lambda\le cN^{\lambda-\mu}\|u\|_\mu\,,\qquad 0\le\lambda\le\mu\,.$$

Here u is the solution of the periodized problem and u_N^k is the kth iteration approximation of the discretized problem, $k\sim\log N$. The $N+1$ parameters of u_N^k can be determined in $\mathcal{O}(N\log N)$ arithmetical operations including the computation of the stiffness matrix of the problem. Actually this is the work to calculate the Fourier coefficients of the interpolant of the right hand term, and all the remaining computations are cheeper.

In [33], two grid iterations were used to solve the discretized problem. An advantage of this method is that a sufficient number of iterations to achieve the optimal accuracy (1.2) is independent of N; a disadvantage is the more complicated iteration formula compared with the conjugate gradient iterations. Numerical experiments in [29] have shown that at least for the Symm integral equation, the conjugate gradients are comparable or even slightly cheeper than rather popular GMRES iterations.

It is well known (see e.g. [28]) that the Fredholm index of problem (1.1) is -1, 0 or 1 depending on the weight $\sigma(x)$ of $L^2_\sigma(-1,1)$ in which the operator defined by the left side of (1.1) is considered. Our considerations in this paper correspond to the case of index 1. The case of index -1 can be treated in a similar way.

First fast solvers of problem (1.1) have been constructed (in the case of Fredholm index 0) by BERTHOLD, HOPPE and SILBERMANN [3] assuming $b_1(x,y) = b_1$ to be constant. They apply the algebraic collocations directly to (1.1) combining with a quadrature method in smaller dimension $M \sim N^{1/3}$; the optimal accuracy in the scale of Sobolev norms is obtained for $\frac{1}{2} < \lambda \le \mu$. Without the setting of fast solving, different analytical and numerical methods to solve problem (1.1) have been studied by many authors [1–4, 7–20, 22–28, 30, 34–36]. Most of these works deal with the cases $b_1(x,y) = 0$, $b_1(x,y) = b_1 = \text{const}$ or $b_1(x,y) = b_1(x)$. The general equation (1.1) has been treated by JUNGHANNS [17] by a fully discrete version of algebraic collocation method. Fast solution of a general class of periodic integral equations of index 0, including the equations of this paper, are discussed in [31–33, 38–39]. The method presented in this paper is new and we describe in details its matrix form.

2. Periodization

By the change of variables

$$
(2.1)\qquad
\begin{aligned}
x &= x(t) = -\cos 2\pi t\,, & 0 < t < \frac{1}{2}\,,\\
y &= x(s) = -\cos 2\pi s\,, & 0 < s < \frac{1}{2}\,,
\end{aligned}
$$

integral equation (1.1) takes the form

$$
\int_0^{1/2} \left\{ \frac{1}{\pi}\frac{1}{x(t)-x(s)} + b_1(x(t),x(s)) \log|x(t)-x(s)| + b_2(x(t),x(s)) \right\} v(x(s))x'(s)ds
$$
$$
= g(x(t))\,, \qquad 0 < t < \frac{1}{2}\,.
$$

For $g(x(t))$ and $b_i(x(t),x(s))$, $i = 1,2$, we will use the natural 1-periodic extensions to $\mathbb{R}$ and $\mathbb{R} \times \mathbb{R}$, respectively, allowing t and s in (2.1) to change in $\mathbb{R}$. Clearly these extensions are even in t and s. On the other hand, for $v(x(s))$ we do not use this natural even extension but the odd extension $w(t)$ defined by

$$
w(t) = \begin{cases} v(x(t))\,, & 0 < t < \frac{1}{2}\,,\\ -v(x(-t))\,, & -\frac{1}{2} < t < 0 \end{cases}
$$

and extended 1-periodically from $(-\frac{1}{2}, \frac{1}{2})$ to $\mathbb{R}$. Thus $w(s)x'(s)$ is an even 1-periodic function, and due to the parity properties, (1.1) is equivalent to the 1-periodic integral equation

$$\frac{1}{2}\int_{-1/2}^{1/2}\left\{\frac{1}{\pi}\frac{1}{x(t)-x(s)}+b_1(x(t),x(s))\log|x(t)-x(s)|\right.$$
$$\left.+b_2(x(t),x(s))\right\}w(s)x'(s)ds = \quad g(x(t))\,, \quad t\in\mathbb{R}\,.$$

We consider

$$u(s) = w(s)x'(s) \tag{2.2}$$

as an 1-periodic unknown function which we will look for. Multiplying both sides of (2.2) by $x'(t)$ and taking into account that due to parity properties $\int_{-1/2}^{1/2}\frac{x'(s)}{x(t)-x(s)}u(s)ds = 0$, we obtain

$$\frac{1}{2}\int_{-1/2}^{1/2}\left\{\frac{1}{\pi}\frac{x'(t)-x'(s)}{x(t)-x(s)}+x'(t)b_1(x(t),x(s))\log|x(t)-x(s)|\right.$$
$$\left.+\,x'(t)b_2(x(t),x(s))\right\}u(s)ds = x'(t)g(x(t))\,, \qquad t\in\mathbb{R}\,.$$

Here $\dfrac{x'(t)-x'(s)}{x(t)-x(s)} = 2\pi\cot\pi(t+s)$, $\log|x(t)-x(s)| = \log 2+\log|\sin\pi(t-s)|+\log|\sin\pi(t+s)|$, and we easily reduce the equation to the form (cf. [33])

$$\int_{-1/2}^{1/2}\left\{\cot\pi(t-s)+a_1(t,s)\log|\sin\pi(t-s)|+a_2(t,s)\right\}u(s)ds = f(t) \tag{2.3}$$

where

$$f(t) = g(x(t))x'(t)\,, \qquad t\in\mathbb{R}\,,$$
$$a_1(t,s) = b_1(x(t),x(s))x'(t)\,, \qquad t,s\in\mathbb{R}\,,$$
$$a_2(t,s) = \frac{1}{2}\left[b_2(x(t),x(s))+(\log 2)b_1(x(t),x(s))\right]x'(t)\,, \qquad t,s\in\mathbb{R}\,.$$

Notice that $g\in C^m[-1,1]$ implies $f\in C^m(\mathbb{R})$, and f is 1-periodic and odd. Similarly $b_j\in C^m([-1,1]\times[-1,1])$, $j=1,2$, imply $a_j\in C^m(\mathbb{R}\times\mathbb{R})$, $j=1,2$, and a_j are 1-biperiodic, odd in t and even in s. We will assume that

$$f\in H^\mu,\ \ \mu>\frac{1}{2},\ \text{is 1-periodic and odd,} \tag{2.4}$$

(2.5) $a_j \in C^\infty(\mathbb{R} \times \mathbb{R})$, $j = 1, 2$, are 1-biperiodic and odd in t and even in s,

Here H^λ, $\lambda \geq 0$, is the Sobolev space of 1-periodic functions u having a finite norm

$$\|u\|_\lambda = \Big(\sum_{k \in \mathbb{Z}} \underline{k}^{2\lambda} |\hat{u}(k)|^2\Big)^{1/2}$$

where $\underline{k} = \max\{1, |k|\}$ and $\hat{u}(k)$ are the Fourier coefficients of u with respect to $\{e^{ik2\pi t}\}$,

$$\hat{u}(k) = \int_{-1/2}^{1/2} u(s) e^{-ik2\pi s} ds \qquad (k \in \mathbb{Z}).$$

Notice that $H^\lambda \subset C(\mathbb{R})$ for $\lambda > \frac{1}{2}$. Thus assumption (2.4) implies the continuity of f.

Let us analyse the periodized problem (2.3). Denote its operator by $\mathcal{A} = A_0 + A_1 + A_2$ where

$$(A_0 u)(t) = \int_{-1/2}^{1/2} \cot \pi(t-s) u(s) ds \tag{2.6}$$

is the Hilbert operator,

$$(A_1 u)(t) = \int_{-1/2}^{1/2} a_1(t,s) \log|\sin \pi(t-s)| u(s) ds\,, \tag{2.7}$$

$$(A_2 u)(t) = \int_{-1/2}^{1/2} a_2(t,s) u(s) ds\,. \tag{2.8}$$

It is well known that $A_0 \in \mathcal{L}(H^\lambda)$ is a Fredholm operator of index 0 whereas $A_1, A_2 \in \mathcal{L}(H^\lambda)$ are compact. Thus $\mathcal{A} \in \mathcal{L}(H^\lambda)$ is a Fredholm operator of index 0, and if problem (2.3) is solvable then (2.4) implies that the solution u also belongs to H^μ. According to (2.2), the smoothness $u \in C^m(\mathbb{R})$ implies the smoothness $v \in C^m(-1,1)$ for the solution of problem (1.1) but v and its derivatives may have (and usually have) special singularities at $x = \pm 1$, namely

$$v(x) \sim \frac{u(0)}{2\pi\sqrt{1-x^2}} \quad \text{as } x \to -1\,, \qquad v(x) \sim \frac{u(\frac{1}{2})}{2\pi\sqrt{1-x^2}} \quad \text{as } x \to 1\,.$$

Solving (2.3) we exploit the parity properties of f, u and a_j, $j = 1, 2$. Denote

$$H^\lambda_{\text{ev}} = \{u \in H^\lambda : \; u(-t) = u(t), \; t \in \mathbb{R}\} = \{u \in H^\lambda : \hat{u}(-k) = \hat{u}(k), \; k \in \mathbb{Z}\}\,,$$

$$H^\lambda_{\text{od}} = \{u \in H^\lambda : \; u(-t) = -u(t), \; t \in \mathbb{R}\} = \{u \in H^\lambda : \hat{u}(-k) = -\hat{u}(k), \; k \in \mathbb{Z}\}\,.$$

These are closed subspaces of H^λ, and $H^\lambda = H^\lambda_{\mathrm{ev}} \oplus H^\lambda_{\mathrm{od}}$. Orthogonal bases of H^λ_{ev} and H^λ_{od} are given, respectively, by $\cos(k2\pi t)$ $(k = 0, 1, 2, \ldots)$ and $\sin(k2\pi t)$ $(k = 1, 2, \ldots)$. Since

$$(2.9) \qquad A_0 \cos(k2\pi t) = \begin{cases} 0, & k = 0 \\ \sin(k2\pi t), & k = 1, 2, \ldots \end{cases}$$

$$(2.10) \qquad A_0 \sin(k2\pi t) = -\cos(k2\pi t), \qquad k = 1, 2, \ldots$$

we see that $A_0 \in \mathcal{L}(H^\lambda_{\mathrm{ev}}, H^\lambda_{\mathrm{od}})$ is Fredholm operator, $\mathrm{ind}\,(A_0) = 1$. It is easy to see also that A_1 and A_2 map H^λ_{ev} into H^λ_{od} and remain to be compact between these spaces. Thus

$$(2.11) \quad \mathcal{A} \in \mathcal{L}(H^\lambda_{\mathrm{ev}}, H^\lambda_{\mathrm{od}}) \text{ is Fredholm operator}, \quad \mathrm{ind}\,(\mathcal{A}) = 1 \ \text{ for } \lambda \geq 0\,.$$

More precisely, $A_1 \in \mathcal{L}(H^\lambda_{\mathrm{ev}}, H^{\lambda+1}_{\mathrm{od}})$, $A_2 \in \mathcal{L}(H^\lambda_{\mathrm{ev}}, H^{\lambda+r}_{\mathrm{od}})$ with any $r > 0$. A consequence is that the solutions of the homogeneous equation $\mathcal{A}u = 0$ are C^∞-smooth. We obtain the following result.

Theorem 2.1. *Assume* (2.4) *and* (2.5) *and let the homogeneous problem* $\mathcal{A}v = 0$ *have in* $\bigcap_{\lambda \geq 0} H^\lambda_{\mathrm{ev}}$ *at most one linearly independent solution. Then integral equation* (2.3) *has a parameter depending family of solutions in* H^μ_{ev}, *and dim* $\mathcal{N}(\mathcal{A}) = 1$, $\mathcal{N}(\mathcal{A}) \subset \bigcap_{\lambda \geq 0} H^\lambda_{\mathrm{ev}}$.

We will look for the solutions of (2.3) of minimal $\|\cdot\|_0$ norm. Under conditions of Theorem 2.1, it exists and is given by $u = A^+ f$ where $A^+ \in \mathcal{L}(H^0_{\mathrm{od}}, H^0_{\mathrm{ev}})$ is the Moore-Penrose inverse to $\mathcal{A} \in \mathcal{L}(H^0_{\mathrm{ev}}, H^0_{\mathrm{od}})$.

3. Interpolation of even and odd functions

Here we recall some results about interpolation. For $n \in \mathbb{N}$, denote

$$\mathcal{T}^{\mathrm{ev}}_n = \mathrm{span}\,\{1, \cos(2\pi t), \ldots, \cos(n2\pi t)\}\,,$$
$$\mathcal{T}^{\mathrm{od}}_n = \mathrm{span}\,\{\sin(2\pi t), \ldots, \sin(n2\pi t)\}\,.$$

By P^{ev}_n and P^{od}_n we denote the corresponding orthogonal projection operators, and by Q^{ev}_n and Q^{od}_n the corresponding interpolation operators defined by conditions

$$Q^{\mathrm{ev}}_n u \in \mathcal{T}^{\mathrm{ev}}_n, \quad (Q^{\mathrm{ev}}_n u)(\tfrac{j}{2n+1}) = u(\tfrac{j}{2n+1}), \quad j = 0, 1, \ldots, n\,,$$
$$Q^{\mathrm{od}}_n u \in \mathcal{T}^{\mathrm{od}}_n, \quad (Q^{\mathrm{od}}_n u)(\tfrac{j}{2n+1}) = u(\tfrac{j}{2n+1}), \quad j = 1, \ldots, n\,.$$

The following estimates are known (see [33]):

$$(3.1) \qquad \|u - P^{\mathrm{ev}}_n u\|_\lambda \leq (n+1)^{\lambda-\mu} \|u\|_\mu \quad (\lambda \leq \mu) \ \text{ for } \ u \in H^\mu_{\mathrm{ev}}\,,$$

$$\|u - P_n^{\text{od}}u\|_\lambda \le (n+1)^{\lambda-\mu}\|u\|_\mu \quad (\lambda \le \mu) \ \text{ for } \ u \in H^\mu_{\text{od}}\,, \tag{3.2}$$

$$\|u - Q_n^{\text{ev}}u\|_\lambda \le \gamma_\mu(n+\frac{1}{2})^{\lambda-\mu}\|u\|_\mu \ (0 \le \lambda \le \mu) \text{ for } u \in H^\mu_{\text{ev}}\,, \ \mu > \frac{1}{2}\,, \tag{3.3}$$

$$\|u - Q_n^{\text{od}}u\|_\lambda \le \gamma_\mu(n+\frac{1}{2})^{\lambda-\mu}\|u\|_\mu \ (0 \le \lambda \le \mu) \text{ for } u \in H^\mu_{\text{od}}\,, \ \mu > \frac{1}{2} \tag{3.4}$$

where $\gamma_\mu = (1 + 2\sum_{j=1}^\infty j^{-2\mu})^{1/2} < \infty$ for $\mu > \frac{1}{2}$. Explicit formulae for the interpolation are given by

$$Q_n^{\text{ev}}u = \sum_{j=0}^n u\Big(\frac{j}{2n+1}\Big)\varphi^{\text{ev}}_{n,j}\,, \qquad Q_n^{\text{od}}u = \sum_{j=1}^n u\Big(\frac{j}{2n+1}\Big)\varphi^{\text{od}}_{n,j} \tag{3.5}$$

where

$$\varphi^{\text{ev}}_{n,j}(t) = \begin{cases} \dfrac{2}{2n+1}\Big(\dfrac{1}{2} + \displaystyle\sum_{k=1}^n \cos(k2\pi t)\Big)\,, & j = 0 \\ \dfrac{4}{2n+1}\Big(\dfrac{1}{2} + \displaystyle\sum_{k=1}^n \cos\Big(kj\dfrac{2\pi}{2n+1}\Big)\cos(k2\pi t)\Big)\,, & j = 1,\dots,n\,, \end{cases}$$

$$\varphi^{\text{od}}_{n,j}(t) = \frac{4}{2n+1}\sum_{k=1}^n \sin\Big(kj\frac{2\pi}{2n+1}\Big)\sin(k2\pi t)\,, \quad j = 1,\dots,n\,,$$

are the fundamental polynomial satisfying

$$\varphi^{\text{ev}}_{n,j}\Big(\frac{k}{2n+1}\Big) = \delta_{jk} \quad (j,k = 0,1,\dots,n)\,,$$
$$\varphi^{\text{od}}_{n,j}\Big(\frac{k}{2n+1}\Big) = \delta_{jk} \quad (j,k = 1,\dots,n)\,.$$

For $v_n \in \mathcal{T}_n^{\text{ev}}$ we have two representations

$$v_n = \sum_{k=0}^n c_k \cos(k2\pi t) = \sum_{j=0}^n v_n\Big(\frac{j}{2n+1}\Big)\varphi^{\text{ev}}_{n,j}(t)\,.$$

The vectors $\underline{c}_n = (c_0, c_1, \dots c_n)^\top$ and $\underline{v}_n = \big(v_n(0), v_n(\frac{1}{2n+1}), \dots, v_n(\frac{n}{2n+1})\big)^\top$ are related by the discrete cosine Fourier transformations

$$\underline{v}_n = \mathcal{C}_n\underline{c}_n\,, \qquad \underline{c}_n = \tilde{\mathcal{C}}_n\underline{v}_n \tag{3.6}$$

where

$$\mathcal{C}_n = \Big(\cos\Big(kj\frac{2\pi}{2n+1}\Big)\Big)_{j,k=0}^n\,, \quad \tilde{\mathcal{C}}_n = \frac{4}{2n+1}D_n\mathcal{C}_nD_n\,, \quad D_n = \text{diag}\{\tfrac{1}{2}, 1, \dots, 1\}$$

are $(n+1)\times(n+1)$ matrices.

For $w_n \in \mathcal{T}_n^{\text{od}}$ we have the representations

$$w_n = \sum_{k=1}^{n} d_k \sin(k2\pi t) = \sum_{j=1}^{n} w_n\Big(\frac{j}{2n+1}\Big)\varphi_{n,j}^{\text{od}}(t)\,.$$

The vectors $\underline{d}_n = (d_1, \dots, d_n)^\top$ and $\underline{w}_n = \big(w_n\big(\frac{1}{2n+1}\big), \dots, w_n\big(\frac{n}{2n+1}\big)\big)^\top$ are related by the discrete sine Fourier transformations

$$\underline{w}_n = \mathcal{S}_n \underline{d}_n\,, \qquad \underline{d}_n = \tilde{\mathcal{S}}_n \underline{w}_n \tag{3.7}$$

where

$$\mathcal{S}_n = \Big(\sin\Big(kj\frac{2\pi}{2n+1}\Big)\Big)_{j,k=1}^{n}\,, \qquad \tilde{\mathcal{S}}_n = \frac{4}{2n+1}\mathcal{S}_n \quad \text{are } n \times n\text{-matrices.}$$

By FFT, an application of $\mathcal{C}_n$ and $\mathcal{S}_n$ costs $\mathcal{O}(n \log n)$ arithmetical operations instead of $\mathcal{O}(n^2)$ operations by the usual matrix- to-vector multiplication. Notice also the symmetry of matrices $\mathcal{C}_n$, $\tilde{\mathcal{C}}_n$, $\mathcal{S}_n$ and $\tilde{\mathcal{S}}_n$.

Formulae (3.6) and (3.7) enable fast changes of the representation forms of trigonometric polynomials. They also enable a fast calculation of Fourier coefficients of interpolants $Q_n^{\text{ev}} v$ and $Q_n^{\text{od}} w$.

4. Approximation of the problem

We approximate the problem $\mathcal{A}u = f$ (the integral equation (2.3)) by a problem

$$\mathcal{A}_N u_N = Q_N^{\text{od}} f \tag{4.1}$$

where $\mathcal{A}_N \in \mathcal{L}(H_{\text{ev}}^\lambda, H_{\text{od}}^\lambda)$ has the following properties:

$$\mathcal{A}_N \mathcal{T}_N^{\text{ev}} \subset \mathcal{T}_N^{\text{od}}\,, \qquad \mathcal{A}_N u \in \mathcal{T}_N^{\text{od}} \Rightarrow u \in \mathcal{T}_N^{\text{ev}}\,, \tag{4.2}$$

$$\|\mathcal{A}_N - \mathcal{A}\|_{\lambda,\mu} := \|\mathcal{A}_N - \mathcal{A}\|_{\mathcal{L}(H_{\text{ev}}^\mu, H_{\text{od}}^\lambda)} \le cN^{\lambda-\mu} \quad \text{for} \quad 0 \le \lambda \le \mu\,, \tag{4.3}$$

$$\|\mathcal{A}_N - \mathcal{A}\|_{\lambda,\lambda} \to 0 \quad \text{as} \quad N \to \infty \quad \text{for} \quad \lambda \ge 0 \tag{4.4}$$

and, moreover, (4.1) has a "cheap" matrix form. The last condition will be made precise in the next section. Here we give the construction of $\mathcal{A}_N$, check the properties (4.2)–(4.4) for it and prove the error estimate

$$\|\mathcal{A}_N^+ Q_N^{\text{od}} f - \mathcal{A}^+ f\|_\lambda \le cN^{\lambda-\mu}\|f\|_\mu \qquad (0 \le \lambda \le \mu) \quad \text{for} \quad f \in H_{\text{od}}^\mu\,; \tag{4.5}$$

number $\mu > \frac{1}{2}$ originates from (2.4).

Introduce natural numbers m, M and n such that

$$(4.6)\qquad \begin{aligned} 2m \le M \le n \le N\,, \quad m \sim N^{\varrho},\ M \sim N^{\sigma},\ n \sim N^{\tau},\\ 0 < \varrho \le \sigma \le \frac{1}{2},\ \frac{\mu}{\mu+1} \le \tau \le 1 \end{aligned}$$

where $n \sim N^{\tau}$ means that there are two positive constants c_1 and c_2 such that $c_1 \le nN^{-\tau} \le c_2$ as $N \to \infty$. Define

$$(4.7)\qquad \mathcal{A}_N = A_0 + Q_M^{\rm od}(A_1^{(M)} + A_2^{(M)})P_m^{\rm ev} + Q_n^{\rm od} A_1^{[d]}(P_n^{\rm ev} - P_m^{\rm ev})$$

where $A_j^{(M)} \in \mathcal{L}(H_{\rm ev}^{\lambda}, H_{\rm od}^{\lambda})$ defined by

$$(A_1^{(M)}u)(t) = \int_{-1/2}^{1/2} \log|\sin\pi(t-s)|Q_{M,s}^{\rm ev}(a_1(t,s)u(s))ds\,,$$

$$(A_2^{(M)}u)(t) = \int_{-1/2}^{1/2} Q_{M,s}^{\rm ev}(a_2(t,s)u(s))ds$$

are product integration approximations of operators A_1 and A_2 (cf. (2.7) and (2.8)) of dimension M, and

$$(4.8)\qquad A_1^{[d]} = \sum_{j=0}^{d-2} A_{1,j} \in \mathcal{L}(H_{\rm ev}^{\lambda}, H_{\rm od}^{\lambda})\,, \qquad d \ge \frac{1-\varrho}{\varrho}\mu\,,$$

with

$$(A_{1,j}u)(t) = \left(\frac{1}{2\pi i}\frac{\partial}{\partial s}\right)^j a_1(t,s)\Big|_{s=t} \int_0^1 \kappa_j(t-s)u(s)ds\,,$$

$$\hat{\kappa}_j(l) = \frac{1}{2}(-l)^{-1-j}\operatorname{sign}(l)\,, \qquad 0 \ne l \in \mathbb{Z}\,, \quad \hat{\kappa}_j(0) = 0$$

presents an asymptotic approximation of A_1 (see [37, 33]). We may put $d = 1$, $A_1^{[d]} = 0$ if $\frac{1-\varrho}{\varrho}\mu \le 1$, i.e. if $\mu \le 1$ and $\varrho \le \frac{1}{2}$ is taken sufficiently large. Since

$$\int_{-1/2}^{1/2} \kappa_j(t-s)e^{il2\pi s}ds = \frac{1}{2}(-l)^{-1-j}\operatorname{sign}(l)e^{il2\pi t}, \quad 0 \ne l \in \mathbb{Z}\,,$$

$$\int_{-1/2}^{1/2} \kappa_j(t-s)ds = 0\,,$$

we have

$$
\begin{aligned}
&A_1^{[d]} 1 = 0\,,\\
(4.9)\quad &A_1^{[d]} \cos(k2\pi t) = \sum_{j=0}^{d-2} b_j(t) k^{-1-j} \left\{ \begin{array}{ll} \cos(k2\pi t), & j \text{ even} \\ \sin(k2\pi t), & j \text{ odd} \end{array} \right\}, \quad k \geq 1\,,\\
&b_j(t) = \left\{ \begin{array}{ll} (-1)^{(j+2)/2}, & j \text{ even} \\ (-1)^{(j-1)/2}, & j \text{ odd} \end{array} \right\} \frac{1}{2(2\pi)^j} \left(\frac{\partial}{\partial s}\right)^j a_1(t,s)\Big|_{s=t}, \quad j \geq 0\,.
\end{aligned}
$$

Notice that b_j is odd for even j and b_j is even for odd j.

Property (4.2) is clear from (4.7) and (2.9), (2.10). Let us prove (4.3) and (4.4). We will locally use simplified designations $P_n = P_n^{\text{ev}}$, $Q_n = Q_n^{\text{od}}$. Representing $\mathcal{A} = A_0 + A_1 + A_2$,

$$
\begin{aligned}
A_1 &= A_1 P_m + A_1(P_n - P_m) + A_1(I - P_n)\\
&= Q_M A_1 P_m + (I - Q_M) A_1 P_m + Q_n A_1 (P_n - P_m) + (I - Q_n) A_1 (P_n - P_m)\\
&\quad + A_1 (I - P_n)\\
A_2 &= A_2 P_m + A_2 (I - P_m) = Q_M A_2 P_m + (I - Q_M) A_2 P_m + A_2 (I - P_m)
\end{aligned}
$$

we have

$$
\begin{aligned}
\mathcal{A} - \mathcal{A}_N &= Q_M (A_1 - A_1^{(M)}) P_m + Q_n (A_1 - A_1^{[d]})(P_n - P_m) + (I - Q_M) A_1 P_m\\
&\quad + (I - Q_n) A_1 (P_n - P_m) + A_1 (I - P_n) + Q_M (A_2 - A_2^{(M)}) P_m\\
&\quad + (I - Q_M) A_2 P_m + A_2 (I - P_m)\,.
\end{aligned}
$$

Using the inequality $2m \leq M$ it is easy to check that (see [33], lemma 8.1, for details in a similar situation)

$$
\| Q_M (A_j^{(M)} - A_j) P_m \|_{\lambda,\lambda} \leq c_r (M - m)^{-r} \quad \forall r > 0\,,\ \lambda \geq 0\,, \quad j = 1, 2\,.
$$

Similarly

$$
\| (I - Q_M) A_j P_m \|_{\lambda,\lambda} \leq c_r (M - m)^{-r} \quad \forall r > 0\,,\ \lambda \geq 0\,, \quad j = 1, 2\,.
$$

Further, since $A_1 \in \mathcal{L}(H_{\text{ev}}^{\lambda}, H_{\text{od}}^{\lambda+1})$, $A_2 \in \mathcal{L}(H_{\text{ev}}^{\lambda}, H_{\text{od}}^{\lambda+r})$ with any $r > 0$, we have due to (3.1) and (3.4)

$$
\| (I - Q_n) A_1 \|_{\lambda,\lambda} \leq c n^{-1}, \quad \| A_1 (I - P_n) \|_{\lambda,\lambda} \leq c n^{-1}, \quad \lambda \geq 0
$$

$$
\| (I - Q_n) A_1 \|_{\lambda,\mu} \leq c n^{-1+\lambda-\mu}, \quad \| A_1 (I - P_n) \|_{\lambda,\mu} \leq c n^{-1+\lambda-\mu}, \quad 0 \leq \lambda \leq \mu\,,
$$

$$
\| A_2 (I - P_n) \|_{\lambda,\lambda} \leq c n^{-r} \quad \forall r > 0\,,\ \lambda \geq 0\,.
$$

Finally, $A_1 - A_1^{[d]}$ is a periodic pseudodifferential operator of order $-d$, hence $A_1 - A_1^{[d]} \in \mathcal{L}(H_{\text{ev}}^{\lambda}, H_{\text{od}}^{\lambda+d})$, $\lambda \geq 0$ (see, e.g., [38]). Due to (3.1) and (3.4) we obtain

$$
\| Q_n (A_1 - A_1^{[d]})(P_n - P_m) \|_{\lambda,\lambda} \leq c m^{-d}, \qquad \lambda \geq 0\,,
$$

$$\|Q_n(A_1 - A_1^{[d]})(P_n - P_m)\|_{\lambda,\mu} \le cm^{-d+\lambda-\mu}, \qquad 0 \le \lambda \le \mu .$$

Combining these estimates we obtain (4.4) and (4.3):

$$\|\mathcal{A} - \mathcal{A}_N\|_{\lambda,\lambda} \le c(n^{-1} + m^{-d}) \to 0 \ \text{ as } \ N \to \infty , \ \lambda \ge 0 ,$$

$$\begin{aligned}\|\mathcal{A} - \mathcal{A}_N\|_{\lambda,\mu} &\le c(n^{-1+\lambda-\mu} + m^{-d+\lambda-\mu}) \le c'(N^{-\tau(1+\mu-\lambda)} + N^{-\varrho(d+\mu-\lambda)}) \\ &\le 2c' N^{\lambda-\mu}, \qquad\qquad 0 \le \lambda \le \mu .\end{aligned}$$

On the last step we used inequalities $\tau \ge \mu/(\mu+1)$, $d \ge \mu(1-\varrho)/\varrho$, see (4.6) and (4.8).

Theorem 4.1. *Assume the conditions of Theorem* 2.1 *and* (4.6). *Then there is a* N_0 *such that* $\mathcal{A}_N H^\lambda_{\mathrm{ev}} = H^\lambda_{\mathrm{od}}$ $(\lambda \ge 0)$ *and* $\dim \mathcal{N}(\mathcal{A}_N) = 1$ *for* $N \ge N_0$. *Further, the error estimate* (4.5) *holds true where* $\mathcal{A}^+ f \in H^\mu_{\mathrm{ev}}$ *is the solution of integral equation* (2.3) *of minimal* $\|\cdot\|_0$ *norm, and* $\mathcal{A}_N^+ Q_N f \in \mathcal{T}_N^{\mathrm{ev}}$ *is the solution of approximating problem* (4.1) *of minimal* $\|\cdot\|_0$ *norm.*

Proof. The first two assertions concerning the mapping properties of $\mathcal{A}_N$ are consequences of similar properties of $\mathcal{A}$ (see Theorem 2.1) and (4.4). Let us prove (4.5) for Moore- Penrose inverses $\mathcal{A}^+, \mathcal{A}_N^+ \in \mathcal{L}(H^0_{\mathrm{od}}, H^0_{\mathrm{ev}})$ of $\mathcal{A}, \mathcal{A}_N \in \mathcal{L}(H^0_{\mathrm{ev}}, H^0_{\mathrm{od}})$. Denote by $\mathcal{N}(\mathcal{A})^\perp$ the orthogonal complement to $\mathcal{N}(\mathcal{A})$ in H^0_{ev}, then

$$\mathcal{A}^+ g = \left(\mathcal{A}\Big|_{\mathcal{N}(\mathcal{A})^\perp}\right)^{-1} g \quad \text{for } \ g \in H^0_{\mathrm{od}} .$$

Since $\mathcal{N}(\mathcal{A}) \subset \bigcap_{\lambda \ge 0} H^\lambda_{\mathrm{ev}}$, $\dim \mathcal{N}(\mathcal{A}) = 1$ (see Theorem 2.1), we have for $\lambda > 0$ the direct sum $H^\lambda_{\mathrm{ev}} = \mathcal{N}(\mathcal{A}) \oplus (H^\lambda_{\mathrm{ev}} \cap \mathcal{N}(\mathcal{A})^\perp)$ where $H^\lambda_{\mathrm{ev}} \cap \mathcal{N}(\mathcal{A})^\perp$ is a closed subspace of H^λ_{ev} of codimension 1. The operator $\mathcal{A}$ is one-to-one on $H^\lambda_{\mathrm{ev}} \cap \mathcal{N}(\mathcal{A})^\perp$ and has the full range H^λ_{od}, therefore the inverse given by $\mathcal{A}^+$ is bounded. We established that $\mathcal{A}^+ \in \mathcal{L}(H^\lambda_{\mathrm{od}}, H^\lambda_{\mathrm{ev}})$ for $\lambda \ge 0$.

Due to (4.4), $\mathcal{A}^0_N := \mathcal{A}\Big|_{\mathcal{N}(\mathcal{A})^\perp} : \mathcal{N}(\mathcal{A})^\perp \to H^0_{\mathrm{od}}$ is one-to-one and onto for sufficiently large N, say $N \ge N_0$, consequently $(\mathcal{A}^0_N)^{-1}$ exists, and again by (4.4), $\|(\mathcal{A}^0_N)^{-1}\|_{0,0} \le$ const for $N \ge N_0$. Since $\|\mathcal{A}^+_N f\|_0 \le \|(\mathcal{A}^0_N)^{-1} f\|_0$, we have

$$\|\mathcal{A}^+_N\|_{0,0} \le c \qquad (N \ge 1) .$$

Further, for $\mathcal{A}, \mathcal{A}_N \in \mathcal{L}(H^0_{\mathrm{ev}}, H^0_{\mathrm{od}})$, $\mathcal{A}^+, \mathcal{A}^+_N \in \mathcal{L}(H^0_{\mathrm{od}}, H^0_{\mathrm{ev}})$ we have

$$\mathcal{A}\mathcal{A}^+ = I , \quad \mathcal{A}^+\mathcal{A} = I - P_0 , \quad \mathcal{A}_N\mathcal{A}^+_N = I , \quad \mathcal{A}^+_N\mathcal{A}_N = I - P_{0,N} \quad (N \ge N_0)$$

where P_0 and $P_{0,N}$ are the orthogonal projection operators in H^0_{ev} to $\mathcal{N}(\mathcal{A})$ and $\mathcal{N}(\mathcal{A}_N)$, respectively. Using these relations we find

$$\mathcal{A}^+_N(\mathcal{A}_N - \mathcal{A})\mathcal{A}^+ = (\mathcal{A}^+_N\mathcal{A}_N)\mathcal{A}^+ - \mathcal{A}^+_N(\mathcal{A}\mathcal{A}^+) = (I - P_{0,N})\mathcal{A}^+ - \mathcal{A}^+_N ,$$

whereby

$$\mathcal{A}^+ - \mathcal{A}_N^+ = \mathcal{A}_N^+(\mathcal{A}_N - \mathcal{A})\mathcal{A}^+ + (P_{0.N} - P_0)\mathcal{A}^+ ,$$

$$\|\mathcal{A}^+ - \mathcal{A}_N^+\|_{0,\mu} \le \|\mathcal{A}_N^+\|_{0,0}\|\mathcal{A}_N - \mathcal{A}\|_{0,\mu}\|\mathcal{A}^+\|_{\mu,\mu} + \|P_{0.N} - P_0\|_{0,0}\|\mathcal{A}^+\|_{0,\mu} .$$

Let $v_0 \in \mathcal{N}(\mathcal{A})$, $v_{0,N} \in \mathcal{N}(\mathcal{A}_N)$, $\|v\|_0 = \|v_{o,N}\| = 1$. Then (see [21])

$$\|P_{0,N} - P_0\|_{0,0} = \max\{\|v_0 - P_{0,N}v_0\|_0 \, , \; \|v_{0,N} - P_0 v_{0,N}\|_0\} .$$

Further,

$$\begin{aligned} \|(I - P_{0,N}v_0\|_0 &= \|\mathcal{A}_N^+\mathcal{A}_N v_0\|_0 = \|\mathcal{A}_N^+(\mathcal{A}_N - \mathcal{A})v_0\|_0 \\ &\le \|\mathcal{A}_N^+\|_{0,0} \; \|\mathcal{A}_N - \mathcal{A}\|_{0,\mu} \; \|v_0\|_\mu \le c' N^{-\mu} \end{aligned}$$

implying also $\|(I - P_0)v_{0,N}\|_0 \le c'' N^{-\mu}$. Hence $\|P_{0,N} - P_0\|_0 \le cN^{-\mu}$ and

$$\|A^+ - A_N^+\|_{0,0} \le cN^{-\mu} .$$

Using also (3.4) we now have

$$\|\mathcal{A}_N^+ Q_N^{\text{od}} f - \mathcal{A}^+ f\|_0 \le \|\mathcal{A}_N^+\|_{0,0} \; \|Q_N^{\text{od}} f - f\|_0 + \|(\mathcal{A}_N^+ - \mathcal{A}^+)f\|_0 \le cN^{-\mu}\|f\|_\mu .$$

This is (4.5) for $\lambda = 0$. Now it is easy to obtain (4.5) also for $0 \le \lambda \le \mu$:

$$\begin{aligned} \|\mathcal{A}_N^+ Q_N^{\text{od}} f - \mathcal{A}^+ f\|_\lambda &\le \|P_N^{\text{ev}}(\mathcal{A}_N^+ Q_N^{\text{od}} f - A^+ f)\|_\lambda + \|(I - P_N)A^+ f\|_\lambda \\ &\le N^\lambda \|\mathcal{A}_N^+ Q_N^{\text{od}} f - A^+ f\|_0 + (N+1)^{\lambda-\mu}\|A^+ f\|_\mu \le cN^{\lambda-\mu}\|f\|_\mu . \quad \square \end{aligned}$$

5. Matrix form of the method

By (4.2) the solutions u_N of approximating problem (4.1) belong to $\mathcal{T}_N^{\text{ev}}$. Moreover, representing

$$u_N = \sum_{k=0}^{N} c_k \cos(k2\pi t) , \quad Q_N^{\text{od}} f = \sum_{k=1}^{N} d_k \sin(k2\pi t) ,$$

we have on the basis (2.9) and (2.10)

$$c_k = d_k \quad \text{for} \quad n < k \le N ,$$

so we have to complete only $c_0, c_1, \dots, c_n$. For

$$v_n := P_n^{\text{ev}} u_N = \sum_{k=0}^{n} c_k \cos(k2\pi t)$$

we obtain from (4.1) equation $\mathcal{A}_N v_n = w_n$ with $w_n = P_n^{\mathrm{od}} Q_N^{\mathrm{od}} f = \sum_{k=1}^{n} d_k \sin(k2\pi t)$, or

$$A_0 v_n + Q_M^{\mathrm{od}}(A_1^{(M)} + A_2^{(M)}) P_m^{\mathrm{ev}} v_n + Q_n^{\mathrm{od}} A_1^{[d]}(P_n^{\mathrm{ev}} - P_m^{\mathrm{ev}}) v_n = w_n \,. \tag{5.1}$$

So the dimension of the problem is reduced from N to n: we solve (5.1) and put

$$u_N = v_n + \sum_{k=n+1}^{N} d_k \cos(k2\pi t)\,; \tag{5.2}$$

notice that the minimal $\|\cdot\|_0$ norm solution u_N of (4.1) corresponds to the minimal $\|\cdot\|_0$ norm solution of (5.1). In sequel of this solution we present the matrix form

$$\mathbb{M}_n \underline{c}_n = \underline{d}_n \tag{5.3}$$

of equation (5.1) where $\underline{c}_n = (c_0, c_1, \ldots, c_n)$, $\underline{d}_n = (d_0, d_1, \ldots, d_n)$ and $\mathbb{M}_n$ is a $n \times (n+1)$-matrix. The formulae contain some "trivial" operations (e.g. omitting of a part of coordinates of vector $\underline{c}_n$) but we present also these operations as matrix multiplications. This enables us to write the formula also for $\mathbb{M}_n'$, the Hermite dual of $\mathbb{M}_n$.

(i) Due to formula (2.9), A_0 is given by $n \times (n+1)$ matrix

$$\mathbb{A}_0 = \begin{pmatrix} 0 & \mathbb{I}_n \end{pmatrix} = \begin{pmatrix} 0 & 1 & 0 & \ldots & 0 \\ 0 & 0 & 1 & \ldots & 0 \\ \ldots & \ldots & \ldots & \ldots & \ldots \\ 0 & 0 & 0 & \ldots & 1 \end{pmatrix}.$$

(ii) Representing $Q_{M,s}^{\mathrm{ev}}(a_1(t,s)v(s))$ via (3.5), changing twice the representation form of polynomials (see (3.6), (3.7)) and taking into account that

$$\int_{-1/2}^{1/2} \log|\sin \pi(t-s)| \cos(k2\pi s)ds = \begin{cases} -\log 2\,, & k = 0 \\ -\frac{1}{2k}\cos(k2\pi t)\,, & k \geq 1 \end{cases}$$

we find that the operator $Q_M^{\mathrm{od}}(A_1^{(M)} + A_2^{(M)})P_m^{\mathrm{ev}}$ is given by $n \times (n+1)$ matrix

$$\mathbb{I}_{n,M} \tilde{\mathcal{S}}_M (\mathbb{A}_1^{(M)} + \mathbb{A}_2^{(M)}) \mathcal{C}_M \mathbb{P}_{M,m,n}$$

with $\mathcal{C}_M$ and $\tilde{\mathcal{S}}_M = \frac{4}{2M+1}\mathcal{S}_M$ defined in Section 3,

$(M+1) \times (n+1)$ matrix $\mathbb{P}_{M,m,n} = \begin{pmatrix} \mathbb{I}_{m+1} & 0 \\ 0 & 0 \end{pmatrix}$,

$n \times M$ matrix $\mathbb{I}_{n,M} = \begin{pmatrix} \mathbb{I}_M \\ 0 \end{pmatrix}$,

$M \times (M+1)$ matrixes $\mathbb{A}_1^{(M)} = (a_{j',j}^{(1)})$ and $\mathbb{A}_2^{(M)} = (a_{j',j}^{(2)})$,

$$a^{(1)}_{j',j} = -\frac{\theta_j}{2M+1}(\gamma_{j'+j} + \gamma_{|j'-j|})a_1\Big(\frac{j'}{2M+1}, \frac{j}{2M+1}\Big),$$

$$a^{(2)}_{j',j} = \frac{2\theta_j}{2M+1}a_2\Big(\frac{j'}{2M+1}, \frac{j}{2M+1}\Big), \quad 1 \le j' \le M, 0 \le j \le M,$$

$$\theta_j = \begin{cases} 1/2 & j = 0 \\ 1, & j = 1, \ldots, M, \end{cases}$$

$$\gamma_l = \log 2 + \sum_{k=1}^{M} \frac{1}{k} \cos\Big(kl\frac{2\pi}{2M+1}\Big), \; 0 \le l \le M,$$

$$\gamma_{M+l} = \gamma_{M+1-l}, \; 1 \le l \le M.$$

Notice that $(\gamma_0, \ldots, \gamma_M)$ is the image of $(\log 2, 1, \frac{1}{2}, \ldots, \frac{1}{M})$ by $\mathcal{C}_M$. Clearly the entries of $\mathbb{A}_1^{(M)}$ and $\mathbb{A}_2^{(M)}$ can be computed in $\mathcal{O}(M^2)$ arithmetical operations.

(iii) Due to (4.9), the operator $Q_n^{\text{od}} A_1^{[d]}(P_n^{\text{ev}} - P_m^{\text{ev}})$ is given by $n \times (n+1)$ matrix

$$\tilde{\mathcal{S}}_n \sum_{j=0}^{d-2} \mathbb{B}_n^{(j)} \left\{ \begin{array}{ll} \mathcal{S}_n \mathbb{J}_n, & j \text{ even} \\ \mathbb{J}_n \mathcal{C}_n, & j \text{ odd} \end{array} \right\} \mathbb{G}_n^{(j,m)}$$

with

$$(n+1) \times (n+1) \text{ matrix } \mathbb{G}_n^{(j,m)} = \begin{pmatrix} 0 & 0 \\ 0 & \mathbb{H}_{n-m}^{(j)} \end{pmatrix},$$

$$\mathbb{H}_{n-m}^{(j)} = \operatorname{diag}\{(m+1)^{-(j+1)}, \ldots, n^{-(j+1)}\},$$

$$n \times (n+1) \text{ matrix } \mathbb{J}_n = \begin{pmatrix} 0 & \mathbb{I}_n \end{pmatrix},$$

$$n \times n \text{ diagonal matrix } \mathbb{B}_n^{(j)} = \operatorname{diag}\Big\{b_j\Big(\frac{1}{2n+1}\Big), \ldots, b_j\Big(\frac{n}{2n+1}\Big)\Big\}.$$

(iv) Finally, having grid values $\underline{f}_N = \Big(f\Big(\frac{1}{2n+1}\Big), \ldots, \Big(\frac{1}{2n+1}\Big)\Big)$, we compute $\underline{d}_N = \tilde{\mathcal{S}}_N \underline{f}_N$ and $\underline{d}_n = \mathbb{P}_{n,N}\underline{d}_N$ with

$$n \times N \quad \text{matrix} \quad \mathbb{P}_{n,N} = \begin{pmatrix} \mathbb{I}_n & 0 \end{pmatrix}.$$

So we have

$$\begin{aligned} \mathbb{M}_n = \; & \mathbb{A}_0 + \mathbb{I}_{n,M}\tilde{\mathcal{S}}_M(\mathbb{A}_1^{(M)} + \mathbb{A}_2^{(M)})\mathcal{C}_M\mathbb{P}_{M,m,n} \\ & + \tilde{\mathcal{S}}_n \textstyle\sum_{j=0}^{d-2} \mathbb{B}_n^{(j)} \left\{ \begin{array}{ll} \mathcal{S}_n \mathbb{J}_n, & j \text{ even} \\ \mathbb{J}_n \mathcal{C}_n, & j \text{ odd} \end{array} \right\} \mathbb{G}_n^{(j,m)}, \end{aligned} \tag{5.4}$$

$$\begin{aligned} \mathbb{M}'_n = \; & \mathbb{A}_0^\top + \mathbb{P}_{M,m,n}^\top \mathcal{C}_M(\mathbb{A}_1^{(M)} + \mathbb{A}_2^{(M)})'\tilde{\mathcal{S}}_M\mathbb{I}_{n,M}^\top \\ & + \Big(\textstyle\sum_{j=0}^{d-2} \mathbb{G}_n^{(j,m)} \left\{ \begin{array}{ll} \mathbb{J}_n^\top \mathcal{S}_n, & j \text{ even} \\ \mathcal{C}_n \mathbb{J}_n^\top, & j \text{ odd} \end{array} \right\} \overline{B}_n^{(j)}\Big)\tilde{\mathcal{S}}_n. \end{aligned} \tag{5.5}$$

Most of occuring matrices are of diagonal type but not $\mathcal{C}_n$, $\mathcal{S}_n$ and $\mathbb{A}_1^{(M)} + \mathbb{A}_2^{(M)}$. The last one is of reduced dimension; using the FFT technique for $\mathcal{C}_n$ and $\mathcal{S}_n$, the application of $\mathbb{M}_n$ to a $(n+1)$- vector and the application of $\mathbb{M}'_n$ to a n-vector costs $\mathcal{O}(M^2 + n\log n) = \mathcal{O}(N^{2\sigma} + N^{\tau}\log N)$ arithmetical operations, see (4.6). This is $\mathcal{O}(N)$ for $\sigma = \frac{1}{2}$ and $o(N)$ for $\sigma < \frac{1}{2}$; of course, the choice $\varrho = \sigma = \frac{1}{2}$ is preferable from the point of view of approximation accuracy. The computation of $\underline{d}_N$ costs $\mathcal{O}(N\log N)$ arithmetical operations.

Notice that $\mathbb{M}_n$ is real if functions $a_j(t,s)$, $j = 1,2$, are real.

6. Conjugate gradients

We first recall the algorithm of the conjugate gradient method for symmetrized equation and a convergence result. After that we apply this method for equations (5.1) and (5.3).

1. Let H_1 and H_2 be Hilbert spaces, $A \in \mathcal{L}(H_1, H_2)$, and $A^* \in \mathcal{L}(H_2, H_1)$ the adjoint operator of A. To solve the equation $Ax = f$ we first symmetrize it,

$$A^*Ax = A^*f\,,$$

and then apply the following algorithm of classical conjugate gradient iterations (see [5,6]).

Step 0: $x^0 = 0$, $r^0 = -A^*f$, $y^0 = -f$.

For $k = 0, 1, 2, \ldots$:

(1) if $r^k = 0$ then terminate;

(2) if $r^k \neq 0$ then proceed step $k+1$, and compute

$$z^k = \begin{cases} -r^k + \dfrac{\|r^k\|^2}{\|r^{k-1}\|^2} z^{k-1} & \text{if } k \geq 1 \\ -r^0 & \text{if } k = 0 \end{cases}$$

$$\begin{aligned} x^{k+1} &= x^k + w_k z^k\,, \qquad & w_k = \frac{\|r^k\|^2}{\|Az^k\|^2}\,, \\ y^{k+1} &= y^k + w_k A z^k\,, & \\ r^{k+1} &= r^k + w_k A^* A z^k\,. & \end{aligned}$$

Under condition $\mathcal{R}(A) = H_2$ for $Ax^k - f = y^k$, the following estimate is valid (see [5]):

$$\|Ax^k - f\| \leq 2q^k\|f\| \quad (k = 1, 2, \ldots)\,, \ q = \frac{\kappa - 1}{\kappa + 1}\,, \ \kappa = \|A\|\,\|A^+\|\,.$$

According to this estimate, the residual level

$$\|Ax^k - f\| \le \varepsilon \, \|f\| \tag{6.1}$$

with a given $\varepsilon \in (0,1)$ is guaranteed for $2q^k \le \varepsilon$, i.e. $k \ge \ln(2/\varepsilon)/|\ln q|$. Since $|\ln q| = \ln(\kappa+1) - \ln(\kappa-1) \ge 2/\kappa$, a sufficient condition on k guaranteeing residual level (6.1) is given by

$$k \ge \frac{\kappa}{2} \ln \frac{2}{\varepsilon}. \tag{6.2}$$

It follows from (6.1) that, for those k

$$\|x^k - A^+ f\| \le \|A^+\| \, \|f\| \, \varepsilon. \tag{6.3}$$

2. Now apply these results to equation (5.1), i.e.

$$\mathcal{A}_N v_n = w_n, \quad \mathcal{A}_N \in \mathcal{L}(H^0_{\mathrm{ev}}, H^0_{\mathrm{od}}), \quad w_n = P^{\mathrm{od}}_n Q^{\mathrm{od}}_N f.$$

We terminate the conjugate gradient iterations on the first k satisfying

$$\|\mathcal{A}_N v^k_n - w_n\|_0 \le \|w_n\|_0 \, \delta N^{-\mu} \tag{6.4}$$

with a given $\delta > 0$. Then (see (6.3))

$$\|v^k_n - \mathcal{A}^+_N w_n\|_0 \le \|\mathcal{A}^+_N\|_{0,0} \, \|w_n\|_0 \, \delta N^{-\mu}$$

implying

$$\|u^k_N - \mathcal{A}^+_N Q^{\mathrm{od}}_N f\|_\lambda = \|v^k_n - \mathcal{A}^+_N w_n\|_\lambda \le \|\mathcal{A}^+_N\|_{0,0} \, \|Q^{\mathrm{od}}_N f\|_0 \, \delta N^{\lambda-\mu} \quad (0 \le \lambda \le \mu)$$

where (cf. (5.2))

$$u^k_N := v^k_n + \sum_{k=n+1}^{N} d_k \cos(k 2\pi t). \tag{6.5}$$

Since $\|\mathcal{A}^+_N\|_{0,0} \to \|\mathcal{A}^+\|_{0,0}$, $\|Q^{\mathrm{od}}_N f\|_0 \to \|f\|_0$ as $N \to \infty$, together with (4.5) we obtain an optimal order estimate for u^k_N:

$$\|u^k_N - \mathcal{A}^+ f\|_\lambda \le c N^{\lambda-\mu} \|f\|_\mu \qquad (0 \le \lambda \le \mu). \tag{6.6}$$

The residual level (6.4) is guaranteed (cf. (6.2)) for

$$k \ge \frac{\kappa_N}{2}\left(\mu \ln N + \ln\left(\frac{2}{\delta}\right)\right) \sim \frac{1}{2} \kappa \mu \ln N \quad \text{as} \quad N \to \infty$$

where $\kappa_N = \|\mathcal{A}_N\|_{0,0} \, \|\mathcal{A}^+_N\|_{0,0} \to \kappa = \|\mathcal{A}\|_{0,0} \, \|\mathcal{A}^+\|_{0,0}$. Let us summarize the results.

Theorem 6.1. *Under conditions of Theorem 4.1, for $N \geq N_0$, the conjugate gradient method applied to equation (5.1) of reduced dimension n, with $\mathcal{A}_N^* \in \mathcal{L}(H^0_{\text{od}}, H^0_{\text{ev}})$ corresponding to $\mathcal{A}_N$ as an operator from H^0_{ev} onto H^0_{od}, with the termination on the first k satisfying (6.4), provides via (6.5) an approximation u_N^k of the optimal order accuracy (6.6), and k is of order not exceeding $\frac{1}{2}\kappa\mu \ln N$ where $\kappa = \|\mathcal{A}\|_{0,0}\,\|\mathcal{A}^+\|_{0,0}$.*

3. Introduce the bijections $V_n \in \mathcal{L}(\mathcal{T}_n^{\text{ev}}, \mathbb{C}^{n+1})$ and $W_n \in \mathcal{L}(\mathcal{T}_n^{\text{od}}, \mathbb{C}^n)$ defined by

$$V_n \sum_{l=0}^{n} c_l \cos(l2\pi t) = (c_0, c_1, \ldots, c_n)\,,$$

$$W_n \sum_{l=1} d_l \sin(l2\pi t) = (d_1, \ldots, d_n)\,.$$

Using in $\mathcal{T}_n^{\text{ev}}$ and $\mathcal{T}_n^{\text{od}}$ the scalar product $(\cdot,\cdot)_0$ and in $\mathbb{C}^{n+1}$ and $\mathbb{C}^n$ the scalar products

$$\begin{aligned} (\underline{c}_n, \underline{c}'_n) &= c_0\overline{c'_0} + \frac{1}{2}\sum_{l=1}^{n} c_l\overline{c'_l}\,, \\ (\underline{d}_n, \underline{d}'_n) &= \frac{1}{2}\sum_{l=1}^{n} d_l\overline{d'_l}\,, \end{aligned} \tag{6.7}$$

operators V_n and W_n are isometric, $V_n^*V_n = I$, $V_nV_n^* = I$, $W_n^*W_n = I$, $W_nW_n^* = I$. Futher,

$$\mathcal{A}_N\Big|_{\mathcal{T}_n^{\text{ev}}} = W_n^*\mathbb{M}_nV_n\,, \quad \mathcal{A}_N^*\Big|_{\mathcal{T}_n^{\text{od}}} = V_n^*\mathbb{M}_n^*W_n$$

where $\mathbb{M}_n^* = D_n\mathbb{M}'_n$ is the adjoint to $\mathbb{M}_n$ with respect to scalar products (6.7), $\mathbb{M}'_n$ is the Hermite dual to $\mathbb{M}_n$ (see (5.4), (5.5)), and $D_n = \text{diag}\,(\frac{1}{2}, 1, \ldots, 1)$. The algorithm of the conjugate gradient method for the equation (5.1) can be reformulated in terms of coefficients vectors $\underline{c}_n^k$ of $v_n^k = \sum_{l=0}^n c_l^k \cos(l2\pi t)$ and $\underline{d}_n$ of $P_n^{\text{od}}Q_N^{\text{od}}f = \sum_{k=1}^n d_k \sin(k2\pi t)$ as follows.

Step 0: $\underline{c}_n^0 = \underline{0}$, $\underline{r}_n^0 = -D_n\mathbb{M}'_n\underline{d}_n$, $\underline{y}_n^0 = -\underline{d}_n$.

For $k = 0, 1, 2, \ldots$:

(1) if $\|\underline{y}_n^k\| \leq \|\underline{d}_n\|\delta N^{-\mu}$ then terminate;

(2) if $\|\underline{y}_n^k\| > \|\underline{d}_n\|\delta N^{-\mu}$ then proceed step $k+1$, and compute

$$\underline{z}_n^k = \begin{cases} -\underline{r}_n^k + \dfrac{\|\underline{r}_n^k\|^2}{\|\underline{r}_n^{k-1}\|^2}\underline{z}_n^{k-1}\,, & k \geq 1\,, \\ -\underline{r}_n^0\,, & k = 0\,, \end{cases}$$

$$\underline{c}_n^{k+1} = \underline{c}_n^k + w_k \underline{z}_n^k, \qquad w_k = \frac{\|\underline{r}_n^k\|^2}{\|\mathbb{M}_n \underline{z}_n^k\|^2},$$
$$\underline{y}_n^{k+1} = \underline{y}_n^k + w_k \mathbb{M}_n \underline{z}_n^k,$$
$$\underline{r}_n^{k+1} = \underline{r}_n^k + w_k D_n \mathbb{M}_n' \mathbb{M}_n \underline{z}_n^k.$$

In this algorithm, the norms in $\mathbb{C}^{n+1}$ and $\mathbb{C}^n$ are induced by (6.7). We have incorporated the termination rule (6.4) with a parameter $\delta > 0$ into the algorithm.

As we saw in Section 5, an application of $\mathbb{M}_n$ and $\mathbb{M}_n'$ costs $\mathcal{O}(N)$ arithmetical operations for $\sigma = \frac{1}{2}$ and $o(N)$ for $\sigma < \frac{1}{2}$. An iteration step of conjugate gradients contains one application of $\mathbb{M}_N$ and one application of $\mathbb{M}_n'$, and the optimal accuracy (6.6) is achieved in $\mathcal{O}(\log N)$ iteration steps (see Theorem 6.1). Thus the amount of arithmetical operations to solve (5.1) is $\mathcal{O}(N \log N)$ in case $\sigma = \frac{1}{2}$ and $o(N)$ in case $\sigma < \frac{1}{2}$. Together with the computation of $Q_N^{\mathrm{od}} f$, the amount of total work is $\mathcal{O}(N \log N)$ arithmetical operations in both cases. If f has known Fourier coefficients, we can use $P_N^{\mathrm{od}} f$ instead of $Q_N^{\mathrm{od}} f$, and the total amount of arithmetical and logical operations to compute u_N^k of optimal order accuracy (6.6) is $N + o(N)$ for $\sigma < \frac{1}{2}$.

References

[1] S.M. Belotserkovsky and I.K Lifanov, *Method of Discrete Vortices*, CRC Press, London, 1993.

[2] D. Berthold, W. Hoppe and B. Silbermann, *The numerical solution of the generalized airfoil equation*, J. Integral Equations Appl. **4** (1992), 309–336.

[3] ———, *A fast algorithm for solving the generalized airfoil equation*, J. Comput. Appl. Math. **43** (1992), 185–219.

[4] D. Berthold and P. Junghanns, *New error bounds for the quadrature method for the solution of Cauchy singular integral equations*, SIAM J. Numer. Anal. **30** (1993), 1351–1372.

[5] J.W. Daniel, *The Approximate Minimization of Functionals*, Prentice-Hall, Engelwood–Cliffs, 1 edition, 1971.

[6] G.H. Golub and C.F. vanLoan, *Matrix Computations*, 2 edition, The Johns Hopkins Univ. Press, Baltimore, 1989.

[7] M.R. Capobianco, *The stability and the convergence of a collocation method for a class of Cauchy singular integral equations*, Math. Nachr. **162** (1993), 45–58.

[8] G. Chiocchia, S. Prössdorf and D. Tordella, *The lifting line equation for a curved wing in oscillatory motion*, Z. angew. Math. Mech. **77** (1997), no. 4, 295–315.

[9] D. Elliott, *The classical collocation method for singular integral equations*, SIAM J. Numer. Anal. **19** (1982), 816–832.

[10] ———, *A comprehensive approach to the approximate solution of singular integral equation over the arc* (-1,1), J. Integral Equations Appl. **2** (1989), 59–94.

[11] F. Erdogan, G.D. Gupta and T.S. Cook, *Numerical solution of singular integral equations*, Mech. Fractures **1** (1973), 368–425.

[12] S. Fenyö and H.W. Stolle, *Theorie und Praxis der Linearen Integralgleichungen*, Deutscher Verlag der Wissenschaften, Berlin, 1984.

[13] J.A. Fomme and M.A. Golberg, *Numerical solution of a class of integral equations arising in two-dimensional aerodynamics*, J. Optim. Theory Appl. **24** (1978), 169–206.

[14] ———, *On the L^2-convergence of collocation for the generalized airfoil equation*, J. Math. Anal. Appl. **71** (1979), 271–286.

[15] I. Gohberg and N.Y. Krupnik, *Einführung in die Theorie der eindimensionalen singulären Integraloperatoren*, Birkhäuser Verlag, Basel, 1979.

[16] ———, *One-dimensional linear singular integral equations*, Vol. 1, Birkhäuser Verlag, Basel, 1992.

[17] P. Junghanns, *Product integration for the generalized airfoil equation*, Beiträge zur Angewandten Analysi und Informatik (E. Schock, ed.), Shaker Verlag, 1994, 171–188.

[18] P. Junghanns and B. Silbermann, *Zur Theorie der Näherungesverfahren für singuläre Integralgleichungen auf Intervallen*, Math. Nachr. **103** (1981), 199–244.

[19] ———, *Numerical analysis of the quadrature method for solving linear and nonlinear singular integral equations*, Wiss. Schr. Tech. Univ. Karl-Marx-Stadt **10** (1988).

[20] A.I. Kalandiya, *Mathematical methods of two-dimensional elasticity*, Mir Publishers, Moscow, 1973 (Russian).

[21] M.A. Krasnoselskii, G.M. Vainikko, P.P. Zabreiko, Y.B. Rutitskii and V.Y. Stetsenko, *Approximate Solution of operator Equations*, Wolters-Nooedhoff Publ., Groninger, 1972.

[22] I.K. Lifanov, *Singular Integral Equations and Discrete Vortices*, VSP, Utrecht, 1996.

[23] I.I. Lifanov and I.K. Lifanov, *Antenna-diffraction problems and singular solutions of singular integral equations*, J. Electromagn. Waves and Appl. **10** (1996), 925–937.

[24] G. Mastroianni and S. Prössdorf, *A quadrature method for Cauchy integral equations with weakly singular perturbation kernel*, J. Integral Equations Appl. **4** (1992), 205–228.

[25] S.G. Mikhlin and S. Prössdorf, *Singular integral operators*, Springer-Verlag, Berlin, 1986.

[26] G. Monegato and S. Prössdorf, *Uniform convergence estimates for a collocation and discrete collocation method for the generalized airfoil equation*, Contributions in Numerical Mathematics (R.P. Agarwal, ed.), Vol. 2, World Scientific Series in Applicable Analysis (1993), 285–299.

[27] S. Okada and D. Elliott, *The finite Hilbert transform in L^2*, Math. Nachr. **153** (1991), 43–56.

[28] S. Okada and S. Prössdorf, *On the solution of the generalized airfoil equation*, J. Integral Equations Appl. **9** (1997), 71–98.

[29] R. Plato and G. Vainikko, *On the fast and fully discretized solution of integral and pseudodifferential equations on smooth curves* (submitted).

[30] S. Prössdorf and B. Silbermann, *Numerical analysis for integral and related operator equations*, Akademie-Verlag, Berlin, 1991; Birkhäuser Verlag, Basel, 1991.

[31] J. Saranen and G. Vainikko, *Trigonometric collocation methods with product integration for boundary integral equations on closed curves*, SIAM J. Numer. Anal. **33** (1996), 1577–1596.

[32] ———, *Fast solution of integral and pseudodifferential equations on closed curves*, Math. Comp. **67** (1998), 1473–1491.

[33] ———, *Fast collocation solvers for integral equations on open arcs*, J. Integral Equations Appl. **11** (1999), 57–102.

[34] M. Schleiff, *Über eine singuläre Integralgleichung mit logaritmischem Zusatzkern*, Math. Nachr. **42** (1969), 79–88.

[35] ———, *Singuläre Integraloperatoren in Hilbert-Räumen mit Gewichtsfunktion*, Math. Nachr. **42** (1969), 145–155.

[36] P.S. Theocaris, *On the numerical solution of Cauchy-type singular integral equations*, Serdica **2** (1976), 252–275.

[37] V. Turunen and G. Vainikko, *On symbol analysis of periodic pseudodifferential operators*, J. Anal. Appl. **17** (1988), 9–22.

[38] G. Vainikko, *Periodic Integral and Pseudodifferential Equations*, Research report C13, Helsinki Univ. Technol., 1996.

[39] ———, *Trigonometric Galerkin fast solvers for periodic integral equation of the first kind*, Funct. Differ. Equations **4** (1997), 419–441.

Helsinki University of Technology
Institute of Mathematics
Otakaari 1M
02150 Espoo, Finland

1991 Mathematics Subject Classification: Primary 65R20; Secondary 45L10

Submitted: 25.4.2000

Operator Theory:
Advances and Applications, Vol. 121

On Plane Potential Flow Past a Porous Circular Cylinder

LOTHAR VON WOLFERSDORF

A mathematical model for the plane potential flow of an inviscid incompressible fluid around and through a circular cylinder of porous material is presented. The flow inside the cylinder obeys Darcy's law. The flow problem is reduced to a nonlinear boundary value problem of Poincaré type and this one in turn to a nonlinear infinite sytem of algebraic equations and a related nonlinear integral equation.

Introduction

Recently the problem of steady plane potential flow of an inviscid incompressible fluid past a circular cylinder with porous surface has been dealt with in papers by the author [5] jointly with E. Wegert [4] and K. Heier [1], respectively, and separately by R. Kühnau [2] (cf. also Wegert's monograph [3]). In the case of a linear filtration law on the boundary of the cylinder the solution of this flow problem is given in closed from.

In the present paper the related steady plane potential flow of an inviscid incompressible fluid past a cylinder of porous material is considered where the flow inside the cylinder is assumed to obey the linear Darcy law. The flow problem is reduced to a quadratic boundary value problem of Poincaré type for the normed pressure function inside the unit circle. The Poincaré problem is equivalent to an infinite system of quadratic equations for the Fourier coefficients of the pressure function from which the velocity and pressure distributions of the flow on the cylinder surface and the flow pattern inside and outside the cylinder can be calculated. Further the flow problem is reduced to a quadratic integral equation with a positive solution.

For the details of the analysis of the flow problem and numerical results of flow patterns see a forthcoming joint paper of the author with W. Mönch [6].

1. Mathematical model

We consider the plane steady irrotational flow of an incompressible fluid with density ρ past a circular cylinder of porous material. Let the circle $G : |z| < a$ with radius a be the cross-section of the cylinder in the complex z-plane, where $z = x + iy$, and $\Gamma : |z| = a$ the boundary of G. Outside G we look for the velocity field $\underline{v} = (v_x, v_y)$ of the flow with the speed $q = (v_x^2 + v_y^2)^{1/2}$. The flow at infinity is parallel to the x-axis and has the speed q_∞:

$$v_x(\infty) = q_\infty > 0, \quad v_y(\infty) = 0 . \tag{1.1}$$

The pressure field in the flow is denoted by $p(z)$ and the pressure at infinity by p_∞.

In G the flow obeys Darcy's law. We have

$$\underline{q} = \gamma \underline{v}, \quad \underline{v} = -k \operatorname{grad} p \tag{1.2}$$

for the seepage velocity $\underline{q}$ with porosity $\gamma \in (0,1)$ and permeability $k \in (0,\infty)$

$$\underline{q} = -\kappa \operatorname{grad} p \tag{1.3}$$

with the (horizontal) filtration coefficient $\kappa = \gamma k \in (0,\infty)$. Assuming no sources and sinks in the flow, we have div $\underline{q} = 0$ and $\Delta p = 0$ in G by (1.3).

On Γ pressure and mass flux (normal component of seepage velocity) are continuous

$$p^- = p^+ , \quad v_\nu^- = \gamma \, v_\nu^+ \tag{1.4}$$

where ν denotes the inner normal of Γ and $\pm$ mark the inner and outer limit values on Γ. By Bernoulli's theorem we have

$$p^- = c - \frac{1}{2}\rho(q^-)^2 , \quad c = p_\infty + \frac{1}{2}\rho \, q_\infty^2 , \tag{1.5}$$

and by Darcy's law (1.2)

$$v_\nu^+ = -k \, \frac{\partial p^+}{\partial \nu} . \tag{1.6}$$

For the complex velocity function $W(z) = v_x - iv_y$ of the flow in the domain G_1 outside Γ we make the ansatz

$$W(z) = q_\infty (1 - \frac{a^2}{z^2}) + \frac{a}{z} \, w(z), \quad z \in G_1 , \tag{1.7}$$

where w is a holomorphic function in G_1 satisfying $w(\infty) = 0$. On $\Gamma : z = t = ae^{is}$ we have

$$v_\nu^- = - \operatorname{Re} \, w(t), \quad v_\sigma^- = - \operatorname{Im} \, w(t) - 2q_\infty \sin s \tag{1.8}$$

and $(q^-)^2 = (v_\nu^-)^2 + (v_\sigma^-)^2$, where σ is the arc length of Γ.

By (1.5), (1.6) and (1.8) the conjugacy conditions (1.4) take the form

$$p^+ = c - \frac{\rho}{2}[(\text{ Re } w(t))^2 + (\text{ Im } w(t) + 2q_\infty \sin s)^2] \tag{1.9}$$

$$\frac{\partial p^+}{\partial \nu} = \frac{1}{\kappa} \text{ Re } w(t) \tag{1.10}$$

with a real constant c.

The flow problem is reduced to the transmission problem (1.9), (1.10) for $\{p(z), z \in G; w(z), z \in G_1\}$, where the pressure function p is determined apart from an additive constant.

By (1.10) for the boundary values of the holomorphic function w in G_1 with $w(\infty) = 0$ it follows

$$\text{Im } w(t) = -\kappa \frac{\partial p^+}{\partial \sigma} . \tag{1.11}$$

Inserting (1.10) and (1.11) into (1.9), we arrive at the boundary condition

$$p = c - \frac{\rho}{2}\kappa^2 \left[\left(\frac{\partial p}{\partial \nu}\right)^2 + \left(\frac{\partial p}{\partial \sigma} - \frac{2q_\infty}{\kappa} \sin s\right)^2 \right]$$

for p on Γ.

We further transform the problem to the unit circle by $\zeta = z/a = re^{is}$ and introduce a normed (dimensionless) pressure function P by

$$p - c = -M\, P, \quad M = 2q_\infty a/\kappa . \tag{1.12}$$

Then the harmonic function $P = P(r, s)$ in $r < 1$ fulfills the quadratic boundary condition of Poincaré type

$$\left(\frac{\partial P}{\partial r}\right)^2 + \left(\frac{\partial P}{\partial s} + \sin s\right)^2 = \lambda P \quad \text{on } r = 1 \tag{1.13}$$

with the dimensionless parameter $\lambda = a/\rho q_\infty \kappa > 0$. The solutions $P = P(r, s)$ of (1.13) are positive and even functions of s. The limits $\lambda \to 0$ and $\lambda \to \infty$ correspond to free flow and flow around an impermeable cylinder, respectively. Due to the assumption (1.2) of a linear Darcy law the stated mathematical model of the flow problem is expected to be appropriate at best for not too small λ.

2. Reduction to an infinite system of algebraic equations and an integral equation

We represent the solution of (1.13) in form of the series

$$P(r, s) = C_0 + \frac{1}{2} r \cos s + \sum_{n=1}^{\infty} \frac{p_n}{n} r^n \cos ns, \quad r \le 1 , \tag{2.1}$$

with a positive constant $C_0 > 0$ and a square summable sequence $\{p_n\}$. Then the Poincaré condition (1.13) yields the system of equations for p_n, $n \geq 1$:

$$\lambda\left(\frac{1}{4}+\frac{p_1}{2}\right) = \sum_{k=1}^{\infty} p_k p_{k+1}$$

$$\lambda\frac{p_n}{n} = p_{n-1} + 2\sum_{k=1}^{\infty} p_k p_{k+n},\ n \geq 2$$

and the formula for C_0

$$C_0 = \frac{1}{\lambda}\left(\frac{1}{4}+\sum_{k=1}^{\infty} p_k^2\right) .$$

In addition we put $p_0 = -\lambda/2$ and obtain the system of equations

(2.2) $$\lambda\,\frac{p_n}{n} = p_{n-1} + 2\sum_{k=1}^{\infty} p_k p_{k+n}\ ,\ n \geq 1$$

as equivalent to the Poincaré problem.

For given $p_n, n \geq 1$, the complex velocity functions $W(z)$ in G_1 and $Q(z) = \gamma(v_x - iv_y)$ in G are determined by the series

(2.3) $$W(z) = q_\infty\left(1+2\sum_{n=1}^{\infty} p_n(\frac{a}{z})^{n+1}\right),\quad |z| > a$$

(2.4) $$Q(z) = q_\infty\left(1+2\sum_{n=1}^{\infty} p_n(\frac{z}{a})^{n-1}\right),\quad |z| < a\,.$$

The value of the constant C_0 has no influence on the flow parameters.

The system of equations (2.2) with (2.3) and (2.4) is the basis for the numerical solution of the flow problem. For a theoretical investigation of this system a reduction to an integro-differential equation is useful. Making the hypothetical ansatz

(2.5) $$p_n = \frac{\lambda}{2}\int_0^b \rho^n f'(\rho)d\rho,\quad n = 0, 1, \ldots$$

with some $b \in (0,1)$ and a smooth function f satisfying $f(0) = 1, f(b) = 0$ the system (2.2) is fulfilled iff f is a solution of the integro-differential equation

(2.6) $$f'(\xi) + \lambda\,\xi^2 f'(\xi)\int_0^b \rho f'(\rho)\frac{d\rho}{1-\rho\xi} + \lambda\,f(\xi) = 0, \qquad \xi \in (0,b)\,.$$

By the substitution $\varphi = -f'$ with

$$f(\rho) = 1 - \int_0^\rho \varphi(\xi)d\zeta$$

we obtain the quadratic integral equation for φ

$$\varphi(\xi) = \lambda[1 - \int_0^\xi \varphi(\rho)d\rho] + \lambda\xi^2\varphi(\xi)\int_0^b \rho\varphi(\rho)\frac{d\rho}{1-\rho\xi}\ , \ \xi \in (0,b) \tag{2.7}$$

together with the condition for $b \in (0,1)$

$$\int_0^b \varphi(\xi)d\xi = 1\ . \tag{2.8}$$

So the flow problem is reduced either to the infinite quadratic system of equations (2.2) for the Fourier coefficients p_n or to the quadratic integral equation (2.7) with (2.8) for the function φ and the parameter $b \in (0,1)$.

3. Further investigation of the Poincaré problem (1.3)

We introduce the holomorphic functions $\Phi(\zeta)$ in $|\zeta| < 1$ defined by $P = Re\Phi$ and $Im\Phi(0) = 0$, i.e. by (2.1)

$$\Phi(\zeta) = C_0 + \frac{1}{2}\zeta + \sum_{n=1}^{\infty}\frac{p_n}{n}\zeta^n \quad \text{in } |\zeta| < 1 \tag{3.1}$$

and

$$\Psi(\zeta) = \zeta^2\Phi'(\zeta) + \frac{1}{2}(1-\zeta^2)\ ,$$

i.e.

$$\Psi(\zeta) = \frac{1}{2} + \sum_{n=1}^{\infty} p_n\zeta^{n+1} \quad \text{in } |\zeta| < 1\ . \tag{3.2}$$

Then the formulas (2.3) (2.4) write

$$\begin{aligned} W(z) &= 2q_\infty\Psi(1/\zeta) \quad \text{in } |z| > a \\ Q(z) &= 2q_\infty\Phi'(\zeta) \quad \text{in } |z| < a \end{aligned} \qquad (\zeta = z/a) \tag{3.3}$$

The Poincaré condition (1.13) takes the complex form

$$|\Psi(\zeta)|^2 = \lambda \ \mathrm{Re}\, \Phi(\zeta) \qquad \text{on } |\zeta| = 1 \tag{3.4}$$

together with Im $\Phi(0) = 0$. This condition is equivalent to the integral differential equation for Φ

$$(1-\zeta^2)\Phi'(\zeta) + \lambda\, \Phi(\zeta) = \frac{1}{2}(1-\zeta^2) + S[|\Phi'|^2](\zeta) \quad \text{in } |\zeta| < 1 \tag{3.5}$$

together with Im $\Phi(0) = 0$, where S denotes the Schwarz integral

$$S[\varphi](\zeta) = \frac{1}{2\pi i} \int\limits_{|\tau|=1} \varphi(\tau)\, \frac{\tau+\zeta}{\tau-\zeta}\, \frac{d\tau}{\tau}\,.$$

We are looking for solutions Φ of (3.5) possessing a continuous derivative Φ' in $|\zeta| \le 1$.

For physical reasons we make two assumptions on the functions Φ and Ψ.

Assumption A:

$\Phi'(-1) > 0$, Φ' strictly decreasing in $[-1, 1]$,

Ψ strictly increasing in $[-1, 0]$ and decreasing in $[0, 1]$.

Assumption B:

Φ' and Ψ have only one zero in $|\zeta| \le 1$ which lie in $|\zeta| < 1$.

Then Φ' has a simple zero $c \in (0,1)$ and Ψ has a simple zero $b \in (0,1)$, where $c < b$. By (3.3) the point $z_1 = a \cdot c$ is the stagnation point of the flow inside the cylinder and $z_2 = a/b$ is the stagnation point of the flow outside (behind) the cylinder.

In view of (3.5) the function Φ can be analytically continued to the whole ζ-plane slitted along the real axis from $R = 1/b$ to ∞ and satisfies the differential functional equation

$$\begin{aligned}(1-\zeta^2)\Phi'(\zeta) + (1-\frac{1}{\zeta^2})\Phi'(\frac{1}{\zeta}) + \lambda[\Phi(\zeta) + \Phi(\frac{1}{\zeta})] \\ = 1 - \frac{1}{2}(\zeta^2 + \frac{1}{\zeta^2}) + 2\Phi'(\zeta)\Phi'(\frac{1}{\zeta})\,.\end{aligned} \tag{3.6}$$

A discussion of the behaviour of Φ at $R = 1/b$ and at infinity and the jump condition on the slit $L = (R, \infty)$ yield again the integro-differential equation (2.6), respectively, the integral equation (2.7), (2.8) with the additional information

$$\varphi(\xi) = -f'(\xi) > 0\,, \qquad \xi \in (0, b) \tag{3.7}$$

implying the correctness of the ansatz (2.5) with

$$p_n < 0 \quad \text{and} \quad |p_n| \quad \text{decrease} \quad \text{for all } n \,. \tag{3.8}$$

Indeed, it can be shown that the $|p_n|$ behave asymptotically like $n^{-\alpha}b^n$ with some positive α and the $b \in (0,1)$ from above.

So, for the physically relevant solution of the flow problem we have to look either for a solution vector $\{p_n\}$ with (3.8) of the system of equations (2.2) or for a positive continuous solution φ in $(0, b)$ with $b \in (0,1)$ of the integral equation (2.7) and the additional condition (2.8).

References

[1] Heier, K.; von Wolfersdorf, L.: Numerical evaluation of potential flows past a circular cylinder with porous surface. Z. angew. Math. Mech. **70** (1990), 65-66.

[2] Kühnau, R.: Zur ebenen Potentialströmung um einen porösen Kreiszylinder. J. Appl. Math. Phys. **40** (1989), 395-409.

[3] Wegert, E.: Nonlinear Boundary Value Problems for Holomorphic Functions and Singular Integral Equations. Akademie Verlag, Berlin 1992.

[4] Wegert, E.; von Wolfersdorf, L.: Plane potential flow past a cylinder with porous surface. Math. Meth. Appl. Sci. **9** (1987), 587-605.

[5] von Wolfersdorf, L.: Potential flow past a circular cylinder with permeable surface. Z. angew. Math. Mech. **68** (1988), 11-19.

[6] von Wolfersdorf, L.; Mönch, W.: Potential flow past a porous circular cylinder, Z. angew. Math. Mech. **80** (2000), 465-479.

Lothar von Wolfersdorf, Fakultät für Mathematik und Informatik, TU Bergakademie Freiberg, D 09596 Freiberg, Germany

1991 Mathematics Subject Classification: Primary 76B99, 76S05; Secondary 30E25, 31A25, 45G10

Submitted: 4.4.2000